FIBER OPTICS COMMUNICATIONS

HAROLD KOLIMBIRIS

Seneca College

Upper Saddle River, New Jersey
Columbus, Ohio

Editor in Chief: Stephen Helba
Assistant Vice President and Publisher: Charles E. Stewart, Jr.
Assistant Editor: Mayda Bosco
Production Editor: Alexandrina Benedicto Wolf
Production Coordination: Karen Fortgang, bookworks
Design Coordinator: Diane Ernsberger
Cover Designer: Deb Warrenfeltz
Cover art: Digital Vision
Production Manager: Matt Ottenweller
Marketing Manager: Ben Leonard

This book was set in Times Roman by Carlisle Communications, Ltd. It was printed and bound by Courier, Kendallville. The cover was printed by Phoenix Color Corp.

Pearson Education Ltd.
Pearson Education Singapore, Pte. Ltd.
Pearson Education Canada, Ltd.
Pearson Education—Japan

Pearson Education Australia Pty., Limited
Pearson Education North Asia Ltd.
Pearson Educación de Mexico, S.A. de C.V.
Pearson Education Malaysia, Pte. Ltd.

10 9 8 7 6 5 4 3 2 1
ISBN 0-13-015883-6

To
STAVROULA
and
PANAGIOTIS

PREFACE

The last two decades have seen a phenomenal increase in the transmission of electronic information—a trend that is now, thanks to the Internet, growing exponentially. Traditional communication links are incapable of satisfying the demand for this information, due to the limitations imposed by their bandwidth constraints. Data transmission limitations of traditional links have been overcome since the introduction of optical fiber networks. Studies have demonstrated that the information carrying capacity of optical fiber networks has increased 50% per year over the last decade. Current and projected demands for high capacity links and their anticipated economic benefits have triggered a flurry of activity among researchers. It is the purpose of this text to present a comprehensive account of both the theoretical and the practical aspects of optical communications.

This volume is intended for several different audiences. It can be used by undergraduates beginning their study of fiber optics communications and also by advanced graduate students. It can also be used as a convenient reference source for academics and practicing engineers. The text covers a broad range of integrated materials relevant to optical fiber communications and can be used systematically or, because each chapter has been written independently, selectively.

The text is organized into four main parts: Introductory Concepts, Electro-optics, Optics, and Systems. The fundamentals of optics are presented in Chapters 1–4. In Chapter 1, the elements of geometric, physical, and quantum optics are introduced as preparation for the more advanced topics in subsequent chapters. Chapter 2 introduces the basic concepts of atomic and semiconductor theory. Because the key devices of optical fiber systems are based on semiconductors, it is essential that topics such as energy bands, Fermi-Dirac distribution, nondegenerative and highly degenerative semiconductor materials, the PN-junction, and biasing techniques be discussed in depth. The theory and operating characteristics of two types of optical sources, those of the LED and laser diodes, are discussed in Chapter 3. The structure and performance characteristics of edge emitting and surface emitting LEDs are also discussed. The chapter concludes with a detailed discussion of Fabry-Perot lasers, distributive feedback lasers, and multi-quantum well vertical cavity service emitting lasers. Chapter 4 is primarily devoted to a discussion of the photodetection process and the examination of

the theory and performance characteristics of PIN and avalanche photodiodes. The structure and operating characteristics of advanced photodetecters, including resonance cavity enhanced heterojunction are also presented in this chapter.

The concept of optical amplification and its implementation is presented in Chapter 5. This chapter is devoted to the theory and performance characteristics of available bandwidth, dynamic range, and signal to noise ratios of semiconductor amplifiers and erbium doped fiber amplifiers. Optical transmitters and the fundamental building blocks are presented in some detail. Several types of transmitters developed by leading manufacturers, design philosophy, operating characteristics, and intended applications are discussed, along with a brief description of Lucent Technologies' uncooled laser transmitter. Chapter 7 begins with a description of basic data patterns and their differences and similarities, and continues with a description of the role of optical receivers in a fiber communications system. The classification of optical receivers is also discussed, followed by the presentation of several transimpedance preamplifiers and postamplifiers developed by industry as essential blocks of optical amplifiers. The function within a fiber optics link, the building blocks, and operating characteristics of optical transreceivers is the topic of Chapter 8. Classification of optical transreceivers in terms of their optical source is discussed, followed by a discussion of a number of LED and laser based transreceivers from several manufacturers. The theory, fabrication, and operating characteristics of optical fibers are the subjects of Chapter 9. Fundamental laws governing the transmission of light through fibers and the classification of fibers as single-mode or multimode fibers are presented. Limitations of standard fiber transmission capabilities are identified, and the incentives for the development of advanced fibers such as non-zero dispersion shifted and large area fibers are discussed. The chapter concludes with a detailed description of the adverse effects of four wave mixing, cross phase modulation, and self phase modulation in dense wavelength division multiplexing systems.

Chapter 10 presents the theory and operating characteristics of external modulators, an essential element of high speed optical communications systems. The chapter begins with an introduction of the Mach Zehender interferometer and leads to a discussion of the Mach Zehender $LiNbO_3$ optical modulator. The chapter concludes with a presentation of several modulator driving circuits and a description of the operation and performance characteristics of electroabsorption modulated lasers. Chapter 11 discusses the concept of multiplexing in detail. Frequency division multiplexing, time division multiplexing, code division multiplex access, and wavelength division multiplexing are introduced, and dense wavelength division multiplexing and add/drop multiplexing are discussed in considerable detail. Chapter 12 is dedicated to the design of optical systems and begins by outlining the rules and procedures used in the design process. Wave polarization and dispersion, and their negative effect on system performance, are discussed. Design examples from analog, hybrid, and digital systems, as well as systems that employ spectral inversion, are explained. A number of experimental systems that are designed to operate in the Tb/s range are also discussed. The chapter concludes with an introduction to soliton and soliton transmission in transoceanic systems. Chapter 13 briefly introduces the elements of optical networks. It begins with the fundamental components of data communication links and continues with a discussion of local area networks, network transport architecture, Gb/s Ethernet, and a synchronous transfer mode, and concludes with a description of optical synchronous networks.

ACKNOWLEDGMENTS

I wish to express my thanks and deepest appreciation to all the people who contributed to the development of this book.

Technical reviews were done by Matthew Frank, RSoft; Byron Todd, Tallahassee College; Richard L. Windley, ECPI College of Technology; Fred Seals, Blinn College; Costas Vassiliadis, Ohio University; Robert Borns, Purdue University; Lee Rosenthal, Farleigh Dickinson University; Chris Derventzis and Owen Moorhouse, Seneca College; and Bill Hessmiller, Editors and Training Associates. Language reviews were done by: David Phillips, Seneca College; and Laurence Orenstein and Andrew Kenneth Tamblyn, Shanghai College.

In particular, I would like to thank James Moran for his unending patience. Kenneth Fernandez did a fantastic job on the graphics. Jason Cousin and Todd Julien of Seneca College provided excellent computer support.

My thanks are also extended to the following companies for providing up-to-date technical information: Analog Devices, Corning Inc., Hewlett Packard, Lucent Technologies (TriQuint Semiconductors), Motorola, Nortel Networks, Philips Semiconductors, and Vitesse Semiconductors.

I would like to thank the Prentice Hall team: Mayda Bosco, Alex Wolf, Karen Fortgang, Pat Wilson, and especially Tricia Rawnsley for her excellent work of editing the complete manuscript. Special thanks to RSoft Inc. for providing the CD-ROM to accompany the text. Finally, my thanks go to Ms. Jiang Wen for her undiminished encouragement and support.

Harold Kolimbiris

CONTENTS

INTRODUCTION

THE NEED FOR IMPROVEMENTS

In a communication system, the volume, speed, and clarity of information processed are key parameters that determine its performance level. With the advent of digital communications in the mid-1960s, research and development has concentrated efforts on increasing the volume and improving the quality of information processed through the system.

Because electromagnetic waves are almost always used as a primary means of transmitting information electronically, the increasing demand for a high volume of processed information has generated the need for a corresponding increase in carrier frequencies and bandwidths.

Carrier frequencies present certain limitations in higher ranges when applied to terrestrial systems. These limitations generated the incentive to search for new ways to increase the carrier frequencies, thus improving the information carrying bandwidth. Optical fibers are the most suitable media, because they present theoretically unlimited possibilities.

OPTICAL COMMUNICATIONS

A great interest in optical communication was triggered in the 1960s, with the development of optical sources. These sources were capable of generating frequencies of about 5×10^{34} Hz, with corresponding bandwidths that have the potential to increase the information capacity of the communications system by 100,000%. Early development of this discovery proceeded slowly, due to technical limitations related to optical fiber manufacturing techniques.

A Brief History

Before proceeding with a further discussion of optical fiber communications, it is appropriate to review major discoveries in physics and applied science that directly or indirectly contributed to fiber optics communications systems development.

In 1820, Hans Christian Oersted discovered the relationship between electric current and magnetic field. In 1831, Michael Faraday substantiated the relationship between magnetic field and electric current. The combined discoveries of Oersted and Faraday led to the development of the telegraph system in 1837 by Samuel Morse in the United States and William Cooke and Sir Charles Wheatstone in the U.K. The American system was somewhat simpler than the British and was based on a newly developed code (dot-dash). It was widely accepted, and achieved unprecedented commercial success. In the United States, the first transcontinental telegraph line was developed by Western Union. It was fully operational by 1861. Five years later, Lord Kelvin was instrumental in establishing the first transatlantic telegraph line.

In 1874, Alexander Graham Bell envisioned the conversion of sound waves into electric current through a small magnet. In 1876, he patented the first primitive telephone set. By 1876, the first telephone exchange was fully operational. In 1877, Thomas Edison and Elisha Gray developed a telephone that performed better than Alexander Graham Bell's.

Western Union produced approximately 50,000 of Gray's telephone and placed them into regular service. Although the telephone set exhibited relative success for short distance voice communications, it was found to be totally impractical for long distance applications.

In 1864, James Maxwell formulated the theoretical interpretation of electromagnetic wave propagation, and in 1884, Heinrich Hertz succeeded in demonstrating the applicability of Maxwell's theory. Although Hertz's demonstration was very promising, his findings could not be applied to long distance voice transmission because of the large attenuation that the voice signal suffered during transmission.

In order to improve the quality of long distance transmission of voice signals, a new knowledge of electromagnetic propagation was essential. Advances in transmission line theory were crucial if improvements were to be made in the quality of voice transmission through wired lines. Based on work done by Oliver Heaviside in 1887, a substantial improvement in transmission line signal attenuation could be accomplished if discrete series inductors are added to the line at intervals equal to one tenth of the wavelength of the traveling wave.

Although this innovation substantially enhanced the distance of voice transmission, further improvement in both distance and received signal quality could not be achieved without the implementation of repeaters at regular intervals on the transmission line. Such amplifiers, which had high gain and wide dynamic range, were impossible at that time because the electron had not yet been discovered.

Discovery of the Electron and Signal Amplification

In 1897, Joseph J. Thompson at Cambridge conducted extensive studies of cathode ray discharge, which ultimately led to the discovery of the electron. In 1904, John Ambrose Fleming produced the electron vacuum tube, and in 1907, Lee deForest invented the triode vacuum tube—an extension of Fleming's invention.

Both devices were initially used as signal detectors, not as signal amplifiers. In 1913, H. D. Arnold at AT&T succeeded in developing the first vacuum tube amplifier. The combination of transmission line theory implementation and the use of vacuum tube amplifiers as repeaters greatly enhanced the performance of the telephone system and enabled wireless voice transmission. Vacuum tubes contributed to a substantial increase of system capacity, with a simultaneous decrease in system costs, through the development of circuits such as high power amplifiers, oscillators, multiplexers, demultiplexers, and filters.

In 1918, William Schottky predicted thermal noise, an inherited destructive component in any electronic device. By 1925, J. B. Johnson substantiated Schottky's predictions by actually observing the phenomenon of thermal noise. At the same time (during the 1920s), Europe experienced a phenomenal growth in quantum mechanics research.

Semiconductor Devices

In 1928, Felix Bloch developed the quantum theory of metals, which eventually led to the development of the fundamental principles of semiconductor physics by Peierls in 1929. During the 1930s, substantial progress in solid state physics had shown promising results in the area of semiconductor devices for communications systems. The application of semiconductor materials, such as silicon and germanium, as detector devices that operate in the microwave range was very successful during World War II.

In late 1947, Walter Brittain and John Bardeen successfully demonstrated a PNP semiconductor device used as a voice signal amplifier. This successful demonstration opened new horizons in electronic and communication developments.

By the 1950s, semiconductor devices began to replace vacuum tubes in practically all the electronic and communications circuits and systems. The first transistorized communications repeaters, developed by Bell labs, became fully operational by 1959. Continuous improvement in semiconductor device design and fabrication substantially enhanced transistor device performance characteristics such as gain, noise reduction, and dynamic range.

Parallel to semiconductor device development in 1947, another concept in communication development began to take shape: digital voice transmission. The implementation of such a concept was only possible with the invention of the transistor device and the eventual design and implementation of integrated circuits (Jack S. Kilby of TI in 1958). Integrated circuits were able to facilitate the complex functions required by Pulse Code Modulation (PCM) systems.

In 1962, the first PCM digital system, capable of digitally transmitting 24 voice channels with a total bit rate of 1.544 Mb/s was developed. This system was called the Bell T1 digital system.

Computer and Integrated Circuit Technology

Parallel advancement in computer and integrated circuit technology propelled the communication industry to previously unimaginable heights. Until the middle 90s, long distance communications systems were defined as those systems capable of transmitting voice signals for relatively short distances via line of sight or for long distances mainly via satellite transmission.

Modern Communication Technology

Today, the communications industry is undergoing another revolution. The explosive demand for Internet services has completely redefined the communications industry as we know it. As mentioned previously, the present and future demand for combined voice, video, and data transmission, makes optical fiber communications systems the most desirable. Optical fiber communications systems differ from the traditional systems because they use glass waveguides instead of free space or wire pairs, multichannel modulators (WDM and DWDM) instead of single analog or digital voice channels, and coherent light sources that exhibit low noise, high optical power, and low loss when traveling through optical fiber.

Important Discoveries

At this point, it is necessary to review some of the major discoveries that eventually led to the development of optical fiber communications systems. The inception of such systems goes back to 1917, when Albert Einstein first developed the concept of stimulation emission.

In 1955, Towns observed stimulation emission when experimenting with ammonia beam MASERs (microwave amplification by stimulation emission of radiation).

In 1956, Bloembergen achieved population inversion in MASER amplifiers, and in 1957, the first solid state MASER amplifier was developed by Bell Labs. The primary application of such a low noise and high gain amplifier was in the Telstar series of communications satellites.

In 1960, T. Maiman demonstrated the first pulse ruby laser at Hughes research labs. In 1962, various research facilities such as IBM, GE, and Lincoln Labs, working independently, succeeded in developing the first semiconductor laser. This device was found to be unsuitable for implementation in optical fiber communications systems because of its inability to operate continuously at room temperature. It was not until 1970 that such a device was developed for full system implementation.

The second major element required for long distance transmission with fiber systems is the optical fiber. Early development proceeded rather slowly because of technical limitations related to fiber manufacturing techniques. For example, early fibers measured transmission losses of 100 dB/km. Such losses were considered impractical for system implementation.

In 1970, Felix Kapron and his team at Corning Labs managed to reduce attenuation of an optical fiber that operates at 850 nm to 20 dB/km, a definite improvement over previous fibers.

By 1976, a dramatic improvement in optical fiber manufacturing techniques reduced fiber losses from 20 dB/km to 1 dB/km at the 1300 nm wavelength. This breakthrough in optical fiber manufacturing made the commercial implementation of the first fiber optics communications systems possible in 1978.

Following the relative success of the first system, a total of 40,000 km of optical fiber line was in operation by 1982. Continuous improvements in fiber manufacturing technology further decreased fiber attenuation, and by 1985, fibers could be manufactured with attenuation as low as 0.5 dB/km at 1550 nm.

The combination of low loss optical fibers that operate at the 1300 to 1550 nm wavelength range with laser diodes made the installation of the first transatlantic system (TAT-8) possible in 1988, operating at a wavelength of 1300 nm.

In terms of information carrying capacity, it is estimated that optical fiber communications systems are doubling every two years. Today's optical fiber communications systems that employ TDM schemes have reached their limits in terms of information carrying capacity and path length. For practical purposes, such systems are divided into two major groups: electrical and optical.

Although the information superhighway (optical) theoretically has unlimited data transmission capability, standard electrical interface schemes, such as synchronous optical networks (SONET) and synchronous digital hierarchies (SDH), impose limits on the speed and overall performance of the optical network. The modules from which they are composed (amplifiers, multiplexers, demultiplexers, and regenerators) exhibit limited data handling capabilities.

To elevate the system performance limits imposed by the various electrical layers of the optical network, the system must be capable of transmitting the optical signal as far as possible into the traveling path before it is recovered and converted back to electrical signals. This can only be achieved by the implementation of dense wavelength division multiplexing (DWDM) schemes.

FUTURE OPTICAL SYSTEMS

Future optical systems will include fewer electrical layers in comparison to today's systems. The combination of ATM and IP protocols will be replaced by a combination of STM and optical layer over DWDM, encompassing the service performance of SONET/SDH, and enhanced by optical layer capabilities.

Metropolitan optical networks (MONETs) can significantly benefit from DWDM layers. High bandwidth demands for such networks can easily be accommodated by wavelength routing. Optical rings can perform reliably as demonstrated by the SONET/SDH, while greatly reducing system complexity. Future optical networks require a new family of active components such as broad bandwidth erbium doped fiber amplifiers (EDFAs) and exotic fibers. The ever increasing demand for higher capacity can be satisfied by further technological advancement in areas such as optical sources, optical amplifiers, modulators, and optical fibers.

Future optical networks utilizing DWDM schemes require highly sophisticated optical sources capable of selective wavelength transmission. The advantage of selective wavelength transmission by an optical source is that it provides an optical system with the capability of utilizing any available wavelength for data transmission. Wavelength conversion and wavelength utilization upon availability and demand greatly contribute to bandwidth optimization and traffic management.

Research and development in the above areas has led to a new development such as quantum well (QW) laser technology, nonlinear optics, and advanced optical fibers. A QW based laser diode is a double heterojunction structure composed of different band gap semiconductor materials with a very thin $(100 \overset{\circ}{A})$ layer of lower band gap in the middle. The confined carriers dominate the properties of the thin center layer. A team of Bell Labs scientists observed promising evidence of QW lasers in 1975.

An extension of quantum well technology is the multiple quantum well (MQW) process. The performance characteristics of MQW based laser diodes were combined by incorporating a laser diode with an MQW modulator in the same semiconductor structure. The overall performance of the combined structure qualified it for full implementation in optical systems that employ DWDM.

Further development in QW based laser structure led to the development of the QW cascade laser diode. In this device, the QW states are used for the generation of the desired wavelength emissions instead of the traditional band gap.

Such an optical device has been tested by Bell Labs with very positive results. It incorporates a number of DFB lasers, with each laser device attached to its own modulator, creating a much desired multiwavelength laser source.

The capability and performance of this optical source led to another form of multiwavelength transmission: chirp pulse wavelength division multiplexing (CPWDM). In this scheme, a fiber doped laser, operating as a single-mode optical source, transmits ultra short pulses over a large wavelength band. A dispersive fiber is used to enlarge the width of the pulse to the nanosecond range. A TDM modulator is used to further divide the pulses into time slots, each of which places data into different wavelengths. It is, therefore, evident that a large number of wavelength channels are generated from a single laser source. Experimental work has shown that, by applying this concept, one hundred optical wavelengths, each transmitting 10 Gb/s of error free data, could be optically transmitted to a distance of 400 kilometers. The other device critical to the implementation of a DWDM scheme in an optical fiber communications system is the optical amplifier.

Erbium Doped Fiber Amplifiers (EDFAs), in conjunction with gain flattening filters, have demonstrated a significant and even amplification over the 80 nm bandwidth. This bandwidth can be translated into a significant and even amplification of 100 optical channels at a bit rate of 10 Gb/s per channel. Experimental optical amplifiers operating at the 1400 nm WDM layer demonstrate significant gain over the entire spectrum. The design philosophy of such an amplifier is based on multiple high power lasers that produce the required gain over a 93 nm bandwidth. Bell Labs' research results have shown that fiber graded lasers are capable of providing a phenomenal optical power output at the 1100 nm wavelength, in comparison to traditional lasers.

It is projected that, in the near future, such optical amplifiers will extend their operation to the 400 nm range, and plans are underway for such devices to approach 1000 channels. EDFAs and Raman amplifiers use fiber gratings in their fabrication process. Such structures are composed of lengths of fibers capable of changing their refractive index periodically. This periodic refractive index change is embedded in the core of the fiber by ultraviolet light intended to generate an almost arbitrary combination of reflection and transmission light spectra.

In the area of filter applications, fiber grated based filters are capable of separating group wavelengths into individual bands. Furthermore, wavelength selectivity enables the optical system to add or drop single wavelengths spaced at just a few tenths of a nanometer.

EDFAs capable of providing parallel amplification of several optical channels, in conjunction with MQW laser devices, made the implementation of DWDM schemes in optical communications systems possible.

In the early stages of the implementation of DWDM schemes (the 1990s), series problems were encountered, which were directly related to third order nonlinear effects. Up to this point it was common practice to adjust the optical fiber chromatic absorption to zero in order to minimize the problem.

Zero fiber chromatic absorption was necessary to eliminate the possibility of pulse spreading due to the pulse traveling through the fiber, which ultimately was attributed to an optical source's finite spectral width. Pulse spreading generates unwanted wave mixing products and contributes directly to the degradation of overall system performance. In practice, for long distance transmission, the difficulty was solved through the introduction of a new type of optical fiber by Bell Labs in 1993. In such a fiber, a small amount of dispersion was allowed in order to discourage the formation of wave mixing. This improvement in fiber design greatly contributed to the implementation of DWDM.

CONCLUSIONS AND RECENT BREAKTHROUGHS

It is therefore evident that improvement in laser diodes, modulators, optical amplifiers, and optical filters dramatically enhanced the overall optical fiber communications system performance, measured in terms of distance, capacity, and low noise.

Engineers at Bell Labs, being in the forefront of optical research, were successful in integrating microelectronic and photonic technologies into a single chip. A two-by-ten array of vertical cavity surface emitting laser (VCSEL) was successfully incorporated into a CMOS structure. The device exhibited a driving capability of 1 Gb/s per laser device, low power dissipation, and low cost.

Collaboration between Bell Labs and Lucent Technologies in the area of commercial optical communications systems has driven system capacities over the 3 Tb/s range, through the development of the Wave Star system. This optical line system incorporates eight fibers with eighty wavelength channels per fiber. Each channel is capable of transmitting 5 Gb/s, with a total of 3.2 Tb/s system capacity.

The Wave Star system also incorporates the add/drop channel function while conforming to SONET/SDH standards for 2.5 Gb/s and 10 Gb/s formats. It also provides cross connections, asynchronous/synchronous conversion, performance monitoring, IP routing, and switching, and low level cross connect.

In this book, an attempt will be made to thoroughly explore the chemical and physical properties of the various materials incorporated into the fabrication of semiconductor components, which are utilized in a complete optical fiber communications system. Such devices are optical sources, amplifiers, multiplexers (modulators), fiber detectors, and filters.

Integrating the special functions and performance characteristics of all the above devices, optical fiber communications system design will satisfy various parameters and system specifications. Today's optical fiber communications systems are capable of simultaneously transmitting up to 40,000 voice channels over a distance of 200 km, without the need for interim signal amplification.

Further developments in cooled laser technology, and the ability to integrate optical sources with optical modulators in the same microchip, yielded the electroabsorption modulated isolated laser module (EMILM) device developed by Lucent Technologies. This module exhibits a superior performance in terms of transmission distance, signal degradation, and chirp level. Employed in optical communications systems that utilize DWDM schemes, the device is capable of supporting transmission rates of 2.5 Gb/s per channel for 38 optical channels (ITU-T standard) to a total of 95 Gb/s, which translates into simultaneous transmission of approximately 1.48 million voice channels.

1

ELEMENTS OF OPTICS AND QUANTUM PHYSICS

OBJECTIVES

1. Define the fundamental concepts of optics

2. Establish the differences among geometric, physical, and quantum optics

3. Discuss in detail the theory of geometric optics, including the concepts of reflection and refraction

4. Fermat's principle

5. Discuss some fundamental principles of physical optics, including wave velocity, intensity, frequency and wavelength, absorption, scattering, dispersion, and polarization of optical waves

6. Explain some of the basic concepts of quantum optics, including the photoelectric effect, radiation from a black body, Planck's radiation law and wave particle duality

KEY TERMS

aberration
absorption
absorption coefficient
acoustic wave
acoustic wave photon
Ampere's law
angle of deviation
angle of incidence
angle of incident
angle of reflection
angle of refraction
angular dispersion
anti-Stokes photon
birefringent
Bohr's atomic model
black body

Brewster's law
Brillouin scattering
Cauchy's equation
circular polarization
Compton scattering
cut off frequency
cut off wavelength
degree of polarization
diffraction
dipole moment
direction of propagation
dispersion
elastic scattering
elastic theory
electric field
electromagnetic wave

electromotive force
electron mass
electrostriction
elliptical polarization
emission
energy density
Faraday's law
Fermat's principle
frequency
general absorption
geometric optics
Helmholtz constant
Huuygens' principle of
   reflection
incident wave
incoherent scattering

inelastic scattering
intensity
interference
inverse square law
Kerr effect
kinetic energy
light rays
magnetic field
Malus' law
medium
optical path
optics
particle natural vibrating
   frequency
period
phase difference

1.1 INTRODUCTION

Because light is the medium of information transmission in an optical communication system, it is imperative that the nature and behavior be briefly examined. The study of light, commonly referred to as **optics,** is divided into three main categories: **geometric optics, physical optics,** and **quantum optics.** Geometric optics treats light as if it were composed of individual light rays, physical optics treats light as a wave phenomenon, and quantum optics treats light as quanta. (It studies the interaction of light with the atomic particles of matter.) The division of the study of light into geometric and physical optics emerged as a result of the failure to isolate a single light ray. It was observed that when a **point source** (a very small light source) was interrupted by an opaque diaphragm with a hole larger than the light source, the generated light spot projected onto a screen is propagated in straight lines. These lines are referred to as light rays (see Figure 1–1).

If the hole of the diaphragm is progressively reduced as shown in Figure 1–2a and Figure 1–2b, the observation is the same as Figure 1–1. That is, light rays propagate in

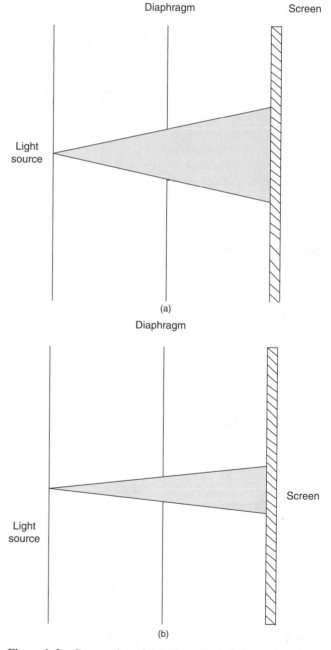

(a)

(b)

Figure 1–2. Propagation of light through a hole larger than the point source.

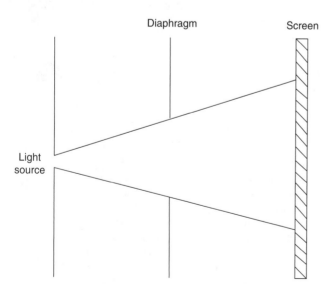

Figure 1–1. Propagation of light through a hole larger than the point source.

straight lines. Based on these observations, scientists once thought that further reducing the size of the diaphragm hole would isolate single ray. However, when the diaphragm hole was reduced to a very small size (see Figure 1–3), the projected light spot actually increased significantly. The inability to isolate a single light ray was the result of two factors. One was the non-ideal sharpness of the edges of the diaphragm hole and the other was the wave nature of the light. The combined effect is referred to as the "**diffraction** of light."

In applications where wide beams of light are used, diffraction is not a problem. However, it becomes a significant problem in applications where a very narrow beam of light is required. The interaction of light waves can be explained satisfactorily by wave theory. However, **wave theory** does not explain the phenomena resulting from the interaction of light with matter (emission, absorption). For example, the observed energy distribution curve reaches a maximum at a **wavelength** determined by a **black body** temperature, and approaches zero as the wavelengths increase without limit and as they approach zero. Classical theory does a fairly good job of predicting the shape of this curve for long wavelengths but cannot predict the energy distribution for short wavelengths. In fact, according to classical theory, the radiated power should increase without limit as the wavelength approaches zero. The observation that the rate of emission of a black body approaching zero could only be explained if it were assumed (as in classical theory) that the energy of each mode of vibration of the radiation could not take on all possible values from zero to infinity. Instead, it must be assumed that each mode could only take certain discrete values of the energy, namely $0, \varepsilon_v, 2\varepsilon_v, 3\varepsilon_v \ldots \varepsilon_v = hv$, where v = the **frequency** of the vibrational mode and h = **Planck's constant.** This fundamental energy level is referred to as **quantum** energy and has a value equal to 6.6254×10^{-34} J/s. This was de-

termined by Planck in 1900, and is now referred to as Planck's constant. Another phenomenon, which demonstrated the fact that electromagnetic radiation has particle-like properties, is the **photoelectric effect.** When a weak light wave was used to eject **photoelectrons** from a metallic surface, the energy of the generated photoelectrons was larger than the energy of the impeding light wave. This observation led Einstein, in 1905, to propose the existence of **photons.** Further studies and experimentation confirmed that some light phenomena could be explained satisfactorily by the wave theory, while others can only be explained by quantum theory. In this chapter, the nature of light as a wave and as quantum will be briefly examined.

1.2 ELEMENTS OF GEOMETRIC OPTICS

The fundamental laws establishing the basis for geometric optics are those of **reflection** and **refraction.** Both these laws were extrapolated from experimental results.

Reflection and Refraction

When an optical ray impedes upon a transparent surface that separates two *media* through which the light is propagating with different velocities, the optical ray is divided into two rays: reflected rays and refracted rays (see Figure 1–4).

Where ϕ_1 is the **angle of incidence,** ϕ_2 is the **angle of reflection,** and ϕ_3 is the **angle of refraction.** In Figure 1–4, the reflected angle ϕ_2 is equal to the incident angle ϕ_1. Therefore, when a ray is impeding upon a transparent surface that separates two media that are propagating light with different velocities, the angle of reflection is equal to the incident angle (see Equation (1–1)).

$$\phi_2 = \phi_1 \qquad \textbf{(1–1)}$$

In 1621, Willebrord Snell formulated the law of refraction, which states that the refractive ray lies on the same plane as that of the incident ray and the ratio of the sin ϕ_1 to the sin ϕ_3 maintains a constant value (see Equation (1–2)).

$$\frac{sin\ \phi_1}{sin\ \phi_3} = \frac{n_2}{n_1} = \text{k (constant)} \qquad \textbf{(1–2)}$$

where n_1 is the refractive index of medium 1 and n_2 is the refractive index of medium 2. The refractive index can be established experimentally by measuring the angles ϕ_1 and ϕ_3. Based on the above, Snell's law can be formulated for two media as shown in Equation (1–3).

$$n_1 \sin\ \phi_1 = n_2 \sin\ \phi_3 \qquad \textbf{(1–3)}$$

For small angles, an acceptable approximation can be established (see Equation (1–4)).

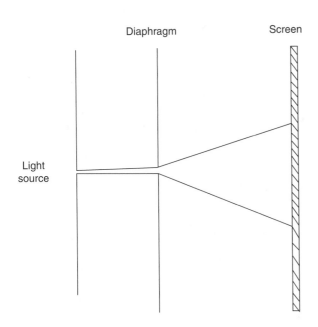

Figure 1–3. Propagation of light through a small hole.

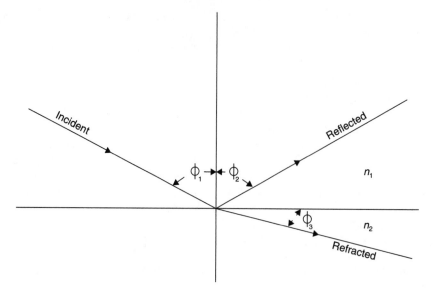

Figure 1–4. Reflected and refracted optical rays.

$$\frac{\phi_1}{\phi_3} = \frac{n_2}{n_1} \qquad \text{(1–4)}$$

The angle between the incident ray and the refracted ray is referred to as the angle of deviation (β). Equation (1–5) expresses this angle of deviation.

$$\beta = \phi_1 - \phi_3 \qquad \text{(1–5)}$$

where β is the angle of deviation, ϕ_1 is the angle of incident, and ϕ_3 is the angle of refraction.

Example 1–1

Calculate the arrival time difference between two monochromatic optical beams that are emitted simultaneously from a common reference point and are traveling an equal distance (d). However, one optical beam travels the distance (d) through air, while the other travels the same distance through air and through another medium with a refractive index of 1.5 and length equal to 5 cm.

Solution
i) Calculate the time (t_1) required for the first optical beam to travel the distance (d) through air.

$$t_1 = \frac{d}{c}$$

where c is the velocity of light (3×10^8 m/s).

ii) Calculate the time (t_2) required for the second beam to travel the distance (d) composed of air and another medium.

$$t_2 = \frac{d - 5 \; cm}{c} + \frac{5 \; cm}{v_2}$$

The velocity of the second beam through the other medium expressed as follows:

$$v_2 = \frac{c}{n_2}$$

where v_2 is the velocity of light through the second medium and n_2 is the refractive index of the second medium (see Equation (1–5)).

$$
\begin{aligned}
t_2 &= \frac{d - 5}{c} + \frac{5}{v_2} \\
&= \frac{d - 5}{c} + \frac{5}{\dfrac{c}{n_2}} \\
&= \frac{d - 5}{c} + \frac{5n_2}{c} \\
&= \frac{(d - 5) + 5(1.5)}{c} \\
&= \frac{d - 5 + 7.5}{c}
\end{aligned}
$$

Therefore, $t_2 = \dfrac{d + 2.5 \; cm}{c}$.

iii) Calculate the time difference (Δt).

$$\Delta t = t_2 - t_1 = \frac{d + 2.5 - d}{c}$$

$$= \frac{2.5}{c} = \frac{2.5 \times 10^{-2} \; m}{3 \times 10^8 \; m/s}$$

$$= 0.83 \times 10^{-10} \; s$$

Therefore, $\Delta t = 83$ ps.

Example 1–2

An optical ray of wavelength of 620 nm is traveling through the air. It is striking a medium with a refractive index equal to 1.333, at an angle of 42°.

Calculate:

- the angle of refraction (θ_2)
- the velocity of the optical ray (v_2) in the second medium
- the wavelength of the optical ray (λ_2) in the second medium

Solution

i) Calculate the angle of refraction, θ_2. Applying Snell's law we have:

$$n_1 \sin \theta_1 = n_2 \sin \theta_2$$

Solve for θ_2.

$$\sin \theta_2 = \frac{n_1 \sin \theta_1}{n_2}$$

or

$$\sin \theta_2 = \frac{(1) \sin 428}{1.333} = 0.5$$

Therefore, $\sin \theta_2 = 0.5$, or $\theta_2 = 30°$.

ii) Calculate the velocity (v_2) of the optical ray through the second medium.

$$v_2 = \frac{c}{n_2} = \frac{3 \times 10^8 \, m/s}{1.333} = 2.25 \times 10^8 \, m/s$$

Therefore, $v_2 = 2.25 \times 10^8$ m/s.

iii) Calculate the wavelength (λ_2) of the optical ray inside the medium.

$$\lambda_2 = \left(\frac{n_1}{n_2}\right)\lambda_o = \left(\frac{1}{1.333}\right)(620 \text{ nm}) = 465.1 \text{ nm}$$

Therefore, $\lambda_2 = 465.1$ nm.

Example 1–3

An optical ray in the air is incident upon a medium with a refractive index equal to 1.56 at an angle of 60°. Calculate:

- The angle of refraction (ϕ_3)
- The angle of deviation (β)

Solution

DATA:

$n_1 = 1$ (refractive index of air)
$n_2 = 1.56$ (refractive index of the medium)
$\phi_1 = 60°$ (angle of incident)

i) Use Snell's law ($n_1 \sin \phi_1 = n_2 \sin \phi_3$) and solve for ϕ_3.

$$\sin \phi_3 = \frac{n_1 \sin \phi_1}{n_2}$$

$$\sin \phi_3 = \frac{(1) \sin (608)}{1.56} = 0.555$$

or

$$\phi_3 = \sin^{-1}(0.555) = 37°$$

Therefore, $\phi_3 = 37°$.

ii) Using Equation (1–5):

$$\beta = 60° - 37° = 23°$$

Therefore, $\beta = 23°$.

The Principle of Reversibility

In Figure 1–4, if the direction of the reflected or refracted (the **direction of propagation**) ray is reversed, each will retrace the optical path. That is, n_1 and n_2 are not altered. This process is referred to as "the *principle of reversibility*."

Optical Path

An **optical path** is defined as the product of the distance (d) traveled by a ray through a medium and the refractive index of that medium (see Equation (1–6)).

$$\text{optical path } [d] = (d)(n) \quad \textbf{(1–6)}$$

where d is the distance and n is the refractive index. The optical path can also be defined for a ray traveling $d_1, d_2, \ldots d_k$ through media with refractive indices $n_1, n_2, \ldots n_k$ respectively (see Equation (1–7)).

$$\text{optical path } [d] = d_1 n_1 + d_2 n_2 + d_3 n_3 + \ldots + d_k n_k$$

or

$$\text{optical path } [d] = \sum_{k=1}^{k} d_k n_k \quad \textbf{(1–7)}$$

The concept of optical path formulates a more general interpretation of the reflection and refraction principle.

Fermat's Principle

In reference to optical path, **Fermat's principle** is the path taken by a light ray traveling from one point to another through any set of media such as to render its optical path equal in the first approximation to other paths closely adjacent to the actual one. Furthermore, the time required by the light to travel the path is a minimum and the optical path is a measure of this time. Both laws of reflection and refraction follow Fermat's principle. The mathematical verification of Fermat's principle as applied to the laws of reflection and

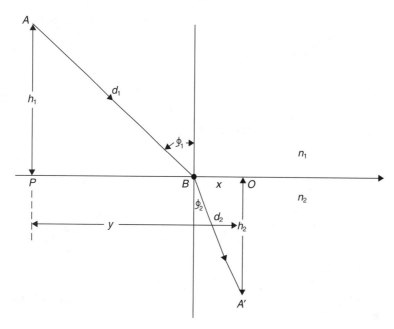

Figure 1–5. Fermat's principle.

refraction can be demonstrated as follows in Figure 1–5, assuming an incident ray falls upon a surface separating two media with refractive indices n_1 and n_2. The optical path between points A and A', passing through point B and two media with refractive indices n_1 and n_2, is expressed by Equation (1–8).

$$d = n_1 d_1 + n_2 d_2 \qquad (1\text{–}8)$$

where d is the optical path, d_1 is the distance AB, d_2 is the distance BA', n_1 is the refractive index of medium 1, and n_2 is the refractive index of medium 2. Let h_1 and h_2 be perpendiculars from A and A' to the refractive surface, and let y be the distance between OP. Applying the Pythagorean theorem, we have Equation (1–9) and Equation (1–10).

$$d_1^2 = h_1^2 + (y - x)^2 \qquad (1\text{–}9)$$

$$d_2^2 = x^2 + h_2^2 \qquad (1\text{–}10)$$

Substituting Equation (1–9) and Equation (1–10) into Equation (1–8) yields Equation (1–11).

$$d = n_1 [h_1^2 + (y - x)^2]^{1/2} + n_2 [x^2 + h_2^2]^{1/2} \quad (1\text{–}11)$$

In accordance with Fermat's principle, the value of d (the optical path) must be a maximum or a minimum. Differentiating Equation (1–11) in reference to y, the slope of a corresponding graph can be obtained. By setting the equation as equal to zero, the point at y can be established where the slope of the curve is equal to zero (see Equation (1–12)).

$$\frac{d(d)}{dy} = \frac{0.5 n_1}{[h_1^2 + (y - x)^2]^{1/2}}(-2y + 2x) + \frac{0.5 n_2}{(h_2^2 + x^2)^{1/2}}(2x) = 0$$

$$(1\text{–}12)$$

or

$$(n_1)\frac{y - x}{[h_1^2 + (y - x)^2]^{1/2}} = (n_2)\frac{x}{(h_2^2 + x^2)^{1/2}}$$

or

$$(n_1)\frac{y - x}{d_1} = (n_2)\frac{x}{d_2}$$

From Figure 1–5,

$$\frac{y - x}{d_1} = \sin \phi_1 \text{ and } \frac{x}{d_2} = \sin \phi_2$$

or

$$n_1 \sin \phi_1 = n_2 \sin \phi_2.$$

Same as Equation (1–3).

Therefore, Fermat's principle of the optical path has been mathematically verified.

Example 1–4

If a light ray in a medium with a refractive index (n_1) equal to 1.05 is incident upon the surface of a medium with a refractive index (n_2) of 1.5, with corresponding distances h_1

and h_2 of 12.5 cm each, and the distance BD equal to 5 cm, calculate the optical path (d). See the following diagram.

Solution

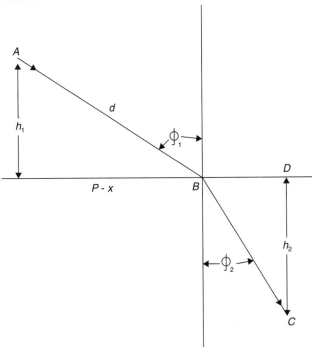

Schematic for Example 1–4.

Calculate optical path (d).

$$(d) = \sum_{k=1}^{k} d_k n_k$$

From Equation (1–11),

$$(d) = d_1 + d_2 = n_1 \left[h_1^2 + (p - x)^2\right]^{1/2} + n_2 \left(h_2^2 + x^2\right)^{1/2}$$

Calculate ($p - x$).

From the schematic for Example 1–4,

i) $\tan \phi_2 = \dfrac{(BD)}{(DC)} = \dfrac{5}{12.5} = 0.4$

or

$$\phi_2 = \tan^{-1}(0.4) = 21.8°$$

$$\therefore \phi_2 = 21.8°$$

and

$$\sin \phi_2 = 0.37$$

ii) Applying **Snell's law,** solve for ϕ_1.

$$n_1 \sin \phi_1 = n_2 \sin \phi_2$$

$$\sin \phi_1 = \frac{n_2 \sin \phi_2}{n_1} = \frac{(1.5)(0.37)}{1.05} = 0.53$$

$$\sin \phi_1 = 0.53$$

or

$$\phi_1 = \sin^{-1}(0.53) = 32°$$

$$\therefore \phi_1 = 32°$$

$$\tan \phi_1 = \frac{(p - x)}{12.5}$$

or

$$p - x = (12.5) \tan 32° = 7.8 \text{ cm}$$

$$(d) = n_1 \left[h_1^2 + (p - x)^2\right]^{1/2} + n_2 \left(h_2^2 + x^2\right)^{1/2}$$

$$(d) = 1.05[12.5^2 + 7.8^2]^{1/2} + 1.5(12.5^2 + 5^2)^{1/2}$$

$$= 15.47 \text{ cm} + 20.19 \text{ cm}$$

$$= 35.66 \text{ cm}$$

Therefore, the optical distance (d) = 35.66 cm.

1.3 ELEMENTS OF PHYSICAL OPTICS

Physical optics deals with the nature of light. That is, it deals with the interaction of light waves with matter (**emission** and **absorption**). If these optical phenomena are to be completely and accurately interpreted, a knowledge of **quantum mechanics** is necessary. The depth and breadth of quantum mechanics is far beyond the scope of this book. However, elements of quantum optics will be presented at the end of this chapter.

Light as an Electromagnetic Wave

The **wave theory of light** will be used to interpret the interaction of a wave with another wave. In 1814, Fresnel attempted to explain the diffraction and interference of light using wave theory, and he achieved some success. Fresnel's so-called **elastic theory** could not be fully supported because the knowledge of light as a **transverse electromagnetic wave** was unavailable before 1864. That year J. C. Maxwell (1831–1879) published a paper with the title *A Dynamic Theory of the Electromagnetic Field* in which he stated that light is a transverse wave with the fluctuating magnetic and electric fields perpendicular to each other and perpendicular to the direction of propagation. Maxwell presented his theory in the form of four equations, which summarizes the relationship between electricity and magnetism. These equations, expressed in differential Equations (1–13), (1–14), (1–15), and (1–16) were the starting point for further investigations into the wave nature of light.

First set:

$$\frac{1}{c} \frac{\partial E_x}{\partial t} = \frac{\partial H_z}{\partial y} - \frac{\partial H_y}{\partial z} \qquad \textbf{(1–13)}$$

$$\frac{1}{c}\frac{\partial E_y}{\partial t} = \frac{\partial H_x}{\partial z} - \frac{\partial H_z}{\partial x} \qquad \textbf{(1–14)}$$

$$\frac{1}{c}\frac{\partial E_z}{\partial t} = \frac{\partial H_y}{\partial x} - \frac{\partial H_x}{\partial y} \qquad \textbf{(1–15)}$$

Equations (1–13), (1–14), and (1–15) state that an **electric field** generates a **magnetic field.** This is known as **Ampere's law,** while Equations (1–16), (1–17), and (1–18) also include **Faraday's law** of induced **electromotive force.**

Second set:

$$-\frac{1}{c}\frac{\partial E_x}{\partial t} = \frac{\partial H_z}{\partial y} - \frac{\partial H_y}{\partial z} \qquad \textbf{(1–16)}$$

$$-\frac{1}{c}\frac{\partial E_y}{\partial t} = \frac{\partial H_x}{\partial z} - \frac{\partial H_z}{\partial x} \qquad \textbf{(1–17)}$$

$$-\frac{1}{c}\frac{\partial E_z}{\partial t} = \frac{\partial H_y}{\partial x} - \frac{\partial H_x}{\partial y} \qquad \textbf{(1–18)}$$

Third:

$$\frac{\partial E_x}{\partial x} + \frac{\partial E_y}{\partial y} + \frac{\partial E_z}{\partial z} = 0 \qquad \textbf{(1–19)}$$

Fourth:

$$\frac{\partial H_x}{\partial x} + \frac{\partial H_y}{\partial y} + \frac{\partial H_z}{\partial z} = 0 \qquad \textbf{(1–20)}$$

The partial differential equations relate the electric field (E) and magnetic field (H) of an electromagnetic wave in space and time (t). More specifically, the right-hand sides of the equations define the magnetic field distribution in space while the left-hand sides define the rate of change in time of the electric field. In accordance with the equations, a change in the electric field in time produces a magnetic field. Likewise, a change in the magnetic field in time generates an electric field. Equation (1–19) expresses the idea that no free electric charges are present in a vacuum. This concept is substantiated by Equation (1–20), which expresses the impossibility of a free magnetic pole.

Wave Motion

An electromagnetic wave propagating through free space is illustrated in Figure 1–6. The electric field is perpendicular to the magnetic field, and both are perpendicular to the direction of propagation. A two-dimensional transverse wave traveling in the x direction and vibrating in the $x y$ plane can be expressed by Equation (1–21).

$$y = f(x) \qquad \textbf{(1–21)}$$

Assuming that the wave propagates in the x direction with a constant velocity (u), the value of y will also change with time. If two values, y_1 and y_2, are taken, these values can be represented as follows in Equations (1–22) and (1–23).

$$y_1 = f(x - ut) \qquad \textbf{(1–22)}$$

Because distance is $x = ut$, then $\Delta x = u\Delta t$, or $y_2 = f[(x + u\Delta t) - u(t + \Delta t)]$

$$y_1 = f[x + u\Delta t - ut - u\Delta t]$$
$$= f(x - ut)$$

Therefore,

$$y_1 = y_2 \qquad \textbf{(1–23)}$$

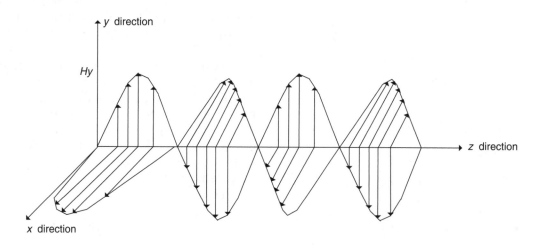

Figure 1–6. Wave motion in free space.

From Equation (1–23), it is evident that a transverse wave in a plane can be represented as follows in Equation (1–24).

$$y = f(x \pm ut) \tag{1–24}$$

where $(+)\rightarrow$ represents forward wave propagation and $(–)\rightarrow$ represents reverse wave propagation.

Equation (1–24) is referred to as the **wave equation.** The solution of Equation (1–24) is applied to waves traveling along the x-axis and perpendicular to y-axis as follows in Equation (1–25).

$$\frac{\partial^2 y}{\partial t^2} = u^2 \frac{\partial^2 y}{\partial x^2} \tag{1–25}$$

The solution of $y = f(x \pm ut)$ yields Equation (1–26).

$$\frac{dy}{dx} = -uf'(x - ut)$$

and

$$\frac{\partial^2 y}{\partial t^2} = u^2 f''(x - ut) \tag{1–26}$$

Equation (1–26) expresses the acceleration of a particle at time (t). Differentiating with respect to x yields Equation (1–27).

$$\frac{\partial^2 y}{\partial x^2} = f''(x - ut) \tag{1–27}$$

Equation (1–27) represents the wave curvature progression with time t. At time $t + \Delta t$, the waveform will be at the same displacement of distance ut. That is, the particles of the medium through which the wave is traveling are not moving along with the wave but are oscillating around their position of equilibrium.

The Velocity of Light

Because optical waves of defined frequency will travel through a medium with a finite velocity, electromagnetic waves (light waves) travel through space with the same velocity for all frequencies. The velocity of light in space is defined by the constant c. The first attempt to compute this very important constant in nature was made by Röemer (1644–1710), a Danish astronomer, in 1676. Using the orbits of Earth and Jupiter and the distance between the Earth and the Sun, he was able to compute the velocity of light as approximately 292,000 km/s. In 1727, the English astronomer James Bradley (1693–1764) used **aberration,** the motion of the stars, which he believed to be caused by the earth's orbital motion. Bradley found star phase displacement to be equal to $\pi/2$ in relation to the Earth's motion. His calculations placed the velocity of light at 299,714 km/s, a very close approximation. In 1849, a French physicist named H. L. Fizeau (1819–1896) measured the velocity of light through terrestrial means. His method set the velocity of light at 313,200 km/s, higher than the speed previously calculated. Refinement of Fizeau's experiment by Young and Forbes established c at 301,400 km/s. In 1850, J. L. Faucoult (1814–1868) calculated the velocity of light to be equal to 298,000 km/s. In 1926, Michelson, experimenting at Mt. Wilson observatory, calculated the velocity of light in a vacuum to be 299,776 km/s. In the 1950s, Bol, Essen, Aslakson, and Froome conducted further experiments establishing the velocity of light at 299,792 km/s. In all practical calculations for this book, the velocity of light will be considered 300,000 km/s.

The Sinewave

A *sinewave* is a periodic waveform (repeats at regular intervals) described by frequency, **period,** and peak amplitude. See Figure 1–7. The mathematical expression of a sinewave is given in Equation (1–28).

$$y = a \sin \frac{2\pi x}{\lambda} \tag{1–28}$$

where a is the amplitude (maximum displacement) and λ is the wavelength. If the wave travels in the $x\rightarrow$ direction, in time t, the Equation (1–28) is modified as follows in Equation (1–29).

$$y = a \sin \frac{2\pi x}{\lambda}(x - ut) \tag{1–29}$$

Equation (1–29) indicates that the contour of the wave is displaced in the $x\rightarrow$ direction with velocity u (see Equation (1–30)), while any point along the contour will perform harmonic oscillations. The number of vibrations per second of any point along the contour, defined within a full period, is referred to as the frequency (f) of the wave (see Equation (1–31)). The period (T) of the wave is equal to the reciprocal of the frequency shown in Equation (1–32). The velocity of the propagation of the wave in terms of frequency and wavelength is expressed in Equation (1–33).

$$u = \frac{\lambda}{T} \tag{1–30}$$

where u is the velocity and λ is the wavelength.

$$f = \frac{1}{T} \tag{1–31}$$

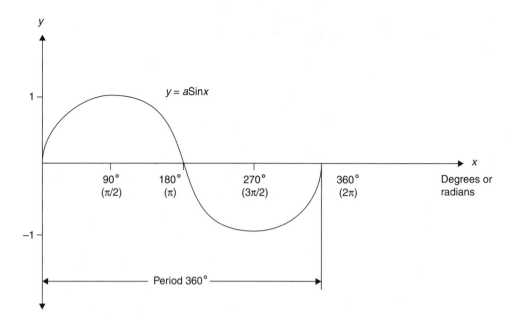

Figure 1–7. The sinewave.

or

$$T = \frac{1}{f} \qquad (1\text{–}32)$$

Therefore,

$$u = \lambda f \qquad (1\text{–}33)$$

The complete expression of a sinewave is given by Equations (1–34) and (1–35).

$$y = a \sin (\omega t - kx) \qquad (1\text{–}34)$$

or

$$y = a \cos(\omega t - kx) \qquad (1\text{–}35)$$

where $\omega = 2\pi f$ (angular velocity) and $k = \dfrac{2\pi}{\lambda}$ (number of waves per unit distance of 2π).

Wave Phase and Phase Difference

A closer examination of Equations (1–34) and (1–35) indicates that the phase of the wave varies proportionally to the x direction. For a complete oscillation, the phase is shifted by 2π. The **phase difference** between two points at x_1 and x_2 along the x direction is expressed as follows in Equation (1–36).

$$\delta = k(x_2 - x_1) = \frac{2\pi}{\lambda}(x_2 - x_1)$$

Because $k = \dfrac{2\pi}{\lambda}$, then

$$\delta = \frac{2\pi}{\lambda}(x_2 - x_1) \qquad (1\text{–}36)$$

where δ is the phase difference. Knowledge of the phase difference between two light waves is essential in several applications, including fiber optics communications. If a beam of monochromatic light is divided into two beams, which both travel different paths but are then recombined at some predetermined point, the intensity of the resulting beam is determined by the phase difference of the two beams at the point of recombination. However, the phase difference is also subject to the optical distance traveled by the two beams. This fact is widely used in optical modulation. It is essential to emphasize once more the difference between a geometric path and an optical path. For an optical path, the distance is the product of the geometric path and the refractive index of the medium through which the wave is traveling.

Phase Velocity

The velocity with which a wave crest moves along the x axis is called the wave velocity or phase velocity. This velocity is the same as that identified in Equations (1–30) and (1–33) and represents the rate of change of the x coordinate under the assumption of a constant phase ($\omega t - kx$ constant). The phase velocity of a traveling wave through the x direction is derived as follows in Equation (1–37).

Assuming $\omega t - kx$ = constant, then

$$u = \frac{dx}{dt} = \frac{\omega}{k}$$

Because $\omega = 2\pi f$ and $k = \dfrac{2\pi}{\lambda}$, then

$$u = \frac{2\pi f}{\dfrac{2\pi}{\lambda}} = \lambda f$$

Therefore,

$$u = \lambda f \qquad \textbf{(1–37)}$$

Equation (1–37) is identical to Equation (1–33) (wave propagation velocity).

However, the ratio $\dfrac{2\pi f}{\lambda}$ for a specific wave depends on the physical properties of the medium and the frequency of the wave traveling through that medium.

Example 1–5

Calculate the phase velocity (u) of an optical ray of wavelength 830 nm traveling through a medium with a refractive index of 1.5.

Solution

i) Using Equation (1–37), compute λ.

$$\lambda = \frac{830 \text{ nm}}{1.5} = 553 \text{ nm}$$

Therefore, $\lambda = 553$ nm.

ii) Compute f.

$$f = \frac{c}{\lambda} = \frac{3 \times 10^8 \text{ m/s}}{553 \times 10^{-9} \text{ m}} = 542 \text{ THz}$$

Therefore, $f = 542$ THz.

iii) Compute u.

$$u = \lambda f = (553 \times 10^{-9} \text{ m})(542 \times 10^{12} \text{ c/s}) = 2.997 \; 10^8 \text{ m/s}$$

Therefore, phase velocity: $u = 2.997 \times 10^8$ m/s.

Wave Intensity

Waves are energy transporters. The amount of energy flowing through a unit area per unit time is referred to as the *intensity* of the wave. If the wave flows with a constant phase velocity (u), a defined energy density per unit area exists. The intensity of a wave is then given as the product of phase velocity and **energy density.** If a point is considered along the contour of a sinewave, and is moving along the contour, it will possess both potential and **kinetic energies.** However, at any point along the contour, the sum of both kinetic and potential energies remains constant. For example, if the kinetic energy of the point is at its maximum, its potential energy is equal to zero, and vice versa. Considering Equation (1–32), the displacement of the point as a function of time is expressed by Equation (1–34).

The velocity of the point is calculated as follows in Equation (1–38).

From Equation (1–34),

$$\frac{dy}{dt} = \omega a \cos(\omega t - k) \qquad \textbf{(1–38)}$$

For $y = 0$, the cosine takes a maximum value with a velocity equal to $-\omega a$ and kinetic energy given by Equation (1–39).

$$E = \frac{1}{2}m\left(\frac{dy}{dt}\right)^2 = \frac{1}{2}m\omega^2 a^2 \qquad \textbf{(1–39)}$$

$\dfrac{1}{2}m\omega^2 a^2$ represents the sum of both potential and kinetic energies per unit volume. Therefore, the energy density is proportional to $\omega^2 a^2$.

Frequency and Wavelength

An oscillating source generates a wave that has the same frequency as that source. The wavelength of a wave in a medium is a function of the velocity of the wave in that medium. The relationship of the frequency, wavelength, and velocity of a wave in a medium is expressed by Equation (1–40).

$$\lambda = \frac{c}{f} \qquad \textbf{(1–40)}$$

where λ is the wavelength, c is the velocity, and f is the frequency. If a wave passes from one medium to another, its wavelength will change at a rate that is proportionate to its velocity in the new medium. Because wavelengths are proportional to velocity, the ratio of the wavelength of a wave traveling in a vacuum to the wavelength of the same wave traveling through another medium is expressed by Equation (1–41).

$$\frac{\lambda}{\lambda_m} = \frac{c}{u} = n \qquad \textbf{(1–41)}$$

where λ is the wavelength in vacuum, λ_m is the wavelength in a medium, u is the velocity in a medium, n is the refractive index, and c is the velocity in vacuum. From this list, it can be stated that the optical path of an optical ray can also be expressed as follows in Equation (1–42).

$$\left(\frac{\lambda}{\lambda_m}\right)d = nd \qquad (1\text{–}42)$$

where nd is the optical path, as in Equation (1–6). The wavelength of visible light covers the range between 4×10^{-5} cm (violet), to 7.2×10^{-5} cm (red).

Example 1–6

A light ray of wavelength 720 nm is traveling through a medium with a refractive index of 1.5.

Calculate the wavelength (λ_m) of the light ray in the medium and the velocity (u) of the light ray in the medium.

Solution

i) Using Equation (1–41), calculate and solve for λ_m.

$$\lambda_m = \frac{\lambda}{n} = \frac{720 \text{ nm}}{1.5} = 480 \text{ nm}$$

Therefore, $\lambda_m = 480$ nm.

ii) Calculate u, using Equation (1–41). Solve for u.

$$u = \frac{c}{n} = \frac{3 \times 10^8 \text{ m/s}}{1.5} = 2 \times 10^8 \text{ m/s}$$

Therefore, $u = 2 \times 10^8$ m/s.

Example 1–7

A light ray with a wavelength of 640 nm is traveling through a medium at a velocity of 2.2×10^8 m/s.

Calculate the wavelength (λ_m) of the light ray in the medium and refractive index (n) of the medium.

Solution

i) Using Equation (1–41), compute and solve for λ_m.

$$\lambda_m = \frac{u\lambda}{c} = \frac{(2.2 \times 10^8 \text{ m/s})(640 \text{ nm})}{3 \times 10^8 \text{ m/s}} = 496.3 \text{ nm}$$

Therefore, $\lambda_m = 496.3$ nm.

ii) Using Equation (1–41), compute n.

$$n = \frac{\lambda}{\lambda_m} = \frac{640 \text{ nm}}{496.3 \text{ nm}} = 1.289$$

Therefore, $n = 1.289$.

1.4 ABSORPTION

The intensity and velocity of light waves traveling through solid, liquid, or gaseous matter are altered to a degree determined by the physical and molecular properties of that mat-

ter. A decrease in intensity is caused by absorption and **scattering,** while the change in the wave velocity is caused by **dispersion.** Loss of intensity is defined as the decay of the intensity of the incident beam traveling through matter. That is, the intensity of all the wavelengths in the beam is reduced by the same amount. This process is referred to as **general absorption.** However, through the same process, a reduction in the intensity of some wavelengths composing the light wave can also be achieved. This method is referred to as **selective absorption.**

The relationship between input and output intensities is expressed by Equation (1–43).

$$I_o = I_{in}e^{-ad} \qquad (1\text{–}43)$$

where I_o is the intensity of the optical beam at the output, I_{in} is the intensity of the optical beam at the input, a is the absorption coefficient, and d is the distance between input and output.

The **absorption coefficient** is a measure of the relative decrease in intensity per unit length of the light beam. However, a closer examination of the intensity loss mechanism reveals that the decrease of the light intensity depends on two factors. One factor is the conversion of the light energy into thermal energy and the other is on the scattering of light waves by the molecules of the matter in any direction other than that of the forward direction of the light beam. Therefore, a more accurate representation of the previous expression is given by Equation (1–44).

$$I_o = I_{in}e^{-(a_a + a_s)d} \qquad (1\text{–}44)$$

where a_a is the absorption coefficient due to thermal motion and a_s is the absorption coefficient due to scattering.

Example 1–8

Calculate the ratio of input/output intensities (I_o/I_{in}) of the optical beam traveling in parallel to the center axis of a 25 cm tube filled with hydrogen gas.

Solution

i) Determine ($a_a + a_s$), using Equation (1–44). From graphs: The combined attenuation due to absorption (a_a) and scattering (a_s) of the hydrogen gas is equal to 6.3. (Please research and obtain related graphs for various gases.)

ii) Calculate the ratio I_o/I_{in}.

$$\frac{I_o}{I_{in}} = e^{-(a_a + a_s)d} = e^{-(6.3)(0.25m)} = 0.207$$

Therefore, $\dfrac{I_o}{I_{in}} = 0.207$, or $I_o = 0.207I_{in}$.

That is, only 20% of the optical beam intensity at the input of the tube will emerge at the output.

Example 1–9

If the ratio of the output to the input intensities I_o/I_{in} of an optical beam traveling parallel to the main axis of a tube filled with helium gas is 0.15, compute the length (d) of the tube.

Solution

Solve for d, using Equation (1–44).

$$\frac{I_o}{I_{in}} = e^{-(a_a + a_s)d}$$

or

$$-[(a_a + a_s)d] = \ln\left(\frac{I_o}{I_{in}}\right)$$

$$-63.1d = -1.9$$

or

$$d = \frac{1.9}{63.1} = 3 \text{ cm}$$

Therefore, the length of tube $d = 3$ cm.

Absorption and Radiation

When light waves strike matter containing bound charged particles capable of vibrating with frequencies equal to those of the incident light wave, the electric field component of the impeding wave will act upon the charges with force eE. If the electric field component changes with a frequency equal to the natural frequency of the charged particles, resonance will be observed. The charged particle will radiate electromagnetic radiation of the same frequency. However, if charged particles are in close proximity to each other, the particles will interact with each other as well as with the electric field of the incident light ray. The reemitted radiation from all charged particles will be in phase. The generated wavefront will constitute the reflective wave with an angle of reflection equal to the angle of incidence (**Huygens' principle of reflection**). Selective reflection results from the phenomenon of resonance occurring at wavelengths almost equal to the vibrating frequency of the charged particles composing the matter.

1.5 SCATTERING

Scattering is the phenomenon whereby the frequency, direction, or polarization of an incident wave changes with a random change in energy distribution. This occurs when the wave is interacting with the atomic or molecular structure of the matter through which it travels. The interaction of the electric field component (E) of an **incident wave** with a charged particle results in particle motion. If the frequency of the incident wave is close to that of the natural frequency of the particle, **resonance** will occur. Through this process, substantial optical energy will be lost to absorption. On the other hand, if the frequency of the incident wave is close but not equal to the **particle natural vibrating frequency,** optical energy will be lost to scattering. This results from the forced vibrations upon the charged particles. These vibrations will have the same frequency and direction as those of the electric field component of the incident wave, while the phase of the two waves will be different. However, the amplitude of the scattered signal will be very small compared to the resonant signal in the previous case. It is evident, then, that the velocity of light will be different in a medium other than free space. In 1871, Lord Rayleigh investigated the phenomenon of scattering. He was followed by John Tyndall (1820–1893) and V. C. Raman.

RAYLEIGH SCATTERING. When optical waves travel through a medium, they normally follow a forward path. However, a small part of the optical energy is scattered by the media **refractive index** inhomogeneities. **Rayleigh scattering** is referred to as the scattering of light caused by the material structure's inhomogeneities (material nonuniformity). Small refractive index fluctuations caused by the materials nonuniformity scatter light in all directions without altering the frequency. Rayleigh scattering is illustrated in Figure 1–8.

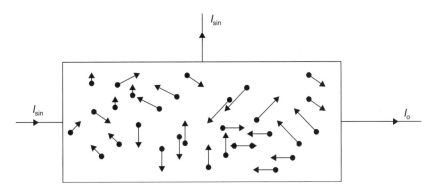

Figure 1–8. Rayleigh scattering.

The intensity of scattered light in a medium surrounded by another medium that has a different refractive index is proportional to the intensity of the incident wave and also proportional to the square of the volume of the scattered particles when the particle linear dimensions are smaller than those of the wavelength of the incident wave. The relationship between the intensity of scattered waves and the wavelength of incident waves is illustrated in Figure 1–9.

From Figure 1–9, it is evident that the relative intensity of the scattered wave from charged particles with specified linear dimensions will be higher for shorter wavelengths than for those scattered wave intensities of longer wavelengths. Rayleigh established the mathematical relationship between the intensity of the scattered wave and wavelength as follows in Equation (1–45).

$$I_{sw} \propto \frac{1}{\lambda^4} \qquad \textbf{(1–45)}$$

where I_{sw} is the intensity of the scattered wave and λ is the wavelength of the incident wave.

That is, the intensity of the scattered wave is proportional to $(1/\lambda^4)$ of the incident wave.

RAMAN SCATTERING. When optical energy is scattered from a molecule, the majority of the photons are **elastic scattering.** That is, they have the same frequency and wavelength as the incident photons. However, some photons (one in every ten million) are scattered at lower frequencies than those of the incident photons. This process, leading to **inelastic scattering,** is referred to as the **Raman effect.** When a photon strikes a molecule, the interaction between the photon and the molecule alters the electric field of that molecule. Classical mechanics interprets this as a change in

the molecule's electric field. Quantum mechanics interprets it as a change of a molecule's vibrational state from a higher to a lower excitation level. **Raman scattering** occurs within the time frame of 1×10^{-14} seconds. The energy levels of Raman scattering are illustrated in Figure 1–10.

The energy difference between the initial and final vibrational levels of the molecule are expressed in Equation (1–46).

$$\bar{\nu} = \frac{1}{\lambda_{\text{inc.}}} - \frac{1}{\lambda_{\text{sct.}}} \qquad \textbf{(1–46)}$$

where $\bar{\nu}$ is the vibrational energy level difference, λ_{inc} is the wavelength of the incident photon, and λ_{sct} is the wavelength of the scattered photon.

Energy levels are not dissipated, but energy is. An increase in the overall temperature of the material could be anticipated. However, because it is very small, it does not cause any *measurable* temperature increase. The vibrational energy is a function of molecular parameters such as mass, molecular geometry, bond order and so on. These factors affect the vibrational constant, and consequently, the vibrational energy. Based on classical electromagnetic theory, when an external electric field (E) is applied to a molecule, a dipole moment (P) proportional to the applied electric field will be induced (see Equation (1–47)).

$$P = aE \qquad \textbf{(1–47)}$$

where P is the **dipole moment,** E is the electric field, and a is the constant of proportionality (polarizability of the molecule). Polarizability of the molecule defines the ease with which the electronic cloud of the molecule can be distorted.

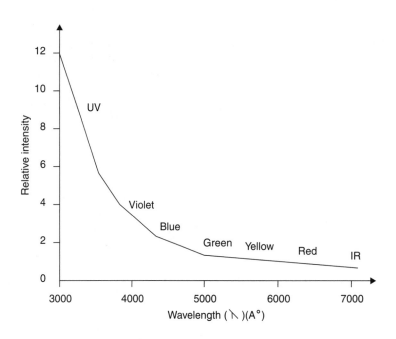

Figure 1–9. Relative intensity of scattered waves in relation to the wavelengths of incident waves.

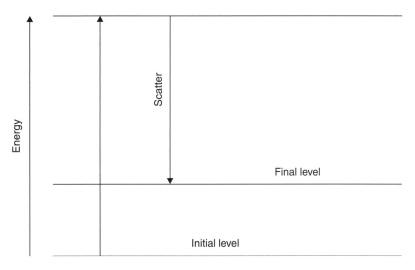

Figure 1–10. Raman scattering energy levels.

The induced dipole is capable of scattering light energy at the same frequency as that of the incident wave or at frequencies usually lower than the incident frequency. Therefore, Raman scattering is the result of increased molecular vibrations caused by an external electric field capable of altering the molecular polarizability $\partial a/\partial Q$, Q being the normal coordinate of the molecular vibration. For a change of the polarizability of a molecule to occur, the following relationship must be satisfied (see Equation (1–48)).

$$\frac{\partial a}{\partial Q} \neq 0 \qquad \textbf{(1–48)}$$

That is, the polarizability of the molecule must be a number other than zero for the normal coordinate under the influence of an external electric field, while the intensity of Raman scattering is expressed as the square of the molecular polarizability (see Equation (1–49)).

$$\left(\frac{\partial a}{\partial Q}\right)^2 \qquad \textbf{(1–49)}$$

It is clear that if the vibration caused by the application of an external electric field to the molecule does not significantly change the polarizability of the molecule, Raman scattering intensity will be very low.

BRILLOUIN SCATTERING (SPONTANEOUS). Before discussing **Brillouin scattering,** a brief description of acoustic waves is essential. In practically all dielectric materials—including silica—used in the fabrication of optical fibers, the refractive index changes under the influence of a strong electric field. When an optical wave travels through a medium composed of silica, its index of refraction will increase proportionally to the intensity of the optical wave. This increase of the material's index of refraction, referred to as **electrostriction,** contributes partially to the **Kerr effect**—two

optical waves of different frequencies traveling through a silica medium in opposite directions. In this process, the superposition of the two waves will generate **interference** fringes composed of high intensity and low intensity electric fields. Because of the varying refractive index within the medium (electrostriction), compression zones will be created with velocities equal to the frequency difference of the opposite traveling optical waves. By appropriately selecting the frequencies of these waves, the frequency difference can fall within the frequency range of sound waves. These waves are referred to as acoustic waves. The process of acoustic wave generation is illustrated in Figure 1–11.

If an acoustic wave travels through a transparent material, it will induce refractive index variations, resulting in the scattering of a small portion of the incident wave. This scattered wave has a frequency that is slightly shifted. This phenomenon is referred to as **spontaneous Brillouin scattering.** Applying the concept of quantum mechanics, Brillouin scattering can be described with the assistance of Figure 1–12.

From Figure 1–12, a photon traveling through a transparent medium is transformed to a lower frequency. This is referred to as a **Stokes photon** or an **acoustic wave photon.** Although the output energy is equal to the input energy, the frequency shift in the Stokes optical wave is determined by the characteristics of the photons that compose the optical wave.

STOKES AND ANTI-STOKES PHOTONS. When photons of the incident light interact with the vibrational and rotational energy in the molecules composing the matter, a new photon of light that has less energy than the original photon will be released. However, if the molecules interacting with the incident photons were previously excited, the new photon (an **anti-Stokes photon**) will have more energy than the incident photon.

BRILLOUIN SCATTERING (STIMULATED). When a light wave travels through an electrostrictive material, it will

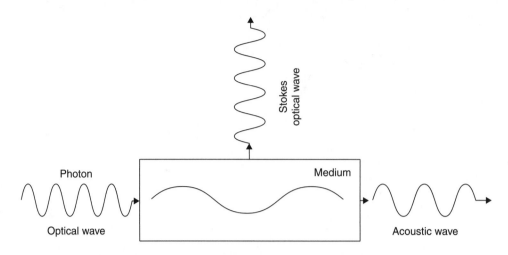

Figure 1–11. The generation of an acoustic wave.

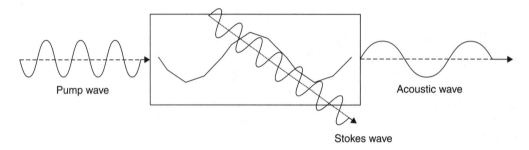

Figure 1–12. Brillouin scattering (spontaneous).

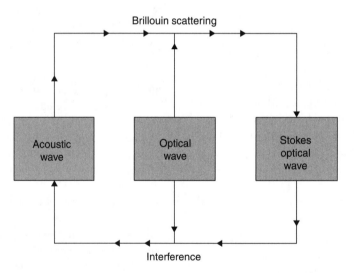

Figure 1–13. The process of stimulated Brillouin scattering.

interact with the acoustic noise generated by the molecular thermal noise. Through this interaction, a portion of the optical wave will be backscattered as Stokes light, and travel in a direction opposite from the incident wave. This Stokes light will generate an acoustical wave that stimulates further Brillouin scattering (see Figure 1–13).

    Stimulated Brillouin scattering can be effectively observed by watching the moment that two optical waves of different frequencies are allowed to travel in an optical fiber in opposite directions (see Figure 1–14).

The gain curve exhibits a bandwidth of between 30 MHz and 50 MHz with a maximum amplitude from 10 GHz to 12 GHz, depending on the wavelength of the traveling waves and fiber material composition. By this method, optical fiber characteristics such as temperature changes, material density, and doping composition can be effectively established.

THE RELATIONSHIP BETWEEN SCATTERING AND THE REFRACTIVE INDEX.   A direct consequence of scattering is the change of the velocity of light in the medium. The mole-

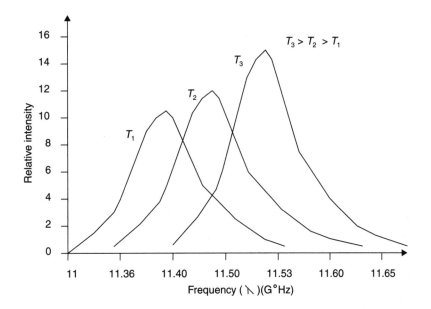

Figure 1–14. Stimulated Brillouin optical gain curves.

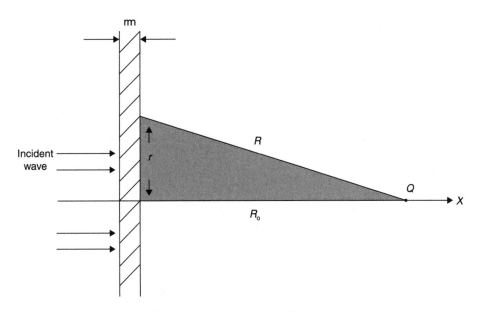

Figure 1–15. An optical beam scattered by a small, optically-transparent membrane.

cules that compose the matter scatter some of the optical energy, and the resulting light waves interfere directly with the original wave, altering its phase and velocity. Assume an optical beam is intercepted by an optically-transparent membrane that is very thin when compared to the wavelength of the optical beam (see Figure 1–15).

The electric vector component of the original optical wave, assuming unity amplitude, is expressed by Equation (1–50).

$$E = e^{ikx} \qquad \textbf{(1–50)}$$

where k is an integer.

From Figure 1–15, at a point (Q) along the x axis the total wave intensity is the sum of the original wave intensity

and the intensity of the scattered wave. In Equation (1–44), the intensity of the optical beam was expressed.

This equation is a measure of optical intensity loss due to absorption and scattering when the light passes through a medium of thickness (d). This scattering intensity loss is proportional to the thickness of the transmission medium (see Equation (1–51)).

$$-\frac{dI}{I} = a_s d \propto I_s \qquad \textbf{(1–51)}$$

where I is the intensity of the wave, a_s is the **scattering coefficient,** d is the thickness of the optically transparent medium, and I_s is the intensity of the scattered wave.

Because there are Nm atoms per unit area, the intensity of a single atom is expressed by Equation (1–52).

$$I_1 \propto \frac{am}{Nm} \qquad \textbf{(1–52)}$$

The amplitude of the scattering atom is expressed by Equation (1–53).

$$E_1 \propto \left(\frac{a_s}{N}\right)^{1/2} \qquad \textbf{(1–53)}$$

Equations (1–52) and (1–53) represent the intensity and amplitude of **incoherent scattering.** For coherent scattering (Rayleigh scattering), the intensities are replaced by amplitudes. Therefore, the coherent amplitude of a scattered wave is expressed by Equation (1–54).

$$E_s \propto Nm\sqrt{\frac{a_s}{N}}$$

or

$$E_s \propto m\sqrt{\frac{a_s N^2}{N}}$$

or

$$E_s \propto m\sqrt{a_s N} \qquad \textbf{(1–54)}$$

Integrating Equation (1–54), and adding the amplitude of the original signal, yields the composite wave, including the original and the scattered signals at point Q (see Figure 1–15 and Equation (1–55)).

$$E_0 + E_s = e^{ikR_0} + m\sqrt{a_s N}\int_0^\infty \frac{2\pi rdr}{R}e^{ikR} \qquad \textbf{(1–55)}$$

The fraction $1/R$ represents the **inverse square law.** Because

$$R^2 = R_0{}^2 + r^2 \text{ and } rdr = RdR$$

Then,

$$\int_0^\infty \frac{2\pi}{R}e^{ikx}\,rdr$$

$$= 2\pi\int_{R_0}^\infty e^{ikR}\,dR$$

$$= \frac{2\pi}{jk}\left[e^{ikR}\right]_{R_0}^\infty$$

The scattering $R \to \infty$ does not contribute to a coherent wave.

Therefore,

$$E + E_s = e^{ikR_0} - m\sqrt{a_s N}\left(\frac{\lambda}{i}\right)e^{ikR_0}$$

$$= e^{ikR_0} + m\sqrt{a_s N}.i\lambda.e^{ikR_0}$$

$$= e^{ikR_0}\left(1 + i\lambda m\sqrt{a_s N}\right)$$

or

$$E + E_0 = e^{i(kR_0 + m\lambda\sqrt{a_s N})} \qquad \textbf{(1–56)}$$

From Equation (1–56), it is evident that the phase of the original wave at point Q has been shifted by $m\lambda\sqrt{a_s N}$.

Furthermore, when an optical wave travels through a medium with thickness (m) and refractive index (n), a phase shift results. This phase shift is expressed in Equation (1–57).

$$m\lambda(a_s N)^{1/2} = \frac{2\pi}{\lambda}(n-1)m$$

Therefore,

$$\frac{\lambda^2}{2\pi}(a_s N) = n - 1 \qquad \textbf{(1–57)}$$

Equation (1–57) is in compliance with Rayleigh's law of scattering, in which the scattering wave intensity is proportional to the scattering coefficient (a_s) and inversely proportional to λ^4, assuming the refractive index of the medium is independent of the wavelength of the optical beam. These relationships hold true only for optical beams whose wavelengths are not close to the absorption wavelengths of the optically transparent medium.

1.6   DISPERSION

Dispersion refers to the behavior of light traveling through a medium of refractive index (n), in relation to its wavelength (λ). Because the velocity of light in a medium other than free space is expressed as ratio of the velocity of light in free space (3×10^8 m/s) to the refractive index of the medium through which it travels (c/n), the velocity of light in any medium with a higher refractive index will be smaller.

Angular Dispersion

When an optical beam travels through a prism, the angle of emergence (θ) can be measured for various wavelengths (Figure 1–16).

The rate of change of the emerging angle to the rate of change of the wavelength is referred to as the **angular dispersion** of the prism and is expressed by Equation (1–58).

$$\frac{d\theta}{d\lambda} = \frac{d\theta}{dn} \times \frac{dn}{d\lambda} \qquad \textbf{(1–58)}$$

where θ is the angle of emergence, n is the refractive index, and λ is the wavelength of the incident beam. Equation (1–58) is composed of two parts: $d\theta/dn$ and $dn/d\lambda$. The first part is indicative of the characteristic properties of the matter composing the prism, and the second part is indicative of the dispersion of the prism. Some of the properties of the prism can be calculated using these properties.

Snell's law of refraction (see Equation (1–59)) states that

$$\frac{\sin\theta}{\sin\phi} = n \qquad \textbf{(1–59)}$$

If $\sin\phi$ is assumed constant, then differentiating Equation (1–59) yields Equation (1–60).

$$\frac{d\theta}{dn} = \frac{\sin\phi}{\cos\theta} \qquad \textbf{(1–60)}$$

For a minimum deviation, there exist equal deviations at both faces of the prism with a total rate of change expressed by Equation (1–61).

$$\frac{d\theta}{dn} = \frac{2s\sin(a/2)}{s\cos\theta}$$

$$= -\frac{B}{b}$$

Therefore,

$$\frac{d\theta}{dn} = -\frac{B}{b} \qquad \textbf{(1–61)}$$

The geometric factor is the ratio of the base of the prism to the width of the optical beam. Therefore, the angular dispersion can be expressed by Equation (1–62).

$$\frac{d\theta}{d\lambda} = \frac{B}{b}\left(\frac{dn}{d\lambda}\right) \qquad \textbf{(1–62)}$$

Equation (1–62) shows that the **angle of deviation** is related to the refractive index of the material that composes the prism and also to the wavelength of the optical beam.

Normal Dispersion

The second part on the right-hand side of Equation (1–58) is indicative of the relationship between the refractive index (n) and wavelength (λ). When plotting the refractive index in relation to wavelength for prisms composed of different refractive indices, a set of curves can be obtained, as illustrated in Figure 1–17.

These curves reflect normal dispersion. They exhibit a common shape, but have different dispersion values. Normal dispersion is characterized by the following:

1. The refractive index increases with a decrease of the wavelength of the optical beam in a non-linear fashion.
2. For materials with higher index of refraction and at a constant wavelength, the dispersion curves exhibit steeper characteristics.

Cauchy's Equation

The mathematical interpretation of normal dispersion was performed by Cauchy in 1836. He related the refractive index of the material composing a prism to the operating wavelength of the incident optical beam as follows in Equation (1–63).

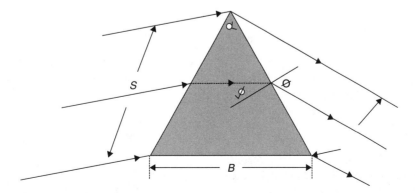

Figure 1–16. A light beam traveling through a prism.

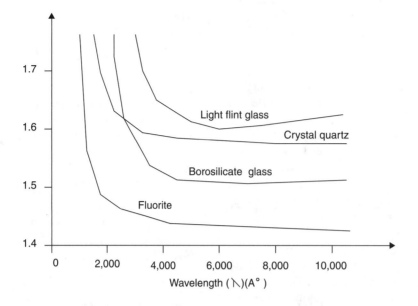

Figure 1–17. Dispersion for various materials in relation to refractive index and wavelength.

$$n = A + \frac{B}{\lambda^2} + \frac{C}{\lambda^4} \qquad (1\text{–}63)$$

where A, B, and C are characteristics of the matter composing the medium.

The values of the parameters (A, B, and C) can be determined if the refractive index (n) is established for two or more values of λ. However, if only two of the material characteristics are used (A and B), the modified **Cauchy's equation** is expressed in Equation (1–64).

$$n = A + \frac{B}{\lambda^2} \qquad (1\text{–}64)$$

Differentiating Equation (1–64) yields Equation (1–65).

$$\frac{dn}{d\lambda} = -\frac{2B}{\lambda^3} \qquad (1\text{–}65)$$

Equation (1–65) shows that the dispersion $dn/d\lambda$ varies as a function of the inverse cube of the wavelength $(1/\lambda^3)$. Cauchy's equation, although inefficient, should be considered an empirical relationship and can be used when describing normal dispersion.

SELLMEIER'S RELATIONSHIP. The inability to apply Cauchy's equation to both normal and anomalous dispersion (large discontinuity in the dispersion curve crossing an absorption bound) prompted Sellmeier, in 1871, to review and improve Cauchy's equation. He assumed that the transmission medium is composed of particles bound with elastic forces capable of vibrating with defined frequency (v_o) without the influence of an external electric field. This frequency is referred to as the particle natural frequency. An optical beam with

wavelength (λ) (close to but not equal to the particles natural frequency), passing through a medium composed of particles exhibiting a specified number of natural frequencies with relatively small amplitudes, will force additional vibrations upon the particles. However, if the wave frequency approaches that of the particle natural frequency, the amplitude of the vibrations will increase. When the frequencies are equal (v equal to v_o), resonance will occur. The large amplitudes of particle vibrations at resonance will interfere with the traveling wave, altering its velocity within the medium. **Sellmeier's equation** is shown by Equation (1–66).

$$n^2 = 1 + \frac{A\lambda^2}{\lambda^2 - \lambda_o{}^2} \qquad (1\text{–}66)$$

where n is the refractive index, λ is the wavelength of the traveling optical beam, and λ_o is the wavelength corresponding to the natural frequency of the matter. The relationship between natural frequency and the corresponding wavelength is expressed in Equation (1–67).

$$\lambda_o = \frac{c}{v_o} \qquad (1\text{–}67)$$

where c is the velocity of light in free space (3×10^8 m/s). A more general expression (Equation (1–68)) has been developed to account for a series of different natural frequencies.

$$n^2 = 1 + \frac{A_o\lambda^2}{\lambda^2 - \lambda_o{}^2} + \frac{A_1\lambda^2}{\lambda^2 - \lambda_1{}^2} + \frac{A_2\lambda^2}{\lambda^2 - \lambda_2{}^2}$$

$$+ \frac{A_3\lambda^2}{\lambda^2 - \lambda_3{}^2} + \ldots + \sum \frac{A_i\lambda}{\lambda^2 - \lambda_i{}^2} \qquad (1\text{–}68)$$

where λ_o, λ_1, λ_3, ... λ_i = corresponding wavelengths to natural frequencies v_o, v_1, v_2, v_3, ... v_i.

THE RELATIONSHIP BETWEEN DISPERSION AND ABSORPTION. Sellmeier's equation, although valid for wavelengths away from the absorption bands of the medium, develops definite weaknesses at wavelengths very close to the absorption wavelengths. Sellmeier's deficiency was rectified by Helmholtz, a German physicist, who assumed that the oscillatory particles must continuously draw energy (absorption) from the traveling wave in order to maintain their motional status. He introduced a constant (k_o) that relates the absorption coefficient (a) to the wavelength (λ) (see Equation (1–69)).

$$k_o = \frac{a\lambda}{4\pi} \qquad (1\text{–}69)$$

where k_o is the **Helmholtz constant,** λ is the wavelength, and a is the absorption coefficient. The constant (k_o) signifies a decrease in the intensity of the wave from its initial value while traveling a distance (λ). This decrease in intensity is expressed by Equation (1–70).

$$\frac{1}{e^{4\pi k_o}} = \frac{1}{e^{4\pi\left(\frac{a\lambda}{4\pi}\right)}} \qquad (1\text{–}70)$$

Therefore, intensity decrease $= \dfrac{1}{e^{a\lambda}}$.

The final expression formulated by the Helmholtz theory is expressed by Equations (1–71) and (1–72).

$$n^2 - k_o{}^2 = 1 + \sum_i \frac{A_i\lambda^2}{(\lambda^2 - \lambda_i) + \left(\dfrac{g_i\lambda^2}{\lambda^2 - \lambda_i{}^2}\right)} \qquad (1\text{–}71)$$

$$2nk_o = \sum \left[\frac{A_i\sqrt{g_i}\lambda^3}{(\lambda^2 - \lambda_i{}^2) + g_i\lambda^2}\right] \qquad (1\text{–}72)$$

where g_i is the strength of the particle frictional force. Both Equations (1–71) and (1–72) hold true for absorption wavelengths as well. This is a definite improvement over the Cauchy's equation.

1.7 ELECTROMAGNETIC THEORY AND THE DIELECTRIC MEDIUM

The velocity of light in free space was calculated as approximately equal to 3×10^8 m/s. Because optical waves are electromagnetic waves, the velocity through a transparent medium other than free space will be smaller and inversely proportional to the refractive index of the medium (c/n).

When an electromagnetic wave traverses a transparent med-ium, the electric field component of the wave will enforce a small displacement upon the charged particles of the matter composing the medium, thus polarizing them. This displacement from their original position, although very small, diminishes in the absence of an electric field (see Equation (1–73)).

$$D = \varepsilon E \qquad (1\text{–}73)$$

where D is the electric displacement, ε is the dielectric constant, and E is the electric field.

Applying Maxwell's wave theory, elements of which were discussed in Section 1.3, four sets of equations incorporating the medium dielectric constant can be expressed by Equations (1–74), (1–75), (1–76), and (1–77).

$$\left.\begin{aligned}
\frac{\varepsilon}{c}\frac{\partial E_x}{\partial t} &= \frac{\partial H_z}{\partial y} - \frac{\partial H_y}{\partial z} \\[4pt]
\frac{\varepsilon}{c}\frac{\partial E_y}{\partial t} &= \frac{\partial H_x}{\partial z} - \frac{\partial H_z}{\partial x} \\[4pt]
\frac{\varepsilon}{c}\frac{\partial E_z}{\partial t} &= \frac{\partial H_y}{\partial x} - \frac{\partial H_x}{\partial y}
\end{aligned}\right\} \qquad (1\text{–}74)$$

$$\left.\begin{aligned}
-\frac{1}{c}\frac{\partial H_x}{\partial t} &= \frac{\partial E_z}{\partial y} - \frac{\partial E_y}{\partial z} \\[4pt]
-\frac{1}{c}\frac{\partial H_y}{\partial t} &= \frac{\partial E_x}{\partial z} - \frac{\partial E_z}{\partial x} \\[4pt]
-\frac{1}{c}\frac{\partial H_z}{\partial t} &= \frac{\partial E_y}{\partial x} - \frac{\partial E_x}{\partial y}
\end{aligned}\right\} \qquad (1\text{–}75)$$

$$\left.\begin{aligned}
\varepsilon\left(\frac{\partial E_x}{\partial x} + \frac{\partial E_y}{\partial y} + \frac{\partial E_z}{\partial z}\right) &= 0 \\[4pt]
\frac{\partial H_x}{\partial x} + \frac{\partial H_y}{\partial y} + \frac{\partial H_z}{\partial z} &= 0
\end{aligned}\right\} \qquad (1\text{–}76)$$

Differentiating the equations for plane waves yields

$$\left.\begin{aligned}
\frac{\partial^2 H_z}{\partial t^2} &= \frac{c^2}{\varepsilon}\frac{\partial^2 H_z}{\partial x^2} \\[4pt]
\frac{\partial^2 E_y}{\partial t^2} &= \frac{c^2}{\varepsilon}\frac{\partial E_y}{\partial x^2}
\end{aligned}\right\} \qquad (1\text{–}77)$$

From the previous examples, it is evident that the velocity of a wave in a medium is equal to $c/(\varepsilon)^{1/2}$. Because the index of refraction composing the medium is c/u, where u is the velocity, the refractive index in relation to the dielectric constant of the material is calculated as follows in Equation (1–78).

$$u = \frac{c}{\sqrt{\varepsilon}} \quad \text{(a)}$$

or

$$n = \frac{c}{u} \quad \text{(b)}$$

Substituting (a) into (b) yields:

$$n = \frac{c}{\frac{c}{\sqrt{\varepsilon}}} = \sqrt{\varepsilon}$$

or

$$n = \sqrt{\varepsilon} \qquad \textbf{(1–78)}$$

Therefore, the index of refraction is equal to the square root of the dielectric constant of the material. For a monochromatic plane wave, the instantaneous amplitude of the electric and magnetic fields can be expressed as follows in Equations (1–79) and (1–80).

$$E_y = A \sin(\omega t - kx) \qquad \textbf{(1–79)}$$

$$H_z = \sqrt{\varepsilon} A \sin(\omega t - kx) \qquad \textbf{(1–80)}$$

The ratio of the above expressions yields Equation (1–81).

$$\frac{E_y}{H_z} = \frac{1}{\sqrt{\varepsilon}}$$

or

$$H_z = \sqrt{\varepsilon} E_y \qquad \textbf{(1–81)}$$

For $\varepsilon > 1$, the magnetic field component is larger than the electric field component by a ratio equal to that of the refractive index (n), because n is equal to $\sqrt{\varepsilon}$. See Equation (1–78). The intensity of the electromagnetic wave in a dielectric medium is expressed by Equation (1–82). Because the instantaneous density of the electric field $\varepsilon E_y^2/8\pi$ is equal to the instantaneous density of the magnetic field $H_z^2/8\pi$, the field intensity (I) is the product of the instantaneous density of the electric field, the magnetic field components, and the velocity of light.

$$I = \left(\frac{\varepsilon E_y^2}{8\pi}\right)\left(\frac{\varepsilon H_z^2}{8\pi}\right)(u)$$

$$= \frac{2(\varepsilon E_y^2)(u)}{8\pi}$$

$$= \frac{\varepsilon E_y^2(u)}{4\pi}$$

Because $u = \dfrac{c}{n}$ and $n = \sqrt{\varepsilon}$

Therefore,

$$I = \frac{c}{n}\frac{\varepsilon E_y^2}{4\pi}$$

$$= \frac{c}{n}\frac{n^2 E_y^2}{4\pi}$$

$$= \frac{c n E_y^2}{4\pi}$$

Therefore,

$$\text{field intensity: } I = \frac{c n E_y^2}{4\pi} \qquad \textbf{(1–82)}$$

where E_y is the root mean square (RMS) of the electric vector. Equation (1–82) indicates that the intensity of an electromagnetic wave in a dielectric medium is proportional to the velocity of light in free space, the refractive index of the medium, and the instantaneous densities of both the electric and magnetic field components of the transverse wave.

When an electromagnetic wave impedes upon an atom or a molecule, the periodic electric field component of the wave exerts a force upon the atom or the molecule, setting it into a vibrational motion with a frequency equal to that of the impeding wave. However, the phase of the induced particle motion is defined by the difference between the frequency of the impeding wave and the natural frequency of the particle. The forced vibrations upon the particles will alter the velocity of the electromagnetic wave traversing the medium. Dispersion of optical energy within the medium has its source in the secondary waves generated by the forced vibrations upon the particles. In order to better explain the effect of dispersion on the wave velocity traveling through a transparent medium, the concept of scattering must be recalled. When an optical beam traverses a transparent medium, primary and secondary waves are generated. The primary waves are very small in amplitude. Because they are 180° out of phase compared to the principal waves, they are canceled out. However, the secondary waves traveling in parallel to the direction of the principal waves do not cancel out and combine to form other waves also traveling parallel to the principal waves. Based on the concept of the superposition of two waves, the principal waves and the secondary waves will be added together. In other words, the secondary waves will interfere with the principal waves altering their phase velocity. Because the phase of the secondary waves depends on the velocity of the principal wave, it can be said that the velocity of an electromagnetic wave transversing a transparent medium is subject to the frequency of the principal wave.

1.8 WAVE POLARIZATION

Electromagnetic waves are three dimensional waves. That is, the electric field and magnetic field vectors are perpendicular to each other and they are both perpendicular to the direction of propagation (see Figure 1–18). Optical waves are electromagnetic waves occupying a narrow band of the electromagnetic spectrum (see Figure 1–19). Usually such waves, generated by natural or man made sources are referred to as **unpolarized waves.** Unpolarized waves are those with magnetic and electric field components that appear on a number of planes. However, in numerous applications, the electric field vector component of the optical ray must occupy only one plane. In order to achieve this, the three dimensional electromagnetic wave must be transformed into a two dimensional wave. This process is referred to as **polarization.** **Wave polarization** is accomplished through the interaction of an unpolarized wave with matter. Therefore, polarization of light is referred to as the process whereby the electric field vector component vibrates in a single plane, producing a sinewave laying on the plane of polarization.

Three types of wave polarization exist. They are **plane polarization, circular polarization,** and **elliptical polarization.**

Plane Polarization

A plane polarized wave is defined as a wave whose the amplitude vector of the electric field component always travels the same direction (see Figure 1–20).

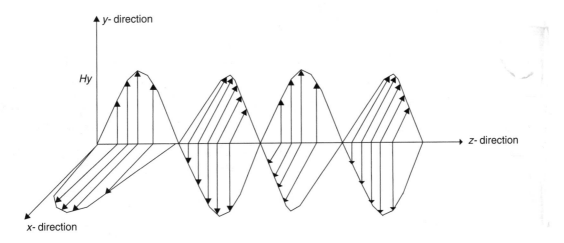

Figure 1–18. An electromagnetic wave.

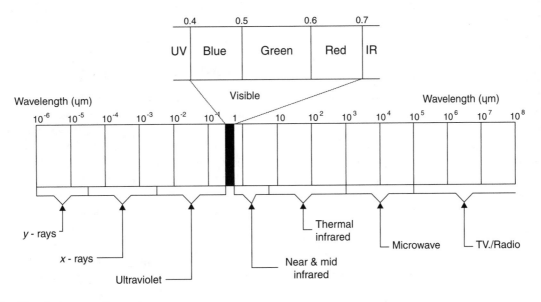

Figure 1–19. The electromagnetic spectrum.

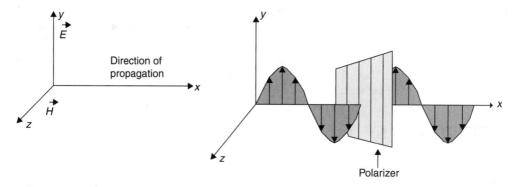

Figure 1–20. Plane wave polarization.

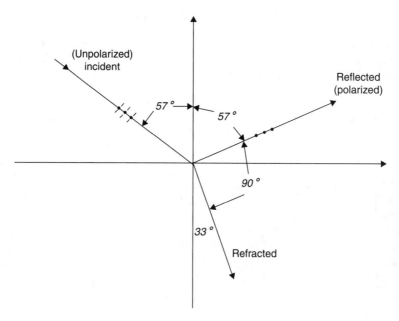

Figure 1–21. Wave polarization by reflection (Brewster's law).

The general mathematical expression of a plane polarized wave is given by Equation (1–83).

$$\vec{E} = \vec{E}_m \sin (kz - \omega t) \quad \textbf{(1–83)}$$

where $\vec{E}_m$ is the amplitude constant in time and pointing in the same direction, k is the $2\pi/\lambda$, ω is the $2\pi v$, v is the frequency, λ is the wavelength equal to u/v or equal to $(c/v)/n$, n is the refractive index of the medium through which the wave is traversing, u is the velocity of the wave in a medium, and z is the direction of propagation.

A more detailed mathematical expression of a plane polarized monochromatic ray is given by Equation (1–84).

$$\vec{E}(z, t) = E_x \sin(kz - \omega t) + E_y \sin(kz - \omega t + \delta) \quad \textbf{(1–84)}$$

where E_x is the amplitude of the vector component in the x direction, E_y is the amplitude of the vector component in the

y direction, and δ is the phase difference between the above two components.

Several methods of plane wave polarization exist, some are polarization by reflection, by absorption, by scattering, and by double reflection.

POLARIZATION BY REFLECTION. When an unpolarized optical wave impedes upon a reflecting surface plane, polarization (shown in Figure 1–21) can be achieved. The reflected light is polarized when the electric field vector component changes parallel to the reflecting surface. In a plane polarized optical wave, the reflected ray and the refracted ray are at an angle of 90° (**Brewster's law**).

In his drive to achieve maximum plane polarization of an optical ray, Brewster observed that when an unpolarized optical ray is incident upon a reflecting medium of refractive index (n) with an **angle of incident** (ϕ_1), a reflected ray and a refracted ray will appear. However, when the angle of incident was set at 57°, maximum plane polarization was ob-

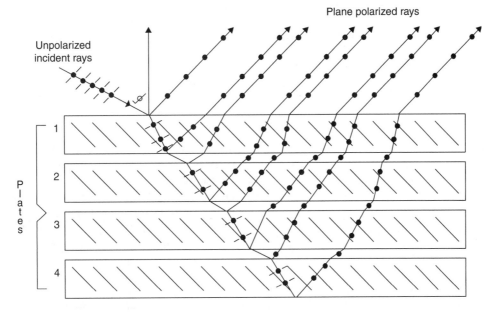

Figure 1–22. Polarization with reflecting plates.

served. At this point it was also observed that the reflected and refracted rays were at 90° angles, as shown in Figure 1–22. This observation made by Brewster, using an ordinary glass as the reflected medium, is referred to as Brewster's law. In essence, this law interrelates the polarization of an optical ray with the index of refraction of the reflected medium as follows in Equation (1–85).

From Figure 1–21,

$$\frac{\sin \phi_1}{\sin \phi_2} = n$$

Because

$$< xyz = 90°,$$
$$\sin \phi_2 = \cos \phi_1$$

or

$$\frac{\sin \phi_1}{\sin \phi_2} = \frac{\sin \phi_1}{\cos \phi_1} = n$$

Because

$$\frac{\sin \phi_1}{\cos \phi_1} = \tan \phi_1$$

$$\tan \phi_1 = n \qquad \textbf{(1–85)}$$

Brewster's law establishes the relationship between the angles of incidence as a function of the index of refraction for a maximum plane polarization of a monochromatic optical ray incident upon a reflective medium with an angle of incident of 57°.

POLARIZATION WITH REFLECTING PLATES. During the process of plane polarization, not all the energy of the incident optical ray emerges as the polarization beam, because a portion of it is reflected. In order to increase the **degree of polarization** (*P*), several reflective plates can be used simultaneously as shown in Figure 1–22.

Figure 1–22 shows that the larger the number of the reflecting plates, the higher the degree of plane polarization. P. Desains and F. Provosstaye formulated the mathematical interpretation of this concept in 1850, as shown in Equation (1–86).

$$P = \frac{I_p - I_s}{I_p + I_s}$$

$$= \frac{m}{m + \left(\dfrac{2n}{1 - n^2}\right)^2}$$

Therefore,

$$P = \frac{m}{m + \left(\dfrac{2n}{1 - n^2}\right)^2} \qquad \textbf{(1–86)}$$

where P is the degree of polarization, I_p is the intensity of the parallel component, I_s is the intensity of the perpendicular component, m is the number of reflecting plates, and n is the index of refraction.

Equation (1–86) does not take into account the effect of absorption, therefore it is a conservative formula establishing the degree of polarization of an optical ray as a function of the refractive index and the number of reflecting plates.

Example 1–10

In order to increase the degree of polarization (P), five reflective plates are used simultaneously with refractive indices of 1.33 each. Calculate the degree of polarization.

Solution

From Equation (1–86),

$$P = \frac{m}{m + \left(\dfrac{2n}{1 - n^2}\right)^2} = \frac{5}{5 + \left[\dfrac{2(1.33)}{1 - 1.33^2}\right]^2} = 0.3$$

Therefore, the degree of polarization is 0.3.

Example 1–11

Determine the number of reflective plates required to achieve a degree of polarization of 0.45. The reflective plates are made of material with a refractive index of 1.5.

Solution

Solve for m, using Equation (1–86).

$$P\left[m + \left(\frac{2n}{1 - n^2}\right)^2\right] = m$$

$$Pm + P\left(\frac{2n}{1 - n^2}\right)^2 = m$$

or

$$m - Pm = P\left(\frac{2n}{1 - n^2}\right)^2$$

Substituting for values,

$$m - 0.45m = 0.45\left(\frac{2(1.5)}{1 - 1.5^2}\right)^2$$

or

$$0.55m = 2.6$$
$$m = 4.73$$

Therefore, it will require **five** reflective plates to achieve a degree of polarization equal to 0.45.

THE LAW OF MALUS. E'tienne Malus (1775–1812), a French engineer, accidentally discovered the polarization of light by reflection while observing a beam of light reflected from a window through a calcite crystal. Through his observation, Malus was able to establish the relationship between the **intensity** of a fully polarized optical ray produced by a polarizer and the intensity of the same optical beam that traveled through an analyzer (Figure 1–23).

More specifically, **Malus' law** states that the intensity of a fully polarized optical ray is a function of the square of the cosine of the angle (θ) between the polarizer and the analyzer. That is, when a plane polarized optical ray enters an analyzer, the amplitude of the wave is composed of two components (A_1 and A_2). From these two components, only A_1 will pass through, while A_2 is eliminated by the analyzer. This amplitude is expressed by Equation (1–87).

$$A_1 = A \cos \theta \qquad \textbf{(1–87)}$$

where A_1 is the amplitude of the optical component parallel to the analyzer, A is the amplitude of the polarized ray entering the analyzer, and θ is the angle between the polarizer and the analyzer. The corresponding intensity (I_1) of the optical component parallel to the analyzer is expressed by Equations (1–88) and (1–89).

$$I_1 = A_1^2 = A^2 \cos^2 \theta \qquad \textbf{(1–88)}$$

or

$$I_1 = I_o \cos^2 \theta \qquad \text{(Malus' law)} \qquad \textbf{(1–89)}$$

where I_o is the intensity of the impeding ray on the analyzer optical ray. Therefore, the law of Malus states that the intensity of the optical component of a fully polarized optical ray

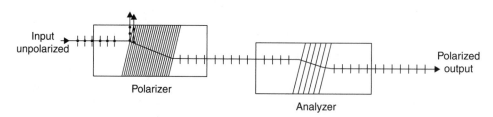

Figure 1–23. An unpolarized optical ray, incident upon a polarizer, then an analyzer.

parallel to the analyzer varies with the square of the cosine of the angle between the polarizer and the analyzer.

Example 1–12

Calculate the ratio of the optical ray's intensity parallel to the analyzer and the intensity of the optical ray approaching the analyzer (I_1/I_o) if the angle (θ) between the polarizer and the analyzer is $30°$.

Solution
From Equation (1–89),

$$\frac{I_1}{I_o} = \cos^2 \theta = \cos^2 30° = 0.75$$

Therefore, $\dfrac{I_1}{I_o} = 0.75$.

Example 1–13

If the ratio of the intensity of the optical ray parallel to the analyzer and the intensity of the optical ray entering the analyzer (I_1/I_o) is 0.55, calculate the angle (θ) between the polarizer and the analyzer.

Solution
Use Equation (1–89) to solve for θ.

$$\cos \theta = \sqrt{\frac{I_1}{I_o}} = \sqrt{0.55} = 0.74$$

Therefore, $\theta = 42°$.

Polarization by Selective Absorption

Polarization by selective absorption occurs when unpolarized optical waves enter an anisotropic material that has characteristics that allow it to absorb optical energy in one direction and to transmit optical energy parallel to the crystallographic axis. Unpolarized optical waves entering materials such as biotite and tourmaline are divided into two waves perpendicular to each other. The sum of the intensities are equal to the intensity of the entering wave. However, if the geometry satisfies certain conditions, one of the two optical rays will be fully absorbed by the medium while the other will emerge vertically polarized. It is therefore evident that plane polarization can be accomplished through selective absorption.

Elliptical and Circular Polarization. If the vector magnitude of the electric field component varies while rotating with time, the wave will be elliptically polarized (see Figure 1–24). On the other hand, if the vector magnitude of the electric field component is constant while rotating with time, the wave will be circularly polarized (see Figure 1–25).

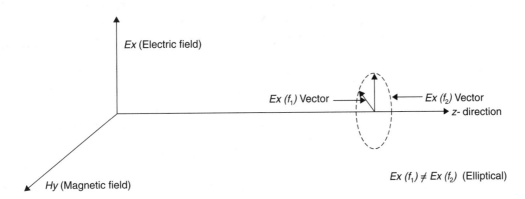

Figure 1–24. Elliptical wave polarization.

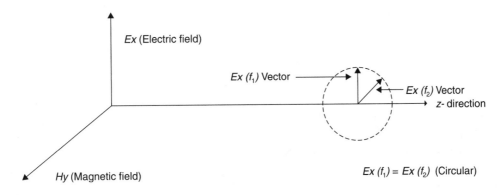

Figure 1–25. Circular wave polarization.

Circular wave polarization is a modified version of elliptical polarization. When two plane polarized waves are combined to form a new polarized wave, and if the phase difference between them is 90°, then the new wave is circularly polarized. Circular wave polarization can be achieved with **birefringent** material.

Birefringent (double refraction) materials are crystals that allow an optical ray to travel in a special direction through the crystal. This is referred to as the *optical axis*. A plane polarized optical ray travels through a birefringent material parallel to the optical axis with a different velocity than those that travel perpendicular to the optical axis. Because two plane polarized optical rays travel within the crystal with two different velocities, it is necessary to associate these velocities with two different material indices (n_o and n_e). n_o is the refractive index corresponding to the material traveled by the ordinary wave parallel to the optical axis, and n_e is the refractive index corresponding to the material traveled by the extraordinary wave perpendicular to the optical axis. A birefringent material may naturally have or can be forced to have its optical axis along the plane of its surface. When an optical ray is incident upon such a crystal with an angle of incidence of 0°, it is anticipated that the angle of refraction will also be 0°. Therefore, the two angles will emerge at the exit as a single ray. However, because the two rays travel at two different velocities within the crystal, they will emerge with a phase difference relevant to both traveling times. This phase difference can be adjusted to a predetermined level by manipulating the thickness of the crystal. If the phase difference is made equal to $\pi/2$, the combined wave at the exit of the crystal will be circularly polarized. Assuming a birefringent material with thickness (T), the time required for the optical ray to travel this thickness is expressed by Equation (1–90).

$$t = \frac{nT}{c} \tag{1–90}$$

where t is the time required to travel the crystal, T is the thickness of the crystal, n is the refractive index, and c is the velocity of light in space. Because the crystal exhibits two different refractive indices for the two polarized components of the incident optical ray, there will be a time difference (Δt) between the plane polarized waves emerging at the exit. This time difference is expressed by Equation (1–91).

$$\Delta t = |n_e - n_o|\left(\frac{T}{c}\right) \tag{1–91}$$

where n_o is the refractive index of the path traveled by the ordinary ray (parallel to the optical axis) and n_e is the refractive index of the path traveled by the extraordinary ray (perpendicular to the optical axis). Converting the time difference Δt to phase difference $\Delta\phi$ yields Equation (1–92).

$$\Delta\phi = \omega\Delta t \tag{1–92}$$

where $\Delta\phi$ is the phase difference, Δt is the time difference, and ω is the angular velocity ($2\pi f$).

Substituting Equation (1–91) into Equation (1–92) yields Equation (1–93).

$$\Delta\phi = |n_e - n_o|\left(\frac{\omega T}{c}\right)$$
$$= 2\pi T\left|\frac{1}{\lambda_e} - \frac{1}{\lambda_o}\right| \tag{1–93}$$

When a plane polarized wave is incident upon a birefringent material with an angle of incident (θ) relative to the crystal optical axis, the vector magnitude of the electric field component at a distance (x) and time (t) can be expressed as follows in Equation (1–94).

$$\vec{E}_m = |E_m|\sin(kx - \omega t)(\cos\theta\hat{o} + \sin\hat{e}) \tag{1–94}$$

where k is the $\frac{2\pi}{\lambda}$, $\hat{o}$ is the wave direction parallel to the optical axis, and $\hat{e}$ is the wave direction perpendicular to the optical axis.

Because the average over a period of the square of a sine function is $^1/_2$, the intensity of the electric field is then expressed by Equation (1–95).

$$I_{\vec{E}m} = \frac{1}{2}\vec{E}_m \tag{1–95}$$

where $I_{\vec{E}m}$ is the intensity of the electric field and $\vec{E}_m$ is the vector magnitude of the electric field component.

At the exit of the crystal, the two polarized optical rays will combine to form a single ray. However, because both of them have traveled at different times, they will emerge at the exit of the crystal with a phase difference ($\Delta\phi$). A general relationship between the magnitude of the combined electric field component ($\vec{E}_o$) and phase difference ($\Delta\phi$) is expressed by Equation (1–96).

$$\vec{E}_o = |\vec{E}_m|\cos\theta.\sin(kx-\omega t_o)\hat{o} + |\vec{E}_m|\sin\theta.\sin(kx-\omega t+\Delta\phi)\hat{e}$$
$$\tag{1–96}$$

where x_o is the distance traveled (at the exit of the crystal), t_o is the time required for the ray to transverse the crystal, $\Delta\phi$ is the phase difference between the two rays, and $\vec{E}_o$ is the magnitude of the combined wave at the exit of the crystal.

If the phase difference $\Delta\phi$ is 90°, the crystal is capable of performing circular polarization as shown in Equation (1–97).

$$\Delta\phi = \frac{\pi}{2}$$

or

$$\Delta\phi = \frac{\pi}{2} = 2\pi T \left| \frac{1}{\lambda_e} - \frac{1}{\lambda_o} \right| \qquad \textbf{(1–97)}$$

where λ_o is the wavelength of the polarized optical ray parallel to the optical axis and λ_e is the wavelength of the polarized optical ray perpendicular to the optical axis. Solving for the thickness (T) of the crystal yields Equation (1–98).

$$2\pi T \left| \frac{1}{\lambda_e} - \frac{1}{\lambda_o} \right| = \frac{\pi}{2}$$

or

$$4\pi T \left| \frac{1}{\lambda_e} - \frac{1}{\lambda_o} \right| = \pi$$

Therefore,

$$T = \frac{1}{4} \left| \frac{1}{\lambda_e} - \frac{1}{\lambda_o} \right|^{-1} \qquad \textbf{(1–98)}$$

Equation (1–98) indicates that the thickness of a crystal required to perform circular polarization of a monochromatic wave must be equal to $\frac{1}{4}$ of its wavelength.

Example 1–14

Calculate the time difference (Δt) and phase difference ($\Delta\phi$) of two plane polarized components emerging at the output of a birefringent material with refractive indices (n_e) of 1.4, (n_o) of 1.25, and thickness (T) of 2×10^{-5} m. The wavelength of the optical ray at the input of the birefringent material is equal to 740 nm.

Solution

i) Using Equation (1–91), solve for Δt.

$$\Delta t = |n_e - n_o| \left(\frac{T}{c} \right) = |1.4 - 1.25| \left(\frac{2 \times 10^{-5} \text{ m}}{3 \times 10^8 \text{ m/s}} \right)$$

$$= 0.01 \times 10^{-12} \text{ s}$$

Therefore,

$$\Delta t = 0.01 \text{ ps}$$

ii) From Equation (1–9), solve for $\Delta\phi$.

$$\Delta\phi = 2\pi T \left| \frac{1}{\lambda_e} - \frac{1}{\lambda_o} \right|$$

$$\lambda_o = \frac{\lambda}{n_o} = \frac{740 \text{ nm}}{1.25} = 592 \text{ nm}$$

$$\therefore \lambda_o = 592 \text{ nm}$$

$$\lambda_e = \frac{\lambda}{n_e} = \frac{740 \text{ nm}}{1.4} = 528 \text{ nm}$$

$$\therefore \lambda_e = 528 \text{ nm}$$

$$\Delta\phi = 2\pi T \left| \frac{1}{\lambda_e} - \frac{1}{\lambda_o} \right| = 2\pi (2 \times 10^{-5} \text{m}) \left| \frac{1}{528 \text{ nm}} - \frac{1}{592 \text{ nm}} \right| = 2.5 \text{ rad}$$

Therefore, $\Delta\phi = 2.5$ rad.

1.9 ELEMENTS OF QUANTUM OPTICS

Introduction

In 1921, Albert Einstein was awarded the Nobel Prize in physics for his contribution to quantum theory. In 1905, Einstein published a paper on the photoelectric effect. From then on, he was the principal defender of the existence of light quanta or photons in the scientific community. However, Einstein had major reservations about the theory of quantum mechanics developed by Schroedinger and Hesenberg between 1905 and 1926. In 1909, Einstein introduced the concept of **wave particle duality.** This theory explains that light can be considered both a wave and a discrete particle. Neil Bohr, a Danish physicist, borrowed some of Einstein's ideas and developed his own model of the atom. He suggested that electrons occupy a definite orbit around the nucleus of an atom. Therefore, when absorbing a discrete quantum of energy, these electrons can move from an orbit of lower energy to an orbit of higher energy. In 1916, Einstein was able to interpret the black body radiation suggested by Max Planck by integrating the concept of the photon and Neil **Bohr's atomic model.** However, the scientific community was reluctant to accept Einstein's quantum theory and Bohr's atomic model due to a lack of experimental verification. In 1923, the American physicist Arthur Compton succeeded in measuring the transfer of energy from photons to electrons, thus providing the required proof and establishing the factuality of the quantum theory. A critical date for the wider acceptance of the quantum theory was 1926, when an alternative interpretation of quantum theory was developed, based on De Broglie's concept that matter can behave as waves in a manner similar to electromagnetic waves behaving as particles. De Broglie's theory found support from prominent physicists of the time such as Einstein, Planck, and Schroedinger. The light properties of reflection, refraction, and interference can be interpreted by wave theory because these concepts reflect the interaction of optical waves with other optical waves. However, if one attempts to explain the phenomena (absorption of scattering), resulting from the interaction of a wave with matter, wave theory is insufficient. Attempting to explain the phenomena of absorption and scattering through wave theory was unsuccessful because experimental results differed markedly from the theoretical assumptions.

The Photoelectric Effect

In 1905, Albert Einstein referred to the photon concept in his attempt to explain the photoelectric effect (see Figure 1–26).

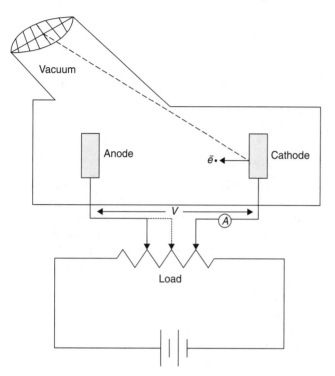

Figure 1–26. The photoelectric effect.

Figure 1–26 shows that when light is incident upon a metallic surface, electrons can be released when they are attracted by (+) of the potential difference applied across the plates. The generated electrons constitute the photocurrent flown through the circuit. Reducing the potential difference across the plates, the photocurrent can be reduced to zero. That is, the kinetic energy of the photoelectrons can be reduced to zero. The kinetic energy of photoelectrons is expressed by Equation (1–99).

$$E_{\text{kinetic}} = eV_o$$

or

$$E_{\text{kinetic}} = \frac{1}{2} mV_{\text{max}}^2 \qquad \textbf{(1–99)}$$

where m is the **electron mass** and V_{max} is the voltage applied across the plates (max).

Through experimentation, it was observed that photoelectron generation was impossible if the wavelength of the impeding light was below a certain level. This minimum wavelength was referred to as the **cut off wavelength** (λ_c). This phenomenon could not be explained by wave theory, which stipulates that the energy of the incident light used for the photoelectron generation is independent of the wavelength (frequency). Einstein was able to explain this phenomenon by elaborating Planck's photon theory. He postulated that the incident photons have energy equal to the energy between two adjacent electron orbital levels and that it is proportional to the frequency of the incident wave, see Equation (1–100).

$$E = hv \qquad \textbf{(1–100)}$$

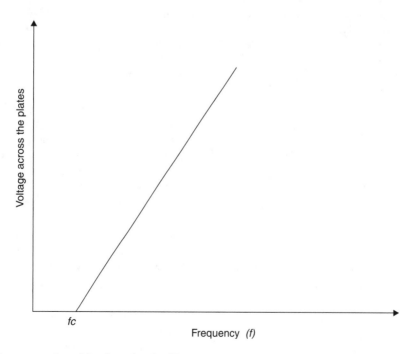

Figure 1–27. Graphical representation of the photoelectric effect.

where h is Planck's constant ($6.628 \times 10^{-38} \, J - s$) and v is the wave frequency. An electron that requires a specific amount of energy in order to become a free electron absorbs the energy of the photon incident upon the metallic surface. The energy required to release an electron from its orbital bond and to become a free electron (photoelectron) is referred to as the **work function** (ϕ) of the metal. If the work energy is smaller than the photon energy, then the excess energy is converted to kinetic energy ($E_{kinetic}$), expressed by Equation (1–101).

Therefore,

$$E_{kinetic} = eV_o$$

or

$$E_{kinetic} = hv - \phi$$

Because

$$v = \frac{c}{\lambda}$$

then

$$E_{kinetic} = \frac{hc}{\lambda} - \phi \qquad \textbf{(1–101)}$$

where c is the velocity of light in vacuum (3×10^8 m/s).

A more common representation of the wave frequency is f. The **cut off frequency** is represented as f_c.

Therefore, f_c is the cut off frequency.

Or

$$f_c = \frac{\phi}{h}$$

where ϕ is the constant that is characteristic of the metal used for the cathode of the photoelectric apparatus. The graphical representation of the photoelectric effect is shown in Figure 1–27.

Example 1–15

Compute the wavelength of a quantum of electromagnetic radiation with energy equal to 3 KeV.

Solution

From Planck's relationship (Equation (1–100)),

$$E = hv = \frac{hc}{\lambda}$$

Solve for λ.

$$\lambda = \frac{hc}{E} = \frac{(6.63 \times 10^{-34} \, \text{J/s})(3 \times 10^8 \, \text{m/s})}{3 \times 10^3 \, eV}$$

Because

$$1eV = 1.6 \times 10^{-19} \, J$$

then,

$$\lambda = \frac{hc}{E} = \frac{(6.63 \times 10^{-34} \, \text{J/s})(3 \times 10^8 \, \text{m/s})}{(3 \times 10^3)(1.6 \times 10^{-19} \, \text{J})} = 4.14 \times 10^{-10} \, \text{m}$$

Therefore,

$$\lambda = 0.414 \text{ nm}.$$

Example 1–16

When an optical monochromatic ray with a wavelength of 670 nm is incident upon a metallic surface, it generates photoelectrons at a potential difference of 0.75 V.

Compute the constant ϕ (characteristic of the metal) and cut off frequency (f_c).

Solution

i) Using Equation (1–101), solve for ϕ and compute ϕ.

$$\phi = \frac{hc}{\lambda} - E_{kinetic} = \frac{(6.63 \times 10^{-34} \, \text{J/s})(3 \times 10^{17} \, \text{nm/s})}{670 \text{ nm}} - 0.75 \text{ eV}$$

$$= 1.85 \text{ eV} - 0.75 \text{ eV} = 1.1 \text{ eV}$$

Therefore, $\phi = 1.1$ eV.

ii) Compute f_c.

$$f_c = \frac{\phi}{h} = \frac{1.1 \text{ eV}}{6.63 \times 10^{-34} \, \text{J/s}} = \frac{(1.1)(1.6 \times 10^{-19} \, \text{J})}{6.63 \times 10^{-34} \, \text{J/s}}$$

$$f_c = 0.265 \times 10^{15} \text{ Hz}$$

Therefore, $f_c = 265$ THz.

From the cut off frequency, the cut off wavelength (λ_c) can also be calculated.

$$\lambda_c = \frac{c}{f_c} = \frac{3 \times 10^{17} \text{ nm}}{265 \times 10^{12} \text{ Hz}} = 1132 \text{ nm}$$

Therefore, $\lambda_c = 1132$ nm.

Example 1–17

Calculate the voltage required to accelerate an electron to the point at which a collision with a target is capable of producing a photon with a wavelength equal to 0.45 nm.

Solution

The energy acquired by an electron when accelerated by a voltage (V) is eV. However, if this energy is sufficient to produce a photon of energy (E) and wavelength (λ), then

$$E = eV$$

Because $E = \dfrac{hc}{\lambda}$ then $eV = \dfrac{hc}{\lambda}$.

i) Evaluate E.

$$E = \frac{hc}{\lambda} = \frac{(6.63 \times 10^{-34}\ \text{J/s})(3 \times 10^{17}\ \text{nm/s})}{0.045\ \text{nm}}$$

$$E = 442 \times 10^{-17}\ \text{J}$$

Because $1\ \text{eV} = 1.6 \times 10^{-19}\ \text{J}$,

$$E = \frac{442 \times 10^{-17}\ \text{J}}{1.6 \times 10^{-19}\ \text{J}} = 276 \times 10^{2}\ \text{eV}$$

ii) Evaluate V, the required voltage.

$$V = \frac{E(\text{J})}{e(\text{C})} = \frac{276 \times 10^{2}\ \text{eV}(1.6 \times 10^{-19}\ \text{J})}{1.6 \times 10^{-19}\ \text{C(eV)}} = 276 \times 10^{2}\ \text{V} = 27.6\ \text{KV}$$

The Compton Effect

In 1923, Arthur Compton illustrated the photon nature of light with an experiment that involved the scattering of X-rays. In his honor, this effect has since been referred to as **Compton scattering** (Figure 1–28).

    With this experiment, Compton proved that the wavelength of scattered light is different from the wavelength of incident light. This cannot be explained by wave theory. According to wave theory, the wavelength of scattered light remains the same as that of incident light. However, if the light is considered as a photon with energy

$(E = hv)$, and momentum (p), where p is expressed by Equation (1–102),

$$p = \frac{h}{\lambda} \qquad \textbf{(1–102)}$$

where p is the momentum, h is Planck's constant, and λ is the wavelength. From Figure 1–28, the incident X-ray is scattered by a scatterer at an angle θ. From the scattered light, a very narrow beam is isolated through a double slit aperture. The narrow beam then falls onto a crystal (A), which diffracts the X-ray to a photographic plate. Comparing the spectral lines of the original X-ray to the spectral lines of the scattered rays, it shows that the scattered ray wavelength is shifted slightly toward a longer wavelength. This shift in wavelength progresses with an increase in the angle (θ). The wavelength difference $(\Delta\lambda)$ is expressed by Equation (1–103).

$$\Delta\lambda = \frac{c}{v'} - \frac{c}{v} = \frac{h}{m_o c}(1 - \cos\theta) \qquad \textbf{(1–103)}$$

where $\Delta\lambda$ is the wavelength difference, v' is the frequency of the scattered X-ray, v is the frequency of the incident X-ray, h is Planck's constant, m_o is the electron mass at rest (9.11×10^{-31} kgr), c is the velocity of light in vacuum (3×10^{8} m/s), and θ is the angle of scattering. Equation (1–104) expresses the wavelength of the scattered wave.

$$\lambda = \lambda_o + \frac{h}{m_o c}(1 - \cos\theta)$$

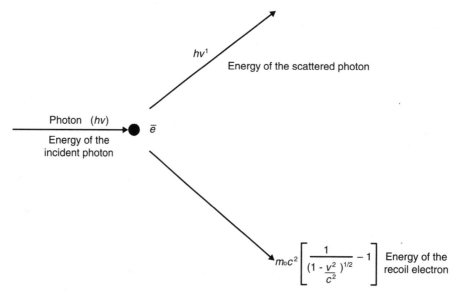

Figure 1–28. Compton scattering.

or

$$\lambda \approx \lambda_o + 0.024(1 - \cos \theta) \qquad \textbf{(1–104)}$$

From Equation (1–104), it is evident that the wavelength of the scattered wave is larger than the wavelength of the incident wave by $0.024(1 - \cos \theta)$, a phenomenon that is contrary to wave theory. Therefore, the phenomenon of scattering of X-rays can be satisfactorily explained by wave particle duality. In order to obtain the difference between the wavelength of the incident wave and the wavelength of the scattered wave ($\Delta\lambda$), the momenta of the photons must be equal to $h\nu/c$. That is, the photons must travel with a velocity of c in any medium. It is evident that the velocity of photons is different than the velocity of particles in matter, and may have velocities less than the velocity of light in a vacuum. However, observations have shown that the velocity of light (photon velocity) in matter appears to be smaller than in free space. In fact, the combined phase velocity does not change the photon velocity, which still remains at c. The only way that the velocity of a photon can be slowed down is by complete photon annihilation.

The interaction of waves with matter is summarized under a specific set of rules referred to as radiation laws. These laws are as follows:

1. **Planck's radiation law**

2. **Stefan-Boltzmann law**

3. **Wien displacement law**

When a body is heated enough, it emits radiation with a continuous spectrum whose amount and distribution of radiation is a function of wavelength. In accordance with Kirchhoff's law of radiation, the ratio of the emitted radiation to the absorption is the same for all bodies heated at the same temperature, while absorption is a measure of the amount of light energy disappearing at a single reflection. By definition, a black body is one that exhibits perfect absorption and emission characteristics when heated. The distribution of energy in the spectrum of a black body at three different temperatures is illustrated in Figure 1–29.

Figure 1–29 shows that the energy emitted at a particular wavelength increases with temperature.

PLANCK'S RADIATION LAW. One of the principal laws governing the radiation from a black body is the Planck law of radiation. This law presents the intensity of radiation as a unit surface area emitted from a black body when heated at a specific temperature as a function of wavelength (see Figure 1–29). The energy distribution shows a maximum at a particular wavelength. It shifts to shorter wavelengths with an increase of temperature, with a corresponding increase of the area under the curve. The shape of the curve of a black body heated at a temperature (T) is defined by Equation (1–105).

$$Wd\lambda = \frac{c_1}{\lambda^5}(e^{c_2/\lambda T} - 1)^{-1}\, d\lambda \qquad \textbf{(1–105)}$$

Where e is the base of natural log (2.718). c_1 and c are constants, depending on wavelength (λ).
For λ in cm,

$$c_1 = 3.7413 \times 10^{-5} \text{ erg.cm}^2/\text{sec}$$

$$c_2 = 1.4388 \text{ cm.deg}$$

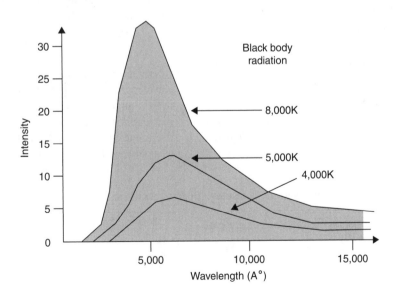

Figure 1–29. Energy distribution of a black body.

The intensity of radiation from a black body as a function of temperature (T) and wavelength (λ) is expressed by Equation (1–106).

$$E(\lambda, T) = \frac{2hc^2}{\lambda^5} \frac{1}{e^{hc/\lambda kT} - 1} \qquad \textbf{(1–106)}$$

where k is the Boltzman's constant (1.38×10^{-23} J/K), c is the velocity of light in vacuum (3×10^8 m/s), and h is the Planck's constant (6.628×10^{-38} J/s).

At long wavelengths, the curves were in compliance with the theory of energy equipartition and electromagnetic theory. However, for shorter wavelengths, experimental results showed that the curves kept increasing indefinitely instead of peaking to a maximum and falling back to zero. This observation led to the assumption that the amplitudes and the energy levels of the oscillating sources could not exist in a continuum, but only at definite points with energy levels being a multiple of specific values referred to as quantum levels (derived by Planck in 1900). Planck also stated that when a molecule goes from a higher energy level to a lower energy level, a packet of energy is emitted. This packet of energy is referred to as *photon energy*. Applying the concept of the photon, the black body radiation curves can be interpreted for both long and short wavelengths.

Black body temperatures and characteristic wavelengths for various regions of the spectrum are listed in Table 1–1.

Two other laws were derived from Planck's radiation law: the Stefan-Boltzmann law and the Wien displacement law.

STEFAN-BOLTZMANN LAW. Josef Stefan (1835–1893) and Ludwig Boltzmann (1844–1906) developed an equation that governs the total energy (E) emitted by a black body at all wavelengths. Equation (1–107) expresses this relationship, referred to as the Stefan-Boltzmann law.

$$E = \sigma T^4 \qquad \textbf{(1–107)}$$

Where E is the total radiated energy from a black body (energy under the Planck curve), σ is the Stefan-Boltzmann constant (5.6705×10^{-5} erg cm^2 K^{-4} sec^{-1}), and T is the black body temperature.

In essence, the Stefan-Boltzmann law quantifies the energy emitted by a black body under the Planck curve. One characteristic curve indicating the radiated intensity of a black body heated at 3179°K is illustrated in Figure 1–30, and another one heated at 18423°K is illustrated in Figure 1–31.

THE WIEN DISPLACEMENT LAW. William Wien, a German physicist who was awarded the Nobel Prize in 1911 for his work on radiation and optics, formulated the relationship between the peak energy distribution at wavelength (λ) and the temperature (T) of black body radiation. Equation (1–108) expresses the Wien law.

$$\lambda_{max} = \frac{3 \times 10^7}{T} \qquad \textbf{(1–108)}$$

where λ_{max} is the wavelength at peak distribution radiation ($A°$) and T is the temperature of the black body in Kelvin.

The Wien law peak distribution radiation wavelength for three different sources is illustrated in Figure 1–32.

WAVE PARTICLE DUALITY AND THE UNCERTAINTY PRINCIPLE. It is important to emphasize the dual character of light. From the previous discussion, it is evident from the results of such experiments that black body radiation and scattering can be satisfactorily interpreted by assuming the light is a particle, while diffraction and interference can be interpreted by assuming that light is a wave. This led to the establishment of the dual character of light as a wave and a particle. However, there is a drawback to the assumption of the dual character of light. This is referred to as the **uncertainty principle.** According to this principle, developed by Heisenberg, when the position and momentum of a particle needs to be known, the particle must be seen by exposing it to a beam of light of specified intensity and wavelength (λ). Assuming the light as a wave, the particle position can be determined by Equation (1–109).

$$\Delta x \to \lambda \qquad \textbf{(1–109)}$$

where Δx is the difference in particle position and λ is the wavelength of the illuminating light.

Equation (1–109) indicates that the position of the particle is not certain (it is uncertain). If the beam of light illuminating the particle under observation is considered as a photon, the particle should give up part or all of its mo-

TABLE 1–1 Black Body Temperatures and Characteristic Wavelengths.

Spectral Region	Energy (EV)	Wavelength (cm)	Black Body Temperature (K)
Radio	$< 10^{-5}$	> 10	< 0.03
Microwaves	$1 \times 10^{-5} - 1 \times 10^{-2}$	$1 \times 10^1 - 1 \times 10^{-2}$	$3 \times 10^{-2} - 3 \times 10^1$
Infrared	$1 \times 10^{-2} - 2 \times 10^1$	$1 \times 10^{-2} - 7 \times 10^{-5}$	$3 \times 10^1 - 41 \times 10^2$
Visible	$2 - 3$	$7 \times 10^{-5} - 4 \times 10^{-5}$	$4.1 \times 10^3 - 7.3 \times 10^3$
Ultraviolet	$3 \times 10^0 - 1 \times 10^3$	$4 \times 10^{-5} - 1 \times 10^{-7}$	$7.3 \times 10^3 - 3 \times 10^6$
X-rays	$1 \times 10^3 - 1 \times 10^5$	$1 \times 10^{-7} - 1 \times 10^{-9}$	$3 \times 10^6 - 3 \times 10^8$
Gamma rays	$> 10^5$	$> 10^{-9}$	$> 3 \times 10^8$

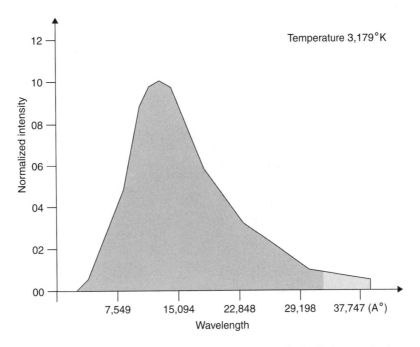

Figure 1–30. Black body energy intensity integrated over a wavelength of 3179°K (Stefan-Boltzmann law).

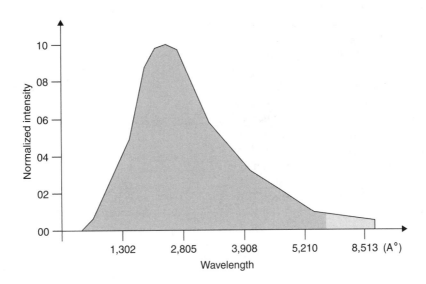

Figure 1–31. Black body energy intensity integrated over wavelength at 18423°K (Stefan-Boltzmann law).

mentum to the particle under observation. Because the momentum transferred to the particle is not precisely known, there is uncertainty in the momentum expressed by Equation (1–110).

$$\Delta p \to \frac{h}{\lambda} \qquad \textbf{(1–110)}$$

where Δp is the difference in the particle momentum, h is Planck's constant, and λ is the wavelength of the illuminating light. Multiplying Equations (1–109) and (1–110) yields Equation (1–111).

$$\Delta x\, \Delta p \to \lambda\left(\frac{h}{\lambda}\right) = h$$

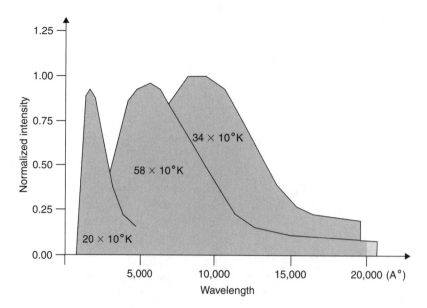

$$\text{Figure 1–32.} \quad \text{Wien law for three different body temperatures.}$$

Therefore,

$$\Delta x\, \Delta p \rightarrow h \qquad \qquad \textbf{(1–111)}$$

The product of the difference of position and momentum is independent of the wavelength of the illuminating light. In principle, there is a limit to the accuracy for establishing both the position and momentum of a particle. Decreasing the wavelength of the illuminating light while the uncertainty of the particle momentum increases proportionally can reduce the level of uncertainty for the position of the particle. However, if the wavelength of the illuminating light increases, the opposite will occur. The inability to reduce the uncertainty for both the position and momentum of the particle to zero simultaneously is a direct consequence of the dual character of light as a wave and a photon. Heisenberg refined the uncertainty principle for both the position and momentum of the particle through a precise relationship expressed by Equation (1–112).

$$\Delta x\, \Delta p \geq \frac{h}{4\pi} \qquad \qquad \textbf{(1–112)}$$

where $P = \dfrac{m}{\nu}$.

Although in a macroscopic world the principle of uncertainty can be marginalized, this is not the case when it is applied in the microscopic world. For example, the principle of uncertainty plays an important role at the atomic level.

Example 1–18
Calculate the uncertainty of the electron velocity when light of a wavelength equal to 620 nm is used to determine the position of the electron within the optical wavelength.

Solution
i) Applying Heisenberg's uncertainty principle (Equation (1–112)), solve for Δp.

$$\Delta p = \frac{h}{4\pi \Delta x}$$

or

$$\Delta(m\nu) = \frac{h}{4\pi \Delta x}$$

Solve for Δv.

$$\Delta \nu = \frac{h}{4\pi m \Delta x}$$

where m = mass of the electron (9.11×10^{-31} kg) and Δx = difference in the particle momentum (6.2×10^{-7} m).

ii) Calculate Δv.

$$\Delta \nu = \frac{h}{4\pi m \Delta x} = \frac{(6.63 \times 10^{-34}\ \text{J/s})}{4\pi(9.11 \times 10^{-31}\ \text{kg})(6.2 \times 10^{-7}\ \text{m})} = 93\ \text{m/s}$$

Therefore, the uncertainty of the electron velocity:
$\Delta v = 93$ m/s

SUMMARY

Chapter 1 discussed the fundamental theorems of optics and quantum physics. The chapter started with an introduction to optics, defining geometric, physical, and quantum optics. The discussion of geometric optics included concepts such as reflection, refraction, Fermat's principle, and optical distance.

The discussion of physical optics included the concepts of wave motion, velocity of light, phase velocity, wave intensity, absorption, scattering, dispersion, and wave polarization. Finally, the concept of quantum optics was introduced, including topics such as the photoelectric effect, Compton's effect, radiation from a black body, Planck's radiation law, the Stefan-Boltzmann law, and the wave particle duality concept. All topics were demonstrated by numerous examples.

QUESTIONS

Section 1.2

1. List the three main categories of optics.
2. Define *geometric optics*.
3. Define *physical optics*.
4. Define *quantum optics*.
5. Briefly describe reflection and refraction.
6. What is Snell's law? Formulate Snell's law for two media.
7. Define *refractive index*.
8. What is the principle of reversibility?
9. Define *optical path*.
10. What does Fermat's principle state? Give a brief explanation.

Section 1.3

11. Define *electromagnetic wave*.
12. Give and translate the wave equation.
13. List some of the important dates and people involved in the measurement of the velocity of light.
14. What are the main components of a sinewave? Use a formula and explain the relationships among them.
15. Describe the phase difference between two light waves and its importance to optical communications.
16. What is the difference between optical and geometric paths?
17. Define *wave phase velocity* and derive the corresponding relationship.
18. Define *wave intensity*.

Section 1.4

19. Briefly describe the concepts of light intensity, wave absorption and selective absorption.
20. Give the mathematical expression of lightwave intensity loss and determine the main factors contributing to this loss.
21. Describe Huygens' principle of reflection.

Section 1.5

22. Define *light wave scattering*.
23. Define *Rayleigh scattering*. What are the main components contributing to it?
24. Define *Raman scattering*. Define *elastic* and *inelastic scattering*.
25. What is a dipole moment? Give its mathematical relationship.
26. Describe Brillouin spontaneous scattering.
27. Briefly describe the *Kerr effect*.
28. With a simple schematic diagram, describe the generation of an acoustic wave.
29. What are *Stokes photons* and *anti-Stokes* photons?
30. With the assistance of a schematic diagram, explain Brillouin stimulation scattering.
31. Briefly describe the relationship between the refractive index and scattering.

Section 1.6

32. Define *light wave dispersion*.
33. List the characteristics of *normal* dispersion.
34. Interpret Cauchy's equation in terms of its material characteristics.
35. What is Sellmeier's relationship? How does it relate to Cauchy's equation?
36. How is Helmholtz' theory an improvement of Cauchy's equation?

Section 1.7

37. List Maxwell's four sets of equations and briefly explain their relevance to wave theory.
38. Describe the mathematical relationship between the refractive index of a dielectric material and its dielectric constant.
39. Describe the field intensity of an electromagnetic wave in terms of its electric field component, magnetic field component, and velocity of light.

Section 1.8

40. Define *light polarization*.
41. List the three types of polarization of light.
42. Name the four methods of plane wave polarization.
43. Briefly describe Brewster's law.
44. With a mathematical expression, translate the P. Desains-F. Provosstaye degree of polarization.

45. What does Malus' law state?

46. When does polarization by selective absorption occur? Explain in some detail.

47. Define *elliptical wave polarization*.

48. Define *circular wave polarization*.

49. What is a birefringent material?

Section 1.9

50. Briefly describe the evolutionary process of quantum theory.

51. Describe Einstein's concept of photoelectric effect.

52. Give the mathematical expression relating the photoelectron kinetic energy in terms of electron mass and velocity.

53. With the assistance of a schematic diagram, describe the Compton effect.

54. What is a black body?

55. Describe Planck's radiation law.

56. List the fundamental components incorporated in the relationship defining radiation from a black body.

57. Describe the Stefan-Boltzmann law.

58. What is the Wien displacement law and how does it relate to black body radiation?

59. Briefly explain the wave particle duality and the uncertainty principle.

60. How does Heisenberg define the uncertainty principle?

PROBLEMS

1. Two monochromatic optical beams are emitted simultaneously from a common reference point and travel the same distance (d). Beam A travels the distance (d) through the air, and beam B travels part way through the air and the rest of the distance through a medium with a refractive index of 1.4. If the beam arrival time difference is 50 ps, calculate the length (d_2) of the second medium.

2. An optical ray with λ_o of 640 nm traveling through another medium measures a wavelength of 427 nm. Calculate:

   i) The index of refraction of the other medium (n_2)

   ii) The velocity of the ray in the other medium (v_2)

   iii) The angle of incident (θ_1) if the angle of refraction (θ_2) is equal to 33°

3. A light ray travels from one medium with a refractive index (n_1) of 1.25 to another medium with a refractive index (n_2) of 1.15. If the optical distance (d) is 30 cm, h_1 and h_2 are equal to 10 cm, and BD is equal to 7 cm, compute the angle of refraction (ϕ_2) (see schematic of Example 1–4).

4. A monochromatic ray (λ_o) equal to 840 nm travels through a medium other than air with a phase velocity (u) of 2.994 m/s. If the frequency (f) of the optical ray in that medium is 550 THz, compute the index of refraction (n_1).

5. An optical beam travels through a glass tube filled with helium gas. If the combined attenuation due to absorption and scattering $(a_a + a_b)$ is equal to 7.2, and the ratio of the output/input (I_o / I_{in}) intensities is 0.354, compute the length (d) of the tube.

6. An optical beam traveling through a number of reflective plates with refractive indices of 1.5 has achieved a degree of polarization equal to 0.55. If the degree of polarization is to be increased to 0.3, calculate the number of the required additional reflective plates.

7. Compute the intensity ratio of two rays I_1 and I_2 parallel to two sets of analyzers and polarizers with corresponding angles between polarizer and analyzer of 35° and 70°, respectively.

8. A monochromatic optical ray of wavelength λ_o equal to 680 nm is incident upon a birefringent material with refractive indices n_o equal to 1.25, n_e equal to 1.5, and thickness of the material equal to 3.1×10^{-5} m. Calculate the time difference between Δt and phase difference $\Delta \phi$ of the two plane polarized components emerging at the output of the birefringent material.

9. An optical monochromatic ray of 940 nm wavelength is incident upon a metallic surface with a characteristic constant ϕ of 0.25 eV. Compute:

   i) kinetic energy $(E_{kinetic})$ required for photoelectron generation

   ii) cut-off frequency (f_c) and cut-off wavelength (λ_c)

10. Calculate the uncertainty of electron velocity when a monochromatic optical ray of 820 nm wavelength is used to determine the position of an electron within the defined wavelength. Assume particle momentum difference (Δx) equal to 4.8×10^{-7} m.

2

FUNDAMENTALS OF SEMICONDUCTOR THEORY

OBJECTIVES

1. Give a historical perspective of the development of the atomic theory that lead to semiconductor theory

2. Discuss the fundamental concepts of semiconductor theory

3. Explain *P*-type and *N*-type semiconductor materials

4. Describe the biasing techniques for *PN* semiconductor devices

5. Explain optical emission

KEY TERM

absolute temperature	carrier density	electron-volt	heterojunction
acceptor	carrier mobility	element emission	hole concentration
acceptor doping	carrier velocity	coefficiency	honojunction
Aniso-type	conduction band	energy band	indirect band gap insulator
atom	conductivity	equilibrium	intrinsic semiconductors
atomic number	conductor	Fermi efficiency	iso-type
atomic weight	crystal momentum	Fermi energy	Maxwell-Boltzmann statistics
band gap	depletion region	Fermi integral of half-order	molecule
band gap width	direct band gap	Fermi-Dirac distribution	neutron
barrier potential	donor	Fermi-Dirac functions	nondegenerate
barrier region	donor doping	field intensity	semiconductors
barrier voltage	drift current	force of attraction	*N*-type semiconductor
Bohr's atomic model	drift mobility	forward biasing	nuclear force
Boltzmann's constant	electromagnetic force	forward biasing current	nucleus
breakdown current	electron	free electron	optical emission
breakdown voltage	electron concentration	gravitational force	periodic table of elements

2.1 SEMICONDUCTOR THEORY

Elements of Atomic Theory

In the middle of the fifth century B.C., the Greek philosopher Democritus developed the hypothesis that all matter in nature is composed of **atoms,** discrete units that are not further divisible. In Greek, *atom* means single, non-divisible. His philosophical interpretation of matter was carried on by the Epicurian School of the third century B.C. For almost seventeen centuries, these original ideas and philosophies were buried intentionally by the Romans, and later covered by the shroud of medieval darkness until the seventeenth century A.D., when they were brought back to light by Galileo, Newton, Boyle, and others.

All matter in nature is composed of basic building blocks called **molecules.** In Latin, *molecule* means small object. Furthermore, each molecule is composed of a number of discrete units non-further divisible by chemical means called "atoms." If all the atoms that compose the molecules are the same, then the resultant matter is called an **nonelement.** If the atoms that compose the molecules are different, then the resultant matter is called a **substance.** Therefore, the atom is the smallest unit in an element, retaining all the fundamental chemical characteristics of that element. The atom itself is also a microcosmic system resembling, in some ways, that of a macrocosmic planetary system. It is composed of an extremely massive central core called the **nucleus,** which is surrounded by a predetermined number of orbiting **electrons.** The number and arrangement of electrons within the atomic structure determines the interaction of atoms, and also determines the overall physical and chemical properties of these elements or substances. The nucleus of the atom is not homogeneous, but is composed of subatomic particles called **protons** and **neutrons.** In Greek, proton means first. Protons are positively charged particles with an electrical charge equal and opposite to that of the orbiting electron ($+1.602 \times 10^{-19}$C), and a mass equal to 1.643×10^{-27} kg. The neutron is a particle that carries no electric charge and has a mass almost equal to that of the proton. In 1897, the English scientist J. J. Thomson discovered the electron. Ernest Rutherford discovered the proton in 1914, and James Chadwick discovered the neutron in 1932. Around 1911, Rutherford suggested that practically all the mass of the atom is concentrated in the nucleus. While the diameter of the atom is approximately 10^{-10} m, the diameter of the nucleus is equal to

10^{-15} m. Thus, protons and neutrons in the nucleus are virtually in contact with each other. This observation defied the law of electromagnetic force whereby the like charges repel each other and unlike forces attract each other. Until 1935, the two known forces in nature were the **electromagnetic force,** with a relative strength equal to 1 and gravitational force with a relative strength equal to 10^{-37}. Employing the principle of **gravitational force** to interpret why like charges such as protons are held together at the nucleus of an atom was unsuccessful, because gravitational force is too weak. On the other hand, the electromagnetic force, although strong, reacts only to electric charges in terms of repulsion or attraction and is also unable to provide solutions to the question. In 1935, the Japanese scientist H. Yucawa introduced the theory of **nuclear force.** He proposed that when protons and neutrons are in very close proximity, almost in contact with each other, a nuclear force holds them together. The same nuclear force diminishes almost to zero for a distance of separation larger than 5×10^{-15} m. On the other hand, the electrons separated from the protons by a relatively large distance and carrying opposite charges are subject to electromagnetic field influence and thus are kept in orbit around the nucleus. For the outermost electron orbit in the atom of a conducting material, the electromagnetic force is quite weak. A relatively small amount of energy absorbed by this electron will be enough for it to break away from its predetermined orbit and become a **free electron.** Free electrons and their random behavior are the main source of electronic noise.

All matter in nature is composed of atoms. Each atom is composed of protons, neutrons, and electrons. The atom is the smallest (non-further divisible by chemical means) particle in an element. All 109 natural elements are composed of different atoms. Therefore, the structure of each atom is characteristic of the element.

In 1813, Neil Bohr modeled the atom as having a spherical shape with the nucleus composed of protons and neutrons, at the center and electrons orbiting around the nucleus in predetermined orbits. **Bohr's atomic model** is illustrated in Figure 2–1.

The number of protons and electrons within an atom is characteristic of the element to which the atom belongs. Protons carry positive electric charges, while electrons carry negative electric charges. The sum total of all the positive charges in an atom is equal to the sum total of all the negative charges, therefore the atom is perceived to be electrically

Figure 2–1. Bohr's atomic model.

neutral. The natural and man-made elements are all classified in terms of their total number of electrons. This number is referred to as the *atomic number,* and it was used as the basis for the construction of the **periodic table of the elements,** starting with hydrogen. Because the hydrogen atom is composed of one proton at the nucleus and one orbiting electron, it occupies the first place in the periodic table of the elements shown in Figure 2–2.

Another way the elements are arranged is by their **atomic weight,** which is defined by the sum total of the protons and neutrons in the nucleus of the atom.

Bohr's atomic model also shows that electrons are orbiting at discrete distances from the nucleus of the atom, carrying discrete energy levels referred to as **energy bands,** or shells. The number of energy bands or shells is different in each atom, and this difference is a characteristic of the element. There is also a difference in energy levels between shells. The shells closest to the nucleus have less energy than the outer shells. The shells in the atom are classified and labeled with the letters K, L, M . . . and so on. K is the orbit closest to the nucleus. Sub-shells that carry different energy levels within each shell also exist. The electron at the bottom of the shell carries the lowest energy level, and the electron at the top of the shell carries the highest energy level (see Figure 2–3). In accordance with Bohr's atomic model, an increase of the atomic number of the atom will reflect an increase of the emitted energy from that atom because a larger number of positive charges will attract a larger number of negative electrons. Furthermore, ionization energies should drop at the start of each level because they are located at longer distances from the nucleus of the atom. However, it was observed that the ionization rise and fall energy patterns were more complex than Bohr's model predicted. This fact promoted the hypothesis that there may be sub-shells within each shell. Furthermore, the energy line spectra of individual shells were not single lines but rather multiple lines close to each other and with almost equal energy levels distinguished only by the difference of their degree of brightness and line width. For example, a sharp

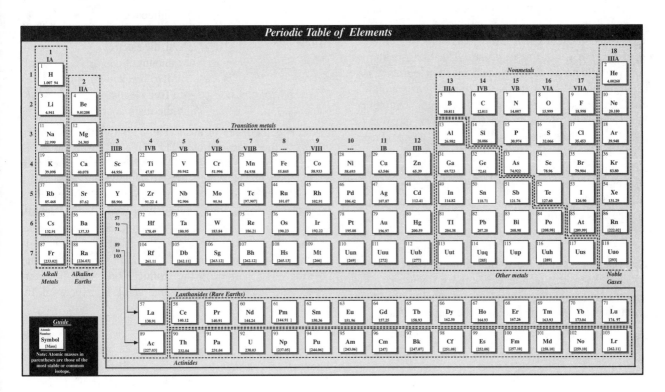

Figure 2–2. The periodic table of elements.

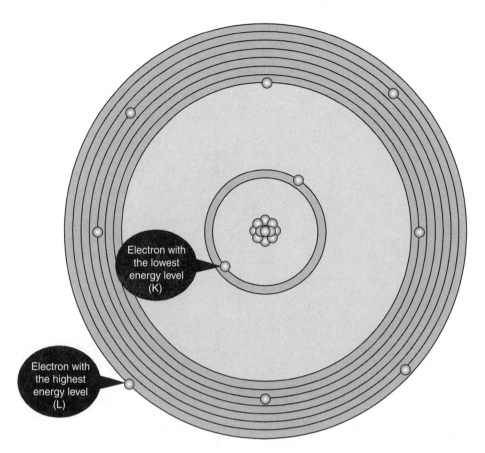

Figure 2–3. An atomic model illustrating sub-shell energy levels.

line is a characteristic of the sub-shell *s,* a bright line, referred to as principle line, is characteristic of the sub-shell *p*, a diffused line is characteristic of the sub-shell *d* and finally, a fundamental line is characteristic of the sub-shell *f*. The electrons in an atom, according to Schroedinger, are governed by four **quantum numbers.** The first quantum number, referred to as the principle quantum number, is identified by the letter (*n*) and is that of the Bohr's atomic model. The second is referred to as the orbital quantum number, is identified by the letter (*l*), and corresponds to a sub-shell with values (*l*) equal to 0, 1, 2, 3, . . . (*n*−1). That is, for *n* = 1, *l* = 0. For *n* = 2, *l* = 0, 1. For *n* = 2, *l* = 0, 1, 2 . . . The third quantum number is referred to as the orbital quantum number (*m*) reflecting the orbital alignment of the electron under the influence of an externally applied magnetic field. This quantum number takes values as follows: $m_1 = -l, (-l - 1), (-l - 2), \ldots, -1, 0, +1, \ldots, (l - 1)$, *l*. That is, for *l* = 1, *m* = −1, 0, +1 with *p* sub-shell composed of three orbits. The fourth quantum number (m_s) describes electron spin direction under the influence of an applied magnetic field. For example, for *n* equal to one, the shell contains one orbit with one or two electrons. That is,

n = 1	*n* = 2
l = 0	*l* = 0, 1
$m_1 = 0$	*l* = 0 ($m_1 = 0$, $m_s = -x, +x$)
$m_s = -x, +x$	*l* = 1 ($m_1 = -1, 0, +1$, $m_s = -x, +x$)

Thus the third sub-shell contains eighteen electrons and the fourth contains thirty-two electrons, and so on.

The **force of attraction** between electrons and protons within the atom is given by Equation (2–1).

$$F = \frac{q_n q_p}{r^2} \tag{2–1}$$

where *F* is the force of attraction, q_n is the electron mass, and q_p is the **proton mass.** The force of attraction is inversely proportional to the square of the distance between the two opposite charged particles. Therefore, the electrons that occupy the outermost orbiting shell carry the most energy. The outermost shell is referred to as the **valence shell,** and the electrons that occupy this shell are referred to as **valence electrons.** Valence electrons determine the chemical properties of the element. The relationship between the position of each shell in the atomic model and the number of electrons occupying that shell is expressed by Equation (2–2).

$$N_e = 2n^2 \tag{2–2}$$

where N_e is the maximum number of electrons and *n* is the position of the shell in the atom (the first closest to the nucleus).

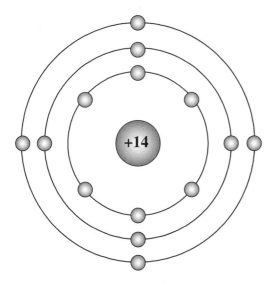

Figure 2–4. An atomic model of silicon (Si).

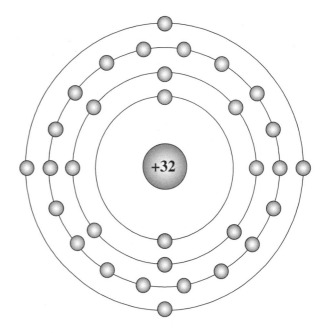

Figure 2–5. An atomic model of germanium (Ge).

Example 2–1

Calculate the maximum number of electrons occupying each of the four shells in a balanced atom.

Solution

First shell:

$$N_e = 2n^2$$
$$= 2(1)^2$$
$$= 2$$

Therefore, the maximum number of electrons occupying the first energy orbit is two.

Second shell:

$$N_e = 2(n)^2$$
$$= 2(2)^2$$
$$= 8$$

Therefore, the maximum number of electrons occupying the second orbit is eight.

Likewise, the third orbit is occupied by eighteen electrons and the fourth orbit is occupied by thirty-two electrons. All of the inner shells of the atom must be complete. That is, they must carry all the required number of electrons. The only shell that may have electrons less than the maximum is the valence shell. Figure 2–4 illustrates the atomic model of silicon (Si), and Figure 2–5 illustrates the atomic model of germanium (Ge). The atomic models of these two elements are presented because they constitute the very foundation of semiconductor devices.

Figure 2–4 shows the Si atom composed of fourteen protons and fourteen neutrons at the nucleus, surrounded by fourteen electrons distributed among three orbiting shells. It is evident that the first and the second shells are complete (they have the required number of electrons), while only four electrons occupy the valence shell.

Figure 2–5 shows the atomic model of germanium, composed of thirty-two protons in the nucleus and thirty-two electrons distributed among four orbiting shells and sub-shells. The electron distribution among shells is listed in Table 2–1.

Within the germanium atom, electrons belonging to the same shell occupy different energy levels. The shells in the atomic model of germanium are therefore divided into sub-shells in order to facilitate the different electron energy levels (see Table 2–2).

TABLE 2–1   Element: Germanium (Ge)

Shell	Required Electrons: $2(n)^2$ Max.	Available
K ($n = 1$)	2	2
L ($n = 2$)	8	8
M ($n = 3$)	18	18
N ($n = 4$)	32	4
		32 Total

TABLE 2–2   Element: Germanium (Ge)

Shell	Sub-shells	Required Electrons: $2(n)^2$ Max.	Available
K ($n = 1$)	s	2	2
L ($n = 2$)	s	2	2
	p	6	6
M ($n = 3$)	s	2	2
	p	6	6
	d	10	10
N ($n = 4$)	s	2	2
	p	6	6
	d	10	10
	f	14	14
			32 Total

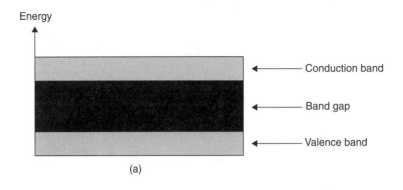

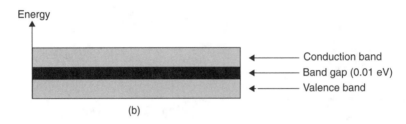

Figure 2–6. Energy gaps: **(a)** Insulator band gap, **(b)** Conductor band gap.

At a specified temperature, the atom of an element is said to be neutral. That is, all the electrons are at their predetermined orbits and their combined negative charge is balanced by the positive charges of the nucleus. If the atom absorbs additional thermal energy, the electrons in the valence shell absorb energy. If this energy level is higher than a predetermined threshold energy, the electron can leave the atom altogether and become a free electron. In this case, the remaining atom is deprived of one negative charge and therefore has become a positive ion, while the free electron carrying the fundamental negative charge has become a negative ion.

Principles of Semiconductor Theory: Insulators, Conductors, and Semiconductors

ENERGY BANDS. All elements in nature, in terms of their conductivity, are divided into two basic categories. They are either **insulators** or **conductors.**

In order to establish the difference between an insulator material (not able to conduct electric current) and a conductor material (able to conduct electric current), we must first define *energy bands*. From the atomic model, the outermost shell of an atom is referred to as the valence shell. In some elements, if sufficient thermal energy is absorbed by the atom, a valence electron may be able to achieve enough kinetic energy to escape from the valence shell and become a free-electron. Free electrons are capable of traveling between atoms, thus significantly contributing to the conducting properties of the element. It is therefore evident that the electron has been transferred from the **valence band** to the **conduction band** through thermal energy absorption.

The amount of absorption energy required by an electron to be able to transfer from the valence band to the conducting band is referred to as the *energy gap*. The distinct levels of energy gaps are what classify elements as either conductors or insulators (see Figure 2–6).

BAND GAP WIDTH. The **band gap energy** (E_g) of semiconductor materials, such as Si, Ge, or GaAs, is temperature dependent, and is expressed by Equation (2–3).

$$E_{g_{(T_x)}} = E_{g_{(T_0)}} - \frac{\alpha T^2_x}{\beta + T_x} \qquad \textbf{(2–3)}$$

where $E_{g_{(T_x)}}$ is the band gap energy at T_x, $E_{g_{(T_0)}}$ is the band gap energy at $0°$K, and (α, β) are coefficients. Table 2–3 shows that the band gap energy is higher in silicon than in germanium, and that both are lower than GaAs. It is also evident from Table 2–3 that with an increase in the operating temperature, the band gap energy correspondingly decreases, thus increasing the probability that electrons from the valence band will transfer to the conducting band to become free electrons. The band gap energy of an insulator is much larger than that of a conductor. The band gap energy levels defining insulators, conductors, and semiconductors are as follows: The band gap energy of a conductor is approximately 0.1 eV, the band gap energy of an insulator is larger than 5 eV, and the band gap energy of a semiconductor is approximately 1 eV.

TABLE 2–3 Band Gap Energy

Semiconductor	$E_g/0°$K (eV)	$E_g/300°$K (eV)	α (eV/K)	β/K
Germanium	0.7437	0.66	4.774×10^{-4}	235
Silicon	1.170	1.12	4.73×10^{-4}	636
GaAs	1.519	1.42	5.405×10^{-4}	204

Example 2–2

Calculate the band gap energy (eV) of silicon, germanium, and GaAs for temperatures ranging from 0°K to 400°K, and plot the corresponding graphs.

Solution

At $T_o = 50°/K$,

$$E_{g_{(50)}} = E_{g_{(0°)}} - \frac{\alpha T^2}{\beta + T}$$

$$= 1.17 - \frac{4.73 \times 10^{-4} \times (50)^2}{636 + 50}$$

$$= 1.161$$

Therefore, $E_g = 1.161$ eV.
At $T_o = 100°K$,

$$E_{g_{(100)}} = E_{g_{(0)}} - \frac{\alpha T}{\beta + T}$$

$$= 1.17 - \frac{4.7 \times 10^{-4} \times 100}{636 + 100}$$

$$= 1.16 \text{ eV}$$

Therefore, $E_g = 1.163$ eV.

Applying the same relationship for different temperatures gives the corresponding band gap energy levels in Table 2–4.

THE FERMI-DIRAC DISTRIBUTION. Through the **Fermi-Dirac distribution** statistics, hole-density in the valence band and electron-density in the conducting band in a semiconductor material is expressed by Equations (2–4) and (2–5).

$$dn(E) = P_c(E)f_n(E) \qquad (2–4)$$

$$dp(E) = P_v(E)f_p(E) \qquad (2–5)$$

where $P_c(E)$ is the density of state function in the conduction band, $P_v(E)$ is the density of state function in the valence band, $f_n(E)$ is the Fermi-Dirac distribution function of electrons in the conduction band, and $f_p(E)$ is the Fermi-Dirac distribution function of holes in the valence band. Furthermore, the density of state functions for both conducting and valence bands are expressed by Equations (2–6) and (2–7).

$$P_c(E) = \frac{8\sqrt{2}\pi(m_e)^{\frac{3}{2}}\sqrt{E - E_c}}{h^3} \qquad (2–6)$$

$$P_v(E) = \frac{8\sqrt{2}\pi(m_h)\sqrt{E_v - E}}{h^3} \qquad (2–7)$$

where E_v is the energy at the top of the valence band, E_c is the energy at the bottom of the conducting band, m_e is the density of state effective mass of electrons, and m_h is the density of state effective mass of holes. Equations (2–8) and (2–9) express the **Fermi-Dirac functions.** Equation (2–8) expresses the probability that an electron occupies an available state with energy (E), and Equation (2–9) expresses the probability that a hole occupies an available state with energy (E).

$$f_n(E) = \frac{1}{1 + e^{\frac{E - E_{F_n}}{kT}}} \qquad (2–8)$$

$$f_p(E) = \frac{1}{e^{\frac{E_{F_p} - E}{kT}}} \qquad (2–9)$$

where E_{F_n} is the **Fermi energy** for the electron, E_{F_p} is the Fermi energy for the hole, k is Boltzmann's constant, and T is the **absolute temperature.** Although these two energy levels are not exactly the same, for simplicity they are considered to be equal (see Equations (2–10) and (2–11)).

$$f_p(E) = 1 - f_n(E) \qquad (2–10)$$

$$f_n(E) = 1 - f_p(E) \qquad (2–11)$$

From Equations (2–10) and (2–11), it is evident the sum of the two probabilities is equal to one.

$$f_p(E) + f_n(E) = 1 \qquad (2–12)$$

In reference to (kT), it is a common practice that this product is expressed in eV as follows in Equation (2–13).

$$(kT) = \frac{T°K}{11605} \text{ eV} \qquad (2–13)$$

THE CONCENTRATION OF ELECTRONS AND HOLES IN INTRINSIC SEMICONDUCTORS. The electron and hole distribution in an intrinsic semiconductor material is defined as the product of the density of state and the Fermi distribution function, shown in Figure 2–7.

TABLE 2–4 Band Gap Energy Levels for Si, Ge, and GaAs at Different Temperatures

Semiconductors	Temperature °K							
	50	100	150	200	250	300	350	400
Si (eV)	1.168	1.163	1.156	1.147	1.137	1.125	1.111	1.097
Ge (eV)	0.7395	0.7295	0.7158	0.700	0.678	0.663	0.644	0.624
GaAs (eV)	1.514	1.501	1.485	1.465	1.445	1.422	1.399	1.376

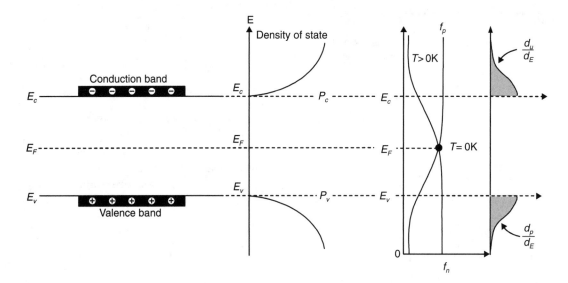

Figure 2–7. Electron and hole distribution.

ELECTRON CONCENTRATION IN THE CONDUCTING BAND. The **electron concentration** in the conducting band of an intrinsic semiconductor material is derived as follows, from the Fermi-Dirac density of state distribution function (Equation (2–14)).

$$dn(E) = p_c(E)f_n(E) \qquad (2\text{--}14)$$

Integrate

$$\int dn(E) = \int_{E_c}^{\infty} pc(E)f_n(E)\,dE$$

because

$$p_c(E) = \frac{8\sqrt{2}\pi(m_c)^{3/2}}{h^3}\sqrt{E - E_c}.$$

Substituting this into Equation (2–14) and solving yields Equation (2–15).

$$n = \frac{8\sqrt{2}\pi(m_c)^{3/2}}{h^3}\int_{E_c}^{\infty}\sqrt{E - E_c}\,f_n(E)\,dE. \qquad (2\text{--}15)$$

From Fermi-Dirac,

$$f_n(E) = \frac{1}{1 + e^{\left(\frac{E - E_{f_n}}{kT}\right)}}$$

Substituting this into Equation (2–15) yields Equation (2–16).

$$n = \frac{8\sqrt{2}\pi(m_c)^{3/2}}{h^3}\int_{E_c}^{\infty}\sqrt{E - E_c}\left(\frac{1}{1 + e^{\frac{E-E_c}{kt}}}\,dE\right) \qquad (2\text{--}16)$$

If the ratio $\dfrac{E - E_c}{kT}$ is defined as the efficiency (η), then by differentiating both parts of the efficiency equation, we have Equation (2–17).

$$d\eta = \frac{dE}{kT}$$

or

$$dE = kTd\eta \qquad (2\text{--}17)$$

Substituting Equation (2–17) into Equation (2–16) yields Equation (2–18).

$$n = \frac{8\sqrt{2}\pi(m_c)^{3/2}}{h^3}(kT)^{3/2}\int_{0}^{\infty}\frac{\sqrt{\eta}}{1 + e^{(\eta - \eta_F)}}\,dn \quad (2\text{--}18)$$

where (η_F) is the **Fermi efficiency,** expressed by Equation (2–19).

$$\eta_F = \frac{E_F - E_c}{kT} \qquad (2\text{--}19)$$

Equation (2–18) can also be expressed by Equation (2–20).

$$n = \frac{2}{\sqrt{\pi}}N_cF_{0.5}(\eta_F) \qquad (2\text{--}20)$$

where $F_{0.5}(\eta_F)$ is the **Fermi integral of half-order,** expressed by Equation (2–22) and N_c is expressed by Equation (2–21).

$$N_c = \frac{8\sqrt{2}\pi(m_c)^{3/2}}{h^3} \qquad (2\text{--}21)$$

$$F_{0.5}(\eta_F) = \int_{0}^{\infty}\frac{\sqrt{t}}{1 + e^{(t - \eta_F)}}\,dt \qquad (2\text{--}22)$$

Because the integral in Equation (2–22) is not closed, and therefore cannot be solved, an approximate solution can be derived for values of η_F much larger than one, or much smaller than one (see Equations (2–23) and (2–24)).

$$F_{0.5}(\eta_F) = \frac{\sqrt{\pi}}{2} e^{\eta_F} \qquad (2\text{--}23)$$

where $\eta_F \ll -1$, or

$$F_{0.5}(\eta_F) = \frac{2}{3} \eta_F^{3/2} \qquad (2\text{--}24)$$

where $\eta_F \gg 1$.

FOR NONDEGENERATE SEMICONDUCTORS: FERMI LEVEL WITHIN THE BAND GAP. By applying first approximation ($\eta_F \ll -1$), and because $\eta_F = \dfrac{E_F - E_C}{kT}$, yields Equation (2–25).

$$E_F - E_c \ll -kT \qquad (2\text{--}25)$$

Equation (2–25) indicates that, for **nondegenerate semiconductors,** the Fermi level must lie within the band gap and also be away from the bottom of the conducting band by several (kT).

By applying the second approximation, for $\eta_F \gg 1$, and because $\eta_F = \dfrac{E_F - E_c}{kT}$, yields Equation (2–26).

$$E_F - E_c \gg kT \qquad (2\text{--}26)$$

From Equation (2–26), it is evident that the Fermi level lies well within the conducting band and, therefore, that the semiconductor's electrical properties are very similar to those of a conductor. Because this book's intent is to describe the electrical properties of semiconductor materials, non-degenerate semiconductors will be considered. The first approximation solution reflects the properties of an **intrinsic semiconductor.** Therefore, the electron concentration in the conduction band is given by Equation (2–27) (based on **Maxwell-Boltzman statistics**).

$$n = 2 \left[\frac{2\pi(kT)m_c}{h^2} \right]^{3/2} e^{\left(\frac{E_F - E_c}{kT} \right)} \qquad (2\text{--}27)$$

HOLE CONCENTRATION IN THE VALENCE BAND. In order to determine the **hole concentration** in the valence band, the same process as that with the electrons in the conduction band is applied.

Integrating the Fermi-Dirac distribution function (Equation (2–5)) from $-\infty$ to E_v in the valence band yields Equation (2–28).

$$\int dp(E) = \int_{-\infty}^{E_v} P_v(E) f_p(E) dE$$

or

$$p = \int_{-\infty}^{E_v} p_v(E) f_p(E) dE \qquad (2\text{--}28)$$

where

$$p_v(E) = \frac{8\sqrt{2}\pi(m_v)^{3/2}}{h^3}$$

$$f_p(E) = \frac{1}{1 + e^{\frac{E_{F_p} - E}{(kT)}}}$$

Substituting $p_v(E)$ and $f_p(E)$ into Equation (2–28) yields Equation (2–29).

$$p = \frac{8\sqrt{2}\pi(m_v)^{3/2}}{h^3} \int_{-\infty}^{E_v} \sqrt{E_v - E} \left(\frac{1}{1 + e^{\frac{E_F - E}{(kT)}}} dE \right) \qquad (2\text{--}29)$$

Because $\dfrac{E_v - E}{(kT)} = \eta$ (*efficiency*), by differentiating both sides we have

$$dn = -\frac{dE}{(kT)}$$

or

$$dE = -(kT)dn$$

Substituting the above into Equation (2–29) yields Equation (2–30).

$$p = \frac{8\sqrt{2}\pi(m_v)^{3/2}}{h^3} (kT)^{3/2} \int_{0}^{\infty} \frac{\sqrt{\eta}}{1 + e^{\eta - \eta_F}} \qquad (2\text{--}30)$$

where (η_F) is the Fermi efficiency, equal to $\dfrac{E_F - E_v}{(kT)}$.

Equation (2–30) can also be expressed as follows in Equation (2–31).

$$p = \frac{2}{\sqrt{\pi}} N_v F_{0.5}(\eta_F) \qquad (2\text{--}31)$$

$$N_v = 2 \left[\frac{2\pi(kT)m_v}{h^2} \right]^{3/2} \qquad (2\text{--}32)$$

where N_v is the effective density of states in the valence band.

$$E_{(0.5)}(\eta_F) = \int_{0}^{\infty} \frac{t}{1 + e^{t - \eta_F}} dt \qquad (2\text{--}33)$$

Equation (2–33) is the Fermi integral of half-order.

Because Equation (2–33) is not closed, and therefore cannot be solved, an approximate solution can be derived for values of (η_F) much larger than one, or much smaller than one (see Equations (2–34) and (2–35)).

For $\eta_F \ll -1$,

$$F_{0.5}(\eta_F) = \frac{\sqrt{\pi}}{2} e^{\eta_F} \qquad (2\text{--}34)$$

For $\eta_F \gg 1$,

$$F_{0.5}(\eta_F) = \frac{2}{3}\eta_F^{3/2} \qquad (2\text{--}35)$$

FOR NONDEGENERATE SEMICONDUCTORS: FERMI LEVEL WITHIN THE BAND GAP. By applying the first approximation solution ($\eta_F \ll -1$), and because

$$\eta_F = \frac{E_\nu - E_F}{(kT)},$$

$$E_\nu - E_F \ll -kT \qquad (2\text{--}36)$$

Equation (2–36) indicates that the Fermi level must be within the band gap and also that it must be away from the top of the valence band by several (kT). Therefore, the hole concentration in the valence band is expressed by Equation (2–37).

$$p = 2\left[\frac{2\pi(kT)m_\nu}{h^2}\right]^{3/2} e^{\left(-\frac{E_F - E_\nu}{kT}\right)} \qquad (2\text{--}37)$$

From Figure 2–6a, it is evident that the energy gap of an insulator is very large and, therefore, that a very high energy in the order of 3 eV will have to be absorbed by an electron in order to elevate it from the valence band to the conducting band. Furthermore, all the energy states in the valence band of insulators are fully occupied by electrons while the conducting band is absolutely free of the presence of electrons. Figure 2–6b indicates that, in conductors, some electrons move freely from the valence band to the conducting band under room temperature.

In 1947, a new material was invented. The new material exhibited electrical properties of both conductors and insulators. The band gap energy level of this new material is higher than a conductor and lower than an insulator (see Figure 2–8).

Semiconductor materials were instrumental in launching a new industrial revolution—the microchip revolution. The basic semiconductor materials used in the design and fabrication of semiconductor devices are silicon (Si) and germanium (Ge). The silicon atom consists of fourteen protons in the nucleus and fourteen electrons distributed among three orbital shells. The first shell (closest to the nucleus) contains two electrons, the second shell contains eight, and the third shell, the valence shell, contains only four electrons (intrinsic material).

The germanium atom consists of 32 protons at the nucleus and 32 electrons distributed among four orbital shells. The first shell (closest to the nucleus) contains two

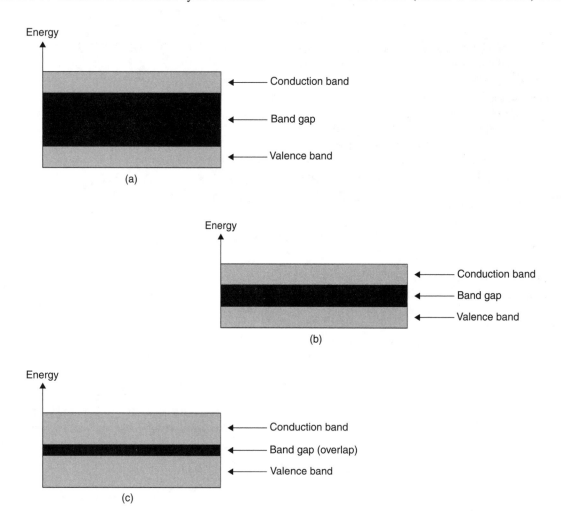

Figure 2–8. Band gap energy of (a) an insulator material, (b) a semiconductor material, and (c) a conductor material.

electrons, the second shell contains eight electrons, the third contains eighteen, and the fourth (valence) shell contains only four electrons (intrinsic material).

From the atomic models of silicon and germanium, it is evident that in a germanium atom, electrons occupying the valence shell are at higher energy levels because they are farther away from the nucleus than electrons of a silicon atom also occupying the valence shell, which are closer to the nucleus. It is, therefore, obvious that the valence electrons of a germanium atom need to absorb less energy in order to become free electrons than do valence electrons of a silicon atom. This fact makes germanium somewhat more susceptible to thermal noise and silicon the preferred choice in the fabrication of semiconductor devices.

The atomic structure of silicon has been examined. Bonding silicon atoms together forms crystalline silicon. These atoms are held together in a crystalline form by covalent bonding, shown in Figure 2–9, achieved when the electrons occupy the valence shells of other atoms (see Figure 2–10).

Figure 2–10 illustrates silicon in crystalline form. Each atom shares an electron with an adjacent atom, thus establishing the required chemical stability within the intrinsic crystalline structure. When an electron at the valence band acquires enough thermal energy or is influenced sufficiently by an electric field, it can transfer from the valence band to the conducting band, thus becoming a free electron. This free electron is the carrier of a negative electric charge. The empty space, which the electron leaves behind at the valence band, is referred to as a *hole* and represents a unit of positive charge.

In the same way, if the free electron loses its acquired thermal energy or is deprived of the influence of the electric field, a necessary condition for maintaining free electron status, it can fall back to the valence band, thus making the atom electrically neutral again. This process will continue as long as the right conditions external to the atom continue to exist. The process of electron-hole generation (shown in Figure 2–11) is referred to as electron-hole recombination. At the absolute temperature of 0°K, the crystalline silicon exists in a neutral state electrically. That is, no free electrons or positive holes exist. However, at room temperature, a number of free electrons and holes do exist and they will be present as long as the temperature is maintained equal or higher to the room temperature. These very properties make silicon and germanium elements ideal for the design and fabrication of semiconductor devices. Recently, other compounds with similar crystalline properties, such as GaAs and InP, have been used in the design and fabrication of semiconductor devices.

From Figure 2–11, it seems that electrons move in one direction and holes move in the exact opposite direction. At the state of equilibrium, the concentration of electrons and holes is equal, and the semiconductor is referred to as an *intrinsic* semiconductor.

Basic semiconductor materials such as silicon (Si) and germanium (Ge), in their intrinsic state, possess a very small number of free electrons and positive holes. Therefore, they have very little use in the fabrication of semiconductor

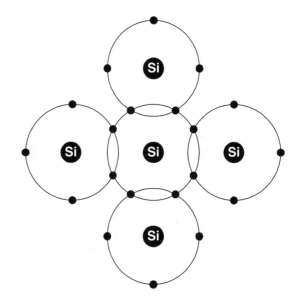

Figure 2–9. Covalent bond of Si atoms.

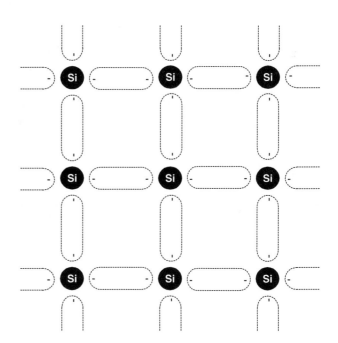

Figure 2–10. Silicon in crystalline form.

devices. The great potential of these two elements was realized when certain impurities were injected into the molecular structure of the element. This controlled molecular modification enabled silicon and germanium to substantially enhance their conducting properties, which makes them extremely valuable in the fabrication of semiconductor devices. Through the process of doping, the intrinsic state of silicon or germanium was altered, resulting in a modified semiconductor material exhibiting extrinsic properties.

The careful selection of the type of impurities and the level of doping produced two new types of semiconductor

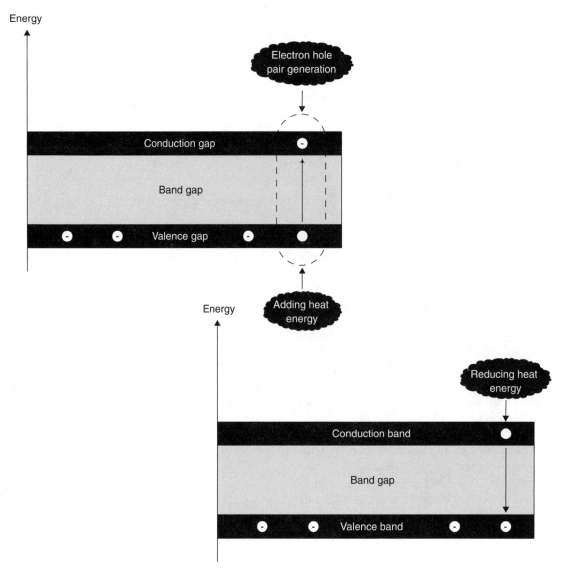

Figure 2–11. Electron hole recombination.

materials, namely the *N*-type and the *P*-type with a varying degree of conductivity.

2.2 N-TYPE SEMICONDUCTORS

To convert an intrinsic semiconductor material such as silicon or germanium into an extrinsic material, pentavalent atoms of such elements as phosphorus (P), arsenic (As), or bismuth (Bi) are induced as impurities into the crystalline structure of an intrinsic silicon or germanium element. The interaction between an arsenic atom and four silicon atoms is illustrated in Figure 2–12.

Four arsenic valence electrons are used to establish a covalent bond with each silicon atom, while the remaining fifth electron of the arsenic atom becomes a free electron.

Therefore, each covalent bond of one arsenic atom and four silicon atoms produces a free electron. This free electron directly contributes to the enhancement of the conducting properties of the semiconductor material. Because the interaction of silicon with arsenic produces a free electron, the arsenic is referred to as the **donor** atom and the silicon atoms as **acceptors.** The resultant semiconductor material is classified as *N*-type because electrons are the majority carriers and holes the minority carriers.

It is evident in the *N*-type semiconductor material that the ratio of the majority carriers (electrons) to minority carriers (holes) is directly related to the degree of doping. Light doping produces a small ratio of electrons to holes in an *N*-type material and holes to electrons in a *P*-type material while heavy doping produces a high ratio of majority to minority carriers in both *N*-type and *P*-type semiconductor materials. The product of electron and hole

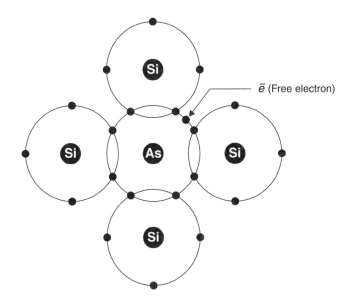

Figure 2–12. Molecular structure of an *N*-type semiconductor.

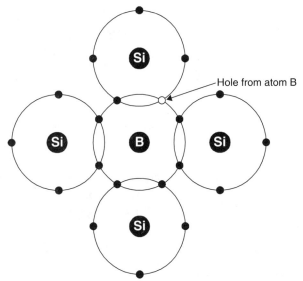

Figure 2–13. Molecular structure of a *P*-type semiconductor.

densities in an intrinsic semiconductor is given by Equation (2–38).

$$np = N_c N_v e^{\frac{-(E_c - E_v)}{(kT)}} \qquad (2\text{–}38)$$

In an intrinsic semiconductor n is equal to p, therefore the Equation (2–38) can be expressed by Equation (2–39).

$$n^2 i = N_c N_v e^{\frac{-(E_c - E_v)}{(kT)}} \qquad (2\text{–}39)$$

Furthermore, in an intrinsic semiconductor, $E_c - E_v = E_g$, therefore the above is expressed by Equation (2–40).

$$n^2 i = N_c N_v e^{\frac{-E_g}{(kT)}} \qquad (2\text{–}40)$$

Where E_c is the conducting band energy level, E_v is the valence band energy level, and E_g is the band gap energy level. Equation (2–40) indicates that in an intrinsic semi-conductor, the product of electron/hole density is independent of the Fermi level and dependent on the operating temperature.

For an intrinsic Si material at room temperature (300° K), the electron and hole densities are approximately 2.4×10^{17} carriers/m³. For an intrinsic Ge material, the electron and hole densities are approximately 1.5×10^{14} carriers/m³.

It is imperative to remember that in intrinsic semiconductors, the number of electrons is equal to the number of holes.

2.3 P-TYPE SEMICONDUCTORS

In order to produce a *P*-type semiconductor material, a trivalent atom must exist (three electrons in the valence shell). Atoms of such elements as aluminum (Al), boron (B), indium (In), or gallium (Ga) are induced into the crystalline structure of silicon (Si) or germanium (Ge) (see Figure 2–13).

In *P*-type semiconductor material, positive holes or the absence of electrons are the majority carriers. To convert an intrinsic semiconductor material such as silicon or germanium to an extrinsic semiconductor, where holes are the majority carriers and electrons the minority carriers, a trivalent atom such as boron (B) is induced as impurity into the crystalline structure of silicon or germanium.

Figure 2–13 illustrates that the trivalent boron is interacting with four silicon atoms. The three valence electrons of boron are interacting with three silicon atoms, while the fourth silicon atom is deprived of the presence of one electron. This space, unoccupied by an electron, is considered to be equivalent to the presence of a positive hole.

The resultant compound is a *P*-type semiconductor material in which the majority carriers are holes and the minority carriers are electrons. In this case, the donor material is silicon and the acceptor is boron. Controlling the acceptor level, the conducting properties of the *P*-type semiconductor material can be brought into preestablished specifications.

Carrier Mobility

If an external voltage source is applied across a **PN junction,** negative charges (electrons) and positive charges (holes) will move in opposite directions and defuse to the opposite type of materials (**carrier mobility**). The cumulative charge flow is referred to as drift current, and is a function of operating temperature, type of semiconductor material, and charge mobility. It is identified as (μ_n) for electrons and (μ_p) for holes.

For Si: $\mu_n = 0.14$ m²/Vs, and $\mu_p = 0.05$ m²/Vs

For Ge: $\mu_n = 0.8$ m²/Vs, and $\mu_p = 0.18$ m²/Vs

Drift mobility ($\mu_{n,p}$) is a measure of **carrier velocities** in (m/s) and **field intensity** in (V/m) (see Equation (2–41)).

$$\mu_{n,p} = \frac{\text{Carrier velocity (m/s)}}{\text{Field intensity (V/m)}} \quad \text{(2-41)}$$

Therefore, the unit for drift current is expressed as $\left(\dfrac{m^2}{Vs}\right)$.

The electron hole velocities in a semiconductor material are expressed by Equations (2–42) and (2–43).

$$\nu_n = E\mu_n \quad \text{(2-42)}$$

$$\nu_p = E\mu_p \quad \text{(2-43)}$$

where $\nu_{n,p}$ are the electron/hole velocities (m/s), E is the electric field intensity (V/m), and $\mu_{n,p}$ is the **drift mobility** (m^2/Vs). The electron hole velocity units are derived as follows in Equation 2–44.

$$\nu_{n,p} = E\mu_{n,p} = \text{V/m} \times m^2/\text{Vs} = \text{m/s} \quad \text{(2-44)}$$

Current Density (J-A/m^2)

In a semiconductor material, current density (J) is defined as the sum total of the electron and hole current densities (see Equation (2–45)).

$$J_T = J_n + J_p \quad \text{(2-45)}$$

where $J_n = nq_n\mu_n E$ (n−Electron density: Number of carriers/m^3), $J_p = pq_p\mu_p E$ (p−Hole density: Number of carriers/m^3), q_n is the electron charge (1.6×10^{-19}C), q_p is the hole charge (1.6×10^{-19}C), μ_n is the electron mobility (m^2/Vs), μ_p is the hole mobility (m^2/Vs), E is the electric field intensity (V/m), ν_n is the electron velocity (m/s), and ν_p is the hole velocity (m/s). In Equation (2–45), substitute for J_n, J_p with their equivalent (see Equation (2–46)).

$$J = J_n + J_p = nq_n\mu_n\overline{E} + pq_p\mu_p\overline{E} \quad \text{(2-46)}$$

For intrinsic semiconductors,

$$n_iq_n\overline{E}(\mu_n + \mu_p) = p_iq_p\overline{E}(\mu_n + \mu_p)$$

or

$$n_iq_n(\overline{E}\mu_n + \overline{E}\mu_p) = p_iq_p(\overline{E}\mu_n + \overline{E}\mu_p)$$

since

$$\overline{E}\mu_n = \nu_n \text{ and } \overline{E}\mu_p = \nu_p$$

then Equation (2–47) must be true.

$$n_iq_n(\nu_n + \nu_p) = p_iq_p(\nu_n + \nu_p) \quad \text{(2-47)}$$

The unit for current density is determined as follows.

$$J = nq\overline{E}\mu = \text{carriers/}m^3 \times \text{coulombs/carriers(V/m)}(m^2/\text{Vs})$$

Therefore, the current density is equal to amps per square meter.

$$J = \frac{A}{m^2}$$

Example 2–3

Compute the carrier velocities and the current density of an intrinsic silicon semiconductor with physical dimensions as follows:

- length: $l = 10 \times 10^{-3}$m
- width: $w = 2 \times 10^{-3}$m
- height: $h = 2 \times 10^{-3}$m

Assume a potential difference of 12 V is applied across the semiconductor.

Solution

For a meaningful solution, two assumptions must be made:
(1) The electric field must be uniform across the entire material.
(2) The current flows in the direction of the electric field.

(i) Carrier velocities: Use $\nu_n = \overline{E}\mu_n$ to compute $\overline{E}$.

$$\overline{E} = \frac{V}{m} = \frac{12V}{10 \times 10^{-3}m} = 1.2 \text{ KV/m}$$

Compute (ν_n).

$$\nu_n = \overline{E}\mu_n = 1.2 \times 10^3 \text{ V/m} \times 0.14 \ m^2/\text{Vs} = 0.168 \times 10^3 \text{ m/s}$$

Therefore, electron velocity is equal to: $\nu_n = 0.168$ Km/s.

Compute (ν_p).

$$\nu_p = \overline{E}\mu_p = 1.2 \times 10^3 \text{ V/m} \times 0.05 \ m^2/\text{Vs} = 0.06 \text{ km/s}$$

Therefore, hole velocity is equal to: $\nu_p = 0.06$ km/s.

Si: Electron mobility ($\mu_n = 0.14 \ m^2$/Vs)

Hole mobility ($\mu_p = 0.05 \ m^2$/Vs)

Compute current density (J).

$$J = J_n + J_p$$

Evaluate:

$$J_n = n_iq_n\nu_n = (2.4 \times 10^{19})(1.6 \times 10^{-19})(0.168 \times 10^3) = 645.12 \text{ A/}m^2$$

Therefore, $J_n = 645.12 \text{ A/}m^2$.
Evaluate:

$$J_p = p_iq_p\nu_p = (2.4 \times 10^{19})(1.6 \times 10^{-19})(0.06 \times 10^3) = 230.4 \text{ A/}m^2$$

Therefore, $J_p = 230.4 \text{ A/}m^2$.

Substitute J_n and J_p into J.

$$J = J_n + J_p = 645.12 \text{ A/}m^2 + 230.4 \text{ A/}m^2 = 875.52 \text{ A/}m^2$$

Therefore, $J = 875.52 \text{ A/}m^2$.

The total current in the Si semiconductor material is calculated as follows:

$$I_T = J \times A_{surface-area} = 875.52 \text{ A/}m^2 88 \times 10^{-6} \ m^2 = 77 \times 10^{-3} \text{ A}$$

Therefore, $I_T = 77$ mA.

Conductivity

The conductivity of a material is the opposite of resistance. It is expressed as $\left(\dfrac{1}{R}\right)1/\Omega$.

The resistivity of a conductor is expressed by Equation (2–48).

$$R = \rho\frac{l}{A}\,\Omega \qquad\qquad (2\text{--}48)$$

where, ρ is the resistivity. Characteristic of the material (Ωm), l is the length of the conductor (m), and A is the cross section area of the conductor (m^2).

THE CONDUCTIVITY OF EXTRINSIC SEMICONDUCTORS (σ). The conductivity of a semiconductor material is given by Equation (2–49).

$$\sigma = n\mu_n q_n + p\mu_p q_p \qquad\qquad (2\text{--}49)$$

where, σ is the semiconductor conductivity (S/m), n is the number of majority carriers in an N-type semiconductor, μ_n is the electron mobility (0.14m^2/Vs), q_n is the electron charge (1.6×10^{-19} C), p is the number of majority carriers in a P-type semiconductor, μ_p is the hole mobility (0.05m^2/Vs), and q_p is the hole-charge (1.6×10^{-19} C).

Example 2–4

The electron density of a silicon semiconductor material in the intrinsic state is doped with trivalent boron atoms. The hole density of the resultant extrinsic semiconductor material is equal to 5.7×10^{20} holes/m^3.

Assuming an electron density of silicon in the intrinsic state of 2×10^{17} electrons/m^3, determine

- the electron density (extrinsic)
- type of semiconductor (N-type or P-type)
- extrinsic semiconductor material conductivity

Solution

(i) The relationship between electron density and hole density in an extrinsic semiconductor is expressed by

$$n^2 i = n \times p$$

Solve for (n).

$$n = \frac{n^2 i}{p} = \frac{(2 \times 10^{17})^2}{5.7 \times 10^{20}} = \frac{4 \times 10^{34}}{5.7 \times 10^{20}} = 0.7 \times 10^{14}$$

Therefore, $n = 70 \times 10^{12}$ electrons.

(ii) It is evident that the number of holes is much higher than the number of electrons. Therefore the resultant extrinsic semiconductor is a P-type material. The conductivity of an extrinsic semiconductor is shown in Equation (2–49).

Substituting for values we have:

$$\sigma = n\mu_n q_n + p\mu_p q_p = (70 \times 10^{12})(0.14)(1.6 \times 10^{-19}) +$$

$$(5.7 \times 10^{20})(0.05)(1.6 \times 10^{-19}) = 20.24 \text{ S/m}$$

Therefore, $\sigma = 20.24$ S/m.

2.4   THE PN JUNCTION

P-type and N-type semiconductors are used extensively in the fabrication of practical semiconductor devices, and are used exclusively by the electronics industry.

When a PN junction semiconductor is formed, the joint region of the two materials is referred to as the PN junction. To form a PN junction, a fabrication process is used by which a gradual transition from P-type to N-type material within the same crystalline structure is achieved. Figure 2–14 is an energy diagram of a PN junction at the moment of formation.

In Figure 2–14, it appears that the two types of semiconductors are discretely jointed together. Although this is incorrect, it simplifies the explanation.

In P-type materials, holes are the majority carriers, generated by the diffusion of acceptor atoms into the crystalline structure, while electrons are the minority carriers. In N-type materials, the majority carriers are electrons, generated by the diffusion of donor atoms into the crystalline structure, and minority carriers are holes.

The diffusion process takes place at the junction of the two materials. That is, some electrons from the N-region absorb sufficient thermal energy and will be diffused through the junction to occupy the same number of holes in the P-region. At the same time, a number of holes from the P-region will be diffused through the junction to the N-region and will recombine with the same number of electrons (**drift current**). This diffusion process results in the following.

When an electron leaves the N-region (very close to the junction), it deprives the donor atom of a negative charge. Therefore, the donor atom becomes a positive ion. Likewise, when a hole leaves the P-region (very close to the junction) and diffuses into the N-region, it deprives the acceptor atom of a positive charge. Therefore, it becomes a negative ion.

Across the junction, a small electric field is developed, due to an accumulation of opposite charges, negative at the P-region and positive at the N-region. The small space that is free of charges is referred to as the **depletion region.** It is illustrated in Figure 2–15.

In a PN junction, the product of electron/hole concentration is constant at a specific temperature and is expressed by Equation (2–50).

$$n_i^2 = n_o p_o \qquad\qquad (2\text{--}50)$$

where n_i is the concentration of carriers in an intrinsic semiconductor, n_o is the electrons in **equilibrium,** and p_o is the holes in equilibrium. After the initial state of diffusion, the established positive charges at the N-region prevent holes from further diffusing from the P-region to the N-region. Likewise, electrons in the P-region prevent more electrons from diffusing from the N-region to the P-region. It is therefore evident that the diffusion of charges to the opposite regions is temporary and has very short duration. However, other charges, such as minority carriers, exist. Their contribution to the overall semiconductor material performance is quite significant.

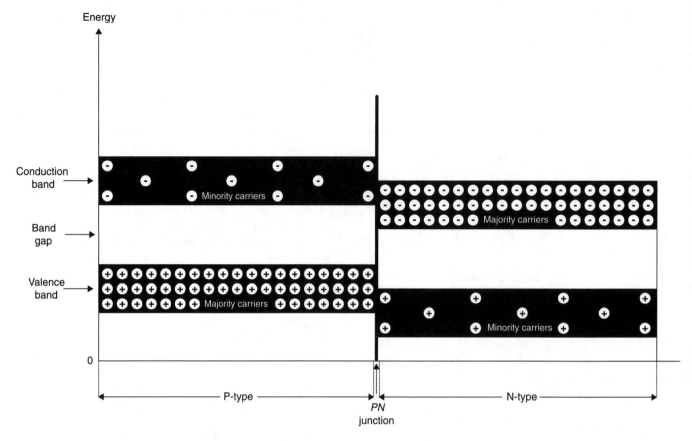

Figure 2–14.   A *PN* junction energy diagram.

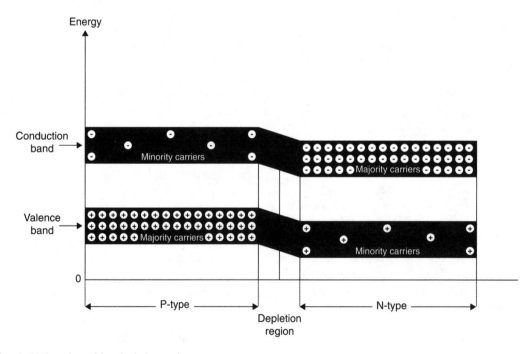

Figure 2–15.   A *PN* junction with a depletion region.

As stated previously, in a *P*-type material, majority carriers are holes and minority carriers are electrons, while in an *N*-type material, majority carriers are electrons and minority carriers are holes. During the diffusion process, minority holes from the *N*-region and minority electrons from the *P*-region are diffused to the opposite regions under the influence of the small electric field established across the junction by the diffusion process. The small current, based on minority carriers, is referred to as **reverse current**. The total sum of the diffusion and drift currents at equilibrium is equal to zero. The region close to, and alongside of the *PN*-junction at equilibrium is free of charges and therefore is referred to as the depletion or **barrier region**. The already-established electric field, through the initial majority carrier diffusion process, opposes further majority carrier diffusion.

The width of the depletion region is about 1×10^{-18} m and it is directly dependent on the degree of doping for both *P*-type and *N*-type semiconductors while the barrier voltage depends of three factors:

a. level of doping for both *N*-type and *P*-type materials

b. operating temperature

c. intrinsic semiconductor material used

From Figure 2–15, the potential difference across the depletion region due to the opposite charges is referred to as the **barrier potential** or **barrier voltage**. Barrier voltage is expressed by Equation (2–51).

$$V_B = \frac{(KT)}{q} \ln\left(\frac{N_A N_D}{n_i^2}\right) \tag{2–51}$$

where V_B is the barrier voltage (V), K is the **Boltzmann's constant** (1.38×10^{-23} J/K), T is the operating temperature ($K = 273° + °C$), q is the electron charge (1.6×10^{-19} C), N_A is the **acceptor doping** (*P*-type), N_D is the **donor doping** (*N*-type), and n_i is the intrinsic density. It is evident from Equation (2–51) that the key variable capable of inducing barrier voltage variations is the operating temperature. The first part of Equation (2–51) is referred to as the thermal voltage (V_T), and is expressed by Equation (2–52).

$$V_T = \frac{(kT)}{q} \tag{2–52}$$

Therefore Equation (2–51) can also be expressed as Equation (2–53).

$$V_B = V_T \ln\left(\frac{N_A N_D}{n_i^2}\right) \tag{2–53}$$

Example 2–5

Compute the barrier voltage of a silicon *PN* junction doped with impurities as follows:

- *P*-type (1×10^{22} acceptors/m³)
- *N*-type (1.2×10^{21} donors/m³)
- Operating temperature (25° C)

Solution
From Equation (2–51),

$$V_B = V_T \ln\left(\frac{N_A N_D}{q}\right)$$

Solve for (V_T).

$$V_T = \frac{kT}{q} = \frac{1.38 \times 10^{-23} \times (273 + 25)}{1.6 \times 10^{-19}} = 257 \times 10^{-4} \ V$$

Therefore, $V_T = 25.7$ mV.
For Si: $n_i = p_i = 2.4 \times 10^{17}$ carriers/m³

$$n_i^2 = (2.4 \times 10^{17})^2 = 5.76 \times 10^{34}$$

$$V_B = V_T \ln\left(\frac{N_A N_D}{n_i^2}\right) = (25.7 \times 10^{-3})$$

$$\times \ln\left(\frac{1 \times 10^{22} \times 1.2 \times 10^{21}}{5.76 \times 10^{34}}\right) = 0.492 \ V$$

Therefore, $V_B = 492$ mV.

PN Junction Forward Biasing

PN junction biasing is referred to as the process whereby an external DC voltage source is connected across the junction with the positive of the source connected to the *P*-type material and the negative of the source connected to the *N*-type material, as shown in Figure 2–16. In such a biasing arrangement, an external resistor (current limiting device) must be inserted in series with a power supply to protect the *PN* junction from an excess current flow.

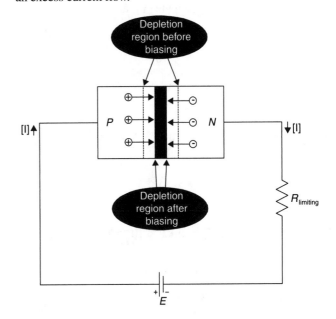

Figure 2–16. *PN* junction biasing.

In a **forward biasing** arrangement, the externally applied electric field counteracts the space charge (barrier voltage). If the biasing voltage starting from zero is progressively increasing at the barrier voltage level, the depletion region will be substantially reduced (not fully eliminated), and the diffusion current will be increased. A further increase in the biasing voltage will progressively increase the current flow through the junction and from the P-type to the N-type material (here the conventional current flow is assumed).

The biasing of a PN junction, promoting current flow composed of majority carriers, is referred to as forward biasing. In practical applications, the PN junction is referred to as the **semiconductor diode.**

The **forward biasing current** flow through the diode in relationship to the externally supplied voltage is given in Equation (2–54).

$$I = I_s\left(e^{\frac{v}{\eta V_T}} - 1\right) \qquad (2\text{--}54)$$

where I is the forward biasing current (A), I_s is the **saturation current** (A), η is the **emission coefficient** ($1 \le \eta \le 2$), and V_T is the **thermal voltage** ($\frac{kT}{q}$). Thermal voltage at room temperature is approximately equal to 0.026 V.

Example 2–6

Calculate the current through a silicon diode when forward biased and operating at 100°C with the following data:

- $I_s = 0.12$ pA at 20°C
- $V = 0.6$ V

Solution
Use Equation (2–54) to compute (V_T) at 20°C.

$$V_T = \frac{kT}{q} = \frac{(1.38 \times 10^{-23})(273 + 20)}{(1.6 \times 10^{-19})} = 0.2527 \text{ V}$$

Therefore, $V_{T(20°C)} = 0.2527$ V.
Compute (V_T) at 100°C.

$$V_T = \frac{kT}{q} = \frac{(1.38 \times 10^{-23})(273 + 100)}{1.6 \times 10^{-19}} = 0.03217 \text{ V}$$

Therefore, $V_{T(100°C)} = 0.03217$ V.

$$I = I_s\left(e^{\frac{V}{\eta V_T}} - 1\right) = 0.12 \times 10^{-12}(e^{\frac{0.6}{1.25 \times 0.03217}} - 1)$$

$$= 0.36 \text{ μA.}$$

Therefore, $I_{(100°C)} = 0.36$ μA.

The relationship between the saturation current (I_s) and operating temperature (T_x) is expressed by Equation (2–54).

$$I_{s(T_x)} = I_{s(T_0)}\left(2^{\frac{T_x}{10}}\right)$$

where $I_{s(T_x)}$ is the saturation current at operating temperature, $I_{s(T_0)}$ is the saturation current at room temperature (20°C), and T_x is the operating temperature (°C).

Example 2–7

A silicon diode exhibits a saturation current of 100 nA at 20°C. Compute the saturation current at 100°C.

Solution

$$I_{s(100°C)} = I_{s(20°C)}\left(2^{\frac{100}{10}}\right) = 100 \times 10^{-12} \times 1024 = 10.24 \times 10^{-6} \text{ A}$$

Therefore, $I_{s(100°C)} = 10.24$ μA.

PN Junction Reverse Biasing

Reverse biasing is the process whereby an external DC source is connected across a PN junction (diode). The positive of the source is connected to N-type material (cathode of the diode) and the negative of the source is connected to the P-type material (anode of the diode) (see Figure 2–17).

The externally applied electric field adds onto the space-charge field, enlarging the depletion region. The positive charges of the external source attract electrons (majority carriers) from the N-region and the negative charges of the external field attract positive charges from the P-region, depriving the junction of carrier charges and further enhancing the depletion region. Although the majority carrier flow is zero, a small current consisting of minority carriers (electrons in the P material and holes in

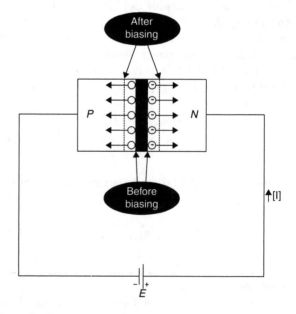

Figure 2–17. Reverse biasing of a PN junction.

the *N* material) does exist and is referred to as the **reverse biasing current.** If the reverse biasing voltage increases beyond a predetermined level characteristic of the device electrical parameters, the minority carriers will gain enough kinetic energy to cross the depletion region and interact with atoms in the opposite region. If the kinetic energy is high enough, electrons occupying the valence band will travel the band gap and transfer to the conduction band, thus becoming free electrons.

These newly generated free electrons, under the influence of a high external electric field, will generate more free electrons. The number of free electrons in the conduction band is much higher than the holes, and although some of them are lost through the recombination process, the rest of them contribute to the generation of a large current referred to as the **breakdown current.** The process is referred to as *Avalanch-breakdown* and eventually leads to the destruction of the device.

In a reverse biasing condition, the reverse current is very small until the externally applied voltage reaches a certain level referred to as **breakdown voltage** (V_{BR}).

The relationship of the reverse current, very close to the breakdown voltage, is expressed by Equation (2–55).

$$I_{BR} = \frac{I_s}{1 - \left(\dfrac{V}{V_{BR}}\right)^K} \qquad (2\text{--}55)$$

where I_{BR} is the breakdown current (A), I_s is the saturation current (A), V is the applied voltage (V), V_{BR} is the breakdown voltage (V), and K is the constant characteristic of the material ($2 \le K \le 6$). One of the critical operating characteristics of any semiconductor device is the **power dissipation.** Power dissipation is the ability of the device to dissipate the internally generated heat and protect itself from the so called *thermal runaway effect.* The power dissipation of a semiconductor diode is expressed in Equation (2–56).

$$P_{D(\text{max})} = V \times I_R \qquad (2\text{--}56)$$

where $P_{D(\text{max})}$ is the power dissipation (W), V is the applied voltage (V), and I_R is the reverse current (A). If the P_D exceeds the recommended maximum, the device will be unable to dissipate the thermal energy generated internally by the excessive current flow, resulting in permanent damage. To prevent this, it is recommended that a current limiting resistor be placed between the external voltage source and the diode.

2.5 OPTICAL EMISSION

The fundamental theory of semiconductors, the formation of *P*-type and *N*-type materials, and the behavior of the *PN* junction in forward and reverse biased states have been dealt-with quite extensively.

The depletion region and the barrier voltage of the *PN* junction are subject to carrier concentration and externally applied voltage. In forward biasing, both the depletion region and the barrier voltage are substantially reduced (close to but never zero). Therefore, holes from the *P*-region and electrons from the *N*-region can easily flow to the *N*-region and *P*-region, respectively.

Given the appropriate materials composing the semiconductor diode, the minority carrier recombination process in both types of materials, under the influence of the externally applied voltage, can lead to emission of light.

Spontaneous Emission

When electrons from the *N*-region are diffused into the *P*-region, some of them are recombined with holes, or when holes from the *P*-region are diffused into the *N*-region they are recombined with electrons. Through the recombination process, if the semiconductor materials exhibit special characteristics, the recombination process is followed by emission of light (see Figure 2–18).

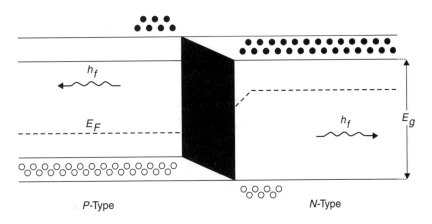

Figure 2–18. Spontaneous emission of light.

The energy released through the recombination process is estimated to be equal to the band gap energy (E_g) and it can take the form of thermal energy (heat dissipation) or photon energy (hf) of a specific wavelength (see Equation (2–57)).

$$E_g = hf \qquad \textbf{(2–57)}$$

Because $f = \dfrac{c}{\lambda}$, then Equation (2–58) must be true.

$$E_g = \frac{hc}{\lambda} \qquad \textbf{(2–58)}$$

where E_g is the band gap energy (eV), c is the velocity of light in vacuum (3×10^8 m/s), h is the **Planck's constant** (6.63×10^{-34} J/s), and λ is the **Wavelength** (m). From Equation (2–58), solving for (λ) yields Equation (2–59).

$$\lambda = \frac{hc}{E_g} \qquad \textbf{(2–59)}$$

Electron/hole recombination, resulting in the emission of light, is subject to the materials used in the formation of the PN junction. In this respect, semiconductor materials are divided into two basic categories: (1) direct band gap semiconductors and (2) **indirect band gap** semiconductors.

DIRECT BAND GAP SEMICONDUCTORS. **Direct band gap** semiconductor are materials that have electrons in the conduction band and holes in the valence band that exhibit the same level of **crystal momentum.** The crystal momentum is expressed by Equation (2–60).

$$C_m = 2\pi hK \qquad \textbf{(2–60)}$$

where C_m is the crystal momentum, h is Planck's constant (6.63×10^{-34} J/s), and K is the **wave vector momentum.** Direct band gap spontaneous emission is shown in Figure 2–19.

From Figure 2–19, because both the electrons in the conduction band and the holes in the valence band exhibit the same crystal momentum, direct recombination can occur. During the recombination process, the energy release is equal to band gap energy (E_g). In Equation (2–57), it was stated that $E_g = hf$ where hf is equal to photon energy. Therefore, during the direct band gap **spontaneous emission** process, photon generation can be achieved.

The direct method of photon generation is an efficient method because the minority carrier lifetime is very short ($\approx 1 \times 10^{-8} - 1 \times 10^{-10}$ s).

INDIRECT BAND GAP. In indirect bandgap semiconductors, the electrons in the conduction band exhibit higher values of crystal momentum than the holes in the valence band. Therefore, for electron/hole recombination to occur, the level of electron crystal momentum must be reduced to the level of the highest hole-crystal momentum in the valence band. This energy transition requires additional time, thus allowing the minority carriers a larger lifetime inside the bandgap before recombination. A larger minority carrier lifetime substantially reduces the probability of photon generation. The **radiative minority carrier lifetime** (τ_r) is expressed by Equation (2–61).

$$\tau_r = \frac{1}{B_r(P + N)} \qquad \textbf{(2–61)}$$

where τ_r is the radiative minority carrier lifetime (s), B_r is the recombination coefficient (see Table 2–5), P is the hole concentration in P-type material, and N is the electron concentration in N-type material. It is evident the recombination coefficients of GaAs and InAs are better than Si and Ge, thus making GaAs and InAs better suited for optical semiconductor devices.

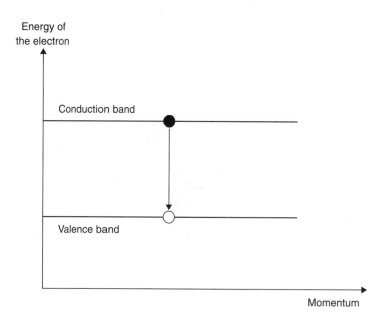

Figure 2–19. Direct band gap spontaneous emission.

TABLE 2–5   The Recombination Coefficient (B_r)

Semiconductor material	Recombination coefficient
Si	1.79×10^{-15}
Ge	5.25×10^{-14}
GeAs	7.21×10^{-10}
InAs	8.5×10^{-11}

Example 2–8

Compute the radiative minority carrier lifetime for Ge, Si, GeAs, and InAs and compare them. Assume electrons as minority carriers into the P-type material with a majority carrier concentration of $20 \times 10^{20}/cm^3$.

Solution

For Ge:

$$\tau_r = \frac{1}{B_r(P + N)} = \frac{1}{5.25 \times 10^{-14}(2 \times 10^{20}/cm^3)}$$

$$= \frac{1}{10.5 \times 10^6} = 0.095 \times 10^{-6} \text{ s}$$

Therefore, $\tau_r = 0.095$ μs.

For Si:

$$\tau_r = \frac{1}{Br(P + N)} = \frac{1}{1.79 \times 10^{-15}(2 \times 10^{20})}$$

$$= \frac{1}{3.58 \times 10^5} = 2.79 \times 10^{-6} \text{ s}$$

Therefore, $\tau_r = 2.79$ μs.

For InAs:

$$\tau_r = \frac{1}{B_r(P + N)} = \frac{1}{8.5 \times 10^{-11}(2 \times 10^{20})}$$

$$= \frac{1}{17 \times 10^9} = 0.059 \times 10^{-9} \text{ s}$$

Therefore, $\tau_r = 59$ ps.

For GaAs:

$$\tau_r = \frac{1}{7.2 \times 10^{-10}(2 \times 10^{20})} = \frac{1}{14.4 \times 10^{10}} = 0.07 \times 10^{-10} \text{ s}$$

Therefore, $\tau_r = 7$ ps.

It is evident that the radiative minority carrier lifetime is much smaller for InAs and GaAs than for Si and Ge. Therefore, InAs and GaAs are better suited for the fabrication of semiconductor diodes than Si or Ge are.

Stimulation Emission of Radiation

The fabrication of laser diodes requires the utilization of semiconductor materials that exhibit special characteristics, such as the ability to form PN junctions, high quantum efficiency, and a wavelength of the light generated within the 800 nm to 1800 nm range.

Heterojunction

The optical performance characteristics of a single PN junction laser diode can be enhanced through the implementation of several layers of extrinsic semiconductor materials of different band gaps, thus enclosing the original PN junction diode. The extrinsic semiconductor layers, used as interfaces between two **homojunction** materials, are referred to as a **heterojunction.** In order to fully understand the characteristics of a heterojunction structure, we must first fully understand the fundamental deficiencies of a simple PN junction laser diode. The deficiencies of such a laser diode are its inability of complete carrier confinement within the active layer and excessive absorption of light at the edges of the absorption region.

Confining the laser diode between heterojunction layers can compensate for these deficiencies. Heterojunction semiconductors belong to one of two categories, they are either (1) **iso-type,** or (2) **aniso-type.**

An iso-type (equal) semiconductor material is in the form of (*n-n*) or (*p-p*), while an aniso-type semiconductor material is in the form of (*n-p*) or (*p-n*). The contribution of an iso-type heterojunction layer alongside a laser diode is an improvement in the minority carrier confinement, resulting in a reduction of the diffusion length and ultimately leading to a reduction of the minority carrier lifetime. Furthermore, iso-type semiconductor interfaces, because of their unique transparent properties, reduce light absorption in the active layer.

Aniso-type heterojunction semiconductor layers, used as dielectric isolators with band gaps larger than those of the active layers, contribute to the enhancement of quantum efficiency. Another major advantage of heterojunction structures incorporated in the design and fabrication of laser diodes is the substantial reduction of the forward current required for laser action to be initiated, and therefore a reduction in the forward biasing voltage.

The implementation of heterojunction materials in the fabrication of LEDs and laser diodes goes as far back as the 1960s. With the introduction of direct band gap (III-V) GaAs, GaP, and GaAsp semiconductor materials, laser diodes were able to generate light covering the 850 nm to 910 nm optical spectrum with a reasonably satisfactory quantum efficiency while operating at low temperatures. Because system applications were demanding higher performance optical sources (higher optical power with less biasing current), new fabrication techniques such as the liquid-phase-epitaxy (LPE) were developed for single heterojunction laser diodes (GaAs/AlGaAs) structures.

As a matter of natural laser device evolution, the above technique was employed in the fabrication of double heterojunction structures composed of As/GaAs/AlGaAs and Al/Ga/ compounds.

Double heterojunction sources were capable of generating optical carrier waves covering the range between 800 nm and 1700 nm at room temperature and with relatively small biasing currents.

To further improve laser diode performance, allowing for full control of band gap width and lattice match, quaternary alloys such as InGaAs and Ga/AlAsSb were introduced. The quaternary alloys employed in the fabrication of laser semiconductor diodes allow for full band gap control, lattice parameter flexibility, and wave generation covering the 1000 nm to 1700 nm optical spectrum. InAs is also used for the active layer, exhibiting band gap energies in the range of 1.37 eV.

SUMMARY

In this chapter, the fundamental concepts of atomic and semiconductor theory were discussed in some detail, including Bohr's atomic model and the atomic structures of silicon and germanium. The principles of semiconductor theory were also discussed, including such topics as energy bands, Fermi-Dirac distribution, nondegenerative, and highly degenerative semiconductors. *P*-type and *N*-type semiconductor materials were dealt with, as well as *PN* junction biasing techniques (including the voltage-current relationship). The chapter concluded with the introduction of spontaneous and stimulated optical emission.

QUESTIONS

Section 2.1

1. List, in chronological order, the main discoveries contributing to the formulation of the atomic theory.
2. What is a *free electron*?
3. Describe Neil Bohr's atomic model.
4. What does *fourth quantum number* indicate?
5. Sketch the atomic models of silicon and germanium, indicating the number of protons and electrons in the outer orbit.
6. Describe insulator and conductor materials in terms of their atomic structure and band gap width.
7. Describe electron density in the conducting band and hole density in the valence band through Fermi-Dirac distribution statistics.
8. What are the fundamental parameters incorporated in the Fermi-Dirac function?
9. What are *intrinsic semiconductors*?
10. Describe the relationship of the Fermi efficiency in terms of its fundamental components.
11. In terms of the Fermi level within a band gap, describe nondegenerative semiconductors.
12. What does Maxwell-Boltzmann's statistic represent?

Section 2.2

13. With the assistance of a diagram, briefly describe *N*-type semiconductor material.
14. Briefly describe why the product of electron/hole density in an intrinsic semiconductor is independent of the Fermi level.

Section 2.3

15. What is a *P-type semiconductor*?
16. Sketch the molecular structure of a *P*-type semiconductor and explain its principle function.
17. What is *drift current*?
18. Define *drift mobility* in terms of carrier velocity and field intensity.
19. List the components that determine current density in a semiconductor material.
20. Define *conductivity* of a material.
21. Give the mathematical expression of semiconductor conductivity and list the contributing parameters.

Section 2.4

22. What is a *PN semiconductor*?
23. With the assistance of an energy diagram describe the operation of a *PN* semiconductor.
24. Define *minority* and *majority carriers*.
25. Define *reverse current*.
26. List the all the components incorporated into the relationship that determines the forward current through a *PN* junction.
27. Describe the relationship between saturation current and the operating temperature of a *PN* junction.
28. What are the factors contributing to breakdown current? Use the appropriate mathematical formula and explain.
29. Define *thermal runaway effect*.

Section 2.5

30. Describe *optical emission*.

31. With the assistance of a schematic diagram, describe the concept of *spontaneous emission of light*.

32. Give the mathematical expression of the relative mobility carrier lifetime and list its determining parameters.

33. Define *stimulation emission of radiation*.

34. What is a *heterojunction semiconductor structure*?

35. List and define the two categories of heterojunction semiconductors.

PROBLEMS

1. Calculate the number of electrons occupying each of the five shells and the maximum number of electrons of a balanced atom.

2. Compute and compare the band gap energies in (eV) of silicon, germanium, and gallium arsenide for the following operating temperatures: $0°\,K$, $120°\,K$, $240°\,K$, and $350°\,K$.

3. A potential difference of 8.5 V is applied across an intrinsic silicon semiconductor material with length 12 mm, width 3 mm, and height 1.5 mm. Compute carrier velocities and current density.

4. An extrinsic semiconductor exhibits material conductivity of 75 S/m. If the hole density is 6.3×10^{12} charges/m^3, compute electron density and determine the type of semiconductor material.

5. The impurity content of a silicon *PN* junction is 1.5×10^{23} acceptors/m^3 for the *P*-type and 0.85×10^{21} donors/m^3 for the *N*-type semiconductor materials. If the device is operating at 35°C, calculate the barrier voltage (V_B).

6. Calculate the operating temperature of a silicon diode if when it is forward biased with a biasing voltage of 1.2 V, it generates a forward current of 20 µA. The device operating characteristics are as follows:

  (i) saturation current = 0.32 pA at 20°C

  (ii) emission efficiency = 1.25

3

OPTICAL SOURCES

1. Establish the need for semiconductor optical sources
2. Determine the differences between LEDs and laser diodes in terms of operating characteristics and performance
3. Explain, in detail, the structure and operating characteristics of LEDs
4. Discuss the fundamental principles of edge emitting LEDs and surface emitting LEDs
5. List and discuss the most important methods of coupling LEDs into optical fibers
6. Explain the principle of operation of lasers
7. Describe in detail the types, structure, and operating characteristics of laser diodes, with emphasis on the latest development in laser technology, including vertical cavity surface emitting (VCSEL) and multi-quantum well (M-QW) lasers

KEY TERMS

absorption
active layer
active layer thickness
aluminum gallium arsenide (AlGaAs)
back reflections
band gap
base failure rate
Butt direct coupling method
carrier lifetime
carrier recombination
cavity length

cavity refractive index
coupling efficiency
distributed feedback laser (DFB)
distribution trace (DT)
doping density
edge emitting LED
electro-luminescence
emission
epitaxial growth
Fabry-Perot laser
fall time

forward biasing voltage
forward current
frequency domain
full width at half maximum (FWHM)
gain average lifetime
gallium arsenide (GaAs)
half-length
half power point
heterojunction laser
injection efficiency
internal cavity losses

laser
LED bandwidth
lens method of coupling
light emitting diode (LED)
mean wavelength
mirror losses
modal gain
mode spacing (MS)
modulation bandwidth
noise intensity
nonlinear process
nonradiative recombination

3.1 INTRODUCTION

The primary optical source of a fiber optics transmission system is the semiconductor **light emitting diode (LED)**. The selection of this device as the primary optical source was based on its ability to provide optical power ranging from 0.05 mW to 2 mW over optical fibers several kilometers in length. Semiconductor optical sources are very reliable devices with long operating life projections. Another advantage of the devices is that for ordinary system applications they require no modulating circuitry because the optical power of the device can be altered in accordance with the input current variations.

In the early 1970s the basic device structure was composed of **gallium arsenide (*GaAs*)** and **aluminum gallium arsenide (*AlGaAs*)** materials that were capable of generating wavelengths of 800 nm to 900 nm.

These LED devices were limited to short transmission optical links because the optical fibers' large signal attenuation at high data rates. Continuous improvement in optical fiber manufacturing technology, especially in the 1700 nm to 1500 nm wavelengths, significantly improved the overall data transmission capability of LED devices. The continuous drive to satisfy an ever-increasing system performance demand generated the need for the development of improved LED devices such as indium gallium arsenide phosphorus (InGaAsP). These devices were capable of generating light power covering the 920 nm to 1650 nm spectrum.

Today, two basic types of semiconductor optical sources exist. The LEDs we have just briefly described, and **laser diodes**. Laser diodes are used for long distance, high data rate transmissions, while LED devices are used for shorter distance, lower data rate transmission. The progressive evolution of LED technology led to the development of two types of LED devices: edge emitting LEDs (ELEDs), and surface emitting LEDs (SLEDs).

Edge emitting LEDs are used for both single-mode and multi-mode operations with bit rates in excess of 400 Mb/s. **Lasers** made their appearance in the early 1960s. However, these optical sources had to wait for another ten years for full implementation in optical communications systems. This was essential if improvements in areas such as photon confinement, excessive device degradations, improved stability, and forward current biasing requirements were to be made.

In the early 1970s, AlGaAs **heterojunction lasers** emitting at wavelengths between 800 nm and 4000 nm were already operational. To be compatible with optical fibers operating at wavelengths between 1000 nm and 2000 nm, laser diodes composed of AlGaAsSb/GaSb and InGaAsP/InP alloy structures were fabricated. Other diode structures, employing such alloys as GaAlInN and MgZnSSe, are already in the advanced state of development and are capable of emitting at blue and green wavelengths.

The demand for high bandwidths and long optical link distances accelerated the development of sophisticated laser structures. The evolution of advanced optical fibers, erbium doped fiber amplifiers (EDFAs), optical add/drop multiplexers (OADMs), Mach-Zehner modulators, and dispersion compensation modules in the middle of 1990s, in conjunction with advanced laser diode designs made possible the development of **dense wavelength division multiplexing (DWDM)** optical systems. Laser diodes that required relatively low forward current to achieve lasing, and that exhibited characteristics such as high **optical efficiency,** higher optical power, very narrow **spectral linewidth, operating wavelengths** between 630 nm and 1550 nm, and direct modulation capability were absolutely essential for the development of DWDM long distance optical links.

Since the 1990s, laser diode technology has dramatically improved device performance characteristics through the introduction of **vertical cavity surface emitting lasers (VCSEL)** and **quantum well (QW) lasers.** Such diodes exhibit a superior optical efficiency (conversion from electrical to optical power), tunable capabilities for a wide wavelength area, extremely high modulating rates (GHz), and very narrow line widths down to (KHz) range—all necessary parameters for implementation in DWDM optical systems.

Surface emitting LEDs are better suited to multi-mode, medium range optical fiber transmission. The fundamental theory for the design and construction of LED and laser optical sources is based on electron hole recombination within a semiconductor material. This electron hole recombination results in a generation of photons that have frequencies that are determined by the physical parameters of the semiconductor material. The two phenomena derived from the interaction between matter and light are **emission** and **absorption.** Emission can also be divided into two basic

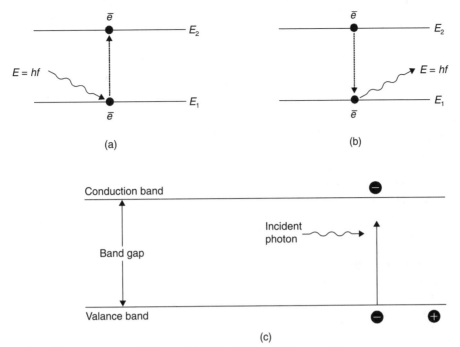

Figure 3–1. (a) Absorption, (b) spontaneous emission, and (c) stimulated emission.

categories: **stimulation emissions** and **spontaneous emissions.** Both stimulation emission and spontaneous emission can be better explained with the assistance of the following oversimplified atomic-model (Figure 3–1).

An electron of energy E_1 can be elevated to level E_2 by absorbing a photon of energy $E = hf$. An electron that is already at energy level E_2 can decay to energy level E_1 by releasing a photon (spontaneous emission). The third phenomenon is observed when an electron already at energy level E_2 absorbs a photon and decays to energy level E_1. Through this process, two photons that have the exact same phase direction and energy levels are simultaneously released. If continuous radiation is maintained, the generation of photons will also be continued (stimulation emission).

LED operation is based on spontaneous emission, while the laser operation is based on stimulated emission. In general, for both devices, the principle of operation is based on the interaction of light and matter within a semiconductor material.

3.2 LIGHT EMITTING DIODES (LEDS)

Perhaps the best representative of the optical source device family is that of the light emitting diode (LED). LEDs are classified as **electro-luminescent** devices that are composed of semiconductor materials that are capable of generating light when they are forward biased by a current source. Basic semiconductor materials that compose LED structures are GaAsP, GaAlAs, GaAs, and GaP. Gallium arsenide phosphate (GaAsP) LEDs generate light between 640 nm and

700 nm, with a peak optical power at 660 nm. Gallium aluminum arsenide (GaAlAs) LEDs produce light between 650 nm and 700 nm, with a peak optical power at 670 nm. Gallium phosphate (GaP) LEDs generate light between 520 nm and 570 nm, with a peak optical power at 550 nm. Indium gallium arsenide phosphorus devices are recent developments that present advantages over GaAsP semiconductor devices. These devices can generate optical wavelengths between 930 nm and 1650 nm. The selection of semiconductor materials and their proportional contribution to optical device fabrication is indicative of the wavelengths required. The optical energy obtained from semiconductor material combination is measured in electron volts (eV), as follows:

Ge = 0.7 eV	GaP = 2.2 eV
Si = 1.1 eV	GaAs = 1.4 eV
Cds = 2.4 eV	GaAlP = 0.8 − 2.0 eV

The center wavelength (λ_c) of an LED device is determined by the **band gap** energy (E_g) in eV at the active layer, given by Equation (3–1).

$$\lambda_c = \frac{hc}{E_g} \tag{3–1}$$

where h is Planck's constant (6.63×10^{-34} J/s), c is the velocity of light in vacuum (3×10^8 m/s), and E_g is the band gap (eV).

Edge Emitting LEDs

Edge emitting LEDs were first introduced in the mid-1970s. Their basic structure closely resembles the laser diode, with one fundamental difference. For laser diodes, positive feedback is promoted in order to enhance stimulated emission, while with edge emitting LEDs, the feedback mechanism is suppressed for the exact opposite reason, to prevent the device from going to a saturated emission mode of operation. The **active layer** (of *n*-AlGaAs) **thickness** of 0.05 μm is confined by two layers of *p* and *n* semiconductor materials, such as *p*-AlGaAs and *n*-AlGaAs with a thickness of 0.115 μm. External to these two optical guiding layers are another two layers of p^+-GaAs and n^+-GaAs with a corresponding thickness of 3.5 μm. The optical guide layers confine the generated light into the **active layer.** This represents an advantage when coupling the source to the optical fiber. The basic device structure is shown in Figure 3–2.

An external DC source, connected across the device, will provide the necessary **forward biasing voltage**. The ejected electrons from the *n*-AlGaAs and holes from the *p*-AlGaAs layers will recombine at the thin *n*-AlGaAs layer. During the recombination process, a certain number of photons will escape through the edge of the active layer. By reducing the active layer, self-absorption is kept at a minimum, thus eliminating the possibility of stimulation emission. A typical voltage/current characteristic curve of an edge emitting LED (E-LED) is shown in Figure 3–3.

Figure 3–3 shows the current increasing exponentially beyond the threshold biasing voltage, which is characteristic of the device. The current (*I*) generated by the **forward biasing voltage** of the E-LED device is used to determine the optical power generated by that device (see Figure 3–4).

LED Characteristics

One of the most important characteristics of an LED source is that of power efficiency. Power efficiency is subdivided into two categories: **internal power effi-**

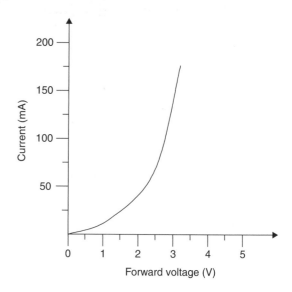

Figure 3–3. A typical voltage/current characteristic curve.

ciency, also called quantum efficiency (η_g), and **external power efficiency** (η_c).

Internal power efficiency is defined as the ratio of photons generated to the number of electrons induced into the active layer of the device (see Equation (3–2)).

$$\eta_g = \frac{N_{ph}}{N_{e^-}} \qquad (3\text{--}2)$$

where η_g is the **quantum efficiency** (%), N_{ph} is the number of photons, and N_{e^-} is the number of electrons. **External power efficiency** (η_c) is defined as the ratio of the optical power coupled into the fiber to the electrical power applied by the optical device (see Equation (3–3)).

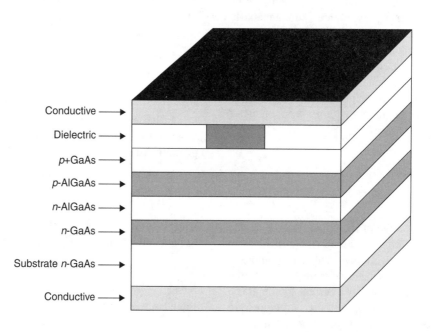

Figure 3–2. Edge emitting LED device structure.

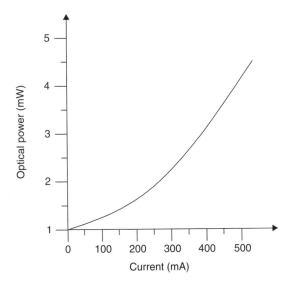

Figure 3–4. Current/optical power output characteristic curve.

$$\eta_c \% = \frac{P_F}{P_{in}} \times 100 \qquad (3\text{–}3)$$

where η_c is the external power efficiency, P_F is the optical fiber power, and P_{in} is the input power.

Optical fiber power (P_F) is only a fraction of the power generated internally by the optical device. This optical power loss is relevant to the device-optical fiber coupling efficiency expressed by Equation (3–4).

$$\eta_c = (NA)^2 \qquad (3\text{–}4)$$

where *NA* is the **numerical aperture** of the optical fiber.

Example 3–1

A light source generating an optical power output equal to 1 µW is coupled into an optical fiber with a cross sectional area larger than that of the active area of the light source.

Determine the power coupled into the fiber with a fiber $\theta°$ equal to 15°.

Solution

$$\theta = 15°$$

Therefore, $\sin\theta° = NA$.

$$\sin(15°) \cong 0.26 = NA$$

Coupling efficiency: $\eta_c = (NA)^2 = (0.26)^2 = 0.0676$

Power coupled into the fiber: $\eta_c = \dfrac{P_F}{P_{in}}$

or $P_F = \eta_c \times P_{in} = 0.0676 \times 1\ \mu W = 67.6\ nW.$
Therefore, $P_F = 67.6\ nW.$

It is evident from the above example that only a small fraction of the optical power generated in the active region of the optic device will be coupled into the optical fiber.

Example 3–2

Calculate the optical power coupled into the fiber, generated by an optical source with a bias current of 20 mA and forward voltage of 1.5 V. The source internal efficiency is 2% and the fiber $\theta° = 20°$.

Solution
Power input to the optical device:

$$P_{in} = I_F \times V_F = 1.5\ V \times 20\ mA = 30\ mW$$
$$\therefore P_{in} = 30\ mW$$

The power output of the optical source is given by: $\eta_{lin} = \dfrac{P_o}{P_{in}}$

or $P_o = \eta_{lin} \times P_{in} = 0.02 \times 30\ mW = 0.6\ mW.$
Therefore, $P_o = 0.6\ mW.$
This power output (P_o) of the optical source becomes the power input to the optical fiber.

$$P_{in} = 0.6\ mW$$

From $\eta_c = \dfrac{P_F}{P_{in}}$, solve for P_F.

$$P_F = \eta_c \times P_{in}$$

and $\eta_c = (NA)^2$ where $NA = (\sin\theta)^2$.
Therefore, $\eta_c = (\sin 20°)^2 = 0.116.$
Substituting into P_F, $P_F = \eta_c \times P_{in} = 0.116 \times 0.6\ mW$
$= 0.0696\ mW.$
Therefore, the optical power coupled into the optical fiber is:
$P_F \cong 70\ \mu W.$

A more detailed relationship establishing the maximum power coupled into the optical fiber is given by Equation (3–5).

$$P_{Fmax} = I_o A_{min} \pi (NA)^2 \qquad (3\text{–}5)$$

where P_{Fmax} is the maximum power coupled into the fiber, A_{min} is the minimum cross section area (selected between the active region of the optical source, or the cross section area of the fiber), *NA* is the numerical aperture of the fiber, and I_o is the ratio of the source optical power output and active area.

LED Spectral Bandwidth at the Half Power Point (Δλ)

The LED **spectral bandwidth (LED bandwidth)** determined at the **half power point** (50%) of the spectral density in reference to wavelength is illustrated in Figure 3–5.

LED optical devices exhibit a spectral bandwidth of between 20 nm and 200 nm at the half power point. This spectral bandwidth is translated to the pulse broadening as it travels through the optical fiber per kilometer of fiber length. LEDs emitting at peak wavelength of 800 nm exhibit a pulse broadening of 5 ns/km. This disadvantage of the LED device can be controlled by shifting the peak wavelength from

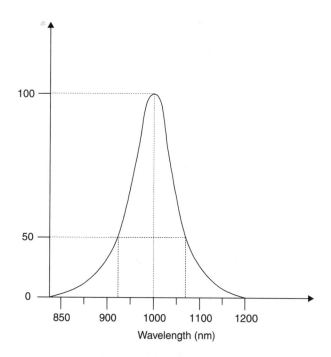

Figure 3–5. An LED spectral bandwidth curve.

800 nm to 1300 nm. At this wavelength, a smaller dispersion is encountered.

LED Bandwidth

Light emitting diodes are intensity modulated devices. That is, the input current can directly affect the output intensity of the device. In digital transmission, the LED device is turned ON and OFF in accordance with the input binary data. Ideally, turning ON and OFF the device must occur simultaneously with the input binary data. In reality there is a time delay between the bias current changes and turning ON and OFF the LED. This delay is caused by the **rise time** (t_r) and **fall time** (t_f) of the LED source (Figure 3–6).

Rise time (t_r) is measured between 10% and 90% of the power output. That is the time it takes the **output power** to increase from 10% to 90%. Fall time (t_f) is the time it takes the output to decrease from 90% to 10%. Rise time and fall time are the two most significant factors contributing to LED bandwidth limitations, which in turn establish the maximum data bit rate the device is capable of handling. This total time delay is the result of factors such as carrier recombination, time and space change capacitance inherent to the LED device physical dimensions, and semiconductor properties. The equation that establishes total optical bandwidth at the half power point is given by Equation (3–6).

$$B_W = \frac{1}{2\pi r} \qquad (3\text{–}6)$$

where B_W is the bandwidth and r is the **carrier lifetime** or **carrier recombination.**

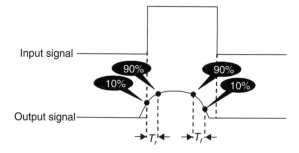

Figure 3–6. Time delay.

A more practical formula that establishes optical bandwidth is given by Equation (3–7).

$$B_W = \frac{0.35}{t_r} \qquad (3\text{–}7)$$

where t_r is the rise time.

Example 3–3
Determine the bandwidth of an LED source with a rise time equal to 10 ns.

Solution

$$B_W = \frac{0.35}{t_r} = \frac{0.35}{10 \times 10^{-9}\,\text{s}} = 35 \times 10^6\,\text{Hz}$$
$$B_W = 35\,\text{MHz}$$

Therefore, 35 MHz is the maximum operating bandwidth of this LED source.

The **modulation bandwidth** of an LED device can be increased by increasing the carrier concentration in the active region, with a simultaneous decrease of the carrier lifetime. This, of course, has a negative effect on the LED overall optical output power. Therefore, a compromise between modulation bandwidth and power output must be reached during the design process.

Surface Emitting LED

The design of a surface emitting LED (SLED) was based on a massive electron injection into a thin, optically transparent layer of *p*-material. This thin layer, confined between two other layers with larger band gaps, secures the confinement of the injected carriers, thus promoting a higher degree of recombination and a larger number of photon generations (see Figure 3–7).

In contrast to edge emitting LEDs, the optical radiation of the surface emitting LED takes place from the surface of the active layer. An examination of an SLED cross section reveals the following: The SiO_2 layer acts as an insulator between the GaAs *p*-layer and metallic conduct. The other two AlGaAs and GaAs *p*-materials are performing dual functions: light

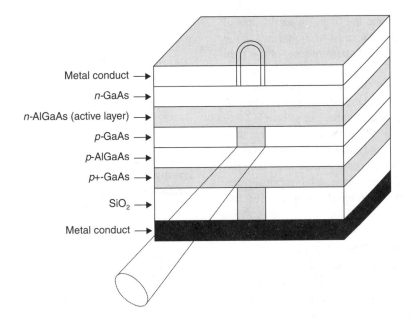

Figure 3–7. SLED cross section.

confinement and minimization of the recombination process close to the *p*-GaAs and *n*-AlGaAs junction.

The active region substrate of the SLED device is etched away in a well-type manner. This reduction of the active region dramatically reduces the recombination process, while the well-shaped area enhances the focus of the emitted light into the optical fiber. Another structure of an SLED semiconductor device is shown in Figure 3–8.

The SLED optical source shown in Figure 3–8 is designed to generate optical power in the 1.3 μm range. It uses an InP substrate because InP is transparent to this wavelength. The other four layers are grown **epitaxially** on this substrate, with varying doping levels and thicknesses in order to facilitate the device's design objectives (maximization of optical power and modulation speed). The first layer grown on the wafer substrate is a buffer composed of *n*-InP substance with an average thickness of 3 μm and a **doping density** of $2 \times 10^{18}/\text{cm}^3$.

The second layer is the much thinner active layer of *p*-I$_n$PGaAsP substance, with an average thickness of 1 μm doped with Zn to an average density of $2 \times 10^{18}/\text{cm}^3$.

The third layer is the cladding layer, which is somewhat thicker than the first layer, with an average thickness of 2.5 μm and, similarly, doped with Zn to an average density of $2.75 \times 10^{18}/\text{cm}^3$. The fourth and final layer is composed of a very thin (0.25 nm) layer of *p*-ZnGaAsP substance, which is heavily *p*-doped and is intended to minimize conduct resistivity.

Coupling SLED Devices into Optical Fiber

One of the major problems inherent to optical fiber communication systems is that of the coupling of optical power, that is generated by the optical source, into the optical fiber. The ratio of the optical power coupled into the fiber to the power generated by the optical source is called the **coupling efficiency** and is given by Equation (3–8).

$$\eta_c = \frac{P_F}{P_s} \qquad \textbf{(3–8)}$$

where η_c is the coupling efficiency, P_F is the optical power coupled into the fiber, and P_S is the optical power generated by the source.

The inability to transfer all the generated optical power of the source into the fiber is caused by the different physical

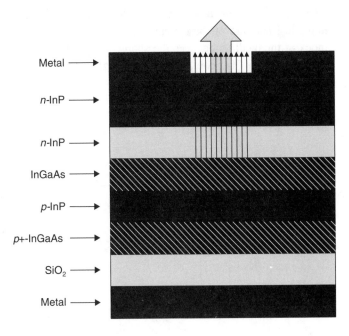

Figure 3–8. SLED device structure.

characteristics of the source and the fiber. Either the optical area of the source is larger than the cross-sectional area of the core of the fiber, or the numerical aperture (*NA*) of the light beam is larger than that of the optical fiber (*NA*). In the first case, when the optical area of the source is larger than the cross-sectional area of the optical fiber, only a fraction of the generated optical power of the source can be coupled into the fiber. Similarly, when the *NA* of the beam is larger than the *NA* of the core of the fiber, only some of the optical power will be coupled into the fiber. In order to maximize the coupling efficiency between optical source and the optical fiber, two methods have been implemented: the Butt method and the lens method.

THE BUTT METHOD. The simplest method of coupling optical power, generated by the SLED device, into the fiber is the **Butt direct coupling method**. The fundamental requirement for efficient Butt coupling is that the core cross-sectional area of the fiber must be at least equal to the optical emission area of the optical source. If the source area is larger than the core area, only a fraction of the generated optical power will be coupled into the cladding of the fiber, resulting in a quick attenuation of that optical power. Attenuation of the power coupled into the cladding section of the optical fiber can be considered optical power loss. Figure 3–9 illustrates the Butt direct coupling method.

The coupling efficiency obtained by the Butt method for a uniformly exited guiding source (lambertian) such as the SLED devices varies with step index and graded index fibers. For $y > x$, the coupling efficiency is given by Equation (3–9).

$$\eta = T \left(\frac{NA}{n_o} \right) \left(\frac{x}{y} \right)^2 \left(\frac{\alpha}{\alpha + 2} \right) \quad \textbf{(3–9)}$$

where T is the **transmissivity** of the medium, NA is the numerical aperture of the optical fiber, n_o is the refractive index of the medium between the source and the fiber, x is the radius of the fiber core, y is the **half length** of the optical

source, and α is the fiber core refractive-index parameter. For step index: $\alpha = \infty$ for graded index: $\alpha = 2$.

For $y < x$, the coupling efficiency is given by Equation (3–10).

$$\eta_c = \frac{T \left(\frac{NA}{n_o} \right)^2 \left[\alpha + \left[1 - \left(\frac{y}{x} \right)^2 \right] \right]}{\alpha + 2} \quad \textbf{(3–10)}$$

Example 3–4

Determine the coupling efficiency of an optical source coupled into the fiber, given the following data: $T = 1$ (air), $n_o = 1$, $NA = 0.3$, and $x = y$.

Solution

For step index: ($\alpha = \infty$)

$$\begin{aligned}
\eta_c &= T \left(\frac{NA}{n_o} \right)^2 \left(\frac{x}{y} \right)^2 \left(\frac{\alpha}{\alpha + 2} \right) \\
&= 1 \left(\frac{0.3}{1} \right)^2 \left(\frac{1}{1} \right)^2 (1) \\
&= 0.09
\end{aligned}$$

Therefore, $\eta_c = 9\%$.

For graded index: ($\alpha = 2$)

$$\begin{aligned}
\eta_c &= \frac{T \left(\frac{NA}{n_o} \right)^2 \left[\alpha + \left[1 - \left(\frac{y}{x} \right)^2 \right] \right]}{\alpha + 2} \\
&= \frac{1 \left(\frac{0.3}{1} \right)^2 \left[2 + \left[1 - (1)^2 \right] \right]}{2 + 2} \\
&= \frac{0.09(2 + 0)}{4} = \frac{0.18}{4} = 0.045
\end{aligned}$$

Therefore, $\eta_c = 4.5\%$.

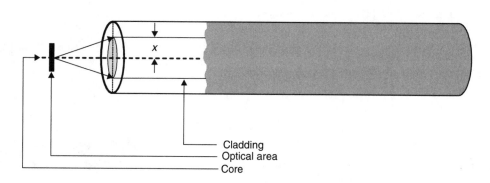

Figure 3–9. The Butt direct coupling method.

Example 3–5

Determine the coupling efficiency (η_c) when an optical source is coupled into a fiber, given the following data: $T = 1$ (air), $n_o = 1$, $NA = 0.3$, $y = 0.75\ x$, and $x > y$.

Solution:

For step index:

$$\eta_c = T\left(\frac{NA}{n_o}\right)^2\left(\frac{x}{y}\right)^2\left(\frac{\alpha}{\alpha + 2}\right)$$
$$= 1\left(\frac{0.3}{1}\right)^2\left(\frac{1}{0.75}\right)^2\left(\frac{2}{2 + 2}\right)$$
$$= (0.09)(1.8)(0.5)$$
$$= 0.08$$

Therefore, $\eta_c = 8\%$.

For graded index ($\alpha = 2$):

$$\eta_c = \frac{T\left(\frac{NA}{n_o}\right)^2\left[\alpha + \left[1-\left(\frac{y}{x}\right)^2\right]\right]}{\alpha + 2}$$
$$= \frac{1\left(\frac{0.3}{1}\right)^2[2 + [1-(0.75)^2]]}{2 + 2}$$
$$= \frac{0.09[2 + 0.4375]}{4}$$
$$= 0.055$$

Therefore, $\eta_c = 5.5\%$.

It is evident from the above examples that, for both step index and graded index fibers, the Butt coupling efficiency is very small. It is also evident that the coupling efficiency is higher when the optical area of the source is smaller than the cross-sectional area of the optical fiber.

The Lens Method

It is evident from the previous examples that only a small fraction of the generated optical power from optical source can be coupled into the narrow fiber angle of acceptance. In order to improve the coupling efficiency of optical power into the fiber, a lens is inserted between the radiating area of the source and the fiber core cross-sectional area (the **lens method**). The objective of such an insertion is to equalize the optical area of the source to that of the fiber core cross-sectional area. This lens scheme allows for a maximum coupling efficiency when the ratio of the radius of the power optical sources and the radius of the fiber core (*x/y*) become equal to the magnification factor of the inserted lens (Figure 3–10).

Figure 3–10 shows that when the optical source radiating area is smaller than the core cross-sectional area of the fiber, the solid optical angle of the source is larger than the solid angle of acceptance of the core. The lens's physical properties, and the precise location between the source and the core, achieve an equalization of the two solid angles (θ_1 equal to θ_2), resulting in an optimum coupling efficiency.

There are two methods of lens coupling. One requires that the radiating surface of the optical source placed behind the focal point of the lens (Figure 3–11), and the second requires that the source be placed in front of the focal point (Figure 3–12).

Figure 3–11 illustrates the image of the optical source concentrated at the fiber core surface area, resulting in an improved coupling efficiency.

In Figure 3–12, the optical source is placed between the lens and the focal point. In this arrangement, the optical solid angle can be made equal or smaller to the numerical aperture of the fiber core, thus enhancing coupling efficiency.

Several methods of incorporating a lens into the source to fiber assembly exist. The three most commonly used methods are shown in Figure 3–13.

In Figure 3–13a, the lens is embedded at the input of the fiber core. In Figure 3–13b, the lens is an integral part with the optical area of the source. In Figure 3–13c, the lens is between the optical source and the fiber core. These methods of lens coupling result in a significant improvement in efficiency over Butt coupling.

LED Reliability and Operational Lifetime

LED operational lifetime is an important factor in determining optical link long term reliability. LED operational lifetime is determined by the reliability factor, based on a model

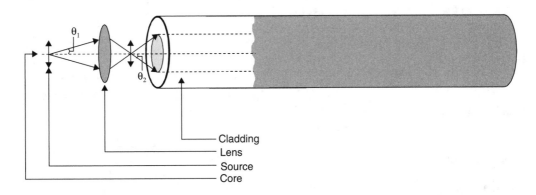

Figure 3–10. The lens coupling method.

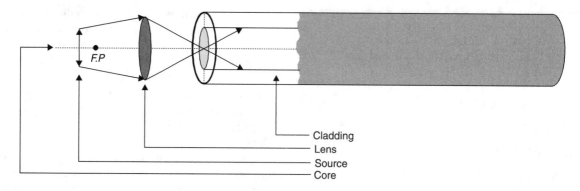

Figure 3–11. The optical source concentrated at the fiber core.

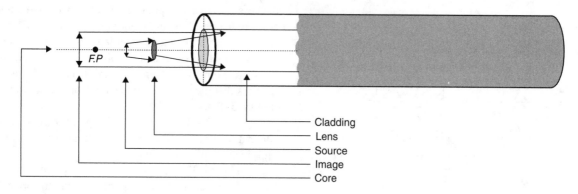

Figure 3–12. The optical source is placed between the lens and the focal point.

developed to calculate the device maximum operational time between failures. Maximum LED operational time is expressed by (Equation (3–11)).

$$R_F = B_F T_F E_F Q_F \times \frac{\text{failure}}{1 \times 10^6 \text{ hr}} \quad \text{(3–11)}$$

where R_F is the **reliability factor** (part failure rate model), B_F is the **base failure rate**, T_F is the **temperature factor**, E_F is the **environmental factor**, and Q_F is the **quality factor**. The mean time between failures (MTBF) is the inverse of the reliability factor (Equation (3–12)).

$$\text{MTBF} = \frac{1}{R_F} \quad \text{(3–12)}$$

The base failure rate (B_F) of a single LED is expressed by Equation (3–13).

$$B_F = 6.5 \times 10^{-4} \quad \text{(3–13)}$$

To calculate the MTBF of discrete optical fiber components, the above individual parameters must either be calculated or be obtained from tables provided by the component manufacturers. The individual parameters are calculated as follows.

TEMPERATURE FACTOR (T_F). The temperature factor of an LED device is expressed by Equation (3–14).

$$T_F = 8.01 \times 10^{12} \, e^{-\left(\frac{8111}{T_J + 273}\right)} \quad \text{(3–14)}$$

where T_J is the operating junction temperature (°C), and is expressed by Equation (3–15).

$$T_J = T_A + \theta_{JA} P_d \quad \text{(3–15)}$$

where T_A is the ambient temperature (°C), θ_{JA} is the **thermal resistance** $\left(\dfrac{°C}{W}\right)$, and P_d is the device **power dissipation** (W).

LED power dissipation is expressed by Equation (3–16).

$$P_d = I_F \times V_F \quad \text{(3–16)}$$

where I_F is the forward current (mA) and V_F is the forward biasing voltage (V). For DC operation, use $I_{F(\text{max})}$, and for AC operation use $I_{F_{\text{Average(max)}}}$.

Example 3–6
Compute the MTBF of an LED device operating at room temperature (25°C) with forward current (I_F) of 85 mA and forward voltage of 2.5 V.

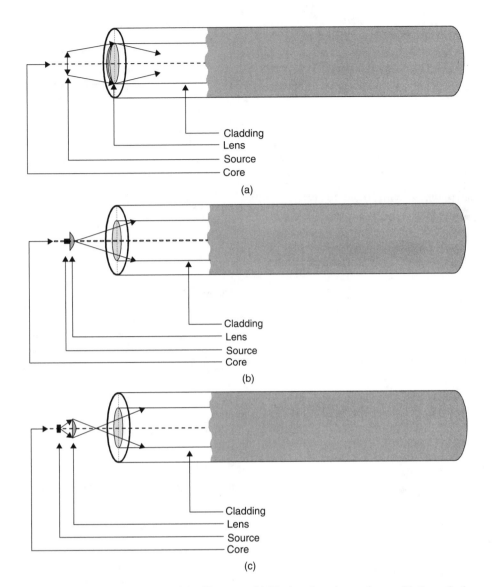

(a)

(b)

(c)

Figure 3–13. (a) The lens is embedded at the input of the fiber core. (b) The lens is an integral part with the optical area of the source. (c) The lens is imbedded between the optical source and the fiber core.

Solution

The MTBF is given by the expression

$$\text{MTBF} = \frac{1}{R_F}$$

where $R_F = B_F\, T_F\, E_F\, Q_F \times \dfrac{\text{failure}}{1 \times 10^6\ \text{hr}}$

i) Calculate (T_F).

$$T_F = 8.01 \times 10^{12}\ e^{-\left(\frac{8111}{T_J + 273}\right)}$$

ii) Calculate (T_J).

$$T_J = T_A + \theta_{JA} P_d$$

where $T_A = 25°C$, $\theta_{JA} = 150\left(\dfrac{°C}{W}\right)$ (For a hermetic LED—from the manufacturer's tables), and

$P_d = I_F \times V_F = 85\ \text{mA} \times 2.5\ \text{V} = 212.5\ \text{mW}$.

Substituting,

$$T_J = 25 + 150\left(\frac{°C}{W}\right)(0.2125\ \text{W}) = 56.9°C$$

Therefore, $T_J = 56.9°C$.

Substitute the value of (T_J) into the value of (T_F):

$$T_F = 8.01 \times 10^{12}\ e^{-\left(\frac{8111}{56.9\ +\ 273}\right)} = 168°C$$

Substitute (T_F) for (R_F). Also assume an indoor operation with E_F equal to 1, and Q_F equal to 0.5 (from manufacturer's tables).

$$R_F = B_F\, T_F\, E_F\, Q_F\, \frac{\text{failure}}{1 \times 10^6\ \text{hr}}$$

$$= 6.5 \times 10^{-4}\,(168)(1)(0.5)\frac{\text{failure}}{1 \times 10^6} = 5.46 \times 10^{-8}$$

iii) Calculate MTBF.

$$\text{MTBF} = \frac{1}{R_F} = \frac{1}{5.46 \times 10^{-8}} = 18 \times 10^6$$

Therefore, MTBF $= 8 \times 10^6$ hours.

Example 3–7

Compute the MTBF of an LED device operating at 80°C and with electrical parameters as follows:

- Operating frequency: 50 MHz with 50% duty cycle
- Peak drive current: 125 mA
- Forward voltage: 1.8 V

Solution

i) Compute (R_F).

$$R_F = B_F \, T_F \, E_F Q_F \frac{\text{failure}}{1 \times 10^6}$$

where $B_F = 6.5 \times 10^{-4}$, $E_F = 0.75$ (from tables), and $Q_F = 0.2$ (from tables).

ii) Compute (T_F).

$$T_F = 8.01 \times 10^{12} \, e^{-\left(\frac{8111}{T_j + 273}\right)}$$

iii) Compute (T_J).

$$T_J = T_A + \theta_{JA} \, P_d$$

iv) Compute (P_d).

$$P_d = I_F \times V_F$$
$$= (0.5)(120 \text{ mA})(1.8 \text{ V})$$
$$= 0.108 \text{ W}$$
$$T_J = 75 + \left(150 \frac{^\circ C}{W}\right)(0.108 \text{ W})$$
$$= 91.2 ^\circ C$$

Therefore,

$$T_J = 91.2 ^\circ C$$
$$T_F = 8.01 \times 10^{12} \, e^{-\left(\frac{8111}{91.2 + 273}\right)}$$
$$= 1704$$
$$R_F = B_F \, T_F \, E_F \, Q_F \frac{\text{failure}}{1 \times 10^6 \text{ hr}}$$
$$= (6.5 \times 10^{-4})(1704)(0.75)(0.2)\frac{\text{failure}}{1 \times 10^6 \text{ hr}}$$
$$= 0.166 \times 10^{-6}$$

Therefore, $R_F = 0.166 \times 10^{-6}$.

Finally, $\text{MTBF} = \dfrac{1}{R_F} = \dfrac{1}{0.166 \times 10^{-6}} = 6 \times 10^6$

Therefore, MTBF $= 6 \times 10^6$ hours.

3.3 LASER DIODES

Laser is an acronym for **L**ight **A**mplification by **S**timulated **E**mission of **R**adiation. Laser devices were first introduced in 1961, and their operations were based on stimulated emission instead of spontaneous radiation emission. Stimulated emission of radiation is the process whereby photons are used to generate other photons that have the exact phase and wavelength as the parent photons (Figure 3–14).

Fabry-Perot Lasers

Laser diodes are semiconductor devices capable of generating highly directional optical beams of particular wavelengths. Another important characteristic of laser diodes is that they can be modulated by very high rates, perhaps as high as 10 *GHz*.

These two very important characteristics make laser diodes ideal for applications in optical communications systems. However, laser diodes suffer from three inherited but very fundamental problems when employed as source devices in optical communications systems:

a) laser diode temperature sensitivity

b) **back reflections**

c) susceptibility to optical interference

The **Fabry-Perot laser** structure is a simpler laser diode in comparison with LEDs capable of producing optical beams that have substantially wider spectral bandwidths. The only drawback of the Fabry-Perot structure is that it exhibits a higher degree of chromatic absorption, an unwanted component, that imposes limits on the overall performance of optical communication systems.

Emphasizing once more the basic difference between LED and Fabry-Perot diodes; LEDs generate optical power through spontaneous emission of radiation while Fabry-Perot structures generate optical power through stimulation emission of radiation.

Photon generation within a Fabry-Perot device starts with spontaneous emission. The initial photons trigger the generation of more electron-hole recombination, which in turn generates additional photons, and so on. Examining the stimulation emission of radiation reveals that a laser diode performs two basic functions: It initiates optical power generation of a specific wavelength and at the same time it amplifies this optical power to a level determined by the semiconductor materials used in the fabrication of the device and its physical characteristics. To achieve optical amplification, Fabry-Perot devices employ reflective mirrors on both sides of the photon traveling path (Figure 3–15).

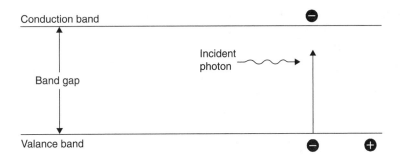

Figure 3–14. Stimulated emission of radiation.

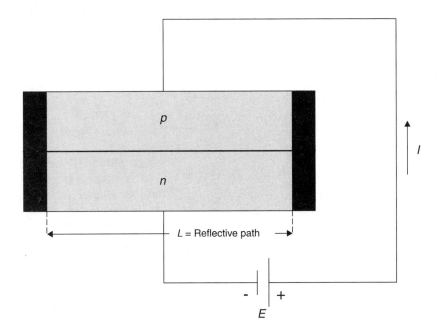

Figure 3–15. Simplified Fabry-Perot semiconductor structure.

These mirrors reflect the photons through the band gap, thus allowing for additional electron-hole recombination and larger photon generation. Not all the photons traveling between the reflective mirrors will contribute to optical amplification of the laser structure. These photons, which are in phase at a specific wavelength, will contribute to light amplification. Other photons, at different wavelengths, will contribute (to a lesser extent) to light amplification of other wavelengths. Again, the physical dimensions of the Fabry-Perot structure will determine the principal wavelength, achieving the highest optical gain. That is, if the mirror spacing is an integer number of half wavelength ($\lambda/2$), then the maximum optical amplification will occur in that wavelength while lesser optical amplification will be observed at the wavelength adjacent to the main. The optical spectrum of a Fabry-Perot resonator is illustrated in Figure 3–16.

The number of wavelengths that can be generated by a Fabry-Perot laser diode is given by Equation (3–17).

$$f_{res} = \frac{mc}{2ln} \qquad (3\text{–}17)$$

where c is the velocity of light (3×10^8 m/s), m is the integer, l is the length between mirrors, and n is the **cavity refractive index**, and $\lambda = \dfrac{\lambda_o}{n}$

where n is the refractive index of the material. The spectral optical power density of Figure 3–16 shows that there is a separation between the different wavelengths around the center wavelength. This modal separation is a direct function of the distance between the two mirrors and is expressed by Equations (3–18) and (3–19).

$$\text{Spacing}_{\text{Modal}} = \frac{c}{2ln} \text{ (Hz) } (\textbf{frequency domain}) \quad (3\text{–}18)$$

where c is the velocity of light (3×10^8 m/s), l is the length between mirrors (**cavity length**), and n is the cavity reflective index, or

$$\text{Spacing}_{\text{Modal}} = \frac{\lambda^2}{2ln} \text{ (m) } (\textbf{time domain}) \quad (3\text{–}19)$$

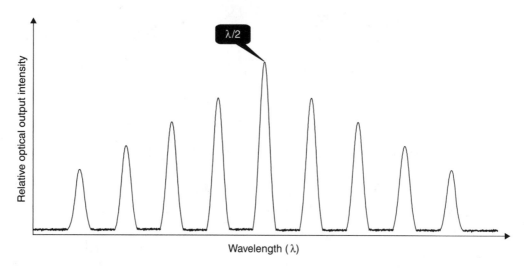

Figure 3–16. Fabry-Perot resonator optical spectrum.

where λ is the wavelength of the center mode. To evaluate the performance of a Fabry-Perot laser structure, certain measurements and calculations must be performed. Therefore, we will define and discuss them in some detail.

TERMS AND DEFINITIONS. *Total optical power output* (P_{TO}). The optical power output of a Fabry-Perot laser device is defined as the total sum of all the spectral components displayed by the optical spectral analyzer and set forth by **peak excursion** criteria (Equation (3–20)).

$$P_{TO} = \sum_{i}^{N} P_i \qquad \textbf{(3–20)}$$

where N is the number of spectral components and P_i is the power of each spectral component.

Full Width at Half Maximum: For a continuous Gaussian power distribution (CGPD), **full width at half maximum (FWHM)** is defined as the optical power spectral density at half of the peak amplitude.

Mean wavelength (MW) defines the center of all the spectral components and is calculated by incorporating all the modal components in terms of their wavelengths and optical power. Mean wavelength is expressed by Equation (3–21).

$$\overline{\lambda}_{MW} = \sum_{i}^{N} P_i \left(\frac{\lambda_i}{P_O} \right) \qquad \textbf{(3–21)}$$

where $\overline{\lambda}$ is the mean wavelength, P_i is the power of individual modes, P_O is the total optical output power, and λ_i is the modal wavelength. **Sigma (σ)** is defined as the Fabry-Perot laser root-mean-square of the spectral width, assuming Gaussian spectral distribution. It is calculated as follows in Equation (3–22).

$$\sigma = \sqrt{\frac{\sum_{i}^{N} P_i(\lambda_i - \lambda)}{P_O}} \qquad \textbf{(3–22)}$$

where σ is the root-mean-square, P_O is the total optical output power, and λ_i is the wavelength of the individual modal component.

Peak amplitude (PA) defines the Fabry-Perot laser's center modal component amplitude.

Mode spacing is the average wavelength spacing between individual spectral modes.

Peak wavelength (PW) spacing in a Fabry-Perot laser diode is the wavelength with the highest amplitude within the total spectrum.

Peak function (PF): A single line can represent each component of the displayed spectrum. The display of this line indicates whether an adjustment is required for the peak excursion value (see peak excursion).

The **distribution trace (DT)** displays a more complete picture of the Fabry-Perot laser diode. This curve is Gaussian, and it represents a continuous approximation of the discrete spectrum that encompasses all the individual components, such as total optical output power, modal wavelengths, modal spacing, and mean wavelengths (Figure 3–17).

The **peak excursion (PE)** function of an optical spectrum analyzer allows for the adjustment of the base level of all the discrete spectral components. This is a critical adjustment because if it is not carefully set, it will result in either failure to detect useful spectral components near the noise level, or it will induce unwanted noise components in the useful spectrum.

Early in their development stage, laser devices exhibited a substantial increase in optical power output, confined to a very narrow beamwidth, in comparison with standard LED devices. Some of these early structures were based on heavily pumped (2×10^{18} e⁻/cm³), thin (0.2 μm) GaAs alloy forming the laser cavity and sandwiched between four **epitaxially grown** layers of AlGaAs, providing carrier and optical power confinement (see Figure 3–18).

For stimulation emission to occur, the number of electrons in the conducting band must exceed the number of electrons in the valence band. This increase can be accom-

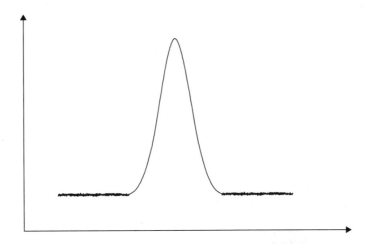

Figure 3–17. A distribution trace for a Fabry-Perot laser diode.

plished through the process of **pumping.** Pumping, or the elevation of electron from the valence band to the conducting band, is achieved by passing a sufficient amount of current through the active region of the laser device. Stimulated emission of radiation will start to take place at a minimum current level at the active region of the laser structure, called the *threshold current*. Below this threshold current, stimulated emission does not take place because radiation and absorption losses occurring inside the active region offset the additional photon generation below the threshold current. The relationship between threshold current and optical power output of a laser device is illustrated in Figure 3–19.

Figure 3–19 shows that at approximately 50 mA of injected current into the active region of the laser device, stimulated emission of radiation begins. Below 50 mA, the device operates under the spontaneous emission of radiation mode. The early laser structures suffered from a lack of efficiency, that is, the electron-to-photon conversion ratio was low. In addition, they required very high currents, around 50 KA/cm^2 of active area, for stimulation emission to take place. These deficiencies of early laser diodes were addressed with the introduction of the double heterostructure laser device. A very thin low band gap, high refractive index AlGaAs active layer is confined between n-AlGaAs and p-AlGaAs layers with higher band gaps and lower refractive indexes (Figure 3–20).

In 1975, a threshold current of 0.5 KA/cm^2 was achieved in a laser device with an active layer of 0.1 μm thickness. This device was capable of generating optical power in the range of 800 nm to 900 nm. The optical wavelength generated by the laser device is related to the alloy composing the structure, and is expressed by Equation (3–23).

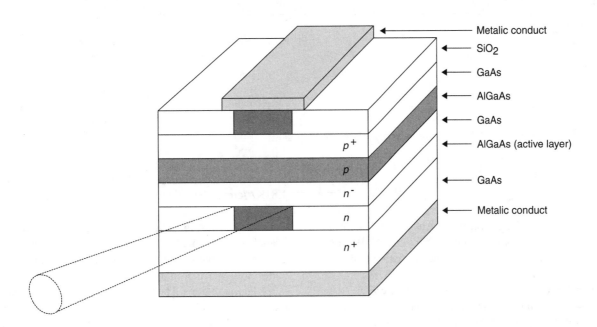

Figure 3–18. Fabry-Perot AlGaAs/GaAs laser diode.

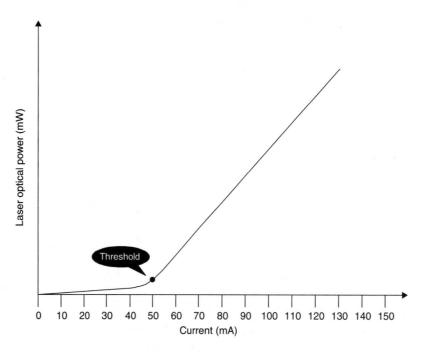

Figure 3–19. Laser optical power output v. biasing current

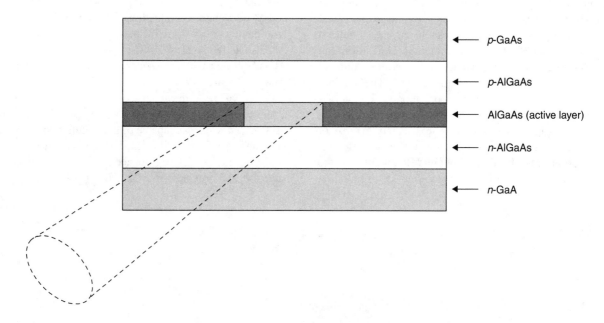

Figure 3–20. Fabry-Perot thin layer (AlGaAs/GaAs) laser diode.

$$\lambda = \frac{hc}{E_g} \qquad (3\text{--}23)$$

where λ is the wavelength (nm), h is the Planck's constant $(6.63 \times 10^{-34}$ J/s$)$, c is the velocity of light $(3 \times 10^8$ m/s$)$, and E_g is the band gap energy of the active region (eV).

For the InGaAsP alloy, $E_g = 0.74$ to 1.13 eV.

For the AlGaAs alloy, $E_g = 1.42$ to 1.61 eV.

For operating wavelengths in the spectral region between 1300 nm and 1500 nm, substances such as InGaAsP were introduced for the fabrication of laser semiconductor structures that required a threshold current of 0.7 KA/cm² to 1.8 KA/cm². The schematic diagram of such a device is shown in Figure 3–21.

If a forward biasing voltage between 1.5 V and 2.0 V is applied across the laser diode, a carrier concentration will gradually be built up into the active region. This carrier concentration is denser at the center of the active region than in the lateral regions. The high carrier density at the center of the active region far exceeds the transparency concentration, which results in an overall optical gain. The lesser lateral carrier concentration is far below the transparency level, thus

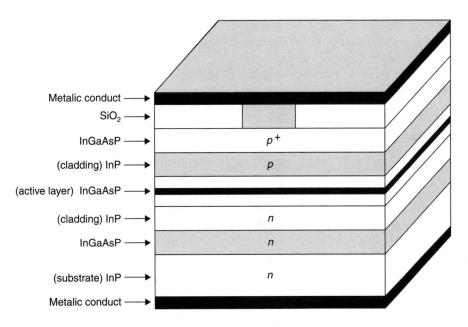

Figure 3–21. InGaAsP laser diode structure.

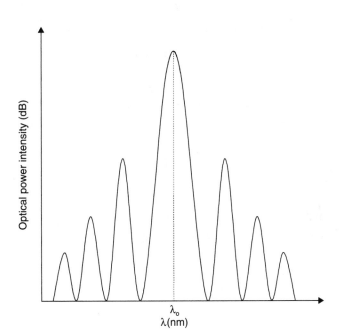

Figure 3–22. Spectral density.

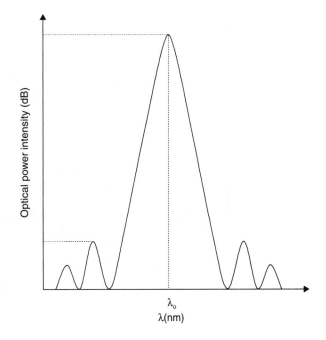

Figure 3–23. Laser diode spectral line.

an overall optical power loss is encountered in the lateral regions of the active layer. The double heterostructure laser is a gain guided multi-mode device. The spectral density of such a device is shown in Figure 3–22.

The multi-mode behavior of a gain guided laser device can be remedied by the reduction of the active region's width. Width reduction permits the formation of a substantial number of discrete dielectric levels, which allows for the development of a fundamental mode instead of a multi-mode spectral output, and results in a guided index laser. Both multi-mode and single-

mode devices have been utilized in optical fiber communications systems that can take advantage of their fundamental properties. That is, for multi-mode transmission systems, the multi-mode laser device is used as the optical power source. For single-mode transmission systems, the single-mode optical source is utilized. The implementation of the single-mode optical source in such a transmission system complies with fundamental system design requirements such as low optical modal noise. Laser diodes exhibit a single spectral line with sidelines reduced by at least 25 dB (Figure 3–23).

In optical fiber communications systems, direct optical modulation of the optical source is an established practice. At very high rates of modulation, the behavior of a single mode optical source begins to change, resulting in an output spectrum that shows more than one spectral line. In order to maintain a single mode spectral output, a modified version of the single-step index laser device is fabricated, that is, a **distributed-feedback laser (DFB)**.

Distributed Feedback Laser Structures

Careful examination of the spectral display of a Fabry-Perot laser diode reveals that it is composed of a central (highest amplitude) modal component and several modal components that have progressively lower amplitudes.

This is a definite disadvantage when used in optical communications systems because its wide spectral width significantly increases chromatic absorption, with a consequent reduction of the usable transmission bandwidth. Ideally, such applications will require a laser diode that exhibits an output spectrum that is composed of a central wavelength component of set amplitude. All the other wavelengths around the center have amplitudes that are significantly reduced.

In DFBs (see Figure 3–24), the main design objective is to generate a single line spectrum at the output, under high data rates of modulation. This is achieved by incorporating a corrugated layer below the active layer of the DFB device. The hills and valleys generate a constant change of the refractive index, which contributes to the device's feedback

mechanism, so that a single mode is produced and undesirable modes are suppressed.

In order for the structure to operate as a DFB device, the grading period must satisfy the relationship exhibited in Equation (3–24).

$$g_P = \frac{\lambda_{\text{mode}}}{n} \qquad \textbf{(3--24)}$$

where g_P is the grading period, λ_{mode} is the operating wavelength, and n is the refractive index of the effective mode. Typical operating characteristics of a DFB laser diode are shown in Table 3–1.

DISTRIBUTED BRAGG REFLECTORS. Another single mode feedback laser is the distributed Bragg Feedback device shown in Figure 3–25.

The fundamental structural difference between DFB and DBR lasers is their grading mechanisms. In a DFB device, the grading is at the bottom of the active layer, while in a DBR device the grading is at both ends of the active region. When it is at both ends, it can act as a perfect optical mirror, because of the difference between the constant refractive index of the active layer and the continuously changing refractive index of the grading layer. This structural arrangement provides the required feedback mechanism for optical power generation and spectral purity. DBR devices require a higher threshold current than DFB structures. DBR devices also exhibit a higher degree

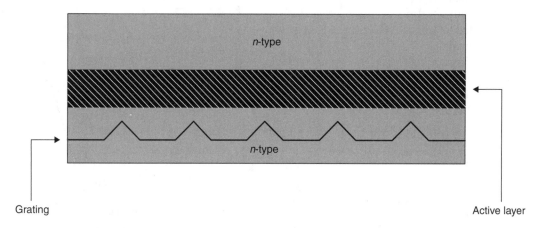

Figure 3–24. Distributed feedback laser.

TABLE 3–1 Typical Operating Characteristics of DFB Lasers

Operating wavelength	1300 nm
Output power (max)	5×10 mW
Threshold current	40 mA
Temperature coefficient (for threshold current)	1.3 mA/°K
Modulation bandwidth	800 MHz
Temperature coefficient	0.07 nm/°K
Spectral bandwidth	20 MHz

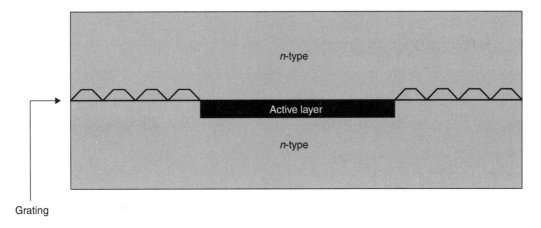

Figure 3–25. A distributed Bragg feedback reflector.

of susceptibility to temperature variations, and their line width is more sensitive to the rate of modulation than DFB laser devices.

LASER DIODE CHARACTERISTICS. There is a spectral difference between gain guided and index guided laser devices. As we mentioned previously, gain guided lasers are multi-mode devices, while index guided lasers are single-mode devices. This is because the spontaneous emission component within index guided structures is smaller than that of the gain guided structure. Under modulated conditions, the single-mode, index guided laser can become a multi-mode device with additional broadening of the modal line width. This is as a result of a small change of the modal frequency, and is attributed to carrier density variations due to pulse modulation. Maintaining a narrow modal line width is crucial when laser devices are used as optical sources for long-distance communications systems that are designed to process high data rates while operating under the most stringent system noise restrictions.

Temperature v. Optical Power

Experimental results have shown that lasers are temperature dependent devices. The relationship of the threshold current and optical power output are subject to operating temperature conditions. This relationship between threshold current and temperature is given by Equation (3–25).

$$I_{th} = I_o e^{\frac{T}{T_o}} \qquad (3\text{–}25)$$

where I_o is the threshold current at room temperature (characteristic of the laser), I_{th} is the current at operating temperature, T_o is the room temperature (characteristic of the laser device), and T is the operating temperature (characteristic of the laser device).

T_o is an intrinsic value and it is different for GaAs/AlGaAs and InGaAsP/InP laser structure. For example, in InGaAsP devices, T_o has an average value of 65 °K, while for AlGaAs devices, the average T_o is observed to have a value

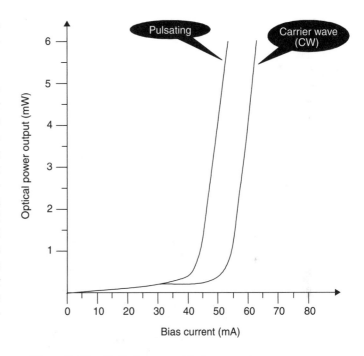

Figure 3–26. Optical power v. biasing current.

of approximately 125 °K. The lower T_o value for InGaAsP devices is attributed to various phenomena occurring inside the device, such as heterobarrier carrier leakage and bond absorption.

When laser diodes are used as optical sources in optical fiber communications systems, the generated output optical power is different for the two basic modes of operations: carrier wave and pulsating. This difference is shown in Figure 3–26.

Figure 3–26 shows that under pulsating modulation, the laser diode requires a smaller threshold current for laser action to take place, while its optical power output is considerably higher in comparison to the carrier wave mode of operation. The threshold current variations v. different operating temperatures are shown in Figure 3–27.

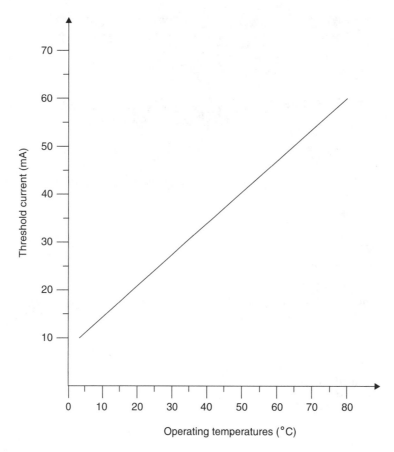

Figure 3–27. Threshold current *v.* temperature.

Figure 3–27 illustrates a proportional increase of the threshold current in relationship to a corresponding increase of the operating temperature.

These temperature variations also have an effect on the operating wavelength (λ) of the laser device. An increase of 1°C in the operating temperature shifts the operating wavelength by 0.3 nm. This wavelength change is attributed to device cavity expansion due to the increase of the operating temperature.

Laser Bandwidth

One of the fundamental advantages of an optical fiber communications system is its ability to directly modulate the optical source at a very high data rate. Although these modulating rates are very high (approx. 20 GHz), there are limits beyond which the laser diode cannot respond. Figure 3–28 shows the output frequency response of a typical laser diode under modulating conditions.

Figure 3–28 illustrates that under relatively low modulating frequencies, spectral output intensity is constant. When modulating frequency reaches a certain level, the carriers injected into the device cavity interact with the generated photons, enhancing a self-oscillatory process and thus sharply reducing the output spectral density. Another serious problem is **noise intensity**. There is a direct relation-

ship between the noise spectral density and the injected carriers into the active region. The result is an increase of the threshold current beyond the level required for laser action to take place. This spectral noise density is observed to take a maximum value at maximum modulating frequencies. The self-oscillating frequency of a laser device is expressed by Equation (3–26).

$$f_{so} = \frac{1}{2\pi}\sqrt{\frac{AP_o}{\tau}} \qquad \textbf{(3–26)}$$

where f_{so} is the self-oscillating frequency, A is the gain, τ is the **gain average lifetime** of the photon inside the cavity, and P_o is the **photon density**.

It is evident from Equation (3–26) that in order to increase the range of the oscillating frequency, and consequently to increase the modulation rate and enhance output stability, both photon density and gain must be increased with a simultaneous decrease of the photon (cavity) lifetime. Stabilizing the device operating temperature can be achieved by using a **thermoelectric aperture** (heat sinking), while output power stabilization can be achieved by incorporating a feedback mechanism that can increase the differential gain through a PIN diode. Because the lifetime of photons inside the cavity depends on the physical characteristics of the laser structure, a decrease of the length of the cavity will result in a corresponding decrease

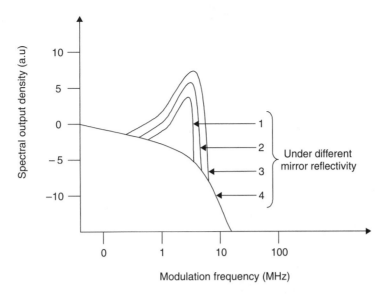

Figure 3–28. Output spectral density *v.* modulation frequency.

of the (τ). This, of course, has its drawbacks. For example, a decrease in the length of the cavity will increase the carrier density and increase temperature to beyond a certain level.

Laser Device Reliability

Laser devices present a number of challenging problems that result from their fabrication processes. Two common problems are control of the physical dimensions and control of the heterobarrier lattice growth. Various techniques, such as *Liquid Phase Epitaxial* (*LPE*) growth, have been utilized in the fabrication of laser devices. This technique represents certain difficulties related to uniform reproduction of larger areas. The introduction of the *Vapor Phase Epitaxial* (*VPE*) growth method has eliminated some of the problems encountered by the LPE fabrication technique. Fabrication techniques, as well as operating conditions, are key factors in determining a device's reliability and life span. Both these factors are very important when the optical device is considered for utilization in an optical fiber communications system. The operating life span of an optical source is the period for which the device is capable of delivering specified optical power at a predetermined maximum threshold current. Over time, internal degradation based on operating temperatures, crystal defects, facet, and conduct damages limit a device's life span.

3.4 QUANTUM WELL LASER DIODES

The basic objective in the design of laser diodes is to obtain the highest optical gains at the lowest possible carrier densities, ultimately reflecting much lower threshold current requirements. Conventional double **heterojunction laser** diodes are unable to fully satisfy performance requirements because of the relatively thick active layer. If the active layer is divided in sublayers with widths of the order of a few

nanometers, while separated by equally thin barrier layers, then the carrier movement across the individual thin active layers will be somewhat restricted and the kinetic energy will appear to be quantized to discrete energy levels. Because of the transformation of the forward kinetic energy to quantum energy levels in the active region, these devices are referred to as quantum well (QW) laser diodes. The energy bands of such a structure are illustrated in Figure 3–29.

Figure 3–29 illustrates the optical spectrums of four single quantum well (SQW) laser diodes with different active layer thicknesses. It also illustrates a slight peak wavelength shift at an active layer thickness of less than 17 A°.

The discrete energy states within the active region alter the optical and electronic properties of the beam, promoting higher optical gains at much lower threshold currents in comparison to double heterojunction laser diodes.

Multi Quantum Well Laser Diodes

To further enhance laser diode operating performance, the quantum well concept was extended beyond the SQW concept, to a new semiconductor structure incorporating more than one quantum well, referred to as multi quantum well laser diodes (MQW). The energy bands of an MQW diode are illustrated in Figure 3–30.

MQW lasers employ a multilevel active region that is separated by barrier layers. In such a device, the barrier energy level and the cladding energy level are equal. A different energy level between the barrier and the cladding layers results in a modified version of the MQW structure, referred to as a modified multi quantum well laser diode (M-MQW*).* M-MQWs have no significant changes in their performance characteristics.

QW semiconductor structures allow carrier energy states to be elevated above the band gap level by confining the carriers in quantum wells within the active layer in a super lattice

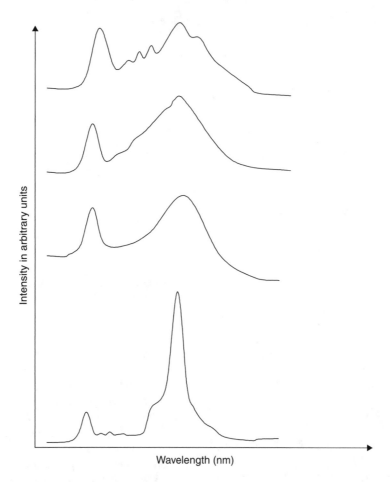

Figure 3–29. Energy band levels of single quantum well laser structures composed of four different well sizes.

structure. Laser action can be initiated by employing radiative transitions among the confined carriers in the quantum wells.

MQW lasers are fabricated from nitride materials grown on MgAlO substrates. These materials exhibit characteristics such as high thermal conductivity, very large heterojunction offsets, high melting temperatures, and physical hardness. That is, an InGaN MQW active layer is sandwiched between two GaN optical guides and AlGaN cladding layers. Multi quantum well technology is today applied in the fabrication of both LED and laser diode devices that operate at very short wavelengths, down to the 410 nm range.

InGaN compounds that have direct bandgaps between 1.95 eV and 3.4 eV at room temperature can be used as active layers in MQW LED and laser diode structures that operate in the blue wavelength region. Figure 3–31 illustrates a cross section of an MQW laser diode.

The device in Figure 3–31 is capable of emitting at very short wavelengths (~400 nm), the shortest ever achieved by a semiconductor laser diode. For Figure 3–31, a buffer layer of GaN is grown on a $M_qAl_2O_4$ substrate. The active region is composed of an InGaN MQW layer (20-periods), the emitting light from which is confined by a p-type and n-type AlGaN guiding layer, while n-type InGaN is used as buffer to protect the AlGaN from damaging the very thin film. The p-type Al-GaN is used to maintain the bond between the InGaN layers of

the active region during the p-type layer growth process. The p-type and the n-type GaN on both sides of the active region are functioning as light guiding layers. The electrical characteristics of an InGaN MQW device is illustrated in Figure 3–32 (see page 87).

Figure 3–32 illustrates InGaN MQW laser diode polarized output intensity versus a pulse modulated biasing current. Close examination of Figure 3–32 shows that for a biasing current of 300 mA, no stimulation of emission is observed. At 310 mA, stimulation of emission begins to form (threshold current), and at 400 mA of biasing current, a maximum optical power output of 35 mW is obtained. Figure 3–33 (see page 87) illustrates the optical spectrums of an InGaN MQW laser diode for three different biasing current levels.

Line (a) illustrates the optical spectrum of a laser diode at a biasing current level of 0.6 A. The measured full wave at half maximum (FWHM) is 22 nm. The optical output power is the result of the spontaneous emission of radiation and it peaks at ~404 nm wavelength. Line (b) illustrates the optical spectrum at a biasing current of 1.2 A. The peak wavelength has been shifted to 410 nm and the optical power output (much higher), indicates the starting of the stimulation of emission process. Line (c) illustrates the optical spectrum at a biasing current above the threshold level. It is evident that the FWHM is approximately 1/10 (21 nm) that of line (a).

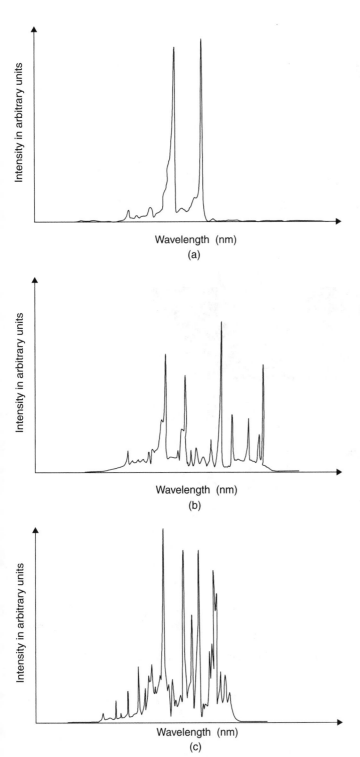

Figure 3–30. Energy band levels of MQW laser structure for different biasing current levels.

InGaN MQW laser diodes exhibit very low FWHM and high optical output power at very low operating wavelengths. A decrease of the thickness of the active layer will further shift the operating wavelengths. If the emission energy increases, the FWHM will also increase, resulting in a corresponding decrease of quantum well luminescence.

3.5 SURFACE EMITTING LASERS

The ever-increasing demand for higher transmission capacities, longer distances between amplification, optical interconnections, and optical computing generated the incentive for the development of surface emitting lasers (SEL). SELs are classified into four major categories:

i)   **V**ertical **C**avity **S**urface **E**mitting **L**asers (VCSEL)
ii) **F**olded **C**avity (FCSEL)
iii) 45°C **R**eflecting **M**irror (45°C-RMSEL)
iv) **G**rating **C**oupled (GCSEL)

The fundamental advantages of SELs are that they can be massively fabricated by monolithic processing, they are able to vertically harvest the generated optical power, and they are able to be tested and have their performance evaluated before separated in individual devices. The fact that they can be fabricated in large numbers of arrays leads to the generation of relatively high optical laser power (Figure 3–34, see page 88).

3.6 VERTICAL CAVITY SURFACE EMITTING LASERS

The fundamental difficulty with standard SEL diodes lies in their inability to operate at room temperature because of short gain path relevant to insufficient mirror reflectivity. To improve the p-side mirror reflectivity, a ring electrode was introduced, while on the n-side, a multilayer reflector was introduced to improve mirror reflectivity. However, in order to improve the second important performance parameter (that is, the substantial reduction of the threshold current (I_{th})), a thin circular heterostructure active layer was induced in the optical confinement waveguide. The first CW-VCSEL diode operating at room temperature was fabricated in 1988, and was composed of GaAlAs/GaAs compounds. By 1989, CW-VCSEL diodes lasing at 2 mA threshold current at room temperature were realized. Full commercial applications of such diodes began in early 1996. A three dimensional visualization of a VCSEL diode is illustrated in Figure 3–35 (see page 88).

Figure 3–35 shows the active layer embedded in a smaller bandgap material. Therefore, the injected carriers are confined within the active region. The surface area of the active region πD^2 defines the base of the cylindrical vertical waveguide. In order for the optical of a dominant mode to be detected at the output of the cylindrical waveguide, the modal gain must be at least equal to the total optical losses. The relationship between the modal gain and the optical losses is expressed by Equation (3–27).

$$KG_{th} = (\alpha_m + \alpha_T) \tag{3–27}$$

where K is the **optical energy confinement factor,** G_{th} is the **modal gain** at threshold current, α_m represents **mirror**

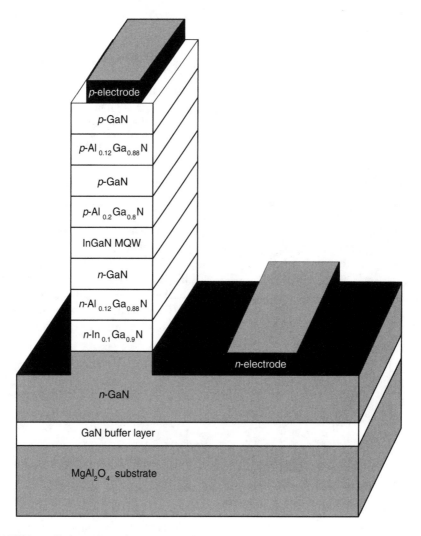

Figure 3–31. InGaN -MQW laser diode cross section.

losses, and α_T represents total **internal cavity losses.** Mirror losses (α_m) are defined by Equation (3–28).

$$\alpha_m = \frac{1}{L} \ln\left[\frac{1}{(R_F R_R)^{1/2}}\right] \qquad (3-28)$$

where α_m represents mirror losses, R_F is the front mirror reflectivity, and R_R is the rear mirror reflectivity. Total internal cavity loss (α_T) is defined by Equation (3–29).

$$\alpha_T = \alpha_D + \alpha_A \qquad (3-29)$$

where α_D is the diffraction loss (subject to guide length), and α_A is the absorption loss. The absolute threshold modal gain (G_{th}) can be expressed by Equation (3–30).

$$G_{th} = A_o (N_{th} - N_T) \qquad (3-30)$$

where A_o is the differential gain close to threshold (dG_{th}/dN), N_{th} is the current density at threshold, and N_T is the transparency current density. Solving Equation (3–30) for N_{th} yields Equation (3–31).

$$N_{th} = N_T = \frac{G_{th}}{A_o} \qquad (3-31)$$

Combining Equation (3–27) with Equation (3–30) yields Equation (3–32).

$$N_{th} = N_T + \frac{1}{KA_o} (\alpha_T + \alpha_m) \qquad (3-32)$$

Furthermore, the current density at threshold is expressed by Equation (3–33).

$$N_{th} = \frac{\tau_s \, n_i \, (J_{th})}{ed} \qquad (3-33)$$

where n_i is the **injection efficiency,** τ_s is the carrier lifetime, e is the electron charge, d is the active layer thickness, and J_{th} is the **threshold current density.** Solving Equation (3–33) for threshold current density yields Equation (3–34).

$$J_{th} = \frac{ed}{n_i \tau_s} \qquad (3-34)$$

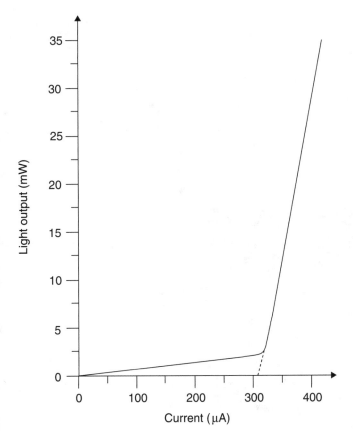

Figure 3–32. Electrical characteristics of an InGaN MQW laser diode.

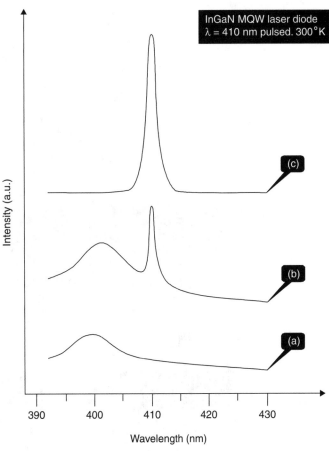

Figure 3–33. InGaN MQW laser diode optical spectrums of three different biasing currents.

The carrier lifetime is expressed by Equation (3–35).

$$\tau_s = \frac{1}{A + BN + CN^2} \quad \text{(3–35)}$$

Where A is the **non-radiative recombination**, BN is **radiative recombination**, and CN^2 is the **non-linear process**. If $BN \gg CN^2$, then Equation (3–35) is approximated by Equation (3–36).

$$\tau_s = \frac{n_{\text{spont}}}{B_{\text{eff}} N_{\text{th}}} \quad \text{(3–36)}$$

Where n_{spont} is the efficiency of spontaneous emission, and B_{eff} is the coefficient of effective radiative recombination. Substituting Equation (3–35) into Equation (3–33) yields Equation (3–37).

$$J_{\text{th}} = \frac{edB_{\text{eff}} N_{\text{th}}^2}{n_i\, n_{\text{spont}}} \quad \text{(3–37)}$$

Early laser diodes exhibited a threshold current density of 30 KA/cm^2 with mirror reflectivity up to 95%. An active layer thickness reduction combined with an increase of mirror reflectivity can significantly reduce the threshold current density.

VCSEL Threshold Current (I_{th})

VCSEL threshold current is expressed by Equation (3–38).

$$I_{\text{th}} = \pi \left(\frac{D}{2}\right)^2 J_{\text{th}} \quad \text{(3–38)}$$

Substituting Equation (3–37) into Equation (3–38) yields Equation (3–39).

$$I_{\text{th}} = \pi \left(\frac{D}{2}\right)^2 \left(\frac{edB_{\text{eff}} N_{\text{th}}^2}{n_i n_{\text{spont}}}\right) \quad \text{(3–39)}$$

where $\pi \left(\dfrac{D}{2}\right)^2 d$ = active region volume (V).

or

$$V = \pi \left(\frac{D}{2}\right)^2 d \quad \text{(3–40)}$$

Substituting Equation (3–40) into Equation (3–39) yields Equation (3–41).

$$I_{\text{th}} = \frac{eVB_{\text{eff}} N_{\text{th}}}{n_i\, n_{\text{spont}}} \quad \text{(3–41)}$$

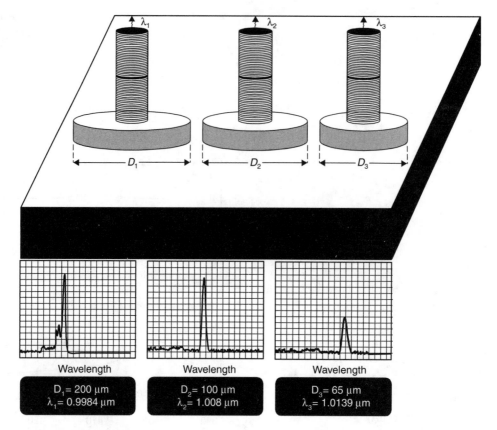

Figure 3–34. An array of surface emitting lasers.

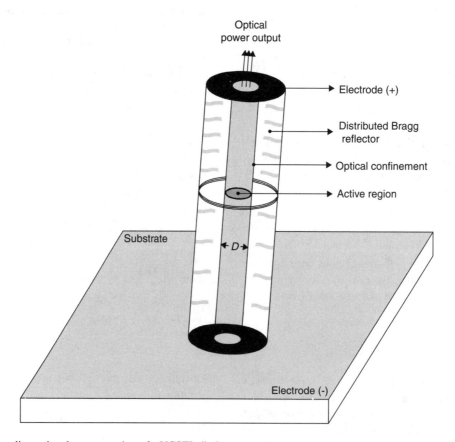

Figure 3–35. A three dimensional representation of a VCSEL diode.

TABLE 3–2 VCSEL Operating Characteristics.

Parameters	VCSEL
Mirror reflectivity (R_M)	99% to 99.9%
Length of cavity (L)	300 nm
Thickness of active layer (d)	80 A°
Confinement factor (K) optical	6%
Active region area (S)	25 µm²
Carrier lifetime (τ_s)	1×10^{-13} s

From Equation (3–41), it is evident that the threshold current (I_{th}) can be reduced by effectively reducing the volume of the active region and the injection carrier density or by a simultaneous increase of one or both injection and spontaneous emission efficiencies. Table 3–2 illustrates operating characteristics of VCSELs.

Difficulties associated with VCSEL diodes are optical power loss minimization, optical field gain overlap heat sinking for high optical power operation, and electrode resistivity minimization at high efficiency levels. For a typical GaAs VCSEL diode, threshold current versus different active region diameters and mirror reflectivity is illustrated in Figure 3–36.

Laser diodes are classified based on their application as either short wavelength or long wavelength device.

Short Wavelength Diode Characteristics

Laser diodes that emit in the red to blue optical spectrum are required for display and disk applications. The fabrication of VC-SELs composed of VII-V compounds such as GaInAlP/GaAs have already been realized, despite extreme difficulties encountered during the design and fabrication process. At the wavelength range between 340 nm and 470 nm, a combination of such elements as In, As, B, Ga, and N have exhibited potentials for implementation in VCSEL diode fabrication. For standard CW laser diodes emitting in the 500 nm wavelength window, ZnSe is normally used, while GaN is more appropriate to be used in the fabrication of VCSEL devices. For applications such as DVDs and laser printers, systems that utilize plastic optical fibers that operate in wavelength windows of 650 nm to 670 nm, AlGaInP/GaAs VCSEL devices have been developed. They are capable of generating optical power at low mW with threshold currents in the µA range.

VCSEL diodes operating in the 780 nm wavelength and capable of generating optical power in the mW range with approximately 250 µA of threshold current have already been developed for commercial applications.

Long Wavelength Diode Characteristics

The emergence of optical interconnect, parallel optical processing, and long distance all optical systems require laser diodes capable of emitting in the 1300 nm and 1550 nm wavelength window. CW-VCSEL diodes emitting in these wavelengths at room temperature encounter significant difficulties such as high threshold current and resonant intravalence band absorption (mainly at 1550 nm). Because GaInAsP/InP based compounds are used for the fabrication of such diodes, and the index of refractivity difference between these two groups is very small, fabrication of the devices requires a relatively large number of compound pairs ($\lambda/4$) that have a very thin active

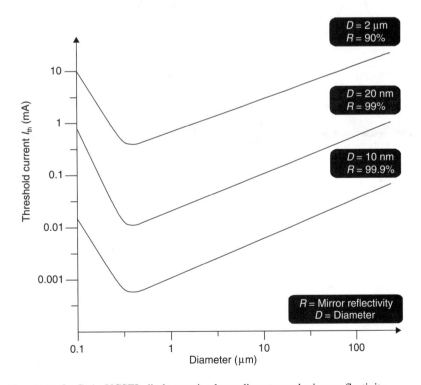

Figure 3–36. Threshold current of a GaAs VCSEL diode *v.* active layer diameter and mirror reflectivity.

layer (few μm). A typical VCSEL diode cross section is illustrated in Figure 3–37 with a corresponding I v. L curve illustrated in Figure 3–38.

VCSEL Diodes Composed of GaAlAs/GaAs

A VCSEL diode composed of GaAlAs/GaAs compounds fabricated through the metal organic chemical vapor deposition process is illustrated in Figure 3–39.

The laser diode pictured in Figure 3–40 is composed of a 3 μm thick active layer grown at a temperature of 780°C and under atmospheric pressure. The circular active layer is buried in a 10 μm diameter cylindrical mask composed of a Si_3N_4 compound surrounded by a p-cladding GaAlAs compound. The current blocking layer is composed of a 0.7 μm thick n-type and a 0.3 μm thick p-type GaAs compound, while the length of the vertical waveguide is 5 μm. At both ends, SiO_2/TiO_2 reflective mirrors and 40 μm and 10 μm diameter ring electrodes are inserted. Operational characteristics of VCSEL laser diodes are as follows:

i) operating temperature: room temperature

ii) threshold current: (I_{th}) 20 mA

iii) threshold current density: (J_{th}) 260 μA/μm²

iv) differential quantum efficiency: 10%

v) CW optical power output: 2 mW(max)

vi) spectral line width: 1 A°(above threshold)

vii) modal spacing: 170 A°

iii) single mode suppression ratio: 35 dB $\left(\text{at } \frac{I_{th}}{J_{th}} = 1.25 \right)$

VCSEL Diodes Composed of GaInAs/GaAs

In the quest for higher laser gain, VCSEL diodes employing GaInAs/GaAs compounds were fabricated. These structures utilize a GaAs substrate, on top of which multiple layers of GaInAs/GaAs are grown, while GaAlAs/AlAs multilayer compounds are used as optical reflectors. Such standard diode structures have exhibited very low threshold currents—around 1 mA at carrier-wave (CW) operations. However, VCSEL structures have exhibited lasing action at threshold currents below the mA level, at the 980 nm wavelength window. Furthermore, experimental results have shown that, with current confinement and CW operation at room temperature, threshold current levels around 10 μA are possible. An innovative VCSEL diode structure is illustrated in Figure 3–40.

The above diode is composed of a three QW 80A⁰ GaInAs active layers and twenty-four λ/4 stacks of GaAs/AlAs Bragg reflector layers. The current v. optical output power graph of such a device is illustrated in Figure 3–41.

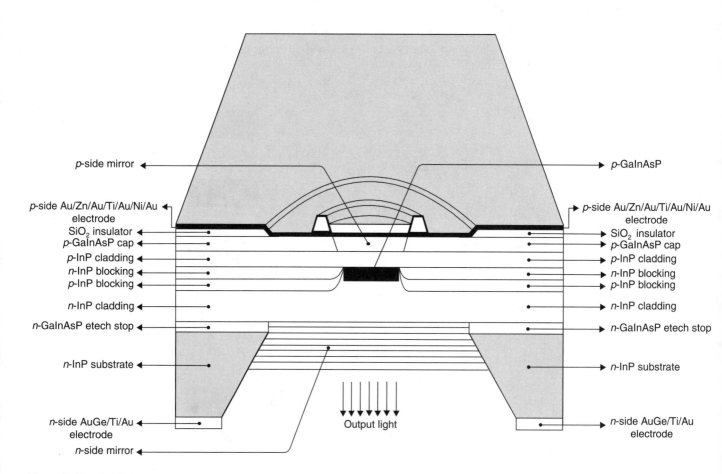

Figure 3–37. VCSEL cross section.

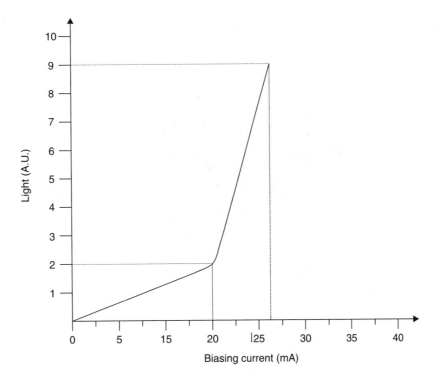

Figure 3–38. Injected diode current *v.* optical output power (I-L).

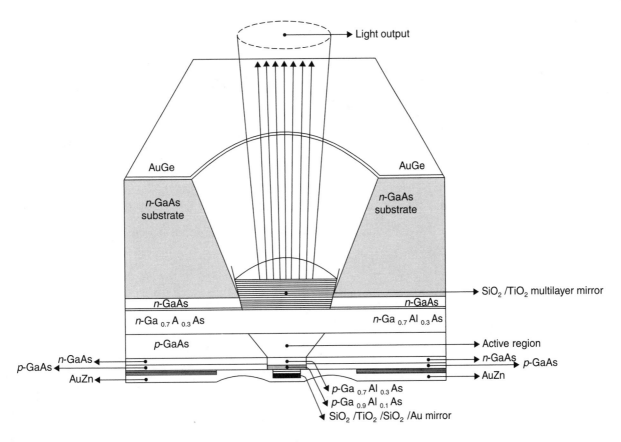

Figure 3–39. Cross section of a GaAlAs/GaAs VCSEL diode.

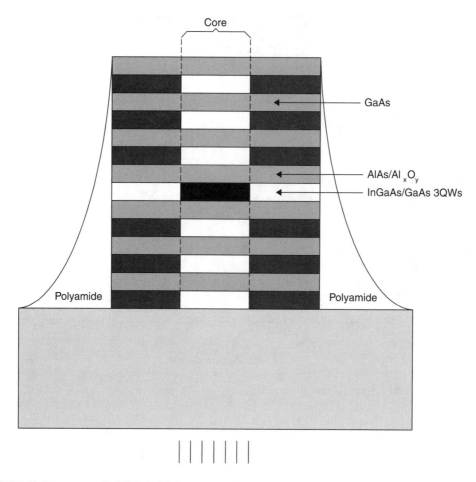

Figure 3–40. VCSEL diodes composed of GaInAs/GaAs compounds.

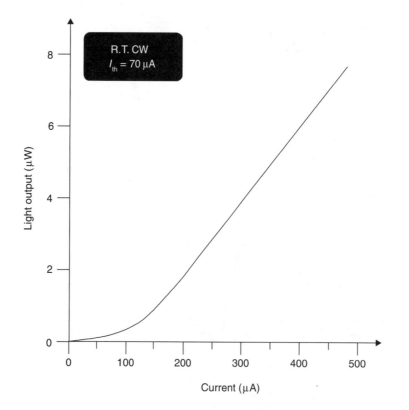

Figure 3–41. L *v*. I of a VCSEL diode.

From Figure 3–41, it is evident that lasing action takes place at an approximate threshold current of 80 mA at room temperature with carrier wave operation. Subsequent research has shown that optical power around 50 mW to 200 mW with efficiencies up to 50% can be achieved.

Performance Characteristics of VCSEL Diodes

The VCSEL diode performance characteristics can be described as follows:

THRESHOLD CURRENT. One of the most critical parameters of a VCSEL diode structure is the threshold current (I_{th}). Threshold currents as low as 10 μA can be achieved. Further reduction of threshold current is only possible through a proportional reduction of the device physical dimensions, electrode resistivity, and heat sinking improvement.

EFFICIENCY. Another very important device characteristic is efficiency. The ratio of the applied biasing current to photon generation can be substantially improved by improving the current confinement of the VCSEL diode structure with a parallel overall loss reduction. Improvement of current confinement enables VCSEL diodes to form array structures.

MODULATION CAPABILITIES. The third characteristic is high rate modulation capability, a very desirable parameter in optical communications system applications. Such systems can substantially benefit from lasers that exhibit high modulation capabilities at low threshold currents, operating at room temperatures.

NOISE. Tests conducted with VCSEL diodes to establish noise performance showed noise levels around 150 dB/Hz and side mode suppression ratios of 35 dB.

WAVELENGTH DRIFT v. TEMPERATURE. VCSEL diodes exhibit a wavelength drift in reference to temperature variations around 70 pm/K.

SPECTRAL LINE WIDTH. In optical communications systems employing DWDM schemes, laser diode spectral line width is very important. VCSEL diodes exhibit very impressive spectral line width around 50 MHz at an optical power output of approximately 1.5 mW. Because spectral line width is subject to cavity length and mirror reflectivity, a reduction of the cavity length and an increase of mirror reflectivity will further reduce the diode spectral line width.

RELIABILITY. VCSEL diode operational lifetime is a critical factor in establishing overall optical link reliability. Similarly, with LEDs, VCSEL diode lifetime can be expressed by the MTBF factor. Applying the same criteria, the VCSEL reliability factor was found to be compatible with that of the LED diodes.

SUMMARY

In Chapter 3 the need for and developmental processes of the two major types of optical sources, those of light emitting diodes (LEDs) and laser diodes, were presented. The chapter began with introductory concepts, including the classification of optical sources into LEDs and lasers, then continued with a detailed discussion of edge emitting and surface emitting LEDs, emphasizing differences in their performance and operating characteristics. The two most important methods of coupling optical power generated by an LED into the optical fiber were discussed at some length. A major part of the chapter was devoted to the presentation of laser sources, an essential component for high bit rate long distance optical fiber transmission systems. Fabry-Perot and distributed feedback (DFB) lasers were discussed in terms of their physical structure and performance characteristics, followed by quantum well (QW), multi quantum well (MQW), and vertical cavity surface emitting lasers (VCSEL).

QUESTIONS

Section 3.1

1. Explain why LEDs are primary sources in optical fiber communications systems.

2. Name and describe the advantages and disadvantages of the two types of LEDs.

3. List the critical components that led to the development and implementation of DWDM.

4. What does VCSEL mean? List the advantages of such a device.

5. With the assistance of a simple diagram, describe the processes of spontaneous and stimulation emissions.

Section 3.2

6. What is an electro-luminescent device? Explain.

7. Name the various types of LEDs, based on their material composition, and list their corresponding performance characteristics.

8. Briefly describe the basic structure and operating mechanism of an edge emitting LED.

9. Draw the current/optical power output curve of an LED and explain its relationship.

10. With the assistance of a formula, describe *internal power efficiency* of an LED.

11. Define *external power efficiency*. Give the corresponding formula and explain it in some detail.

12. What is the spectral bandwidth of an LED? Draw the spectral bandwidth curve and explain.

13. Describe the relationship between LED bandwidth and carrier recombination.

14. Briefly describe the basic structure and operating mechanism of a surface emitting LED.

15. What is *coupling efficiency*? With the assistance of a formula, briefly explain.

16. Explain, in detail, the Butt method of SLED coupling into an optical fiber. Give an example of coupling efficiency.

17. Explain in detail the Lens method of SLED coupling into an optical fiber.

18. List the various methods of incorporating a lens into the source-to-fiber assembly.

19. What are the parameters that determine the reliability factor of LEDs? Give the corresponding formula.

Section 3.3

20. What does the acronym LASER mean?

21. Describe the Fabry-Perot laser diode and list its operating characteristics.

22. Give the mathematical expression of *mean wavelength* and define its associated parameters.

23. Draw the current/optical power output curve of a Fabry-Perot laser and explain the relationship.

24. How was the low efficiency of early laser diodes overcome? Explain.

25. Name the parameters involved in the establishment of the laser diode generated wavelength.

26. With the assistance of diagram, describe DFB operation.

27. What is the critical relationship which must be satisfied in order for a laser to operate as a DFB?

28. Describe Bragg reflector structure and compare it to the structure of a DFB.

29. Describe the relationship between the threshold current and operating temperature of a laser. Briefly explain.

30. List the components that influence the self-oscillating frequency of a laser diode.

31. How can the modulation rate of a laser diode be increased? Explain.

Section 3.4

32. What is a quantum well laser and how does it differ from a conventional double heterojunction laser?

33. Describe the difference in semiconductor structure and performance between QW and MQW lasers.

Section 3.5

34. What is a surface emitting laser and how does it compare to conventional double heterojunction and QW lasers?

Section 3.6

35. What was the principal incentive for the development of vertical cavity surface emitting lasers?

36. Give the mathematical expression relating modal gain and optical losses in a VCSEL.

37. Derive the expression of threshold current for a VCSEL.

38. List the major difficulties associated with VCSELs.

39. Compare the characteristics of a laser diode for both short wavelength and long wavelength operations.

40. List the operating characteristics of VCSELs composed of GaAlAs/GaAs.

PROBLEMS

1. An LED device composed of GaAlP semiconductor material is emitting at a wavelength of 650 nm. Calculate the band gap energy for the material.

2. Calculate the coupling efficiency and optical power launched into the fiber if the generated power by the optical source is 1.7 μW, the angle of incident is 22.5°, and the active area is smaller than that of the optical fiber.

3. If an optical source coupled to a fiber is forward biased with a biasing voltage of 1.75 V, it will generate a forward current of 24 mA. Assuming the diode's internal efficiency is 1.5% and has a coupling angle of 30°, calculate the coupling efficiency and the optical power coupled into the fiber.

4. The Butt method is used to couple an optical source into both a step index and graded index fiber. Compute the coupling efficiency for both cases for the given operating parameters:

i) transmissivity of the medium: 0.8

ii) refractive index of the medium between source and fiber: 1.2

iii) radius of the fiber core: $x = 0.9\,y$

iv) fiber numerical aperture: 0.45

5. When an LED source operating at 30°C is forward biased with 2 V, it generates 50 mA of forward current. If the source is to operate at 45°C, calculate the MTBF.

4

OPTICAL DETECTORS

1. Describe the photodetection process and state the importance of photodetector operating characteristics in the design of optical receivers.

2. Explain in detail the nature and performance characteristics of PIN photodetectors in terms of response time, dark current, reliability, and operational lifetime.

3. Discuss, in detail, the theory of avalanche photodetectors, including semiconductor material composition, optical gain, noise, and response time.

4. Establish the need for advanced optical detectors required in the design of modern sophisticated optical receivers.

5. Describe, in detail, resonant cavity enhanced (RCE) photodetector diodes in terms of material composition, wavelength selectivity, and quantum efficiency.

6. Describe the operation and performance characteristics of resonant cavity enhanced heterojunction photodetectors (RCE-HP).

K E Y   T E R M S

absolute temperature
absorption coefficient
absorption layer
active layer length
avalanche effect
avalanche photodetectors
   (APD)
background current
base failure rate
Boltzmann's constant
breakdown voltage

confinement layer
cut off wavelength
dark current
decay losses
depletion area
doping density
electron velocity
environmental factor
excess noise factor
free spectral range
Johnson noise

junction capacitance
junction voltage
kinetic energy
linearity
mean square avalanche gain
mean square spectral density
mean time between failures
   (MTBF)
mirror effective length
mirror reflection coefficient
multiplication factor

noise voltage
photocarriers
photocurrent
photocurrent gain
photodetection
photodetector
photon energy
photon lifetime
PIN
primary photocurrent
propagation constant

quality factor
quantum efficiency
relative permittivity
reliability factor

resonant cavity enhanced
(RCE)
response time
responsivity

saturation velocity
shot noise
shot noise power
temperature factor

transit time
tunneling current

4.1 PHOTODETECTION

Photodetection is the process whereby optical power is detected and then converted to electrical power. Photodetector devices (optical detectors) perform **photodetection.** Optical detectors perform the opposite function of that of optical sources: they convert electric power into optical power.

In any optical fiber communications system, the optical source is part of the transmitter section, while optical detectors are part of the receiver section. The performance of an optical detector incorporated into the receiver section of an optical fiber communications system can be determined by its ability to detect the smallest optical power possible (detector sensitivity) and generate a maximum electric power at its output with an absolute minimum degree of distortion (low noise). Optical detectors must also exhibit a comparatively wide bandwidth and sharp response to accommodate the high bit-rate required by such a system. Other criteria for selecting a particular photo diode for implementation into an optical fiber communications system are the ability to interface with optical cables, a long operating life, and cost.

Although there are several types of **photodetectors,** not all of them are suitable for use in optical fiber communications systems. In such systems, the optical detector device, which is almost always utilized, is the semiconductor photodiode.

Photodetector design criteria are set forth by system parameters such as size, sensitivity, bandwidth, and degree of tolerance to temperature variations.

The two photodetector devices most commonly used in optical fiber communications systems are PIN and APD devices.

4.2 PIN PHOTODETECTORS

The principal theory on which a *P*-region, I-intrinsic, *n*-region (**PIN**) photodetector device is based is illustrated in Figure 4–1.

When a photon is incident upon a semiconductor photodetector device, and it has more energy than the band gap energy of that device, the energy of the photon is absorbed by the band gap and an electron-hole pair is generated across the band gap. The energy of the incident photon is given by Equation (4–1).

$$E_{ph} = \frac{hc}{\lambda} \qquad \textbf{(4–1)}$$

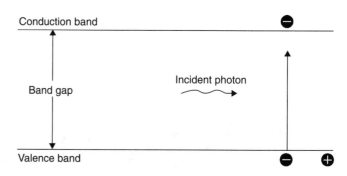

Figure 4–1. Photodetector device.

where E_{ph} is the energy of the photon, h is Planck's constant $(6.62 \times 10^{-34}\ \text{Ws}^2)$, c is the velocity of light $(3 \times 10^8\ \text{m/s})$, λ is the wavelength (m), E_g is the band gap energy.

It is evident from Equation (4–1) that the photon energy (E_{ph}) is inversely proportional to the wavelength (λ). Therefore, a wavelength, at which the photon energy becomes equal to the band gap energy, exists. At this photon energy level, electron-hole generation will occur. The wavelength at which the photon energy becomes equal to band gap energy is called the **cut off wavelength** (λ_c). From Equation (4–1), solving for (λ) we have Equation (4–2).

$$\lambda = \frac{hc}{E_{ph}}$$

Because $E_{ph} = E_g = eV$

then $\lambda = \dfrac{hc}{E_g}$.

Substituting for: $h\ (6.62 \times 10^{-34}\ \text{Ws}^2)$
$\qquad\qquad\qquad c\ (3 \times 10^8\ \text{m/s})$

Therefore,

$$\lambda_c = \frac{1.24\ \mu m}{E_g} \qquad \textbf{(4–2)}$$

Semiconductor materials employed in the fabrication of photodetectors are the same with the materials employed in the fabrication of optical sources.

Each individual element or substance is classified by a band gap energy level (E_g) characteristic of that element or substance. Therefore, different materials exhibit different cut off wavelengths. Some materials, with their corresponding band gap energy levels (eV), are listed in Table 4–1.

Applying these energy gap levels to Equation (4–2) yields the following cut off wavelengths.

For Ge: $\lambda_c = \dfrac{1.24}{0.67} = 1.85\ \mu m$

TABLE 4–1 Band Gap Energy Levels

Elements/Substances		Band Gap Energy (eV)
Germanium	Ge	0.67
Silicon	Si	1.11
Indium gallium arsenide	InGaAs	0.77
Indium gallium arsenide phosphorus	InGaAsP	0.89

Therefore, the cut off wavelength for Ge is 1.85 μm.

For Si: $\lambda_c = \dfrac{1.24}{1.11} = 1.11$ μm

Therefore, the cut off wavelength for Si is 1.11 μm.

For InGaAs: $\lambda_c = \dfrac{1.24}{0.77} = 1.61$ μm.

Therefore, the cut off wavelength for InGaAs is 1.61 μm.

For InGaAsP: $\lambda_c = \dfrac{1.24}{0.89} = 1.4$ μm

Therefore, the cut off wavelength for InGaAsP is 1.4 μm.

A cross section of a silicon PIN diode is shown in Figure 4–2. When a photon enters the photodetector, the low band gap absorption layer absorbs the photon, and an election-hole pair is generated. This election-hole pair is called a **photocarrier.**

These photocarriers, under the influence of a strong electric field that was generated by a reverse bias potential difference across the device, are separated and form a **photo-current** intensity proportional to the number of incident photons. The DC biasing of a PIN diode photo detector is shown in Figure 4–3.

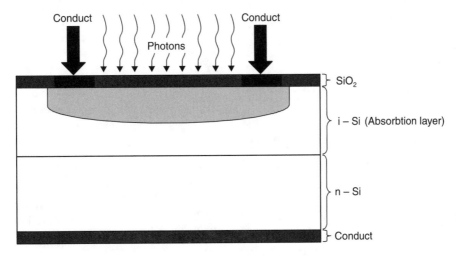

Figure 4–2.   A cross section of a silicon PIN diode.

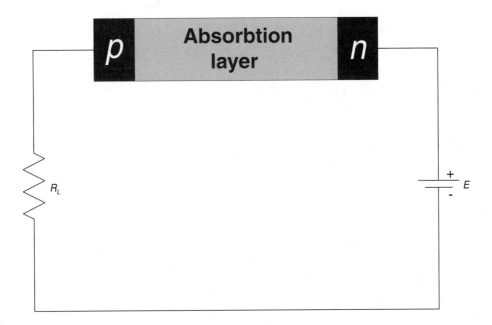

Figure 4–3.   Diode biasing.

OPTICAL DETECTORS

The photocurrent generated from the PIN photodetector device develops a potential difference across the load resistance (R_L), with a frequency calculated as follows in Equation (4–3).

$$f = \frac{E_{ph}}{h} \qquad (4\text{–}3)$$

where E_{ph} is the **photon energy** (*eV*), *h* is Planck's constant 6.62×10^{-34} Ws^2, and *f* is the frequency. Because a cut off wavelength for each substance used in the fabrication of a PIN photodetector exists, a cut off frequency for each element or substance also exists. The cut off frequency can be calculated as follows.

For Ge: $f_c = \dfrac{E_{ph}}{h}$

First, the E_{ph} must be converted from *eV* to Joules: 1 eV $= 1.6 \times 10^{-19}$ J

For germanium, Ge $= 0.67$ eV
In Joules, $1.6 \times 0.67 \times 10^{-19}$ J $= 1.07 \times 10^{-19}$ J.

$$f_c = \frac{1.07 \times 10^{-19}\,\text{J}}{6.62 \times 10^{-34}\,\text{W/s}^2} = 161 \times 10^{12}\,\text{Hz}$$

Therefore, the cut off frequency for Ge is 161 THz.
For silicon, Si $= 1.11$ eV.
In Joules, $1.11 \times 1.6 \times 10^{-19}$ J $= 1.776 \times 10^{-19}$

$$f_c = \frac{1.776 \times 10^{-19}}{6.62 \times 10^{-34}} = 268 \times 10^{12}\,\text{Hz}$$

Therefore, the cut off frequency for Si is 268 THz.
For indium gallium arsenide, InGaAs $= 0.77$ eV.
In Joules, $0.77 \times 1.6 \times 10^{-19} = 1.232 \times 10^{-19}$ J

$$f_c = \frac{1.232 \times 10^{-19}}{6.62 \times 10^{-34}\,\text{W/s}^2} = 186 \times 10^{12}\,\text{Hz}$$

Therefore, the cut off frequency for InGaAs is 186 Thz.
For indium gallium arsenide phosphorus, InGaAsP $= 0.89$ eV.
In Joules, $0.89 \times 1.6 \times 10^{-19} = 1.424 \times 10^{-19}$ J

$$f_c = \frac{1.424 \times 10^{-19}\,\text{J}}{6.62 \times 10^{-34}\,\text{W/s}^2} = 215 \times 10^{12}\,\text{Hz}$$

Therefore, the cut off frequency for InGaAsP is 215 THz.
The combination of different semiconductor alloys that are all operating at different wavelengths allows the selection of material capable of responding to the desired operating wavelength. For example: GaAs/AlGaAs substances operate at wavelengths between 800 nm and 900 nm, while photodetector devices composed of InGaAs/InP alloys operate at wavelengths between 1000 nm and 1600 nm.

PIN Photodetector Characteristics

The fundamental PIN photodiode operational characteristics are quantum efficiency (η), **responsivity** (*R*), speed, and **linearity.**

Quantum efficiency (η) is defined by the number of electron-hole pairs generated per photon (Equation (4–4)).

$$\eta = \frac{N(e^-, p^+)}{N_{ph}} \qquad (4\text{–}4)$$

where N (e^-, p^+) is the number of generated electron-holes, N_{ph} is the number of photons, and η is the **quantum efficiency.** The number of generated electron-hole pairs is translated to current by using Equation (4–5).

$$I_p = q \times N_{e^-} \qquad (4\text{–}5)$$

To determine the number of electrons, solve Equation (4–5) for N_{e^-}, yielding Equation (4–6).

$$N_{e^-} = \frac{I_p}{q} \qquad (4\text{–}6)$$

where I_P is the photocurrent (mA), *q* is the electron charge (1.6×10^{-19} C), and N_{e^-} is the number of electrons. Consequently, the number of incident photons is translated to light power through Equation (4–7).

$$P_o = N_{ph} \times h\nu \qquad (4\text{–}7)$$

To determine the number of photons, solve Equation (4–7) for N_{ph}, yielding Equation (4–8).

$$N_{ph} = \frac{P_o}{h\nu} \qquad (4\text{–}8)$$

where P_o is the light power, N_{ph} is the number of photons, *h* is Planck's constant (6.628×10^{-38} J/s), ν is the velocity of light, and λ is the wavelength.
Substituting Equations (4–6) and (4–8) into Equation (4–4) yields Equation (4–9).

$$\eta = \frac{N_{e^-}}{N_{ph}} = \frac{\dfrac{I_P}{q}}{\dfrac{P_o}{h\nu}} = \frac{I_P\,h\nu}{q\,P_o}$$

Substituting for $\nu = \dfrac{c}{\lambda}$,

$$\eta = \frac{I_p hc}{qP_o \lambda}$$

Because $E_{ph} = \dfrac{hc}{\lambda}$,

therefore,
$$\eta = \frac{I_p E_{ph}}{qP_o} \qquad (4\text{–}9)$$

From the quantum efficiency equation, it is evident that the efficiency of a PIN photodetector is proportional to the photon energy absorbed by the **absorption layer** of the device. Larger photon energy requires a thicker absorption layer, allowing longer time for electron-hole pair generation to take place.

Response Time (Speed)

Response time or speed of a photodetector is referred to as the time required by the generated carriers within the absorption region to travel that region under reverse bias conditions.

The main factor that determines this time is the thickness of the absorption region. The thicker the absorption region, the longer the time. Here, there is a conflict between photodetector efficiency and response time. For higher efficiency, a thicker absorption region is needed, while for higher speed, a thinner absorption region is required. In practice, trade-offs between efficiency and speed are made to accommodate design objectives. The response time (t_r) of a photodetector is given by the relationship

$$t_r = \frac{\text{thickness of the absorption layer}}{\text{saturation velocity}}$$

Saturation velocity (V) for a typical InGaAs alloy is 10^7 m/s. Given the device absorption layer thickness, the response time can be calculated.

Example 4–1
Compute the response time of PIN photodetector composed of InGaAs with 5 μm of absorption layer thickness.

Solution

$$t_r = \frac{5 \times 10^{-6} \text{ m}}{1 \times 10^7 \text{ m/s}} = 5 \times 10^{-13} \text{m}$$

or $t_r = 0.5$ ps.

The key parameter for determining photodetector device performance is responsivity. Responsivity is defined by the ratio of the current generated in the absorption region per-unit optical power incident to the region. Responsivity is closely related to quantum efficiency and is expressed by Equation (4–10).

$$R = \eta \frac{q}{E_{\text{ph}}} \qquad \textbf{(4–10)}$$

where R is the responsivity, η is the quantum efficiency, q is the electron charge (1.59×10^{-19} C), and E_{ph} is the energy of the photon (hv). Substituting Equation (4–4) (quantum efficiency) into Equation (4–10) yields Equation (4–11).

$$R = \frac{I_p E_{\text{ph}}}{P_o} \times \frac{q}{E_{\text{ph}}} = \frac{I_p}{P_o}$$

Therefore,

$$R = \frac{I_p(\mu\text{A})}{P_o(\mu\text{W})} \qquad \textbf{(4–11)}$$

The responsivity of a PIN photodiode is the ratio of the generated photocurrent per incident of unit-light power.

A graphical representation of quantum efficiency (η) and responsivity is shown in Figure 4–4.

Figure 4–4 illustrates the fundamental difference between responsivity and quantum efficiency. For different semiconductor materials, the responsivity is linear up to a particular wavelength, then it drops quickly. Beyond this point, the photon energy becomes less than the energy required for electron-hole generation.

Dark Current

Dark current (I_d) is defined as the reverse leakage current of a photodetector device in the absence of optical power entering the photodetector device. Dark current is an unwanted element caused by factors such as current recom-

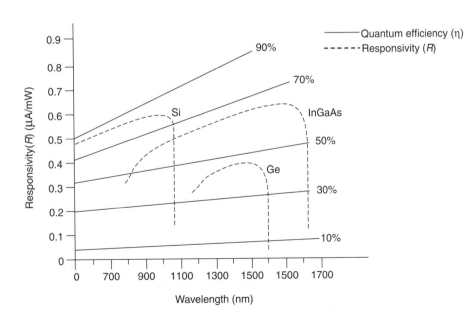

Figure 4–4. A graphical representation of quantum efficiency (η) and responsivity (R) in reference to operating wavelength.

bination within the depletion region and surface leakage current. The negative effects of such unwanted currents contribute to thermal **shot noise** first observed by W. Schottky. While experimenting with vacuum tubes he observed that certain spontaneous fluctuations of the DC anode current were periodically occurring. Further observations and studies concluded that these fluctuations were the result of the particle nature of free electrons moving randomly under the influence of a potential difference applied across the anode and the cathode of the tube.

Shot Noise

In semiconductor devices, shot noise is the result of electron-hole recombination and majority carrier random diffusion. The power spectral density of shot noise is proportional to the dark current and is expressed by Equation (4–12).

$$P_n = 2I_d q B_w \qquad (4\text{–}12)$$

where P_n is the **shot noise power** (W), I_d is the dark current (A), q is the electron charge $(1.59 \times 10^{-19}$ C$)$, and B_W is the operating bandwidth. Shot noise-voltage (V_n) is expressed by Equation (4–13).

$$V_n = 2I_d B_W \qquad (4\text{–}13)$$

where V_n is the **noise voltage** and B_W is the receiver operating bandwidth.

Thermal (Johnson) Noise

Thermal noise is the result of thermally agitated free-electron motion within any conducting material. Thermally agitated electrons within a conductor collide with the molecules of that conductor, thus starting a chain reaction with all other free electrons. The average of thermal noise is zero, at which point the material is said to be in thermal equilibrium. Therefore there is no DC component. The voltage probability is Gaussian with a mean square voltage given by Equation (4–14).

$$V_n = 4KT \int_{f_1}^{f_2} R(f) P(f)\, df \qquad (4\text{–}14)$$

$P(f)$ is defined as follows in Equation (4–15).

$$P(f) = \frac{hf}{KT}(e^{-hf/KT} - 1)^{-1} \qquad (4\text{–}15)$$

where T is the absolute temperature (290°K), K is **Boltzmann's constant** $(1.38 \times 10^{-23}$ J-K$)$, f is the frequency (Hz) (bandwidth of the observed voltage), h is Planck's constant $(6.62 \times 10^{-34}$ J/s$)$, and R is the impedance (Ω). For a temperature of $T = 290°$K and frequency $f \geq 100$ GHz, $P(f)$ takes a value between 0.992 and 1 $(0.992 < P(f) < 1)$; therefore, $P(f) \cong 1$ and noise voltage becomes Equation (4–16).

$$V_n = \sqrt{4KTB_W R} \qquad (4\text{–}16)$$

The thermal noise power $P_{n(\text{thermal})}$ is expressed by Equation (4–17).

$$P_{n(\text{thermal})} = KTB_W \qquad (4\text{–}17)$$

Shot-noise-voltage (V_n) is expressed by Equation (4–18), which is the same as Equation (4–13).

$$V_n = 2I_d B_w \qquad (4\text{–}18)$$

where V_n is the **noise voltage** and B_W is the receiver operating bandwidth.

Signal to Noise Ratio

The signal to noise ratio (SNR) expresses (in decibels) the difference between the signal power and noise power at the input or output of an electronic device or circuit. This ratio is perhaps the most important criteria used to establish the performance of electronic devices or circuits. The SNR redundant is given by Equation (4–19).

$$SNR_{dB} = 10\log\frac{P_s}{P_n} \qquad (4\text{–}19)$$

where P_s is the power of the signal (W) and P_n is the power of the noise (W) (total noise power). Based on Equation (4–19), the SNR can be thought of as the logarithmic ratio of the signal power to the sum total of the noise power (shot noise and thermal or **Johnson noise**).

Equation (4–13) indicates that an increase of dark current will decrease the overall receiver operation bandwidth by maintaining a constant noise voltage level. Although it can be reduced considerably by proper material selection and controlled fabrication techniques, the dark current cannot be totally eliminated. This unwanted current is also temperature dependent. That is, it increases with an increase in the operating temperature. Therefore, proper control of materials fabrication techniques and operating temperatures are key factors for the reduction of dark current and, consequently, for enhancing the operating performance of the PIN photodetector device.

4.3 AVALANCHE PHOTODETECTORS

Avalanche photodetectors (APD) are very similar to PIN diodes, with only one exception: the addition to the APD device of a high intensity electric field region. In this region, the primary electron-hole pairs generated by the incident photons are able to absorb enough **kinetic energy** from the strong electric field to collide with atoms present in this region, thus generating more electron-hole pairs. This process of generating more than one electron-hole pair from one incident photon through the ionization process is referred to as the **avalanche effect.**

It is apparent that the photocurrent generated by an APD photodetector device exceeds the current generated by a PIN device by a factor referred to as the **multiplication factor (M).**

Because the current generated by a PIN device is expressed as $I = qN_{e-}$, the generated photocurrent is expressed by Equation (4–20).

$$I_p = (qN_{e-})M \qquad (4\text{–}20)$$

where I_P is the generated photocurrent, q is the electron charge (1.59×10^{-19} C), N_{e-} is the carrier number, and M is the multiplication factor. The multiplication factor depends on the physical and operational characteristics of the photodetector device. Operational characteristics include the width of the avalanche region, the strength of the electric field, and the type of semiconductor material employed. The cross section of a short wavelength silicon APD device is shown in Figure 4–5. This structure is composed of a $p^+p^-pn^+$ semiconductor material. A light doped p^- region is epitaxially grown on a heavily doped p^+-type substrate.

When a reverse biased voltage is gradually applied across the diode, an electric field develops across the avalanche region. Its strongest intensity is measured at the pn^+ junction. As the reverse biasing voltage gradually increases, the corresponding electric field across the region also increases and, as a consequence, there is an expansion of the depletion region. If the electric field intensity increases just below the avalanche breakdown point, the depletion region has almost reached the total width of the p^- layer. Under these circumstances, if a photon is incident upon the device, it will be absorbed by the p^- layer. The energy released by the incident photon causes the generation of the first electron-hole pair. Under the influence of the strong electron field, the generated electron is guided from the p^- intrinsic region closer to the pn^+ junction. At this point, the electric field intensity is at its maximum.

Under the influence of the strong electric field, the entering electron collides with other atoms, thus generating a new electron pair. The secondary electron, still under the influence of the strong electric field, generates another electron-hole pair, and so on (avalanche effect).

The number of the secondary electron-hole pairs that are generated is proportional to the carrier distance traveled through the avalanche region and to the semiconductor materials used for the fabrication of photodetector devices. Figure 4–6 shows the relationship between electric field strengths, and ionization rates for different alloys employed in the fabrication of photodetector diodes.

The equivalent circuit of an avalanche-photodetector diode is shown in Figure 4–7. When reverse biased, an electronic field develops across the depletion region with its maximum field strength across the p-n junction. The avalanche effect will occur only when the depletion region has reached its maximum or is fully developed.

A cross section of an InGaAs photodetector device is shown in Figure 4–8. This is a double heterostructure device incorporating an InGaAs low bandgap absorption layer, while the InP is used as a confinement layer with a band gap higher than that of the absorption layer. This higher band gap of InP ensures that no photon absorption will occur in the confinement layer.

One of the fundamental operating characteristics of a photodetector device is its available bandwidth. The number of holes present in the device limits available bandwidth. The InGaAP layer is epitaxially grown on the absorption layer with a band gap somewhere between the absorption and **confinement layers.** This arrangement removes the bandwidth limiting holes, and thus enhances the overall bandwidth characteristics of the photodetector device. The absorption layer is elevated to a higher band gap region, resulting in the first electron-hole pair generated by the incident photon. The generated electron under the influence of the very strong electric field is guided to the p^- InP ring. At the pn^+ junction, the electric field is at maximum, with a maximum depletion region. Through this region, impact ionization takes place, resulting in photocurrent multiplication. Another basic performance characteristic of an APD photodetector device is the ratio of the electron and hole ionization rates to the different semiconductor materials used during the fabrication process. This electron-hole ionization ratio is a key factor, as it determines gain bandwidth product and noise performance characteristics. For example, let us suppose that electrons are the primary carriers in an APD structure. A higher electron-to-hole ratio reflects a low noise and higher gain bandwidth product, while a low electron-to-hole ratio reflects a higher noise and lower gain bandwidth product.

Gain

The photocurrent gain in an APD device is a function of several elements such as: the wavelength of the incident photons, the electric field strength as a result of the reverse bias voltage, the width of the depletion region, and the types of semiconductor materials used for the fabrication of the APD device. The relationship of the photocurrent gain to biasing

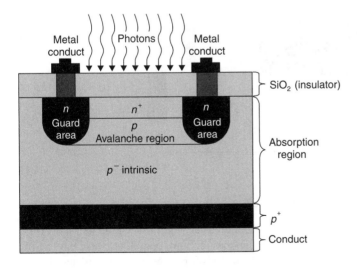

Figure 4–5. Silicon APD device.

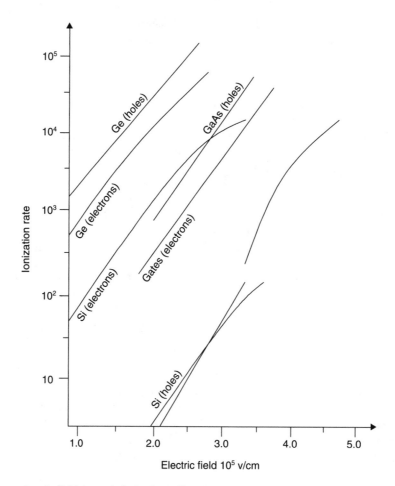

Figure 4–6. Ionization rate *v.* electric field strength for various alloys.

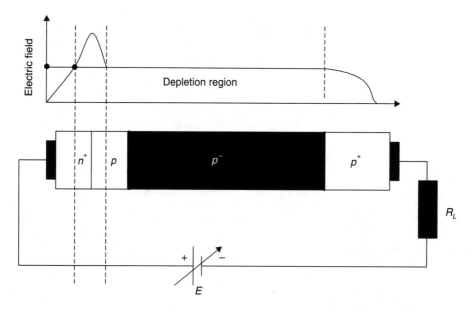

Figure 4–7. Equivalent circuit for APD (minimum electric field required for avalanche effect).

voltage for different wavelengths is shown in Figure 4–9. It is evident that higher photocurrent gain is observed at higher wavelengths at specific reverse biasing voltages.

The function of the guard rings in an APD structure is to prevent edge breakdown around the avalanche region.

When InGaAsP materials are used in the fabrication of APD devices, these devices exhibit operating wavelengths of between 900 nm and 1600 nm; when silicon materials are used, they exhibit operating wavelengths of between 400 nm and 900 nm.

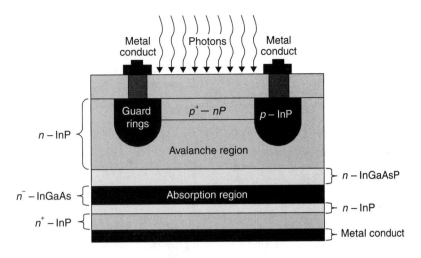

Figure 4–8. A cross section of an InGaAs photodetector device.

Photocurrent gain, an important parameter of an APD device, is also temperature dependent. Figure 4–10 shows an increase of the multiplication factor with a corresponding increase in the operation temperature at a constant reverse biasing voltage. Current fluctuation with temperature is an undesirable phenomenon and must be confined to a tolerable level or completely eliminated. To maintain a constant multiplication factor for a wide range of operating temperatures, any increase in the generated photocurrent due to an increase in the operating temperature must be compensated for by a proportional decrease of the reverse biasing voltage. The equation that determines the (current-gain) multiplication factor (M), in reference to reverse biasing voltage and operating temperature, is given by Equation (4–21).

$$M = \frac{1}{1 - \left[\dfrac{V_a - I_p R_L}{V_{BK}[1 + \alpha(T_1 - T_o)]} \right]} \quad \textbf{(4–21)}$$

where V_a is the reverse biasing voltage (V), I_P is the multiplied photocurrent (mA), R_L is the device resistance plus load resistance, V_{BK} is the **breakdown voltage** at room temperature, T_O is the room temperature, T_1 is the operating

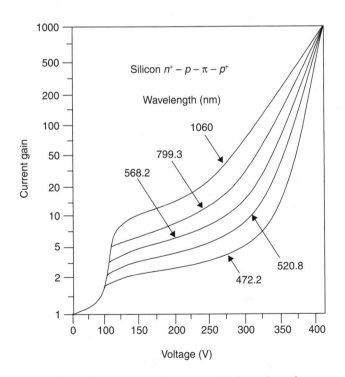

Figure 4–9. Photocurrent gain *v.* reverse biasing voltage for different wavelengths.

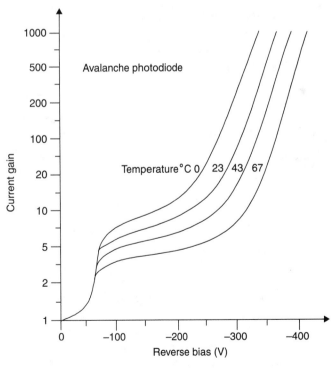

Figure 4–10. Photocurrent gain *v.* reverse biasing voltage at different operating temperatures.

temperature, and α is the factor determined from gain v. temperature graphs.

Photodetector Noise

Avalanche photodetectors exhibit higher noise levels than PIN devices. This is a result of the ionization and photocurrent multiplication process that takes place within the APD device. The random nature of the incident photons on the APD device results in a random photocurrent generation at the output of the device. This current fluctuation is classified as shot noise, and can be expressed in Equation (4–22).

$$\frac{d(i_p)^2}{df} = 2qI(M)^2 \tag{4–22}$$

where $(i_p)^2 =$ **mean square spectral density,** $f =$ frequency (Hz), $q =$ electron charge (1.6×10^{-19} C), $I =$ **primary photocurrent,** $(M)^2 =$ **mean square avalanche gain**. The primary photocurrent can be found through the relationship $I = i_P + I_{BR} + I_{dk}$, where i_P is the photocurrent, I_{BR} is the **background current,** and I_{dk} is the dark current.

Equation (4–22) can be further modified to incorporate the gain nonlinearities resulting from the statistical nature of the ionization and avalanche processes (Equation (4–23)).

$$\frac{d(j_p)^2}{df} = 2qI(M)^2F(M) \tag{4–23}$$

where M is the multiplication factor, $F(M)$ is the **excess noise factor,** I is the primary photocurrent, (j_p) is the mean square spectral density, and f is the frequency. The excess noise factor $(F(M))$ is related to the semiconductor material used in the fabrication of the APD device, the electron-to-hole carrier ratio, and the electric field profile across the depletion region. An empirical equation determining the excess noise factor $F(M)$ is given by Equation (4–24).

$$F(M) = 2(1 - K) + K(M) \tag{4–24}$$

where M is the multiplication factor and K is the ratio of the smallest to the largest ionization coefficients

For silicon: $K \cong 0.02$–0.1
For germanium: $K \cong 0.5$
For InGaAsP: $K \cong 0.3$–1

Dark Current

Dark current is referred to as the current present at the photodetector output at the absence of incident light. For an APD device, the dark current is multiplied by the device multiplication factor (M) to determine the reduction in device sensitivity.

The dark current is a nonlinear function of the reverse biased voltage at avalanche breakdown levels, and is sometimes referred to as **tunneling current.** Different semiconductor materials exhibit different levels of tunneling current resulting from different band gap sizes. For example, devices with small band gaps measure small tunneling currents, while large band gap devices measure larger tunneling currents. A practical solution for a substantial reduction of the tunneling current is the fabrication of structures with a separation between the absorption (low band gap) region and the avalanche (high band gap) region.

Response Time

The **response time** of a photodetector device is the time a carrier takes to cross the depletion region. For APD devices, the response time is almost double that of PIN devices. Because the APD structure incorporates a large band gap region produced by the large electric field, the generated photocarriers must travel twice the distance from the low band gap region to the higher band gap region and back, after the multiplication process has taken place. It is, therefore, evident that response time is directly related to **depletion area** width. The larger the width, the larger the response time. If a reduction of the depletion region is attempted in order to reduce response time, inevitably a substantial quantum efficiency reduction will result. Therefore, in APD photodetector devices, a trade-off is necessary between quantum efficiency and response time. A typical response time of 0.5 ns at 800 nm–900 nm has been achieved.

Capacitance

In a photodetector device, internal capacitance is a parasitic component that affects the overall response time of the detector. As with any other capacitance, **junction capacitance** of an APD device is determined by the cross-sectional area and width of its depletion region. It is expressed by Equation (4–25).

$$C = \frac{\varepsilon qAN}{2(V_R + V_j)} \tag{4–25}$$

where C is the **junction capacitance (F),** ε is the dielectric constant, A is the depletion area, N is the **doping density** (depletion region), V_R is the reverse bias voltage (V), V_j is the **junction voltage,** and q is the electron charge.

4.4 PHOTODETECTOR DEVICE CHARACTERISTICS

Conventional PIN Photodetectors

Some characteristics of commercially available PIN photodetector devices are as follows:
Device type: 35PD300-FC (InGaAs photodetector)
Manufacturer: Telcom Devices
Brief description: The 35DP300-FC is an InGaAs photodetector that was designed to operate at a wavelength of 1300 nm, applicable to medium data rate optical communications systems and high sensitivity instrumentation. This device is packaged in a TO-46 header. Some of the basic device characteristics are listed in Table 4–2.

TABLE 4–2 Device Characteristics

Parameter	Minimum	Typical	Maximum	Units
Operating voltage	—	—	–15	V
Responsivity	0.7	0.9	—	A/W
Rise/fall	—	—	3	ns
Capacitance	9	12	—	pF
Frequency response (–3dB)	—	300	—	MHz
Dark current	–5V	—	1	10	nA

The absolute maximum ratings are listed in Table 4–3.

TABLE 4–3 Absolute Maximum Ratings

Parameter	Rating
Reverse voltage	20 V
Forward current	25 mA
Reverse current	5 nA
Operating temperature	–40°C–+85°C
Storage temperature	–40°C–+85°C

Device type: 35PD300LDC-S (InGaAs photodetector)
Manufacturer: Telcom Devices
Brief description: The 35PD300LDC-S is a PIN InGaAs photodetector device, designed to operate in the 1300 nm– 1500 nm wavelength rage. It is mostly applicable to laser back facet monitoring, medium data rate optical communications systems, and high sensitivity instruments. This device also exhibits a very low dark current. Some of the important device characteristics are listed in Table 4–4.

TABLE 4–4 Device Characteristics

Parameter	Minimum	Typical	Maximum	Units
Operating voltage	—	—	–15	V
Responsivity				
(i) 1300 nm	0.7	0.9	—	A/W
(ii) 1500 nm	—	1.0	—	A/W
Rise/fall time	—	—	3	ns
Capacitance (–5V)	—	4	12	pF
Dark current (–5V)	—	—	0.51	nA

The absolute maximum ratings are listed in Table 4–5.

TABLE 4–5 Absolute Maximum Ratings

Parameter	Rating
Reverse voltage	20 V
Forward current	25 mA
Reverse current	5 mA
Operating temperature	–40°C – +85°C
Storage temperature	–40°C – +85°C
Soldering temperature	250°C

Device type: 131-/type long wavelength
Manufacturer: Lucent Technologies
Brief description: This is a high reliability, pigtailed device that exhibits high performance, optical coupling stability, low capacitance, and wide dynamic operating range (1100 nm – 1600 nm wavelength). The device operating characteristics are listed in Table 4–6.

TABLE 4–6 Device Characteristics

Parameter	Minimum	Typical	Maximum	Units
Responsivity	0.75	0.85	—	A/W
Rise/fall time	—	≤0.5	—	ns
Capacitance	—	0.7	—	pF
Dark current	—	1.0	5	nA

The absolute maximum ratings are listed in Table 4–7.

TABLE 4–7 Absolute Maximum Ratings

Parameter	Minimum	Maximum	Units
Reverse voltage	—	30	V
Photocurrent	—	4	mA
Operating temperature	–40	85	C°
Storage temperature	–40	90	C°

Specific Applications
Analog systems:
    CATV trunk and loop
    Microwave

Telecommunications:
    Fiber in the loop (FITL)
    Broadband

Military:
    Microwave systems
    Remote antennae
    Tactical command

Digital systems:
    Telecommunications
    Fiber in the loop (FITL)
    SONET/SDH transmission systems
    Digital cellular

Data communications:
    Local Area Network (LAN)
    1 Gb/s fiber channel

Military
    Microwave systems
    Remote antennae
    Tactical command

Conventional APD–InGaAs photodetectors

Some commercially available APD photodetector devices and their operating characteristics are listed as follows:

Device type: 126 A InGaAs avalanche photodetector.
Manufacturer: Lucent Technologies
Brief description: A high performance optical detector, responding to 1300 nm and 1500 nm wavelengths. This device exhibits a high sensitivity, wide bandwidth and it can operate at a maximum data rate of 2.5 Gb/s. The device operating characteristics are listed in Table 4–8.

TABLE 4–8   Device Characteristics

Parameter	Minimum	Typical	Maximum	Units
Breakdown voltage	55	65	95	V
Maximum gain	30	—	—	—
Bandwidth	130	1500	—	nm
Responsivity	10	10.7	—	A/W
Dark current	—	50	100	nA
Capacitance	—	0.3	0.4	pF
Noise factor	—	5	6	—

The absolute maximum ratings are listed in Table 4–9.

TABLE 4–9   Absolute Maximum Ratings

Parameter	Minimum	Maximum	Units
Reverse current	—	1.0	mA
Operating temperature	−40	80	°C
Storage temperature	−55	100	°C
Soldering temperature	—	275/20	°C/s

Photodetector Reliability and Operational Lifetime

As with optical sources, photodetector operational lifetime is an important factor in determining optical link long-term reliability. The reliability factor is determined by Equation (4–26).

$$R_F = B_F T_F E_F Q_F \frac{\text{failure}}{1 \times 10^6 \text{ hr}} \qquad (4\text{–}26)$$

where R_F is the **reliability factor** (part failure rate model), B_F is the **base failure rate,** T_F is the **temperature factor,** Q_F is the **quality factor,** and E_F is the **environmental factor.** The mean time between failures (MTBF) is expressed by Equation (4–27).

$$\text{MTBF} = \frac{1}{R_F} \qquad (4\text{–}27)$$

The base failure rate for a single photodetector device is expressed by Equation (4–28).

$$B_F = 1.1 \times 10^{-3} \qquad (4\text{–}28)$$

To compute the MTBF of a single photodetector device, the individual components of Equation (4–26) must either be calculated or be obtained from tables provided by the component manufacturers. The individual parameters are calculated as follows.

The temperature factor of a single photodetector device is expressed by Equation (4–29).

$$T_F = 8.01 \times 10^{12} e^{-\left(\frac{8111}{T_J + 273}\right)} \qquad (4\text{–}29)$$

where the operating junction temperature (T_J, in °C) expressed by Equation (4–30).

$$T_J = T_A + \theta_{JA} \cdot P_d \qquad (4\text{–}30)$$

where T_A is the ambient temperature (°C), θ_{JA} is the thermal resistance $\left(\frac{°C}{W}\right)$, and P_d is the device power dissipation (W).

Example 4–2

Compute the MTBF of a photodetector diode (standard hermetic) mounted in a receptacle operating in an office environment at a temperature of 25°C. The maximum power dissipation of the device is 1.15 mW.

Solution
The MTBF is given by the Equation (4–27), where

$$R_F = B_F T_F E_F Q_F \times \frac{\text{failure}}{1 \times 10^6}$$

i) Compute (T_F).

$$T_F = 8.01 \times 10^{12} e^{-\left(\frac{8111}{T_J + 273}\right)}$$

ii) Compute (T_J).

$$T_J = T_A + \theta_{JA} P_d$$

where $T_A = 25°C$, $\theta_{JA} = 200 \left(\frac{°C}{W}\right)$,

and $P_d = 1.15$ mW.
Substituting into the equation,

$$T_J = T_A + \theta_{JA} P_d$$

$$T_J = 25°C + 200 \left(\frac{°C}{W}\right)(0.00115 \text{ W}) = 25.23°C$$

Therefore, $T_J = 25.23°C$.
Substitute T_J into T_F.

$$T_F = 8.01 \times 10^{12} e^{-\left(\frac{8111}{25.23 + 273}\right)} = 12.36$$

Therefore, $T_F = 12.36°C$.
Substitute T_F into R_F.

$$R_F = (1.1 \times 10^{-3})(12.36)(1)(0.5) \times \frac{failure}{1 \times 10^6}$$

$$= 6.8 \times 10^{-9}$$

Therefore, $R_F = 6.8 \times 10^{-9}$.

iii) Compute MTBF.

$$\text{MTBF} = \frac{1}{R_F} = \frac{1}{6.8 \times 10^{-9}} = 147 \times 10^6$$

Therefore, MTBF = 147×10^6 hours.

4.5 ADVANCED OPTICAL SEMICONDUCTOR DEVICES

High demand optical networks require high performance optical devices. One way to improve the performance of such solid state devices is through the **resonant cavity enhancement (RCE)** method developed by (Fabry and Perot).

The utilization of the resonant microcavity principle for the design and fabrication of optical devices enhances the wavelength selectivity and resonant optical field, ultimately leading to improved quantum efficiency at the operating resonant wavelength. Furthermore, an enhanced optical field allows for a reduction of the physical dimensions of the optical device active region, and consequently increases the operating speed. Substantial improvement of operating characteristics such as high speed and wavelength selectivity qualifies optical devices for system implementation that employs Wavelength Division Multiplexing (WDM) schemes.

RCE Detector Quantum Efficiency

Photodetector efficiency (η), or quantum efficiency, is defined as the ratio of the generated current flux to the incident photon flux, or the ratio of the power absorbed to the incident power. The increase of quantum efficiency for RCE photodetectors can be interpreted with the assistance of Figure 4–11.

An RCE device is structured with a selection of insulator and semiconductor materials such as AlAs/GaAs and InAlAs/InGaAs combined in such a way as to generate the required index differences. The absorption layer (active layer), with an absorption coefficient (α) and thickness (d), is sandwiched between two mirrors that are separated by distances l_1 and l_2. The field reflection coefficients for both reflective mirrors are denoted as $\sqrt{R_1} \exp^{-j\theta_1}$ and $\sqrt{R_2} \exp^{-j\theta_2}$, where θ_1 and θ_2 are the phase shifts of the light penetrating the reflective mirrors.

The electric field component of the incident light wave traveling forward is composed of both the transmitted and the feedback fields. The feedback component is the result of the cavity mirror internal reflections. The wave **propagation constant** is given by Equation (4–31).

$$\beta = \frac{2n\pi}{\lambda_o} \qquad \textbf{(4–31)}$$

where β is the propagation constant, n is the refractive index, and λ_o is the wavelength in vacuum. It is evident from

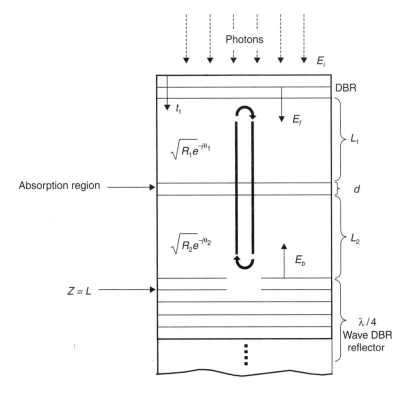

Figure 4–11. Optical detector semiconductor structure.

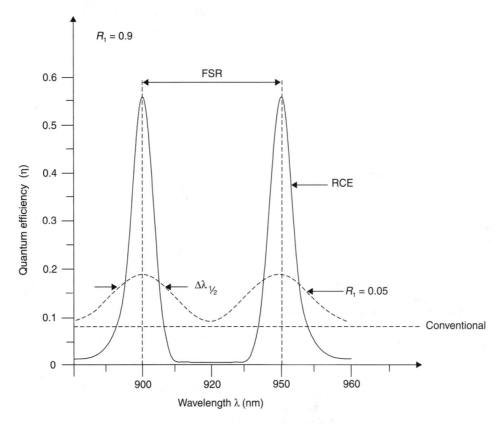

Figure 4–12. Quantum efficiency comparison between conventional optical detectors.

Equation (4–31) that the propagation constant (β) is wave-length dependent, and the refractive index (n) is periodic to the inverse wavelength (λ_o). Figure 4–12 compares the quantum efficiency (η) of both conventional and RCE photodetector devices. Photodetectors with very-close-to-unity quantum efficiency are required for specialized applications such as astronomical measurements. High quantum efficiency in photodetectors can be accomplished by a combination of high reflectivity of the bottom mirror and an absorption layer of moderate thickness. High **mirror reflection coefficients** for RCE structures are very difficult to achieve.

For conventional devices, an absorption coefficient of only a few percent can be obtained across the entire operating wavelength. For RCE devices, selecting the appropriate device parameters for the required design specifications can increase the absorption coefficient.

Compounds such as GaAs, used as bottom mirrors, exhibit a nonimpressive ninety-four percent reflectivity at approximately 1000 nm wavelength in vacuum, while metals such as Au exhibit a ninety-eight percent reflectivity under the same conditions. A combination of metal film and 1/4 wavelength substances such as AlAs/GaAs can achieve an almost unity reflective coefficient at a desired wavelength.

The high quantum efficiency of an RCE detector device is mainly due to a substantial increase of the electric field intensity within the resonance cavity (high-Q), result-ing in a higher energy absorption in the active region. This increased electric field intensity translates into a corresponding increase in much needed optical power.

The power enhancement factor of an RCE photodetector diode is shown in Figure 4–13.

Figure 4–13 shows that the power enhancement factor varies in accordance with mirror reflectivity and cavity absorption coefficients. For low cavity loss and a combination of mirror reflectivity the enhancement factor can exceed fifty percent, while for high cavity absorption and low mirror reflectivity the enhancement factor is substantially smaller.

In high Q cavities, the dramatic decrease of the enhancement factor is attributed to the higher absorption coefficient within the cavity. The thicker active layer absorbs most of the optical power. The optical power that remains is insufficient to reach the bottom mirror and is thus unable to trigger the required feedback mechanism for the enhancement process to take place.

Figure 4–14 illustrates the quantum efficiency in reference to the absorption coefficient for various levels of mirror reflectivity.

For conventional photodetector diodes with the bottom mirror reflective coefficient equal to zero, the quantum efficiency in relation to the absorption coefficient is the top line in Figure 4–14, while the RCE with the bottom mirror reflective coefficient of 0.99 is the bottom line. It is evident that the quantum efficiency of a selected wavelength is better for the RCE device than for conventional photodetector devices.

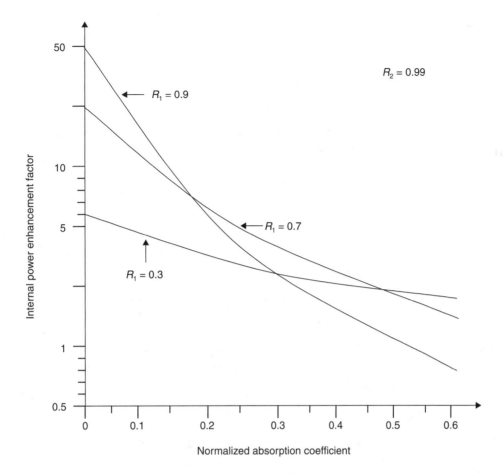

Figure 4–13. Enhanced factors for RCE photodetector diodes with $R_2 = 0.99$ and various R_1 mirror reflections.

RCE Photodetector Wavelength Selectivity

In an RCE photodetector structure, the quantum efficiency is dramatically reduced at wavelengths outside the resonant wavelength. This is due to the mutual cancellations of both the forward and the reverse traveling waves within the resonant cavity. The spectrum between resonant peaks, referred to as the **free spectral range,** is expressed by Equation (4–32).

$$f_{sr} = \frac{\lambda^2}{2(L + L_{eff_1} + L_{eff_2})n_{eff}} \qquad \textbf{(4–32)}$$

where f_{sr} is the **free spectral range,** λ is the wavelength, L_{eff_1}, L_{eff_2} is the **mirror effective lengths,** n_{eff} is the mirror-refractive index, and L is the **active layer length.** The mirror reflective length is wavelength dependent. The length of a quarter wave mirror stack composed of GaAs/AlGaAs is around 700 nm.

Wavelength selectivity of an RCE photodetector device can be measured as the free spectral-to-$\Delta\lambda_{1/2}$ ratio, and is given by Equation (4–33).

$$\frac{f_{sr}}{\Delta\lambda_{1/2}} = \frac{\pi(R_1 R_2)^{1/4} e^{-\frac{ad}{2}}}{1 - \sqrt{R_1 R_2}\, e^{-ad}} \qquad \textbf{(4–33)}$$

where R_1, R_2 is the mirror refractive index and ad are **absorption coefficients.** From Equation (4–33), it is evident that

RCE photodetector wavelength selectivity greatly depends on high mirror reflectivity and thin layers of active (absorption) region.

With typical mirror reflectivity indexes of 0.985 for the top mirror, and 0.99 for the bottom mirror, and an absorption coefficient of one percent, the ratio of f_{sr} to $\Delta\lambda_{1/2}$ can be as high as one hundred (high wavelength selectivity).

Figure 4–15 illustrates the wavelength dependency of the mirror length for a GaAl/AlGaAs mirror stack.

Material Requirements for RCE Photodetectors

From the brief description of RCE photodetectors, it is evident that their performance depends strongly on minimum cavity loss.

For minimum cavity loss, cavity materials must be selected from those that exhibit zero absorption coefficients. Both reflective mirrors must also exhibit zero absorption coefficients as well as high mirror reflectivity. Another critical element in the design and fabrication of RCE photodetectors is the mirror thickness. The thickness of a stack of mirrors, of a total width of $\lambda/4$, must not exceed a few micrometers.

The lattice structure of cavity materials must be absolutely perfect in order to have a defect-free active layer. Furthermore, the fabrication complexity of RCE devices can

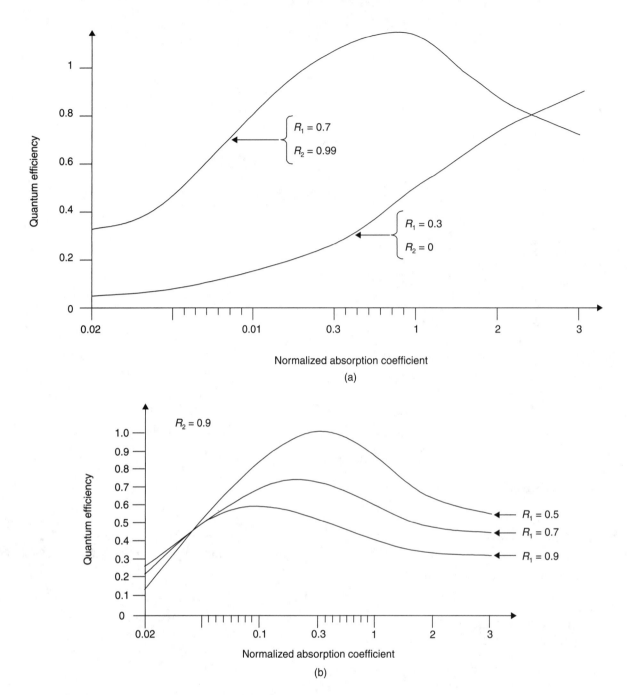

Figure 4–14. (a) Quantum efficiency *v.* absorption coefficient for a conventional device ($R_1 = 0.3$, $R_2 = 0$) and an RCE device ($R_1 = 0.7$, $R_2 = 0.99$). (b) Quantum efficiency *v.* absorption coefficient for different levels of mirror reflectivity.

be somewhat simplified if the number of mirror periods is kept to an absolute minimum. This can be accomplished by selecting a large refractive index difference among the materials that compose the reflecting mirrors.

Both cavity and mirror materials must have a larger band gap than the active region. There is, however, a low band gap limit for the active layer beyond which extraction of the photo-generated carriers will be inhibited by the heterojunction band effects. The absorption coefficient of the active layer must be between 1×10^3 cm $\leq$ and $\leq 5 \times 10^4$ cm across the operating wavelength. For larger absorption coef-

ficients, a very thin active layer is required. The negative effects of such a thin active layer are undesired electric field standing waves and a substantial reduction in the breakdown voltage. Thicker active layers reflect higher transit times and, consequently, lower device speeds.

For RCE photodetector diodes, to achieve low noise for high speed applications, low surface recombination and high current saturation velocities must be considered. High carrier velocities can reduce transit time delays. High carrier velocities are the result of higher biasing voltages. Low recombination rates result in corresponding reduction of the

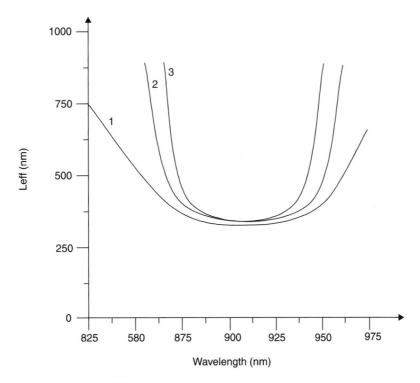

Figure 4–15. Wavelength dependency of a GaAl/GaAs stack.

dark current and detector noise. On the other hand, in order to achieve high operating bandwidths, the surface area of the device must be reduced. Surface reduction increases surface recombination and, consequently, higher dark current and detector noise. Therefore, it is imperative that a balance be maintained between surface area of the device, dark current, and detector noise and operating bandwidth.

All the criteria outlined above for the successful design and fabrication of RCE photodetector diodes can be satisfied by the combination of different semiconductor materials that operate at wavelengths of 1550 nm and beyond. A short list of such combined materials and their properties are shown in Table 4–10.

High Speed Comparison of Conventional and RCE Photodetectors

Speed limitations for heterojunction PIN photodetector devices are caused by several factors such as charging and discharging time of parasitic and inherent capacitances, charge trapping at the heterojunction, and **transit time** across the depletion region.

Improved fabrication techniques have successfully reduced the impacts of charge trapping at the heterojunction and diffusion time. The most serious problem—transit time and parasitic capacitance charging and discharging time—must, therefore, be used to compare the high speed performances of conventional and RCE photodetectors.

HIGH SPEED PERFORMANCE OF CONVENTIONAL DEVICES. For a conventional photodetector device (thin active layer), transit time is given by Equation (4–34).

$$t_r = 0.45\frac{v_e}{W} \qquad (4\text{–}34)$$

where t_r is the transit time(s), v_e is the **electron velocity** (m/s), and W is the width of the depletion region (m). When a GaAs substance with a hole velocity (v_e) of 6×10^6 cm/s is the active layer, it will require a depletion width (W) of 0.6 μm. It is evident that a decrease in the thickness of the active layer will increase the parasitic capacitance and, consequently, limit the device operating bandwidth.

The operating bandwidth of a conventional photodetector device, based on its physical characteristics (surface area) and parasitic capacitance, is given by Equation (4–35).

$$B_w = \frac{CW}{2\pi\varepsilon_r\varepsilon_o R_t A} \qquad (4\text{–}35)$$

where W is the depletion width, ε_r is the **relative permitivity**, ε_o is the permitivity of free space, R_t is the total load and conduct resistance, A is the surface area of the device, and C is the parasitic capacitance. It is evident from Equations (4–34) and (4–35) that, for each combination of device surface area (A) and total resistivity (R), there can be an optimum depletion layer width (W) at a predetermined operating bandwidth (B_w).

Figure 4–16 shows the peak operating bandwidth of two conventional photodetector devices with dimensions 5 μm × 5 μm and 10 μm × 10 μm and with different surface areas and different depletion layer thicknesses.

Figure 4–16 shows that by maintaining constant total resistivity of 50 Ω for both devices, the smaller detector, which has a depletion layer thickness of 150 nm, achieves

TABLE 4–10   Properties of Selected Semiconductor Materials

Combined Materials	Properties
AlGaAs/GaAs/InGeAl	1. Good electronic properties 2. Low current recombination rates 3. Excellent lattice matching to AlAs 4. Good refractive index
AlGaAs	Easily incorporated as a wide band gap conduct layer
AlAs	Good refractive index allows for mirror of almost unity reflectivity (with a twenty period wave)
InGaAs	1. Active layer material 2. With GaAs/AlAs mirrors, can extend spectral wavelength to 1550 nm 3. Active layer thickness detrimental to detector performance 4. Important for testing and prototypes 5. Limited commercial application
InP/InGaAs/InAlAs	1. The InGaAs alloy (lattice) matches the InP substrate perfectly 2. Exhibits excellent electrical properties 3. Covers the wavelength spectrum between 1300 nm and 1550 nm 4. Exhibits a poor refractive index between InAlAs and InGaAs 5. To achieve near unity mirror reflectivity, approximately 35 periods are required 6. InP/InGaAlAs very promising alloys for future commercial applications
AlAs/GaAs/Ge	1. The use of Ge as an active layer extends the operating wavelength beyond the 1550 nm range 2. Ge exhibits moderate absorption coefficients beyond the 1000 nm range 3. Ge exhibits very low recombination rates 4. Very difficult to grow GaAs on Ge
Si/SiGe	1. Si RCE photodetectors are commonly used at wavelengths below 1000 nm 2. Si as active layer, in conjunction with SiGe alloys, is used for longer wavelengths 3. Quarter mirrors exhibit good potentials at wavelengths below 1000 nm 4. Moderate absorption rates at 1000 nm 5. Operate with optimum performance at 1500 nm
Si/AlP/GaP	1. Operate successfully over the visible wavelength span 2. On the negative side, the refraction index between AlP and GaP is somewhat smaller than that of GaAs/AlAs. Increasing the number of mirror periods and using thinner mirrors for shorter wavelength applications can compensate for the negative effect

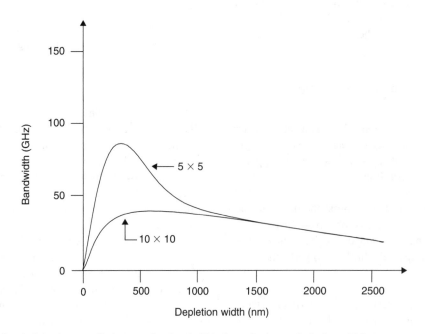

Figure 4–16. Conventional photodetector diode operating bandwidth dependency on depletion width.

a peak bandwidth of 86 GHz. The larger device, which has a depletion layer thickness of 300 nm achieves a peak bandwidth at 43 GHz. The peak bandwidth is directly proportional to the depletion layer thickness.

HIGH SPEED PERFORMANCE OF RCE PHOTODETECTORS.
The improvement factor of the limiting bandwidth performance of RCE photodetector devices is attributed mainly to innovative design and improved fabrication techniques.

In such a PIN semiconductor structure, a small band gap absorption layer with thickness (D) is placed in the depletion region (Figure 4–17).

Figure 4–17 shows that the carriers (electrons and holes) must travel a smaller distance in an RCE than in a conventional structure. The transit time for both carriers is expressed by Equations (4–36a) and (4–36b).

$$t_e = \frac{W_1}{\nu_e} \qquad \textbf{(4–36a)}$$

$$t_h = \frac{W_2}{\nu_h} \qquad \textbf{(4–36b)}$$

where W_1 is the distance traveled by the electrons, W_2 is the distance traveled by the holes, ν_e is the velocity of the electron, and ν_h is the velocity of the hole. The limited bandwidth (B_W) for an RCE photodetector is given by Equation (4–37).

$$B_w = 0.45\frac{\nu_e + \nu_h}{W + D} \qquad \textbf{(4–37)}$$

when $D < W$ and $\nu_e > \nu_h$ in a GaAs substance, electron velocity (ν_e) is 1×10^7 cm/s.

Comparing Equations (4–36) and (4–37), it is evident that the limited operating bandwidth of an RCE photodetector is significantly higher than the limited bandwidth of a conventional device.

Photon Lifetime

Another important factor contributing to RCE photodetector diode bandwidth limiting performance is the **photon lifetime** (t_{ph}). Photon lifetime is defined as the time required for the building or decaying of the optical field within the resonant cavity, and is expressed by Equation (4–38).

$$t_{ph} = \frac{t_{rt}}{L_d} \qquad \textbf{(4–38)}$$

where t_{ph} is the photon lifetime (s), t_{rt} is the total photon round trip (s), and L_d is the **decay losses** during the round trip. Experimental results have shown that for a GaAs cavity with a length of 1000 nm, the total photon round trip is approximately 23 fs (23×10^{-15} s). Decay losses (L_d) are given by Equation (4–39).

$$L_d = 1 - R_1 R_2 e^{-2ad} \qquad \textbf{(4–39)}$$

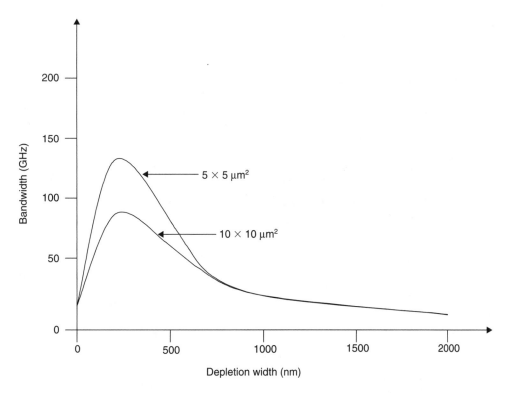

Figure 4–17. RCE photodetector diode bandwidth dependency on depletion width.

where L_d is the decay loss, R_1, R_2 is the mirror reflectivity, and ad are absorption coefficients.

Example 4–3

Compute the photon lifetime inside an RCE photodetector diode with the following characteristics:

$$R_1 = 0.7$$
$$R_2 = 0.99$$
$$ad = 0.1$$

Solution

i) Compute decay losses (L_d).

$$L_d = 1 - R_1 R_2 e^{-2ad}$$

$$= 1 - (0.7)(0.99) e^{-2(0.1)}$$

$$= 0.4326$$

Therefore, $L_d = 0.4326$.

ii) Compute photon lifetime (t_{ph}).

$$t_{ph} = \frac{t_{rt}}{L_d} = \frac{40\,fs}{0.4326} = 92.46\,fs$$

Therefore, $t_{ph} = 92.46\,fs$, assuming a total photon round trip within the cavity of 40 fs.

The RCE photodetector device bandwidth limit is based on photon lifetime decay as follows.

$$B_w = \frac{1}{t_{ph}} = \frac{1}{92.46 \times 10^{-15}\text{s}}$$
$$B_w = 10.8\text{ THz}$$

Therefore, $B_w = 10.8$ THz

It is evident that the photon lifetime component is critical only at the THz range.

RCE Schottky Photodiodes

Researchers have been aware, since 1995, that placing a conventional Schottky photodetector semiconductor device (InAlAs) into a Fabry-Perot resonant cavity can greatly enhance the device performance. By employing a thinner absorption region, a significant reduction of the transit time was observed, with a parallel increase of quantum efficiency at resonant wavelength.

Laboratory experimentation has demonstrated that an RCE structure using an InP substrate to grow an InAlAs/InGaAlAs eight-period mirror, with an absorption region composed of a 475 nm InGaAs, a 50 nm AlInAs Schottky layer, and a high reflectivity Al Schottky layer for conduct, an impressive increase of detector photocurrent in comparison to standard Schottky photodiode structures results.

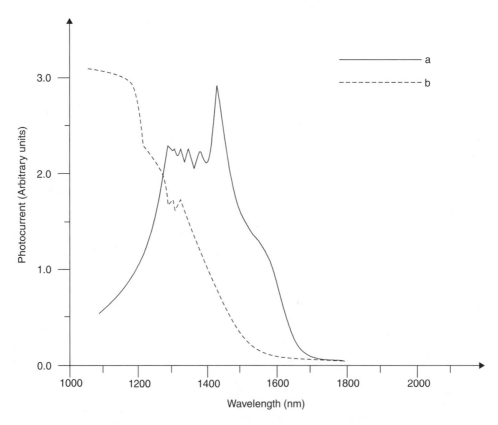

Figure 4–18. RCE and standard Schottky photodetector current performance at resonant wavelength.

Figure 4–18 shows a significant increase of RCE photodetector current at resonant wavelength in comparison to that of a standard Schottky photodetector diode.

RCE Avalanche Photodiodes

Applying the RCE concept to the design and fabrication of an APD photodetector device results in an optimum quantum efficiency bandwidth product. One of the fundamental bandwidth limiting factors of a standard APD photodetector device is the transit time of the secondary electrons that travel through the depletion region. If the transit time is somehow reduced, the overall device operating bandwidth will be substantially increased.

The other component that is crucial in determining device performance is the quantum efficiency. An increase in quantum efficiency will enhance the photodetector quantum efficiency bandwidth product. This increase in quantum efficiency can be achieved by employing a thinner absorption region. A thinner absorption region also means that a larger electric field will be generated across the region, thus allowing for a proportional reduction of the externally applied DC biasing, and consequently, lower power supply voltage. An RCE based APD photodetector is illustrated in Figure 4–19.

Figure 4–19 illustrates an early RCE APD photodetector structure composed of a very thin $(900 \, A^\circ)$ absorption layer that is located in the optical cavity of a fifteen quarter-wave AlAs/GaAs mirror that has an approximate reflectivity factor of ninety percent at a predetermined wavelength. The performance of this device was measured with different mirror reflectivity and the result is shown in Figure 4–20.

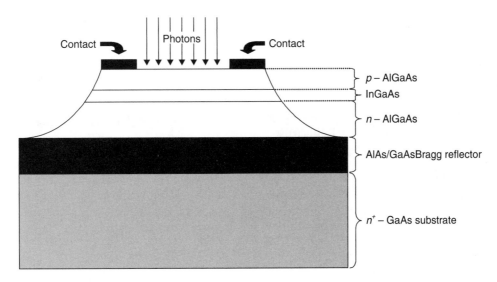

Figure 4–19. An RCE based APD photodetector structure.

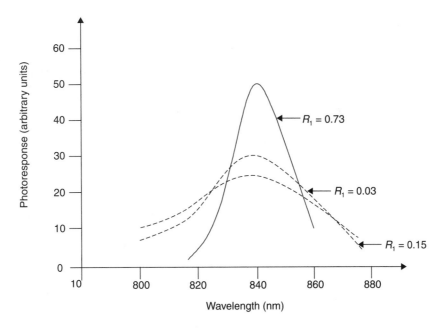

Figure 4–20. RCE APD photoresponse at different mirror reflective indexes.

Figure 4–20 illustrates that maximum quantum efficiency can be achieved at a specific high Q of the resonant cavity and at a high mirror reflective index. A bottom mirror low reflectivity contributes to the quantum efficiency limit. Therefore, a better lower mirror reflectivity can substantially enhance quantum efficiency. In the same device, photodetector gain was measured and found to be very high. This high gain was attributed to the thin absorption layer. Such a thin layer generates a comparatively high electric field across itself as a function of the external DC biasing voltage. A gain of close to two hundred was measured at a DC biasing of nine volts.

RCE PIN Photodetector Devices

For very high-speed optical detectors (above the 100 GHz range) conventional PIN devices exhibit compatible characteristics. The only disadvantage is their low sensitivity. Low sensitivity can be remedied by introducing the RCE PIN photodetector diode.

The first RCE PIN device was composed of an InGaAs/InGaAsP/InP structure with a 200 nm InGaAs absorption layer. This device exhibited a quantum efficiency of 82% at the 1550 nm operating wavelength. Mirror reflectivity was set at $R_2 = 0.95$ and $R_2 = 0.7$. The cross section of this structure is illustrated in Figure 4–21.

Various researchers experimenting with substances such as SiGe based RCE PIN structures noticed a significant reduction in quantum efficiency. The limited degree of quantum efficiency was due to the small refractive index of Si/SiGe mirrors. Further experimentation demonstrated a

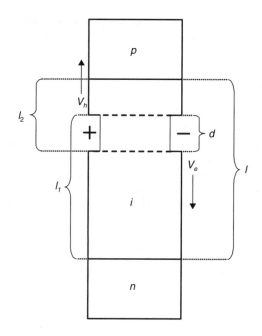

Figure 4–21. An RCE PIN photodetector diode.

significant improvement (over 30%) in quantum efficiency when optimizing the location of the absorption layer in the resonant cavity, while incorporating a thinner absorption layer in an InGaAs/InAlAs photodetector structure significantly reduced dark current.

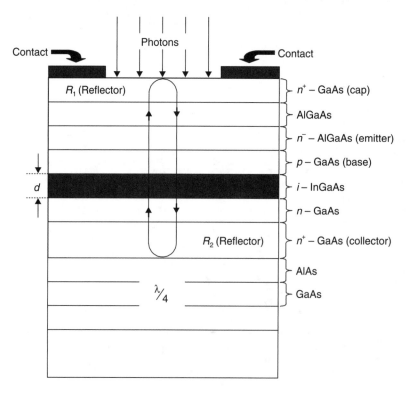

Figure 4–22. An RCE HPT detector/amplifier combination.

RCE Heterojunction Phototransistors

In high density optical fiber communications systems, amplifiers coupled to the output of optical detectors are highly desirable. Such a combination of photodetectors and optical amplifiers, exhibiting high gain and low noise as required by high density optical systems, can be accomplished through the combination of RCE structure with heterojunction transistors.

An RCE HPT detector/amplifier combination that employs AlGaAs/GaAs substances for the heterojunction and InGaAs for the active layer was tested experimentally. The cross section of such a device is illustrated in Figure 4–22.

The early and experimental optical detector/amplifier combination pictured in Figure 4–22 employed a top mirror composed of a GaAs substrate with a 30% reflectivity, a bottom mirror composed of ten AlAs/GaAs substrate stacks with a 90% reflectivity, and an active layer of 100 nm width composed of InGaAs materials.

The high gain of this device was mainly due to nonexisting cavity loss and a large photocurrent generation of 900 nm at the resonant wavelength. The relatively smaller gain shown at the 850 nm wavelength is mainly attributed to small cavity losses. For an improved quantum efficiency and wavelength sensitivity of the RCE HPT device, a larger Fabry-Perot cavity is required. In such a device, operating at the 930 nm wavelength, InGaAl quarter wavelengths are optimally placed in a high Q Fabry Perot cavity with mirror reflectivity of 99% and 70% for R_2 and R_1, respectively. The device exhibited very high gain and excellent photoresponse. Heterojunction phototransistors composed of InGaAs/InAlAs/In and operating at 1300 nm and 1600 nm, with an Au high reflectivity mirror, have exhibited very high optical gain.

SUMMARY

In Chapter 4, the concept of photodetection and its application in the design of optical receivers was presented. A detailed examination of the two major types of photodetector diodes (PIN and avalanche) was performed in terms of fabrication, optical gain, noise, and response time. The chapter concluded with an introduction of advanced photodetector diodes, including the RCE and RCE HPT. In order to reenforce the concepts, numerous examples were given throughout the chapter.

QUESTIONS

Section 4.1

1. Define *photodetection process*.

2. What is the role of the optical detector in a fiber optics communications system?

3. Name the fundamental performance characteristics of optical detectors.

4. What are the basic criteria for the selection of a photodetector diode to be incorporated in the design of an optical receiver?

5. Name the two basic types of photodetector diodes.

Section 4.2

6. What does the acronym *PIN* mean?

7. Briefly describe the principal theory upon which the operation of a PIN photodiode is based.

8. Draw a schematic of a silicon PIN diode and explain its operation.

9. What are the parameters involved in the establishment of the cut off frequency of a semiconductor material employed in the fabrication of a PIN photodetector?

10. List the PIN diode's basic characteristics.

11. Define *quantum efficiency* and its relation to PIN efficiency.

12. Define *response time* of a photodiode.

13. What are the main factors that contribute to response time?

14. With the assistance of a graph, explain the relationship between wavelength, responsivity, and quantum efficiency.

15. What is *dark current?*

16. Describe *shot-noise*.

17. With the assistance of a formula, describe *noise power.*

18. What is *Johnson noise?*

19. Define *signal to noise ratio (SNR)*. Give an example.

20. Describe the negative effect of dark current in the operation of a PIN photodiode.

21. How can the performance of a PIN device be enhanced through the reduction of the level of dark current?

Section 4.3

22. What is an *avalanche photodetector?*

23. How do avalanche photodetectors differ from PIN diodes?

24. What is the *multiplication factor* and what is its dependency?

25. Draw the physical layout of an APD silicon diode and briefly explain its operation.

26. With the assistance of an equivalent circuit, describe the function and operation of an APD device.

27. Describe a method through which the available bandwidth of an APD diode can be enhanced.

28. Define *photocurrent gain.*

29. What elements contribute to photocurrent gain?

30. Comment on the relationship between reverse biasing voltage and photocurrent gain.

31. What is the function of the guard rings in an APD structure?

32. How can undesirable current fluctuations within an APD structure be confined to acceptable levels, or be completely eliminated?

33. Why do APD diodes exhibit higher noise levels compared to PIN diodes?

34. With the assistance of a formula, relate shot noise to avalanche gain of an APD device.

35. Define *excess noise factor* of an APD diode.

36. Describe *dark current* of an APD diode.

37. What are the methods used for the reduction of the *dark* or *tunneling current* of an APD device?

38. Define *response time* of an APD device.

39. What is the relationship between quantum efficiency and response time of an APD photodetector?

Section 4.4

40. Why are advanced optical detectors required?

41. Describe the advantages of RCE devices compared to standard photodetectors.

42. With the assistance of a diagram depicting the semiconductor layout of an RCE diode, describe its function and operation.

43. Explain why a thicker active layer is a detrimental factor to the enhancement process of an RCE photodetector diode.

44. Give reasons why quantum efficiency is substantially reduced outside the resonant wavelength.

45. How can RCE devices achieve minimum cavity loss?

46. Describe method(s) through which transit time can be reduced in an RCE photodiode.

47. List some of the combined materials used in the fabrication of the resonant cavity enhanced devices.

48. Compare the properties of AlGaAs/GaAs/In GeAl and InGaAs. Comment.

49. List the factors that contribute to speed limitation of conventional RCE photodiodes.

50. With the assistance of a formula, describe the parameters that determine the operating bandwidth of a conventional photodetector diode.

51. What design changes are required to convert a conventional RCE photodiode to a high speed device?

52. What are the key parameters that determine transit time?

53. Describe the relationship between the bandwidth and physical characteristics of an RCE photodiode.

54. Explain why photon lifetime is a contributing factor in limiting the available bandwidth of an RCE diode.

55. Describe the principle of operation of an RCE Schottky photodiode.

56. How can the RCE principle employed in the design and fabrication of APD photodiodes achieve optimum quantum efficiency?

57. Sketch the physical layout of an RCE APD device and briefly explain its operation.

58. Describe the operation and performance characteristics of an RCE PIN photodiode.

59. Draw the physical layout and describe the operation and performance characteristics of an RCE HPT photodiode.

PROBLEMS

1. Calculate the photon energy required for electron-hole generation in a semiconductor material with cut off wavelength of 1.4 μm. Name the semiconductor material.

2. A PIN photodiode generates 1 mA of current through electron-hole generation. If the diode operates with a quantum efficiency of 25%, calculate:

  i) the number of electrons (N_{e^-}),

  ii) the number of photons (N_{ph}), and

  iii) optical power (P_O).

3. A PIN photodiode operating at the 940 nm wavelength generates 100 μA of photocurrent. If the quantum efficiency is 20%, calculate the optical power generated by the diode.

4. A PIN photodiode operating at the 1500 nm wavelength generates 15 μA of photocurrent. If the quantum efficiency of the diode is 22%, the active layer thickness is 3 μm, and carrier velocity 1.2×10^7 m/s, calculate:

  i) optical power generated by the diode,

  ii) responsivity, and

  iii) speed.

5. Compute the noise power and noise voltage of a semiconductor material operating with 100 MHz bandwidth and generating a dark current of 24 pA.

6. Calculate the thermal noise power and thermal noise voltage generated by a semiconductor device operating at a temperature of 310°K, bandwidth of 6 MHz, and 100 Ω impedance.

7. If an SNR of 25 dB is to be maintained at the input of a communications receiver that operates with 27 MHz of bandwidth, calculate the maximum noise temperature that the receiver can tolerate at the input for a constant signal level of 10 pW.

8. Calculate the multiplication factor of an APD device generating 65 μm of photocurrent with a carrier number of 10^{14}.

9. Calculate multiplication factor (M) for an APD that operates with the following parameters:

    i)   reverse bias voltage: $V_a = 10$ V
    ii) generated photocurrent: $I_P = 20$ μA
    iii) load and device resistance: $R_L = 150$ Ω
    iv) breakdown voltage: $V_{BK} = 25$ V
    v)   room temperature: $T_O = 20°C$
    vi) operating temperature: $T_1 = 27°C$
    vii) factor: $a = 0.2$

10. Compute the MTBF of an APD, operating at 30°C that has power dissipation of 1.55 mW.

11. Calculate the available bandwidth of a conventional photodetector diode given the following data:

    i)   depletion width: D = 0.5 μm,
    ii) relative permitivity: $\varepsilon_r = 7.5 \times 10^{-7}$ F/m,
    iii) total load and conduct resistance: $R_t = 100$ Ω,
    iv) device surface area: D = 25 μm^2
    v)   parasitic capacitance: $C = 2$ pF.

12. A high speed RCE photodiode is fabricated with a band gap absorption layer of 5 μm. If the electron velocity is 1×10^5 m/s, the hole velocity 0.85×10^5 m/s, and the distance traveled by the electrons is 2 μm, calculate the diode available bandwidth.

13. Calculate the available bandwidth of an RCE photodetector diode based on photon lifetime and for the following parameters:

    i) $R_1 = 0.85$,
    ii) $R_2 = 0.98$,
    iii) ad = 0.11, and
    iv) total round trip = 20 fs.

5

OPTICAL AMPLIFIERS

5.1 TYPES OF OPTICAL AMPLIFIERS

In various stages of an optical fiber communications system, the utilization of a number of amplifiers is absolutely essential. Two basic types of amplifiers are used in optical communications systems, electronic amplifiers, and **optical amplifiers.**

Electronic Amplifiers

The fundamental function of an electronic amplifier/ **repeater** is to amplify the electrical signals that were generated from the conversion of photons to electrons through the optical detector to a voltage level set forward by the system specifications. At the same time, they maintain a constant level of **signal to noise ratio (SNR)** at the receiver input of an analog system, or a predetermined **bit error rate (BER)** at the receiver input for a digital system.

After amplification, electric signals are converted back to photons, and then retransmitted to the next amplification point through the optical fiber (Figure 5–1).

Optical Amplifiers

Optical amplifiers differ from their conventional counterparts in that they do not require conversion from photons to electronic signals, but instead directly amplify the photons and transmit them to the next point of amplification through the optical fiber.

Through the utilization of optical amplifiers, the distance between repeaters is substantially extended. At the same time, the volume of the transmitted data does not inhibit the amplification process.

The basic application of optical fiber communications systems is in the area of long distance and **CATV** transmissions. Both systems require high optical amplification for different reasons. For long distance transmission, high optical amplification is required in order to extend the distance between repeaters as far as possible. For CATV, high optical amplification is required because of the multiple division of the transmitted signal. **Erbium doped fiber amplifiers (EDFAs)** are capable of satisfying the requirements of both systems.

The amplification requirements of optical communications systems can be divided into the following categories:

a) **Power amplifiers**

b) **In-line repeaters**

c) **Optical preamplifiers**

d) **Loss compensators**

Figure 5–2 illustrates the block diagram of an oversimplified optical fiber trans/receiver.

POWER OPTICAL AMPLIFIER. The function of an optical amplifier is to increase the optical power generated by the transmission source to a maximum level required by

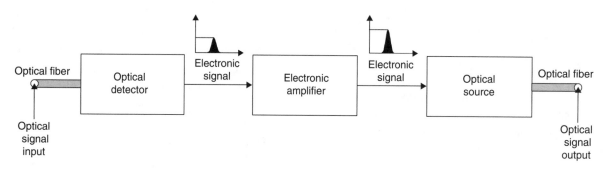

Figure 5–1. A block diagram of an electronic amplifier/repeater.

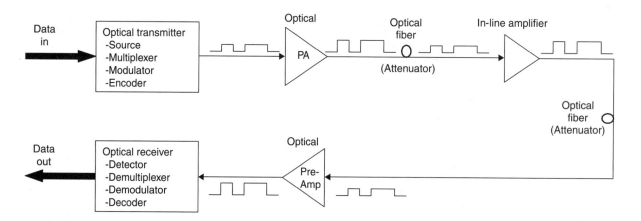

Figure 5–2. A block diagram of a simplified optical fiber.

system specifications. This maximum optical power is absolutely essential in order for the signal to be transmitted as far as possible along the fiber length, thus minimizing the number of in-line repeaters.

Although some noise may be added into the line at this stage of amplification, a reasonably high SNR can be maintained due to the relative high signal amplitude at the input of the amplifier.

IN-LINE REPEATERS. The relatively high optical signal generated by transmitter amplifiers suffers a substantial attenuation while traveling through the optical fiber. In-line optical amplifiers must amplify this weak incoming signal to the specified level before retransmission to the next repeater. EDFAs are most capable of providing the required optical amplification with a minimum level of added noise. The noise factor is very crucial because if it is excessive, it will limit the repeater-to-repeater optical length.

Preamplifier

The ever increasing demands for higher data rates and longer transmission distance increases the demand for higher **receiver sensitivities**. In earlier systems, receiver sensitivities in the –30 dBm range were considered satisfactory. Today, system performance requires receiver sensitivities of –40 dBm and beyond. These high sensitivity demands and high SNRs can only be satisfied by incorporating a preamplifier in front of the system's optical receiver.

EDFAs capable of providing high gain with low added noise levels necessary to satisfy receiver SNR requirements are ideal for such preamplifiers.

Optical Amplifiers as Loss Compensators

Wide bandwidth demand networks require **optical switching** (Figure 5–3). At switching points, it is inevitable that significant loss of optical power will occur. Optical amplifiers must compensate for these losses. The selection of such amplifiers is limited to either the EDFA or the semiconductor optical amplifier (SOA). For some particular applications, SOA amplifiers are best suited because of their small size. In such applications in which large optical signal amplification is a must, and more than one amplifier may be required, the selection of the smallest device will be preferable.

In any optical communications link, system upgrading is a standard practice. This upgrading is required mainly because of customer demands and future technological trends. Generally, system upgrading is difficult and very expensive. In optical fiber systems, upgrading is very easy to implement through the combination of **wavelength division multiplexing (WDM)** schemes, and the utilization of optical amplifiers.

Three types of optical amplifiers exist: Raman, SOA, and EDFA.

Optical amplifiers such as the SOA are capable of operating in the 1300 nm and 1500 nm wavelength range, but are subject to **polarization dependency** and high **coupling losses.** In contrast to SOAs, EDFAs exhibit low coupling losses and low polarization dependency while at the same time providing very high **optical gain** with relatively low noise. The only problem with EDFAs is their inability to operate in the 1300 nm wavelength window.

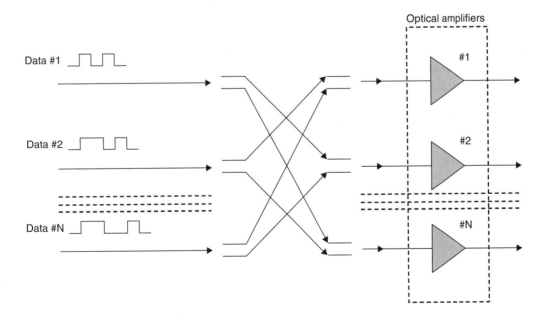

Figure 5–3. Optical switching.

5.2   RAMAN OPTICAL AMPLIFIERS

Raman optical amplifiers utilize the principle of **Raman scattering** for selective wavelength amplification through **stimulation emission** of radiation.

Raman Scattering

When the vibrating atoms in a crystalline lattice interact with optical waves, the vibrating atoms absorb some of the energy of the optical wave. This absorbed energy, plus the vibration energy characteristic of the absorbing atoms within the crystalline lattice, is almost instantaneously reemitted in the form of photons. The reemitted photons combined photon/atom vibration energy is a form of light scattering, which results in wavelength shifting.

The scattering effect of the light traveling through the fiber and wavelength shifting results in optical energy transfer from wavelength to wavelength within the optical fiber. This optical power spillover, or optical power transfer from wavelength to wavelength, becomes the basic for Raman optical amplification. The equivalent principle is used for signal amplification in microwave range by a **traveling wave tube (TWT)**.

In such a structure, an electron beam of a relatively high energy is used to interact through a coil with a forward-traveling electromagnetic wave. From this interaction, some of the energy of the electron beam is transferred to the electromagnetic wave, thus increasing its amplitude to a desired level.

Likewise, Raman scattering occurs when a **laser pump** emits optical power at a predetermined level within the fiber, while at the same time a weak optical signal is traveling through the same fiber. The optical energy of the weak signal initiates atom vibrations within the crystalline lattice, while at the same time additional atoms from the high energy laser pump absorb optical energy. The Raman optical amplification process is illustrated in Figure 5–4. If the wavelength of the weak optical signal is selected to be exactly the same as the wavelength generated by Raman scattering, then optical energy will be transferred from the laser pump into the weak signal, and therefore, optical amplification is observed at the wavelength of the weak signal. It is evident from the above that, in order for Raman amplification to occur, a high power laser pump is absolutely essential.

Raman optical amplification can take place in ordinary semiconductor structures without the necessity of special doping, and it can cover the entire 1300 nm to 1550 nm wavelength range. Raman amplification can be accomplished along the length of the transmission fiber carrying the signal through the addition of a high-power pump. This high-power pump can be inserted either at the front end of the fiber referred to as forward pumping or co-pumping, or at the far end of the fiber referred to as backward pumping or counter pumping. However, in WDM systems where several wavelengths must be amplified simultaneously, an equal number of pumps with corresponding wavelengths will also be required.

This type of amplification is highly desirable in optical communications systems employing Wavelength-Division-Multiplexing (WDM) schemes, submarine optical links, and optical networks. The main disadvantages of Raman amplification are the high laser pump power requirement, usually on the order of 0.5 W, and signal power (loss) to other shifted wavelengths.

5.3   SEMICONDUCTOR OPTICAL AMPLIFIERS

Semiconductor optical amplifiers (SOAs) are designed to amplify weak optical signals of a specific wavelength to a level defined by system requirements and specifications, with a minimum degree of distortion. The principal devices that satisfy all the fundamental requirements for optical signal amplification are slightly modified (reflective mirrors on both sides are removed) semiconductor laser devices. Because these diodes are capable of generating optical wavelengths across the 1300 nm to 1550 nm range and through stimulation of emission are also capable of optical power amplification across the entire operating spectrum, they can be classified as SOAs.

A block diagram of an oversimplified semiconductor optical amplifier is shown in Figure 5–5.

The optical signal from the output of the fiber is coupled into the active layer of the semiconductor amplifier, where it is amplified by the laser diode through stimulation. The required **carrier recombination** in the active layer is provided by an external current biasing source. Such optical amplifiers can be modulated at very high rates through the external biasing current, a definite advantage for optical switching applications.

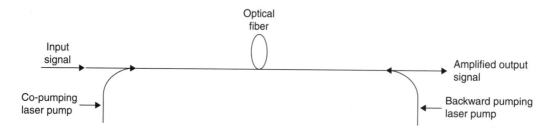

Figure 5–4.   The Raman amplification process.

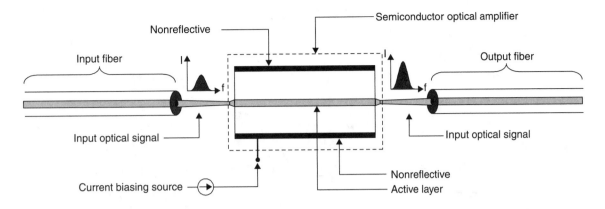

Figure 5–5. A block diagram of a semiconductor optical amplifier.

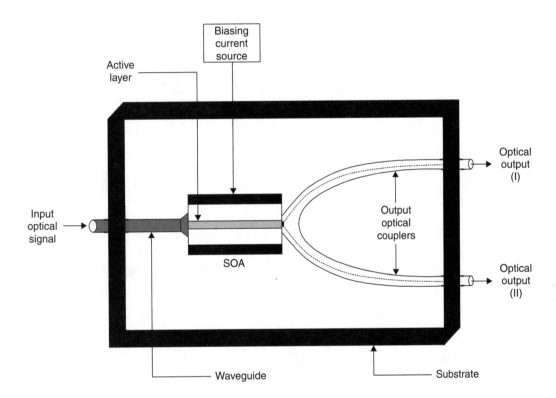

Figure 5–6. Multidevice optical semiconductor structure.

The quick **response time** of the **input signal intensity** variations reflect corresponding variations in the **device gain**. That is, SOAs amplify smaller input optical signals at higher gain and larger input signals with smaller gain, thus inducing undesirable signal distortion. Furthermore, SOAs can be fabricated in the same wafer with other optical semiconductor devices, and are therefore capable of multifunctional applications and enhanced performance (Figure 5–6).

Figure 5–6 illustrates the fact that several optical devices, such as **optical waveguides**, SOAs, **optical couplers,** and **modulators** can all be integrated into a single microchip. One of the main advantages of such an integrated semiconductor device is the minimization of coupling losses. This is

attributed to zero reflections between the output of the amplifier and the optical coupler.

Optical power losses based on **coupling reflections** are unavoidable when SOAs operate as discrete devices, and optical power from their output is coupled into the optical fiber input. Another application of an integrated semiconductor device is illustrated in Figure 5–7.

In Figure 5–7, four semiconductor laser diodes emitting at four different wavelengths are multiplexed and then redistributed to four different outputs. The integrated semiconductor structure is composed of four laser diodes, an optical multiplexer (**star coupler**), a set of eight optical waveguides, and four SOAs.

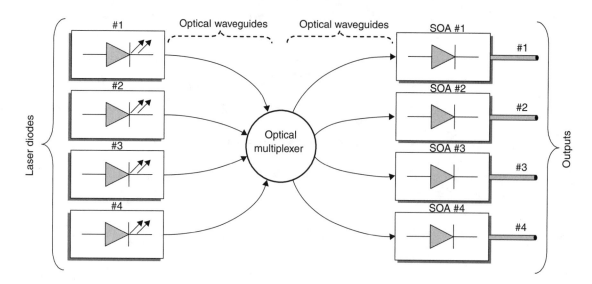

Figure 5–7. Optical multiplexing with integrated SOA devices.

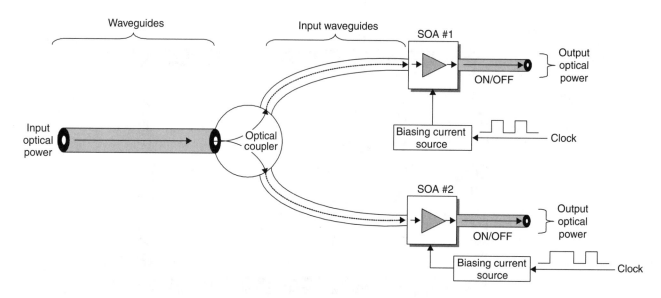

Figure 5–8. Optical power channelling.

The four discrete wavelengths, each generated by a corresponding laser diode, are multiplexed through the optical multiplexer (star coupler). Through this process of multiplexing, substantial optical signal attenuation is observed. In order to maintain a constant signal output for each channel, an **optical amplifier** is required to increase the **signal power** to the level that existed prior to multiplexing.

A fundamental characteristic of SOAs is that carrier recombination occurs through an externally-supplied biasing current. The laser diode responds extremely quickly to driving current changes. This implies that a complete amplifier shut-off will occur if no biasing current is supplied. This principle can be used to channel optical power to different waveguides (Figure 5–8).

The incoming optical signal is divided by the optical coupler and routed through the input waveguides to the corresponding inputs of the SOAs. A separate clock controls the biasing current for each amplifier. The precise timing of the clocks will determine which amplifier will be on and which will be off at any given time. Therefore, optical power can appear at either output upon demand.

5.4   ERBIUM DOPED FIBER AMPLIFIERS

Principles of Operation

Erbium doped fiber amplifiers (EDFAs) are composed of a relatively small amount (a few meters) of optical fiber doped with the rare earth **erbium (Erb^{3+}),** which has several energy levels. Again, a laser pump is used to excite the erbium ions. Initially,

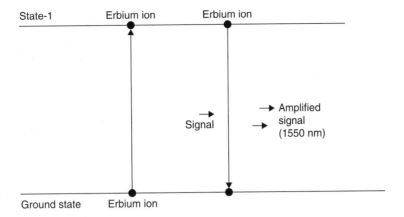

Figure 5–9. Amplification by spontaneous emission.

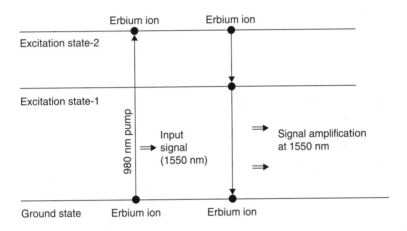

Figure 5–10. Amplification by stimulated emission.

the erbium ions are at ground level. If a laser operating at 1480 nm wavelength is used as a pump, the Erbium ions from ground level are elevated to state-1 (Figure 5–9). These ions will eventually fall back to ground level and, at the same time, release a photon. This process is referred to as **spontaneous emission,** and the accumulation of photons contributes to the process referred to as **amplified spontaneous emission (ASE).**

ASE is an undesirable condition that adds noise to the optical amplifier. However, if a laser pump that operates at the 980 nm wavelength is used to excite erbium ions to state-2, while an optical signal of 1550 nm wavelength is applied at the input, the erbium ions falling from state-2 to state-1 will release a photon of the same wavelength as that of the applied signal, thus adding to the signal strength. This process is referred to as signal amplification through stimulated emission (Figure 5–10).

Device Description

EDFA devices are composed of an erbium doped fiber (EDF), laser pump(s), a WDM module, isolators, and a microcontroller (Figure 5–11). A several-meters-long EDF composes the transmission line within the optical amplifier. A laser diode for a single pump, or two laser diodes for a dual

pump structure, is used to pump the Erb^{3+} ions at either 980 nm or 1480 nm resonant absorption wavelengths.

As the heavily attenuated optical signal travels through the fiber within the amplifier structure at the 1550 nm wavelength, the laser pump action (stimulation emission) excites the Erb^{3+} ions, thus generating the predictable and much-required optical amplification.

The input tap senses the incoming optical signal. **Optical isolators** are used to eliminate optical reflections from the input or the outputs to be induced into the amplification section of the device. Optical amplification is achieved through one or two laser pumps.

The **optical reflector tap** at the output senses whether the output of the device is terminated or disconnected. Because these optical structures generate very high optical power, which may damage the human eye, it is necessary to incorporate a special function by which the optical device automatically reduces its power to a safe level for the human eye if the output is disconnected. This is achieved through the output reflector tap and the microcontroller module.

If the output reflected optical signal is detected to be larger than a predetermined level, the microcontroller automatically adjusts the laser pump biasing current to a safe level. When the output of the amplifier is properly

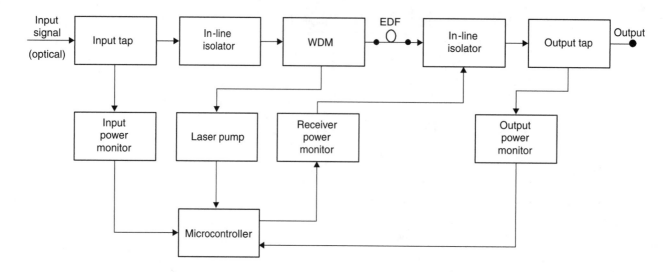

Figure 5–11. A block diagram of an EDFA with a single laser pump.

reconnected, the microcontroller (through the reflector tap) restores the amplifier to its regular mode of operation.

Another important function of the microcontroller module is to sense and adjust the case temperature to a constant level, usually 25°C, through the incorporation of **thermistors** into the amplifier circuit. If the temperature increases beyond the set maximum of 25°C, the thermoelectric cooler adjusts the current flowing through it to the level required to bring the chip temperature back to its normal operating level.

Furthermore, by activating a transistor-transistor-logic (TTL) (logic-1), the microcontroller alerts the system when backdiode current and chip temperature exceed their predetermined maximums. For a specific EDFA (AT&T 1713), monitoring of the various critical operational functions can be performed by analog means. That is, by measuring the voltage levels in reference to ground in various critical points of the device, the operating parameters of the device can be established.

Example 5–1

Assuming the voltage measured at the input power monitor is 0.3 V, calculate the input power, which will be displayed at the input monitor tap.

Solution

$$\text{Input} - \text{power} = \frac{V_{\text{rms}}}{\text{Conversion} - \text{factor}}$$

$$= \frac{0.3 \text{ V}}{0.75 \text{ V/mW}} = 0.4 \text{ mW}$$

Therefore, input power = 0.4 mW

EDFA Characteristics

ELECTRICAL. EDFA devices are designed for applications such as single channel or video power amplifiers, in-line, and preamplifiers. They exhibit high optical gain, low noise figure, low power dissipation, and adjustable output optical power.

Device Characteristics (Electrical)

Parameters	Minimum	Typical	Maximum	Units
Power supply	4.75	5.0	5.25	V
Power supply (current)	−1.0	—	—	A
Power consumption	−5.0	—	—	W

Absolute Maximum Ratings

Parameters	Minimum	Typical	Maximum	Units
Storage temperature	−40	20	70	°C
Operating temperature	0	35	65	°C
Absolute humidity	—	—	0.024	lbs.H_2O/lbs

Device Characteristics (Optical)

Parameters	Minimum	Typical	Maximum	Units
Signal wavelength range	1530	—	1560	nm
Input signal power (preamp)	−30	—	—	dBm
Input signal power (in-line)	−6.0	—	—	dBm
Optical gain (preamp)	30	—	—	dB
Optical gain (in-line)	13	—	—	dB
Noise figure (preamp)	—	—	5.0	dB
Noise figure (in-line)	—	—	5.5	dB
Return loss	—	—	−40	dB
Output power stability	—	±0.2	±0.5	dB

5.5 APPLICATIONS: 155 MB/SEC SONET/OC3-STM TRANSIMPEDANCE AMPLIFIER

The circuit in Figure 5–12 is based on a Philips SA5223 amplifier design in compliance with 155-SONET/OC3-STM

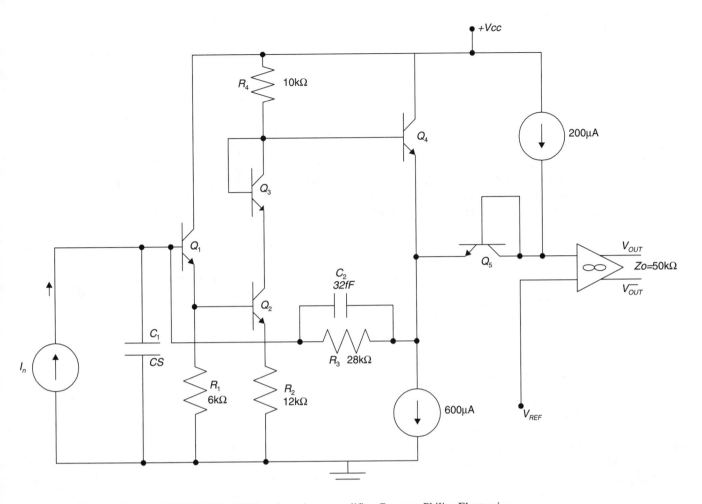

Figure 5–12. A 155 Mb/s SONET/OC3-STM transimpedance amplifier. Courtesy Philips Electronics.

system requirements. It consists of two bipolar transistors, Q_1 and Q_2, at the input in a Darlington pair configuration. The output of Q_3 is fed to the base of Q_4, acting as an emitter follower level detector. Transistors Q_1 to Q_4 and the feedback resistor R_f determine the level of transimpedance. In the first stage, the transimpedance is set at 28 KΩ for a maximum gain of 125, and increased to 100 KΩ by the second stage. The second stage consists of a differential amplifier with a gain of four and low output impedance driving the following post-amplifier.

The transimpedance varies from 98 KΩ to 50 Ω for corresponding input signal variations between 1 μA and 4 mA.

Operating Bandwidth $(B_{W_{-3dB}})$

The two fundamental components determining the bandwidth of the amplifier are the Miller resistance R_{in} and the equivalent input capacitance C_{in}. The value of the Miller resistance R_{in} is expressed by Equation (5–1).

$$R_{in} = \frac{R_f}{1 + A_u} \qquad (5\text{–}1)$$

where (R_{in}) is the Miller resistance (Ω), R_f is the transimpedance of the first stage (28 KΩ), and A_u is the voltage gain of the first stage (125 V).

Compute (R_{in}).

Substitute for values in Equation (5–1).

$$R_{in} = \frac{R_f}{1 + A_u} = \frac{28 \text{ K}\Omega}{1 + 125} = 220 \ \Omega$$

Therefore, $R_{in} = 220 \ \Omega$.

R_{in}: Miller resistance as seen by the photodiode output.

Compute (C_{in}).

The input capacitance C_{in} is expressed by Equation (5–2).

$$C_{in} = (1 + A_u)C_f \qquad (5\text{–}2)$$

where C_{in} is the input capacitance (F), A_u is the gain of the input stage, and C_f is the source capacitance (F). Assuming C_f equal to 32×10^{-15} F, then

$$C_{in} = (1 + 125) \times (32 \times 10^{-15}) = 4 \times 10^{-12} \text{F}$$

Therefore, $C_{in} = 4$ pF.

Compute the operating bandwidth $\left(B_{W_{(-3\,dB)}}\right)$

$$B_{W_{-3\,dB}} = \frac{1}{2\pi R_{in} C_{in}}$$

$$= \frac{1}{2\pi (220 \times 4 \times 10^{-12})}$$

$$= 180 \times 10^6 \text{ Hz}$$

Therefore, $B_{W_{-3\,dB}} = 180$ MHz.

For a realistic operating bandwidth, the photodiode capacitance must be added to the input capacitance. Assuming a photodiode capacitance of 1.2 pF, the **operating bandwidth** is equal to

$$B_{W_{-3\,dB}} = \frac{1}{2\pi(220)(5.2 \times 10^{-12})} = 139 \times 10^6 \text{ Hz}$$

Therefore, $B_{W_{-3\,dB}} = 139$ MHz (actual operating bandwidth).

The calculation of the input system bandwidth is crucial in determining amplifier data transmission capabilities.

Amplifier Input Noise Current (i_{in})

The two main sources that generate noise current at the input of the amplifier are the transistor Q_1 and the feedback resistor R_f. Transistor Q_1 is responsible for the generation of **shot-noise** and feedback resistor R_f is responsible for generating **thermal noise**. Both sources of noise are subject to system operating bandwidth. The mean square noise from both sources is expressed by Equation (5–3).

$$i_{in}^2 = (2qI_B B_{W_n}) + \left(\frac{4KTB_{W_n}}{R_f}\right) \qquad \textbf{(5–3)}$$

where i_{in}^2 is the mean square noise (A), q is the electron charge (1.6×10^{-19}C), I_B is the base current at Q_1 (approximately 1μA), R_f is the feedback resistor (28 KΩ), K is Boltzmann's constant (1.38×10^{-23} J/K), T is the operating temperature (273 + 30°C), and B_W is the **noise bandwidth** (145 MHz).

$$i_{in}^2 = \sqrt{(2 \times 1.6 \times 10^{-19})(1 \times 10^{-6})(139 \times 10^6) + }$$

$$\sqrt{\frac{4 \times 1.38 \times 10^{-23} \times 303 \times 139 \times 10^6}{28 \times 10^3}}$$

$$= 11.3 \text{ nA}$$

Therefore, $i_{in}^2 = 11.3$ nA (based on the entire operating bandwidth).

The spectral noise density can be calculated as follows:

$$i_{in} = \frac{i_{in}^2}{\sqrt{B_W}} = \frac{11.3 \times 10^{-9} \text{ A}}{\sqrt{139 \times 10^6 \text{ Hz}}} = 9.58 \times 10^{-13} \text{ A}/\sqrt{\text{Hz}}$$

Therefore, $i_{in} = 0.958$ pA/$\sqrt{\text{Hz}}$.

Differential Output

The **differential output** of the preceding amplifier significantly contributes to the overall improvement of the system BER.

The optical signal is first converted to an electrical signal, then only the differential mode is utilized for threshold detection at the input of the postamplifier. The advantage of the differential stage is its ability to cancel the common mode noise prior to threshold detection. Both outputs of the emitter follower are set at 3.2 V with a peak-to-peak deviation of ±0.2 V. For a symmetrical signal operation, the DC current of the emitter follower output transistors is set at 2 mA, corresponding with a 1.6 KΩ resistive load for each output. Because each output of the differential mode signal has 50 Ω impedance, they are able to drive a 100 Ω load. This low output impedance enables the system designer to incorporate low impedance LPF (bandwidth limiting) at the differential output in order to improve the system BER.

Automatic Gain Control

Automatic Gain Control (AGC) circuits are incorporated into designs in order to control the output voltage level of the amplifier. The main circuit of the AGC is a peak detector that is composed of an internal capacitor, a buffer, and a threshold comparator. The input of the peak detector is provided by the output of the main differential gain stage, which in turn provides the required control voltage at the gate of the metal oxide semiconductor field effect transistor (MOSFET) that controls its gain level. If the output voltage of the buffer amplifier reaches the peak level of 187 mV (threshold voltage level), AGC action will take place. The input current required to produce the threshold voltage of 187 mV is the result of a function of the optical power at the input of the amplifier and is expressed by Equation (5–4).

$$i_i = \frac{V_{Th}(V)}{R_T(\Omega)} \qquad \textbf{(5–4)}$$

where V_{Th} is the threshold voltage (187 mV) and R_T is the maximum transimpedance (50 KΩ). Substituting for values in Equation (5–4), the input current can be obtained.

$$i_i = \frac{V_{Th}}{R_T} = \frac{187 \text{ mV}}{50 \text{ KΩ}} = 3.74 \times 10^{-9} \text{ A}$$

Therefore, $i_i = 3.74$ nA.

The minimum optical power at the input of the amplifier required to generate the 3.74 nA of input current is calculated as follows in Equation (5–5).

$$P_{o(min)} = \frac{i_i}{\sqrt{2}(R)} \qquad \textbf{(5–5)}$$

where $P_{o(min)}$ is the minimum input optical power (W), i_i is the input current (A), and R is the **photodiode responsivity** (A/W), assuming a PIN photodiode responsivity equal to 0.8 A/W.

Substituting for values in Equation (5–5), we have

$$P_{o(min)} = \frac{i_i}{\sqrt{2} \times (R)} = \frac{3.74 \times 10^{-9}\,A}{\sqrt{2} \times (0.8\,A/W)} = 3.3 \times 10^{-6}\,W$$

Therefore, $P_{o(min)} = 3.3\,\mu W$.

Or in dB_{mW}, $P_{o(min)\,dB_{mW}} = 10\,\log \dfrac{P_{o(min)}}{1\,mW}$

$$= 10\,\log \frac{3.3 \times 10^{-6}}{1 \times 10^{-3}} = -24.8\,dB_{mW}$$

If the input optical power increases beyond the minimum, the input current (i_{in}) will also increase with a proportional decrease of the amplifier transimpedance gain controlled by the internal loop control current (I_{GC}). For an I_{GC} equal to 0 μA, the transimpedance gain is at the maximum. For an I_{GC} equal to 16 μA, the transimpedance gain is at 50% of the maximum.

Amplifier Gain v. Extended Strings of 0s and 1s

In order to determine the relationship between amplifier gain variations and long strings of 0s and 1s, the concept of **AGC loop response time** must be dealt with. The AGC loop response time is determined by a MOSFET transistor that is being used as the peak detector. The loop response time increases proportionally with the voltage above the 187 mV threshold level. The excess voltage level above the threshold charges the (30 pF) hold capacitor. The voltage across the hold capacitor is the gate to source voltage (V_{GS}) for the n-channel MOSFET gain stage used as the driver of the next differential transconductance stage.

The current charging the hold capacitor is proportional to the AC input current. The AGC loop response time for input current with constant amplitude resulting from long string of 1s and 0s is expressed by Equation (5–6).

$$\text{AGC gain change} \approx \pm 1\,dB/ms \qquad \textbf{(5–6)}$$

Having no optical power present at the input of the amplifier will force the AGC circuit to allow for an increase gain. It is therefore evident that if the input binary code is composed of long strings of 1s and 0s, they will be interpreted as DC levels, and the AGC circuit will allow for an increase in the amplifier gain. Equation (5–7) expresses the relationship between the lengths of the same bit string per one dB of gain change.

$$\text{string length} = (1 \times 10^{-3}\,s)\,(\text{bit stream}) \qquad \textbf{(5–7)}$$

Example 5–2
Calculate the pseudo random binary sequence (PRBS) corresponding to 1 dB gain change and determine the impact it may have in the system BER.

Solution
For the input data bit rate of 155 Mb/s,

$$\text{string length} = (1 \times 10^{-3})(155 \times 10^{6})$$

$$= 155 \times 10^{3}\,\text{bits}$$

or, PRBS $= 2^{x} - 1 = \text{string length} = 155 \times 10^{3}\,\text{bits}$

Therefore, PRBS $= 2^{17} - 1$.

Because a 1 dB increase in the gain will occur for each block of $2^{17} - 1$ or higher of the same bits, similarly, the noise will also increase by 1 dB. During the recovery process, time delay systems normally are susceptible to **overshoot.** Most amplifiers are designed to sustain gain changes due to long strings of 1s and 0s up to 10 dB without data loss at the recovery point. Beyond this, it is highly probable that some data may be lost, thus degrading the system BER.

Optical Signal Dynamic Gain

For an BER of 10^{-10}, an SNR of 10.8 dB must be maintained. The SNR is subject to PIN diode operating characteristics and operating bandwidth of the amplifier input stage. Assume maximum input current of 3 mA.

PIN Diode Parameters
a) dark current (5 nA$_{rms}$)
b) input spectral noise current (1.2 pA/Hz)
c) responsivity (0.8 A/W)

Compute Input noise current (i_{in}).

$$i_{in} = \sqrt{(i_{sn})^2 \times B_w}$$

$$= \sqrt{(1.2 \times 10^{-12})^2\,(\times 145 \times 10^{6})}$$

$$= 14\,nA(rms)$$

Therefore, $i_{in} = 14\,nA(rms)$.

Compute Input current $(i_i)_{min}$.

Because SNR is equal to 10.8 dB, then

$$S = Inv.\log(1.08) = 12$$

The input optical current must be twelve times higher than the input noise current.

$$i_i = 12 \times i_n = 12 \times 14\,nA(rms)_{min} = 168\,nA(rms)_{min}$$

Therefore, $i_i = 168\,nA(rms)_{min}$.

Compute Input optical signal power $(P_{in\,(o)})$.

From PIN diode responsivity, $R = 0.8$ A/W, and for an input current of $i_i = 168$ nA(rms), the input optical signal power is (168 nA/0.8 = 210 nW).

Therefore, $P_{in\,(o)} = 210$ nW.

In dB_m, $P_{in(o)_{dBm}} = 10\,\log \left(\dfrac{210 \times 10^{-9}\,W}{1 \times 10^{-3}\,W} \right)$

$$= -36\,dB_m$$

Therefore, $P_{in(o)_{min}} = -36\,dB_m$.

For a maximum input current $(i_{i_{max}} = 3\,mA)$,

$$P_{in(o)} = \frac{3 \times 10^{-3}}{0.8} = 3.75 \times 10^{-3} \text{ W}$$

Therefore, $P_{in(o)} = 3.75$ mW.

In dB$_m$, $P_{in(o)_{dBm}} = 10\log\left(\frac{3.75 \times 10^{-3}}{1 \times 10^{-3}}\right)$

$$= 5.7 \text{ dB}_m$$

Therefore, $P_{in(o)max} = 5.7$ dB$_m$.

The amplifier dynamic range is given as the difference between the maximum input optical power and the minimum input optical power.

Dynamic range $= P_{in(o)max} - P_{in(o)in}$

$$= 5.7 \text{ dB}_m - (-36 \text{ dB}_m)$$

$$= 41.7 \text{ dB}_m$$

Therefore, the amplifier input dynamic range $= 41.7$ dB$_m$.

Output Signal Noise Bandwidth

In some applications, it may be necessary to limit the noise bandwidth at the input of the amplifier. Input signal bandwidth limitation can be accomplished in two ways, through **shunt capacitance** or through a **balanced differential filter**.

INPUT SHUNT CAPACITANCE.   Through this method, a capacitor is connected at the input of the amplifier and ground, and acts as a low pass filter (LPF). The filter limits the input signal noise bandwidth before amplification. This method of noise bandwidth limiting results in the degradation of the input SNR and therefore is not highly recommended.

BALANCED DIFFERENTIAL (RC) FILTER.   A more efficient way to limit noise bandwidth is by inserting a balanced differential low pass filter between the outputs of the preamplifier or the inputs of the postamplifier. The function of the filter is to substantially attenuate the highest frequency components of the data output before detection. In a differential RC low pass filter, the total source resistance (Rs) must be kept reasonably low in order to minimize the thermal noise voltage ($V_{Th} = \sqrt{4kTB_wR}$), while allowing for an appropriate shunt capacitor (Cs) value.

Assuming a preamplifier output signal of 25 mV (approximately what is required for long distance optical fiber transmission),

$$V_{p-p} = 2\sqrt{2} \times V_{rms}$$

$$= 2\sqrt{2} \times (25 \times 10^{-3}) \text{ V}$$

$$= 70 \times 10^{-3} \text{ V}$$

Therefore, $V_{p-p} = 70$ mV.

At the postamplifier, the input current (i_i) is calculated as follows:

$$i_n = \sqrt{(i_{sn})^2 \times B_W}$$

$$= \sqrt{(1.2 \times 10^{-12})^2 \times 145 \times 10^6}$$

$$= 14 \times 10^{-9} \text{ A}$$

Therefore, $i_n = 14$ nA

Assuming an input impedance of 100 K, the post-amplifier noise voltage (V_n) is calculated as follows:

$$V_n = i_n \times R_i = 14 \times 10^{-9} \times 100 \times 10^3$$

$$= 1.4 \times 10^{-3} \text{ V}$$

Therefore, $V_n = 1.4$ mV.

Calculate the SNR at the input of the postamplifier.

The SNR at the input of the post-amplifier is calculated as follows in Equation (5–8).

$$\text{SNR}_{dB} = 20\log\left(\frac{V_S}{V_n}\right) \qquad \textbf{(5–8)}$$

where V_s is the signal (V) and V_n is the noise (V).

For a signal of 25 mV and noise of 1.4 mV, the SNR is equal to: $\text{SNR}_{dB} = 20\log\left(\frac{25 \times 10^{-3}}{1.4 \times 10^{-3}}\right) = 12.5$ dB

Therefore, SNR $= 12.5$ dB.

Because the SNR required to maintain a BER equal to 10^{-9} is set at 10.5 dB, it is evident that the 12.5 dB fully satisfies system specifications.

For accurate signal detection (digital), the binary signal channel must be equal to the highest data frequency at –3 dB. The differential LPF at the input of the postamplifier allows sufficient bandwidth to pass through, and it is in compliance with SONET specifications for data rise/fall times and SNR.

The required noise bandwidth for input data of 155 Mb/s is calculated as follows in Equation (5–9).

$$f_N \geq \frac{1}{2f_s} \qquad \textbf{(5–9)}$$

where f_N is the **Nyquist frequency** (Hz) and f_S is the **system bit rate** (B/s).

or

$$\tau_S = \frac{1}{f_S} = \frac{1}{155 \times 10^6} = 6.45 \times 10^{-9} \text{ s}$$

$$B_{W_{-3dB}}\frac{1}{2\tau_S} = \frac{1}{2(6.45 \times 10^{-9})} = 77.5 \times 10^6 \text{ Hz}$$

Therefore, $B_{W_{-3dB}} = 77.5$ MHz.

LPF design specifications must satisfy the noise bandwidth reduction requirements while reducing the intersymbol interference to a minimum. Both the above requirements can be satisfied with a first order low pass filter that utilizes an external resistor (R_S) with a value

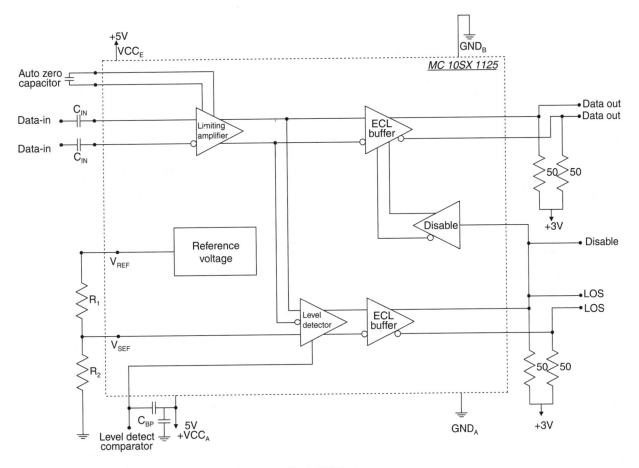

Figure 5–13. A block diagram of an optical fiber amplifier (622 Mb/s).

of 240Ω in series with a 100 Ω internal differential source resistance.

A typical electronics amplifier employed in optical communications systems is illustrated in Figure 5–13.

Typical Amplifiers and Operating Characteristics.

The Motorola SX1125 amplifier is designed for high frequency optical fiber system applications. It operates with a data rate of 622 Mb/s and complies with SONET/SDM-ATM standards.

OPERATING CHARACTERISTICS OF THE MOTOROLA SX 1125.

- Bandwidth: 20 KHz to 550 MHz
- Differential design for noise minimization
- 15 lead SOIC package
- Power supply single +5 V or ECL standard supply
- Programmable input signal detection

The MC10SX1126 is Motorola's enhanced performance amplifier. It is based on the same design philosophy as the MC10SX1125, but with data driven capability of up to 1.25 Gb/s. The operating characteristics of this amplifier are as follows.

DC Characteristics

Parameters	Symbol	Minimum	Typical	Maximum	Units
Input signal voltage	V_{in}	0.008	—	1.5	V_{p-p}
Input offset voltage	V_{os}	—	—	50	μV
Input noise	V_N	—	—	225	μV
Input level detection	V_{TH}	8	—	20	mV_{p-p}
Level detection hysteresis	V_{HYS}	1.5	2.5	7	dB
Input high current disable	I_{IH}	—	—	150	μA
Power supply current	I_{CC}	—	33	45	mA

Input/Output DC Characteristics (–40ºC)

Parameters	Symbol	Minimum	Maximum	Units
Output voltage—high	V_{OH}	1.08	–0.89	V
($V_{CC} = -4.5$ V to –5.5 V)				
($V_{CC} = 5.0$ V)		3.92	4.11	V
Output voltage—low	V_{OL}			
($V_{CC} = -4.5$ V to –5.5 V)		–1.95	–1.65	V
($V_{CC} = 5.0$ V)		3.75	3.35	V
Input voltage—high	V_{IH}			
($V_{CC} = -4.5$ V to –5.0 V)		–1.23	–0.89	V
($V_{CC} = 5.0$ V)		3.77	4.11	V
Input voltage—low	V_{IL}			
($V_{CC} = -4.5$ V to –5.0 V)		–1.95	–1.5	V
($V_{CC} = 5.0$ V)		3.05	3.5	V
Input current—low	I_{IL}	0.5	—	μA

Input/Output DC Characteristics (0˚C)

Parameters	Symbol	Minimum	Maximum	Units
Output voltage—high	V_{OH}			
($V_{CC} = -4.5$ V to –5.5 V)		1.02	–0.84	V
($V_{CC} = 5.0$ V)		3.98	4.16	V
Output voltage—low	V_{OL}			
($V_{CC} = -4.5$ V to –5.5 V)		–1.95	–1.63	V
($V_{CC} = 5.0$ V)		3.05	3.37	V
Input voltage—high	V_{IH}			
($V_{CC} = -4.5$ V to –5.5 V)		–1.17	–0.84	V
($V_{CC} = 5.0$ V)		3.83	4.16	V
Input voltage—low	V_{CC}			
($V_{CC} = -4.5$ V to –5.5 V)		–1.95	–1.48	V
($V_{CC} = 5.0$ V)		3.05	3.52	V
Input current—low	I_{IL}			
($V_{CC} = -4.5$ V to –5.5 V)		0.5	—	μA

Input/Output Characteristics (25ºC)

Parameters	Symbol	Minimum	Maximum	Units
Output voltage—high	V_{OH}			
($V_{CC} = -4.5$ V to –5.5 V)		–0.98	–0.81	V
($V_{CC} = 5.0$ V)		4.02	4.19	V
Output voltage—low	V_{OL}			
($V_{CC} = -4.5$ V to –5.5 V)		–1.95	–1.63	V
($V_{CC} = 5.0$ V)		3.05	3.37	V
Input voltage—high	V_{IH}			
($V_{CC} = -4.5$ V to 5.5 V)		–1.13	–0.81	V
($V_{CC} = 5.0$ V)		3.87	4.19	V
Input voltage—low	V_{IL}			
($V_{CC} = -4.5$ V to –5.5 V)		–1.95	–1.84	V
($V_{CC} = 5.0$ V)		3.05	3.52	V
Input current—low	I_{IL}	0.5	—	μA

OPERATING CHARACTERISTICS OF THE MOTOROLA MC10SX1126.

- Bandwidth: 50 MHz to 900 MHz
- Differential design for noise minimization
- 16 lead SOIC package

AC Characteristics

Parameters	Symbol	Minimum	Typical	Maximum	Units
Bandwidth—lower (–3 dB)	$B_{w_{min}}$	—	—	20	KHz
Bandwidth—upper (–3 dB)	$B_{w_{max}}$	550	—	—	MHz
Pulse width distortion	t_{PWD}	—	—	70	ps
Rise/fall times	t_r/t_f	150	250	650	ps
Auto-zero output resistance	R_{AZ}	200	325	450	KΩ
Level-detect filter resistance	R_F	14	25	41	KΩ
Level-detect time constant	t_{LD}	0.5	—	4	μs

Circuit Performance

Input Voltage (V_{p-p})	BER
6 mV	4×10^{-13}
7 mV	1×10^{-14}
8 mV	5×10^{-15}

- Programmable input signal detection
- Power supply: single +5 V or ELC standard supply

COUPLING CAPACITORS. To accommodate the offset correction function, the data inputs of the amplifier must be AC coupled. The value of the coupling capacitors must be able to pass the lowest input frequency. Therefore,

$$C_{in} = \frac{1}{2\pi R f_L}$$

where C_{in} is the coupling capacitance, R is the input impedance (5 K), and f_L is the lowest frequency.

AUTO-ZERO CAPACITANCE. Any offset voltage of the forward path signal is cancelled through the feedback amplifier, thus allowing the ECL comparator to be at the toggle state without any input signal. The value of the external capacitor required to set the time constant of the cancelling circuit is calculated as follows:

$$C_{AZ} = \frac{150}{2\pi R_{AZ} f_L}$$

where C_{AZ} is the cancelling capacitor, R_{AZ} is the internal driving impedance (290 KΩ), and f_L is the lowest operating frequency.

For a long distance optical transmission system, it is obvious that a large number of electronic amplifiers is required and, consequently, an equal number of conversions between photons to electronic signals and electronic signals to photons is also necessary. This multiple photon to electron and electron to photon conversion, because of its complexity, imposes limits on the overall system performance. Such limits lessen the maximum possible distance between consecutive repeaters and overall system capacity and enhance costs.

To overcome the system limitations imposed by the utilization of electronic amplifiers, a new family of amplifiers, optical amplifiers, was designed to be used in optical communications systems.

OPTICAL AMPLIFIERS

- Device type: P17 fast light PIN preamplifier
- Manufacturer: Lucent Technologies

BRIEF DESCRIPTION. The P17 fast light optical device is composed of a PIN structure coupled into a pigtail multimode fiber and a wideband linear amplifier. The P17 is a high speed, high responsivity and low dark current device, designed to be employed in SONET OC-3/OC-12, SDH, STM-1 and STM-4 telecommunications applications, line terminal equipment, and secure digital data systems.

Device Characteristics (electrical): 155 Mb/s Applications

Parameters	Minimum	Typical	Maximum	Units
Power supply	4.5	5.0	5.5	V
PIN voltage	3.0	5.0	15.0	V
Impedance	45.0	62.5	80.0	KΩ
Capacitance	—	0.65	0.70	pF
Rise/fall time	—	< 0.5	—	ns
Dark current	—	1.0	—	nA

Device Characteristics (optical): 155 Mb/s Applications

Parameters	Minimum	Typical	Maximum	Units
Bandwidth	110	150	—	MHz
Sensitivity	—	−38	−36	dB$_m$
Overload	0	0.5	—	dB$_m$

Device Characteristics (electrical): 622 Mb/s Applications

Parameters	Minimum	Typical	Maximum	Units
Power supply	3	3.3	5.5	V
PIN voltage	3	5.0	15.0	V
Impedance	4.5	6	7.5	KΩ
Capacitance	—	0.65	0.7	pF
Rise/fall time	—	< 0.5	—	ns
Dark current	—	1	5	nA

Device Characteristics (optical): 622 Mb/s Applications

Parameters	Minimum	Typical	Maximum	Units
Bandwidth	460	590	700	MHz
Sensitivity	—	−33	−30	dB$_m$
Overload	0.8	6	—	dB$_m$

SUMMARY

In Chapter 5, the concept of optical amplification and its importance to optical systems was introduced. Optical amplifiers were classified in terms of their functionality and design philosophy. The Raman amplification process was discussed, followed by a detailed examination of EDFAs, including principles of operation and performance characteristics. The chapter concluded with a brief description of transimpedance amplifiers employed in STM/SONET. Key performance characteristics, including available bandwidth, dynamic gain, and SNR of transimpedance amplifiers was also discussed.

QUESTIONS

Section 5.1

1. Name the two basic types of amplifiers that are used in fiber optics systems.

Section 5.2

2. Describe the fundamental function of electronic amplifiers.

Section 5.3

3. Explain the basic difference between electronic and optical amplifiers.

4. What are the advantages of using optical instead of electronic amplifiers in fiber optics systems?

5. List the four signal amplification requirements of optical communications systems.

6. Describe the basic function of an optical power amplifier.

7. Why are in-line repeaters essential in standard optical fiber systems?

8. Justify the need for preamplifiers in high speed, long distance optical links.

9. List the three types of optical amplifiers.

10. List the advantages and disadvantages of SOAs and EDFAs.

Section 5.4

11. Define the principle of operation of Raman amplifiers.
12. Describe, in some length, Raman scattering.
13. Explain why a laser pump is essential for Raman amplification.
14. What is the wavelength range within which Raman amplification can occur?
15. List the advantages and disadvantages of Raman amplification.

Section 5.5

16. Define the principal function of an SOA.
17. With the assistance of a diagram, describe the operation of a semiconductor optical amplifier.
18. How can carrier recombination within the active layer be accomplished?
19. What causes signal distortion in a semiconductor optical amplifier?
20. Give reason for SOA multifunctionality and enhanced performance.
21. What is the main advantage for SOA multifunctionality?
22. Describe how the biasing current or the lack of it can be used for optical power channelling.

Section 5.6

23. Give a short description of an EDFA.
24. Define *spontaneous* and *amplified spontaneous emission (ASE)*.
25. Why is ASE an undesirable condition in the operation of EDFAs?
26. Sketch a simplified diagram of an EDFA, list its main components, and briefly describe its operation.
27. Refer to Figure 5–12. What are the two main components that contribute to amplifier bandwidth limitations?
28. What are the two main sources of noise at the input of the amplifier?
29. Give the mathematical expression of noise bandwidth and comment on the main contributing factors.
30. How can the amplifier BER be improved? Explain.
31. What is the role of AGC in an amplifier? List the main components of the AGC loop and briefly explain its operation.
32. Name two methods used to limit the noise bandwidth at the input of the amplifier.
33. Refer to Figure 5–13. State the purpose of the auto zero capacitor. With the assistance of a formula, list the components that contribute to the establishment of this value.
34. Why do electronic amplifiers impose limitations to system capacity and distance between repeaters? Explain.

PROBLEMS

1. Calculate the available bandwidth of a transimpedance amplifier employed in an SONET/OC-3 STM system. The operating parameters of the amplifier are as follows:
   i) Transimpedance of the first stage: 25 KΩ
   ii) Voltage gain: 100
   iii) Source capacitance: 30 fF
   iv) Photodiode capacitance: 1 pF

2. Calculate the noise current at the input of a transimpedance amplifier employed in a SONET/OC-3 STM system and operating with an available bandwidth of 150 GHz, feedback resistance of 25 K, input transistor base current of 1.1 μA, and operating temperature of 30°C.

3. Calculate the current generated at the input of a transimpedance amplifier with input power of 5 nW and diode responsivity of 0.82 A/W.

4. Calculate the dynamic range of a transimpedance amplifier given the following operating parameters:
   i) SNR: 11.5 dB for a BER of 10^{-11}
   ii) PIN dark current: 4 nA(rms)
   iii) Input spectral noise current: 1 pA/Hz
   iv) Diode responsivity: 0.85 A/W
   v) Bandwidth: 180 MHz
   vi) Input current (max): 2.5 mA

5. In order for a BER of 10^{-9} to be maintained at the input of a postamplifier that is employed in a fiber optics system, an SNR of 11 dB is required. If the signal at the output of the preamplifier is 20 mV, the postamplifier operating bandwidth is 170 MHz, spectral noise current at the input 1.2 pA/Hz, and input impedance 150 kΩ, determine whether or not the given SNR satisfies system requirements.

6

OPTICAL TRANSMITTERS

OBJECTIVES

1. Describe the role of a transmitter in an optical fiber communications system.
2. List and briefly describe the subcircuits in the design of optical transmitters.
3. Describe the function and operating characteristics of the Philips OQ2535HP multiplexer circuit when incorporated into the transmitter module.
4. Describe the VITESSE 2.5 Gb/s multiplexer and clock generator used in optical transmitters.
5. Briefly describe the function and operating characteristics of the Motorola MC10SX1130, Nippon electric NE5300, and Philips 74F5302 LED drivers.
6. Explain the operation of the Philips OQ2545HP, Lucent Technologies LG 1627BXC, Nortel YA08, VITESSE VSC7928 3.2 Gb/s, and Macrocosm CX02066 laser drivers.
7. Describe the effect and management of electronic noise on power supplies used to power optical transmitters.
8. Explain, in some detail, the operation of uncooled laser transmitters developed by Lucent Technologies.
9. Describe the function and operating characteristics of the LCM155EW-64 10 Gb/s DWDM optical transmitter developed by Nortel networks.

KEY TERMS

ambient temperature	common mode impedance	driver current peaking	heat pump current
back facet photodetector	current control loop	driver switching noise	high current sinking
band gap reference	decoupling capacitor	duty cycle distortion	high current sourcing
bias current alarm	decoupling capacitor array	(DCD)	input buffer
clock disable	differential input	electromagnetic interference	input register
clock enable	differential mode impedance	(EMI)	input sensitivity
clock generator	differential pair termination	electro-optical efficiency	input signaling
clock jitter	differential signaling	equivalent inductance	jitter
clock multiplier unit (CMU)	differential tapped termination	extinction ratio	junction temperature

6.1 INTRODUCTION

The main function of an optical transmitter is to accept a number of standard parallel data channels at the input, and through the process of multiplexing, to generate a serial data stream at the input of a laser driver circuit. The driver circuit is then used to modulate a laser diode whose output is coupled into a fiber for optical transmission. A block diagram of an optical transmitter is illustrated in Figure 6–1.

The optical transmitter in Figure 6–1 is composed of a multiplexer, an LED or laser driver, and an LED or laser diode.

6.2 THE MULTIPLEXER

Multiplexer Circuits Based on the Philips OQ2535HP

The function of a multiplexer is to combine a number of standard SONET/SDH data channels and generate a serial data output. The operation and functional characteristics of the following multiplexer are based on the Philips OQ2535HP multiplexer circuit shown in Figure 6–2. This multiplexer combines 32×78 Mb/s data channels into a single 2.5 Gb/s data stream in compliance with OC48/STM-16 format. Some of the most important operating characteristics of the multiplexer are:

1) high **input sensitivity**

2) 5 V TTL clock output

3) low **power dissipation**

4) CML data and clock outputs

5) 3.3 V TTL compatible data inputs

OPERATION. The function of the OQ2535HP multiplexer circuit is divided into two stages. The first stage is composed of four 8:1 multiplexers, each accepting 8×78 Mb/s channels and generating 4×622 Mb/s data streams.

The four data streams are then fed into a 4×1 multiplexer that generates the required 2.5 Gb/s data stream OC-48/STM-16 format. The device incorporates an enabling input $\overline{ENL}$ (active low), which allows the outputs to be switched to DLOOP and DLOOPQ when direct connection to a demultiplexer circuit is required.

The 2.5 GHz reference signal is applied at CIN and CINQ inputs, and then divided into 78 MHz channels that are required for the timing of the input data signals (D0 to D31).

THE $4 \times 8{:}1$ MULTIPLEXER. The low bit rate section of the multiplexer is composed of four 8-bit shift registers, which each perform $8{:}1$ **multiplexing** and generate 4×622 Mb/s data streams, and a synchronization circuit, which generates the load pulse for the shift register module. The phase of the load pulse must be adjusted in accordance to the input data stream. This is accomplished through the SYNSEL1 and SYNSEL2 inputs of the synchronization section, which provides synchronization for both clock and data signals.

THE 4×1 MULTIPLEXER. These 4×622 Mb/s data streams are further multiplexed in two stages through 2×1 (622 Mb/s) multiplexers, that each generate two 1.244 Gb/s data streams, which are then combined to form the required 2.5 Gb/s final data stream. The 2.5 Gb/s data is processed either at the DOUT and DOUTQ outputs or at the DLOOP and DLOOPQ outputs with sequence D31 for MSB and D0 for LSB. The clock and data buffer circuits are internally terminated to 100 Ω resistors that have the ability to drive 50 Ω loads. Furthermore, in order to reduce the power dissipation of the device, the nonfunctional buffers are turned off. The CIN and CINQ inputs to ground are also internally terminated to 50 Ω while the DLOOP and DLOOPQ outputs to ground are internally terminated to 100 Ω and are intended to drive 50 Ω transmission lines.

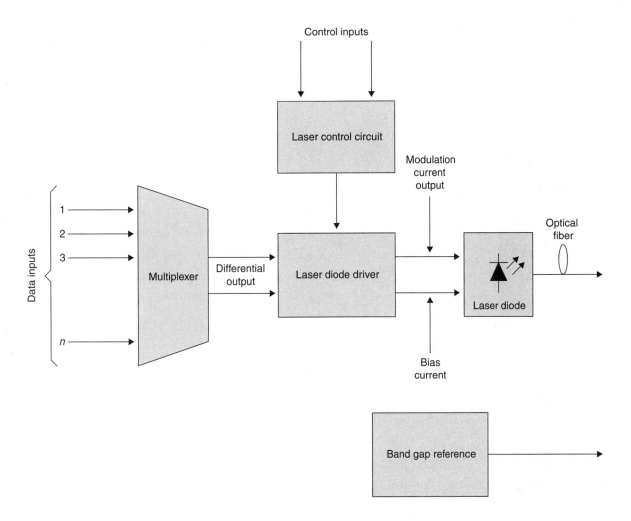

Figure 6–1. A block diagram of an optical transmitter.

All power supply connections must be decoupled with the decoupling capacitors connected as close as possible to the device. All the power supply lines must be filtered with low cut off frequency LC filters. The $V_{cc_{(T)}}$ power supply must be separately filtered by a high cut off LC filter. For ground connections, a relatively large copper fill strip must be connected to a low impedance common ground. All RF connections on the PC board must be short microstrips with 50 Ω characteristic impedance for DOUT and DOUTQ, COUT and COUTQ, DLOOP and DLOOPQ, CLOOP and CLOOPQ and CINQ. For certain inputs such as CIN and CINQ, DOUT and DOUTQ, COUT and COUTQ, the length of the strip lines must not exceed 5 mm, while the length difference between CDIV and D0 to D31 (78 Mb/s interface lines) must not exceed 20 mm.

HEAT SINK REQUIREMENTS. In some applications, a heat sink must be incorporated into the structure. The determination whether or not such a heat sink is required is based on the following:

If $R_{th} > R_{thJ-a}$, then a heat sink is not required.
If $R_{th} < R_{thJ-a}$, then a heat sink is required.
The value of R_{th} can be calculated by using Equation (6–1).

$$R_{th} = \frac{T_J - T_{amp}}{P_T} \qquad (6\text{–}1)$$

where R_{th} is the **thermal resistance** from junction to case, T_J is the **junction temperature,** T_{amp} is the **ambient temperature,** and P_T is the total power dissipation.

Example 6–1
Determine whether or not a heat sink is required for the device just described, given the following data:

- $T_J = 125°C$
- $T_{amp} = 60°C$
- $P_T = 1.8$ W

Assume R_{thJ-a} is equal to 34 K/W.

Solution
Compute R_{th}.

$$R_{th} = \frac{T_J - T_{amp}}{P_T} = \frac{125 - 60}{1.8} = 36 \text{ K/W}$$

Therefore, $R_{th} = 36$ K/W.

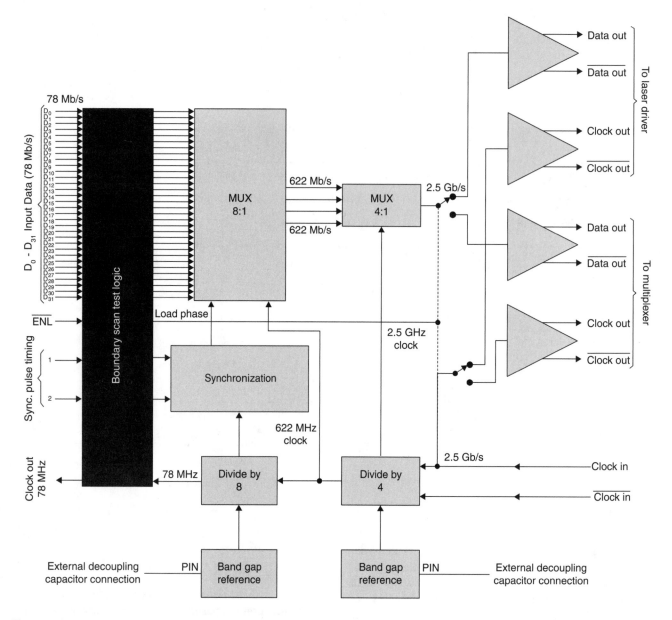

Figure 6–2. The OQ2535HP 32:1 multiplexer circuit.
Reproduced by permission of Phillips Semiconductors.

Because $R_{thJ-a} = 34$ K/W, $R_{th} > R_{thJ-a}$, no heat sink is required.

Example 6–2

Determine whether or not a heat sink will be required, given the following data:

- $T_J = 120°C$
- $T_{amp} = 80°C$
- $P_T = 2.1$ W

Solution

$$R_{th} = \frac{T_J - T_{amp}}{P_T} = \frac{120 - 80}{2.1} = 19 \text{ K/W}$$

Therefore, $R_{th} = 19$ K/W.

$$R_{thJ-a} = 34 \text{ K/W}$$
$$R_{th} < R_{thJ-a}$$

Therefore, a heat sink will be required. The thermal resistance of the heat sink is expressed by Equation (6–2).

$$R_{thH-a} \leq \left(\frac{1}{R_{th}} - \frac{1}{R_{thJ-a}} \right)^{-1} - R_{thJ-c} - R_{thc-H} \quad \textbf{(6–2)}$$

where R_{thJ-c} is the thermal resistance between junction and case, R_{thc-H} is the thermal resistance between case and heat sink, R_{thH-a} is the thermal resistance between heat sink and ambient, and R_{thJ-a} is the thermal resistance between junction and ambient.

Example 6–3

Calculate the heat sink thermal resistance based on the following data:

- $R_{thc-H} = 0.65$ K/W
- $R_{thJ-a} = 33$ K/W
- $R_{thJ-c} = 3$ K/W

Solution

$$R_{thH-a} \leq \frac{1}{\dfrac{1}{R_{th}} - \dfrac{1}{R_{thJ-a}}} - R_{thJ-c} - R_{thc-H} = 34.6$$

Therefore, $R_{thH-a} \leq 34.6$ K/W.

Some of the most important operating parameters (absolute values) of the multiplexer are listed in Table 6–1.

TABLE 6–1 Absolute Values

Parameters	Minimum	Maximum	Symbol	Units
Supply voltage	−0.5	+6.0	V_{cc}	V
Supply voltage	−6.0	+0.5	V_{ee}	V
Supply voltage	−0.5	+5.0	V_{dd}	V
Power dissipation	—	2.35	P_T	W
Junction temperature	−65	+150	T_J	°C

Thermal characteristics of the multiplexer are listed in Table 6–2.

TABLE 6–2 Thermal Characteristics

Parameters	Value	Symbol	Unit
Thermal resistance between junction and ambient	33	R_{thJ-a}	K/W
Thermal resistance between junction and case	2.6	R_{thJ-c}	K/W

Some of the most significant symbols and their descriptions are listed in Table 6–3.

The applications diagram of the Philips OQ2535HP multiplexer circuit is illustrated in Figure 6–3.

6.3   32:1 2.488 Gb/s MULTIPLEXERS WITH CLOCK GENERATORS (VSC8131)

The following multiplexer and **clock generator** circuit is based on the VITESSE VSC8131 2.488 Gb/s device. The circuit is a 32:1 2.488 Gb/s multiplexer and clock generator intended for SONET-48/STM-16 applications, and it operates with a 3.3 V single power supply with a maximum power dissipation of 2.15 W. A block diagram of this device is illustrated in Figure 6–4.

The circuit in Figure 6–4 accepts 32×77.76 Mb/s parallel inputs and generates a PECL 2.488 Gb/s serial output intended for SONET-48/STM-16 applications. The main building blocks of this device are:

TABLE 6–3 Symbols and their Descriptions

Symbols	Description
V_{ee}	Supply voltage (−4.5 V)
CDIV	78 MHz clock output
V_{dd}	Supply voltage (+3.3 V)
V_{cc}(T)	Supply voltage for TTL buffer (+5.0 V)
D0 to D31	78 Mb/s data input channels
SYNSEL1	Selection input 1 for synch pulse timing
SYNSEL2	Selection input 2 for synch pulse timing
$\overline{ENL}$	Loop mode enable (active-low)
DLOOP	Data output to demultiplexer
DLOOPQ	Inverted data output to demultiplexer
CLOOP	Clock output to demultiplexer
CLOOPQ	Inverted clock output to demultiplexer
CIN	Clock input to VCO
CINQ	Inverted clock input to VCO
BGCAP2	Pin for connecting external band gap decoupling capacitor
COUT	Clock output to laser driver
COUTQ	Inverted clock output to laser driver
DOUT	Data output to laser driver
DOUTQ	Inverted data output to laser driver

i)   **clock multiplier unit (CMU)**

ii) 32:1 multiplexer

iii) **timing generator**

iv) **parity register**

v)   **output registers**

vi) **input register**

vii) **input buffers**

Operation

The CMU generates a 2.488 GHz clock output signal (CO$^+$ and CO$^-$). An external reference clock signal of 77.78 MHz is applied at the input of REF.CLK$^+$. This signal is multiplied by the $\times 32$ clock multiplier unit, generating the required 2.488 GHz reference clock signal at the outputs CO$^+$ and CO$^-$. The signal is used to retime the **serial output data** of 2.488Gb/s generated by the 32:1 multiplexer. The timing of the **parallel input data** is achieved through a 77.78 MHz TTL clock signal provided by the CK78OUT. The device also provides parity check of the incoming parallel data through an even or odd parity mode selector. (PAR-MODE–low odd selection and PARMODE–high even selection). The TTL parity error is detected at the PERERR output with a high logic indicating the presence of an error. Furthermore, the device incorporates a **loss of lock indicator (LOL)** to indicate when the reference clock is lost or when the clock multiplier unit has lost lock status.

Data Output

The incoming parallel data stream at D0→D31 inputs, D0 being the least significant bit (LSB) and D31 the most

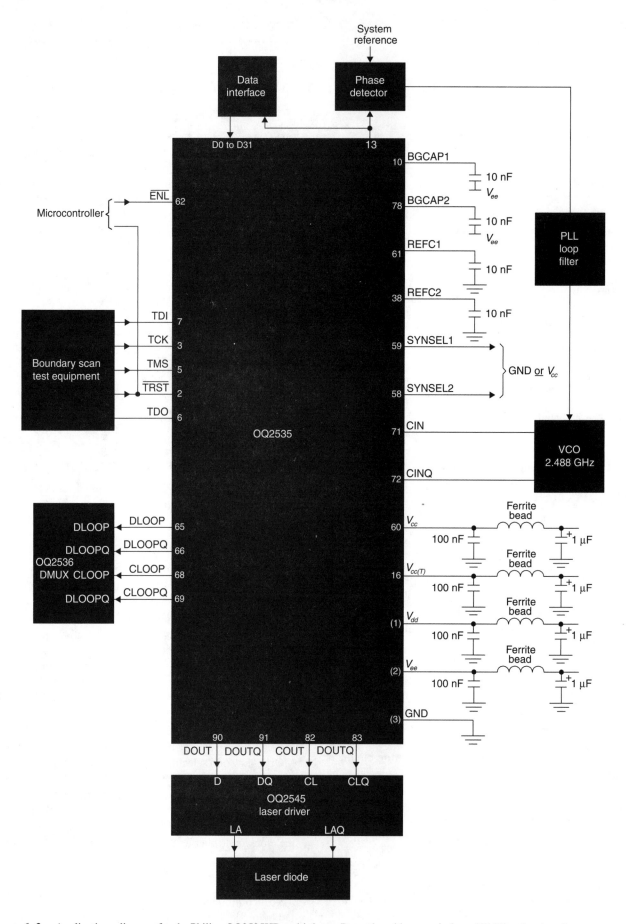

Figure 6–3. Applications diagram for the Philips OQ2535HP multiplexer. Reproduced by permission of Phillips Semiconductors.

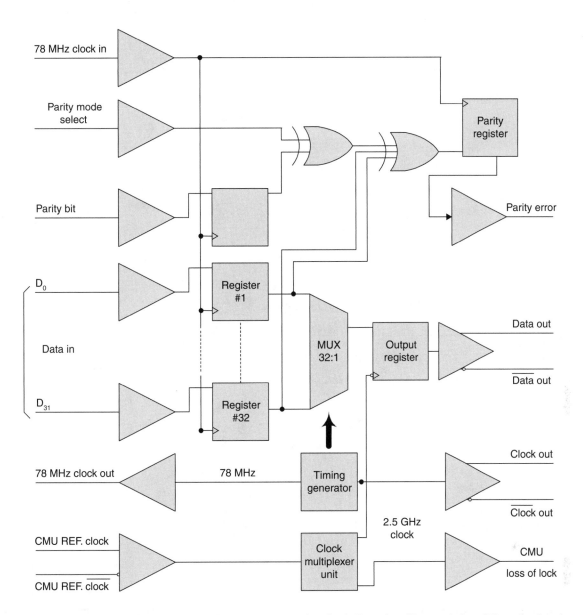

Figure 6–4. A block diagram of a VSC8131 multiplexer and clock generation circuit. Reproduced by permission of Vitese Semiconductor Corp.

significant bit (MSB), will be transmitted to the output through the data and clock driver circuit pictured in Figure 6–5.

The data and clock driver circuit in Figure 6–5 is composed of a **differential pair termination** that is capable of driving 50 Ω transmission lines that are terminated at 100 Ω loads between the differential outputs. To protect against back reflections, both collector resistors of the differential pair are set at 50 Ω. The same circuit is also employed to drive the differential clock output.

The Clock Generator

The required 2.488 GHz clock frequency is generated from a 77.78 MHz reference frequency by a PLL circuit with 2 MHz loop bandwidth in compliance with SONET/SDH specifications. The **voltage control oscillator (VCO)** of the PLL is a low phase noise circuit. The input reference

signal of 77.78 MHz must be noise free. If the noise of the reference clock signal is below the PLL loop bandwidth, it will be propagated to the output in the form of a **clock jitter.** This **noise jitter** must be kept well below the 4 $ps_{(rms)}$ level. In order to maintain the required 4 $ps_{(rms)}$ **jitter** level, the input reference clock signal must be PECL. The internal bias configuration of the reference clock is illustrated in Figure 6–6. In order for the reference signal to be AC coupled without the need for bias resistors, the inputs are internally biased to half of the power supply level ($V_{cc}/2$).

In the case in which the device is to be DC coupled, the input bias voltage must be substantially reduced.

Loss of Lock

The multiplexer also incorporates a loss of lock output to indicate when the clock multiplier unit has lost lock. Two

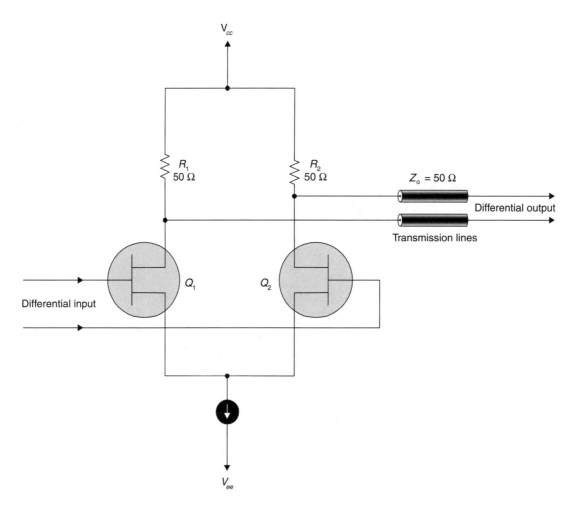

Figure 6–5. Data and clock driver circuit. Reproduced by permission of Vitese Semiconductor Corp.

basic conditions where the CMU might lose its lock exist. One is when the reference clock input is absent, and the other is when the CMU does not lock at the input reference frequency of 77.78 MHz. In lock operating state, the LOL output is high, and in loss of lock state, the LOL output is low.

Power Supply Recommendations

The multiplexer can be powered with either a single 3.3 V power supply (V_{cc}=3.3 V, and V_{ee} to ground) or a V_{ee}= -3.3 V and V_{cc} to ground for ECL applications. For proper operation, it is also necessary that the power supply be decoupled with **decoupling capacitors** of between 0.1 μF and 0.01 μF and be connected as close as possible to the V_{cc} power supply pins. Furthermore, for the analog section of the device, the V_{cc} (analog) must be filtered with a π filter (capacitor-inductor-capacitor) with the inductor value of about 10 μH. The polarities of the decoupling capacitors must be reversed when ECL mode of operation is implemented. The absolute maximum operating values for the device are listed in Table 6–4.

TABLE 6–4   Absolute Maximum Operating Values

Parameters	Minimum	Maximum	Units
Power supply	–0.5	+3.8	V
Input voltage (Diff.)	–0.5	V_{cc} +0.5	V
Input voltage (TTL)	–0.5	+5.5	V
Output voltage (TTL)	–0.5	V_{cc} +0.5	V
Output current (TTL)	—	±50	mA
Output current (Diff.)	—	±50	mA
Case temperature	–55	+125	°C

6.4   LED DRIVERS BASED ON THE MOTOROLA MC10SX1130

Introduction

An **optical driver** is a current source circuit that is capable of driving an LED anode when the input data is switched from low to high or high to low. The basic parameters, which must be taken into consideration when designing optical transmitters using LED devices, are:

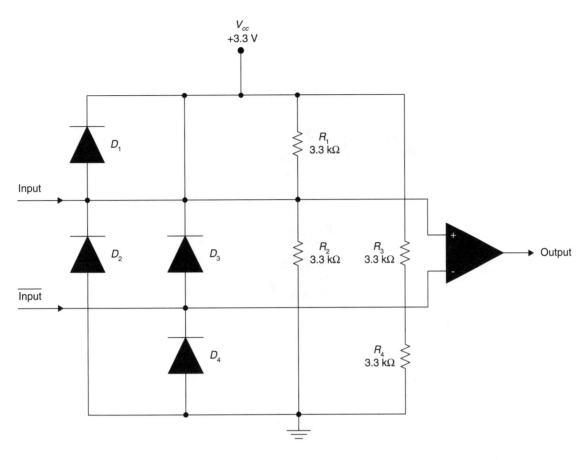

Figure 6–6. A schematic diagram of a reference clock internal bias. Reproduced by permission of Vitese Semiconductor Corp.

i) optical power output

ii) **optical rise time**

iii) **optical fall time**

iv) **pulse width distortion**

v) **optical overshoot**

vi) **optical undershoot**

vii) **optical peaking**

Furthermore, because LEDs are nonlinear devices with a non-linear transfer function and nonuniform low impedance, they exhibit a corresponding nonuniform turn-ON and turn-OFF time. Usually, the turn-ON time is shorter than the turn-OFF time (long-tailed response). Therefore, the transmitter design process must facilitate the optical transmitter requirements and accommodate the LED performance characteristics.

LEDs require forward bias current in order to emit light. The amount of optical power generated by an LED is a linear function of the forward bias current before saturation. However, a critical element that influences the optical power output of an LED device is the junction temperature. That is, an increase of the junction temperature will decrease the LED optical power output at a fixed forward bias current. Another element that must be taken into consideration when LEDs are used in the design of optical transmitters is the uneven (asymmetrical) turn-ON and turn-OFF times of the

output signal. Furthermore, the forward bias current required by the LED to generate the optical power is also used as the modulation current. The LED driver circuit provides this current.

The **LED driver** circuit accepts at its input a binary data stream and generates at the output the required modulated current level capable of driving the LED device. Driver circuits may also incorporate turn-ON and turn-OFF time difference compensating circuitry, a negative optical power output tracking coefficient, and modulation current programming. The block diagram of such an LED driver, circuit, based on the Motorola MC10SX1130 device, is illustrated in Figure 6–7.

The main building blocks of an LED driver circuit are the input line receiver, the pulse stretcher, the bias control circuitry, and the output current switch. The input stage is a differential ECL or PECL used in a dual or single ended arrangement. For single ended applications, a reference voltage source is required when the input is used as AC coupled. The **pulse width** adjust provides two predistortion duty cycles and is controlled by a signal applied at the stretch input. If this input is floating, no correction will be applied to the input waveform. By connecting the stretch input to +5 V (power supply), a pulse correction of 155 ps will occur. If connected to ground, a 310 ps pulse correction will be achieved. The bias control stage controls the voltage

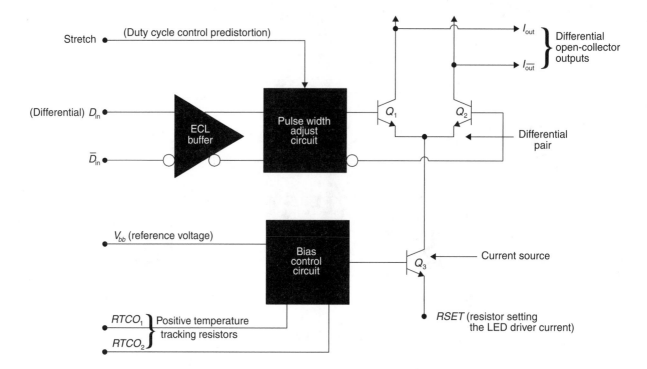

Figure 6–7. A block diagram of an LED driver circuit.

supplied at the output current switch. The LED is modulated through the output stage of the driver circuit. An external resistor, connected in series to the output stage current source, can set the required modulation current. The implementation of a differential input stage that is capable of switching the modulated current from one collector to the other contributes significantly to the minimization of the **driver switching noise.** A detailed treatment of the differential input stage design will follow. The driver circuit can be optimized to facilitate the LED operating characteristics. That is, by controlling key elements through the stages of the driver, the LED modulation current, pulsewidth adjustment, and temperature tracking can be controlled.

Design Examples

SIMPLIFIED DESIGN EXAMPLE.

i) Select the LED. The selected LED generates an optical power output waveform for a desired modulated current while exhibiting a specific optical power output tracking coefficient. That is, an optical power output change of 0.5% will occur with an increase of 1°C in the junction temperature.

ii) Given the set forward voltage, calculate emitter resistor. The external resistor connected in series with the differential emitter current source is expressed by Equation (6–3).

$$R_{\text{emitter}} = \frac{V_{\text{SET}}}{I_{\text{MOD}}} \qquad \textbf{(6–3)}$$

where R_{emitter} is the series resistor, V_{SET} is the voltage set by the external resistor R_{SET} connected between the temperature tracking connectors, and I_{MOD} is the modulation current.

Note: The value of the tracking resistor can be set between 700 Ω and 2 kΩ. For direct connection, the temperature tracking can be set at 1.4 mV/°C, while a 4.9 mV/°C of temperature tracking can be achieved with a maximum resistance of 2 kΩ. The temperature control is determined with Equation (6–4).

$$\text{Temperature control} = \frac{V_{\text{tracking}}}{V_{\text{SET}}} \qquad \textbf{(6–4)}$$

where V_{tracking} can be obtained from data sheets and V_{SET} can be set by the external emitter resistor. The values of V_{SET} are related to the tracking resistor R_{tracking}, and are listed in Table 6–5.

NONSIMPLIFIED DESIGN EXAMPLE. The steps required to determine the values of the external biasing components of the previously described LED driver circuit, and the calculation of the junction temperature (critical to the optical transmitter operation) are as follows:

TABLE 6–5 Temperature Tracking _v._ Tracking Resistor Values

Tracking %/°C	Tracking Resistor R_{tracking} (Ω)
+0.20	0 Ω
+0.52	1 k
+0.89	2 k

Design Steps

i) *Select the LED.*

An LED that generates the required optical power output with a modulation current of 70 mA must be selected. Therefore, $I_{MOD} = 70$ mA.

ii) *Select the output tracking coefficient.*

The output tracking coefficient is selected to be equal to $-0.5\%/°C$.

iii) *LED forward voltage.*

$V_F = 1.5$ V

Basic calculations:

i) *Calculate (R_{SET}).*

Resistor R_{SET} is connected between the emitter of the current source of the driver output stage and V_{ee}, and is used to set the driver circuit current. It also performs voltage compensation. That is, any change in the V_{ee} will be tracked, and the voltage across R_{SET} will be maintained constant. The value of R_{SET} is calculated as follows:

$$R_{emitter} = \frac{V_{SET}}{I_{MOD}}$$

ii) *Determine (V_{SET}).*

Because I_{MOD} is selected at 70 mA, the value of (V_{SET}) is obtained from data sheets as follows: First, select normalized tracking $(\%/°C)$. For $+0.5\%/°C$, the required tracking resistor (R_{TCO}) is 1 kΩ. This resistor is used to control the temperature tracking rate of the voltage across $R_{emitter}$. From data sheets, V_{SET} is equal to 0.65 V operating at 25°C, $R_{TCO} = 1$ kΩ, $V_{cc} = 5$ V, and $V_{ee} = $ GND.

Substitute the above values into Equation (6–3).

$$R_{emitter} = \frac{V_{SET}}{I_{MOD}} = \frac{0.65 \text{ V}}{70 \text{ mA}} = 9 \text{ }\Omega$$

Therefore, $R_{emitter} = 9$ Ω.

iii) *Calculate $R_{\overline{I}out}$.*

A resistor between $\overline{I}_{out}$ and V_{cc} in order to dissipate the power generated by the worst case temperature is required. As a rule of thumb, the voltage across $R_{\overline{I}out}$ is equal to the LED forward voltage (V_F).

$$V_{R_{\overline{I}}} = V_F = 1.5 \text{ V}$$

$$R_{\overline{I}out} = \frac{V_{R_{\overline{I}out}}}{I_{MOD}} = \frac{1.5 \text{ V}}{70 \text{ mA}} = 21 \text{ }\Omega$$

Therefore, $R_{\overline{I}out} = 21$ Ω.

Operating at 85°C, the V_{SET} is equal to 0.855 V (from tables) and the modulation current (I_{MOD}) is equal to:

$$I_{MOD} = \frac{V_{SET}}{R_{emitter}} = \frac{0.855 \text{ V}}{9} = 95 \text{ mA}$$

Therefore, $I_{MOD} = 95$ mA.

Based on the new modulation current requirements, the value of the resistor $R_{\overline{I}out}$ is recalculated.

$$R_{\overline{I}out} = \frac{V_{F_{\overline{I}out}}}{I_{MOD}} = \frac{V_F}{I_{MOD}} = \frac{1.5 \text{ V}}{86 \text{ mA}} = 17 \text{ }\Omega$$

Therefore, $R_{\overline{I}out} = 17$ Ω.

The LED driver and its external bias components operating at 85°C is illustrated in Figure 6–8.

Junction Temperature Considerations

Because LEDs use a relatively large amount of modulation current in order to generate the required optical power, junction temperature becomes a problem for the normal operation of the optical transmitter. Under specific operating conditions, the junction temperature is expressed by Equation (6–5).

$$T_J = P_D + T_A \times \Theta_{J-A} \qquad \textbf{(6–5)}$$

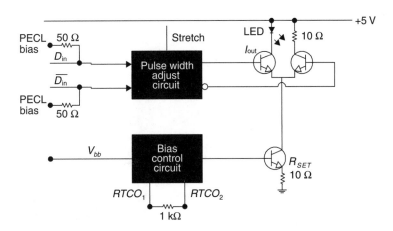

Figure 6–8. LED driver operating at 85°C.

where T_J is the junction temperature, T_A is the ambient temperature, P_D is the power dissipation, and $\Theta_{J\text{-}A}$ is the average thermal resistance between junction and ambient. The value of $\Theta_{J\text{-}A}$ can be obtained from Figure 6–9.

Total power dissipation (P_D) is determined by the sum of static and dynamic (switching) power dissipations, and is expressed by Equation (6–6).

$$P_D = P_{\text{dynamic}} + P_{\text{static}} \qquad \textbf{(6–6)}$$

where $P_{\text{dynamic}} = (V_{cc} - V_F - V_{\text{SET}}) \times I_{\text{MOD}}$ and $P_{\text{static}} = V_{cc} \times I_{cc}$

Example 6–4

Calculate the junction temperature (T_J) of an LED device given the following data:

- $V_{cc} = 5$ V
- $I_{cc} = 24$ mA
- $V_{\text{SET}} = 0.65$ V
- $V_F = 1.5$ V
- $I_{\text{MOD}} = 15$ mA
- $T_A = 25°C$

Solution

i) Calculate the power dissipation under dynamic operation (P_{dynamic}).

$$P_{\text{dynamic}} = (V_{cc} - V_F - V_{\text{SET}}) \times I_{\text{MOD}}$$
$$= (5\text{ V} - 1.5\text{ V} - 0.65) \times 24\text{ mA}$$
$$= 68.4\text{ mW}$$

Therefore, $P_{\text{dynamic}} = 68.4$ mW.

ii) Calculate the power dissipation under static operation (P_{static}).

$$P_{\text{static}} = V_{cc} \times I_{cc}$$
$$= 5\text{ V} \times 24\text{ mA}$$
$$= 120\text{ mW}$$

Therefore, $P_{\text{static}} = 120$ mW.

iii) Calculate the total power dissipation (P_D).

$$P_D = P_{\text{dynamic}} + P_{\text{static}}$$
$$= 68.4 + 120$$
$$= 188.4\text{ mW}$$

Therefore, $P_D = 188.4$ mW.

iv) Calculate the junction temperature (T_J).

$$T_J = T_A + P_D \times \Theta_{J\text{-}A}$$

where $T_A = 25°C$, $\Theta_{J\text{-}A} = 84°C/W$ at air flow of 200l fpm (from Figure 6–9), and $P_D = 188.4$ mW.

$$T_J = T_A + P_D \times \Theta_{J\text{-}A}$$
$$= 25°C + 188.4 \times 10^{-3}\text{ W} \times 84°C/W$$
$$= 25°C + 16°C$$
$$= 41°C$$

Therefore, $T_J = 41°C$.

6.5 LED DRIVERS BASED ON THE NIPPON NE5300

The following circuit description is based on the Nippon NE5300 optical transmitter pictured in Figure 6–10.

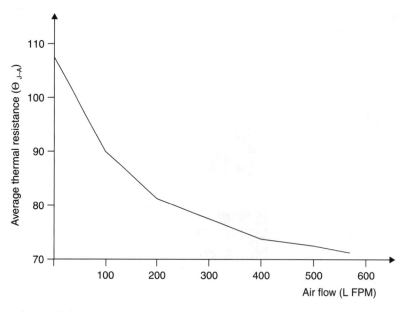

Figure 6–9. $\Theta_{J\text{-}A}$ v. air flow characteristic curve.

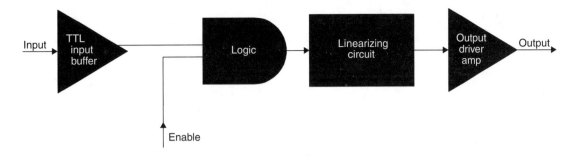

Figure 6–10. A block diagram of an LED driver circuit.

Description of Operation

The data applied at the input of the buffer stage is processed to the output stage through the enabling input. The function of the linearizing circuit is to maintain the rise time (t_{PLH}) and fall time (t_{PHL}) of the input data at a constant. The output stage is capable of sourcing and sinking relatively large currents (up to 160 mA) at low output impedance. The circuit exhibits symmetrical rise times, fall times, and propagation delays while the **duty cycle distortion (DCD)** is kept at a minimum. External precharging can minimize long fall response and overshoots, and prebiasing circuits, designed to comply with the selected LED operating characteristics.

The NE5300 is a typical representative of a TTL interface LED driver circuit. The output of the driver consists of a totem pole stage that is capable of sinking approximately 120 mA of current (low impedance), and of sourcing more than 60 mA with almost equal rise and fall times that do not exceed 2 ns. The circuit exhibits a 23 Ω impedance at an output voltage level of 1.5 V with excellent switching performance. The advantage of the output stage totem pole configuration is that it contributes to the enhancement of the circuit's current sinking capabilities. Furthermore, a voltage reference source incorporated into the design enhances the circuit data handling capability while improving noise immunity. The pull up transistor performs the turn-ON function of the LED, while the pull down transistor of the totem pole stage performs the turn-OFF function. Another benefit of the totem pole arrangement is the ability of the pull down transistor to effectively transfer the charge from the LED anode to the ground through the low impedance. The slight imbalance of the pull up and pull down transistor impedances, with the pull up being the higher of the two, eliminates the possibility of an output power overshoot during turn-ON time. The variable impedance of the totem pole output stage contributes to a significant reduction of pulse width distortion. Furthermore, duty cycle distortion is substantially reduced through propagation delay normalization when the rise and fall times are equal. In addition, two more critical design characteristics further enhance the performance of the driver circuit: prebiasing current and driver current peaking.

Prebiasing Current

Theoretically, at the off state, the LED is deprived of bias current. The absence of bias current completely discharges the junction of the parasitic capacitance so that it will require a larger amount of charges supplied by the driver. Therefore, a **prebiasing current** is applied to the LED at the off state. This current is referred to as prebiasing current.

Driver Current Peaking

Driver current peaking refers to the momentary increase of the forward current provided by the totem pole output stage of the driver circuit to the LED during the rise and fall edges of the modulated current. The time constant of the drive current peaking must be equal to the LED minority carrier lifetime. This momentary current increase will improve the rise and fall times of the driver while minimizing the **optical pulse ringing.**

The optical power at the output of the transmitter must not indicate overshoot or undershoot characteristics induced by LED current precharging (peaking). Excessive peaking, combined with noise at the optical receiver, may cause an increase of the BER beyond a preset threshold level. The required prebiasing current and precharging circuit for an LED driver is illustrated in Figure 6–11.

The values of the passive components R_1, R_2, R_3, and C_1 are expressed by Equations (6–7), (6–8), (6–9), and (6–10).

$$R_1 = \frac{R_o + 10}{2} \tag{6–7}$$

$$R_2 = R_1 - 10 \tag{6–8}$$

$$R_3 = \frac{V_{cc} - V_{F_{ON}}}{I_{F_{ON}}} + \frac{3.2(V_{cc} - V_{F_{ON}} - 1.4)}{I_{F_{ON}}} \tag{6–9}$$

where $V_{F_{ON}}$ is the forward LED bias voltage, $I_{F_{ON}}$ is the forward LED bias current at $V_{F_{ON}}$, and $R_o = R_1 + R_2$.

$$C_1 = \frac{2 \text{ ns}}{R_1} \tag{6–10}$$

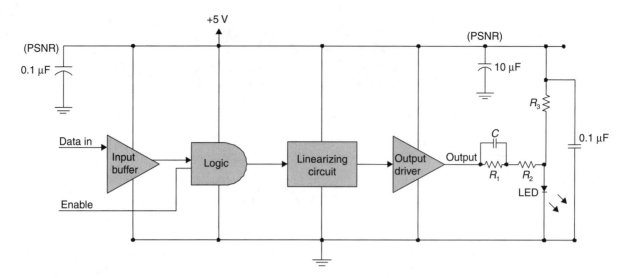

Figure 6–11. Schematic diagram of a 50 Mb/s LED driver.

Example 6–5

Calculate the values of $R_1, R_2, R_3,$ and C_1 for a circuit required to drive an LED with 2 V forward ON voltage, and 120 mA forward ON current at a maximum optical output power of 1 mW.

Solution

i) Compute R_3.

$$R_3 = \frac{V_{cc} - V_{F_{ON}}}{I_{F_{ON}}} + \frac{3.2(5 - 2 - 1.4)}{120 \times 10^{-3}}$$

$$= \frac{3}{120 \times 10^{-3}} + \frac{5.12}{120 \times 10^{-3}}$$

$$= \frac{8.12}{120 \times 10^{-3}} = 68 \ \Omega$$

Therefore, $R_3 = 68 \ \Omega$.

ii) Compute R_o.

$$R_o = \frac{R_3 - 32}{3.2} = \frac{68 - 32}{3.2} = 11 \ \Omega$$

Therefore, $R_o = 11 \ \Omega$.

iii) Compute R_1.

$$R_1 = \frac{R_o + 10}{2} = \frac{11 + 10}{2} \cong 11 \ \Omega$$

Therefore, $R_1 \cong 11 \ \Omega$.

iv) Compute R_2.

$$R_2 = R_1 - 10 = 11 - 10 = 1 \ \Omega$$

Therefore, $R_2 = 1 \ \Omega$.

v) Compute C_1.

$$C_1 = \frac{2 \text{ ns}}{R_1} = \frac{2 \text{ ns}}{11} \cong 0.2 \text{ nF}$$

Therefore, $C_1 = 200 \text{ pF}$.

6.6 LED DRIVERS BASED ON THE PHILIPS 74F5302

Another LED driver circuit incorporated in the design of high speed **optical transmitters** is the Philips 74F5302. A block diagram of this device is illustrated in Figure 6–12. It also consists of a TTL input buffer, a linearizing circuit, and an output driver amplifier.

The input stage accepts TTL data while the linearizing circuit controls the rise time (t_{PLH}) and fall time (t_{PHL}) . The driver circuit in the output stage is used to provide the relatively high source current, usually about 160 mA, at low LED impedance. The design of the output current driver circuit takes into consideration problems associated with the high speed data, with extremely low rise and fall times that must be propagated through transmission lines. Transmitter performance is enhanced through an appropriate impedance matching network between the output of the driver circuit and the LED. Through this, duty-cycle distortion (DCD) is substantially reduced or completely eliminated. Furthermore, the device can be used with other transmitter performance enhancing circuits such as circuits that precharge and prebias the pulse output to produce

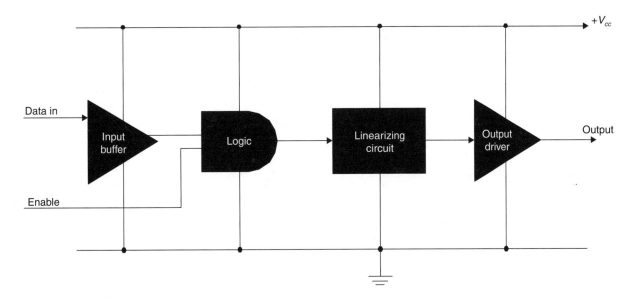

Figure 6–12. A block diagram of the Philips 74F5302 LED and clock driver.
Courtesy Philips Electronics.

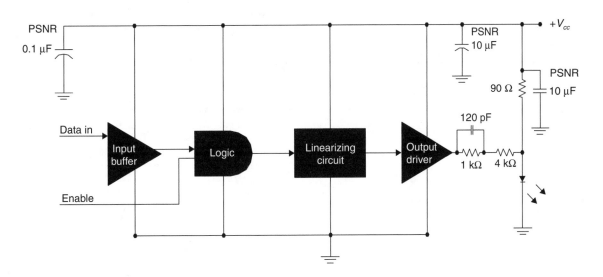

Figure 6–13. A schematic of a 50 Mb/s optical transmitter circuit that utilizes the 74F5302 device.
Courtesy Philips Electronics.

a response specific to the selected LED. The utilization of such circuitry significantly reduces long fall and overshoots of the output pulse. Some of the device's basic operating characteristics follow:

i) output enable control

ii) TTL inputs

iii) **matched propagation delay times** (t_{PHL}, t_{PLH})

iv) **high current sinking and sourcing**

v) symmetrical rise and fall times

vi) single +5 V power supply

The circuit can be used in almost any application, but not in fiber optics transmitters. Applications in which the circuit can be used include:

i) LAN networks

ii) metropolitan networks

iii) HDTV

A schematic diagram of a 50 Mb/s optical transmitter that incorporates the Philips 74F5302 LED and clock driver circuit is illustrated in Figure 6–13.

The DC electrical characteristics of the above LED and clock driver are listed in Table 6–6.

TABLE 6–6   DC Electrical Characteristics

Parameters	Minimum	Typical	Maximum	Symbol	Units
High level output voltage	2.0	3.3	3.9	V_{OH}	V
Low level output voltage	—	0.55	0.8	V_{OL}	V
Input clamp voltage	—	–0.73	–1.2	V_{IK}	V
Input current (at max. input voltage)	—	—	100	I_I	mA
High level input current	—	—	20	I_{IH}	mA
Low level input current	—	—	–0.6	I_{IL}	mA
Supply current at high	—	5.0	12	I_{CCH}	mA
Supply current at low	—	18	25	I_{CCL}	mA

The AC electrical characteristics of the device are listed in Table 6–7.

TABLE 6–7   AC Electrical Characteristics

Parameters	Minimum	Typical	Maximum	Symbol	Units
Propagation delay (L–H)	1.0	2.0	4.5	t_{PLH}	ns
Propagation delay (H–L)	1.0	2.5	5.0	t_{PHL}	ns
Pulsewidth distortion	—	0.8	1.2	D_{tpw}	ns
Rise and fall time skew	—	0.3	1.5	t_{RFS}	ns
Output skew	—	0.9	1.3	t_{sk}	ns

The recommended operating conditions are listed in Table 6–8.

TABLE 6–8   Recommended Operating Conditions

Parameters	Minimum	Typical	Maximum	Symbol	Unit
Supply voltage	4.5	5.0	5.5	V_{cc}	V
High level input voltage	2.0	—	—	V_{IH}	V
Low level input voltage	—	—	0.8	V_{IL}	V
High level output current	—	—	–160	I_{OH}	mA
Input clamp current	—	—	–18	I_{IK}	mA
Low level output current	—	—	160	I_{OL}	mA
Operating free air temperature	0	—	+70	t_{amb}	°C

The device's absolute maximum rating are listed in Table 6–9.

TABLE 6–9   Absolute Maximum Ratings

Parameters	Ratings	Symbol	Unit
Supply voltage	–0.5 to +7.0	V_{cc}	V
Input voltage	–0.5 to +7.0	V_{in}	V
Input current	–30 to +5.0	I_{in}	mA
Voltage output (high)	–0.5 to V_{cc}	V_{out}	V
Current output (low)	240	$I_{out(Low)}$	mA
Operating temperature	0 to +70	t_{amb}	°C
Storage temperature	–60 to +150	t_{stg}	°C

The propagation delay of the output waveform as related to the input waveform is illustrated in Figure 6–14.

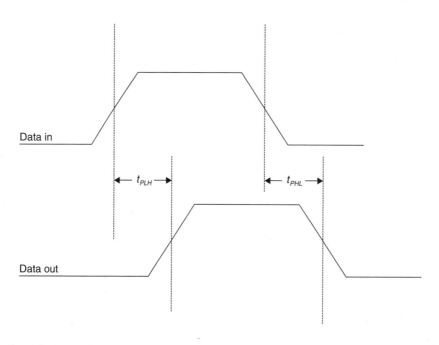

Figure 6–14.   Propagation delay output/input waveforms.

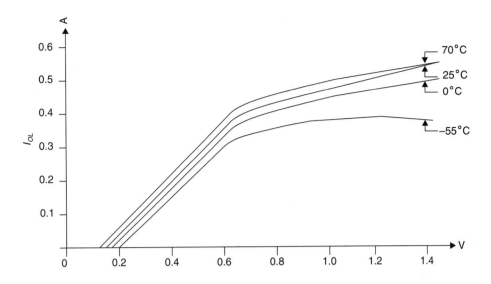

Figure 6–15. V_{OL} v. I_{OL} characteristic curves.

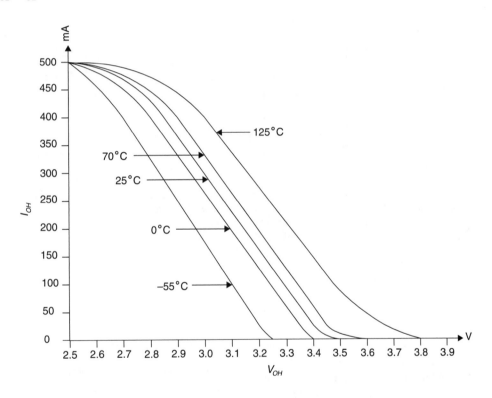

Figure 6–16. V_{OH} v. I_{OH} characteristic curves.

The V_{OL} v. I_{OL} characteristic curves for different operating temperatures are illustrated in Figure 6–15. The V_{OH} v. I_{OH} characteristic curves for different operating temperatures are illustrated in Figure 6–16.

6.7 LASER DRIVERS BASED ON THE PHILIPS OQ2545 HP

The second module of an optical transmitter is the laser diode driver circuit. This circuit accepts an NRZ data stream at its input, to a maximum of 2.5 Gb/s, and generates a single output with enough current to drive a distributed feedback Bragg or a Fabry-Perot laser diode. The block diagram of a typical **laser driver** circuit is illustrated in Figure 6–17.

The main building blocks of the laser driver circuit in Figure 6–17 are an input buffer, a current limiter, a laser control block, and a band gap reference.

The Input Buffer

The CML or PECL multiplexer outputs vary in amplitude between 200 mV and 800 mV (the average is 400 mV) and

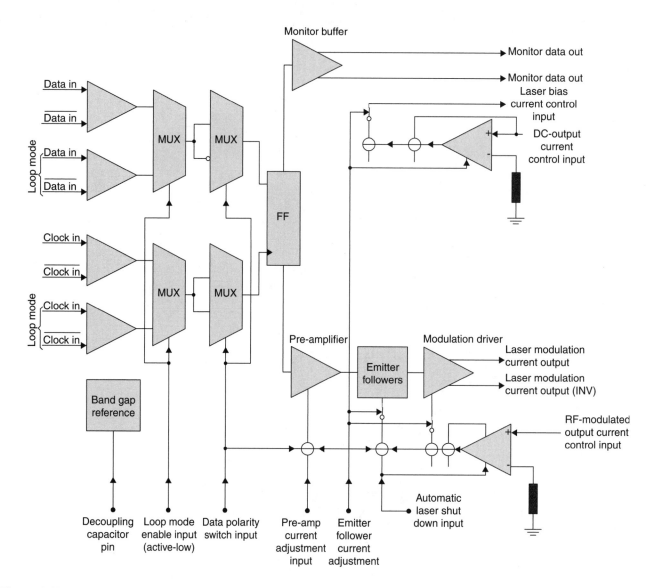

Figure 6–17. A laser diode driver circuit.

are directly connected to the high impedance inputs of the laser driver circuit's buffer. In CML, signals are to be accepted by the buffer inputs, and a 50 Ω pull up external resistor may be required to be connected between the inputs and the power supply. A typical schematic diagram of the buffer differential input stage is illustrated in Figure 6–18. If PECL input signals are used, an equivalent Thevenin circuit, connected at the corresponding inputs, will be required.

The output signal from the buffer circuit is used to drive the current switch that is designed to deliver a maximum current of 60 mA to the laser modulation output and laser modulation output-inverted (Figure 6–19).

In order to maintain the laser output optical power above the light emitting threshold level, a DC bias current of 90 mA (max.) may be required. The laser driver biasing circuit illustrated in Figure 6–20 provides this current.

Laser Current Control

Because a **laser current control** circuit is required, the reference current of such circuit is supplied by a monitor photodiode that is capable of generating an output current of between 0.1 mA and 1 mA. The generated photodiode current is proportional to the laser diode emission and the required reference current for both optical high and optical low is calculated as follows.

For optical high, the reference current is given by Equation (6–11).

$$I_{\text{ref(one)}} = \left(\frac{1}{16}\right) \times I_{\text{MPD(one)}}(\text{A}) \qquad \textbf{(6–11)}$$

where $I_{\text{ref(one)}}$ is the reference current for optical one and $I_{\text{MPD(one)}}$ is the monitor photodiode current for optical high.

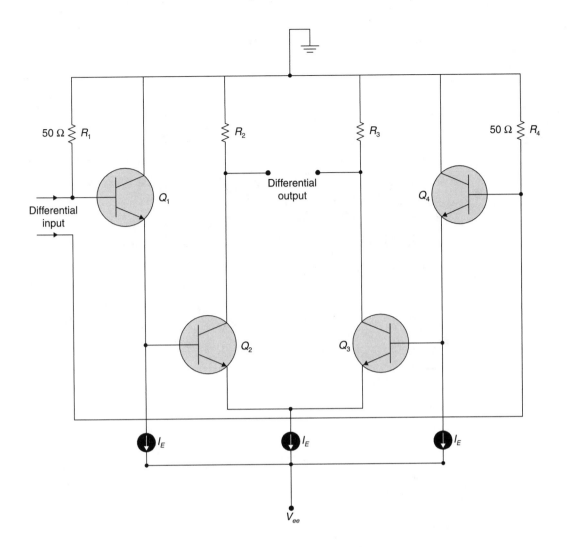

Figure 6–18. The buffer differential input stage.

The value of the required external resistor is given in Equation (6–12).

$$R_{\text{ext(one)}} = \frac{1.5}{I_{\text{ref(one)}}} (\Omega) \qquad \textbf{(6–12)}$$

where $R_{\text{ext(one)}}$ is the external biasing resistance and $I_{\text{ref(one)}}$ is the reference current for optical one. Similarly, the reference current and external bias resistors required for optical low (zero) are expressed by Equations (6–13) and (6–14), respectively.

$$I_{\text{ref(zero)}} = \frac{1}{4} \times I_{\text{MPD(zero)}} \qquad \textbf{(6–13)}$$

$$R_{\text{ref(zero)}} = \frac{1.5}{I_{\text{ref(zero)}}} \qquad \textbf{(6–14)}$$

Example 6–6

A laser diode generates an optical power of 0.3 mW with an **extinction ratio** of 10 dB. Compute the required reference current and external bias resistors for both optical high and optical low, if the monitor photocurrents for optical high and optical low are 250 μA and 25 μA, respectively.

Solution

i) Compute the required reference current ($I_{\text{ref(one)}}$).

$$I_{\text{ref(one)}} = \frac{1}{16} \times I_{\text{MPD(one)}} = \frac{1}{16} \times 250 \ \mu A = 15.625 \ \mu A$$

Therefore, $I_{\text{ref(one)}} = 15.625 \ \mu A$.

ii) Compute the external bias resistor value ($R_{\text{ref(one)}}$).

$$R_{\text{ref(one)}} = \frac{1.5}{I_{\text{ref(one)}}} = \frac{1.5}{15.625 \times 10^{-6} \ A} = 96 \ K\Omega$$

or, $R_{\text{ref(one)}} = \dfrac{24}{I_{\text{MPD(one)}}} = \dfrac{24}{250 \times 10^{-6}} = 96 \ K\Omega$

Therefore, $R_{\text{ref(one)}} = 96 \ K\Omega$.

iii) Compute the required reference current $I_{\text{ref(zero)}}$.

$$I_{\text{ref(zero)}} = \frac{1}{4} \times I_{\text{MPD(zero)}} = \frac{1}{4} \times 25 \ \mu A = 6.25 \ \mu A$$

Therefore, $I_{\text{ref(zero)}} = 6.25 \ \mu A$.

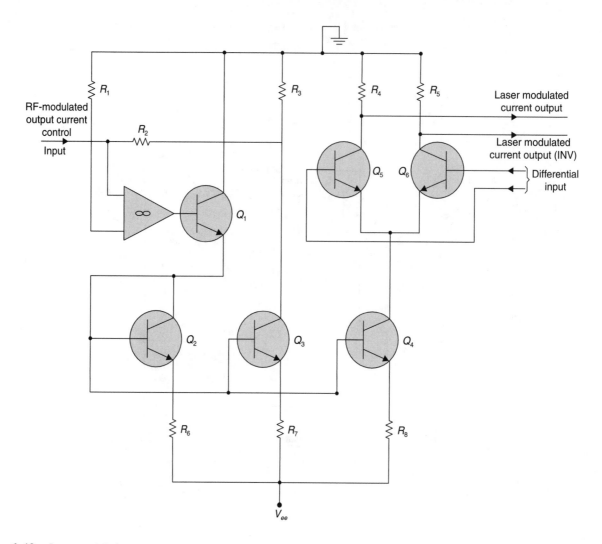

Figure 6–19. Laser modulation output.

iv) Compute the external bias resistor value $R_{ref(zero)}$.

$$R_{ref(zero)} = \frac{1.5}{I_{ref(zero)}} = \frac{1.5}{6.25 \times 10^{-6}} = 240 \text{ k}\Omega$$

Therefore, $R_{ref(zero)} = 240 \text{ k}\Omega$.

Tracking Error

The term **tracking error** refers to the possible difference between the laser optical power output and the monitor photodiode current. Tracking error is impossible to rectify.

Modulation and Current Control Loop

The current control time constant of the laser diode circuit is subject to chip capacitances. If the time constant based only on the chip capacitance is too small, then an external capacitor will be required to increase it to the required level for both optical high and optical low. The time constant (τ) and bandwidth (B_W) of the optical high are expressed by Equations (6–15) and (6–16), respectively.

$$\tau_{one} = (40 \times 10^{-12} + C_{ext(one)}) \times \frac{80 \times 10^3}{\eta_{laser}} (s) \quad \textbf{(6–15)}$$

where $\tau_{(one)}$ is the time constant and $C_{ext(one)}$ is the external capacitance for optical high.

$$B_{W(one)} = \frac{1}{2\pi\tau_{(one)}}(Hz) \quad \textbf{(6–16)}$$

The time constant $\tau_{(zero)}$ and bandwidth $B_{W(zero)}$ for optical low are expressed by Equations (6–17) and (6–18), respectively.

$$\tau_{(zero)} = (40 \times 10^{-12} + C_{(zero)}) \times \frac{80 \times 10^3}{\eta_{laser}} (s) \quad \textbf{(6–17)}$$

$$B_{W(zero)} = \frac{1}{2\pi\tau_{(zero)}}(Hz) \quad \textbf{(6–18)}$$

where η_{laser} is the **laser efficiency.** Laser efficiency (η_{laser}) is defined as the product of the **electro-optical efficiency** and responsivity (Equation (6–19).

$$\eta_{laser} = \eta_{EO} \times R \quad \textbf{(6–19)}$$

Electro-optical efficiency (η_{EO}) is defined as the rate of change of the laser optical power over to the rate of change of the modulation current (W/A) while laser diode respon-

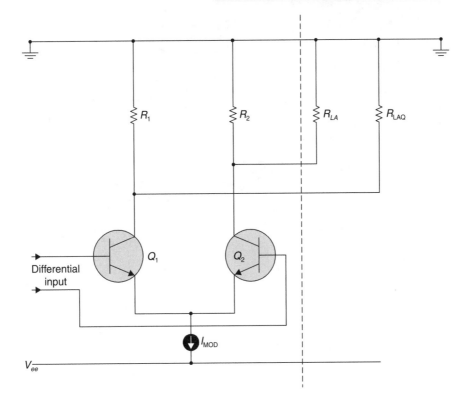

Figure 6–20.   Laser driver bias circuit.

sivity (R) is defined as the monitor photocurrent generation to the laser optical output power (A/W).

$$\eta_{\text{laser}} = \eta_{EO}(W/A) \times R(A/W)$$

Therefore, η_{laser} is dimensionless.

Example 6–7

A laser diode operating above the threshold level ($I_{th} = 30$ mA) generates an optical power output of 1.2 mW. If the electro-optical efficiency (η_{EO}) is equal to 25 mW/A and the responsivity (R) is equal to 400 mA/W, calculate the corresponding bandwidths (B_W) for optical one and optical zero when external capacitance is not required. (B_W refers to the optical modulation current control loop.)

Solution

i)   Compute τ_{one}.

$$\tau_{\text{one}} = (40 \times 10^{-12}) \times \frac{80 \times 10^3}{\eta_{\text{laser}}}$$

ii)   Compute η_{laser}.

$$\eta_{\text{laser}} = \eta_{EO} \times R = 25 \times 10^{-3} \times 400 \times 10^{-3}$$
$$= 1 \times 10^{-2}$$

Therefore, $\eta_{\text{laser}} = 0.01$.

Or,

$$\tau_{\text{one}} = 40 \times 10^{-12} \times 80 \times 10^3 \times \frac{1}{1 \times 10^{-2}}$$
$$= 320 \times 10^{-6}$$

Therefore, $\tau_{\text{one}} = 320$ µs.

iii) Compute $B_{W_{\text{one}}}$.

$$B_{W_{\text{one}}} = \frac{1}{2\pi \times \tau_{\text{one}}} = \frac{1}{2\pi \times 320 \times 10^{-6}} \cong 500 \text{ Hz}$$

Therefore, $B_{W_{\text{one}}} \cong 500$ Hz.

iv) Compute $B_{W_{\text{zero}}}$.

$$B_{W_{\text{zero}}} = 795 \text{ Hz}$$

The equations that determine the reference current (I_{ref}) for both zeros and ones are valid for input data patterns that contain frequent zeros and ones of a predetermined number, that have a time frame of 6 ns. There is a direct relationship between **optical extinction ratio** and bit rate. That is, a higher bit rate generates a higher extinction ratio. It is, therefore, imperative that the input data is used for optical level adjustment.

Modulation Current Variations at Off-State

During normal operation, the modulation current at the output of the laser diode driver circuit alternates between the

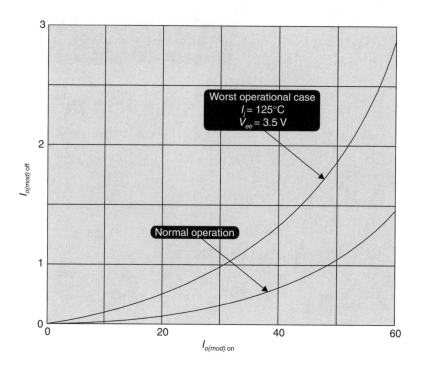

Figure 6–21. Relationship between $I_{o\,(\text{mod})\,\text{off}}$ and $I_{o\,(\text{mod})\,\text{on}}$

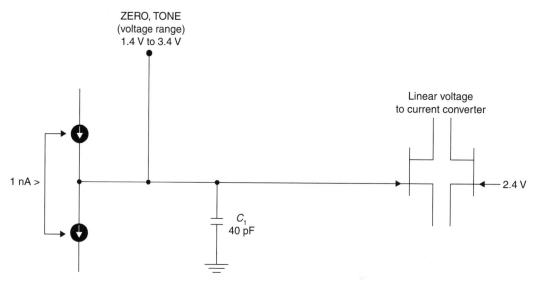

Figure 6–22. Voltage-to-current converter circuit for bias and modulation current monitoring.

two differential outputs. At off state, most of the modulation current appears at the complementary output. Only a small amount appears at the true output. This small amount of current is subtracted from the laser bias current. If the laser diode requires a large bias current for its operation, such a bias current reduction may create operational problems. On the other hand, if the diode operates with a small bias current, the effects will be minimal. The relationship between the modulation current at the off state and modulation current at the on state is illustrated in Figure 6–21.

Because the levels of both bias and modulation currents are so important for the proper operation of laser diodes, both must be monitored in a continuous basis. The continuous monitoring of the bias and modulation current of the laser diode is accomplished through a current-to-voltage converter circuit. Such a circuit is illustrated in Figure 6–22. The voltage applied at ZERO and TONE inputs must be within the range of 1.4 V to 3.4 V. Any circuits connected into the inputs must exhibit extremely high input impedance. Amplifiers that exhibit input impedance on the order of 100 GΩ and have zero (or very close to zero) leakage current must be used. Such current should not exceed the level of a few pA.

Furthermore, for the protection of the optical system, a shut off mechanism must be in place to switch off the laser diode bias current in case of malfunction. A typical shut off circuit is illustrated in Figure 6–23.

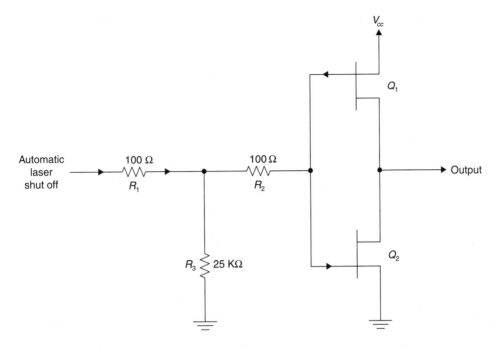

Figure 6–23. Bias current switch off circuit.

A logic high at the input of the protection circuit will disable the automatic laser shut down, while a low at the input will slowly restore the laser diode bias current to the required level. The rate of bias current increase is dictated by τ_{one} and τ_{zero} time constants.

Bias Current Alarm

Beyond the automatic shut off, the driver circuit may be equipped with a **bias current alarm** mechanism that is capable of detecting bias currents outside their predetermined levels. The alarm input accepts currents with a ratio of 1:1500 of that of the bias current (for Philips TZA3031AHL, TZA3031BHL, and TZA3031U laser drivers). Similar circuitry can be implemented for other devices. The level of the alarm current can be set by connecting external resistors between the alarm pins and V_{cc}. The values of these resistors for both optical high and optical low are expressed by Equations (6–20) and (6–21), respectively.

$$R_{\text{alarm}(H)} = \frac{1.5 \times 1500}{I_{o(\text{bias})\text{max}}}(\Omega) \qquad \textbf{(6–20)}$$

$$I_{\text{alarm}(H)} = \frac{I_{o(\text{bias})\text{max}}}{1500}(\text{A})$$

$$R_{(\text{alarm})(L)} = \frac{1.5 \times 300}{I_{o(\text{bias})\text{min}}}(\Omega) \qquad \textbf{(6–21)}$$

$$I_{\text{alarm}(L)} = \frac{I_{o(\text{bias})\text{min}}}{300}(\text{A})$$

Example 6–8

In a laser diode driver circuit, the reference currents required to limit the bias current are set between 5 mA and 80 mA.

Compute the values of the external resistors and the alarm currents for both optical high and optical low.

Solution

i) Calculate alarm resistors.

$$R_{\text{alarm}(H)} = \frac{1.5 \times 1500}{I_{o(\text{bias})\text{max}}} = \frac{2250}{80 \times 10^{-3}} = 28 \text{ k}\Omega$$

Therefore, $R_{\text{alarm}(H)} = 28 \text{ k}\Omega$.

$$R_{\text{alarm}(L)} = \frac{1.5 \times 300}{5 \times 10^{-3}} = 90 \text{ k}\Omega$$

Therefore, $R_{\text{alarm}(L)} = 90 \text{ k}\Omega$.

ii) Calculate alarm currents.

$$I_{\text{alarm}(H)} = \frac{I_{o(\text{bias})\text{max}}}{1500} = \frac{80 \text{ mA}}{1500} = 53 \text{ mA}$$

Therefore, $I_{\text{alarm}(H)} = 53 \text{ mA}$.

$$I_{\text{alarm}(L)} = \frac{I_{o(\text{bias})\text{min}}}{300} = \frac{5 \text{ mA}}{300} = 17 \text{ μA}$$

Therefore, $I_{\text{alarm}(L)} = 17 \text{ μA}$.

Power Dissipation

For the normal operation of a laser driver circuit, knowledge of the maximum power dissipation is important. The terms determining the total power dissipation are expressed by Equation (6–22).

$$P_T = P_{d(V_{cc})} + P_{d(V_{ee1})} + P_{d(V_{ee2})}$$
$$- [P_{d(LD)} + P_{d(LDQ)} + P_{d(\text{bias})}] \qquad \textbf{(6–22)}$$

where P_T is the total power dissipation, $P_{d(V_{cc})}$ is the power consumption of the digital section from V_{cc}, $P_{d(V_{ee1})}$ is the power consumption of the **digital** section from V_{ee1}, and $P_{d(Vee2)}$ is the power consumption of the **analog** section from V_{ee2}. More specifically, the value of $P_{d(Vee2)}$ is expressed by Equation (6–23).

$$P_{d(Vee2)} = V_{ee2} \times I_{ee2} \qquad (6\text{–}23)$$

where V_{ee2} is the power supply for the analog section and I_{ee2} is the current drain of the analog section. The value of I_{ee2} is expressed by Equation (6–24).

$$I_{ee2} = 55 + 1.5I_{mod} + I_{bias} + 3I_{AMP(adj)}55I_{EF(adj)} \qquad (6\text{–}24)$$

where I_{mod} is the modulation current, I_{bias} is the bias current, $I_{AMP(adj)}$ is the amplifier current adjustment, and $I_{EF(adj)}$ is the emitter follower current adjustment. The laser diode power consumption for both differential outputs is expressed by Equations (6–25) and (6-26), respectively.

$$P_{d(LD)} = 0.5\left[I_{LD(ext)} \times V_{LD}\right] \qquad (6\text{–}25)$$

$$P_{d(LDQ)} = 0.5\left[I_{LDQ(ext)} \times V_{LD}\right] \qquad (6\text{–}26)$$

Power dissipation based on bias voltage and bias current is expressed by Equation (6–27).

$$P_{d(bias)} = I_{bias} \times V_{bias} \qquad (6\text{–}27)$$

where I_{bias} is the bias current and V_{bias} is the bias voltage. Because the random data pattern is only ON half the time, the factor 0.5 is incorporated into the formula. V_{LD} and V_{LDQ} represent the voltages at both the laser outputs where the modulation current flows, and are each subject to laser power supply. Therefore, for proper power supply control, external voltages must be completely controlled.

Example 6–9

Compute the total power dissipation of a laser driver circuit that is designed to drive a laser diode that has an optical power output of 0.2 mW and an extinction ratio of 12 dB. Assume bias current (I_{bias}) of 15 mA, laser current (I_{LD}) of 35 mA and laser series resistance 50 Ω.

Solution

The ratio of the laser diode internal to external current is:

$$\frac{I_{LD(int)}}{I_{LD(ext)}} = \frac{100}{50} = 2$$

The total modulation current is:

$$I_{mod} = \left(\frac{100 + 50}{100}\right) \times 35 = 52.5 \text{ mA}$$

Therefore, $I_{mod} = 52.5$ mA.

The current at the complementary output (I_{LDQ}) is:

$$I_{LDQ} = \frac{1.2 \text{ V}}{100 \ \Omega} = 12 \text{ mA}$$

Therefore, $I_{LDQ} = 12$ mA.

The laser voltage (V_{LD}) at optical high is:

$$V_{LD} = -1.2 \text{ V} -50 \ \Omega(15 \times 10^{-3} \text{ mA } + 35 \text{ mA})$$

$$= -3.7 \text{ V}$$

Therefore, $V_{LD} = -3.7$ V.

The laser voltage at optical low (dark) is:

$$V_{LD} = -1.2 \text{ V} -(15 \times 10^{-3} \times 50 \ \Omega) = -1.95 \text{ V}$$

Therefore, $V_{LD} = -1.95$ V.

The laser voltage at the complementary output is:

$$V_{LDQ} = -35 \times 10^{-3} \text{ A} \times 50 \ \Omega = -1.75 \text{ V}$$

Therefore, $V_{LDQ} = 1.75$ V.

The bias voltage V_{bias} is the difference between the laser diode voltage V_{LD} and voltage drop across the RF chock V_{chock}. For an average chock resistance of 5 Ω and bias current 15 mA, the value of V_{chock} is:

$$V_{chock} = -(15 \times 10^{-3} \text{ A} \times 5 \ \Omega) = -0.075 \text{ V}$$

Therefore, $V_{chock} = -0.075$ V.

Bias voltage V_{bias}:

$$V_{bias} = 0.5(V_{LD}-V_{LDQ})-V_{chock}$$

$$= 0.5(-3.7-1.95)-0.075 = -2.9 \text{ V}$$

Therefore, $V_{bias} = -2.9$ V.

Compute total power dissipation P_T:

$$P_T = P_{d(V_{cc})} + P_{d(V_{ee1})} + P_{d(V_{ee2})}-\left[P_{d(LD)} + P_{d(LDQ)} + P_{d(bias)}\right]$$

The total power dissipation is calculated on the following assumptions.

i) For the digital section: 2.5 mA of current is drained from the V_{cc} power supply and 80 mA from the V_{ee1} power supply.

$$P_{d(V_{cc})} = 5 \text{ V} \times 2.5 \text{ mA} = 12.5 \text{ mW}$$

$$\therefore P_{d(V_{cc})} = 12.5 \text{ mW}$$

$$P_{d(V_{ee1})} = 4.5 \text{ V} \times 80 \text{ mA} = 360 \text{ mW}$$

$$\therefore P_{d(V_{ee1})} = 360 \text{ mW}$$

ii) For the analog section: Assuming a total of 160 mA is drained (I_{ee2}) from V_{ee2} then,

$$P_{d(V_{ee2})} = 6.0 \text{ V} \times 160 \text{ mA} = 960 \text{ mW}$$

$$\therefore P_{d(V_{ee2})} = 960 \text{ mW}$$

iii) Laser power dissipation ($P_{d(LD)}$) :

$$P_{d(LD)} = 0.5 \ (V_{LD} \times I_{bias})$$

$$= 0.5(3.75 \text{ V} \times 50 \text{ mA}) = 93.75 \text{ mW}$$

$$\therefore P_{d(LD)} = 93.75 \text{ mW}$$

iv) Laser power dissipation ($P_{d(LDQ)}$) :

$$P_{d(LDQ)} = 0.5 \times 1.95 \text{ V} \times 50 \text{ mA} = 48.75 \text{ mW}$$

$$\therefore P_{d(LDQ)} = 48.75 \text{ mW}$$

v) Power dissipation due to bias voltage and current ($P_{d(bias)}$):

$$P_{d(bias)} = 2.9 \text{ V} \times 15 \text{ mA} = 43.5 \text{ mW}$$

$$\therefore P_{d(bias)} = 43.5 \text{ mW}$$

Substituting the above values into Equation (6–12) yields:

$$P_T = P_{d(V_{cc})} + P_{d(V_{ee1})} + P_{d(V_{ee2})}-\left[P_{d(LD)} + P_{d(LDQ)} + P_{d(bias)}\right]$$

$$= 12.5 + 360 + 960 - (93.75 + 48.75 + 43.5)$$

$$= 1332.5-186 = 1146.5$$

Total power dissipation (P_T) = 1146.5 mW

Maximum Case Temperature

Laser diode driver circuits are designed to operate at maximum case temperatures of 125°C. The case power dissipation is a function of the operating temperature T_{amb} and thermal resistance $R_{th(J-a)}$, expressed by Equation (6–28).

$$P_T = \frac{125 - T_{amb(max)}}{R_{th(J-a)}} \quad \textbf{(6–28)}$$

where $R_{th(J-a)}$ is the thermal resistance and T_{amb} is the ambient temperature. The value of R_{th} is subject to printed circuit board layout, and normally is set at 35 K/W. With a maximum case temperature of 125°C and an ambient temperature of 85°C, the case maximum power dissipation is equal to:

$$P_{T(case)} = \frac{125 - 85}{35} = 1.143 \text{ W}$$

Therefore, $P_{T(case)} = 1.143$ W.

The device maximum power dissipation is expressed by Equation (6–29).

$$P_T = (-V_{cc} \times I_{cc}) + (-V_{cc} + V_{out}) \times I_{mod}$$
$$+ (-V_{cc}V_{bias}) \times I_{bias} \quad \textbf{(6–29)}$$

Example 6–10

For a laser driver operating with the following, compute the maximum case temperature.

- $V_{cc} = -5.0$ V
- $I_{mod} = 35$ mA
- $I_{bias} = 18$ mA
- $V_{bias} = -2.0$ V
- $V_{out} = 2.0$ V
- $T_{J-a} = 30°C$
- $I_{cc} = 140$ mA

Solution

$$P_T = (-V_{cc} \times I_{cc}) + (-V_{cc} + V_{out}) \times I_{mod}$$
$$+ (-V_{cc} + V_{bias}) \times I_{bias}$$
$$= \lfloor -(-5.0) \times 140 \times 10^{-3} \rfloor + (5.0 - 2) \times 35 \times 10^{-3}$$
$$+ (5.0-2) \times 18 \times 10^{-3} = 859 \text{ mW}$$

Therefore, $P_T = 859$ mW.

For $T_{J-a} = 30°C/W$ or $P_T \times T_{J-a} = 0.859 \text{ W} \times 30°C$

$= 25.77$. The maximum case temperature is equal to:

$$T_{case(max)} = 125°C - T_{J-a} = 125°C - 25.77°C \cong 99°C$$

Therefore, $T_{case(max)} = 99°C$.

6.8 LASER DRIVERS BASED ON THE LUCENT TECHNOLOGIES LG1627BXC

The second laser diode driver circuit is based on Lucent Technologies' LG1627BXC design, and is intended for implementation in high speed NRZ direct modulation optical fiber transmission systems. The block diagram of the device is illustrated in Figure 6–24.

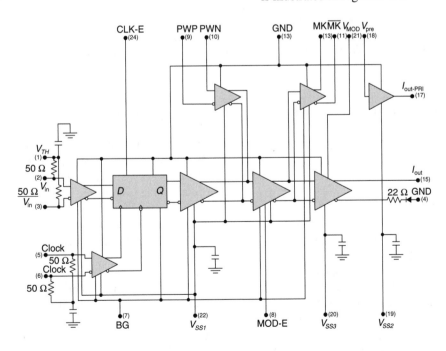

Figure 6–24. A block diagram of the LG1627BXC laser diode driver circuit. Courtesy Agere Systems (Lucent Technologies).

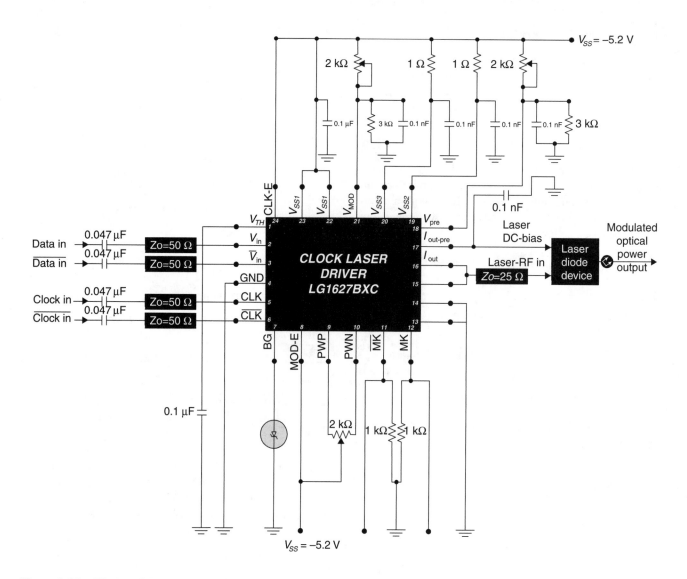

Figure 6–25. Biasing of the LG1627BXC driver circuit. Courtesy Agere Systems (Lucent Technologies).

Some of the important operating characteristics are:

1) maximum operating bit rate: 3 Gb/s

2) single power supply: −5.2 V

3) rise/fall times: 90 ps

4) adjustable output current

The device can also be used in optical test equipment that is intended for SONET/SDH applications. The circuit is composed of three stages in cascade. The first stage is designed to accept differential clock and data ECL inputs. The device is also capable of independently controlling the laser bias and modulation currents through external networks, while an internal 2.5 V band gap reference voltage compensates for supply and temperature variations. Data retiming and jitter minimization are accomplished through internal FFs. Full

biasing and connection to the laser diode are illustrated in Figure 6–25.

For better performance, both data and clock inputs are AC coupled through 0.47 μF capacitors looking into 50 Ω impedances, while the current modulated output is connected to the laser diode input through a 25 Ω impedance. The desired modulation voltage (V_{mod}) and prebias control voltage (V_{pre}) are both set by external voltage divider networks. All the externally connected voltage, dividers as well as those connecting the threshold voltage (V_{th}) and bias output current, should be bypassed by 0.1 μF capacitors that are connected as close as possible between pins and ground. Furthermore, both **mark density** and its complement inputs must be terminated with 1 kΩ pull up resistors. Some of the most important electrical characteristics of the device are listed in Table 6–10.

TABLE 6–10 Electrical Characteristics

Parameters	Minimum	Typical	Maximum	Symbol	Units
Power supply voltage	−4.9	−5.2	−5.5	$V_{ss1}, V_{ss2}, V_{ss3}$	V
Power supply current	100	140	160	I_{ss1}	mA
Voltage control for output mod. current	−4.0	—	−5.5	V_{mod}	V
Output mod. current at ($V_{mod(max)}$)	75	85	—	$I_{mod\ out(high)}$	mA
Output mod. current at ($V_{mod(min)}$)	—	0	2	$I_{mod\ out(low)}$	mA
Control voltage for prebias current	−3.0	—	−5.5	V_{pre}	V
Output prebias current (min)	—	0	0.5	$I_{pre(low)}$	mA
Output prebias current (max)	50	60	—	$I_{pre(high)}$	mA
Data input voltage single ended	0.3	0.6	1.0	V_{in}	V
Mark density (50% duty cycle)	—	−0.5	—	MK	V
Mark density (comp.) (50% duty cycle)	—	−0.5	—	$\overline{MK}$	V
Pulse width adjust (+)	−3.0	−4.2	−5.5	PWP	V
Pulse width adjust (−)	−3.0	−4.2	−5.5	PWN	V
For $I_{mod\ out}$ = 40 mA (Clock enabled)					
Output rise/fall times (20%–80%)	—	90	—	t_R / t_F	ps
Jitter (rms)	—	4	—	–	ps
(Clock disabled)					
Output rise/fall times 20%–80%	—	90	—	t_R / t_F	ps
Jitter (rms)	—	6	—	—	ps
(Clock disabled)					
Output rise/fall times 20%–80%	—	90	—	t_R / t_F	ps
Jitter (rms)	—	6	—	—	ps
For $I_{mod.out}$ = 80 mA (Clock enabled)					
Output rise/fall times (20%–80%)	—	100	—	t_R / t_F	ps
Jitter (rms)	—	4	—	—	ps
(Clock disabled)					
Output rise/fall times 20%–80%	—	100	—	t_R / t_F	ps
Jitter (rms)	—	6	—	—	ps

The absolute maximum ratings are listed in Table 6–11.

TABLE 6–11 Absolute Maximum Ratings

Parameters	Minimum	Maximum	Symbol	Units
Power supply	—	−5.7	V_{ss}	V
Input voltage	GRD	V_{ss}	V_{ss}	V
Power dissipation	—	1.0	P_D	W
Case temperature (operating)	0	100	T_{case}	°C
Storage temperature	−40	125	T_{stg}	°C

The performance of this laser diode driver circuit is evaluated through the eye diagrams in Figure 6–26, which illustrates the eye diagram of a device that generates a modulated current output ($I_{out(mod)}$) of 85 mA under a laser diode biasing current (I_{pre}) of 5 mA.

The characteristic curve that relates the modulated output current ($I_{out(mod)}$) and modulation voltage (V_{mod}) is illustrated in Figure 6–27.

Figure 6–27 shows a very sharp linear increase of the output modulated current between −5.0 V and −4.5 V modulation voltage. Finally, the relationship between the bias output current (I_{pre}) and the bias voltage (V_{pre}) is illustrated in Figure 6–28.

The device pin diagram is illustrated in Figure 6–29, and the description is presented in Table 6–12 (see page 167).

6.9 LASER DRIVERS BASED ON THE NORTEL YA08

The following laser diode driver circuit is based on the NORTEL YA08 device. This device is able to handle data rates of up to 2.5 Gb/s for SONET-48/STM-16 applications. It is a fully integrated, mean power control device, that accepts both CML and PECL **differential inputs**, and is capable of interfacing directly with NORTEL 4:1 or 16:1 multiplexers. A block diagram of the device is illustrated in Figure 6–30 (see page 168).

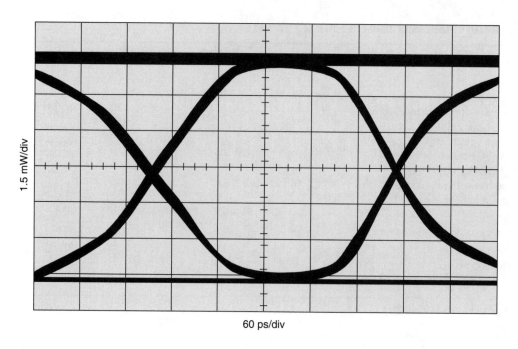

Figure 6–26. Eye diagram for an output modulated current of 85 mA and bias current of 5 mA.

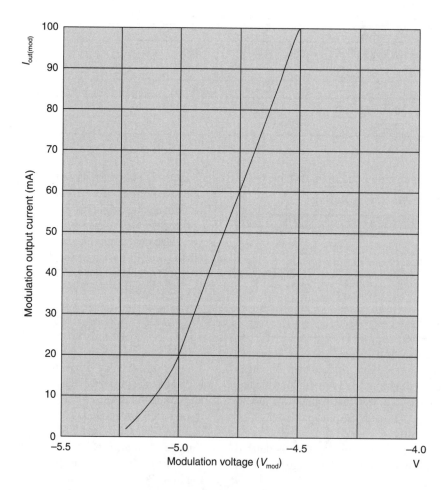

Figure 6–27. Modulated output current ($I_{out(mod)}$) v. Modulation voltage (V_{mod}).

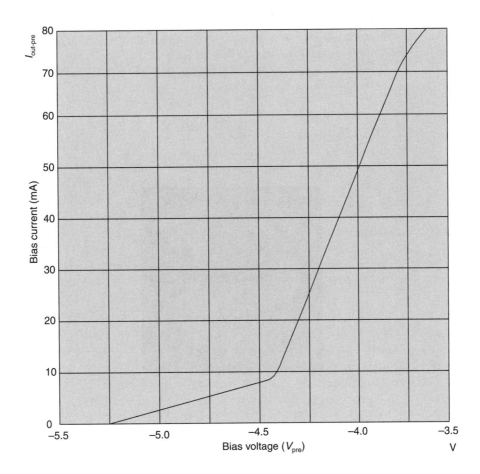

Figure 6–28. Bias current ($I_{out-pre}$) as a function of bias voltage (V_{pre}).

The main building blocks of the above device are:

1) input buffer
2) modulation switch
3) mean power control
4) bias current limiter
5) bias mirror output
6) **laser shutdown input**

The device also provides a built-in reference voltage source for temperature and power supply stabilization.

Description of Operation

Multiplexer differential CML or PECL outputs that have average amplitudes of 400 mV$_{p-p}$ are applied at the input of the buffer stage, the function of which is to provide a modulation current at the output, covering the range between 10 mA and 80 mA, with an average value of 50 mA. The modulation current can be set at the desired level through an external resistor that is connected between the MODSET and GND pins. The complementary output of the driver circuit MODOUTN is connected to the cathode of the laser diode, with the anode connected to the power supply. If laser diode shutdown is required, current is diverted from the modulation output through an appropriate input

pin. It is important to note that the laser diode bias current must be maintained at a constant throughout the device's operational lifetime. The constant bias current is provided by the **mean power control** loop, which senses the laser optical power output through a back facet PIN diode that is connected between the power supply (cathode) and PININ (anode). The generated photocurrent is compared with the current generated by a reference voltage source not exceeding the V_{cc}+0.7 V level. An external resistor, connected between the PINSET input and the bias current generator circuit, sets this reference voltage. The current generated by the reference voltage source can be set at 1 mA at the start of its life, and 100 mA at the end of its life. The device also incorporates a bias current limiting circuit, that limits the laser diode bias current to a threshold level set by an external resistor that is connected between the LIMITSET and GND pins.

Because the maintenance of a constant laser diode bias current is critical to the normal operation of the optical system, a bias current monitoring mechanism is essential. The means for bias current external monitoring is provided through the bias mirror output. The combination of LD and LDSET inputs connected to the EX-OR gate provide the mechanism for laser shutdown. The full biasing arrangement of the YA08 laser diode driver device is illustrated in Figure 6–31 (see page 169).

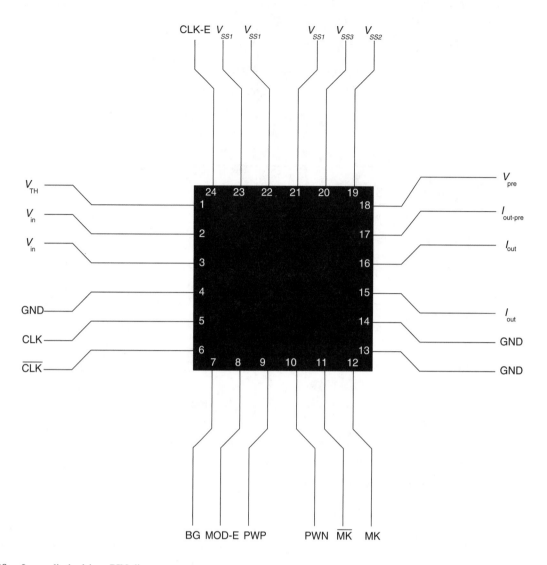

Figure 6–29. Laser diode driver PIN diagram.

The tail current of the modulation circuit driver stage is provided by the external resistor (R_3) connected between the modulation current set (MODSET) and GND pins. This current is diverted either from the modulation complementary output (MODOUTN), or modulation output (MODOUTPUTP), and is subject to applied data at inputs TXDATIP or TXDATIN. If the differential voltage at TXDATIP is positive in reference to TXDATIN, the laser diode will be switched ON. If the differential voltage is zero, the laser diode will be switched OFF. Therefore, the laser diode will be switching ON or OFF in accordance with the binary data at the input of the driver circuit. The laser diode will also be switched OFF when the bias current is diverted to the MODOUTP pin by a potential difference developed across R_6. The voltage drop across resistor R_3 (connected between MODSET and GND) is used as a feedback mechanism, controlling the current through a temperature invariant internal resistor. Through this process, temperature stabilization can be achieved. The ratio of the above two currents is 40:1. The modulation current (I_{mod}) is expressed by Equation (6–30).

$$I_{mod} = 40\left(\frac{1.2\text{ V}}{R_{MODSET}}\right) \quad \text{(6–30)}$$

where $R_{MODSET} = R_3$, 1.2 V is the reference voltage, and I_{mod} is the modulating current.

The value of R_{MODSET} or R_3 is expressed by Equation (6–31).

$$R_{MODSET} = \frac{48}{I_{mod}} \quad \text{(6–31)}$$

Mean Power Controller

The function of the mean power controller circuit, in conjunction with a PIN photodetector diode, is to maintain the bias current of the laser diode between 10 mA and 100 mA. This bias current can be set to the desired level through an external resistor connected between PINSET and GND. The value of the resistor is expressed in Equation (6–32).

$$R_{PINSET(R_1)} = \frac{1.26}{I_{PIN}} \quad \text{(6–32)}$$

TABLE 6–12 Pin Description

Pinnumber	Symbol	Description
1	V_{TH}	Data input reference—capacitor to ground
2	V_{in}	Data input
3	$\overline{V_{in}}$	Data input complementary
4	GRD	Ground
5	CLK	Clock input
6	$\overline{CLK}$	Complementary clock input
7	BG2P5	Band gap reference (–2.5 V)
8	MOD-E	Modulation enable
9	PWP	Pulse width adjust (positive)
10	$\overline{PWN}$	Pulse width adjust (negative)
11	$\overline{MK}$	Complementary mark density out
12	MK	Mark density out
13	GRD	Ground
14	GRD	Ground
15	$I_{mod(out)}$	Output modulation current
16	$I_{mod(out)}$	Output modulation current
17	I_{pre}	Output bias current
18	V_{pre}	Prebias control input
19	V_{ss2}	–5.2 V supply voltage for output prebias
20	V_{ss3}	–5.2 V supply voltage for output modulation
21	V_{mod}	Modulation voltage for current control input
22	V_{ss1}	–5.2 V supply voltage
23	V_{ss1}	–5.2 V supply voltage
24	CLK-E	Clock enable

where I_{PIN} is between 10 mA (start of life) and 100 mA (end of life).

The voltage at PININ developed by the photodetector diode is compared to an internal 2.4 V reference voltage. The bias output current (BIASOUT) is limited to a level above the current that is generated by the PIN photodetector diode, through a current limiting circuit (Resistor R_2), connected between LIMITSET and GND. The value of the current limiting resistor $(R_{LIMITSET})$ can be established by Equation (6–33).

$$\frac{72}{R_{LIMITSET}} < I_{bias(max)} < \frac{80}{R_{LIMITSET}} \qquad \textbf{(6–33)}$$

A block diagram of a mean power controller circuit is illustrated in Figure 6–32.

The AC characteristics of the device are listed in Table 6–13.

TABLE 6–13 AC Characteristics

Parameters	Minimum	Typical	Maximum	Symbol	Units
Data-in	—	—	2.6	$DR\ i$	Gb/s
Rise time	30	—	150	$t_{r(MODoutP/N)}$	ps
Fall time	30	—	150	$t_{f(MODoutP/N)}$	ps
Data-output (mark/ space)	48	—	52	$MODout_{(RMS)}$	%

The DC characteristics of the device are listed in Table 6–14.

TABLE 6–14 DC Characteristics

Parameters	Minimum	Typical	Maximum	Symbol	Units
MODOUTP/N output current setting	10	—	80	I_{MODout}	mA
MODSET bias voltage	1.2	—	1.32	V_{MODSET}	V
Power dissipation (For $I_{MODOUT} = 0$ mA)	—	—	350	$P_{D(0)}$	mW
Power dissipation for: $I_{MODOUT} = 25$ mA $V_{BIAS} = V_{MOD} = 2.5$ V $I_{BIAS} = 60$ mA	—	—	700	$P_{D(wkg)}$	mW
Power dissipation (max)	—	—	1.2	$P_{D(max)}$	W

The recommended operating conditions for the device are listed in Table 6–15.

TABLE 6–15 Operating Conditions

Parameters	Minimum	Typical	Maximum	Symbol	Units
Supply voltage	4.7	5.0	5.3	V_{cc}	V
Input voltage single ended CML-PECL	V_{cc}-2.5	—	V_{cc}+0.2	V_{ICML}	V
Input voltage differential CML-PECL	0.4	—	1.0	V_{IDCML}	V
Compatible input voltage-LOW for CMOS	–0.5	—	0.8	$V_{INLCMOS}$	V
Compatible input voltage-HIGH for CMOS	2.0	—	V_{cc}+0.5	$V_{INHCMOS}$	V
BIAS OUT voltage range	1.2	—	4.0	V_{BIAS}	V
Operating temperature	–40	—	85	T_{amb}	°C

The absolute maximum ratings are listed in Table 6–16.

TABLE 6–16 Absolute Maximum Ratings

Parameters	Minimum	Maximum	Symbol	Units
Power supply	–0.7	6.0	V_{cc}	V
Input voltage Single ended CMOS/ CML/PECL	–0.7	6.0	V_I	V
Differential input voltage	–1.3	1.3	$V_{IDIFF.}$	V
Maximum applied voltage at MODSET/LIMITSET	–0.7	2.2	V_{ISET}	V
Maximum applied voltage at PININ/PINSET	–0.7	V_{cc}+0.7	V_{ISET}	V
Mod. Output current MODOUTP/N	–2.0	120	I_{MODOUT}	mA
Bias output current BIASOUT	–2.0	130	$I_{BIASOUT}$	mA
Output current for ref. voltage	–15	+15	$I_{O.VREF}$	mA

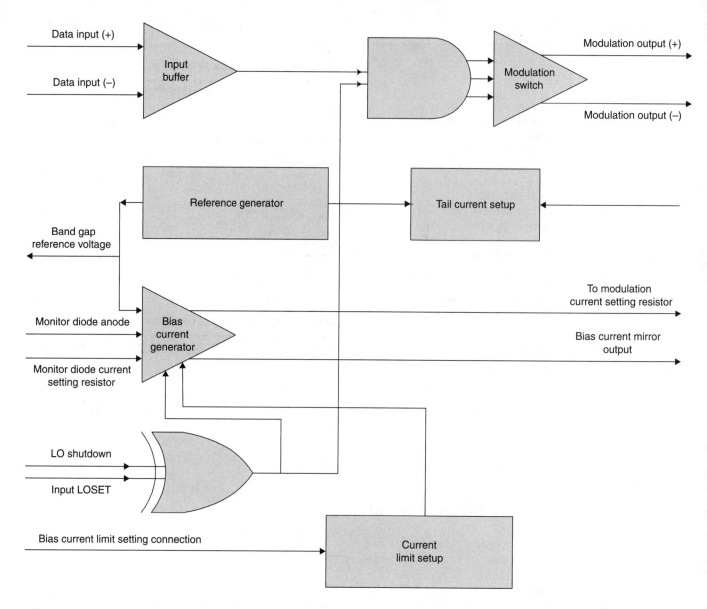

Figure 6–30. A block diagram of the NORTEL YA08 laser diode driver.

The performance of this laser diode driver for a modulation current output of 60 mA, expressed in the form of an eye diagram, is illustrated in Figure 6–33.

6.10 LASER DRIVERS BASED ON THE VITESSE VSC7928 3.2 Gb/s

The VSC7928 in Figure 6–34 is a laser driver device designed and developed by the VITESSE corporation's optoelectronics division. This device has a driving capability of 3.2 Gb/s and direct access to biasing and laser modulation field effect transistors. It operates with a single 5 V power supply. Its differential data and clock inputs are terminated to 50 Ω and the required modulation and biasing currents are precisely controlled by external components. For high unbalanced data input applications, appropriate

laser diode bias is achieved through an external data density adjustment input.

Some of the important operating characteristics of the Vitesse VSC7928 are:

- Input data rate max: 3.2 Gb/s
- Rise time: <100 ps
- Single or differential data and clock inputs

Figure 6–35 illustrates the single ended AC coupling operation, while Figure 6–36 illustrates the differential AC coupling mode of operation. (Figures 6–35 and 6–36 appear on page 172.)

Input Signaling

In high speed optical systems such as fiber channel (1.0625 Gb/s) and OC48 (2.5 Gb/s), the implementation of **differential signaling** is highly desirable. Such technique enables the sys-

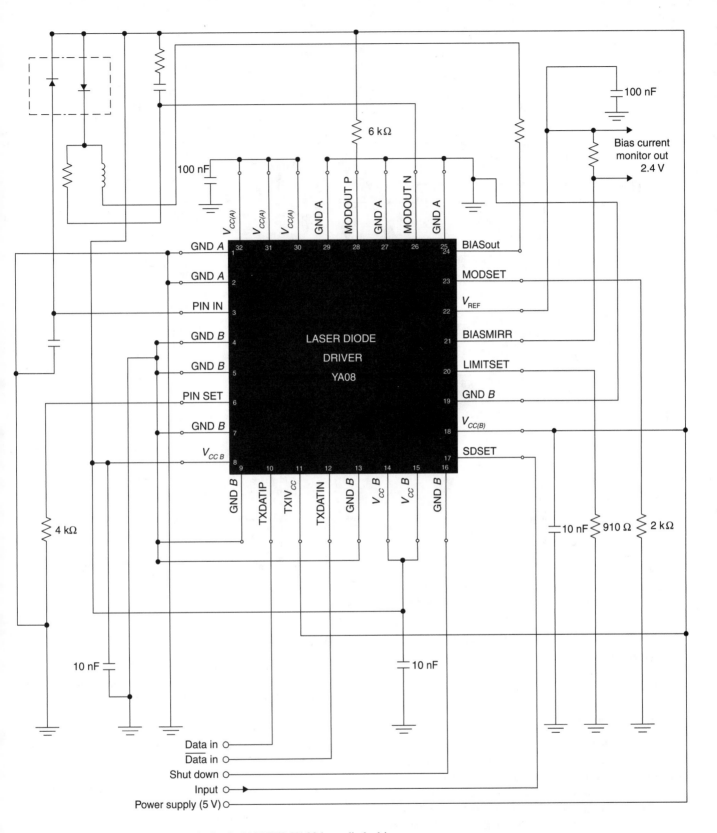

Figure 6–31. Application schematic for the NORTEL YA08 laser diode driver.

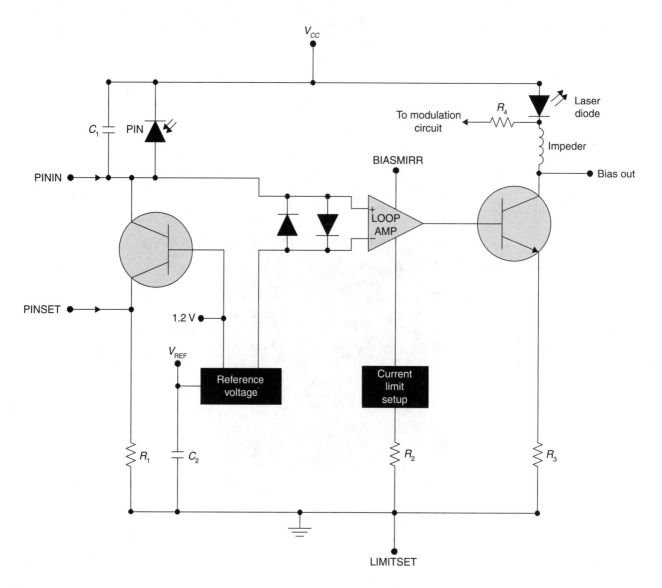

Figure 6–32. A block diagram of a mean power controller.

tem to improve the SNR and the system BER. Before a detailed description of differential **input signaling** takes place, it is appropriate that a brief comparison between single ended and differential ended signaling is presented.

Single Ended Signaling

In **single ended signaling,** the transmitted data is applied to the input of the driver circuit through a single transmission line. At the driver input, the applied signal must be defined as binary one or binary zero. This is accomplished through comparison of the transmitted data to a well defined reference voltage level. It is imperative that the reference voltage be constant if an accurate decision of the input data is to be achieved. In a single ended system, maintaining the reference voltage to a constant level is a challenge. For example, power supply variations, noise (common-mode), crosstalk, and transmission line reflections are some of the sources that cause undesirable variations in the reference voltage.

Differential Signaling

With differential signaling, the input data bit is transmitted through one line and its complement is transmitted through the other. The two signals are applied to a logic circuit, the output of which determines the logic state of the transmitted signal. With differential signaling, no reference voltage is required. Other advantages pertinent to differential signaling are:

a) *Noise immunity*: Parasitic elements such as power supply generated noise will be substantially reduced or altogether eliminated through mutual cancellation.

b) *Precision timing*: With single ended signaling, timing is subject to thermal drifts which affect the reference voltage. With differential signaling, thermal drifts are practically eliminated, therefore allowing for larger timing margins.

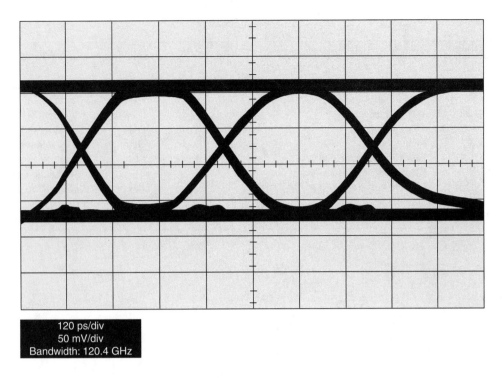

Figure 6–33. Eye diagram for a modulation current output of 60 mA.

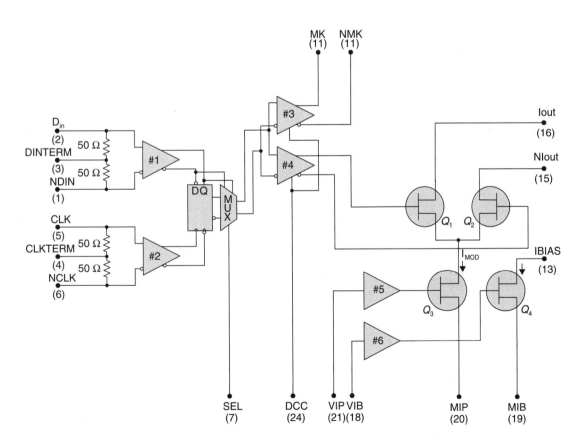

Figure 6–34. VSC7928 Vitesse 3.2 Gb/s laser diode driver circuit. (See page 172 for key.)
Reproduced by permission of Vitesse Semiconductor Corp.

Pin description

Pin Number	Symbol	Description
1	NDIN	Data reference (on chip)
2	DIN	Data input
3	DINTERM	Data reference
4	CLKTERM	Clock reference
5	CLK	Input clock
6	NCLK	Clock reference on chip
7	SEL	Clock-non-clock data select
8, 9, 10, 14, 17	GND	Positive voltage bus
11	NMK	Data density differential output (−)
12	MK	Data density differential output (+)

Pin Number	Symbol	Description
13	IBIAS	Laser bias output
15	NIOUT	Laser modulation current output (−)
16	IOUT	Laser modulation current output
18	VIB	Bias gate node
19	MIB	Bias source node
20	MIP	Modulation source node
21	VIP	Modulation gate node
22, 23	VSS	Negative voltage bus
24	DCC	Duty cycle control

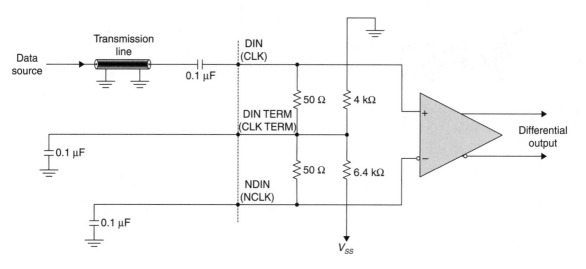

Figure 6–35. Single ended AC coupling input stage.

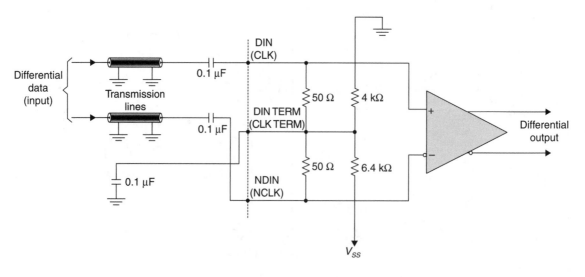

Figure 6–36. Differential AC coupling input stage.

c) *Electromagnetic interference reduction*: Differential signaling is capable of reducing **electromagnetic interference (EMI)** through the reduction of differential skew, a result of proper signal routing and parameter selection.

The Differential Signaling Principle of Operation

Common mode signaling in a two-port network is illustrated in Figure 6–37.

From Figure 6–37, the differential voltage (V_{diff}) is expressed by Equation (6–34).

$$V_{\text{diff}} = V_1 - V_2 \qquad (6\text{–}34)$$

The differential impedance (Z_{diff}) is expressed by Equation (6–35).

$$Z_{\text{diff}} = \frac{V_{\text{diff}}}{i_1}\bigg|_{V_{CM}=0} \qquad (6\text{–}35)$$

where V_{CM} is the common mode voltage. The common mode voltage (V_{CM}) is expressed by Equation (6–36).

$$V_{CM} = \frac{V_1 - V_2}{2} \qquad (6\text{–}36)$$

The **common mode impedance** (Z_{CM}) is expressed by Equation (6–37).

$$Z_{CM} = \frac{V_{CM}}{i_1}\bigg|_{V_{\text{diff}}=0} \qquad (6\text{–}37)$$

Transmission Line Equations

The basic components in a pair of transmission lines, similar to those required in differential signaling, are:

i) L_{11}, L_{22}: self-inductance
   L_{12}, L_{21}: mutual inductance

ii) C_{11}, C_{22}: self-capacitance
   C_{12}, C_{21}: mutual coupling capacitance

iii) Z_{11}, Z_{22}, Z_{12}, Z_{21}: impedance coefficients

iv) V_1, V_2: input voltage and output voltage

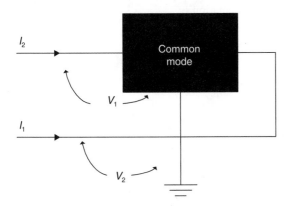

Figure 6–37. Common mode signals.

For symmetric lines, the value of capacitance (C) is calculated as follows.

$$C = \begin{vmatrix} C_{11} & C_{12} \\ C_{21} & C_{22} \end{vmatrix}$$

where $C_{12} = C_{11}$ symmetric lines. Similarly, the value of the inductance (L) is calculated as follows.

$$L = \begin{vmatrix} L_{11} & L_{12} \\ L_{21} & L_{22} \end{vmatrix}$$

where $L_{12} = L_{11}$ symmetrical lines. The value of the impedance (Z) is calculated as follows.

$$Z = \begin{vmatrix} Z_{11} & Z_{12} \\ Z_{21} & Z_{22} \end{vmatrix}$$

where $Z_{12} = Z_{11}$ symmetrical lines. For an infinitely long line with no reflection:

$$\frac{V_1}{V_2} = Z_0 \begin{vmatrix} i_1 \\ 2 \end{vmatrix}$$

Calculate the **differential mode impedance** (Z_{diff}).

Assuming a differential mode voltage (V_{diff}) of 1.5 V and common mode voltage (V_{CM}) of 0 V, then $V_1 = +0.75$ V and $V_2 = -0.75$ V. For an infinitely long line (no reflections),

$$\frac{V_1}{V_2} = Z_0 \begin{vmatrix} i_1 \\ i_2 \end{vmatrix}$$

or

$$\frac{V_1}{V_2} = \begin{vmatrix} Z_{11} & Z_{12} \\ Z_{21} & Z_{22} \end{vmatrix} \begin{vmatrix} i_1 \\ i_2 \end{vmatrix}$$

Solving for i_1 and i_2 yields:

$$i_1 = \frac{1}{2\,(Z_{11} - Z_{12})}$$

$$i_2 = -\frac{1}{2\,(Z_{11} - Z_{12})}$$

Therefore $i_1 = -i_2$.

Calculate Z_{diff}.

From Equation (6–34), and substituting for $V_{\text{diff}} = 1$ V and $i_2 = -\dfrac{1}{2(Z_{11} - Z_{12})}$ then,

$$Z_{\text{diff}} = \frac{1\ \text{V}}{\dfrac{1}{2(Z_{11} - Z_{12})}} = \frac{2(Z_{11} - Z_{12})}{1} = 2(Z_{11} - Z_{12})$$

or

$$Z_{\text{diff}} = 2(Z_{11} - Z_{12})$$

Example 6–11

Assuming microstrip with specific dimensions and weak coupling, calculate Z_{diff} and Z_{CM} for the following values:

- $Z_{11} = 49.95\ \Omega$
- $Z_{12} = 0.15\ \Omega$
- $Z_{21} = 0.15\ \Omega$
- $Z_{22} = 49.95\ \Omega$

Solution

i) Compute (Z_{diff}).

$$Z_{\text{diff}} = 2(Z_{11} - Z_{12}) = 2(49.95 - 0.15) = 98.7$$

Therefore $Z_{\text{diff}} = 98.7\ \Omega$.

By adjusting the differential voltage (V_{diff}) and microstrip physical dimensions, the value of Z_{diff} can be adjusted to required levels.

ii) Compute (Z_{CM}).

$$Z_{CM} = Z_{11} + Z_{12} = 49.95 + 0.15 = 50.1\ \Omega$$

Therefore, $Z_{CM} = 50.1\ \Omega$.

Example 6–12

Assuming microstrip of specific dimensions and strong coupling, calculate Z_{diff} and Z_{CM} for the following values:

- $Z_{11} = 65.4\ \Omega$
- $Z_{12} = 8.2\ \Omega$
- $Z_{21} = 8.2\ \Omega$
- $Z_{22} = 65.4\ \Omega$

Solution

i) Compute (Z_{diff}).

$$Z_{\text{diff}} = 2(Z_{11} - Z_{12}) = 2(65.4 - 8.2) = 114.4\ \Omega$$

Therefore, $Z_{\text{diff}} = 114.4\ \Omega$.

Similarly, adjusting the differential voltage (V_{diff}) and microstrip physical dimensions, the value of Z_{diff} can be adjusted to required levels.

ii) Compute (Z_{CM}).

$$Z_{CM} = Z_{11} + Z_{12} = 65.4 + 8.2 = 73.6\ \Omega$$

Therefore, $Z_{CM} = 73.6\ \Omega$.

Differential Pair Design Guidelines

For side-to-side pair arrangements, the following targets must be identified.

a) intended application (i.e., fiber channel, Ethernet), etc.

b) trace spacing

c) degree of coupling achievable through minimum spacing

d) optimum EMI and cross talk (achievable through minimum spacing)

e) differential impedance requirements (when fiber channel is set at 150 Ω)

In order to calculate the impedance (Z), the trace-to-trace spacing must be continually adjusted until the required differential impedance is produced.

Differential Pair Termination

Termination of the differential pair to their characteristic impedance is a must if negative phenomena such as overshoot and ringing are to be avoided. Several types of differential pair termination exist.

DIFFERENTIAL TAPPED TERMINATION. Common mode signals are usually terminated through the **differential tapped termination** method (Figure 6–38). Through this method, termination noise is reduced to an absolute minimum. Furthermore, in an AC coupling arrangement (PECL), the tap can be used to provide the required receiver input biasing voltage.

PARALLEL TERMINATION. A parallel **termination arrangement** used with open collector drivers can produce efficient common mode current difference as well as providing the required DC biasing (Figure 6–39). Parallel terminators require significantly larger number of components in their implementation.

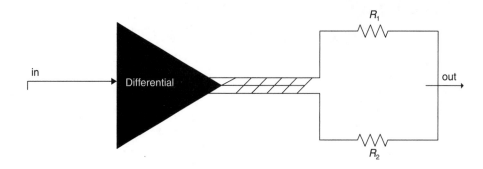

Figure 6–38. A differential tapped terminator.

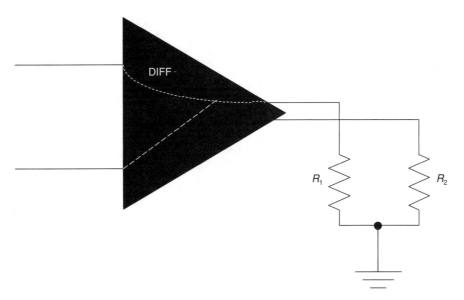

Figure 6–39. Parallel termination.

DIFFERENTIAL TERMINATION. For the implementation of differential terminators, a smaller number of components is required, while push-pull drivers are absolutely essential. Such terminators, although capable of balancing the input current, are unable to provide matching for common mode input signals. Therefore, they are capable of inducing waveshape distortion. The concept of differential termination is illustrated in Figure 6–40.

The implementation of differential signaling in high speed optical systems achieves the much desired high noise immunity, low skew, and significant emission reduction. The main disadvantage with differential termination is circuit complexity and an increase in the signal routing

if target impedance is to be achieved. The differential output stage of the laser driver device is illustrated in Figure 6–41.

Graphical representations of modulation current in reference to modulation voltage and bias current in reference to bias voltage are illustrated in Figures 6–42 and 6–43.

Figures 6–42 and 6–43 show that an exponential increase of both modulation and bias currents occurs above threshold modulation and bias voltages. Therefore, both modulation and bias currents can be set to desired levels by precisely controlling the modulation and bias voltages. The operating characteristics of the device are listed in Table 6–17.

TABLE 6–17   Operating Characteristics (High Speed Input/ECL-Outputs)

Parameters	Minimum	Maximum	Symbol	Unit	Conditions
Input voltage swing (single ended)	300	1500	V_{in}	mV	$V_{CM} = -2.0$ V
Differential input (common mode range)	−2.3	−1.3	V_{CM}	V	$V_{ss} = -5.2$ V
Output voltage (high) ECL	−1.2	—	V_{OH}	V	50 Ω to −2 V
Output voltage (low) ECL	—	−1.6	V_{OL}	V	50 Ω to −2 V
On chip terminators	35	65	—	Ω	—

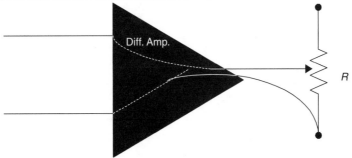

Figure 6–40. Differential termination.

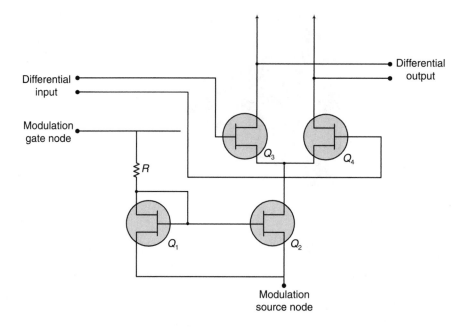

Figure 6–41. The laser diode driver output stage.

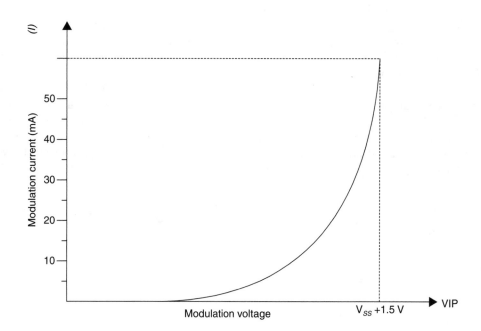

Figure 6–42. Modulation current *v.* modulation voltage.

The operating conditions are listed in Table 6–18.

TABLE 6–18   Operating Conditions

Parameters	Minimum	Typical	Maximum	Symbol	Unit
Power supply voltage	−5.5	−5.2	−4.9	V_{ss}	V
Power supply current	—	80	120	I_{ss}	mA
Power dissipation	—	—	700	P_{total}	mW
Operating temperature	−40	—	85	T_C	°C
Junction temperature	—	—	125	T_J	°C

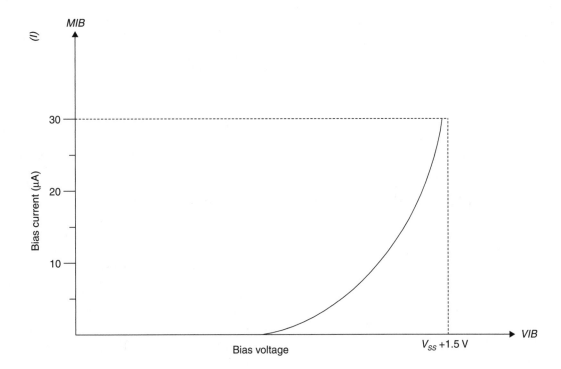

Figure 6–43. Bias current *v*. bias voltage.

The laser diode driver AC-electrical characteristics are listed in Table 6–19.

TABLE 6–19   AC Electrical Characteristics

Parameters	Minimum	Typical	Maximum	Symbol	Unit	Conditions
Output rise time	—	80	100	t_r	ps	Load = 25 Ω I_{mod} (20 mA – 60 mA)
Output fall time	—	80	100	t_f	ps	Load = 25 Ω I_{mod} (20 mA – 60 mA)

The laser diode driver's DC electrical characteristics are listed in Table 6–20.

TABLE 6–20   DC Electrical Characteristics

Parameters	Minimum	Typical	Maximum	Symbol	Unit	Conditions
Laser bias current	2	—	100	I_{bias}	mA	—
Modulation current	2	—	100	I_{mod}	mA	—
Laser bias control voltage	—	—	$V_{ss} + 2.1$	V_{IB}	V	$I_{bias} = 50$ mA
Laser modulation control voltage	—	—	$V_{ss} + 2.1$	V_{IP}	V	$I_{mod} = 60$ mA

Maximum Case Temperature

For efficient operation, the device case-to-junction temperature must not exceed a calculated maximum. The maximum case-to-junction temperature (T_{C-J}), is subject to total power dissipation (P_{total}) which in turn is subject to power supply current (I_{ss}), modulation current (I_{mod}) and bias current (I_{bias}), and thermal rise.

Calculate total power dissipation (P_{total}).

$$P_{total} = (V_{ss} \times I_{ss}) + [(V_{ss} - V_{IP}) \times I_{mod}] + V_{IB} \times I_{bias}$$

Given the following values from tables, the total power dissipation can be established.

Supply voltage: $V_{ss} = -5$ V
Supply current: $I_{ss} = 100$ mA
Bias voltage: $V_{IB} = -3$ V
Mod. voltage: $V_{IP} = -3$ V
Bias current: $I_{bias} = 25$ mA
Mod. current: $I_{mod} = 50$ mA

$$P_{total} = (V_{ss} \times I_{ss} + [(V_{ss} - V_{IP}) \times I_{mod}] + V_{IB} \times I_{bias}$$
$$= (5 \times 100 \times 10^{-3}) + \lfloor (5 - 3) \times 50 \times 10^{-3} \rfloor$$
$$+ 3 \times 25 \times 10^{-3}$$
$$= 675 \text{ mW}$$

Therefore, $P_{total} = 675$ mW.

Calculate thermal rise.
Thermal rise is expressed as the product of junction-to-case thermal resistance $(R_{th(j-c)})$ and total power dissipation (P_{total}). See Equation (6–38).

$$\text{thermal rise} = R_{th(J-c)} \times P_{total} \qquad \textbf{(6–38)}$$

The junction-to-case thermal resistance for a plastic case is typically 15°C/W and for a metal or glass case is typically 32° C/W.

For a plastic case, the thermal rise is equal to $R_{th(J-c)} \times P_{total} = 15 \times 675$ mW = 10.1°C.
The maximum case temperature for a plastic package is equal to 125°C – 10.1°C = 114.9°C.
Therefore, case temperature (max) = 114.9°C.

For a metal or glass case, the thermal rise is equal to $\Theta_{J-C} \times P_{total} = 32°C/W \times 675$ mW = 26.1°C.
The maximum case temperature for a metal or glass package is equal to 125°C – 26.1°C = 98.9°C
Therefore, case temperature (max) = 98.9°C.

6.11 EXTERNALLY MODULATED LASER DIODES

The main building blocks incorporated into indirectly modulated optical transmitters are the multiplexer, the optical modulator, the laser diode, and the EDFA. These components are dealt with in the appropriate chapters. See Chapter 7 for multiplexers, Chapter 8 for modulators, and Chapter 4 for optical sources. This chapter covers optical amplifiers.

6.12 THE EFFECT OF NOISE AND POWER SUPPLY NOISE REJECTION

Introduction

Power supply generated noise and noise rejection mechanisms are critical factors for the normal operation of optical transmitters. There are no established rules for specific plat-

forms that deal with **power supply noise rejection (PSNR).** Calculations and management of power supply generated noise as applied to optical transmitters is based on basic calculations and empirical procedures.

Power Supply Noise Components

Power supply noise can be classified into three groups: **white noise, thermal noise,** and **non-Gaussian (sporadic) noise.**

White and thermal noise are Gaussian in nature. That is, they are random in terms of frequency, phase, and power. In active devices, current surges can be induced by electrostatic induction, and they are inherent to their physical composition. Another form of noise is the delta noise or simultaneous switching noise that results from power supply fluctuations that are generated by multiple output circuit switching. The very fast rise and fall times of high speed signals generate current surges that are capable of completing high inductive paths. Current surges through the inductive paths are sources of electronic noise. Power supply ripple, defined as an unwanted AC voltage of a frequency less than 100 KHz on top of the DC voltage, is not classified as electronic noise because it is not random. In the design of fiber optics systems, thermal noise is dealt with at the circuit (or modal) level, while power supply switching noise generated by surge currents is responsible for impairments that occur at the system level.

The control and confinement of power supply noise voltage to an acceptable level can be achieved by maintaining very low impedance through the noise current path. The value of the desired impedance can be calculated if the noise tolerance of the logic devices employed in the circuit are known, as well as the noise levels generated by the switching speed.

Op-Amp Power Supply Rejection Ratio (PSRR)

The term **power supply rejection ratio (PSRR)** is defined as the change of the input voltage of an operational amplifier resulting from power supply voltage variations, and is expressed by Equations (6–39) and (6–40).

$$\text{PSRR} = \frac{\text{Input} - \text{noise}}{\text{PSnoise}} = \frac{\Delta V_{in}}{\Delta V_{PS}} \qquad \textbf{(6–39)}$$

In dB,

$$\text{PSRR}_{dB} = 20 \log\left(\frac{\Delta V_{in}}{\Delta V_{PS}}\right) \qquad \textbf{(6–40)}$$

Operational amplifiers maintain a power supply rejection ratio (PSRR) between 80 dB and 120 dB and for a specific frequency range. Beyond this frequency, the PSRR decreases by a rate of 20 dB/decade. Because power supply filters limit power supply noise to a very low level, the PSRR applies to noise frequencies below 10 KHz. The schematic diagram of an op-amp PSRR is illustrated in Figure 6–44.

Power Supply Noise Specifications

Power supply noise relevant to fiber optics applications is referred to as the **periodic and random deviation (PARD),**

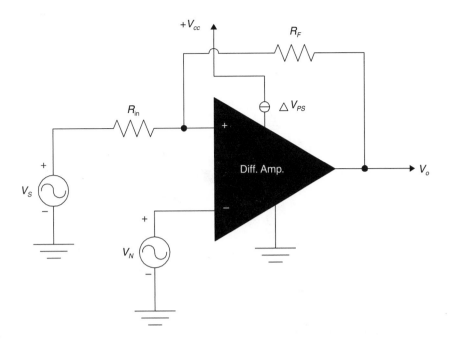

Figure 6–44. An op-amp PSRR.

an undesirable AC component on top of the DC power line. PARD is specified as a peak-to-peak voltage over the 20 Hz to 20 MHz frequency range. Power supply variations below the 20 Hz range are referred to as *drifts*. Power supplies are classified into two major categories: linear and switching.

LINEAR POWER SUPPLIES. With linear power supplies, the total available power is divided between the internal power supply load and the external load. The voltage applied to the external load can be maintained by appropriately adjusting the internal load of the power supply (Figure 6–45).

From Figure 6–45, it is evident that a constant output voltage is maintained through the feedback loop that regulates the power supply internal impedance. Linear power supplies exhibit excellent noise characteristics, and are mainly employed in instruments that are used for RND work.

SWITCHING POWER SUPPLIES. The fundamental difference between a linear power supply and a switching

power supply is that switching power supplies are more efficient. They require smaller heat sink, weigh less, and are more cost effective, but they exhibit higher noise levels. Figure 6–46 illustrates the basic diagram of a switching power supply.

OPERATION OF POWER SUPPLIES. The large value capacitor is charged through the electronic switch, normally a field effect transistor (FET). When the switch opens, the capacitor discharges through the inductance (L). The value of the inductance is selected to be large enough to resist the rapid current change, and to develop a potential difference across it with polarity opposite to V_{DC}. The excess current flows through the forward biased diode, therefore, maintains a constant voltage across the external load. The output of the feedback amplifier is almost a square wave that contains high frequency components. These (undesirable) high frequency components are rejected through special filters, lead length optimization, and

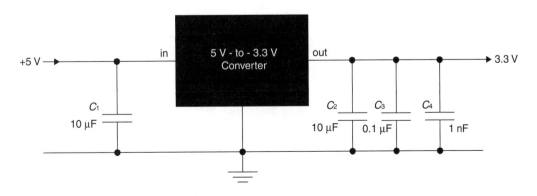

Figure 6–45. Linear power supply.

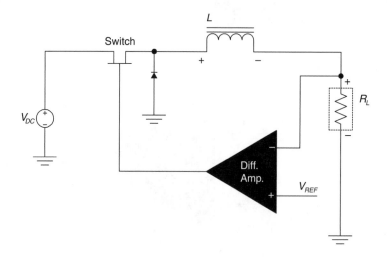

Figure 6–46. A schematic diagram of a switching power supply.

special shielding. The periodic and random deviation of the switching power supply is normally set at a maximum of 50 mV$_{p–p}$.

System Level Noise Suppression

The critical requirement for power supply noise suppression is the control of the noise bandwidth impedance. That is, the lower the impedance, the lower the noise power. The parasitic components that contribute to the generation of the noise power in a power supply and the effect on noise impedance are illustrated in Figure 6–47.

Because noise signals are random, with a frequency band between 20 Hz and several GHz, low system impedance over the after-mentioned frequency range will suppress noise to insignificant levels. The power supply noise level that can be tolerated by the system is subject to power plane impedance (X_{max}), expressed by Equation (6–41).

$$X_{max} = \frac{\Delta V}{\Delta i} \qquad (6–41)$$

where X_{max} is the power plane impedance (Ω), ΔV is the noise voltage (V), and Δi is the **switching current** (A). The maxi-

mum switching current (Δi) is a function of the load parasitic capacitance (C_{load}), noise level (ΔV), and switching time (Δt), and is expressed by Equation (6–42).

$$\Delta i = C_{load} \frac{\Delta V}{\Delta t} \qquad (6–42)$$

From Equation (6–40), it is evident that higher system rates stretch power supply specifications to the limits. The third critical element that establishes power supply noise specifications is the noise voltage in the form of spikes that result from the total **parasitic inductance** (L_{total}), switching current (Δi), and **switching time** (Δt) (Equation (6–43)).

$$\Delta V_{spikes} = L_{total} \frac{\Delta i}{\Delta t} \qquad (6–43)$$

where ΔV_{spikes} is the **parasitic voltage** based on total parasitic inductance, L_{total} is the total parasitic inductance, Δi is the switching current, and Δt is the switching time.

Again, the critical element for power supply noise suppression is the total parasitic impedance. Therefore, control

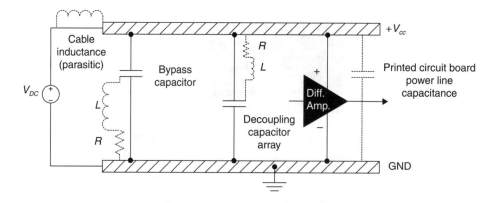

Figure 6–47. Power supply parasitic components.

of the parasitic impedance results in automatic control of the power supply noise. The power supply noise suppression control platform divides the power supply into four sections. Individual control of noise suppression of all four sections reflects the total power supply noise suppression. These four areas are:

a) cable impedance

b) decoupling capacitors

c) bypass capacitors

d) printed circuit board power planes

Power Supply Cable Impedance

The first parasitic element that contributes to the total power supply noise generation is the cable inductive impedance ($X_{L(cable)}$) between the power supply and the printed circuit board (PCB) (Figure 6–48).

The noise voltage generated by the **power supply cable impedance** can be significant for large power supplies used to power high speed optical systems. The noise power contributed to cable impedance is relevant to the noise frequency ($f_{L(cable)}$) and is expressed by Equation (6–44).

$$f_{L(cable)} = \frac{X_{L(cable)}}{2\pi L_{cable}}$$

where

$$X_{L(cable)} = \frac{\Delta V}{\Delta i}$$

or

$$f_{L(cable)} = \left(\frac{\Delta V}{\Delta i}\right)\left(\frac{1}{2\pi L_{cable}}\right) \qquad \textbf{(6–44)}$$

Example 6–13

In an optical fiber transmission system, the power supply maximum generated noise voltage level due to cable parasitic inductance must not exceed 50 mV. Compute the intermediate frequency (IF) noise frequency $f_{L\,(cable)}$ for a power supply switching current of 3 A and cable parasitic inductance of 15 nH.

Solution
Calculate $f_{L\,(cable)}$.

$$f_{L\,(cable)} = \frac{X_{L\,(cable)}}{2\pi L_{cable}} = \left(\frac{\Delta V}{\Delta i}\right)\left(\frac{1}{2\pi L_{cable}}\right)$$

$$= \frac{50 \times 10^{-3}}{3}\left(\frac{1}{2\pi(15 \times 10^{-9})}\right) = 177 \text{ KHz}$$

Therefore, $f_{L\,(cable)} = 177$ KHz.

For noise frequency $f_{noise} \leq f_{L\,(cable)}$, the noise voltage will be maintained to a 50 mV$_{p-p}$ maximum.

Example 6–14

If the maximum IF noise frequency ($f_{L\,(cable)}$) due to the parasitic cable inductance is to be maintained at 120 KHz, calculate the allowed parasitic cable inductance (L_{cable}) of a power supply used in an optical fiber transmission system that requires 4 A of switching current and maintains a maximum noise voltage of 50 mV$_{p-p}$.

Solution
From Equation (6–42) and for the data given, solve for L_{cable}.

$$L_{cable} = \left(\frac{\Delta V}{\Delta i}\right)\left(\frac{1}{2\pi f_{L\,(cable)}}\right)$$

$$= \left(\frac{50 \times 10^{-3}}{4}\right)\left(\frac{1}{2\pi \times 120 \times 10^{3}}\right)$$

$$= 16.5 \text{ nH}$$

Therefore, the maximum allowed parasitic cable inductance (L_{cable}) must not exceed 16.5 nH.

Series Capacitor Reactance

In power supplies used in optical fiber systems, bypass and decoupling capacitors are regularly used to suppress power supply noise by providing very low impedance between the power planes and at the noise bandwidth.

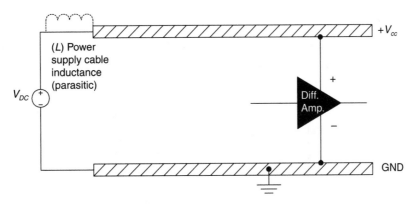

Figure 6–48. Cable impedance.

BYPASS CAPACITORS. Bypass capacitors can suppress noise bandwidths from a few KHz to several MHz by providing low impedance between the power planes and by minimizing (or completely eliminating) the parasitic cable inductance. At low frequencies, bypass capacitors exhibit capacitive reactance. At high frequencies, they exhibit inductive reactance (Figures 6–49 and 6–50).

The noise bandwidth that bypass capacitors can suppress or maintain to a predetermined level is calculated as follows (Equation (6–45).

$$B_{W_{noise}} = f_{H\,(noise)} - f_{L\,(noise)} \qquad \textbf{(6–45)}$$

The high frequency component ($f_{H(noise)}$) of the noise bandwidth is expressed by Equation (6–46).

$$f_{H\,(noise)} = \frac{X_{L(bypass)}}{2\pi L_{bypass}}$$

where $f_{H(noise)}$ is the high frequency component of the noise bandwidth, $X_{L(bypass)}$ is the inductive reactance of the bypass capacitors, and L_{bypass} is the bypass parasitic inductance, or

$$f_{H(noise)} = \frac{X_{L(bypass)}}{2\pi L_{bypass}} = \left(\frac{\Delta V}{\Delta i}\right)\left(\frac{1}{2\pi L_{bypass}}\right) \quad \textbf{(6–46)}$$

where ΔV is the maximum allowed noise voltage (p-p) and Δi is the **system switching current.**

Example 6–15

The number of bypass capacitors used in a power supply to suppress power supply generated noise corresponds to an equivalent 0.5 nH inductance for the high frequency component. If the set noise voltage level is 40 mV$_{p-p}$ and the switching current is 2.5 A, compute the high frequency component ($f_{H(noise)}$) of the noise bandwidth.

Solution

$$f_{H\,(noise)} = \frac{X_{L\,(bypass)}}{2\pi L_{bypass}} = \left(\frac{\Delta V}{\Delta i}\right)\left(\frac{1}{2\pi L_{bypass}}\right)$$

$$= \left(\frac{40 \times 10^{-3}}{2.5}\right)\left(\frac{1}{2\pi(0.5 \times 10^{-9})}\right)$$

$$\cong 5.1 \text{ MHz}$$

Therefore, $f_{H\,(noise)} \cong 5.1$ MHz.

The low frequency component ($f_{L(noise)}$) of the noise bandwidth is expressed by Equation (6–47).

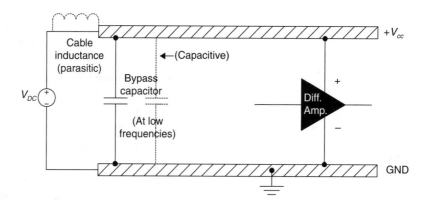

Figure 6–49. Capacitive reactance at low frequencies.

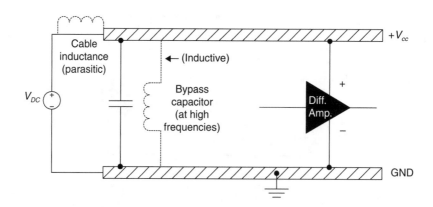

Figure 6–50. Inductive reactance at high frequencies.

$$f_{L\,(noise)} = \frac{1}{2\pi(X_{bypass})C_{bypass}}$$

or

$$f_{L\,(noise)} = \frac{1}{\dfrac{\Delta V}{\Delta i}} \times \frac{1}{2\pi C_{bypass}} \qquad \textbf{(6–47)}$$

Example 6–16
Calculate the low frequency component ($f_{L\,(noise)}$) of the noise bandwidth for a total 220 µF bypass capacitance used for noise voltage suppression in a power supply. The required maximum noise voltage is to be maintained at 50 mV with a switching current of 2.5 A.

Solution

$$f_{L\,(noise)} = \frac{1}{\dfrac{\Delta V}{\Delta i}} \times \frac{1}{2\pi C_{bypass}}$$

$$= \frac{1}{\dfrac{50 \times 10^{-3}}{2.5}} \times \frac{1}{2\pi \times 220 \times 10^{-6}}$$

$$\cong 36\ \text{KHz}$$

Therefore, $f_{L\,(noise)} \cong 36$ KHz.

Example 6–17
Calculate the noise bandwidth ($B_{W_{noise}}$) of a power supply used in an optical fiber system that requires a maximum switching current (Δi) of 4 A. Allow for a maximum noise voltage (ΔV) of 50 mV. The equivalent bypass capacitance (C_{bypass}) is 200 µF reflecting an **equivalent inductance** of 0.2 nH.

Solution

i) Calculate $f_{L\,(noise)}$.

$$f_{L\,(noise)} = \frac{1}{\dfrac{\Delta V}{\Delta i}} \times \frac{1}{2\pi C_{bypass}}$$

$$= \frac{1}{\dfrac{50 \times 10^{-3}}{4}} \times \frac{1}{2\pi \times 220 \times 10^{-6}}$$

$$\cong 67\ \text{KHz}$$

Therefore, $f_{L\,(noise)} \cong 67$ KHz.

ii) Calculate $f_{H(noise)}$.

$$f_{H\,(noise)} = \frac{X_{L\,(bypass)}}{2\pi L_{bypass}} = \left(\frac{\Delta V}{\Delta i}\right)\left(\frac{1}{2\pi L_{bypass}}\right)$$

$$= \left(\frac{50 \times 10^{-3}}{4}\right)\left(\frac{1}{2\pi(0.2 \times 10^{-9})}\right)$$

$$\cong 10\ \text{MHz}$$

Therefore, $f_{H\,(noise)} \cong 10$ MHz.

The effective noise bandwidth for the total bypass capacitance of 200 µF is:

$$B_{W_{noise}} = f_{H\,(noise)} - f_{L\,(noise)}$$

$$= 10\ \text{MHz} - 67\ \text{KHz}$$

$$= 9.93\ \text{MHz}$$

Therefore, $B_{W_{noise}} = 9.93$ MHz.

The bandwidth figured in the preceding example reflects a 50 mV noise level. Any component change that contributes to the increase of the high frequency component ($f_{H(noise)}$) of the noise bandwidth beyond the 10 MHz level will ultimately increase the set 50 mV target noise level.

DECOUPLING CAPACITORS. The main function of decoupling capacitors in a power supply for optical fiber systems is to divert the high frequency noise signal from the V_{cc} line to ground (Figure 6–51).

The effect of decoupling capacitors in noise suppression is relevant to the power supply driver circuit arrangement. For example, there exist two possibilities:

a) when the output of the driver circuit (signal trace) is closer to the V_{cc} line than the ground line

b) when the output of the driver circuit is closer to the ground line than the V_{cc} line

When the driver output trace is closer to V_{cc} than ground, the signal (with very fast rise and fall times) will perceive the

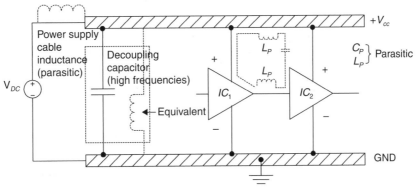

Figure 6–51. Decoupling capacitors for high frequency noise suppression.

connection as a transmission line and the high frequency component will follow the less inductive path. In this case, exactly the same as the V_{cc} line. Similarly, when the driver output trace is closer to ground than to V_{cc}, the output signal of the driver circuit perceives the path between the driver output and the ground also as a transmission line and follows that path, which is less inductive. In both cases, decoupling capacitors provide the less inductive path for the driver output signal. In essence, noise power between V_{cc} and ground is dissipated through decoupling capacitors. When incorporating decoupling capacitors into the power supply circuit design, the following considerations must be taken into account:

a) location of the capacitors

b) number of decoupling capacitors required

c) equivalent source impedance

A **decoupling capacitor array** used for power supply noise suppression is illustrated in Figure 6–52. In a low frequency range, the array behaves as capacitive, while at high frequencies it behaves as inductive. The purpose of decoupling capacitors is to suppress noise signals within the 10 MHz to 100 MHz window. Beyond 100 MHz, noise voltage between V_{cc} and ground is suppressed through the PCB's internal capacitance. The decoupling capacitor effective bandwidth is calculated as follows in Equation (6–48).

$$B_{W_{noise}} = f_{C_{decoupling\,(high)}} - f_{C_{decoupling\,(low)}} \quad \textbf{(6–48)}$$

$f_{C_{decoupling\,(high)}}$ is expressed by Equation (6–49).

$$f_{C_{decoupling\,(high)}} = \frac{X_T}{2\pi L_T} = \left(\frac{\Delta V}{\Delta i}\right)\left(\frac{1}{2\pi L_T}\right) \quad \textbf{(6–49)}$$

where X_T is the total capacitive reactance ($N \times C_{decoupling}$), ΔV is the noise voltage, Δi is the switching current, and L_T is the total inductive reactance.

Example 6–18

Calculate the effective high frequency component of a decoupling capacitor array used for power supply noise suppression. Use the following data:

- Total equivalent inductance: 0.05 nH
- Maximum noise voltage: 40 mV
- Switching current: 3 A

Solution

$$f_{C_{decoupling\,(high)}} = \frac{X_T}{2\pi L_T} = \left(\frac{\Delta V}{\Delta i}\right)\left(\frac{1}{2\pi L_T}\right)$$

$$= \frac{40 \times 10^{-3}}{3}\left(\frac{1}{2\pi \times 0.05 \times 10^{-9}}\right)$$

$$= 42.3 \text{ MHz}$$

Therefore, $f_{C_{decoupling\,(high)}} = 42.3$ MHz.

Similarly, the effective low frequency ($f_{C_{decoupling\,(low)}}$) is expressed by Equation (6–50).

$$f_{C_{decoupling\,(low)}} = \left(\frac{1}{2\pi X_T C_T}\right) = \left(\frac{1}{\frac{\Delta V}{\Delta i}}\right)\left(\frac{1}{2\pi C_T}\right) \quad \textbf{(6–50)}$$

Example 6–19

Calculate the effective low frequency component for a 2.2 µF decoupling capacitor array (pure capacitive reactance) required to maintain a 45 mV$_{p-p}$ noise voltage at a switching current of 2.5 A.

Solution

$$f_{C_{decoupling\,(low)}} = \left(\frac{1}{2\pi X_T C_T}\right) = \left(\frac{1}{\frac{\Delta V}{\Delta i}}\right)\left(\frac{1}{2\pi C_T}\right)$$

$$= \left(\frac{1}{\frac{40 \times 10^{-3}}{2.5}}\right)\left(\frac{1}{2\pi \times 2.2 \times 10^{-6}}\right)$$

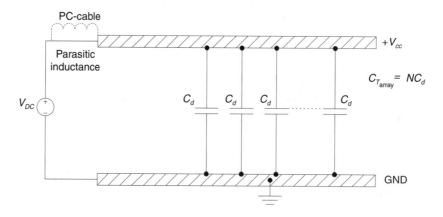

Figure 6–52. A decoupling capacitor array.

The noise bandwidth for the given parameters is:

$$B_{W_{\text{noise}}} = f_{C_{\text{decoupling (high)}}} - f_{C_{\text{decoupling (low)}}}$$

$$= 42.3 \ \text{MHz} - 4 \ \text{MHz}$$

or

$$4 \ \text{MHz} \leq B_{W_{\text{noise}}} \leq 42.3 \ \text{MHz}$$

Noise below 4 MHz and above 42.3 MHz will generate noise levels beyond the set 45 mV, therefore it will be considered unacceptable.

V_{cc} and Ground Lines as Pure Capacitance

In printed circuit board, V_{cc} and ground lines are very close to each other. Increasing the power line area, decreasing the distance between the power lines, or increasing the dielectric constant between the lines can increase the value of the capacitor. Higher capacitance between the power lines is highly desirable in order to decrease the effective lower noise frequency.

V_{cc} and Ground as Transmission Lines

In optical fiber transmission systems, the power lines will behave as transmission lines when the PCB length becomes very close to the noise signal high frequency component and the rise time of the transmitted signal becomes very small (Figure 6–53).

For example, a noise frequency of 0.5 GHz sees the power lines of a 28.5 cm long PCB as transmission lines and can propagate at half wavelength. Power supply noise management is a very complex issue. Several software developers have attempted to address the problem, with various degrees of success, while some circuit designers employ the traditional methods of power supply noise evaluation and management through the use of standard instrumentation.

6.13   UNCOOLED LASER TRANSMITTERS

Uncooled laser transmitters, pioneered by Lucent Technologies, are designed to operate in optical fiber systems and are in compliance with SONET and ITU-T SDH formats. They can be used in systems such as interoffice systems, subscriber loops, metropolitan area networks, and high speed data communications. Some of the most important operating characteristics of uncooled laser transmitters are as follows:

a) they do not require a thermoelectric cooler (small size and less power consumption)

b) operating wavelengths: 1300 nm or 1550 nm

c) input data rates: up to OC48/STM-16 (2.448 Gb/s)

d) optical power output: –11 dBm to 0 dBm

e) ECL differential inputs

f) 5 V single power supply

g) InGaAsP MQW laser technology

h) CMOS based technology

An oversimplified schematic diagram of Lucent Technologies' uncooled laser transmitter is illustrated in Figure 6–54. The basic building blocks of the optical transmitter are:

a) Input data comparator

b) Modulation circuit

c) Band gap reference

d) Automatic control circuitry

e) Laser diode

f) Back-facet detector

The Input Data Comparator

Incoming data is applied to the differential inputs of the comparator stage. Pull down resistors at both inputs of the comparator are used to maintain a reference voltage of 1.3 V below the V_{cc} level. Through this arrangement, the transmitter can be driven with either differential or singe-ended signals. For differential driving, the input voltage levels are not critical because a small differential voltage above a set threshold is required to change the comparator output stage. However, for single ended applications, the input signal must be centered at the V_{cc} –1.3 V level. Through this arrangement, the PWD is very minimal or nonexisting. If the noninverting input is used, the transmitter output optical signal will follow the input data. That is, for a logic high at the input, the laser diode will turn ON and an optical signal will appear at the output of the transmitter. For a logic low at the input, the laser diode will turn OFF, and no optical power will appear at the output of the transmitter. However, if the complementary input is used, the exact opposite will occur.

The Laser Driver

The laser diode is driven by a CMOS based integrated circuit that is capable of providing the reference voltage for the input data signal, for the laser diode bias for the uncooled temperature compensation, and for the modulation current control. The laser diode output optical power is constantly monitored by a **back facet photodetector** diode. A resistor multiplies

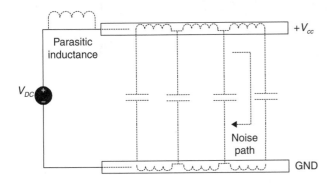

Figure 6–53. V_{cc} and ground planes as transmission lines.

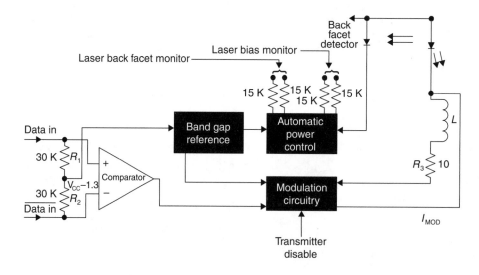

Figure 6–54. A simplified schematic diagram of Lucent Technologies' uncooled laser transmitter.

the photodetector output current and the voltage across it is monitored at the back facet output monitoring pins. The critical laser biasing current for the efficient operation of the transmitter is subject to operating temperature, power supply variations, input data patterns, and laser diode aging. The minimum input data rate for this family of optical transmitters is 1 Mb/s. When designing the transmitter driver circuit interface, special attention is placed on the power supply and data lines. Because the transmitter operates at very high data rates, the input signals perceive the traces as transmission lines and, therefore, they must be kept as short as possible and equal in length when used as a differential pair. For efficient operation of the optical transmitter, the input impedance must be controlled through microstrip or strip line construction.

Connecting the transmitter to power supply is also critical. The transmitter may be connected to either V_{cc} (+5 V) with V_{ee} to ground for positive power supply operation, or V_{ee} (–5 V) and V_{cc} to ground for negative power supply operation. In either case, power supply noise reduction is of

paramount importance. Decoupling and bypass capacitors must be incorporated into the design in order to reduce the power supply generated noise to a level not exceeding 50 mV$_{p-p}$. For a more detailed explanation of power supply noise reduction, see the previous section. Another critical element in the normal operation of the optical transmitter is the maintenance of the transmitter optical output power within a specified range. This optical power output range is subject to power supply operating characteristics and temperature variations, and is set by the manufacturer to a required level under a specific power supply voltage (normally 5 V), and at a specific temperature.

Interfacing the driver circuits to Lucent Technologies' 1241, 1243, and 1245 uncooled optical transmitter family is illustrated as follows: Figure 6–55 illustrates the arrangement for a DC coupled differential input. Figure 6–56 illustrates a DC coupled single ended input. Figure 6–57 illustrates an AC coupled, single ended input.

The electrical characteristics of the uncooled optical transmitter are listed in Table 6–21.

TABLE 6–21   Electrical Characteristics

Parameters	Minimum	Typical	Maximum	Symbol	Units
P. supply voltage	4.75	5.0	5.5	V	V
P. supply current	—	30	130	I_{total}	mA
Input data levels					
high	–1.16	—	–0.88	V_{IH}	V
low	–1.81	—	–1.47	V_{IL}	V
Disable voltage	$V_{cc} - 2.0$	—	V_{cc}	V_D	V
Enable voltage	V_{ee}	—	$V_{ee} + 0.8$	V_{EN}	V
Input transition time	—	$t/4$	—	t_I	ns
Output disable time	—	—	0.2	t_D	μs
Output enable time	—	—	2.0	t_{EN}	μs
Laser diode bias voltage	0.01	0.06	0.70	V_B	V
Laser monitor voltage	0.01	0.05	0.20	V_{BF}	V

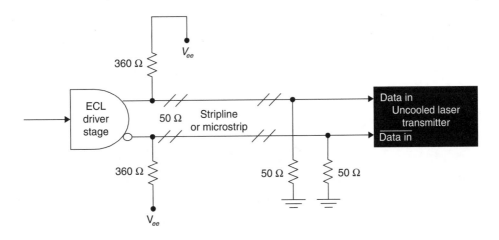

Figure 6–55. DC coupled differential input.

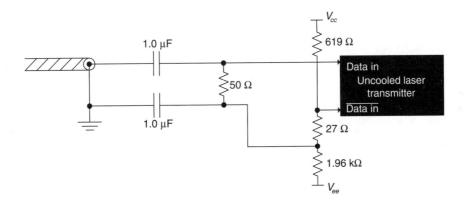

Figure 6–56. AC coupled differential input.

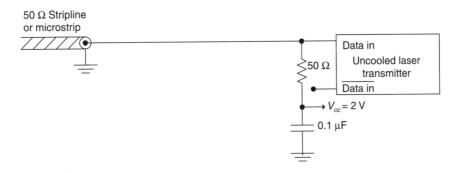

Figure 6–57. DC coupled, single ended input.

6.14   10 Gb/s DWDM OPTICAL TRANSMITTERS

NORTEL has developed a 10 Gb/s DWDM optical transmitter (LCM155EW-64) that incorporates an MQW DFB laser diode and a **Mach Zehnder modulator.** The transmitter module also incorporates a precision thermistor (NTC), thermoelectric heat pumps, and an optical isolator.

Wavelength monitoring and locking to the ITU grid is achieved through a rear facet. Another important feature of this module is the optical output power control. A DC bias voltage applied to the internal attenuator controls the optical power into the modulator while maintaining a constant operating wavelength. A transmitter can be used with **nondispersion shifted fibers (NDSF)** to a maximum distance of 80 km. For dispersion shifted fiber, the link distance can be extended to 400 km. The choice of interfacing the transmitter with either NDSF or DSF is provided by a DC voltage biasing of the Mach Zehnder modulator. The transmitter is configured to provide either positive or negative chirp pulses as required by the appropriately selected optical fiber. The operating characteristics of the LCM155EW-64 transmitter module are listed in Table 6–24.

The transmitter absolute ratings are listed in Table 6–25.

Table 6–22   Optical Characteristics

Parameters	Minimum	Typical	Maximum	Symbol	Units
Power output av.		Various	upon application		
Extinction ratio	10	—	—	r_E	dB
Optical *rise* and *fall* times					
OC-3/STM-1	—	—	1.0	t_R, t_F	ns
OC-12/STM-4	—	—	0.5	t_R, t_F	ns
Fiber Ch.1062.5 Mb/s	—	—	0.37	t_R, t_F	ns
Spectral width	—	—	4	$\Delta\lambda$	nm
Side mode suppression ratio	30	—	—	SMSR	dB

Table 6–23   Absolute Maximum Ratings

Parameters	Minimum	Maximum	Symbol	Units
Supply voltage	—	5.5	—	V
Operating case temperature range	–40	85	T_C	°C
Storage case temperature range	–40	85	T_{str}	°C
Relative humidity	—	85	RH	%
Minimum fiber bend radius	25.4	—	—	mm
Lead soldering temp/time	—	250/10	—	°C/s

TABLE 6–24   LCM155EW-64 Transmitter Operating Characteristics

Parameters	Minimum	Typical	Maximum	Conditions	Units
Threshold current	—	—	40	—	mA
Slope efficiency	3.7	6	—	—	μW/mA
RF input reflection coefficient	10	—	—	to 8 GHz	dB
Laser diode forward voltage	—	—	3.7	—	V
Peak wavelength	1528.77	—	1562.23	—	nm
Side mode suppression ratio	37.5	—	—	—	dB
Optical rise/fall time	—	—	50	20%–80%	ps
Monitor photocurrent	0.15	—	—	at locked-λ	mA
Monitor dark current	—	—	100	—	nA
Thermistor resistance	6.81	—	10.09	at locked-λ	kΩ
Heat pump current	—	—	1.1	T_C 0–70°C	A
Heat pump voltage	—	—	5	T_C 0–70°C	V
M-Z bias voltage (L)	–4	–2	–1.5	at ≤ 12 mA	V
M-Z bias voltage (R)	–5	—	0	at ≤ 12 mA	V
Modulation voltage	3	4	5.5	AC(p-p)	V
Extinction ratio	13	—	—	DC	dB
Attenuator voltage (0–15 dB)	–7.3	—	0.5	at ≤ 30 mA	V
Dispersion penalty	—	—	1	±1500 ps/nm	dB

TABLE 6–25 Absolute Ratings

Parameters	Minimum	Maximum	Units
Operating temperature	0	70	°C
Storage temperature	–50	85	°C
Laser forward current	—	275	mA
Laser reverse voltage	—	2	V
Monitor diode bias	—	15	V
TE-cooler current	—	1.8	A
M-Z bias voltage	–6	0.5	V
Attenuator voltage	–8	0.5	V
Fiber bend radius	30	—	mm

SUMMARY

In Chapter 6, the role of optical transmitters in fiber optics communications systems was defined, followed by a detailed description of the basic building blocks of the design of such transmitters. The chapter continued with a description of the various types of transmitters from different manufacturers such as the VSC8131 2.5 Gb/s multiplexer and clock generator developed by VITESSE, the MC10SX1130 LED driver developed by Motorola, the NE5300 LED driver developed by Nippon Electric, and others by Philips, Lucent Technologies, and Nortel. The intention was to give the reader an opportunity to examine the design philosophies and compare the specifications and operating characteristics of various transmitters intended for specific applications. Finally, the uncooled laser transmitter developed by Lucent Technologies was briefly described.

QUESTIONS

Section 6.1

1. Describe the function of an optical transmitter incorporated into the design of fiber optics communications systems.

2. Draw a block diagram of an optical transmitter and briefly explain the function of each block.

Section 6.2

3. Draw a block diagram of the Philips OQ2535HP multiplexer circuit and list its basic building blocks.

4. Describe briefly the operating function of the OQ2535HP multiplexer circuit.

5. What is the function of the synchronization circuit in the 4 × 8:1 multiplexer?

6. How is power dissipation further reduced in the above device?

7. What are the necessary steps required to maintain proper power supply operation?

8. List the conditions that necessitate heat sinking.

Section 6.3

9. Draw a block diagram of the VSC8131 2.5 Gb/s multiplexer/clock generator device.

10. List the building blocks of the above device and briefly describe the function of each block.

11. Briefly describe the operation of the clock generator circuit.

12. Why is the input reference clock signal PECL?

13. What are the conditions through which the CMU loses lock?

14. Name the conditions under which the power supply operates properly.

Section 6.4

15. Describe the function of an LED circuit.

16. List the fundamental parameters which must be taken into account when designing optical transmitters.

17. Draw the basic LED driver circuit of the MC10SX1130 and briefly explain the function of each block.

18. How can switching noise in an LED driver circuit be reduced?

19. Why is junction temperature of the LED incorporated into a driver circuit problematic for the proper operation of an optical transmitter?

Section 6.5

20. Draw an NE5300 LED driver block diagram and describe the function of each block.

21. What are the critical elements which must be taken into account in order to enhance the performance of the above LED driver circuit?

Section 6.6

22. Draw a block diagram of the Philips 74F5302 LED driver and briefly describe its operation.

23. Why are transmission lines required between the LED and the driver circuit?

24. How can the performance of the optical transmitter be enhanced through impedance matching networks?

25. List the above device basic operating characteristics.

Section 6.7

26. Draw a block diagram of the Philips OQ2545HP laser driver. List its main building blocks and describe, in some detail, the function of each block.

27. What is meant by *tracking error*?

28. How are the time constant and bandwidth of an optical high related? Describe the relationship.

29. What will be required in a case where the chip capacitance of a laser driver circuit is smaller than a set minimum?

30. Describe the difference between electro-optical efficiency and laser diode responsivity.

31. Express the relationship between case power dissipation and thermal resistivity of a laser diode driver circuit.

32. List the parameters incorporated in the determination of the maximum power dissipation of a laser driver circuit.

Section 6.8

33. Describe the basic operating characteristics of the Lucent Technologies LG1627BXC laser diode driver circuit.

Section 6.9

34. Describe the basic operating characteristics of the NORTEL YA08 laser driver circuit.

35. List the main building blocks of the above device and briefly explain their functions.

36. What is the reference voltage in this device?

37. Why is a biasing current monitoring mechanism essential in such a device?

38. How is this biasing current monitoring mechanism accomplished?

39. Describe the function and operation of the mean power controller circuit.

40. Give the relationship establishing the value of the external resistor connected between PINSET and GND, and which is used for bias current control.

Section 6.10

41. Why is the implementation of differential signaling desirable in high speed optical systems?

42. List the causes for drifts in the reference voltage in an AC coupling differential stage.

43. Describe, in detail, the principle of operation of common mode arrangement.

44. Derive the differential mode impedance.

45. In a differential pair terminator, how can termination noise be reduced to a minimum? Explain with the assistance of a diagram.

46. What is the advantage of using parallel termination?

47. List the advantages and disadvantages of differential signaling in high speed optical systems.

Section 6.11

48. List the main building blocks that compose indirectly modulated optical transmitters.

49. Name the two critical factors necessary for the normal operation of optical transmitters.

50. Classify and then define power supply noise components and their impact on optical transmitters.

51. Define the operational amplifier PSRR.

52. Name the two classes of power supplies and describe their differences and similarities.

53. How can power supply noise suppression be achieved? Explain.

54. Define power supply cable impedance.

55. Give the relationship between cable impedance and noise frequency.

56. What is the role of bypass and decoupling capacitors in a power supply configuration?

Section 6.12

57. List the possible applications of Lucent Technologies' uncooled laser transmitters.

58. What are the most important operating characteristics of uncooled laser transmitters?

59. With the assistance of a block diagram, describe the operation of an uncooled laser transmitter.

1. A 32:1, 2.5 Gb/s multiplexer, similar to the Philips OQ2535HP, operates with the following parameters: junction temperature: 130°C, ambient temperature: 55°C, and total power dissipation: 1.5 W. Assuming a junction to ambient thermal resistance of 38 K/W, determine whether or not the device will require a heat sink.

2. Similar to problem 1, for an ambient temperature of 85.3°C and total power dissipation of 2.2 W, determine whether or not the device will require a heat sink, assuming a junction to ambient thermal resistance of 32 K/W. If the answer is affirmative, calculate the thermal resistance between heat sink to ambient, given the following data:

   $R_{thc-H} = 0.6$ K/W, $R_{thJ-a} = 29.5$ K/W, and $R_{thJ-c} = 3.3$ K/W

3. An LED device is operating with the following parameters: $V_{cc} = 5$ V, $I_{cc} = 20$ mA, $V_{SET} = 0.7$ V, $V_F = 1.25$ V, $I_{mod} = 10$ mA, and $T_A = 28$°C. Compute the junction temperature (T_J).

4. If the circuit in Figure 6–11 is used to drive an LED that operates with a forward ON voltage of 1.8 V, forward ON current of 100 mA, and maximum optical output power of 1.5 mW, calculate the biasing components R_1, R_2, R_3, and C_1.

5. Calculate the reference current and external bias resistor of the laser driver circuit (in Figure 6–20) if the generated optical power by the laser diode is 0.5 mW and the extinction ratio is 12 dB. Assume a monitor photocurrent of 200 µA for an optical low, and 20 µA for an optical high.

6. Calculate the bandwidth (B_W) of a laser diode for optical-one and optical-zero, when operating above the threshold level with a threshold current of 25 mA that is capable of generating an optical power of 1.5 mW. Assume laser diode responsivity of 350 mA/W, electro-optical efficiency of 20 mA/W, external capacitor for logic-one 10 pF, and logic-zero 2.2 pF.

7. Refer to Philips TZA3031AHL, TZA3013BHL, and TZ3031U laser drivers (page 159). If the reference currents required to limit the bias current are set between 50 mA and 7 mA, calculate the values of the external resistors and the alarm currents for both an optical low and an optical high.

8. Refer to Section 6-7. A laser driver circuit is designed to provide a bias current of 20 mA. If the laser diode operates with a current of 25 mA, a series resistance of 80 Ω, extinction ratio of 15 dB, and an optical output power of 0.3 mW, calculate the total power dissipation of the driver circuit.

7

OPTICAL RECEIVERS

OBJECTIVES

1. Define the role of the receiver in optical fiber communications systems
2. List and briefly explain the different data patterns that optical receivers are capable of detecting
3. Name the two types of photodetector diodes
4. Briefly explain the fundamental differences between a PIN and an APD photodetector diode
5. Name the three categories of optical receivers and briefly explain their operating characteristics
6. Describe the function of a transimpedance amplifier when it is incorporated into an optical receiver
7. Explain in detail the operation of NORTEL AB52 155 Mb/s, AB53 622 Mb/s, and Philips TZ3032 transimpedance amplifiers
8. Define the role of the postamplifier in an optical receiver
9. Describe the function and operating characteristics of Philips TZ3044B 1.25 Gb/s, Nortel AC10AGC, Motorola MC10SX1127 2.5 Gb/s post amplifiers
10. Describe the function and operating characteristics of Philips OQ2541HP, OQ2514U, Lucent Technologies LG1600KXH, and Nortel YA18 data and clock recovery circuits

KEY TERMS

AC coupled circuit
ambient temperature
amplifier front end
amplifier sensitivity
auto zero function
automatic gain control (AGC)
avalanche effect
avalanche photodetector
band gap energy

band gap reference
biphase data encoding
breakdown voltage
carrier number
clock and data recovery
cut off wavelength
damping ratio
dark current
DC biasing
DC offset

depletion region
dielectric constant
differential input voltage
differential output
differential transimpedance
doping density
dynamic range
ECL buffer
edge detection
error probability

eye diagram
fall time
fixed threshold
input buffer
junction voltage
limiting amplifier
linearity
Manchester encoding pattern (NRZi)
mean square avalanche gain

7.1 INTRODUCTION

The function of an optical receiver is to convert modulated optical power to a binary data stream. A block diagram of a typical optical receiver is illustrated in Figure 7–1.

The optical receiver in Figure 7–1 is composed of a PIN photodetector diode, a transimpedance amplifier, an AGC post amplifier, a clock and data recovery, and a demultiplexer circuit. The function and operating characteristics of these circuits within the optical receiver will be described in some detail in the following sections. First,

however, a brief review of the available data patterns and their characteristics is necessary.

7.2 DATA PATTERNS

The Return to Zero Data Pattern

A binary stream of data is composed of logic 1s and logic 0s. In a **return to zero data pattern (RZ)** (Figure 7–2), for logic 1, the pulse remains high for fifty percent of the bit cell then falls back to zero. For logic 0, the pulse remains at zero level

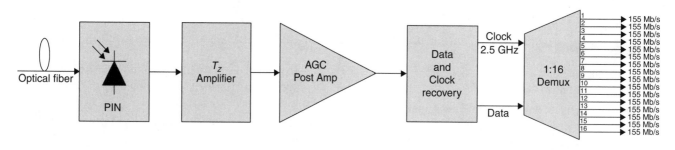

Figure 7–1. A block diagram of a 2.5 Gb/s optical receiver.

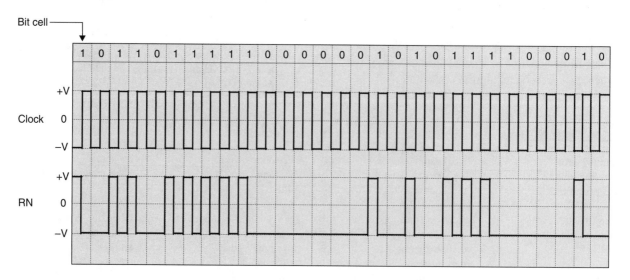

Figure 7–2. A return to zero data pattern.

for the entire bit cell. Because of the presence of long strings of ones and zeros, this represents a definite disadvantage if the system is designed to operate in a synchronous mode (incorporating **clock recovery**). For an **AC coupled circuit**, long strings of ones and zeros can be perceived as DC levels instead of important data, and may therefore be rejected as input offsets. This leads directly to an increase of the system BER.

The Non Return to Zero Data Pattern

In a **non return to zero (NRZ) data pattern,** the logic 1 occupies the entire positive part of the bit cell. Likewise, the logic 0 occupies the entire negative part of the bit cell (Figure 7–3). It is, therefore, evident that the data capacity of a system that employs an NRZ pattern is twice that of the available bandwidth.

This data pattern also presents problems for AC coupled circuits as a result of the long strings of ones and zeros. For both RZ and NRZ data patterns, clock recovery is very difficult to implement. This problem generated the development of a third data pattern, that of the Manchester data pattern.

The Manchester Data Pattern

Manchester, or **biphase, data encoding** is formatted by changes that always occur at the middle of each data cell (Figure 7–4). The logic 1 is represented by a pulse transition from low to high, and the logic 0 is represented by a pulse transition from high to low.

When comparing the three data encoding waveforms, it is evident that the **Manchester encoding pattern (NRZi)** is characterized by frequent pulse transitions from high to low and low to high. This frequent transition allows the system to successfully incorporate clock recovery, a necessary function for synchronous data transmission systems.

The advantage of Manchester encoding is somewhat offset by the higher baseband bandwidth requirements. Through this encoding method, only one half of the data can be transmitted in comparison to RZ and NRZ encoding formats. The question may be asked: Why is AC coupling required? Of course, AC coupling would not have been necessary if all the components of the system had the same **DC biasing.** A common DC biasing requirement for all the components deprives the system engineers of the flexibility of incorporating **preamplifiers** and **postamplifiers** with different DC biasing levels into the optical system. To illustrate the need for AC coupling, we must superimpose the DC average levels on the three encoding schemes previously described (Figure 7–5).

Figure 7–5 illustrates the average DC levels that may be considered significant. RZ and NRZ encoding schemes are capable of inducing problems in the optical system. In an AC coupled receiver, the DC level may cause the switching threshold to move closer to, or away from, the ideal transition point.

If the DC level is away from the ideal transition point, it will require additional DC voltage to reach logic 1. This additional DC voltage requirement translates into a significant reduction of the amplifier sensitivity. On the other hand, if the DC voltage level is closer to logic 1, then a small level of noise voltage could be sufficient to induce a bit error, as a consequence of a lower receiver **signal to noise ratio (SNR).**

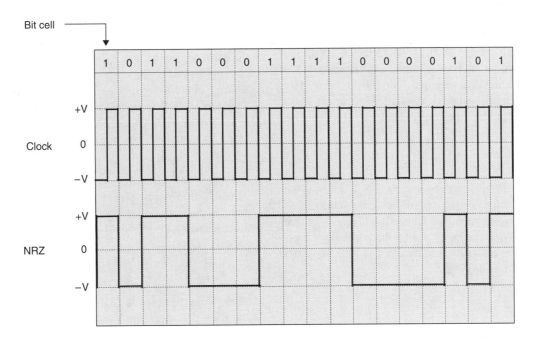

Figure 7–3. A non return to zero data pattern.

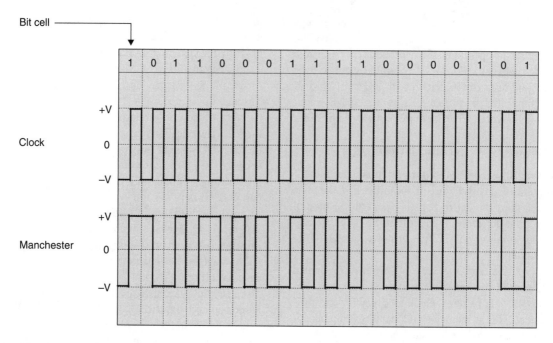

Figure 7–4. Manchester data encoding pattern.

In the Manchester encoding format, the average DC level is very close to zero, thus making this encoding format the ideal candidate for AC coupling.

7.3 PHOTODETECTOR DIODES

Two types of photodetector diodes exist: PIN photodetector diodes and APD photodetector diodes.

PIN Photodetectors

The function of a **PIN semiconductor diode** is to convert the incident photon energy at the photodetector input into electric current. PIN photodetector characteristics are: **quantum efficiency** (η), **responsivity** (R), speed, **linearity, cut off wavelength,** and **dark current.** The cut off wavelength is expressed by Equation (7–1).

$$\lambda_c = \frac{1.24}{E_g} \qquad (7\text{--}1)$$

where λ_c is the cut off wavelength (m) and E_g is the band gap energy (eV). Band gap energy is subject to the semiconductor material used in the device fabrication (Table 7–1). The generated photocurrent from the PIN diode develops a potential difference across the load resistance with a frequency expressed by Equation (7–2).

$$f = \frac{E_{ph}}{h} \qquad (7\text{--}2)$$

where f is the operating frequency (Hz), h is **Planck's constant** ($6.62 \times 10^{-34}\,\text{Ws}^2$), and E_{ph} is the **photon energy** (eV).

The **response time** of a PIN photodiode is referred to as the time required by the generated carriers to travel the absorption region under reverse bias conditions. The response time (t_r) is expressed as the ratio of the thickness of the absorption layer to the saturation velocity. The device responsivity is defined by the ratio of the generated current in the absorption region to the optical power incident to the region, and is expressed by Equation (7–3).

$$R = \eta \frac{q}{E_{ph}} \qquad (7\text{--}3)$$

where R is the responsivity, η is the quantum efficiency, q is the electron charge, and E_{ph} is the photon energy**.** The dark current of a PIN photodetector diode is defined as the **reverse leakage current,** in the absence of optical power, impeding upon the diode. This undesirable current contributes directly to an increase of shot noise. The noise power that is generated, based on shot noise, is expressed by Equation (7–4).

$$P_N = 2I_d q \qquad (7\text{--}4)$$

where P_N is the **noise power** (W), I_d is the dark current (A), and q is the electron charge (1.59×10^{-19} C). Likewise, the noise voltage of a PIN photodetector diode is expressed by Equation (7–5).

$$V_N = 2I_d (B_W) \qquad (7\text{--}5)$$

where V_N is the **noise voltage** (V), I_d is the dark current (A), and B_W is the **operating bandwidth** (Hz). It is evident from Equation (7–5) that the noise voltage is a limiting factor of the receiver operating bandwidth.

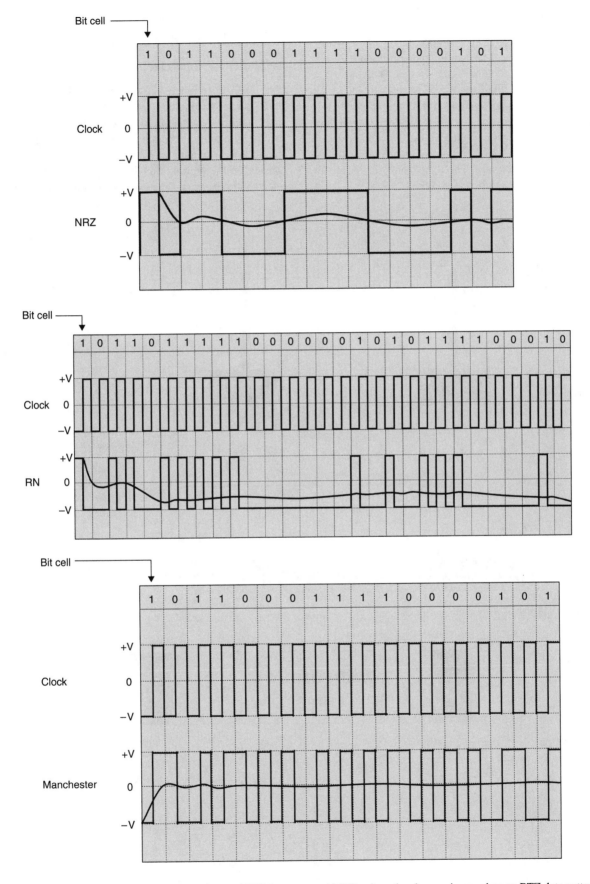

Figure 7–5. a) DC voltage levels superimposed on an NRZ data pattern. b) DC voltage levels superimposed on an RTZ data pattern. c) DC voltage levels superimposed on a Manchester data pattern.

PIN Photodetector Devices

A typical PIN photodetector diode, intended for long distance SONET applications, is illustrated in Figure 7–6.

The PIN photodiode in Figure 7–6 may be coupled to a pigtail (multimode fiber), and because it is intended for SONET long distance applications, it must exhibit very low capacitance and high responsivity. The biasing schematic of this device is illustrated in Figure 7–7.

The photodiode maximum responsivity is achieved under minimum capacitance. The equivalent AC circuit for the capacitance model is illustrated in Figure 7–8.

Finally, a typical frequency response of a PIN photodiode is illustrated in Figure 7–9.

Avalanche Photodetectors

The basic difference between **avalanche photodetectors (APD)** and PIN photodetectors is that APD devices require a high intensity electric field region. This high intensity electric field region is necessary in order to provide the incident photons with enough kinetic energy to collide with the atoms present in the region, thus allowing the generation of more electron-hole pairs. The process whereby more than one

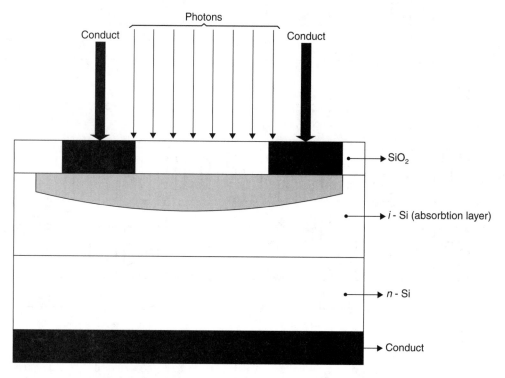

Figure 7–6. A PIN photodetector schematic diagram.

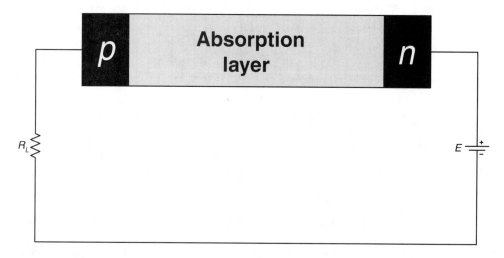

Figure 7–7. PIN photodiode biasing.

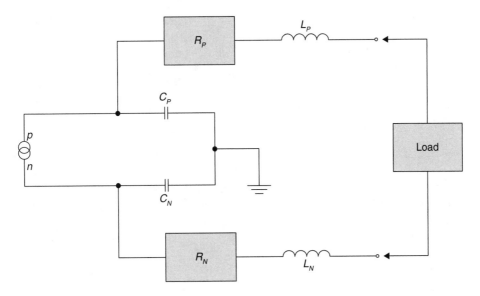

Figure 7–8. PIN photodiode equivalent AC circuit for digital applications.

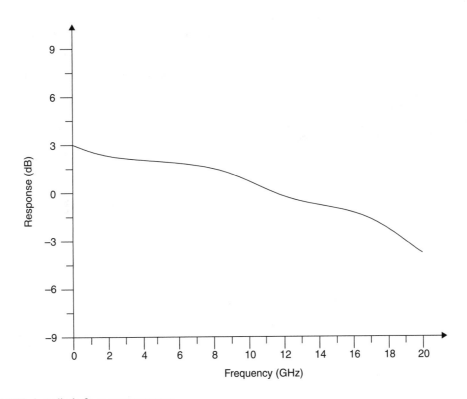

Figure 7–9. Typical PIN photodiode frequency response.

electron-hole pair is generated from a single incident photon is referred to as the **avalanche effect.** The photocurrent gain of an APD is a function of the wavelength of the incident photon, the electric field strength, the width of the **depletion region,** and the type of semiconductor materials used in the fabrication of the APD diode. The current generated by an APD device is expressed by Equation (7–6).

$$I_p = qN_{e^-}M \qquad (7\text{–}6)$$

where I_p = generated **photocurrent** (A), q = electron charge $(1.59 \times 10^{-19}$ C$)$, N_{e^-} = **carrier number,** and M = **multiplication** factor. The multiplication factor (M) depends on the physical and operating characteristics of the photodetector diode. In general, the main operating characteristics of an APD photodetector diode are:

a) photocurrent

b) gain

c) noise

d) dark current

e) response time

f) capacitance

The current gain or multiplication factor (M) of an APD device is expressed by Equation (7–7).

$$M = \frac{1}{\left[\dfrac{V_R - I_P R_L}{V_{BK}[1 + a(T_1 - T_0)]}\right]} \quad (7–7)$$

where M is the multiplication factor, I_P is the photocurrent, a is the factor defined from gain $v.$ temperature graph, T_1 is the **operating temperature,** T_0 is the **ambient temperature,** V_{BK} is the **breakdown voltage** at room temperature, R_L is the device resistance plus load resistance, and V_R is the **reverse bias voltage.**

PHOTODETECTOR NOISE.   The electronic noise of an APD photodetector diode is the result of the ionization and photocurrent multiplication process within the APD device, and is expressed by Equation (7–8).

$$\frac{d(i_p)^2}{df} = 2qI(M)^2 \quad (7–8)$$

where $(i_p)^2$ is the **mean square spectral density,** q is the electron charge, f is the operating frequency, $(M)^2$ is the **mean square avalanche gain,** and I is the **primary photocurrent** ($I = I_{BR} + I_d$).

DARK CURRENT.   The dark current of an ADP photodetector diode is defined as the current present at the device output in the absence of incident light.

RESPONSE TIME.   Response time is the time a carrier takes to cross the depletion region. Response time is a function of quantum efficiency. A typical response time for an APD photodiode is approximately 0.5 ns, at operating wavelengths of 800 nm to 900 nm.

CAPACITANCE.   The internal parasitic capacitance of an APD photodiode significantly affects the overall device response time. This parasitic capacitance is expressed by Equation (7–9).

$$C = \frac{\varepsilon q\, AN}{2(V_R + V_j)} \quad (7–9)$$

where C is the **parasitic capacitance,** ε is the **dielectric constant,** A is the depletion area, N is the **doping density,** V_j is the junction voltage, V_R is the reverse bias voltage, and q is the electron charge.

APD Photodetector Devices

APDs can operate at either 1310 nm or 1550 nm wavelength windows, and are capable of handling data rates of about 2.5 Gb/s. Fabricated by **vapor phase epitaxy (VPE)** and planar structure, they exhibit excellent sensitivity and a high degree of reliability. The performance characteristics of a typical APD photodetector device is illustrated by Figure 7–10, which illustrates a typical diode response as a function of frequency.

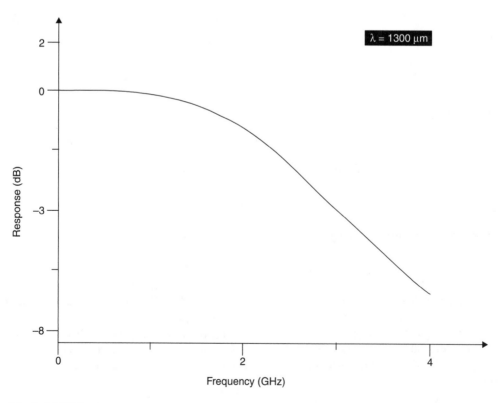

Figure 7–10.   A typical APD frequency response.

Figure 7–10 shows that the photodiode response progressively decreases with an increase in the frequency. Figure 7–11 illustrates the diode responsivity in relation to operating wavelength.

Figure 7–11 shows that diode responsivity remains almost constant at a specific wavelength range but experiences an almost vertical drop outside of the range. Figure 7–12 illustrates the relationship between dark current and reverse bias voltage.

Figure 7–12 shows the almost linear increase of the dark current with a progressive increase of the reverse bias voltage up to a certain level. Beyond that level, avalanche breakdown occurs. The reverse bias voltage level, at which avalanche breakdown takes place, is also temperature dependent. Figure 7–13 illustrates the dark current as a function of reverse bias voltage at different operating temperatures.

From Figure 7–13, the dark current increases progressively with corresponding increases in the operating temperature, while the reverse bias voltage remains constant. Finally, Figure 7–14 presents a graphical comparison of the respective sensitivities of a PIN photodetector diode and an APD diode, both operating at 1300 nm and at bit rates of 1.7 Gb/s.

7.4 CLASSIFICATION OF OPTICAL RECEIVERS

Optical receivers, based on their operating characteristics, can be classified into three main categories:

a) **fixed threshold**

b) **edge detection**

c) **automatic gain control (AGC)**

Fixed Threshold Optical Receivers

Perhaps the simplest of all optical receivers is the fixed threshold receiver (Figure 7–15).

In a fixed threshold optical receiver, the generated current from the PIN photodetector diode is fed into the preamplifier, then into the noninverting input of the comparator circuit, at which point it is compared with the data threshold voltage across R_2. If the voltage at the output of the preamplifier is higher than the data threshold level, a high voltage level will appear at the output of the data comparator. Fixed threshold optical receivers exhibit low performance at high signal levels. The bit detection of the receiver must be set at 50% of the peak signal amplitude. This is required in

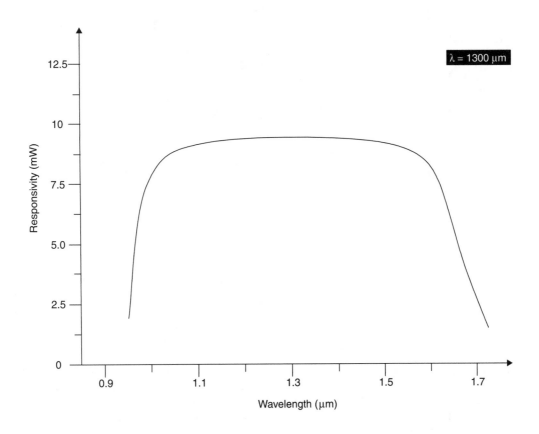

Figure 7–11. APD responsivity *v.* operating wavelength.

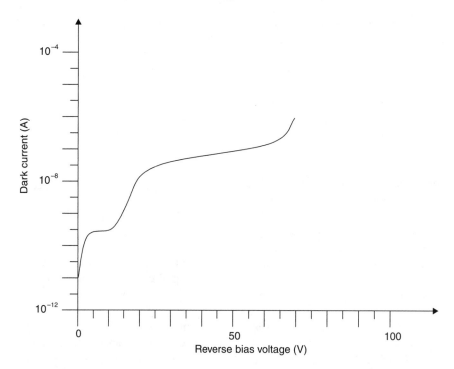

Figure 7–12. APD relationship between dark current and reverse bias voltage.

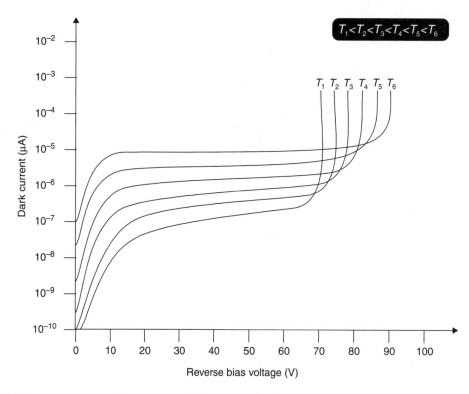

Figure 7–13. APD dark current *v.* reverse bias voltage at different operating temperatures.

order to maintain a constant **signal to noise ratio (SNR)** for both high bit and low bit levels, and also to maintain the least possible **pulse width distortion (PWD)**. If the bit detection is not set at 50% of the peak, a significant PWD will occur, affecting the clock recovery process and directly contributing to an increase in the system BER. The main advantage of the fixed threshold detector is based on the fact that it can receive data of any pulse duration. That is, the receiver is capable of detecting data streams of variable duty cycles. The pulse width distortion is defined as the difference between the transmitted pulse width and the received pulse width, and is expressed by Equation (7–10).

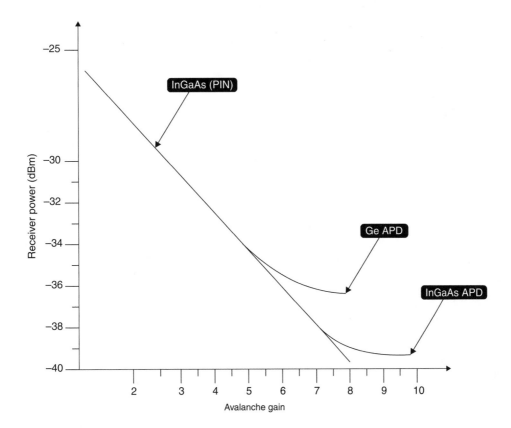

Figure 7–14. PIN and APD receiver sensitivity comparison.

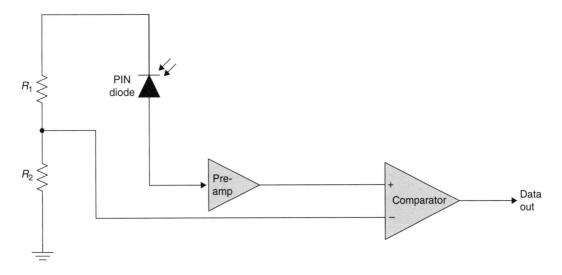

Figure 7–15. A fixed threshold optical receiver.

$$\text{PWD} = \frac{(t_{rec} - t_{trans})}{t_{trans}} \times 100 \qquad \textbf{(7–10)}$$

where PWD is the pulse width distortion, t_{trans} is the transmitted pulse width, and t_{rec} is the received pulse width.

Example 7–1

Compute the PWD of a fixed threshold optical receiver that has a transmitted pulse width of 40 ns and a receiver pulse width of 54 ns.

Solution

$$\text{PWD} = \frac{(t_{rec} - t_{trans})}{t_{trans}} \times 100 = \frac{54-40}{40} \times 100 = 35\%$$

Therefore, PWD=35%.

Edge Detection Optical Receivers

An improved version of the fixed threshold receiver is the edge detection receiver (Figure 7–16).

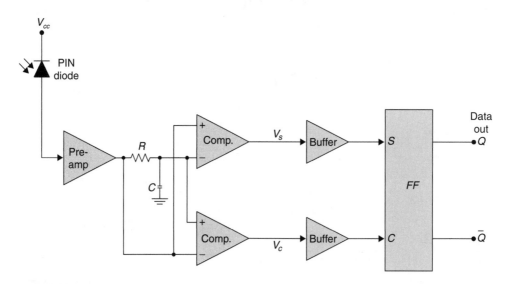

Figure 7–16. An edge detector optical receiver.

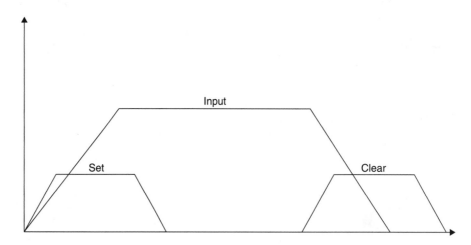

Figure 7–17. A graphical representation of the input/output signal of an edge detection optical receiver.

The edge detection optical receiver in Figure 7–16 is composed of a preamplifier circuit, a differentiator connected at the output of the preamplifier, a positive and a negative comparator, and a set clear *FF*.

OPERATION OF EDGE DETECTION OPTICAL RECEIVERS. The current generated by the PIN photodetector is converted to a voltage level, differentiated, and then fed into the comparator circuit. The edge of the differentiated signal is compared with the fixed threshold detector for a positive or negative slope. The detected edge is then fed into the SET or CLEAR, and the transmitted data is recovered from the Q output of the *FF*. Figure 7–17 graphically illustrates the pulse shape between the input and the output signals of an edge detection optical receiver. From the graph, it is evident that the output pulse is identical to the input pulse, indicating a complete elimination of PWD. Under ideal conditions, that is, for zero PWD, the low to high and high to low transitions must be identical (symmetrical). In practice, slight asymetricity is always present, and it

induces a proportional percentage of PWD. (This is very small in comparison to fixed threshold detection receivers.)

AGC Optical Receivers

The third, and perhaps most efficient, optical receiver is the AGC (Figure 7–18). AGC optical receivers with a data detection threshold set at 50% peak to peak, exhibit minimum error probability and the lowest percentage of pulse width distortion. The only disadvantage is that they are more complex and, consequently, more expensive receivers. A detailed description of the AGC optical receivers of different manufacturers will be presented later in the chapter.

OPERATION OF AGC OPTICAL RECEIVERS. The output of the variable gain amplifier is fed into the input of the buffer and also into the level detector. An increase of the output signal will proportionally increase the voltage drop across R_{AGC} with a corresponding decrease of the differential signal at the input of the amplifier. Consequently, the output

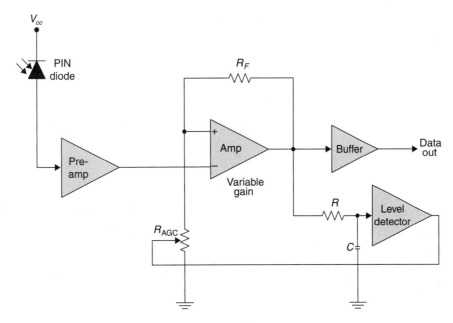

Figure 7–18. An AGC optical receiver.

signal will be reduced to its predetermined level. For proper operation of the receiver, certain conditions must be met. For example, 50% duty cycle is essential in order to ensure that updated information is continuously applied to the receiver.

Furthermore, the packet preamble time must be long enough to allow receiver adjustment to take place for different signal levels. This is imperative because the AGC response time of the amplifier is relatively slow.

7.5 TRANSIMPEDANCE AMPLIFIERS

The function of a **transimpedance amplifier** is to convert the generated current from a photodetector diode into a voltage signal with a level that is capable of driving an AGC postamplifier circuit.

NORTEL AB52 155 Mb/s and AB53 622 Mb/s Devices

A block diagram of a transimpedance amplifier is illustrated in Figure 7–19. The block diagram and the description of this transimpedance amplifier are based on NORTEL AB52 155 Mb/s and AB53 622 Mb/s devices.

The transimpedance amplifier in Figure 7–19 is composed of a front end stage, a driver stage, an AGC stage, and a power supply rejection stage.

OPERATION OF THE NORTEL DEVICE. The generated photocurrent from the PIN or APD photodetector diodes is fed into the input of the transimpedance amplifier front end. The level of this current must not exceed 2.6 mA peak to peak. The input circuit of the front end stage is designed for minimum noise at 15 Mb/s for the AB52 device and

622 Mb/s for the AB53 device. However, in order to further enhance noise performance, separate power pins are used for the front end stage. A dynamic feedback resistor is used to maintain the required constant output voltage.

THE OUTPUT DRIVER STAGE. The function of the output driver stage is to develop a voltage swing of 150 mV peak to peak into a 50 Ω load, or a 300 mV peak to peak differential into a 100 Ω load. This relatively small voltage at the output of the driver stage is the required voltage at the input of an AGC postamplifier stage. The driver output could be either single ended or double ended. When single ended, a positive light at the input of the photodetector will generate a positive voltage at the output of the driver circuit. However, for dual ended operation, a positive light at the input of the photodetector diode will also produce a negative voltage at the output of the driver stage.

THE AUTOMATIC GAIN CONTROL STAGE. The function of the automatic control stage is to constantly monitor the voltage levels at the driver output and to compare them with an internally generated reference voltage. If the voltage at the output of the driver exceeds the internal reference voltage of the AGC control stage, the value of the feedback resistor (R_f) of the front end is automatically reduced through a long time integrated circuit that confines the driver output signal to its predetermined level.

THE POWER SUPPLY REJECTION STAGE. An always-present problem which is capable of reducing system performance is the electronic noise induced by power supply variations. To eliminate the problem, a power supply rejection circuit has been incorporated into the module. This circuit is design to achieve an AC and DC **power supply rejection ratio (PSRR)** of around –40 dB for frequency variations up to

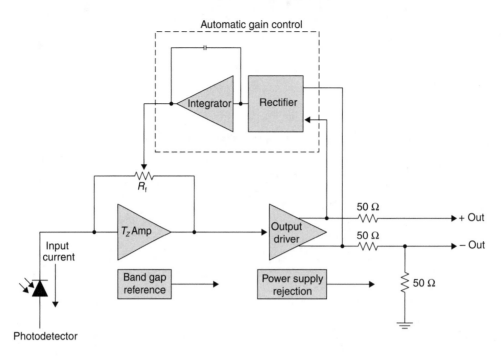

Figure 7–19. A block diagram of the NORTEL-AB52/53 transimpedance amplifier. Reproduced with the permission of Nortel Networks.

TABLE 7–1   Electrical Characteristics (AC and DC)

Parameters	Minimum	Typical	Maximum	Symbol	Unit
Supply current	—	39	58	I_S	mA
Input bias voltage	1.5	2.0	2.6	V_{in}	V
AGC threshold current	4	—	—	$I_{AGC(Th)}$	μA_{p-p}
Transimpedance	11	15	20	T_Z	$k\Omega$
Output bias voltage	2.9	—	3.5	V_{out}	mV
Output resistance	35	50	65	R_{out}	Ω
Bandwidth	150	—	—	$B_W(-3dB)$	MHz
Optical sensitivity	—	−41	—	$P_{opt_{sens}}$	dBm

100 KHz. If parasitic signals of a higher frequency are detected, then an external decoupling circuit will also be required.

The electrical characteristics of a typical transimpedance amplifier are listed in Table 7–1.

The absolute maximum ratings are listed in Table 7–2.

TABLE 7–2   Absolute Maximum Ratings

Parameters	Minimum	Maximum	Symbol	Units
Supply voltage	−0.7	7.0	V_{cc}	V
Junction temperature	−40	120	T_J	°C
Storage temperature	−65	150	T_{STG}	°C

The recommended operating conditions are listed in Table 7–3.

The Philips TZ3023 Transimpedance Amplifier

The transimpedance amplifier in Figure 7–20 is based on the Philips TZ3023 device intended for SONET/SDH, STM4/OC12 digital applications. Some of the most important operating characteristics of the above device are listed as follows:

Operating Characteristics

a) dynamic range: 1 μA

b) **differential transimpedance**: 21 kΩ

TABLE 7–3   Recommended Operating Conditions

Parameters	Minimum	Typical	Maximum	Symbol	Units
Supply voltage	4.7	5.0	5.3	V_{cc}	V
Input current (before overload)	2.6	—	—	I_{in}	mA
Junction temperature	−40	—	95	T_j	°C

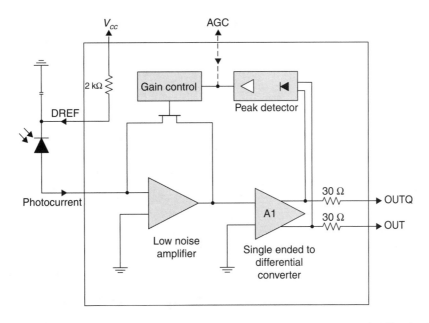

Figure 7–20. A SONET/SDH, STM4/OC12 transimpedance amplifier. Reproduced by permission of Philips Semiconductors.

c) equivalent input noise: $3.5 \text{ pA}/\sqrt{\text{Hz}}$

d) bandwidth: DC—600 MHz

e) on-chip AGC

f) differential output

g) supply voltage: 3.0 V–5.5 V

Applications

a) SONET/SDH

b) STM4/OC-12

c) short, medium, and long haul optical communications systems

OPERATION OF THE PHILIPS TZ3023 TRANSIMPED-
ANCE AMPLIFIER. The transimpedance amplifier in Figure 7–20 exhibits a wide **dynamic range** as a result of the implementation of an internal AGC circuit that operates at input signal levels between 1 μA and 1.5 mA. The receiver sensitivity is greatly increased by a substantial and simultaneous decrease of the electronic noise. The enhancement of the device's dynamic range can be attributed to the internal AGC circuit, while the peak detector monitors the **differential output voltage** of the driver stage. If an increase of the differential output voltage has occurred above a predetermined level, the gain control circuit will decrease the feedback loop dynamic resistance (FET transistor) of the front end stage, automatically decreasing (on demand) the output voltage signal of the driver circuit to its predetermined level. The AGC circuit can be disabled through external biasing arrangements, thus allowing for a maximum receiver gain at a maximum input current. The driver circuit of the transimpedance amplifier is illustrated in Figure 7–21.

The driver circuit converts the single input into a differential output. The key performance characteristics of an optical receiver are subject to PIN photodetector diode and transimpedance amplifier performance. It is, therefore, apparent that the way in which the **PIN photodiode** is connected to the input of the transimpedance amplifier will significantly determine the overall receiver performance as evaluated by such important operating parameters as bandwidth, power supply rejection ratio, and sensitivity. To achieve the highest possible receiver sensitivity, the total capacitance at the input of the PIN diode must be kept to an absolute minimum. In order to maintain the PIN input capacitance to a minimum, the diode must be connected as close as possible to the input of the transimpedance amplifier and the PIN reverse voltage must be kept to a maximum. The PIN diode can be interfaced to the input of the transimpedance amplifier in two ways, which are illustrated in Figures 7–22 and 7–23.

In Figure 7–22, the cathode of the PIN diode is connected to the reference voltage and the anode is connected to the inverted input of the front end stage of the amplifier. The PIN bias voltage is provided by the power supply (V_{cc}) of the T_Z chip through an internal LPF. Both the internal LPF and the external capacitor are intended to reduce the electronic noise generated by the power supply, and to increase the PSRR. A typical value for the external capacitor is 1 nF.

The PIN diode reverse bias voltage is either 2.5 V for a 3.3 V power supply, or 4.2 V for a 5 V power supply. The reverse bias voltage across the PIN diode is a function of the generated signal levels. That is, an increase of the input signal will increase the DC current supplied by the power supply and a larger voltage drop will develop across the resistor (R_1). As a result, a decrease of the voltage across the PIN photodetector will occur. If a negative power supply is

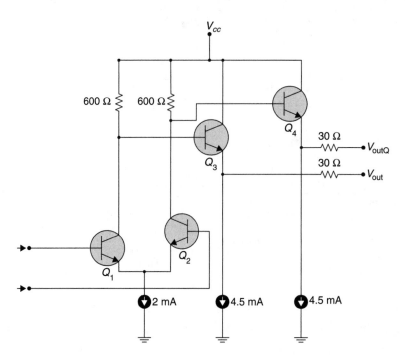

Figure 7–21. A driver circuit for a transimpedance amplifier.

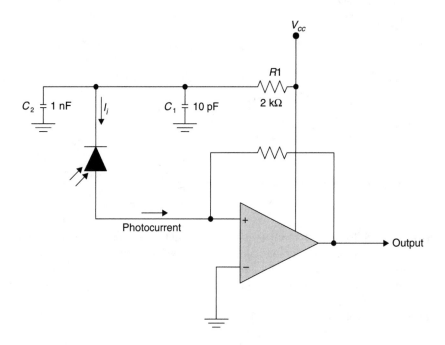

Figure 7–22. PIN-T_Z interface.

available, the anode of the PIN photodetector is connected to the negative power supply.

From Figure 7–23, the cathode of the PIN photodiode is connected to the input of the front end stage of the T_Z amplifier and the anode is connected to the negative supply voltage. An appropriate power supply filter should be incorporated into the design in order to increase the PSRR and maximize the receiver sensitivity.

AGC

The main function of the amplifier is to automatically adjust its transimpedance in order to maintain a constant output voltage, regardless of the current variations induced by the PIN photodetector. That is, if the **differential input voltage** increases then the T_Z must be decreased, and if the input current decreases then the T_Z must be proportionally

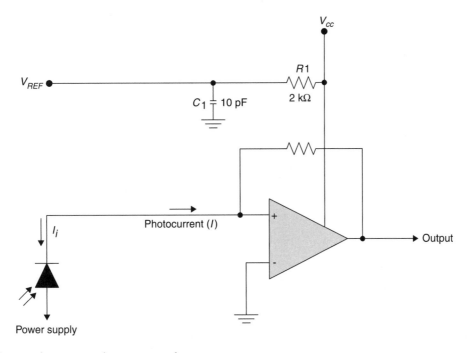

Figure 7–23. PIN connection to a negative power supply.

increased in order to maintain the required constant differential output voltage. Therefore, the gain of the transimpedance amplifier must be a function of the PIN generated photocurrent.

The AGC stage of the transimpedance amplifier is composed of a peak detector, a hold capacitor, and a gain control circuit. The output signal is detected by the peak detector, then stored by a hold capacitor in the form of voltage. This stored voltage is compared to a voltage generated by a threshold input current of 10 µA peak to peak. At input signal levels smaller than 10 µA, the AGC becomes inactive and the T_Z impedance is set at the maximum. If the input signal is larger than the threshold current, the AGC will be activated, forcing the transimpedance of the amplifier to decrease proportionally and, thus, maintaining a constant

output voltage level. A graphical representation of the AGC function is illustrated in Figure 7–24.

A flat differential output voltage is maintained for input signal variations between 10 µA and 200 µA. Above this level, the T_Z impedance of the amplifier is set at the minimum level. However, it is still capable of maintaining linearity up to the 1.5 mA level, beyond which a significant distortion must be anticipated.

7.6 AGC POSTAMPLIFIERS

The third block in an optical receiver, in addition to the PIN or APD photodetector diode and the transimpedance amplifier, is the AGC post amplifier (Figure 7–25). The postamplifier in

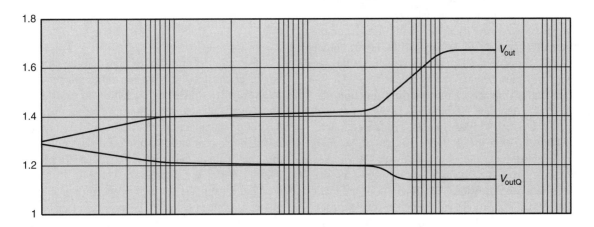

Figure 7–24. A graphical representation of the AGC function.

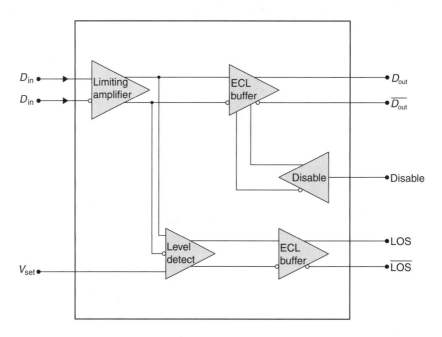

Figure 7–25. A 1.25 Gb/s postamplifer based on the Motorola MC10SX1126 .

Figure 7–25 is composed of a **limiting amplifier,** a level detector, two **ECL buffers,** and a disable circuit.

Although the preamplifier significantly amplifies the electric signals generated by the photodetector diode, the amplitude of these signals is still insufficient. To elevate the signal amplitude to the required levels, a postamplifier is absolutely essential. This amplifier performs a dual function. It amplifies the signal to the required level and also converts them either to the TTL or the ECL logic level, an essential requirement at the input of the clock and recovery module. A detailed diagram of a postamplifier is illustrated in Figure 7–26.

The postamplifier in Figure 7–26 operates with a minimum bandwidth of 600 MHz and an approximate gain of 60 dB. In optical receivers, monitoring the presence of signals applied at the input for levels above the set threshold is essential. This can be accomplished through the postamplifier FLAG output. A FLAG low (TTL level) output indicates that the input optical power is equal to or higher than the set threshold level. Therefore, valid data is anticipated. On the other hand, a FLAG high (TTL level) output indicates that the input optical signal is below the set threshold level and erroneous data may be received.

To take full advantage of this warning function, the FLAG output is connected to both the JAM input and the cathode of an LED driver. For an insufficient input optical signal level (below the set threshold), the FLAG high output connected to the JAM input will automatically disable the postamplifier, preventing it from further processing the input signal (noise amplification). The function of the LED is to provide a visual indication as to the presence or absence of an input optical signal. Under normal operating conditions, the LED connected to the driver output through a

current limiting resistor is ON. If optical power at the input of the receiver is less than a set minimum, or is lost completely, the driver output goes high and the LED is OFF.

The coupling capacitors C_7 and C_8 serve as differentiators, processing the binary data at the transition edge. This guarantees that the output stage of the comparator will not change state at the presence of noise only. Setting a 0.4 V hysteresis of the Schmitt trigger circuit facilitates this.

To prevent the processing of high frequency components present at the input of the receiver, a 100 pF capacitor is connected between the two inputs to act as a low pass filter. The presence of the capacitor between the two inputs increases the overall receiver sensitivity by approximately 3 dB.

The auto zero capacitor (C_3) acts as a DC to low frequency feedback loop, with an operating frequency expressed by Equation (7–11).

$$f_{-3\text{ dB}} = \frac{640}{(2\pi)(1.6\text{ k}\Omega)C_3} \quad (7\text{–}11)$$

The frequency of the applied signal at the input of the postamplifier must be at least ten times higher than the pole frequency determined by the auto zero capacitor (C_3). Also, the value of this capacitor must be larger than the values of the coupling capacitors (C_7 and C_8) by a ratio of 250/1. In addition, experimental results have shown that the value of the coupling capacitors must not exceed 18 pF. If one of the values between the auto zero and coupling capacitors is to be changed during the testing and performance maximization process, it must be that of the auto zero capacitor.

Resistors R_2 and R_3 are used to set the threshold level of the FLAG output and the built-in hysteresis. The value of

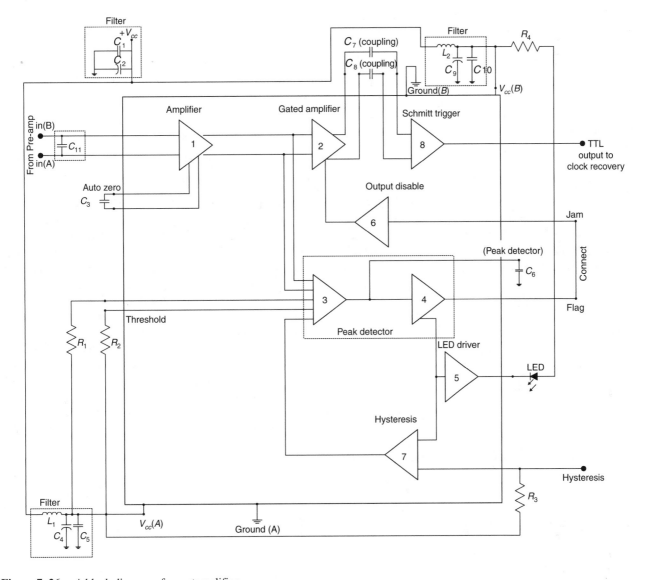

Figure 7–26. A block diagram of a postamplifier.

these resistors can be determined by available graphs. The peak resistor (R_1) and peak capacitor (C_6) are used to set the time constant required by the FLAG in order to change from high to low (high indicating the absence of optical signal and low indicating the presence of optical signal). The time constant is expressed by Equation (7–12).

$$\frac{dV}{dT} = \frac{I_{\text{peak}}}{C_T} \qquad (7\text{–}12)$$

where

$$C_T = C_6 + C_{\text{internal}} \text{ (normally } -10 \text{ pF) and}$$

$$I_{\text{peak}} = \frac{V_{cc} - 0.8 \text{ V}}{67.7 \times 10^3 \times R_1 / 67.7 \times 10^3 + R_1}$$

The combination of a preamplifier and a postamplifier in conjunction with a photodetector diode constitutes the basis for the design of an optical receiver. Upon completion of the receiver design stage, tests must be carried out to measure receiver sensitivity at a constant BER, normally set at 10^{-9} or 10^{-10}, and for various input data and frequency conditions. The external components can be adjusted to satisfy preset performance specifications. The function of the receiver in an optical communications system is to detect an incoming optical signal, amplify it, demodulate/demultiplex it, and generate a SONET/SDH format (high-speed data).

The optical receiver incorporates either an APD or PIN photodetector diode that normally operates in the 1100 nm to 1600 nm wavelength range. More specifically, these detectors are designed to operate at either 1310 nm, or 1550 nm, with WDM modes of operation. Under normal circumstances, optical receivers must comply with SONET/SDH transmission format specifications such as a minimum sensitivity usually set at –30 dBm, and a BER of 1×10^{-10}.

Optical receivers that employ PIN photodetector diodes require supply voltages of +5 V and –5 V, while receivers that employ APD photodetectors require an additional high DC biasing voltage of between 50 V and 60 V at 1 mA current. Furthermore, because of this extra high biasing voltage, a thermistor device is added to the receiver in order to continually monitor the case temperature and to provide the necessary feedback mechanism for temperature compensation.

The Philips TZ3044B 1.25 Gb/s Ethernet Postamplifier

The function of operation and technical characteristics of the Philips TZ3044B 1.25 Gb/s Ethernet postamplifier are based on the Philips TZ3044B 1.25 Gb/s circuit, designed for SONET/SDH, STM4/OC-12 Ethernet operations. The block diagram of the amplifier is illustrated in Figure 7–27.

Operating Characteristics

a) operating bandwidth: 1 KHz–1.25 GHz

b) data handling: 622 Mb/s SONET/SDH and 1.25 Gb/s Ethernet receivers

c) data outputs: PECL adjustable

d) programmable input signal detection

e) single power supply voltage: (3.0 V–5.5 V)

OPERATION OF THE PHILIPS TZ3044B 1.25 GB/S ETHERNET POSTAMPLIFIER. The amplifier receives a maximum of 1.25 Gb/s binary data from the output of the transimpedance amplifier with a voltage swing between 2 mV$_{p-p}$ and 1.5 V$_{p-p}$, then amplifies and limits the detected signal to positive emitter coupled logic (PECL) output voltage levels. The main building blocks of an AGC postamplifier are:

a) input buffer (A1)

b) **DC offset** compensation

c) buffer (A3)

d) buffer (A4)

e) rectifier circuit

f) band gap reference

g) **voltage level comparator**

The incoming data stream looks at a 4.5 kΩ impedance at both inputs of the amplifier (A1). Although both single ended and differential operations are available, the differential operation is highly recommended because it enhances the overall receiver performance. Another crucial element that determines system performance of the postamplifier is the offset voltage. Because the device is a high gain amplifier, a very small offset voltage will dramatically reduce the input sensitivity. It is therefore imperative that a DC offset compensation circuit be incorporated into the design in order to maintain the voltage at the **input buffer** (A3) at a critical threshold level in the absence of input signal. That is, a signal will appear at both outputs of the buffer stage (A3) only if a valid signal is present at the input of buffer (A1) and at a specific SNR. The PECL signal levels of the postamplifier outputs are illustrated in Figure 7–28.

INPUT BIASING. An input reference voltage generator provides the required DC biasing voltage at both inputs of the amplifier. This amplifier can be either DC coupled or AC coupled. When DC coupled, the input voltage source must

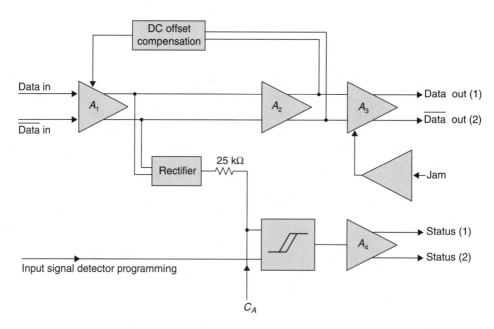

Figure 7–27. A block diagram of a Philips TZ3044B 1.25 Gb/s Ethernet postamplifier.

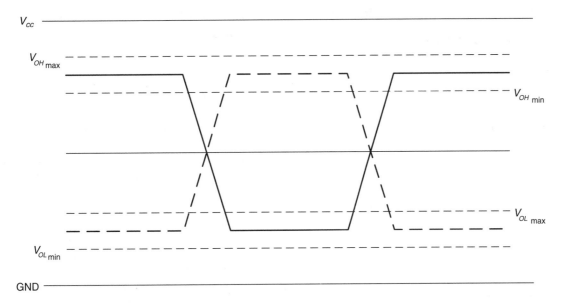

Figure 7–28. PECL logic levels.

not exceed the recommended input signal range, while the **DC offset** voltage must also be maintained below the level determined by the correction range of the compensation circuit.

When the AC coupled mode of operation is used, the coupling capacitor value must be selected so that, in conjunction with the internal resistor, it will be able to pass the lowest input frequency of concern. An AC coupled arrangement imposes limits on the number of consecutive pulses that the amplifier can accurately sense. It is essential that resistor thermal variations, as well as coupling capacitor tolerance, be taken into consideration during the amplifier design process.

DC Offset. The function of the DC offset compensation circuit is to maintain a constant voltage level at the input of the $A3$ buffer stage (threshold level), in the absence of an input signal. The lower cut off frequency of the post-amplifier is determined by the loop time constant.

Input Signal Level Detection. The individual user, through a processing mechanism, can set the input signal detection to its required level. If the input signal is below the minimum set threshold, the PECL output will be disabled. This implies that only the signals that maintain a predetermined SNR at the input of the amplifier will be transmitted, while maintaining the required BER levels. The input signals are amplified by the $(A1)$ amplifier circuit, rectified, and then applied at one of the inputs of the level detector circuit. The other input is connected to the programmable threshold reference input. To protect the level detector from being triggered by noise spikes, an internal filter is incorporated into the microchip while an additional

filter can be externally connected to further secure level detector accidental triggering. The incoming signal is compared with the reference signal at the level comparator, and then amplified by the $A4$ buffer stage. The complementary PECL outputs of $A4$ will indicate whether the input signal is larger or smaller than the threshold level. The **threshold voltage level** can be set through an external resistor connected between the power supply (V_{cc}) and the signal level detection programming input, or by forcing a current of a specific level into the level detection input.

Relationship Between the Detected Input Voltage and Threshold Current. The relationship of the required current (I_{RSET}) at the R_{SET} input to the input voltage is expressed by Equation (7–13).

$$I_{RSET} = 1.8 \times 10^{-3}(V_{DIN} - V_{DINQ}) \qquad \textbf{(7–13)}$$

where ($V_{DIN} - V_{DINQ}$) is the differential input voltage.

The value of the adjust resistor (R_{AD}) is calculated as follows. Because the voltage at the signal level detection is normally set at 1.5 V below V_{cc}, then Equation (7–14) results.

$$R_{ADJ} = \frac{1.5 \text{ V}}{I_{RSET}} \qquad \textbf{(7–14)}$$

Substituting Equation (7–13) into (7–14) yields Equation (7–15).

$$R_{ADJ} = \frac{1.5 \text{ V}}{1.8^{-3}(V_{DIN} - V_{DINQ})} = \frac{833.3}{(V_{DIN} - V_{DINQ})}\Omega$$

Therefore,

$$R_{ADJ} = \frac{833}{(V_{DIN} - V_{DINQ})}\Omega \qquad \textbf{(7–15)}$$

Example 7–2

Compute the value of the resistor (R_{ADJ}) to be connected between V_{cc} and the input of the signal level detector in order to detect differential input voltage levels below 5 mV.

Solution

a) Compute the threshold current (I_{RSET}).

$$I_{RSET} = 1.8 \times 10^{-3}(V_{DIN} - V_{DINQ})$$
$$= 1.8 \times 10^{-3}(5 \times 10^{-3} \text{ V}) = 9 \times 10^{-6} \text{ A}$$

Therefore,

$$I_{RSET} = 9 \text{ μA}$$

b) Compute R_{ADJ}.

$$R_{ADJ} = \frac{V_{RSET}}{I_{RSET}} = \frac{1.5 \text{ V}}{9 \times 10^{-6} \text{ A}} \cong 167 \text{ k}\Omega$$

Therefore,

$$R_{ADJ} \cong 167 \text{ k}\Omega$$

A graphical representation of key performance characteristics of this postamplifier is illustrated in Figure 7–29. The figure illustrates the relationship between the differential input voltage and the differential output voltage.

Figure 7–29 indicates that the differential output voltage is almost constant for a differential input voltage range between 1 mV and 1 V. Beyond this range, a significant decrease of the differential output voltage is encountered.

Figure 7–30 illustrates the relationship between the **rise time** and **fall time** of the differential output voltage and the applied differential input signal.

Figure 7–30 indicates that the lowest rise and fall times occur simultaneously at 3 mV and 100 mV differential input voltage. Therefore, the selection of the differential input voltage can be made to reflect the desired differential output voltage rise and fall times.

Figure 7–31 illustrates the relationship between PWD and differential input voltage.

Figure 7–31 shows a steep increase of PWD below the 10 mV and above the 100 mV range. The lowest PWD is obtained at the differential input range between 10 mV and 100 mV.

Figure 7–32 illustrates the relationship between differential output voltage and junction temperature.

Figure 7–32 shows that a steep decrease of the differential output voltage occurs below +20°C with a minimum reached at –40°C, while an almost constant differential output voltage is obtained at the junction temperature range of between +20°C and +120°C.

Figure 7–33 illustrates the postamplifier overall performance through the eye diagram for a differential input

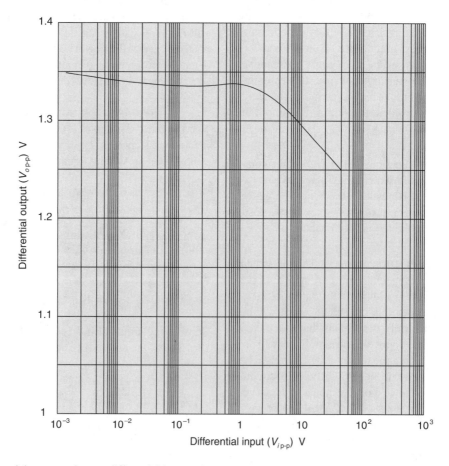

Figure 7–29.   Differential output voltage v. differential input voltage.

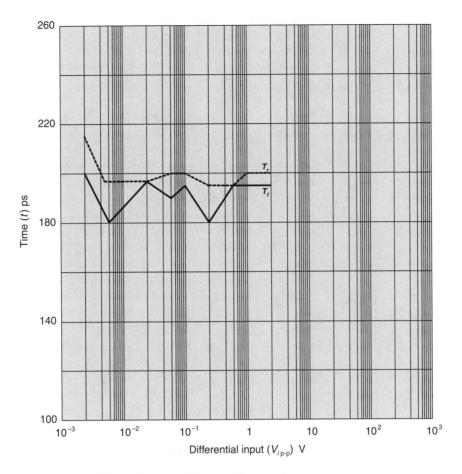

Figure 7–30. Rise and fall time of the differential output voltage *v.* differential input.

voltage of 4 mV. Figure 7–34 illustrates the postamplifier performance for a differential input voltage of 2 V.

Some of the important operating characteristics of the postamplifier are listed in Table 7–4 (see page 218).

A schematic diagram that illustrates a complete gigabit Ethernet optical receiver that incorporates the Philips TZA3043T transimpedance amplifier and the TZ3044 postamplifier is shown in Figure 7–35 (see page 219).

The Nortel AC10 AGC Postamplifier

The AC10 AGC postamplifier, developed by NORTEL NETWORKS, is intended to be used in the design of optical receivers that operate at a maximum data rate of 2.5 Gb/s and are capable of providing the required differential input signals to the **clock and data recovery** stage. A block diagram of the postamplifier is illustrated in Figure 7–36 (see page 220).

The postamplifier of Figure 7–36 is composed of an input stage, two variable gain stages, a rectifier stage, an integrator, a band gap, and a rectifier comparator.

OPERATION OF THE NORTEL AC10 POSTAMPLIFIER. The input stage of the post amplifier accepts either a 150 mV$_{p-p}$ (single-ended), or a 350 mV$_{p-p}$ differential signal. After the initial stage of amplification these input signals are applied at the input of a dual stage variable gain amplifier that has an output

that is required to drive the fixed gain output stage. The output stage generates a fixed 400 mV$_{p-p}$ output signal that is required to drive the next block of the optical receiver, that of the clock and data recovery. Because the output voltage must be maintained at a constant level (400 mV$_{p-p}$), a feedback mechanism, composed of a rectifier and an integrator circuit that is capable of maintaining the required voltage, is incorporated into the design. The rectifier compares the differential output voltage with a 400 mV$_{p-p}$ reference voltage generated by the band gap circuit. If the output voltage exceeds the 400 mV$_{p-p}$ reference level, the integrator is activated, and it supplies an appropriate differential control voltage to both stages of the variable gain control amplifier, forcing it to limit the output voltage to 400 mV$_{p-p}$. The SIGLOSS pin provides the loss of signal output, while the SIGLOS_LEVEL pin is used to provide external control of the loss of signal threshold voltage. Some of the most important performance characteristics of the AC10 AGC postamplifier are listed in Table 7–5.

The operating conditions and electrical characteristics are listed in Table 7–6. The absolute maximum ratings are listed in Table 7–7 (see page 220 for tables).

The Motorola MC10SX1127 2.5 Gb/s Postamplifier

The Motorola MC10SX1127 postamplifier can be incorporated into the design of optical fiber receivers that are

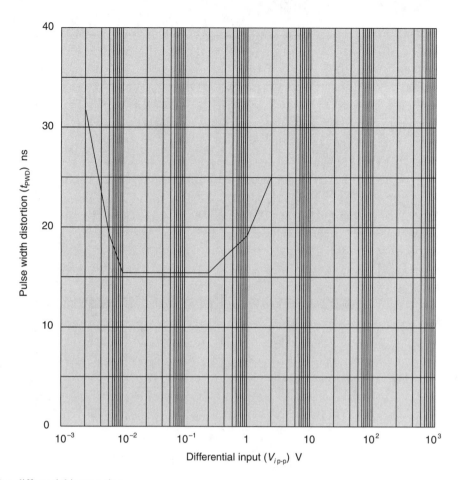

Figure 7–31. PWD *v.* differential input voltage.

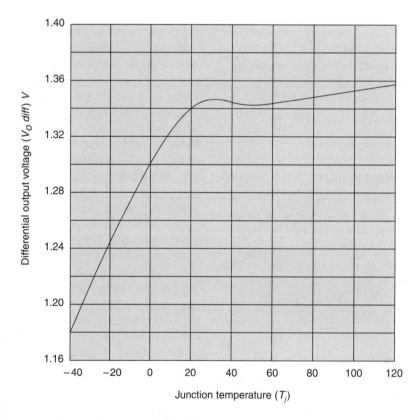

Figure 7–32. Differential output voltage *v.* junction temperature.

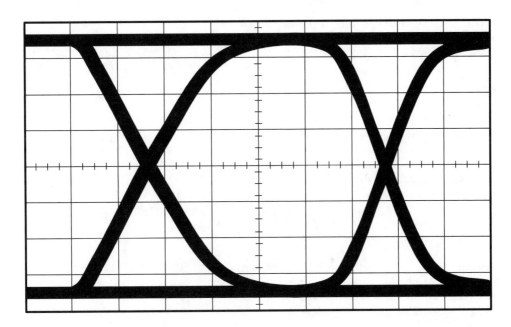

Figure 7–33. Eye diagram for a differential input voltage of 4 mV.

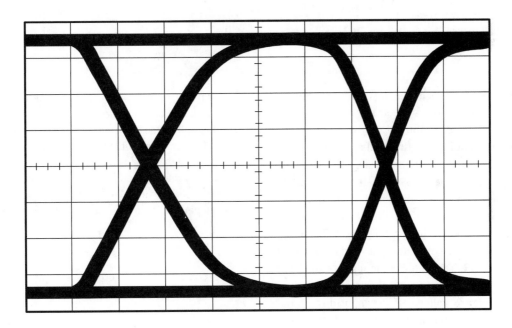

Figure 7–34. Eye diagram for a differential input of 2 V.

employed in applications such as the SONET/SDM, FDDI, and ATM. It is capable of operating at maximum data rates of 2.5 Gb/s (Figure 7–37, see page 221). The device is composed of a limiting amplifier, two ECL buffers, a reference source, a level detector, and a disable stage.

OPERATING CHARACTERISTICS OF THE MOTOROLA MC10SX1127 POSTAMPLIFIER. Some of the basic operating characteristics of the postamplifier are as follows:

Bandwidth range: 50 MHz–1.8 GHz

Single power supply: 3.3 V or 5 V

Programmable input signal detection

Differential design

OPERATION OF THE MOTOROLA MC10SX1127 POSTAMPLIFIER. For efficient operation, the device is intended to operate as AC coupled with coupling capacitors selected to enable the lowest input frequencies. Equation (7–16) gives the computation of the capacitor value.

$$C_{in} = \frac{1}{2\pi R_{in} f_{low}} \qquad (7\text{–}16)$$

TABLE 7–4   Operating Characteristics

Parameters	Minimum	Typical	Maximum	Symbol	Units
Supply voltage	3.0	3.3	5.5	V_{cc}	V
Analog supply current	—	15	24	I_{cca}	mA
Digital supply current	—	18	31	I_{ccd}	mA
Power dissipation	—	110	130	P_t	mW
Junction temperature	–40	—	125	T_j	°C
Ambient temperature	–40	25	85	T_a	°C
Input signal (V_{p-p}) (single ended)	0.002	—	1.5	$V_{Inp(p-p)}$ S.E	V
Input signal (V_{p-p}) (differential)	0.004	—	3.0	$V_{Inp(p-p)}$ Dif	V
Input offset voltage correction	—	3	—	$V_{I.O(corr)}$	mV
Input resistance	2.9	4.5	7.6	R_{in}	kΩ
Input capacitance	—	—	2.5	C_{in}	pF
Equivalent input noise (V_{rms})	—	100	145	$V_{n(rms)}$	μV
Reference current	5	—	60	I_{RSET}	μA
Reference voltage	V_{cc}–1.65	V_{cc}–1.5	V_{cc}–1.4	V_{RSET}	V
Threshold adjust range	2	—	12	$V_{th(p-p)}$	mV
Low level output voltage	V_{cc}–1.84	—	V_{cc}–1.6	V_{OL}	V
High level output voltage	V_{cc}–1.1	—	V_{cc}–0.9	V_{OH}	V
Rise time	—	200	250	t_r	ps
Fall time	—	200	250	t_f	ps
Pulse width distortion	—	—	30	t_{PWD}	ps
Low frequency (–3 dB)	—	0.85	1.5	$f_{-3\,dB(L)}$	KHz
High frequency (–3 dB)	—	1.0	—	$f_{-3\,dB(H)}$	GHz

where C_{in} is the coupling capacitance, R_{in} is the input impedance (5 kΩ), and f_{low} is the lowest frequency.

The differential signal applied at the input of the limiting amplifier is amplified to the level required to drive the input of the buffer circuit, while the same output is fed into the input of the level detector stage. If the differential input signal is less than a set threshold level, the level detector will be activated, and it will disable the switching of the ECL output. Through this process, the ECL output will change state only in the presence of a valid input signal that satisfies a set BER. The signal at the input of the level detector is amplified, rectified, and then compared to a programmable reference voltage. If the input differential signal is higher than the threshold level, the *LOS* output will be activated. If the input signal is below the threshold level, the $\overline{LOS}$ will be activated. To prevent the level detector from activation in the presence of noise, an appropriate filter is incorporated into the design of the postamplifier. The filtering function can be further enhanced by the inclusion of an external capacitor connected between the V_{cc} and the C_{LD} pins. The value of the capacitor is expressed by Equation (7–17).

$$C_{LD} = \frac{t}{R_Z} \qquad (7\text{–}17)$$

where C_{LD} is the external filter capacitor for the level detector comparator, t is the loss of signal filter time constant, and R_Z is the internal impedance (28 kΩ).

THE AUTO ZERO FUNCTION.   For the **auto zero function,** the device uses a feedback amplifier to eliminate any offset voltage along the traveling path of the signal. In this way, the input of the emitter coupled logic (ECL) comparator must be maintained at the toggle point with no input signal present. An external capacitor is, therefore, required to set the time constant. The value of the external auto zero capacitor that is connected between pin-1 and pin-2 is expressed by Equation (7–18).

$$C_{AZ} = \frac{150}{2\pi R_{AZ} f_{low}} \qquad (7\text{–}18)$$

where C_{AZ} is the auto zero external capacitor, R_{AZ} is the internal driving impedance (280 K), and f_{low} is the lowest frequency.

The postamplifier DC operating characteristics are listed in Table 7–8.

The postamplifier AC operating characteristics are listed in Table 7–9.

7.7   DATA AND CLOCK RECOVERY CIRCUITS

The Philips OQ2541HP and OQ2541U Clock and Data Recovery Circuits

The following description is based on the Philips OQ2541HP and OQ2541U clock and data recovery circuits. Both OQ2541HP and OQ2541U devices can be used for OC3/STM1 and OC12/STM4 applications and are capable

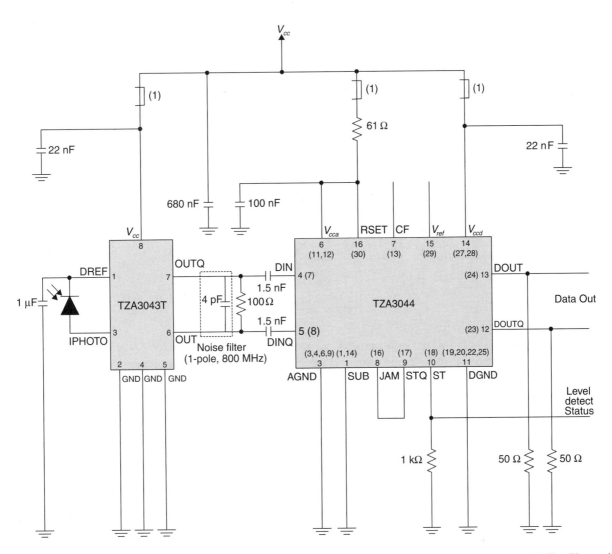

Figure 7–35. A schematic diagram of a Gigabit Ethernet optical receiver developed by Philips electronics. Courtesy Philips Electronics.

List of symbols

Symbol	Explanation
DIN	Differential input complementary to the input DINQ
DINQ	Differential input complementary to input DIN
V_{cca}	Analog supply voltage at the same potential as V_{ccd}
V_{ccd}	Digital supply voltage at the same potential as V_{cca}
AGND	Analog ground at the same potential as DGND
DGND	Digital ground at the same potential as AGND
DOUT	PECL compatible differential output complementary to DOUTQ
DOUTQ	PECL compatible differential output complementary to DOUT
SUB	Substrate pin, must be at the same potential as AGND
CF	Input to connect the capacitor required for setting the level detector time constant
ST	PECL compatible status output of the input signal level detector complementary to pin STQ
STQ	PECL complimentary status output of the input signal detector complementary to pin ST
V_{ref}	Band gap reference voltage
TEST	For tests only, to be left open
RSET	Input signal level detector programming

of extracting clock signals and recovering data from a 2.5 Gb/s (maximum) data stream. The block diagram of the devices is illustrated in Figure 7–38 (see page 222).

CIRCUIT DESCRIPTION. The input stage of the clock and data recovery circuit in Figure 7–38 is used as a buffer for the incoming high speed data stream. The output of the buffer is fed into the input of the phase detector stage, which compares the phase of the incoming signal with an internally generated clock signal. If the phase of the incoming signal is different from the phase of the internally generated clock signal, correction pulses are generated by the phase detector, progressively shifting the phase of the **voltage control oscillator (VCO)** until both the phase of the incoming and the internally generated clock signals are the same. Under ideal (lock) conditions, the data and the clock signals are set as shown in Figure 7–39 (see page 222).

THE PHASE DETECTOR/VCO FUNCTION. Figure 7–39 shows that the data is sampled at three points: At the center of the bit, at the next transition, and at the center of the bit after the transition. If the levels at (x) and (z) are different, and a

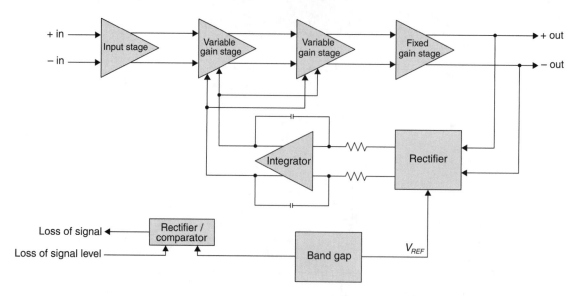

Figure 7–36. A block diagram of the Nortel AC10 postamplifier. Reproduced with the permission of Nortel Networks.

TABLE 7–5 AC10 Performance Characteristics

Parameters	Minimum	Typical	Maximum	Symbol	Units
Input signal level	4	—	300	V_{in}	mV_{p-p}
Output signal level	40	—	—	V_{out}	mV_{p-p}
Bandwidth (–3 dB)	1.9	—	3.9	B_W	GHz
Small signal gain	20	22	24	gain	dB
Noise figure	—	10	12	NF	dB
Rise/fall time	—	100	—	t_r/t_f	ps
Input return loss	—	15	—	S_{11}	dB
Output return loss	—	15	—	S_{22}	dB
Output voltage (AGC)	—	400	—	$V_{out_{age}}$	mV_{p-p}
Low cut off frequency	—	25	—	f_{n_agc}	KHz

TABLE 7–6 Operating Conditions and Electrical Characteristics

Parameters	Minimum	Typical	Maximum	Symbol	Units
Supply voltage	3.1	3.3	3.5	V_{cc}	V
Supply current	—	60	100	I_{cc}	mA
Power dissipation	—	198	350	P_d	mW
Supply ripple voltage	—	—	10	V_s	mV
Operating temperature	–40	—	85	T_{amb}	°C

TABLE 7–7 Absolute Maximum Ratings

Parameters	Minimum	Maximum	Symbol	Units
Supply voltage	–0.7	6.0	V_{cc}	V
Voltage at any input/output	0	$V_{cc}+0.7$	$V_{i/o}$	V
Storage temperature	–65	+135	T_{stg}	°C

transition at (y) has occurred, then the level detector will use the (y) sampling to determine the phase of the data in reference to the phase of the clock. If the level at point (x) is the same as the level at point (y), the clock is too fast and must be slowed down. This can be accomplished through the phase detector and the VCO circuits by increasing the pulse width of the ring oscillator in segments of 1 ps, corresponding to a quarter of one percent of the 400 ps OC48/STM16 mode of operation. This process is repeated until lock conditions have been achieved. If the levels at point (z) and (y) are the same, and are both different from the level at point (x), the clock is too slow and its speed must be increased in segments of 1 ps

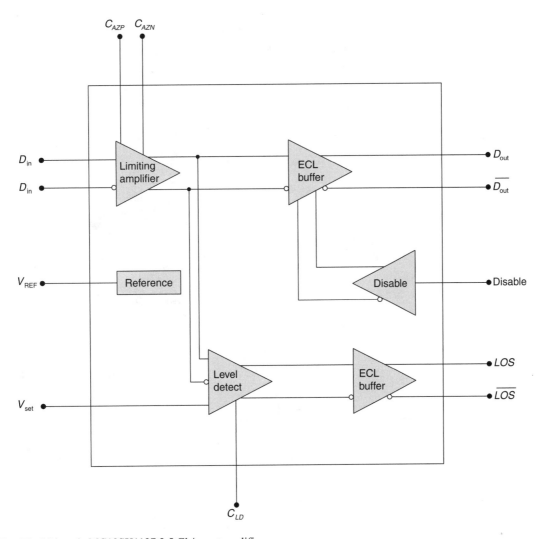

Figure 7–37. The Motorola MC10SX1127 2.5 Gb/s postamplifier.

TABLE 7–8 DC Operating Characteristics

Parameters	Minimum	Typical	Maximum	Symbol	Units
Input signal	—	—	1.5	V_{in}	V_{p-p}
Input noise	—	—	225	V_N	μV_{rms}
Input offset voltage	—	—	50	V_{OS}	μV
Power supply current	—	33	45	I_{cc}	mA
Input level detect	8.0	—	20	V_{TH}	mV_{p-p}

TABLE 7–9 AC Operating Characteristics

Parameters	Minimum	Typical	Maximum	Symbol	Units
Bandwidth (–3 dB)	50 MHz	—	1.8 GHz	B_W	—
Auto zero output resistance	200	325	450	R_{AZ}	$k\Omega$
Level detect filter resistance	14	25	41	R_F	$k\Omega$
Pulse width distortion	—	—	70	t_{PWD}	ps
Rise/fall time	150	250	650	t_r/t_f	ps
Level detect time constant	0.5	—	4.0	t_{LD}	μs

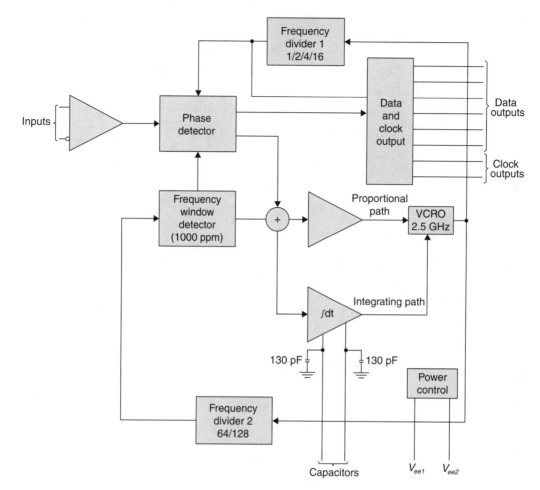

Figure 7–38. A block diagram of a clock and data recovery device.

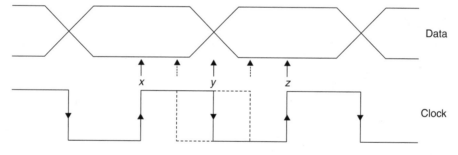

Figure 7–39. Data and clock signal pattern at lock state.

until lock conditions have been achieved. If both the phase and frequency of the voltage control oscillator are different from the incoming data, a number of up-down pulses will be generated by the phase detector and then applied at the inputs of the proportional and the integrated paths. The proportional path will adjust the phase of the clock, while the integrated path will adjust the frequency of the clock through a control voltage generated by the VCO circuit.

FREQUENCY WINDOW DETECTOR FUNCTION. The function of the frequency window stage is to maintain the VCO frequency within the predetermined one thousand parts per million of the operating frequency. The output of the

VCO is fed into the frequency divider ($\div$2). This frequency is compared to the reference frequency applied at the CREF or CREFQ inputs. If the VCO output frequency is outside the 1000 ppm range, the frequency window detector will disable the phase detector and allow the VCO to adjust its center frequency within the set tolerance while enabling the phase detector to acquire phase lock.

THE OUTPUT VOLTAGE SWING. The CML output voltage swing can be controlled by an adjustable external reference voltage and then permanently set by connecting the internal voltage divider (composed of 16 kΩ and 500 Ω resistors through a reference resistor (R_{REF}) between the V$_{ee}$)

and ground. For DC coupling applications, the voltage swing is maintained at 200 mV$_{p-p}$ on a 50 V load, while the reference voltage is set at 100 mV$_{p-p}$ corresponding to 50% of the output voltage swing. For AC coupling applications, Equation (7–19) expresses the required reference voltage.

$$V_{REF} = \frac{R_L + R_O}{R_L} \times 0.5 \, V_{out.Swing} \qquad \textbf{(7–19)}$$

where V_{REF} is the required externally set reference voltage, R_L is the load resistance, R_O is the output resistance, and $V_{out.Swing}$ is the output voltage swing.

Equation (7–20) expresses the value of the reference resistor.

$$R_{REF} = \frac{\left(\dfrac{V_{ee}}{V_{REF}} - 1\right) \times R_1}{1 - \left[\dfrac{R_1}{R_2} \times \left(\dfrac{V_{ee}}{V_{REF}} - 1\right)\right]} \qquad \textbf{(7–20)}$$

where $R_1 = 500 \, \Omega$ and $R_2 = 16 \, \text{k}\Omega$. Assuming a negative power supply of –3.3 V, both V_{REF} and R_{REF} are calculated as follows: For a 200 mV output voltage swing (single ended) with a 50 Ω load, the required reference voltage is:

$$V_{REF} = -100 \text{ mV} \left(\frac{50 + 100}{50}\right) = -300 \text{ mV}$$

Therefore, $V_{REF} = -300$ mV.
Given the preceding data, the value of the reference resistor is calculated as follows:

$$R_{REF} = \frac{\left(\dfrac{V_{ee}}{V_{REF}} - 1\right) \times R_1}{1 - \left[\dfrac{R_1}{R_2} \left(\dfrac{V_{ee}}{V_{REF}} - 1\right)\right]}$$

$$= \frac{\left(\dfrac{3.3}{0.3} - 1\right) \times 500}{1 - \left[\dfrac{500}{16000} \left(\dfrac{3.3}{0.3} - 1\right)\right]} = \frac{5000}{0.6875} = 7273 \, \Omega$$

Therefore, $R_{REF} \cong 7.3$ kΩ.

Example 7–3

Calculate the reference voltage and reference resistor for a required output voltage swing of 450 mV and supply voltage of –3.3 V.

Solution
Calculate V_{REF}.

$$V_{REF} = (-450/2)\left(\frac{50 + 100}{50}\right) = 675 \text{ mV}$$

Calculate R_{REF}.

$$R_{REF} = \frac{\left(\dfrac{3.3}{0.675} - 1\right) \times 500}{1 - \left[\dfrac{500}{16000}\left(\dfrac{3.3}{0.675} - 1\right)\right]} = 2213 \, \Omega$$

Therefore, $R_{REF} \cong 2.2$ kΩ.

In order for an output voltage swing of 450 mV$_{p-p}$ (AC coupled) to be maintained, the required reference voltage must be 0.675 mV and the reference resistance must be equal to 2.2 kΩ. Some of the important operating parameters are listed in Table 7–10.

The performance characteristics for different modes of operation are illustrated in Figure 7–40 (BER as a function of the input signal for an OC3/STM1 (155 Mb/s) mode of operation).

Similarly, Figure 7–41 illustrates the BER in relation to input signal for an OC12/STM-4 (622 Mb/s) mode of operation.

The performance characteristic curve for an OC48/STM-16 (2.5 Gb/s) mode of operation is illustrated in Figure 7–42.

The clock and data output waveforms for two different modes of operation, OC-12/STM-4 (622 Mb/s) and OC-48/STM-16 (2.5 Gb/s) are illustrated in Figures 7–43 and 7–44.

The complete external biasing configuration of the OQ2541 data and clock recovery circuit for an OC-48/STM-16 application is illustrated in Figure 7–45 (see page 227).

TABLE 7–10 Electrical Characteristics

Parameters	Minimum	Typical	Maximum	Symbol	Units
Supply voltage	–3.5	–3.3	–3.1	V_{ee}	V
Supply current	—	105	155	I_{ee}	mA
Power dissipation	—	350	550	P_{total}	mW
Input voltage	7	200	450	$V_{i_{p-p}}$	mV
Input sensitivity	—	2.5	7	$V_{i_{sens_{p-p}}}$	mV
Input impedance	—	50	—	Z_i	Ω

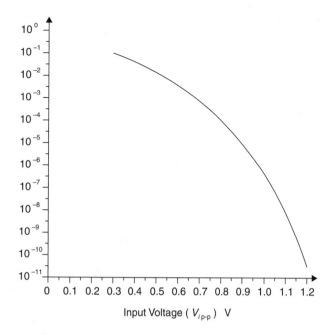

Figure 7–40. BER *v.* V_i for an OC3/STM-1 (155 Mb/s) operating mode.

Some of the device's important operating characteristics are as follows:

Data and clock recovery up to 2.5 Gb/s (it can operate at 155 Mb/s, 622 Mb/s, or 2.5 Gb/s)

Input sensitivity: 2.5 mV$_{\text{p-p}}$ differential data input

50 Ω data clock outputs (differential current mode logic)

Single supply voltage

Power dissipation: 350 mW

Adjustable CML outputs

The Lucent Technologies LG1600KXL Data and Clock Recovery Device

The LG166KXL data and clock recovery circuit is designed to operate in high speed optical fiber systems as well as in high speed digital communications systems. Although the incoming data may be degraded (high intersymbol interference (ISI) and jitter), the device is capable of regenerating the degraded input signals to the level set forth by the receiver specifications through a sophisticated frequency and phase lock loop circuit. The device is also in full compliance with Bellcore/ITU-T jitter tolerance for SONET/SDH modes of operation. Some of the basic operating characteristics of the device are as follows:

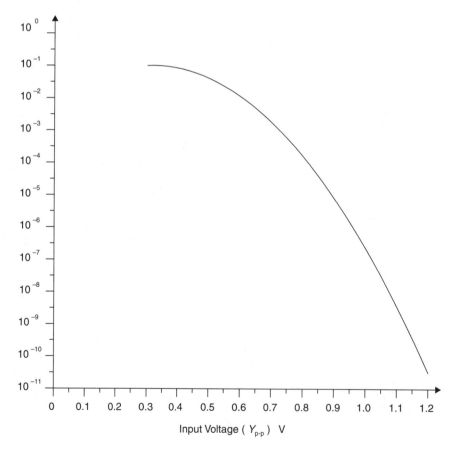

Figure 7–41. BER *v.* V_i for an OC12/STM-4 (622 Mb/s) operating mode.

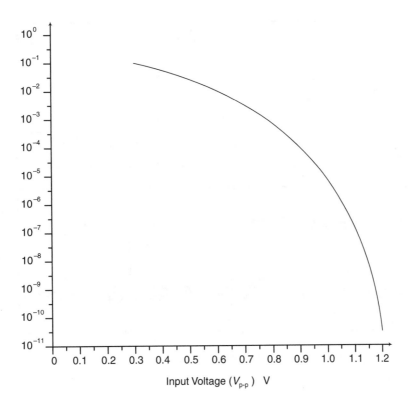

Figure 7–42. BER *v. V_i* for an OC48/STM-16 (2.5 Gb/s) operating mode.

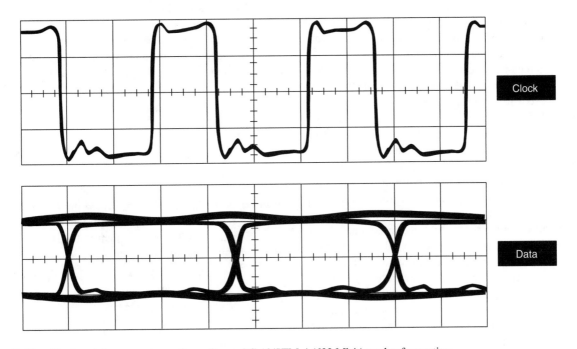

Figure 7–43. Clock and data output waveforms for an OC-12/STM-4 (622 Mb/s) mode of operation.

Operating data rates: 0.5 Gb/s to 5.5 Gb/s

Jitter: $< 5 \times 10^{-3}$ unit intervals (above Bellcore/ITU-T recommendations)

Input/output impedance: 50 Ω

Single ECL power supply

Capable of detecting input signals with a BER up to 10^{-3}

The device can be used in applications such as SONET/SDH, test equipment, and digital video transmission. A block diagram of the device is illustrated in Figure 7–46.

OPERATION OF THE LUCENT TECHNOLOGIES LG1600KXH DEVICE. The function of the data and clock recovery circuit is to regenerate the clock signals and the

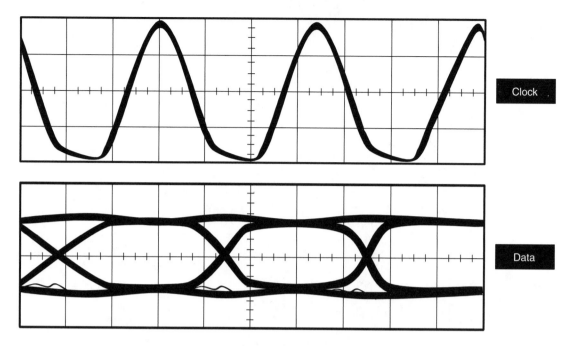

Figure 7–44. Clock and data output waveforms for an OC-48/STM-16 (2.5 Gb/s) mode of operation.

distorted data received at the input to the source level and quality. Because the components of a binary stream of data are the value of the bit and the transition between bits, the data and clock recovery circuit must regenerate the incoming and heavily distorted data by restoring those two components to the original transmitter value. The design of the LG1600KXH data and clock recovery circuit is based on PLL techniques. That is, the PLL filter properties that are determined at low frequencies contribute to a significant reduction of the effect caused by the parasitic elements. Furthermore, the reference frequency is established by the input data rate and not by the properties of the band pass filter. Another important element in optical link design is the receiver jitter function. That is, the overall optical link performance can be predicted well in advance if the receiver jitter transfer function is linear. The LG1600KXH device fulfills extremely well this critical design requirement. The required reference signal for the PLL circuit is obtained from a pulse train extracted from the incoming binary data through the transition detector. That is, the PLL circuit locks into the frequency and phase of the pulse train. A block diagram of the phase and frequency detector is illustrated in Figure 7–47.

A phase and frequency detector circuit is composed of a circulator, an EX-OR circuit, two phase detectors (one In-phase and the other Q-phase), and a logic circuit. If the clocks of both phase detectors are centered at the transition of the incoming pulses, the phase detector output will be zero. If the phase and frequency of the VCO are below or above the phase and frequency of the incoming sampled data, the phase and frequency detector will be adjusted to acquire and maintain lock at the incoming data.

PHASE LOCK LOOP CHARACTERISTICS. The data and clock recovery circuit employs a heavily damped second order phase lock loop circuit (Figure 7–48).

The second order equation expressing the PLL jitter transfer function is given by Equation (7–21).

$$H(s) = \frac{\phi_o}{\phi_i}(s)$$

or

$$\frac{\phi_o}{\phi_i} = \frac{2\zeta\omega_n s + \omega_n^2}{s^2 + 2\zeta\omega_n s + \omega_n^2} \qquad (7\text{–}21)$$

where ϕ_i is the input phase, ϕ_o is the output phase, ζ is the damping ratio, and ω_n is the natural frequency.

Equation (7–22) is the equation that relates the PLL bandwidth and peaking parameters.

$$H(s) = \frac{\omega_b\left(\dfrac{1}{s\tau}\right)}{s = \omega_b\left(1 + \dfrac{1}{s\tau}\right)} \qquad (7\text{–}22)$$

where ω_b is the physical loop gain pole product and τ is the loop filter time constant.

PHASE LOCK LOOP BANDWIDTH ($B_{W_{PLL}}$). The bandwidth of the PLL circuit is expressed by Equation (7–23).

$$B_{W_{PLL}} = K_o K_d R \qquad (7\text{–}23)$$

From Equation (7–23), it is evident that the PLL bandwidth is subject to the external resistor (R).

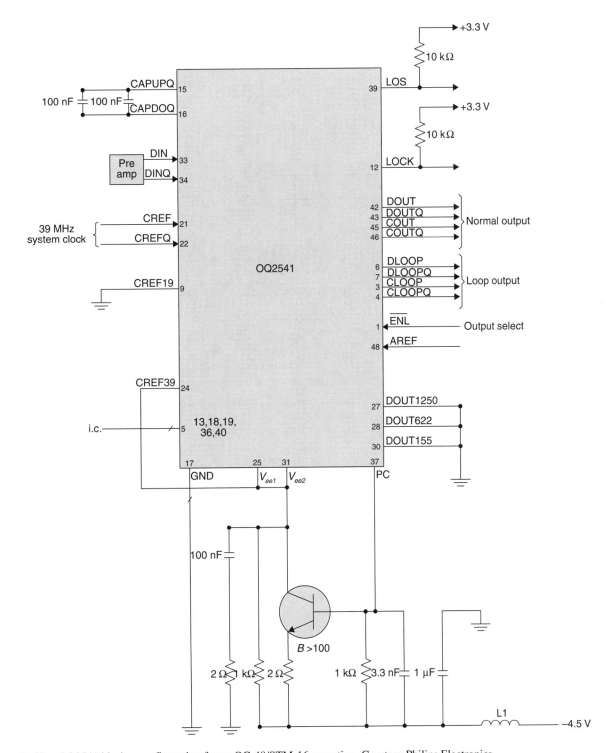

Figure 7–45. OQ2541 biasing configuration for an OC-48/STM-16 operation. Courtesy Philips Electronics.

PLL PEAK. Equation (7–24) determines the phase lock loop peak.

$$H(s) = 1 + \frac{1}{K_o K_d C_T R^2} \qquad (7\text{–}24)$$

where R is the external resistance and C_T is the total loop filter capacitance.

The complete external biasing of the data and clock recovery circuit is illustrated in Figure 7–49.

For an optimum jitter transfer function, the value of the external resistor is approximately 140 Ω. For a DC coupling operation, the outputs are terminated to 50 Ω and are capable of providing a voltage swing to a maximum of –800 mV. AC coupling output format exhibits an excellent return loss. The

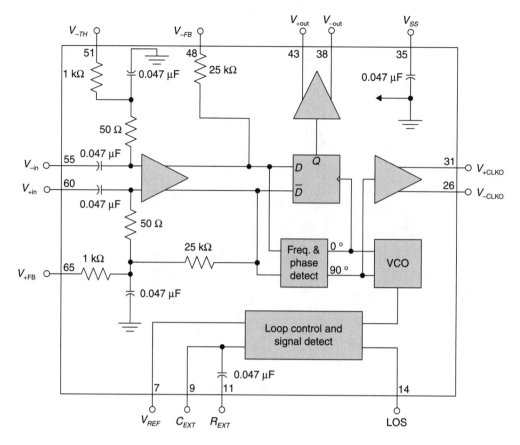

Figure 7–46. Data and clock recovery based on the Lucent Technologies LG166KXL device. Courtesy Agere Systems (Lucent Technologies).

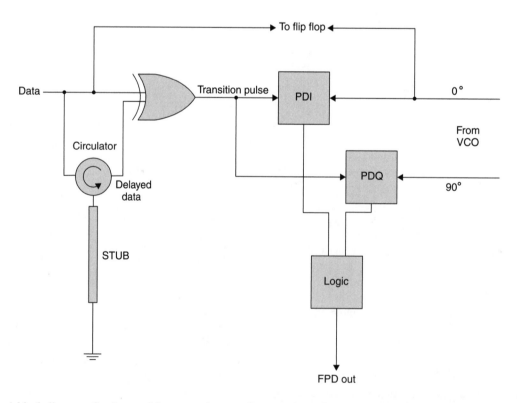

Figure 7–47. A block diagram of a phase and frequency detector. Courtesy Agere Systems (Lucent Technologies).

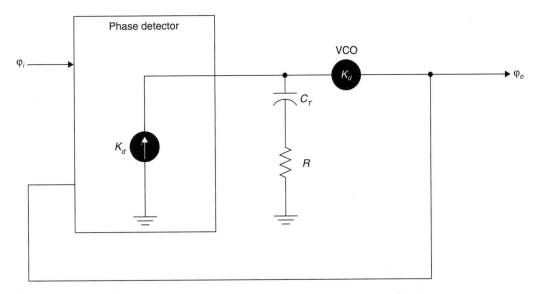

Figure 7–48. A block diagram of a phase lock loop circuit. Courtesy Agere Systems (Lucent Technologies).

performance characteristics of the data and clock recovery circuit are illustrated in Figure 7–50.

Some of the device's electrical characteristics are listed in Table 7–11.

The absolute maximum ratings are listed in Table 7–12.

The NORTEL YA18 2.5 Gb/s Data and Clock Recovery Circuit

The main function of the Nortel YA18 device is to retime the data and to extract a 2.448 GHz clock signal from the high speed input data stream. A block diagram of the device is illustrated in Figure 7–51.

OPERATION OF THE NORTEL YA18 2.5 GB/S DEVICE. The YA18 design incorporates a PLL circuit that extracts a clock frequency from the incoming NRZ 2.5 Gb/s data stream, with the VCO tuned to approximately 2.5 GHz for SONET/SDH applications. Both data inputs are internally terminated to 100 Ω differential CML, and are capable of receiving signals generated with magnitudes of between 60 mV$_{p-p}$ and 350 mV$_{p-p}$ by the postamplifier circuit. When used as AC coupling, the YA18 requires an input voltage swing of between 60 mV and 750 mV peak differential signal, although single ended operations can be implemented through a 50 Ω AC terminator. The input stage of the data and clock recovery circuit is illustrated in Figure 7–52.

The PLL stage is composed of a VCO that is designed to maximize jitter performance and a frequency/phase detector. The output stage, illustrated in Figure 7–53 (see page 233), performs data retiming and generates a 400 mV CML differential output signal that is capable of driving the next module, that of the demultiplexer circuit. The demultiplexer circuit is designed to operate by a 3.3 V single power supply. If a 5 V power supply is considered for use, then an external regulator must be incorporated into the external bi-asing arrangement. The schematic diagram of such a regulator is illustrated in Figure 7–54 (see page 234).

In order to further reduce power supply noise, an external LC filter must be used (Figure 7–55 see page 234). Incorporating the LC filter will limit the supply noise to approximately 50 mV$_{p-p}$ over the 6 KHz to 2 MHz frequency range.

Optical fiber links exhibit definite advantages, such as wide bandwidth, light weight security, and immunity to EMI over the wired-pair.

These advantages have led to the almost exclusive use of optical fibers in all new communications networks. Although optical links are capable of transmission rates well beyond the Gb/s range, they can also be used as low cost, medium rate systems in the range of 100 Mb/s. A block diagram of such a system is illustrated in Figure 7–56 (see page 235).

In Figure 7–56, if the input data is sent as a parallel stream of data at the input of the link, it must be converted to a serial stream of data. The parallel to serial conversion is accomplished through an appropriate parallel to serial converter. This serial data stream must also undergo further formatting in order to maximize bandwidth efficiency and to facilitate clock recovery at the receiver end. Clock recovery is more easily achieved by adapting the Manchester encoding format. Although the Manchester encoding format is ideal for synchronous transmission, it requires higher baseband bandwidth. The data output from the encoder module is converted into an optical signal through the electric to optical converter, and then is transmitted through the optical fiber at the input of the optical receiver.

For a system with such a relatively low data transmission rate, LEDs that operate at the 850 nm wavelength are almost always used as electric to optical converters. If the optical link is meant to accept future updates, that is, to increase its overall system capacity and distance, a laser diode operating at the 1300 nm or 1550 nm wavelengths replaces the LED. At the other end of the optical link, the

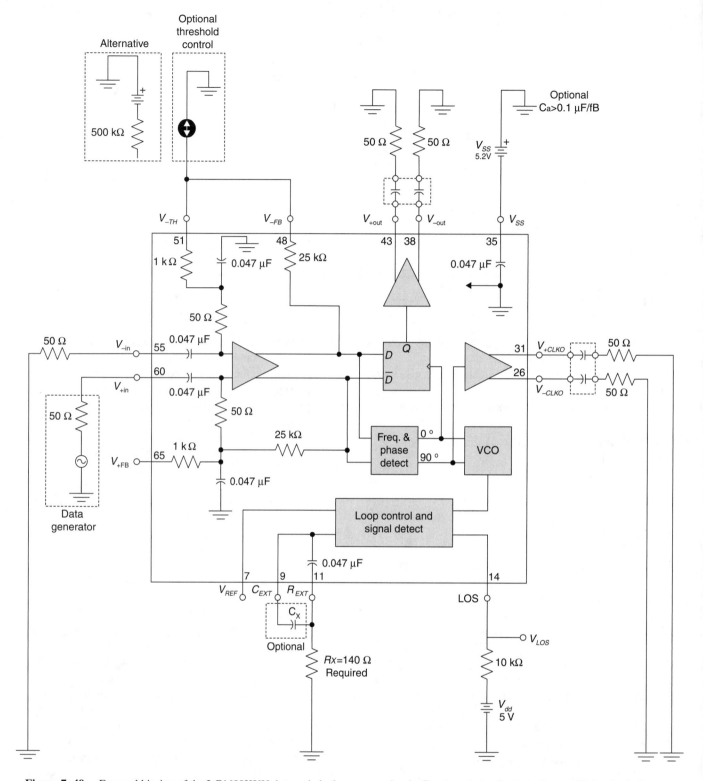

Figure 7–49. External biasing of the LG1600KXH data and clock recovery circuit. Courtesy Agere Systems (Lucent Technologies).

received optical signal is converted to electric through the optical to electric converter. The output of the electric signal is at either the TTL or the ECL logic level.

This complex electric signal, that combines information data and clock signals, is separated by the recovery and data timing module, then converted from serial to parallel data streams, and finally processed to the appropriate destinations. A more detailed explanation of the operation of the optical receiver, with the assistance of Figure 7–57 (see page 235), follows.

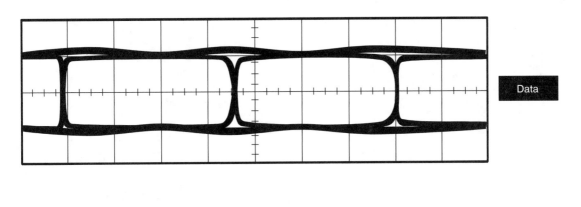

Data

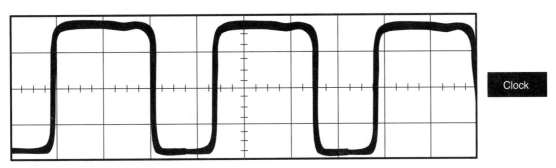

Clock

Figure 7–50. Data and clock recovery performance waveforms. Reproduced with the permission of Nortel Networks.

TABLE 7–11   Electrical Characteristics

Parameters	Minimum	Typical	Maximum	Symbol	Units
Input voltage (single ended)	200	—	800	$V_{in_{single}}$	mV
Input voltage (differential)	200	—	1600	$V_{in_{Diff}}$	mV
Output voltage	750	850	1000	V_{out}	mV
CLK output voltage	750	850	1000	$V_{out_{CLK}}$	mV
CLK output duty cycle	40	—	60	DC_{CLK}	%
Output pulse width relative to the bit period	90	100	110	$P_w\%$	%
BER (max)	1×10^{-3}	—	—	—	—
Jitter	—	0.0025	0.005	J_{Gen}	UI

TABLE 7–12   Absolute Maximum Ratings

Parameters	Minimum	Maximum	Units
Supply voltage	–7	0.5	V
Loss of signal bias voltage (V_{DD})	—	7	V
Power dissipation	—	2	W
Operating temperature	–40	100	°C

The optical signal applied at the input of the receiver is converted to electric current by the photodetector diode, then applied to the input of the preamplifier circuit. The preamplifier circuit of the optical receiver produces a differential output signal of a predetermined level. This differential output is then applied to the input of the postamplifier. The postamplifier further amplifies the signal and generates an output of either TTL or ECL level. The output of the postamplifier, still a complex signal, is fed into the clock recovery module. The function of the clock recovery circuit is to separate the clock from the information data and to use the clock for system synchronization (synchronous transmission). Therefore, the optical receiver generates two outputs: the clock and the information data. The main blocks of an optical receiver are the photodetector diode, the preamplifier, the postamplifier, and the clock recovery circuits. The parameters that determine preamplifier operating characteristics are:

a) bandwidth

b) noise

c) dynamic range

Electronic noise has two sources; one is based on preamplifier equivalent noise temperature (T_0), a key component that determines noise power ($P_n = 4\,KT_0B_w$), and the other is the noise generated by the photodetector diode. The combined noise power at the input of the preamplifier can also be

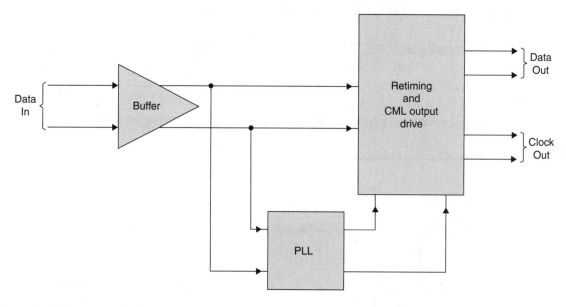

Figure 7–51. A block diagram of the Nortel YA18 data and clock recovery circuit. Reproduced with the permission of Nortel Networks.

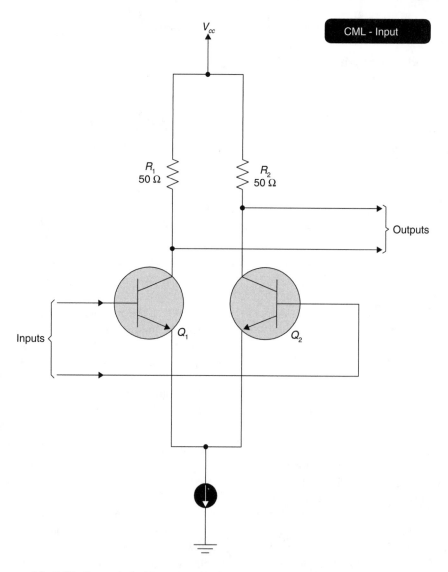

Figure 7–52. Input stage of the DCR (data and clock recovery circuit).

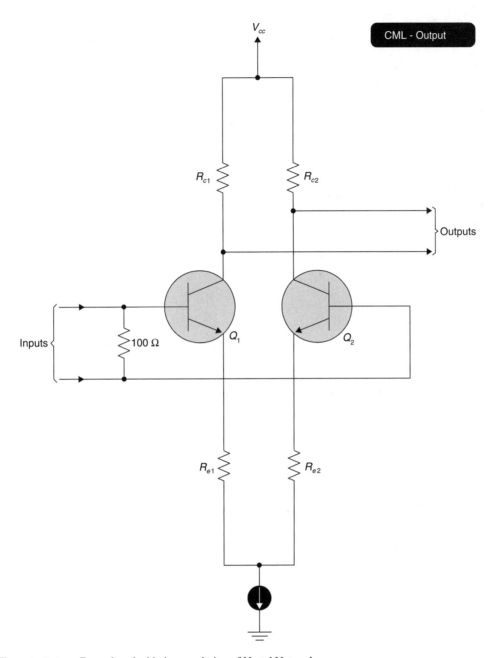

CML - Output

Figure 7–53. The output stage. Reproduced with the permission of Nortel Networks.

considered the optical receiver input noise power. Another important operating characteristic of the optical receiver is sensitivity. Receiver sensitivity is directly related to the type of signal (data) applied at the input.

7.8 OPTICAL RECEIVER PERFORMANCE CHARACTERISTICS

Optical receivers that operate in the 1300 nm to 1500 nm wavelength range must comply with SONET/SDH, OC-3/STM-1 data rates (155 Mb/s), and OC-12/STM-12 (625 Mb/s) rates. They must also comply with ITU-G.954 and G.958 recommendations. The utilization of a PIN InGaAs structure enhances the receiver operating wavelength to the range between 1100 nm to 1600 nm. A block diagram of a typical optical receiver is illustrated in Figure 7–58.

Operation of Optical Receivers

The optical signal is coupled into the receiver input through pigtail multimode optical fiber. The receiver must be sensitive enough to detect and amplify the incoming optical signal.

The power supply filters for both a PIN photodetector and a GaAs amplifier are used to minimize, or perhaps to

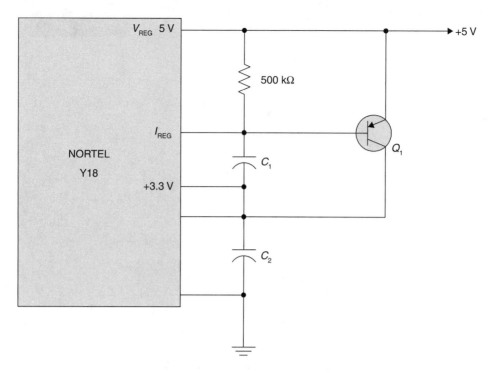

Figure 7–54. The 3.3 V voltage regulator. Reproduced with the permission of Nortel Networks.

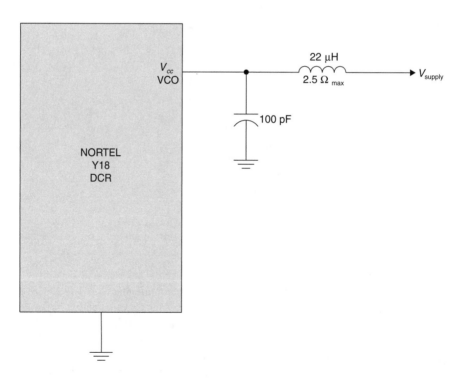

Figure 7–55. A power supply filter. Reproduced with the permission of Nortel Networks.

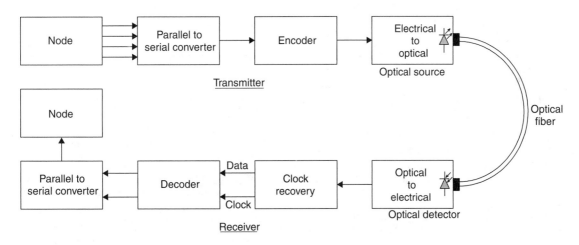

Figure 7–56. A block diagram of a medium data rate optical fiber transmitter/receiver.

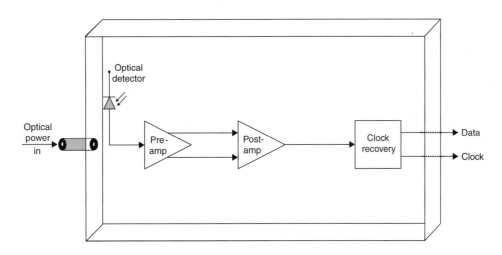

Figure 7–57. An optical receiver.

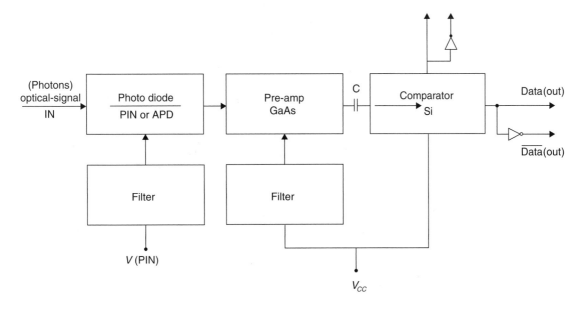

Figure 7–58. A block diagram of an optical receiver.

eliminate, the possibility of electronic noise (which is to be coupled into the receiver) generated by the power supply.

At the output stage of the receiver, *n-p-n* (open-emitter) transistors are usually employed. They are capable of providing 50 mA of load current at 7 Ω output impedance. It is imperative that transmission lines are used for component interconnects, while identical termination for both *DATA* and $\overline{DATA}$ is essential to maintain the required high degree of signal fidelity.

Flag outputs ($\overline{FLAG}$) and ($FLAG$) are required to indicate input optical signal loss. These outputs can be interfaced with CMOS or TTL–logic outputs. Figure 7–59 illustrates the way in which flag outputs can be converted to TTL outputs.

For a satisfactory receiver input optical signal, the TTL true output will produce a value larger than 2.5 V. For a less than satisfactory input signal (or no signal at all), the TTL false output will produce a signal of less than 0.4 V.

Optical Receiver Characteristics (Electrical)

Parameters	Minimum	Typical	Maximum	Units
Power supply	4.75	5.0	5.25	V
Power supply current	—	—	1	mA
Rise/fall time	—	0.7	1.4	ns

Absolute Maximum Ratings

Parameters	Minimum	Maximum	Units
Supply voltage	—	5.5	V
Operating wavelengths	1100	1600	nm
Operating temperature (case)	−40	85	°C
Store temperature	−40	85	°C
Soldering temperature	—	250/10	°C/s

7.9 OPTICAL RECEIVER PERFORMANCE WITH CLOCK RECOVERY

Brief Description

The Lucent Technologies R485 optical receiver with clock receiver (operating at 2.488 Gb/s) employs an APD or PIN photodetector structure (InGaAs based), and operates in the 1200 nm to 1600 nm wavelength range. It also incorporates a clock and data recovery circuit as well as a transimpedance amplifier. The clock and data recovery, a PLL circuit, is capable of extracting and recovering the clock signals embedded in the electrical signals generated by the optical detector.

The receiver also incorporates a flag signal that is used to detect the absence of optical power at the input of the receiver. It also includes a temperature compensation mechanism, necessitated as a result of the high biasing voltage required by the APD device.

The R485 optical receiver is designed to operate at a 2.5 Gb/s rate in applications such as SONET/OC-48 and SDH-STM-16 optical hierarchies and high speed data communications.

Device Characteristics (Electrical)

Parameters	Minimum	Typical	Maximum	Units
Power supply (positive)	4.75	5.0	5.25	V
(negative)	−5.46	−5.2	−4.92	V
Power consumption	—	1.3	2.5	W
Data rates	2.48807	2.48832	2.48857	Gb/s
Jitter generation	—	—	<0.01	UI
Jitter transfer	—	—	<0.1	dB

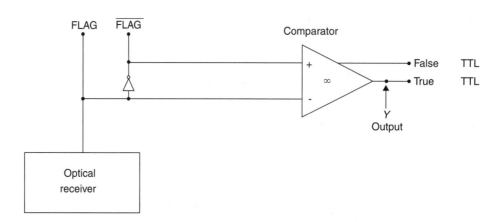

Figure 7–59. A schematic diagram illustrating the conversion of flag outputs to TTL outputs.

Device Characteristics (Optical)

Parameters	Minimum	Typical	Maximum	Units
Sensitivity				
(PIN)	—	−32	−30	dBm
(APD)	—	−23	−21	dBm
Maximum input power				
(PIN)	−3	—	—	dBm
(APD)	−8	—	—	dBm
Flag response time	3	—	1	ms

Absolute Maximum Ratings

Parameters	Minimum	Maximum	Units
Supply voltage	$0\ V_{cc}$	$6.5\ V_{cc}$	V
	$-6.5\ V_{ee}$	$0\ V_{ee}$	V
Operating temperature	−40	85	°C
Storage temperature	−40	85	°C
Lead soldering temperature	—	250/10	°C/s

SUMMARY

In Chapter 7, the role of the optical receiver as part of an optical fiber communications system was defined, followed by a detailed description of the main building blocks of such a receiver. The chapter began with a description of the three fundamental data patterns: Manchester, return to zero, and non return to zero. A brief presentation of the operating characteristics of PIN and avalanche photodiodes (the two basic devices used at the input of an optical receiver) followed. The chapter continued with the introduction of the three most common types of optical receivers: fixed threshold, edge detector, and AGC. The importance of the transimpedance preamplifier and postamplifier circuits incorporated into the design of optical receivers was also examined, followed by an introduction of the Philips TZ3023 transimpedance amplifier, the Motorola MC10SX1126, the Philips TZ3044B, and the Nortel AC10 AGC postamplifiers. The chapter concluded with an introduction of the Philips OQ2541HP, the Lucent Technologies LG1600KXH, and the Nortel YA18 clock and data recovery circuits, important components for the enhancement of the performance of optical receivers.

QUESTIONS

Section 7.1

1. Describe the function of an optical receiver as part of an optical fiber communications system.

2. List the fundamental building blocks of an optical receiver and briefly describe the operation of each block.

Section 7.2

3. Name the three types of data patterns and describe (in detail) their different performance characteristics.

4. Why is it a disadvantage to employ RZ data patterns in systems that are intended to operate in a synchronous mode?

5. What was the motivating factor for the development of the Manchester data pattern?

6. Describe (in detail) the reason why the Manchester data pattern is better suited for synchronous optical systems.

7. Name and explain the main disadvantage of the Manchester data pattern.

8. Explain (in detail) the need for AC coupling in optical receivers.

Section 7.3

9. Name the two types of photodetector diodes.

10. What are the most important PIN photodetector diode characteristics?

11. Define *response time* of a PIN photodiode.

12. List the key parameters that determine diode responsivity.

13. What is dark current and how does it affect the diode performance?

14. Describe the difference between a PIN device and an APD device.

15. List all the parameters that contribute to the photocurrent gain of an APD device.

16. Give the equation that determines the electronic noise of an APD.

17. Define the terms: *dark current, response time,* and *parasitic capacitance* as they relate to an APD.

18. Name the two processes of fabricating avalanche photodiodes.

Section 7.4

19. List the three basic types of optical receivers.

20. With the assistance of a simple schematic diagram, explain the operation and the basic characteristics of a fixed threshold optical receiver.

21. Why must the bit detection of a fixed threshold optical receiver be set at 50% of the peak signal amplitude?

22. Describe the main advantage of a fixed threshold optical receiver.

23. With the assistance of an equation, define *pulse width distortion.*

24. With the assistance of a simple schematic diagram, explain the operation and basic characteristics of an edge detection optical receiver.

25. Describe the operation and performance characteristics of AGC optical receivers.

26. Explain why a 50% duty cycle is essential for the proper operation of an AGC optical receiver.

Section 7.5

27. Define the function of a transimpedance amplifier.

28. With the assistance of a block diagram, describe the operation of the Nortel AB52 transimpedance amplifier.

29. Briefly describe the operation of the Philips TZ3023 transimpedance amplifier and compare its performance characteristics with that of the Nortel AB52.

30. What optical receiver performance parameters will be influenced by properly connecting the PIN photodiode at the input of the transimpedance amplifier?

31. Explain why the total capacitance at the input of the PIN diode should be kept at an absolute minimum.

32. Refer to Figure 7–23. Why must a filter be incorporated into the design of the power supply feeding the transimpedance amplifier?

33. Name the three elements that compose the AGC stage of a transimpedance amplifier, and briefly explain their combined function.

Section 7.6

34. With the assistance of a block diagram, describe the operation of a postamplifier.

35. Refer to Figure 7–26. Monitoring the signal level at the input of an optical receiver is essential. How is this accomplished?

36. In the postamplifier pictured in Figure 7–26, an LED is used as part of the warning circuitry. Explain (in detail) this function.

37. Refer to Figure 7–23. Describe the role and necessity of the out zero capacitor.

38. Optical receivers that employ APD photodetectors require a thermistor to be incorporated into their design. Explain why.

39. Sketch a block diagram of the Philips TZ3044B postamplifier and briefly explain its operation.

40. List operating characteristics of the Philips TZ3044B.

41. What is the function of the DC offset compensation circuit?

42. Describe the relationship between detected input voltage and threshold current of the Philips TZ3044B postamplifier.

43. Sketch a block diagram of the Nortel AC10 postamplifier and list its main building blocks

44. Briefly describe the operation of the Nortel AC10.

45. Sketch a block diagram of the Motorola MC10SX1127 2.5 Gb/s postamplifier and list its main building blocks.

46. Describe (in detail) the operation of the Motorola MC10SX1127.

47. How can the filtering function of the Motorola MC10SX1127 2.5 Gb/s postamplifier be further enhanced?

48. In the above device, offset voltages along the traveling path of the signal must be eliminated. Describe the method and give the equation that determines the value of the external component.

Section 7.7

49. Sketch a block diagram of the Philips OQ2541HP clock and data recovery device, and list its main building blocks.

50. Describe the method through which the output voltage swing of the Philips OQ2541HP can be controlled.

51. Sketch a block diagram of the Lucent Technologies LG1600KXH data and clock recovery device, and list its main building blocks.

52. Describe (in detail) the operation of the Lucent Technologies LG1600KXH.

53. What is the required value of the external resistor (see Figure 7–49) for optimum jitter and transfer function set?

54. Sketch a block diagram of the Nortel YA18 2.5 Gb/s data and clock recovery device, and list its main building blocks.

55. How can power supply noise of the above device be reduced further?

56. Refer to Figure 7–56. List the building blocks of the medium rate system and explain (in detail) its operation.

57. In the Nortel YA18, LED is normally used as the source device. If the system is to be upgraded at some point in the future, explain the method (or methods) through which such upgrading can be accomplished.

58. For the optical receiver in Figure 7–57, describe the upgrading parameters that determine preamplifier operating characteristics.

Section 7.8

59. Refer to Figure 7–58. List the optical receiver building blocks and explain (in detail) its operation.

PROBLEMS

1. A PIN photodetector is fabricated with InGaAsP semiconductor materials that have band gap energy of 0.89 eV. The diode is designed to operate at 300°K. If the diode is connected to a load, and the generated photocurrent develops a potential difference across this load, calculate the diode cut off wavelength, operating frequency, and responsivity. Assume a quantum efficiency of 6%, α coefficient of 5.1×10^{-4}, and β coefficient of 195.

2. If the dark current generated by a PIN photodiode is 2 pA, and the diode bandwidth is 1 KHz, calculate the resulting noise power and noise voltage.

3. An APD generates 1.2×10^{13} carriers. If the diode multiplication factor is 20, calculate the generated photocurrent.

4. Calculate the operating temperature of an APD diode that generates 1 mA of photocurrent with a multiplication factor of 30. The diode is operating with the following parameters: $\alpha = 0.6$, breakdown voltage = 80 V, reverse bias voltage = 50 V, and device resistance plus load resistance = 1 kΩ.

5. If the pulse width at the input of a fixed threshold optical receiver is 32.5 ns, and the pulse width at the output is 35.2 ns, calculate the pulse width distortion of the receiver.

6. A fixed threshold optical receiver exhibits a pulse width distortion of 20%. If the pulse width at the output of the receiver is 28 ns, calculate the pulse width at the input of the receiver.

7. Calculate the value of the adjust resistor required between V_{cc} and input signal adjust programming (Figure 7–27) in order to detect differential input signals below 3.5 mV.

8. If a 330 K external resistor is connected between V_{cc} and input signal adjust programming, calculate the maximum differential input voltage level that can be detected (Figure 7–27).

9. Refer to Figure 7–38. Calculate the value of the external reference resistor required between V_{ee} and ground to maintain an output voltage swing of 0.5 V. Assume a power supply of –3.2 V.

8

OPTICAL
TRANSRECEIVERS

1. Establish the role of transreceivers in optical communications systems

2. Explain the need for LED based transreceivers

3. Describe the operating characteristics of LED driver circuits from leading manufacturers such as Hewlett Packard, Microcosm, Lucent Technologies, and Nortel networks

4. Determine the role of laser based optical transreceivers in fiber optics communications systems

5. Describe, in some detail, the function and operating characteristics of laser based transreceivers such as Philips TZA3005H, HP-HFTC5200, Lucent Technologies TB16 2.5 Gb/s, Texas Instruments SN75FC1000, Microcosm MC4663, and Vitesse VSC7185

6. Establish design guidelines for optical transreceivers

7. Present some of the latest developments in high-speed optical transreceivers, such as the Vitesse VSC7216 and Sumitomo SDG-1201 modules

KEY TERMS ◄┈┈

AC coupling	clock synthesizer	extinction ratio	LED forward current
auto zero	collision function	frame byte	LED forward voltage
band gap reference	common transmit byte clock	hold time	level detector
boundary detection	DC coupling	input data pulse shaper	line loopback
buffer	diagnostic loop back	junction temperature	lock detect
clock and data recovery	differential amplifier	K characters	lock to reference
clock detect	differential input sensitivity	laser bias control	noise bandwidth
clock driver	dynamic range	laser bias driver	optical sensitivity
clock recovery	end of life receiver power	laser modulator	output data jitter
clock synchronizer	dissipation	LED clamping	parallel to serial converter

8.1 INTRODUCTION

Optical transreceivers are fully integrated microchips designed to perform data serialization/deserialization for SONET/ATM optical transmission systems. Two basic categories of optical transreceivers exist: those incorporating LED devices for optical transmission and those incorporating laser diodes. The transmitter section performs data serialization of parallel SONET/ATM signals, while the receiver section performs clock recovery and serial to parallel conversion. A block diagram of an optical transreceiver is illustrated in Figure 8–1.

Figure 8–1 shows that the main components of a typical transreceiver that incorporates an LED as the optical device are the multiplexer and the LED, which composes the transmitter section. Both modules are dealt with extensively in Chapter 7 and Chapter 10. A brief description of LED driver circuits and their operating characteristics will follow. Several manufactures, such as Lucent Technologies, Nortel Networks, Hewlett Packard, Microcosm, and Philips have developed a wide range of LED and laser driver circuits.

8.2 LED TRANSRECEIVERS

LED Transmitters

The simplest form of circuit that drives an LED is illustrated in Figure 8–2. In Figure 8–2a, a single resistor is driving the LED, while in Figures 8–2b and 8–2c, a resistor and a transistor are used to drive the LED. In Figure 8–2a, the LED is connected to the power supply through a resistor in series configuration.

The current (I_f) flowing through the LED pictured in Figure 8–2a is expressed by Equation (8–1).

$$I_f = \frac{V_S - V_f}{R_S} \qquad \textbf{(8–1)}$$

where I_f is the **LED forward current**, V_f is the **LED forward voltage**, V_S is the supply voltage, and R_S is the series resistor. A resistor and a transistor can also drive the LED in a series configuration (Figure 8–2b). The current flow through the LED pictured in Figure 8–2c is expressed by Equation (8–2).

$$I_f = \frac{V_S - V_{CE} - V_f}{R_S} \qquad \textbf{(8–2)}$$

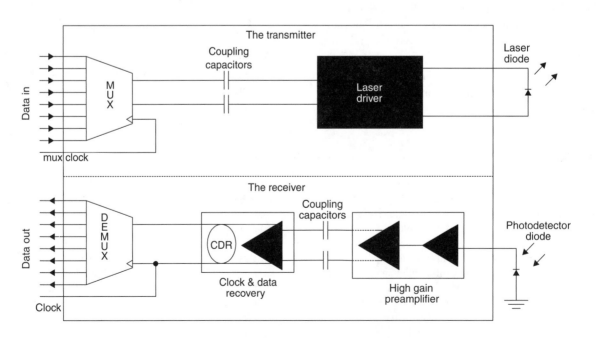

Figure 8–1. A block diagram of an optical transreceiver. Reproduced with the permission of Nortel Networks.

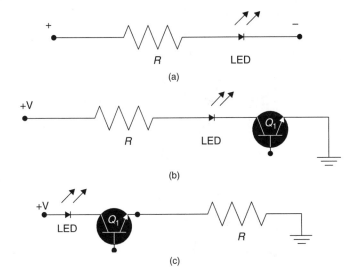

(a)

(b)

(c)

Figure 8–2. (a) A simple resistor driving an LED device. (b) A simple LED driver circuit composed of a resistor and a transistor. (c) A simple LED driver circuit composed of a resistor and a transistor.

where I_f is the LED forward current, V_f is the LED forward voltage, V_S is the supply voltage, R_S is the series resistor, and V_{CE} is the collector to emitter voltage. In these systems, it is important that the LED optical power output be kept constant. This can be achieved by maintaining a constant forward current (I_f), which is also subject to a constant V_f level. LED forward voltage levels can reach 3 V, with corresponding forward currents up to 300 mA. Peak currents up to 1 A can be achieved through very short duty cycles. These high peak currents can only be achieved with an appropriate power supply.

The NAND Gate LED Driver

For high speed optical transmission systems that utilize LEDs, NAND gates that are capable of achieving current peaking can be used as the driver mechanism. In order for the driver circuit with NAND gates to produce the relatively high current required by the LED, special gates such as the 74ACT family of logic gates are used in a shunt configuration (Figure 8–3). Standard NAND gates cannot be used as they are unable to comply with the driver circuit current sinking and current sourcing requirements demanded by the LED during switch-ON and switch-OFF times.

DESCRIPTION OF OPERATION. One of the two inputs of all the NAND gates is permanently connected to the power supply (+5 V). C_1 and C_2 are decoupling capacitors used to provide the required **power supply noise rejection (PSNR).** A logic-0 at the input of gate 1 produces a logic-0 at the combined outputs of gates 2, 3, and 4. That is, the pull down transistors at the output stage of each gate behave as current sinking, allowing the LED to turn ON. A logic-1 at the input of gate 1 produces logic high at the combined outputs of gates 2, 3, and 4, thus turning OFF the LED. The totem pole output stages of each NAND gate are capable of current sourcing and current sinking to a maximum of 100 mA from a low impedance load.

The Microcosm MC2042-4 Driver

Another LED driver circuit, developed by Microcosm Communications, is the MC2042-4. The device is designed to operate at a maximum of 300 Mb/s with an output current and with maximum rise and fall times of 500 ps. It also provides programmable temperature compensation for the LED device, PECL inputs, and optical **pulse width adjust (PWA).** Furthermore it can operate from a +3.3 V or a +5 V single power supply. A block diagram of the device is illustrated in Figure 8–4. The device pictured in Figure 8–4 is composed of the following building blocks:

a) input data pulse shaper

b) current switch and peaking current

c) V_{bb} generator

d) programmable current sink

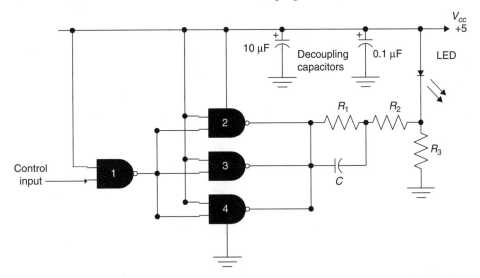

Figure 8–3. A LED driving circuit composed of NAND gates.

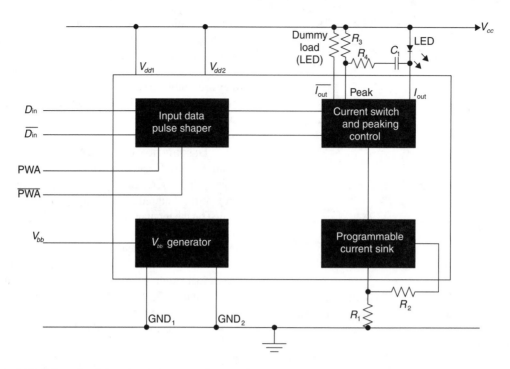

Figure 8–4. An LED driver circuit based on Microcosm Communications, MC2042-4 design.

The device inputs accept differential PECL signals, which ultimately control the LED forward current. In order to minimize the noise generated by the positive power supply, the driver circuit output current flows either through the LED or the dummy load. Furthermore, the LED extinction ratio is improved by **turn OFF time improvement.** The turn OFF time improvement is accomplished by momentarily shorting out the LED through the V_{dd} before the current is diverted to the dummy load. The LED required current is set by the external resistors. R_1 is connected between R_{set1} and ground, and R_2 is connected between R_{set2} and ground. The LED drive current is the current read of the R_{set1} pin. R_1 is used to set the temperature independent component of the LED current, while R_2 is used to set the temperature dependent component of the LED driver current. An internal diode connected to the R_{set2} pin is used for temperature control. If the IC temperature increases, the voltage across the diode decreases with a corresponding increase in LED current.

TURN OFF TIME IMPROVEMENT. The LED turn ON time is improved when a preemphasis circuit is incorporated into the IC design. Preemphasis is implemented through external components R_4 and C_1 connected in series between the I_{out} and Peak pins. At turn ON time, the voltage at the Peak electrode is pulled down very quickly, and the resulting transient voltage coupled through R_4 and C_1 generates a corresponding transient current flowing toward the LED. However, at turn OFF time, the voltage at the Peak electrode is pulled up again very quickly and the resulting transient voltage generates a corresponding transient current away from the LED. The important values of transient cur-

rents (I_{peak}), turn ON time, turn OFF time, and RC decay time (τ) are expressed by Equations (8–3) and (8–4).

$$I_{peak} = \frac{4}{R_4 + 5} \qquad (8-3)$$

$$\tau = \frac{C_1}{R_4 + 5} \qquad (8-4)$$

A typical value for R_4 is 50 Ω, and C_1 is equal to 20 pF.

LED CLAMPING. The MC2042 device incorporates a clamping circuit designed to improve the LED turn OFF time. This improvement is required because the turn OFF time is somewhat longer than the turn ON time. Turn OFF time improvement is accomplished by completely discharging the LED internal capacitance through the V_{dd2} electrode. This has a drawback. At the turn ON time, the LED internal capacitance must be fully recharged before emission begins. This increases the turn ON time, an undesirable characteristic. The preemphasis circuit offsets this slow start. It is, therefore, evident that both turn ON and turn OFF LED times are improved through the preemphasis and clamping circuits.

PULSE WIDTH ADJUST. The linearity of the data pulses coupled into the input of the switch section of the driver circuit must be improved. Pulse linearity improvement is accomplished through a differential voltage at the PWA electrodes of between −500 to +500 ps. PWA linearity is expressed by Equation (8–5).

$$\Delta(PW)_{ps} = KV_{PWA} \qquad (8-5)$$

where $K = 50(\pm 100)$, $V_{PWA} = (^+V_{PWA}) - (^-V_{PWA})$, and, the V_{PWA} maximum range is set at ± 1 V.

The Receiver Module

DESCRIPTION OF OPERATION. At the end of the fiber, the optical signal is converted back to an electrical signal by the photodetector diode. Because the received signal is heavily attenuated, the photodetector diode must be sensitive enough to detect optical signals down to nanowatt levels. The photodiode electric signal generated by the photodetector diode current undergoes further amplification in order to be restored to its original shape and level. Photodetector properties and operating characteristics are dealt with in Chapter 5 (optical detectors). In summary, one of the detector's fundamental properties is **optical sensitivity**—the least level of optical power required to generate the equivalent electrical signal with an absolute minimum degree of distortion. Optical detectors exhibit inherent noise based on the material properties used in the fabrication of the devices. Therefore, the received optical power must be above the detector noise level if an undistorted electrical signal is to be produced. The required level difference between the optical signal and the detector noise level is an operating characteristic called the optical signal to noise ratio (OSNR). Optical detectors are very high speed devices capable of operating over a wide wavelength range. A photodetector diode incorporated into a very simple optical receiver design is illustrated in Figure 8–5.

A typical photodetector, such as the one used in the optical receiver in Figure 8–5, exhibits a switching time of 10 ns under a reverse bias voltage of 10 V and a 50 Ω resistance. The switching time of the photodiode is subject to its internal capacitance, and may vary for different output loads. Internal capacitance of a photodiode is a function of the applied reverse bias voltage (V_r). The resistor, in series with the diode, serves as a current source.

The voltage generated across the series resistor is fed to the input of the preamplifier stage for further amplification. If special preamplifiers, such as the BGA318 or BGA 420 from Infineon are used, the series resistor is no longer required. The preamplifier output is used to drive the next stage, that of the postamplifier, which is intended to generate the required output logic levels. A block diagram of an optical receiver incorporating a postamplifier stage is illustrated in Figure 8–6.

The optical receiver in Figure 8–6, operating in a **DC coupling** mode, exhibits a relatively low sensitivity subject to a critical matching between the preamplifier output and the postamplifier reference voltage (V_{ref}). Similarly, an **AC coupling** optical receiver is illustrated in Figure 8–7.

With an AC coupling mode of operation, the receiver sensitivity is dramatically increased, provided that the applied data is balance coded.

The AD8015 Preamplifier

Among a wide variety of optical preamplifiers is the AD8015, developed by Analog Devices. A block diagram of this device is illustrated in Figure 8–8. AD8015 preamplifier operating characteristics are as follows:

a) **noise bandwidth** (240 MHz)

b) low noise

c) **pulse width modulation** (500 ps)

d) input noise (rms) 26.5 nA

e) wide **dynamic range**

f) optical sensitivity –36 dBm at 155.52 Mb/s

g) power supply +5 V at 25 mA

An optical receiver incorporating a photodetector diode and the SFH551/1 preamplifier developed by Infineon is illustrated in Figure 8–9.

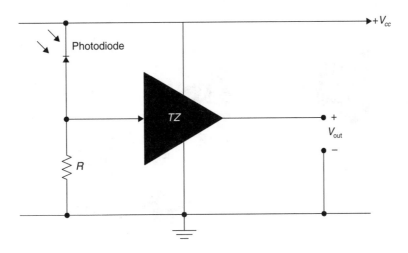

Figure 8–5. A block diagram of a basic optical receiver.

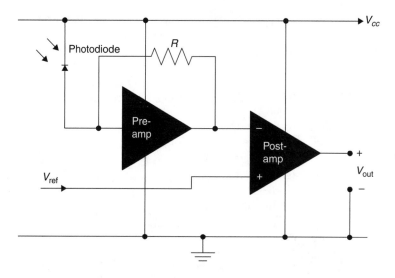

Figure 8–6. A block diagram of an optical receiver incorporating a postamplifier.

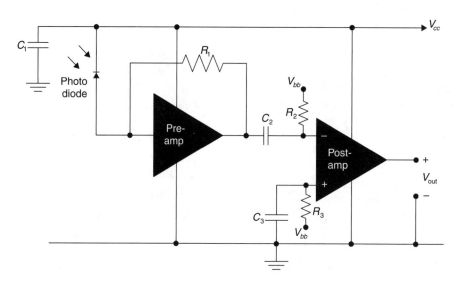

Figure 8–7. An AC coupled optical receiver.

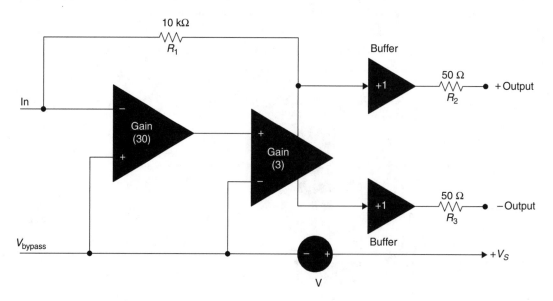

Figure 8–8. A block diagram of an AD8015 preamplifier.

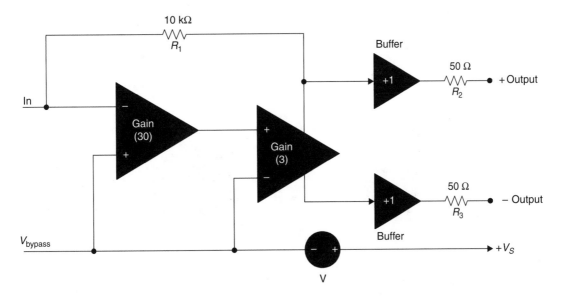

Figure 8–9. An optical receiver incorporating the Infineon SFH155/1, combining a photodiode and preamplifier circuit.

The SIEMENS SFH155/1 Optical Receiver

The SFH155/1 device developed by SIEMENS, which is employed in inexpensive optical transmission links, exhibits very satisfactory noise and dynamic range characteristics. The main components incorporated into the optical receiver are the photodiode, the preamplifier, the comparator, and the Schmitt trigger.

OPERATION OF THE SIEMENS SFH 155/1 OPTICAL RECEIVER. A built-in lens focuses the light from the output of the fiber into the photodetector diode, which converts received light into electric current (photocurrent). The generated photocurrent is converted into a voltage signal by the preamplifier stage. The feedback resistor (R_f) of the preamplifier circuit determines the current to voltage rate of change. One of the preamplifier's critical operating characteristics is that it must exhibit linear characteristics over the entire optical power range. The output of the preamplifier is fed into the inverting input of the comparator circuit, while the noninverting input of the comparator is connected to a reference voltage source (V_{ref}). The preamplifier output is then compared with the reference voltage (V_{ref}) to determine whether or not the comparator output is at a low or high state. However, noise signals with amplitudes very close to the comparator operating threshold can create erroneous states at the differentiator output. To avoid this, a Schmitt trigger is incorporated into the optical receiver, working with two threshold levels. In order for the Schmitt trigger to switch from the OFF state to the ON state, a large signal is required. Therefore, small noise signals are unable to produce erroneous states. The receiver performance, to a large extent, is subject to signal quality at the input of the decision circuit, which in turn is subject to the quality of the optical signal at the input of the preamplifier circuit. Because the preamplifier is designed to operate over a wide bandwidth, its performance characteristics are not critical to the overall receiver operation. However, the received optical signal is subject to transmitter residual emission in the OFF state, to rise and fall time degradation due to transmitter control signals, to optical power losses through the optical link, and to total transmitter optical power. If the optical signal at the input of the receiver is very small, and the generated electrical signal is also very small in comparison to the reference voltage (V_{ref}) at the input of the decision stage, the output of the comparator will be low. If the input signal is high, then the corresponding electrical signal will exhibit a smaller turn ON time than turn OFF time. Therefore, the performance of the optical receiver can only be established if the form and the power level of the optical signal at the input are taken into consideration.

RECEIVER OPERATING LIMITS: DYNAMIC RANGE. The optical receiver dynamic range is defined by the minimum and maximum optical power at the input where the receiver exhibits linear operating characteristics. Switching levels between 2 μW and 4 μW are typical for an error free operation.

The MC2207 Optical Receiver

A complete optical receiver that incorporates an MC2207 low noise transimpedance amplifier and an MC2045 post-amplifier is illustrated in Figure 8–10.

The main building blocks of the preamplifier circuit in Figure 8–10 are as follows.

a) silicon or *InGaAs* photodetector diode

b) voltage regulator

c) DC restore and AGC circuits

d) **band gap reference**

e) **buffer**

f) **differential amplifier**

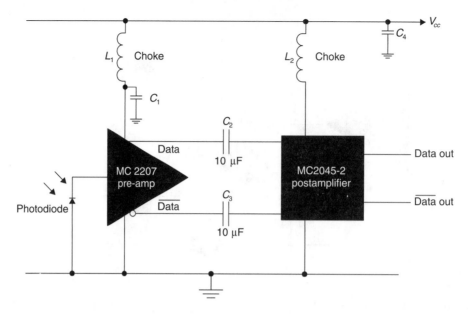

Figure 8–10. A block diagram of a complete optical receiver.

The postamplifier building blocks are as follows:

a) limiting amplifier
b) **level detector**
c) ECL buffer
d) jam buffer
e) buffer

The preamplifier main operating characteristics are:

a) **receiver sensitivity:** –29 dBm at 155 Mb/s
b) bandwidth: 250 Mb/s (max)
c) 16 kΩ differential transimpedance at low level signals
d) AGC providing continuous operation to +6 dBm
e) PSNR >35 dB
f) **Power dissipation:** 100 mW at 3.3 V

THE TRANSIMPEDANCE AMPLIFIER STAGE. The transimpedance, or preamplifier, stage is composed of a CMOS active device capable of maintaining operational stability under all input conditions, and a feedback resistor used to establish the output voltage amplitude. The voltage regulator is used to improve the PSNR up to a few MHz, while power supply noise of higher frequencies is controlled by external decoupling capacitors. The anode of the PIN photodetector is connected to the input of the transimpedance amplifier and the cathode is connected to AC ground. In order to reduce the photodetector internal capacitance, and therefore to increase the diode speed, a reverse bias voltage of 2.6 V is applied at the diode cathode. The AGC stage operates over the +3 dBm to 30 dBm range and for signals larger than –10.5 dBm at 0.45 A/W. For transimpedance variations, a MOS transistor is used.

THE POSTAMPLIFIER. The preamplifier output is fed into the phase splitter and a pair of voltage follower outputs that are designed to drive high impedance loads (500 Ω). The main operating characteristics of a postamplifier circuit are as follows:

a) very wide operating range (1.5 Gb/s)
b) **programmable input signal detect**
c) fully differential
d) CMOS and PECL link status variants
e) BiCMOS process
f) 3.3 V power supply

DESCRIPTION OF FUNCTION. The main function of a postamplifier circuit is to receive the electrical signal from the preamplifier, and to provide the required gain for generation of the PECL logic levels at the output. The device incorporates a programmable **signal level detector** to set the threshold levels based on the application's specific needs, and is intended for fiber channel (1.06 Gb/s) and Ethernet (1.25 Gb/s) applications. A block diagram of the forward gain path is illustrated in Figure 8–11. Because the device is intended for AC coupling operation, two capacitors (C_1 and C_2) are required. **DC coupling** operation can also be accommodated, because the inputs of the first stage are DC coupled through a high input resistance (R_{in} 10 kΩ). Both C_1 and C_2 capacitors must be selected to pass the lowest input frequencies.

THE AUTO ZERO CIRCUIT. The postamplifier also includes an **auto zero** circuit. This circuit is used to cancel the offset voltage inherent in the signal path, thus maintaining the comparator at toggle level. External resistor (R_{az}) and exter-

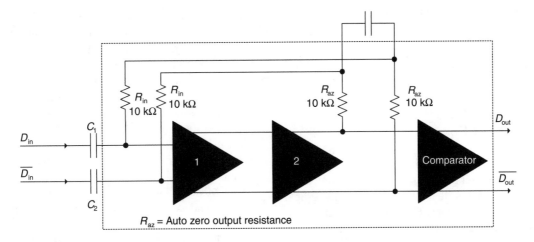

Figure 8–11. A block diagram of the forward gain path.

nal capacitor (C_{az}) are used to set the required time-constant of the autozero circuit. The value of the capacitor C_{az} is normally set at 100 nF.

THE LEVEL DETECTOR. Programmable level detector circuits are designed to set the input signal detection to a specified level, automatically forcing the output data levels to go to a known state if the input signals fall below the set threshold. With the programmable level detector, data is propagated only when the input signals are above the level reflecting a set BER. With this method, changes in the output data will not occur due to the presence of noise. A detailed threshold circuit is illustrated in Figure 8–12.

The generated differential signal of the detection threshold circuit is compared with the input differential peak amplitude. The reference voltage (V_{ref}) is normally set at 0.5 V below the V_{cc}. The external resistors are in parallel with the internal resistors. That is, R_{vset} is in parallel with R_b, and R_{jset} is

parallel to R_c. The voltage controlled current source V_{ccsi} generates a current I_{R_a} expressed by Equation (8–6).

$$I_{R_a} = \frac{0.5}{R_a} \tag{8–6}$$

At V_{set}, the voltage is defined by the I_{R_a}, R_b', and R_c' defined by Equations (8–7) and (8–8).

$$R_b' = R_{vset} \, // \, R_b \tag{8–7}$$

$$R_c' = R_{jset} \, // \, R_c \tag{8–8}$$

The voltage across R_c' is $V_{R_c'}$ and is defined by the voltage controlled current source V_{cc2}. Therefore, the current through R_b generates the V_{det} expressed by Equation (8–9).

$$V_{det} = 0.5 \text{ V} \left(\frac{R_b' R_d}{R_a R_c'} \right) \tag{8–9}$$

where the on chip resistor values are as follows: R_a = 2.8 kΩ, R_b = 5.6 kΩ, and R_d = 60 Ω. V_{det} is equal to 50 mV (differ-

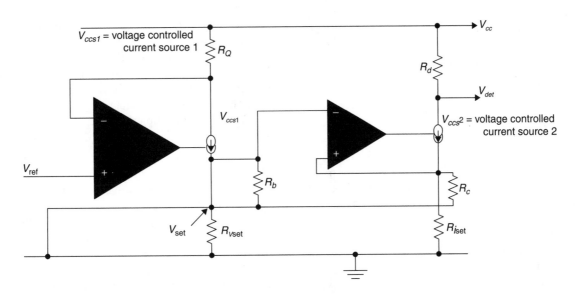

Figure 8–12. A detection threshold circuit.

ential peak), 100 mV (differential peak to peak), or it can be calculated by using Equation (8–10).

$$V_{\text{det}} = 0.025 \text{ V} \left(\frac{R_b / R_{v\text{set}}}{R_a} \right) \qquad (8\text{–}10)$$

8.3 LASER DIODE TRANSRECEIVERS

A large variety of optical transreceivers have been developed by various manufacturers to satisfy the large number of application and design criteria in the rapidly expanding field of fiber optics communications. Examples of optical transreceiver modules from some of the leaders in the fiber optics industry follow.

Optical Transreceivers Based on the Philips TZA3005H

OPERATION OF THE DEVICE. The functional description and operating characteristics of the device are based on those of the Philips TZA3005H optical transreceivers. The TZA3005H is a fully integrated transreceiver that supports OC3/STM1 (155.52 Mb/s) and OC12/STM4 (622.08 Mb/s) formats, and is in full compliance with ANSI, Bellcore, and ITU-T specifications. The module is designed to perform parallel to serial conversion at the transmitter stage, and serial to parallel conversion at the receiver stage, as well as SONET/SDH frame detection. The device complies with Bellcore, ANSI, and ITU-T BER requirements through the implementation of low jitter PECL interface circuitry. A block diagram of the above module is illustrated in Figure 8–13.

The main function of the TZA3005H transreceiver, developed by Philips optoelectronics, is serialization/deserialization, transmission, reception, frame detection, and recovery of data in an optical communications link. The main building blocks of the device are the transmitter, the receiver, and the RF switch box, and the basic operating sequences are as follows: the transmitter section accepts four or eight parallel bits at its input, performs parallel to serial conversion, and produces the required serial data at the output. The receiver accepts the serial data at the input from the optical detector or from the data and clock recovery module. It performs frame detection and serial to parallel conversion, and it generates four- or eight-bit parallel data at the output. The transmitter section, the receiver section, and the clock synthesizer supply the data and clock signals to the RF switch box circuit, which reroutes the signals through the multiplexer circuit to the transmitter, receiver, and clock driver sections. A more detailed description of each of the modules follows.

THE TRANSMITTER. The function of the transmitter within an optical transreceiver is to convert OC3/STM1 or OC12/STM4 (byte serial) input data to a bit serial output stream. For example, four parallel input data lines of 38.88 Mb/s each are converted to a 155.52 Mb/s serial bit stream by the parallel to serial converter of the transmitter module. Likewise, sixteen parallel input data lines of 38.88 Mb/s each are converted to a 622.08 Mb/s serial bit stream. Different input series bytes and the corresponding bit serial outputs of a parallel to serial converter are given in Table 8–1.

The transmitter section provides transmitter to receiver diagnostic loop back and receiver to transmitter line loop back and loop timing. It also incorporates a clock synthesizer circuit that is composed of a phase-lock-loop (PLL) and a driver circuit. The function of the synthesizer is to extract information from the input data signals (19.44 Mb/s, 38.88 Mb/s, 51.84 Mb/s, and 77.76 Mb/s or 155.08 Mb/s) and to synthesize the required high frequency bit clock. A more detailed explanation follows.

The clock synthesizer: The function of the clock synthesizer is to generate an output clock in phase with one of five SONET/SDH input clock frequencies. The output clock frequency is in either 155.52 MHz (OC3/STM1) or 622.08 MHz (OC12/STM4) format. The input reference clock frequency, whether it is 155.52 MHz or 622.08 MHz, is appropriately divided to generate the 19 MHz frequency applied to the PLL **phase detector** input. The very accurate clock frequency required by SONET/SDH formats is achieved through the utilization of a differential PECL crystal oscillator that exhibits an accuracy of better than 4.6 pps (ITU-G.813 option-1) or 20 pps (ITU-G.813 option-2) regulations. In terms of jitter, the reference clock frequency must not exceed 56 ps (rms) for OC3/STM1 or 14 ps (rms) for OC12/STM4 over the 1 KHz to 1 MHz range. The on chip PLL circuit is composed of a phase detector, a loop filter, and a voltage control oscillator (VCO). The phase detector compares the VCO phase with the phase of the divided clock frequency, and the generated output is fed into the loop filter. The DC level filter output is fed into the voltage control oscillator, which lock on to either the 155.52 MHz (OC3/STM1) or 622.08 MHz OC12/STM4 format.

The clock divider: The function of the clock divider is to generate a replica of the serial output clock. The clock divider output transmits the following clock frequencies: 19.44 MHz or 38.88 MHz for OC3/STM1 format and 77.76 MHz or 155.52 MHz for OC12/STM12 format. The selection of one set of frequencies is accomplished through the MODE and BUSWIDTH inputs (Table 8–2).

The clock divider output (SYNCLKDIV) is mainly used for upstream multiplexing, to ensure a stable frequency and phase relationship between the data out and data in.

THE RECEIVER SECTION. The receiver section within the optical transreceiver module performs the exact opposite function to that of the transmitter. That is, it converts the received serial data signal of either 155.52 Mb/s or 622.08 Mb/s to parallel data streams of 19.44 Mb/s or 77.76 Mb/s. It also extrapolates either 38.88 MHz or 155.52 MHz of clock frequency from the received signals.

Frame byte and boundary detection: Because the frame length is 48 bits, the frame and byte detector circuit

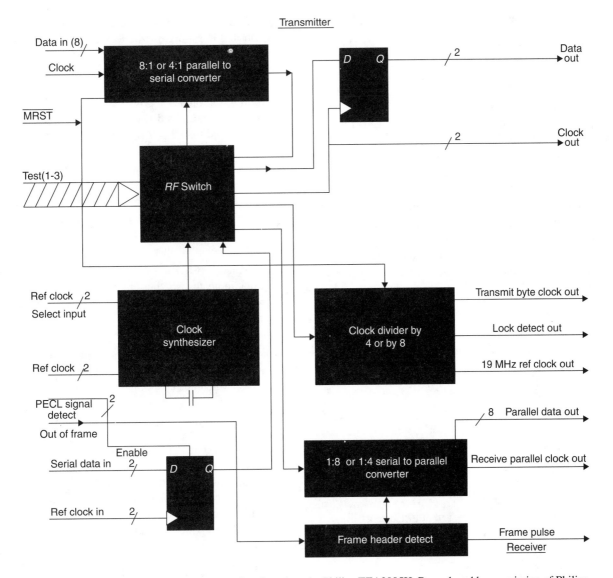

Figure 8–13. A block diagram of an optical transreceiver based on the Philips TZA3005H. Reproduced by permission of Philips Semiconductors.

TABLE 8–1 Conversion Serial Byte to Serial Bit for a 622.08 Mb/s Optical System

Serial Bytes	Data Rate per Input Line	Serial output
32	19.44 Mb/s	622.08 Mb/s
16	38.88 Mb/s	622.08 Mb/s
12	51.84 Mb/s	622.08 Mb/s
8	77.76 Mb/s	622.08 Mb/s
4	155.08 Mb/s	622.08 Mb/s

TABLE 8–2 Frequency Selection

Format	Mode	Buswidth	Frequency
OC3/STM1	0	0	38.88 MHz
	0	1	19.44 MHz
OC12/STM4	1	0	155.52 MHz
	1	1	77.76 MHz

searches for the 48 bit frame pattern that is composed of a sequence of three consecutive A1 (FOH) bytes followed by three consecutive A2 (28H) bytes. Frame and boundary detection can be enabled or disabled through the out of frame enable input. When enabled, the frame pattern is used to identify frame and byte boundaries from the input data stream. Regardless of the speed of the incoming data, the

required time to detect a frame must be less than 250 ns (ITU-G.783 recommendations).

Serial to parallel converter: Through the process of serial to parallel conversion, the converter circuit exhibits a delay between the first bit of the received byte and the start of the parallel output. This time delay is attributed to the time delay between the initial parallel load timing circuit synchronization and the receive parallel clock output (RXPCLK).

Transmitter input signals:

1) *Parallel data inputs* ($D_0 \rightarrow D_7$):   TTL input data word is aligned with the transmit-parallel clock input. D_7 is the MSB transmitted first and D_0 is the LSB transmitted last. All bits $D_0 \rightarrow D_7$ are sampled at the rising edge of the transmit parallel clock input. For a four bit word at the input, D_7 is still the MSB, while D_4 is the LSB for the rest of the unused inputs.

2) *Parallel clock input*:   The parallel clock input is a TTL signal with either a 19.44 MHz, a 38.88 MHz, a 77.76 MHz, or a 155.52 MHz frequency and a 50% duty cycle. One frequency is used to align the corresponding parallel input data. In addition, the parallel clock input is used to transfer the incoming input data to a temporary register located in the parallel to serial converter. The incoming data is sampled at the positive going transition (PGT) of the clock signal.

Transmitter output signals:

1) *Serial data output*:   The transmitter serial data output is a differential PECL signal intended for the input of

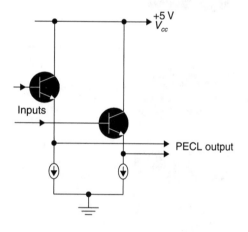

Figure 8–14.   a) A PECL output circuit.

a laser driver circuit. A typical PECL output circuit is illustrated in Figure 8–14a. PECL outputs must be terminated with resistors connected to V_{cc}. The value of these termination resistors is based on the V_{cc} power supply. A differential PECL output termination circuit based on a 5 V power supply is illustrated in Figure 8–14b, and a differential PECL termination circuit based on a 3.3 V power supply is illustrated in Figure 8–14c.

2) *Clock signals*:   Clock signals are also differential PECL signals required in order to retrieve the transmitter output signals. There are only two clock frequencies, 155.52 MHz or 622.08 MHz, depending upon the application. The selection of either one of the frequencies is based on the application.

3) *Lock detect*:   Lock detect is an active-high CMOS signal that indicates whether or not the PLL circuit is locked into the reference clock input signal. Lock conditions are indicated with a high output. A typical CMOS output circuit is illustrated in Figure 8–15.

4) *The 19 MHz clock output*:   The clock synthesizer generates a 19 MHz CMOS clock output, which can be used as a reference clock input for an external clock recovery circuit.

The receiver input signals:

1) *Serial data input*:   The incoming serial data is differential PECL AC coupling, which requires no external biasing, and which is clocked by a serial input clock and its complement. This data may come from the output of a data and clock recovery (DCR) circuit. A typical differential PECL input circuit is illustrated in Figure 8–16.

2) *Serial clock:* The serial clock signals are differential PECL synchronized to the incoming data. These signals are used for framing and deserialization, and can be AC coupled without the need for external biasing.

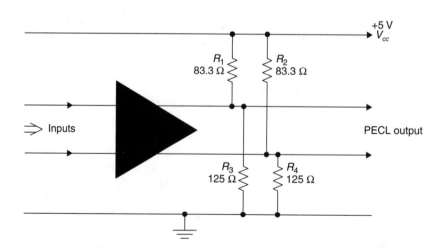

Figure 8–14.   b) A differential PECL output termination circuit where V_{cc} equals 5 V.

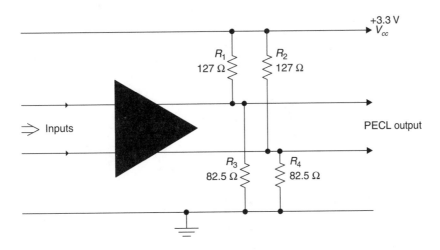

Figure 8–14. c) A differential PECL output termination circuit where V_{cc} equals 3.3 V.

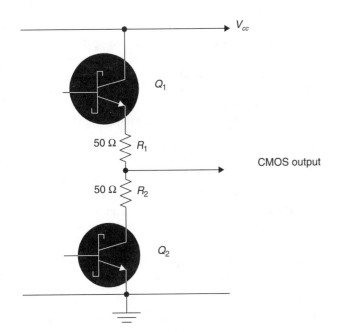

Figure 8–15. A CMOS output circuit.

3) *Optical signal detect*:   This input is used to indicate loss of optical power. It is a single ended PECL input connected to an internal pull down resistor, and it is driven by an external circuit.

4) *Signal detect (TTL)*:   The signal detect input performs the same function as the optical signal detect, but it is single ended TTL connected internally to a pull up resistor instead of to a PECL input. An input TTL circuit is illustrated in Figure 8–17.

Receiver output signals:

1) *Data outputs (parallel)*:   The parallel output bus of the receiver section ($D_0 \rightarrow D_7$) is synchronized with the received parallel clock output. D_0 is the LSR and D_7 is the MSB. If a four bit word is selected, D_7 is still the

MSB and D_0 is the LSB, while the rest of the outputs are disabled by connecting them to logic-0.

2) *Frame pulse*:   Frame pulse is a TTL output signal that indicates the detection of frame boundaries of the incoming serial input data. A TTL output circuit is illustrated in Figure 8–18.

3) *Parallel output clock*:   A parallel output clock is a TTL output clock with frequencies of 19.44 MHz, 77.76 MHz, or 155.52 MHz, 50% duty cycle, and negative going transition (NGT). This signal is used to synchronize the serial byte output signals $D_0 \rightarrow D_7$.

Input signals common to both transmitter and receiver sections:

1) *Buswidth selection*:   Buswidth selection is a TTL input signal used to select either an eight or four bit code of operation. When applying a logic low at this input, the four bit mode is implemented, while with a logic high, the eight bit mode of operation is implemented.

2) *Reference clock inputs*:   Reference clock inputs are differential PECL inputs used by the internal clock synthesizer circuit to synthesize the required clock signals.

3) *Mode-select*:   The mode select is a TTL input used to select the serial transmitter data rate. That is, when a logic low is applied at this input, the 155.52 Mb/s transmission rate is selected, while logic high at the same input results in the selection of the 622.08 Mb/s transmission rate.

4) *Receiver frame alignment*:   Receiver frame and byte boundary detection is enabled on the PGT (positive going transition) of the out of frame (OOF) enabling input, and remains enabled as long as the input remains high. Frame and byte boundary alignment is illustrated in Figure 8–19. Recognition of byte boundaries is accomplished after three consecutive A2 bytes have

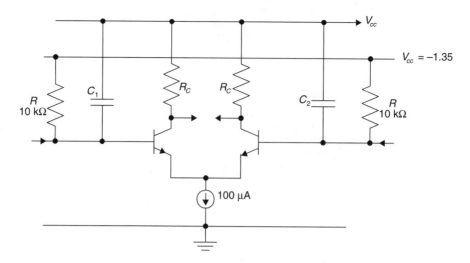

Figure 8–16. A differential PECL input circuit.

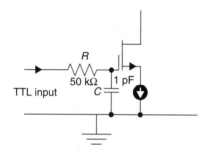

Figure 8–17. A TTL input circuit.

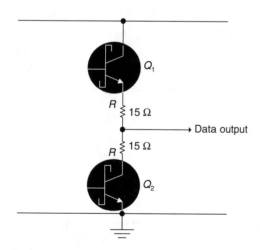

Figure 8–18. A TTL output circuit.

been received. At this time, frame pulse (FP) goes high and remains high for one receive parallel clock period, indicating receipt of the first data byte with correct alignment on the output bus. The rising edge of the OOF signal enables the frame and byte detection circuit. The circuit remains enabled until the FP changes state from low to high, and the OOF pulse goes from high to

low (Figure 8–20). An input signal timing diagram of the transmitter section is illustrated in Figure 8–21, and an output signal timing diagram is illustrated in Figure 8–22. Likewise, an input signal timing diagram of the receiver section is illustrated in Figure 8–23, and the output signal is illustrated in Figure 8–24.

Some of the important operating DC characteristics of the transreceiver are listed in Table 8–3 (see page 257). The AC characteristics of the transreceiver are listed in Table 8–4 (see page 258). The transreceiver absolute maximum values are listed in Table 8–5 (see page 258).

Optical Transreceivers Based on the Hewlett Packard HFCT 5200 Series

Hewlett Packard HFCT 5200 series transreceivers are used in optical fiber links and are in compliance with SONET/SDH operating formats. The transreceivers are composed of two major sections: the transmitter and the receiver.

THE TRANSMITTER SECTION. The transmitter section of the HP HFCT-5200 series optical transreceiver is composed of the following circuits:

a) PECL input

b) **laser modulator**

c) **laser bias driver**

d) **laser bias control**

Operation of the Hewlett Packard HFCT 5200 series: A block diagram of the transmitter section is illustrated in Figure 8–25 (see page 259). It utilizes a diode as the optical source, which is designed to comply with IEC-825 eye safety regulations.

Optical power output generated by the laser diode is monitored by the rear facet photodetector diode and con-

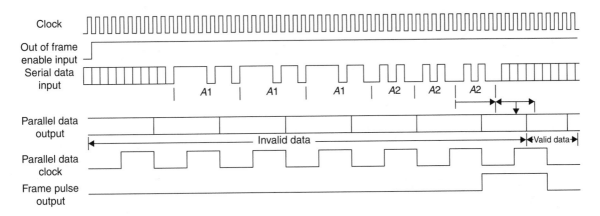

Figure 8–19. Frame and byte boundary alignment.

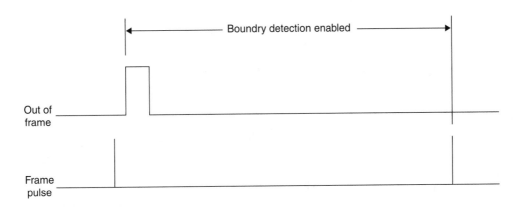

Figure 8–20. OOF timing diagram.

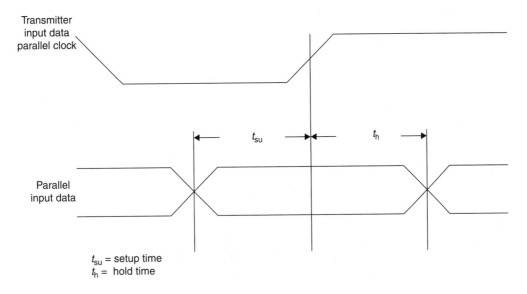

t_{su} = setup time
t_h = hold time

Figure 8–21. Input signal timing diagram of the transmitter section.

trolled by the laser bias current control circuit. The transmitter also provides AC and DC currents, ensuring optical power output under modulation or nonmodulation conditions, while maintaining **extinction ratio** and eye diagram displays within the specified power supply and tempera-ture variations over the entire operating life of the laser diode. The integrated circuit also provides laser diode shut down.

Data inputs: Transmitter inputs accept standard differential PECL signals that have amplitudes as low as

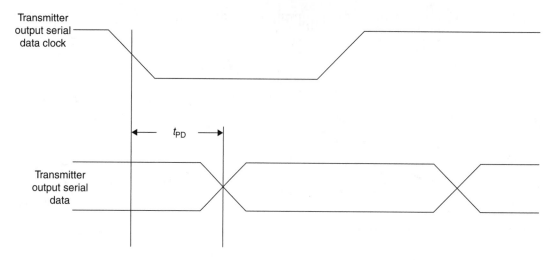

Figure 8–22. Output signal timing diagram of the transmitter section.

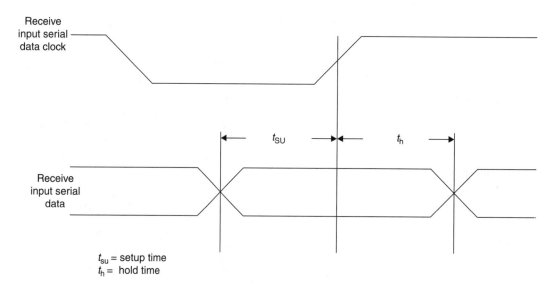

t_{su} = setup time
t_h = hold time

Figure 8–23. Input signal timing diagram of the receiver section.

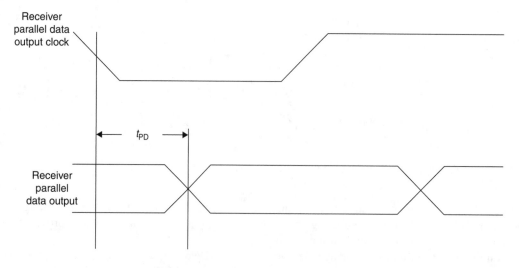

Figure 8–24. Output signal timing diagram of the receiver section.

TABLE 8–3   Transreceiver DC Operating Characteristics (T_{amb}=25° C and V_{cc} = 3.3 V)

Parameters	Minimum	Typical	Maximum	Conditions	Units
Supply voltage (V_{cc})	3.0	3.3	5.5	—	V
Power dissipation (P_{tot})	—	0.9	1.4	Open outputs V_{cc} = 3.45 V	W
	—	—	2.3	V_{cc} = 3.45 V	W
Supply current (I_{cc})	—	272	394	Open outputs V_{cc} = 3.45 V	mA
			420	V_{cc} = 3.45 V	mA
TTL Inputs					
High level input voltage (V_{IH})	2	—	V_{cc}	—	V
Low level input voltage (V_{IL})	0	—	0.8	—	V
High level input current (I_{IH})	–10	—	+10	$V_{IH} = V_{cc}$	μA
Low level input current (I_{IL})	–10	—	+10	$V_{IL} = 0$	μA
Pull up resistor ($R_{P–U}$)	8	10	12		kΩ
Pull-down resistor ($R_{P–D}$)	8	10	12		kΩ
TTL Outputs					
High level output voltage (V_{OH})	2.4	—	—	$I_{OH} = –1mA$	V
Low-level output voltage (V_{OL})	—	—	0.5	$I_{OL} = 4$ mA	V
PECL Inputs					
High level input voltage (V_{IH})	V_{cc}−1.2	—	—		V
Low level input voltage (V_{IL})	—	—	V_{cc}−1.6		V
High level output voltage (V_{OH}) V_{cc}	−1.1	—	V_{cc} − 0.9	50 Ω termination	V
Low level output voltage (V_{OL}) V_{cc}	− 1.9	—	V_{cc} − 1.6	50 Ω termination	V
Differential output voltage (V_{diff})	± 600	—	±900	50 Ω termination	mV
Differential input sensitivity ($V_{diff\,sens}$)	± 100	—	—	PECL inputs AC coupled	mV

250 mV$_{p–p}$. Both inputs are internally terminated to a V_{bb} power supply. If a single ended operation is required, the unused input does not need additional biasing or any other type of connection. However, single ended operation may slightly increase the duty cycle distortion (DCD) because of the imbalance occurring within the input circuit. This problem can be solved by decoupling the unused input to the power supply (V_{cc}) through an 8.6 pF capacitor.

The input capacitor also must be large enough to comply with the low frequency cut off ($f_{L_{cut\,off}}$) input requirements. Equation (8–11) expresses the relationship between low cut off frequency and input capacitor value.

$$f_{L_{cut\,off}} = \frac{1}{2\pi RC} \text{ (Hz)} \qquad (8\text{–}11)$$

where ($f_{L_{cut\,off}}$) is the low frequency cut off, C is the input capacitor, and R is the termination resistor. The low frequency cut off must be set at 1/10 of the lowest frequency content of the transmitted signal. For this application, the input capacitor value must be about 0.1 μF. The differential drive circuit of the transmitter section is illustrated in Figure 8–26 and the single-ended drive is illustrated in Figure 8–27.

For normal circuit operation in which a constant extinction ratio and duty cycle distortion is maintained, the input sig-

TABLE 8–4   Transreceiver AC Operating Characteristics

Parameters	Minimum	Typical	Maximum	Conditions	Units
CLK out serial	155.517	155.52	155.523	MODE-0	MHz
	622.068	622.08	622.092	MODE-1	MHz
Output data jitter	—	0.004	0.006	In lock	UI (rms)
Reference CLK frequency tolerance	−20	—	+20	SONET specs	ppm
Rise/fall times PECL outputs	—	220	450	50 Ω load	ps
Receiver Timing					
TTL output load capacitance	—	—	15		pF
Receiver parallel CLK output duty cycle	40	50	60		%
Propagation delay receiver CLK	−0.5	+1.5	+2.5		ns
Set up time	400	—	—		ps
Hold time	400	—	—		ps
Transmitter Timing					
Serial clock output duty cycle	40	50	60		%
Set up time	−0.5	—	—		ns
Hold time	1.5	—	—		ns
Transmitter serial CLK propagation delay	—	—	440		ps

TABLE 8–5   Transreceiver Absolute Maximum Values

Parameters	Minimum	Maximum	Units
Supply voltage	−0.5	6	V
Input voltage	−0.5	V_{cc} + 0.5	V
Differential PECL inputs	−2	+2	V
PECL signal detect input	V_{cc} − 3	V_{cc} + 0.5	V
Current into any TTL output	−8	+8	mA
Current into any PECL output	−50	+1.5	mA
Power dissipation (total)	—	1.5	W
Storage temperature	−65	+150	°C
Junction temperature	−55	+125	°C
Case temperature under bias	−55	+100	°C

nal must be kept at 50% of the duty cycle. If data is not applied at the input, the transmitter will transmit a specified nominal optical power output. Furthermore, the transmitter is able to function properly with a maximum of 72 consecutive 0s and 1s (PRBS) pattern in compliance with CCITT G.957 (ITU-T) recommendations for OC3/STM1 and OC12/STM4 formats. The transmitter exhibits a 350 ps delay between the time a data bit is applied at the input and the time a corresponding modulation current enters the laser diode.

The Laser Diode Bias Current Monitor: The optical transmitter incorporates a laser bias monitor circuit (Figure 8–28). In the presence of input data, the laser bias current is equal to the laser threshold current. In the absence of input data, the bias current will be equal to the laser forward current required to generate the specified optical power output in the absence of input data. The differential voltage generated across the bias monitor outputs is 10 mV/mA with a common mode voltage equal to V_{cc} − 1.6 V.

The bias current monitor can also be used to detect the end of life (EOL) bias current. A typical EOL bias current of 60 mA will generate a differential voltage of 0.6 V at the bias monitor output, which can drive an EOL indicator alarm. Because the bias current is temperature dependent, a temperature compensating circuit is recommended.

Optical Power Output Monitoring: The optical power generated by the laser diode is very critical to the link operation. A monitoring mechanism must be provided for constant monitoring of the laser power output. Such a circuit is illustrated in Figure 8–29.

The photodetector diode incorporated into the optical power monitoring circuit senses the rear facet power and the generated photocurrent (approximately 1.21 V) at the circuit output. The laser power monitor circuit is used to detect optical power failure by setting the alarm at ±50% of the initial optical level.

THE RECEIVER SECTION.   The receiver section incorporates an InGaAs/InP PIN photodetector diode, a pream-

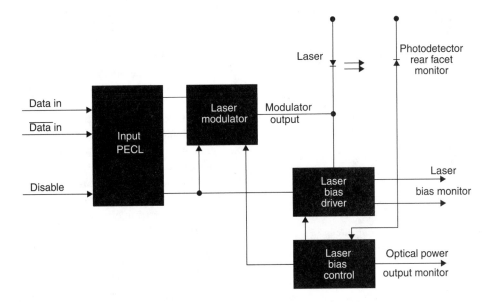

Figure 8–25. A block diagram of the transmitter section.

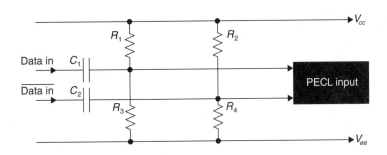

Figure 8–26. Differential drive operation.

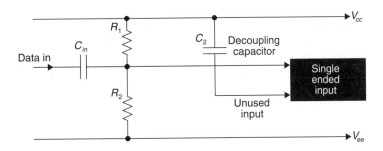

Figure 8–27. Single ended drive operation.

plifier circuit, and a postamplifier timing and recovery circuit. The preamplifier output is AC coupled to the postamplifier inputs. A block diagram of the receiver section is illustrated in Figure 8–30. The value of the coupling capacitor is selected in order to allow the processing of SONET/SDH signal patterns selected at the appropriate speed, without any reduction in receiver performance. If the input data speed is decreased substantially, or if a code with lower frequency content is used, performance charac-

teristics such as jitter pulse distortion and sensitivity will be reduced significantly.

The preamplifier input includes a filter network which limits the bandwidth of the received signal from the preamplifier circuit. By limiting the bandwidth of the preamplifier output signal, a substantial reduction in the noise component is also achieved. The gains achieved in the noise reduction are offset by a proportional reduction in receiver sensitivity. The receiver also incorporates a PSNR circuit. In special

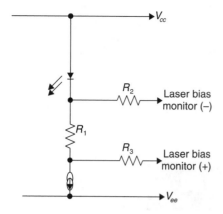

Figure 8–28. A laser diode bias monitoring circuit.

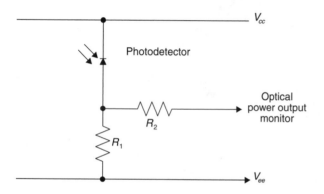

Figure 8–29. An optical power monitoring circuit.

cases, where electromagnetic interference (EMI) is an issue, external filtering is essential. An external filter, designed as a power supply noise reduction network, is illustrated in Figure 8–31. The implementation of the external PSNR filter also enhances receiver sensitivity. The component values of the external filter can be calculated to satisfy a particular application. Typical **end-of-life receiver power dissipation** is 1.443 mW, operating within the power supply range between 4.75 V and 5.25 V.

The Signal Detect Circuit: The peak level of the received signal is compared to a reference signal, resulting in a differential PECL logic output signal. The speed difference between the data path and the signal detect circuit, and consequently the AC noise generated by asymmetric loads, is practically zero.

The Clock and Data Inputs: The receiver section, in addition to acting as the data detect, also provides a clock signal and its complement at the output. Whether or not the clock outputs are used, they must all be appropriately terminated. Some of the termination techniques are illustrated in Figure 8–32. The receiver operates within SONET specified standards for jitter tolerance while maintaining a BER of 10^{-10}.

Lock Clock Signals: The required receiver clock frequency must exhibit zero drift, even in the absence of an input signal. In order to achieve this goal, an external frequency of 19.44 MHz is used, and is then multiplied to obtain the required system operating speed.

Optical Transreceivers Based on the Lucent Technologies TB16 2.5 Gb/s

GENERAL DESCRIPTION. The function of the TB16 2.5 Gb/s optical transreceiver demultiplexer module, developed by Lucent Technologies, is to perform parallel to serial conversion of a parallel 16-bit SONET/SDH format at the transmission section, and to perform serial to parallel conversion, frame and byte boundary detection, and data demultiplexing into OC48/STM-16 16-bit parallel format at the receiver section. A block diagram of this optical transreceiver module is illustrated in Figure 8–33. This module serves as the interface between the SONET/SDH photonic physical layer and the electrical section layer. A block diagram in Figure 8–33 shows that the transreceiver is composed of a 2.488 Gb/s (16 × 155.52 Mb/s) optical transmitter section, a 2.488 Gb/s receiver section, a 155.52 channel demultiplexer circuit, and a clock synthesizer circuit.

THE TRANSMITTER SECTION. The transmitter section of the optical transreceiver incorporates a DFB or Fabry-Perot laser diode as the optical device. Both lasers are capable of

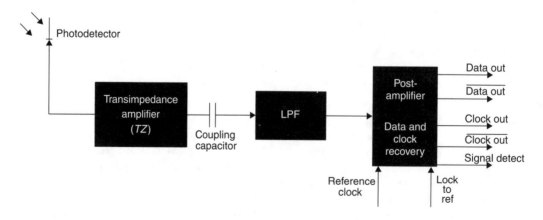

Figure 8–30. A block diagram of the optical receiver section.

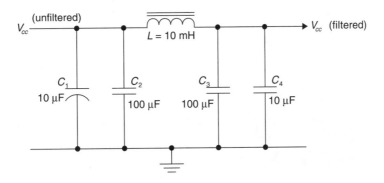

Figure 8–31. A PSNR filter.

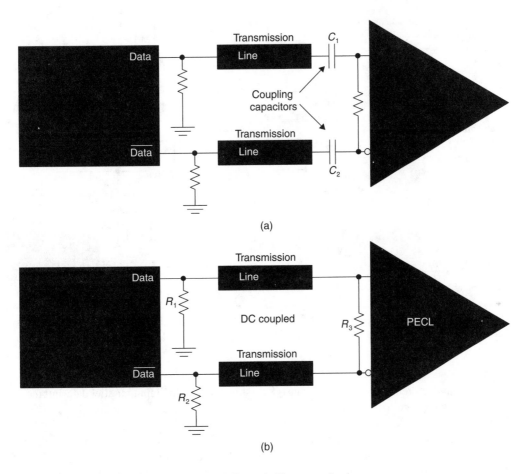

(a)

(b)

Figure 8–32. (a) An AC coupled input termination. (b) A DC coupled input termination.

operating at either 1310 nm or 1550 nm wavelengths. It also includes the parallel to serial converter circuit designed to perform the conversion of 16 (155.52 Mb/s) parallel data channels (serial byte) to 2488.32 Mb/s serial bits. A 155.08 MHz reference signal is used to synthesize a 2488.32 MHz clock that is required for transmitter synchronization.

The Parallel to Serial Converter: The parallel to serial converter in the transmitter section is composed of two shift registers of two byte length. The 16-bit parallel input data is latched into the first 16-bit parallel shift register at the

rising edge of the byte aligned parallel input clock. The second is a parallel input serial output shift register fed by the first register. An internal clock, which is in phase synchronization with the synthesized 2488.32 MHz serial transmitter clock, enables both shift registers. Thus, the parallel input data at the first register is converted to a serial bit stream at the output of the second register.

The Clock Driver and Phase Detector: Clock driver and phase detector circuits are designed to generate the required internal clock signals and the timing for the trans-

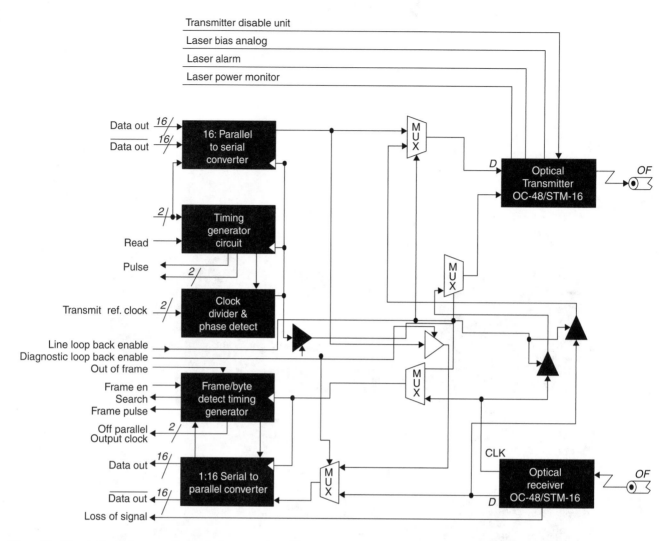

Transmitter disable unit
Laser bias analog
Laser alarm
Laser power monitor

Figure 8–33. A block diagram of an optical transreceiver, based on Lucent Technologies TB16 2.5 Gb/s module. Courtesy Agere Systems (Lucent Technologies).

mitter section. Because these signals are very critical to transmitter performance, the reference clock signal must exhibit very stable operating characteristics and must comply with SONET/SDH jitter requirements. The jitter level set forth by ITU-T recommendations for the SONET/SDH format must not exceed 0.01 unit intervals (UI) over the 12 KH to 20 MHz bandwidth. In order to satisfy these requirements, a crystal oscillator with a frequency stability of 20 ppm is recommended.

The Timing Generator Circuit: Timing generator circuits perform a dual function. They generate a byte-rate signal in synchronization with the 2488.32 MHz serial transmit clock. They also provide the required phase alignment between the data loading clock from the parallel input shift register to the parallel to serial shift register and the byte aligned parallel input clock. The transmitter also includes a parallel byte clock (differential PECL). Dividing the 2488.32 MHz internal clock frequency by sixteen generates this 155.52 MHz byte rate reference clock. This 155.52 MHz reference clock output signal is used for up-stream multi-

plexing, which requires a stable phase-frequency relationship between the input parallel data and the parallel to serial timing and processing circuits.

THE RECEIVER SECTION. The receiver section of the transreceiver module is composed of a PIN or APD photodetector diode, a **clock and data recovery (CDR)** circuit, a serial to parallel converter, and frame and byte detection circuits. The clock and data recovery circuit extracts the 2488.32 MHz clock signal from the detected serial input signal by the receiver section. The recovered clock signal and the retimed serial data signal are fed into the input of the frame and byte control circuit, and also into the input of the 16-bit serial to parallel converter.

The Serial to Parallel Converter: The serial to parallel converter is composed of three 16-bit shift registers. The first register performs the function of serial to parallel conversion, the second is an internal 16-bit shift register which transfers the output data from the first register to byte boundaries, detected by the frame and byte detection circuit.

The parallel data from the second shift register is transferred to the output of the third register at the negative going transition (NGT) of the byte aligned parallel output clock (155.52 MHz).

Frame and Byte Boundary Detection: Frame and byte boundary detection circuits scan the incoming serial data stream for three consecutive A1 data bytes followed by one A2 data byte. The enabling or disabling of the frame and byte detection circuit is accomplished through the frame pattern enable input (high for enabling), and the NGT of the out of frame input. The circuit is disabled when the frame pattern is detected. When the circuit is enabled, the frame and byte boundaries of the incoming signal are locked by the framing pattern. During the search time, the output parallel data is invalid.

The Timing Generator Circuit: The byte boundary located by the frame and byte detector is fed into the timing generator circuit which converts the incoming serial data into blocks of bytes appearing at the output of the 1:16 serial to parallel converter. If a 32-bit pattern matches the set frame pattern, the frame boundary is detected at the frame pulse output.

Loopback Mode of Operation: The optical transreceiver is capable of operating in two modes: **line loopback** and **diagnostic loopback.**

1) *Line loopback:* A receiver to transmitter loopback mode of operation is accomplished by applying a logic low at the line loopback enable input. With the loopback input low, the received serial data and the required 2488.32 MHz serial clock are fed from the optical receiver section directly to the transmitter section, thus establishing a loopback mechanism between the two sections.

2) *Diagnostic loopback:* The diagnostic loopback mode of operation is opposite to line loopback. That is, with a logic low at the diagnostic loopback input, the serial output data stream from the parallel to serial converter, and the clock signal of the transmit section are looped back into the frame and byte detector and into the serial to parallel converter circuit of the receiver section.

Transreceiver Input Termination: Incoming data is fed into the transmitter section through 50 Ω transmission lines. If maximum power transfer is to be achieved, the input lines must also be terminated to 50 Ω loads. One type of input line termination is illustrated in Figure 8–34. In the circuit of Figure 8–34, a 100 Ω resistor is connected between the differential clock and the data lines, and a 240 Ω resistor is connected between each input line and ground. This resistor combination provides the best match between the resistive network and the transmission line's 50 Ω impedance.

In the second case, the clock and input lines are terminated individually with an 82 Ω pulldown resistor to ground, and a 130 Ω pull up resistor to V_{cc}. This arrangement exhibits matching characteristics for the 50 Ω transmission line impedance. Some of the important transmitter section optical characteristics are listed in Table 8–6.

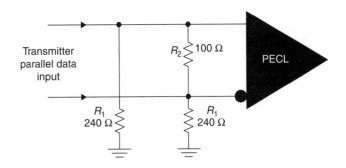

Figure 8–34. Input line termination.

Electrical characteristics of the transmitter section are listed in Table 8–7 (see page 265).

Optical characteristics of the receiver section are listed in Table 8–8 (see page 265).

The electrical characteristics of the receiver section are listed in Table 8–9 (see page 266).

The power supply characteristics are listed in Table 8–10 (see page 266).

Optical Transreceivers Based on the Texas Instruments SN75FC1000

GENERAL DESCRIPTION. The brief description of the following transreceiver is based on the Texas Instruments SN75FC1000 device. This transmitter section is designed to transmit at a maximum data rate of 1.0625 Gb/s and is compatible with ANSI X3T11 10-bit specifications. At the receiver end, it accepts typical 1 V differential PECL input signals terminated to either 50 Ω or 75 Ω impedance. The transreceiver operates with a 3.3 V power supply. This module is capable of accommodating very high speed data transmission (point to point) that complies with ANSI X3T11 fiber channel standard timing and functional requirements for 10-bit interface. The achievable maximum optical link distance through the implementation of this transreceiver depends on optical fiber losses and noise. A block diagram of the module, based on the TI 75F1000 transreceiver, is illustrated in Figure 8–35 (see page 266).

The SN75FC1000 performs the function of parallel to serial conversion for the fiber channel physical layer interface. The maximum operating speed of the module is 1.0625 Gb/s with a corresponding 100 MHz bandwidth. The parallel to serial circuit in the transmitter section accepts 8-bit or 10-bit parallel data and converts it to a serial bit stream. The serial data is transmitted to the input of the receiver section as a differential PECL-NRZ signal. The receiver section performs CDR. That is, it extracts clock and data information from the incoming serial data stream and also performs serial to parallel conversion, generating an 8-bit or 10-bit serial byte (parallel bits) at the output. Although the receiver is capable of automatically locking into the incoming data, it can utilize an external reference clock as the reset function upon demand. The transreceiver also exhibits loopback capabili-

TABLE 8–6   Transmitter Section Optical Characteristics

Parameters	Minimum	Typical	Maximum	Symbol	Unit
Optical Power Out (avg)					
Long haul 1300 nm (DFB) laser	–2	0	2	P_o	dBm
Long haul 1550 nm (DFB) laser	–2	0	3	P_o	dBm
Short haul (DFB) laser	–5	–2	0	P_o	dBm
Interoffice (F-P) laser	–10	–5	–3	P_o	dBm
Optical Wavelength					
Long haul 1300 nm (DFB) laser	1280	—	1335	λ	nm
Long haul 1550 nm (DFB) laser	1500	—	1580	λ	nm
Short haul (DFB) laser	1270	—	1360	λ	nm
Interoffice (F-P) laser	1270	—	1360	λ	nm
Spectral Width					
Interoffice (F-P) laser	—	—	4	$\Delta\lambda$	nm
Short haul (DFB) laser	—	—	1	$\Delta\lambda$	nm
Long haul (DFB) laser	—	—	1	$\Delta\lambda$	nm
Extinction ratio	8.2	—	—	r_e	dB
Side mode suppression ratio (DFB-laser)	30	—	—	SSR	dB
Optical rise and fall times	—	—	200	t_r, t_f	ns
Eye mask of optical power output	In compliance with ITU-T G.957 and GR-253				
Jitter	In compliance with ITU-T G.958 and GR-253				

ties. That is, data can be diverted directly from the transmitter section to the receiver section and vice versa. The purpose of this function is to permit physical layer self-testing.

THE TRANSMITTER SECTION. The incoming 10-bit word encoded data is applied to the input shift register at the rising edge of the 106.25 MHz reference clock. The 106.25 MHz reference clock frequency is multiplied by the 10-bit word, thus generating the 1.0625 Gb/s fiber channel transmission bit rate. The generated serial data output is transmitted bit by bit from D_0 to D_9. The time it takes for the serial byte (10-bit parallel word) at the input of the parallel to serial converter to be transferred to the output (10-bit serial data) is 13 ns and is referred to as time delay.

THE RECEIVER SECTION. The function of the receiver section is to recover the reference clock signal and the encoded data through the CDR circuit, and also to perform the serial to parallel conversion. The clock and data recovery circuit searches the incoming serial data stream for the required data pattern, and aligns it with the 10-bit word boundary. The recovered data and the two byte clocks (RBCs) are fed into the protocol controller circuit for further processing. The two byte clock signals are recovered

from the 531.25 MHz bandwidth that is required to accommodate the 1.0625 Gb/s fiber channel data stream dividing it by the 10-bit word. The resultant 53.125 MHz and its complement (180° out of phase) are the two required clock frequencies. The receiver PLL circuit is able to lock into the incoming 1.0625 Gb/s serial data stream without requiring a lock to reference function. The drift of the received 1.0625 Gb/s data must not exceed 100 ppm if proper operation is to be maintained. The receive byte clocks can deviate from the 53.125 MHz frequency up to a maximum of 60 MHz. This allows word alignment and bus error conditions.

The Phase Lock Loop: The PLL circuit of the receiver section provides an automatic lock for the incoming serial data stream with a critical lock time of 500 µs. A **lock-to-reference** input enables the PLL to lock-on to the external reference clock, thus allowing for PLL set/reset control. The PLL can lose lock due to an arbitrary shift of the incoming data on the order of 200 ppm. However, it can regain lock within a 2.4 µs time frame. This time reflects 2500 serial bit periods. Beyond the 200 ppm, the PLL regains lock within the 500 µs time period.

10-bit Word Recognition and Alignment: 10-bit word recognition and alignment are accomplished through the activation of the synchronous enable input (active high). This enables the synchronous detect circuit to compare ten consecutive bits of serial input data to a synchronization character (K28.5) defined by fiber channel standards. K28.5 consists of a 10-bit standard word, 0011111010, referred to as the "comma character." The comma character is compared to the ten consecutive serial bits received at the input of the receiver section. If the K28.5 bit character falls within the boundary of the received ten bits, proper alignment is achieved. The receive byte clock signal, the synchronization enable signal, and the 10-bit word signal are illustrated in Figure 8–36 (see page 267).

In the event that the received ten consecutive serial bits do not align with the K28.5 pattern, realignment is initiated. Shifting the 10-bit word boundary truncates the character, and alignment can be achieved. The fiber channel requires that three consecutive K28.5 signal patterns between frames must be recognized. The realignment process is illustrated in Figure 8–37 (see page 267).

The time delay between the arrival of the first serial bit at the receiver input and the time the same bit constructs the first bit of the output byte is 21 ns.

Loopback Function: Through the loopback function, serial data from the transmitter section can be rerouted to the receiver section and compared to the receiver parallel output data. Through this process, data verification and transreceiver self-testing can be accomplished. Some of the transreceiver internal circuitry relating to PECL, CMOS inputs, and CMOS outputs are illustrated in Figures 8–38, 8–39, and 8–40 (see page 268).

The differential electrical characteristics of the transmitter section are listed in Table 8–11 (see page 269).

TABLE 8–7 **Transmitter Section Electrical Characteristics**

Parameters	Minimum	Typical	Maximum	Symbol	Unit
Parallel input clock	153.9	155.52	157.0	PICLK	MHz
Parallel input clock duty cycle	40	—	60	—	%
Reference clock frequency tolerance	−20	—	20	TXREFCLK	ppm
Reference clock input duty cycle	30	—	70	TXREFCLK	%
Reference clock rise/fall times	—	—	0.5	t_R/t_f	ns
Input Data Signal Levels					
Input high V_{IH}	$V_{cc} - 1.1$	—	$V_{cc} - 0.7$	TXD(0–15)	V
Input low V_{IL}	$V_{cc} - 1.8$	—	$V_{cc} - 1.5$		V
Input voltage swing ΔV_{in}	150	—	—		mV
Disable input (transmitter)	$V_{cc} - 2$	—	V_{cc}	TXDIS	V
Enable input (receiver)	0	—	0.8	TXEN	V
Laser bias output voltage	0	0.2	1.6	LSRBIAS	V
Laser power monitor output	0.46	0.5	0.54	LPM	V
Line Loopback Enable: Active Low					
Input high (V_{IH})	2.0	—	$V_{cc} + 1.0$ V	LLOOP	V
Input low (V_{IL})	0	—	0.8		V
Diagnostic Loopback Enable: Active Low					
Input high (V_{IH})	2.0	—	$V_{cc} + 1.0$ V	LLOOP	V
Input low (V_{IL})	0	—	0.8		V
Parallel output clock				PCLK	
Output high (V_{OH})	$V_{cc} - 1.3$	—	$V_{cc} - 0.7$		V
Output low (V_{IL})	$V_{cc} - 2.0$	—	$V_{cc} - 1.4$		V
Link status response time	3	—	100		μs
Optical path penalty	—	—	—		dB

TABLE 8–8 **Receiver Optical Characteristics**

Parameters	Minimum	Typical	Maximum	Symbol	Unit
Receiver Sensitivity (avg.)					
PIN	−21	−25	—	P_{Rmin}	dBm
APD	−30	−34	—	P_{Rmin}	dBm
Optical Power (max.)					
PIN	—	0	3	P_{Rmax}	dBm
APD	—	−9	−8	P_{Rmax}	dBm
Link Status Switching Threshold: Decreasing Light Input					
PIN	−30	—	−20	LSTD	dBm
APD	−40	—	−30	LSTD	dBm

The transmitter section differential switching characteristics are listed in Table 8–12 (see page 269).

Transmitter section differential and common mode output voltage waveforms are illustrated in Figure 8–41 (see page 269).

The transmitter section timing requirements are listed in Table 8–13 (see page 269).

The transmitter section interface timing waveform is illustrated in Figure 8–42 (see page 270).

The valid 10-bit parallel input signal is defined from midpoint to midpoint of the reference clock signal.

The differential electrical characteristics of the receiver section are listed in Table 8–14 (see page 270).

The PLL circuit performance characteristics of the receiver section are listed in Table 8–15 (see page 270).

The receiver clock timing requirements are listed in Table 8–16 (see page 270).

The interface timing waveform of the receiver section is illustrated in Figure 8–43 (see page 271).

A typical application for the TI 75FC1000 optical transreceiver is illustrated in Figure 8–44 (see page 272).

TABLE 8–9   Receiver Electrical Characteristics

Parameters	Minimum	Typical	Maximum	Symbol	Unit
Parallel Output Clock				POCLK	
Output high (V_{OH})	$V_{cc} - 1.3$	—	$V_{cc} - 0.7$		V
Output low (V_{OL})	$V_{cc} - 2.0$	—	$V_{cc} - 1.4$		V
Parallel output clock duty cycle	40	—	60		%
Output Data Signal Levels					
Output high (V_{OH})	$V_{cc} - 1.3$	—	$V_{cc} - 0.7$		V
Output low (V_{OL})	$V_{cc} - 2.0$	—	$V_{cc} - 1.4$		V
Output data signal levels rise/fall time	—	—	1		ns
Frame Pulse				FP	
Output high (V_{OH})	$V_{cc} - 1.3$	—	$V_{cc} - 0.7$		V
Output low (V_{OL})	$V_{cc} - 2.0$	—	$V_{cc} - 1.4$		V
Loss of Signal Output					
Output high (V_{OH})	2.4	—	V_{cc}		V
Output low (V_{OL})	0	—	0.4		V
Out of Frame Input					
Input high	2.0	—	$V_{cc} + 1.0$		V
Input low	0	—	0.8		V

TABLE 8–10   Power Supply Characteristics

Parameters	Minimum	Typical	Maximum	Symbol	Unit
Supply voltage	3.13	3.3	3.47	V_{cc}	V
Supply current drain	—	2	—	I_{cc}	A
Power dissipation	—	6.6	—	P_{diss}	W

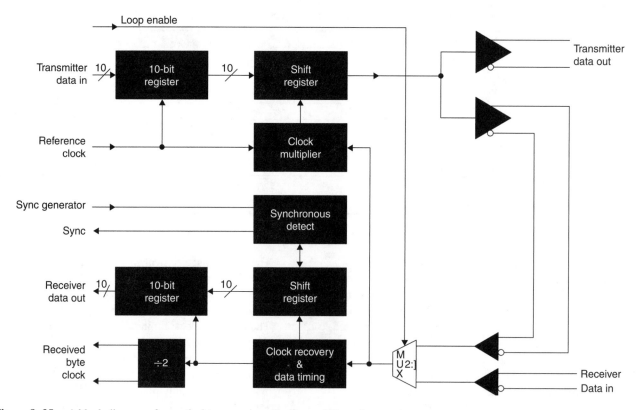

Figure 8–35.   A block diagram of an optical transreceiver. Courtesy of Texas Instruments.

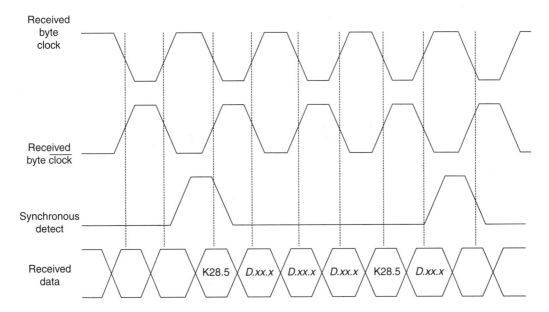

Figure 8–36. 10-bit synchronization waveforms.

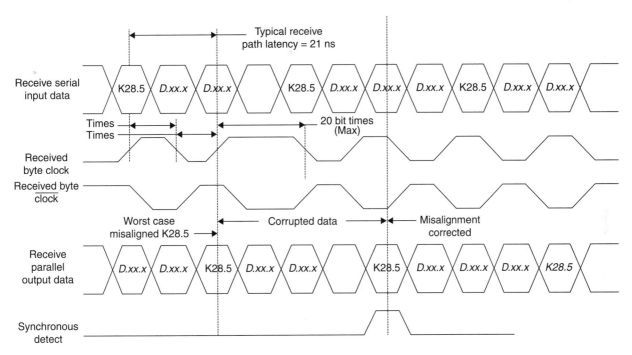

Figure 8–37. Realignment timing waveform.

The Microcosm MC4663 Ethernet Transreceiver

GENERAL CHARACTERISTICS. The MC4663 Ethernet optical transreceiver developed by Microcosm is a fully integrated transreceiver module that employs CMOS technology exclusively. The implementation of CMOS technology in the design and fabrication resulted in high power and an efficient device. The most important operating characteristics of the optical transreceiver are as follows:

a) single chip

b) low cost

c) deep submicron CMOS

d) input sensitivity: 2 mV

e) dynamic range: 62 dB, but if the module is to be interfaced with the MC2003ST photoreceiver, the system will exhibit sensitivity of –36 dBm.

f) resistor programmable 100 mA LED driver allows for low cost LEDs

g) on chip AUI driver/receiver

h) internally generated clock

i) single +5 V power supply

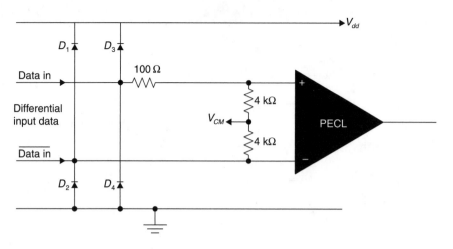

Figure 8–38. PECL input circuitry.

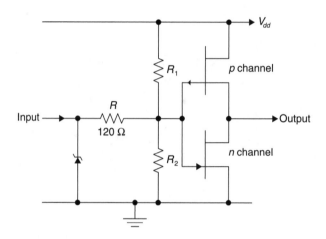

Figure 8–39. CMOS input circuitry.

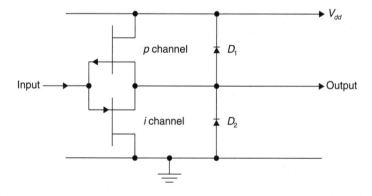

Figure 8–40. CMOS output circuitry.

j) five status LED drivers

k) low cost 10Base-T to FL converter

A block diagram of the Microcosm optical transreceiver is illustrated in Figure 8–45 (see page 274).

THE TRANSMITTER SECTION. The transmitter section of the optical transreceiver module accepts differential data from the AUI interface output through an isolation transformer. The data received at the input of the AUI transmit controls the LED driver current. That is, for a positive

TABLE 8–11   Transmitter Electrical Characteristics

Parameters	Minimum	Typical	Maximum	Symbol	Unit
Differential driver output voltage (p–p)	1.2	—	2.2	V_{OD}	V
Driver common mode output voltage	—	2.1	—	V_{OC}	V

TABLE 8–12   Transmitter Section Differential Switching Characteristics

Parameters	Minimum	Typical	Maximum	Symbol	Unit
Serial data deterministic jitter (p–p)	—	—	75	—	ps
Serial data total jitter (p–p)	—	—	197	—	ps
Differential signal rise time (20% to 80%)	—	—	300	t_r	ps
Differential signal fall time (80% to 20%)	—	—	300	t_f	ps

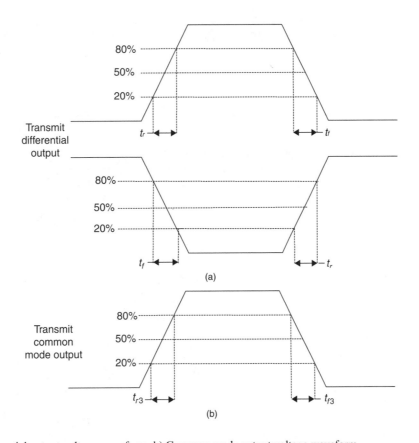

Figure 8–41.   a) Differential output voltage waveform. b) Common mode output voltage waveform.

TABLE 8–13   Transmitter Section Timing Requirements

Parameters	Minimum	Typical	Maximum	Symbol	Unit
$T_{D_0 - D_9}$ set up time (valid to ref. clock)	2	—	—	—	ns
$T_{D_0 - D_9}$ hold time (invalid to ref. clock)	1.5	—	—	—	ns
Parallel to serial time delay	—	13	—	—	ns

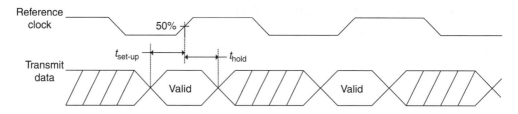

Figure 8–42. Transmitter section interface timing waveform.

TABLE 8–14   Receiver Section Electrical Characteristics

Parameters	Minimum	Typical	Maximum	Symbol	Unit
Differential input voltage	0.2	—	1.3	V_{ID}	V

TABLE 8–15   PLL Performance Characteristics

Parameters	Minimum	Typical	Maximum	Symbol	Unit
Jitter tolerance (input data eye closure)	—	—	70%	—	UI
Data acquisition lock time	—	—	500	—	μs
Data relock time	—	—	2.5	—	μs

TABLE 8–16   Receiver Clock Timing Requirements

Parameters	Minimum	Typical	Maximum	Symbol	Unit
Clock frequency (0)	—	53.125	—	RCLK(0)	MHz
Clock frequency (1) (180° out of phase)	—	53.125	—	RCLK(1)	MHz
Data rise/fall time	0.7	—	2	t_r/t_f	ns
Rise/fall time single-ended output	0.7	—	2	t_r/t_f	ns
Clock duty cycle	40%	—	60%	DC	—
Skew time	8.9	9.4	9.9	t_{skw}	ns
Set-up time $R_{D_0 \to D_9}$ valid: RCLK(0)↑	3	—	—	$t_{set\text{-}up(1)}$	ns
Set-up time $R_{D_0 \to D_9}$ valid: RCLK(1)↑	3	—	—	$t_{set\text{-}up(2)}$	ns
Set-up time $R_{D_0 \to D_9}$ invalid: RCLK(0)↑	1.5	—	—	$t_{set\text{-}up(3)}$	ns
Set-up time $R_{D_0 \to D_9}$ invalid: RCLK(1)↑	1.5	—	—	$t_{set\text{-}up(4)}$	ns
Serial-to-parallel data delay time	—	21	—	—	ns

input voltage $+ VT_x$, the LED will be reverse biased and will therefore be turned off. For negative input voltage, the LED will be forward biased and it will turn on. The differential input squelches undesirable noise, thus preventing it for propagating into the receiver section. The noise suppression (squelch function) is turned on for a differential input signal equal to or larger than 250 mV while remaining at that level for at least 180 ns. The transmitter latency or propagation delay time is 200 ns.

This is the time it requires two data bits of a valid Ethernet packet at the input of the AUI transmit to be transmitted by the LED diode. The output of the transmit circuit is connected to the input of the LED driver circuit, which operates as a current mode switch. The output of the LED driver circuit sinks current through the LED at the ON state and sinks current through T_xV_{cc} at the OFF state. An external resistor sets the required current level through the LED at the ON state.

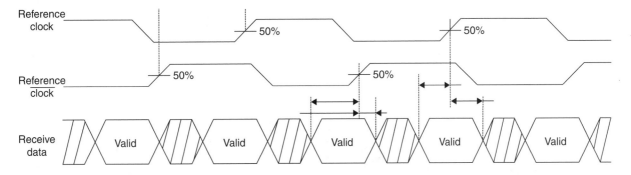

Figure 8–43. Interface timing waveforms of the receiver section.

THE RECEIVER SECTION The received optical signal is converted to an electrical signal by the PIN photodetector diode, and then amplified by the transimpedance amplifier. The differential output of the transimpedance amplifier is AC coupled to the input of the receiver section through two 1 nF capacitors. The input stage of the receiver section is composed of a limiting amplifier and a quantizer circuit. The receiver exhibits an input sensitivity of 2 mV and a dynamic range of 62 dB. It also incorporates a squelch circuit that is capable of rejecting frequencies below the 2.5 MHz range. The receiver also includes an optical monitoring circuit that is able to detect unacceptably low levels at the LED optical power output (low enough compromise the optical link performance integrity).

Loopback: Through the loopback path, data from the transmitter section can be transferred directly to the receiver section for testing and data verification. The loopback function is automatically disabled when link test has failed.

Collision Function: Data collision occurs when data is simultaneously applied to the AUI transmit interface and the AUI receive interface. Data collision is unacceptable beyond the 3.5 Ethernet data bits. At this point, a 10 MHz clock signal will be applied at the AUI collision interface.

Jabber Function: The jabber function is used to prevent the transmitter from transmitting beyond a maximum time. The time length of a transmit packet is set between 20 ms and 150 ms (10BASE-FL standards). The jabber function is set for a packet transmit time of 78 ms. This time is within the specified range. Restoration to the normal function will occur when an idle signal is detected at the input of the AUI transmit interface for a duration of 360 ms.

LED Status Indicators: The transreceiver also incorporates five LEDs as status indicators.

1) *Link monitor.* This LED will be on as long as the link optical power is maintained above the required threshold level.

2) *Receive packet.* This LED will be on during the receive packet process.

3) *Transmit packet.* Similarly, this LED will be on during the transmit packet process.

4) *Jabber active.* This LED will be illuminated during the active jabber.

5) *Collision.* This LED will be illuminated if data collision is detected.

For proper operation, a 510 Ω pull down resistor is required for each LED status indicator that is connected between the driving circuit and the power supply.

Power Supply Decoupling: Due to the highly sensitive circuitry implemented into the transreceiver design, the PSNR must be maximized. For more on PSNR, see Chapter 9.

The device operating characteristics are listed in Table 8–17 (see page 274).

The absolute maximum ratings are listed in Table 8–18 (see page 274).

Some of the important DC characteristics are listed in Table 8–19 (see page 274).

The transmit timing characteristics are listed in Table 8–20 (see page 275).

The receiver timing characteristics are listed in Table 8–21 (see page 275).

The collision timing characteristics are listed in Table 8–22 (see page 275).

Jabber timing characteristics are listed in Table 8–23 (see page 275).

Squelch test timing characteristics are listed in Table 8–24 (see page 275).

The LED indicator outputs are listed in Table 8–25 (see page 275).

Optical Transreceivers Based on the Vitesse VSC7185 Quad Module

GENERAL DESCRIPTION. The Vitesse VSC7185 is a quad fiber channel and Gigabit Ethernet optical transreceiver module designed to accept 8- or 10-bit encoded parallel data and to convert it to a serial data stream at the transmitter end with a maximum speed of 1.36 Gb/s. It incorporates four transmitters with a 5-bit parallel input bus connected to the input of each transmitter by a 272 MHz clock. At the receiver input, the received data is sampled by the clock and data recovery circuit, then converted to 5-bit parallel word and compared to K28.5 character. Clock and data recovery is accomplished through a PLL circuit designed to sample the incoming serial data stream, to extract

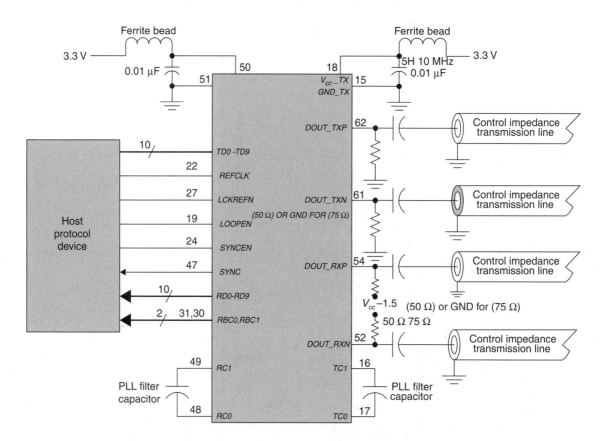

Figure 8–44. Application diagram employing the TI 75FC1000 device. Courtesy of Texas Instruments.

List of Symbols

Symbol	Explanation
DOUT_TXP DOUT_TXN	Differential output transmit: Differential serial outputs interfaced to an optical module, capable of transmitting 1.0625 Gb/s NRZ signals.
DIN_RXP DIN_RXN	Differential input receive: Differential serial input interface from an optical module, which receive 1.0625 Gb/s NRZ signals.
LCKREFN	Lock to reference: With lock to reference low, the PLL circuit of the receiver section locks on to the supplied reference clock signal.
LOOPEN	Loop enable: A logic high at the loop enable input activates the internal loop back path, the serial data stream at the transmitter output is diverted directly to the receiver section for data verification and testing.
RBC1, RBC2	Receive byte clock: The two recovered byte clocks, 53.125 MHz each, are used to synchronize the output 10-bit data $(D_0 \rightarrow D_9)$, producing a valid output at the positive going transition (PGT). Both clocks are adjusted in the middle of the word boundaries and in reference to synchronous detect. During data realignment, both clocks can be expanded but never truncated. The receive byte clock one (RBC1) registers bytes 0 and 2 while clock two registers bytes 1 and 3 of the received data.
RC0, RC1	Receive capacitor: The external capacitor required by the internal filter of the PLL circuit. Receive capacitor recommended value is 2 nF.
$D_0 \rightarrow D_9$	Receive data: Receive data $(D_0 \rightarrow D_9)$ refers to the 10-bit parallel output word at the output of the receiver section of the transreceiver module, referenced to the receive byte clocks RBC1 and RBC2. The received data byte 0 contains the K28.5 character, and is aligned to the rising edge (PGT) of the RBC1.
REFCLK	Reference clock: The reference clock is a 106.25 MHz external clock used for transmitter and receiver synchronization. The reference clock is used by the transmitter to register the input parallel data $(TD_0 \rightarrow TD_9)$ for parallel to serial conversion. It is also used by the receiver as the PLL preset.

continued

Symbol	Explanation
SYNC	Synchronous detect: Synchronous detect output goes high and remains high for 50% of the reference clock (REFCLK) time period upon detection of the K28.5 character from the serial data stream at the receiver input.
SYNCEN	Synchronous function enable: A synchronous function enable high activates the internal synchronization function. By enabling this function, detection of the K28.5 character is accomplished from the serial input data stream. Data realignment on byte boundaries can be initiated if necessary.
TC0, TC1	Transmit capacitor: Similar to the receive capacitor, the transmit capacitor is required by the filter of the transmitter PLL circuit. The capacitor value is 2 nF.
$TD_0 \rightarrow TD_9$	Transmit data: Transmit data is the parallel data output of a protocol interface applied at the input of the parallel to serial converter of the transmitter section, so that it can be converted to a serial data stream for further transmission. The $TD_0 \rightarrow TD_9$ 10-bit parallel data is locked on to the transreceiver on the positive going transition of the reference clock (REFCLK).
V_{cc}_A	Analog power: The analog power supply provides the required reference voltage to the high-speed analog circuits of the transreceiver module.
V_{cc}_CMOS	Digital PECL logic power: The digital PECL logic power supply provides an isolated low-noise power to the logic circuits.
V_{cc}_RX	Receiver power: The receiver power supply provides a low noise reference voltage to the receiver high speed analog circuits.
V_{cc}_TTL	TTL power: The TTL power supply provides the required reference voltage for the TTL circuits of the receiver section.
V_{cc}_TX	Transmitter power: Transmitter power supply provides the required low noise reference voltage for the transmitter high speed analog circuits.
GND_A	Analog ground: The analog ground provides the required ground reference to the high speed analog circuits of the transreceiver module.
GND_CMOS	Digital PECL logic ground: The digital PECL logic ground provides an isolated low noise ground reference to the logic circuits.
GND_RX	Receiver ground: The receiver ground provides a low noise ground reference to the receiver high speed analog circuits.
GND_TTL	TTL ground: The TTL ground provides the required ground reference for the TTL circuits of the receiver section.
GRD_TX	Transmitter ground: Transmitter ground provides the required low noise ground reference for the transmitter high speed analog circuits.

the clock, and to synthesize the transmit clock. A block diagram of the transreceiver module is illustrated in Figure 8–46. Some of the important operating characteristics of the transreceiver module are as follows:

a) a single integrated chip incorporates four transreceivers

b) each channel is capable of speeds from 1.05 Gb/s to 1.36 Gb/s

c) common transmit byte clock

d) common mode detect enable input

e) common serial/parallel loopback controls

f) 1/10 baud rate recovered clocks

g) cable equalization in receivers

h) 3.3 V power supply, 2.5 W power dissipation

The Transmitter Section

THE CLOCK SYNTHESIZER: The baud-rate clock (1.05 GHz to 1.36 GHz) is generated by the clock multiplier unit (CMU).

The loop filter of the PLL circuit is controlled by a 0.1 μF capacitor, which minimizes the effect of common mode noise on the clock multiplier unit (generated mainly by the power supply). It is advisable to use capacitors in order to achieve the best possible noise reduction. This arrangement is illustrated in Figure 8–47 (see page 276).

Parallel to Serial Conversion: The parallel byte (5-bit word) input data and the byte clock are applied to the four parallel to serial converters converting them to four serial high speed data streams as differential outputs. The serialized data is encoded to 8- or 10-bit or equivalent code format.

The Receiver Section

CLOCK RECOVERY: The clock recovery circuit extracts clock information from the high speed serial input data and also retimes the recovered data. A number of equalizer circuits are incorporated into the design to compensate for signal deterioration due to intersymbol interference (ISI). The overall receiver performance can be evaluated through translation of the eye diagram. Encoding the serial bit stream

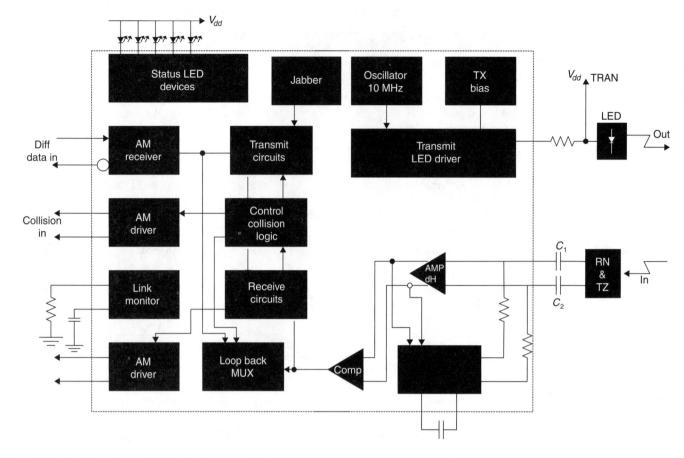

Figure 8–45. A block diagram of the MC 4663 optical transreceiver.

TABLE 8–17 Operating Conditions

Parameters	Minimum	Typical	Maximum	Symbol	Unit
Supply voltage	4.75	5.0	5.25	V^+	V
Current consumption with LED driver current of 55 mA	70	80	90	I_{cc}	mA
Ambient temperature	−40	0	85	T_{amb}	°C

TABLE 8–18 Absolute Maximum Ratings

Parameters	Minimum	Maximum	Symbol	Unit
Power supply voltage	−0.3	6	V_{dd}	V
Input voltage (digital inputs)	0.3	V_{dd}	V_{in}	V
Input current	—	60	I_{in}	mA
Bias current	—	10	I_{bias}	mA

TABLE 8–19 DC Characteristics

Parameters	Minimum	Typical	Maximum	Symbol	Unit
Input voltage (high)	$V_{dd} - 0.5$	—	—	$V_{in(high)}$	V
Input voltage (low)	—	—	0.5	$V_{in(low)}$	V
LED transmit peak output current	80	100	120	$I_{o(LED)}$	mA
AUI drivers common mode output	$V_{dd} - 2.5$	$V_{dd} - 2$	$V_{dd} - 1$	$V_{AUI(CM)}$	V
Differential amplifier gain	40	50	60	A_u	—
Input resistance	1.5	2.0	2.5	R_{in}	kΩ

TABLE 8–20   Transmit Timing Characteristics

Parameters	Minimum	Typical	Maximum	Symbol	Unit
LED ON current	—	—	120	$I_{LED(ON)}$	mA
LED OFF current	—	—	3.6	$I_{LED(OFF)}$	mA
Optical output pulse rise/fall times	—	—	3	t_r/t_f	ns
Transmit steady state delay	—	10	50	t_{Tssd}	ns
Transmit start delay and number of header bits not transmitted	—	—	2	—	bits
Transmit turn OFF time delay from data to idle	0.8	1.1	1.4	$t_{turn\,OFF\,(d-i)}$	μs
Loopback steady state delay	—	20	50	t_{lssd}	ns
Loopback start delay and number of header bits not transmitted	—	—	4	—	bits
Transmit idle frequency	0.85	1	1.25	f_{Ti}	MHz
Transmit mark/space ratio	45	—	55	—	%
Collision occurrence to signal quality error	—	—	350	t_{c-sqe}	ns

TABLE 8–21   Receive Timing Characteristics

Parameters	Minimum	Typical	Maximum	Symbol	Unit
Receiver steady state delay	—	10	50	t_{Rssd}	ns
Receive data to idle delay	—	300	—	t_{Rdid}	ns
Receiver squelch frequency threshold	—	2.5	—	f_{Rsth}	MHz
Receiver start delay and number of header bits not reproduced	—	—	2.5	—	bits
Differential rise/fall times	—	4	—	$t_{r/f(diff.)}$	ns

TABLE 8–22   Collision Timing Characteristics

Parameters	Minimum	Typical	Maximum	Symbol	Unit
Collision occurrence to signal quality error	—	—	350	t_{c-sqe}	ns
Collision occurrence to receive data	—	—	500	t_{c-rd}	ns
End of collision	—	—	500	—	ns

TABLE 8–23   Jabber Timing Characteristics

Parameters	Minimum	Typical	Maximum	Symbol	Unit
Packet transmit length	40	78	125	—	ms
Unjab time	280	360	500	—	ms

TABLE 8–24   Squelch Test Timing Characteristics

Parameters	Minimum	Typical	Maximum	Symbol	Unit
Time from end of transmit to signal quality error test	0.75	0.9	1.2	—	μs
Signal quality error test	0.7	0.8	1.1	—	μs

TABLE 8–25   LED Indicator Outputs

Parameters	Minimum	Typical	Maximum	Symbol	Unit
Status LED stretch time	4	5	6	—	ms
Status LED low time after retrigger	70	100	130	—	ns

is essential if long runs of zeros and ones are to be avoided. The recovered baud rate should be within ±200 ppm of ten times the reference frequency. For the gigabit Ethernet system, a 125 MHz clock that has an accuracy of ±100 ppm must be used.

Serial to Parallel Conversion: The function of the serial to parallel converter circuit is to convert the received

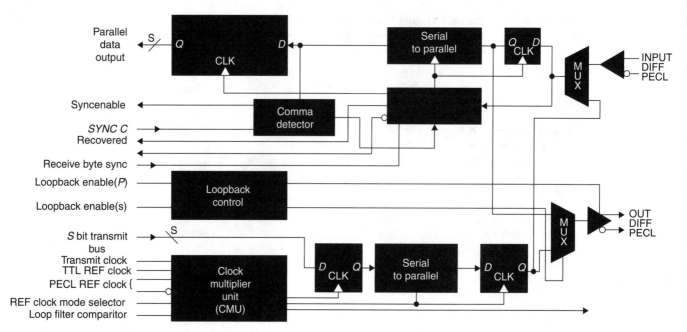

Figure 8–46. A quad optical transreceiver based on the Vitesse VSC7185 module. Reproduced by permission of the Vitesse Semiconductor Corporation.

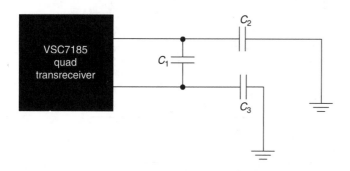

Figure 8–47. A three-capacitor loop filter scheme.

serial data stream into a 5-bit parallel word. The receive section also produces two clock signals of 1/10 or 1/20 of the serial baud rate. The recovered clock is divided by ten or by twenty, while the recovered high speed data is retimed and then converted to a parallel byte at the receiver output. In the event that the incoming serial data does not meet the required baud rate or is missing altogether, the receive section will continue to generate a reference clock signal in order for the rest of the circuitry to function properly.

Word Alignment: The optical transreceiver is cap-able of providing 10-bit comma character and data word alignment. The serial data is converted to 10-bit words, while recognizing a comma character. 10-bit comma characters, such as the K28.1, K28.5, or K28.7 are not contained in any of the 8- or 10-bit coded data characters, and are used only for synchronization purposes. In the event that the "comma character" is not aligned, the recovered clock will be stretched (never slivered) to the point where both the comma character and the received

clock are properly aligned to the 5-bit word. Thus, data word and character synchronization is accomplished. Synchronization waveforms indicating both alignment and misalignment are illustrated in Figure 8–48.

Loopback: The loopback function enables data from the transmitter section to be transferred directly to the receiver section for data verification and evaluation.

Transmitter AC characteristics are listed in Table 8–26.

The receiver AC characteristics are listed in Table 8–27.

The reference clock requirements are listed in Table 8–28.

The DC operating characteristics are listed in Table 8–29.

The absolute maximum values are listed in Table 8–30.

8.4 DESIGN GUIDELINES FOR A FIBER CHANNEL TRANSRECEIVER

The following guidelines are essential for the design of a transreceiver module for a functional and reliable optical link operation.

Reference Clock Jitter

Optical transreceiver performance and overall link performance is largely dependent on the reference clock signals. This external clock provides the required reference signal for the PLL circuit, and multiplied by ten it provides the baud rate clock. Both clock signals (the external and the internally generated baud rate clock signals) must be in phase. Phase lock of the two clock signals is accomplished at the rising edge of the external clock. The

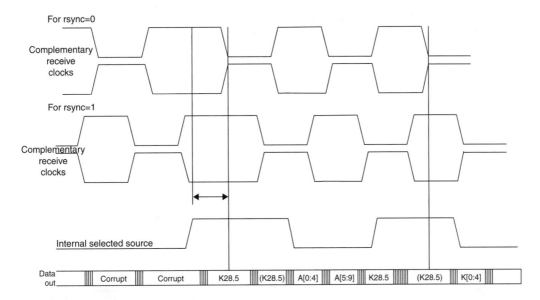

Figure 8–48. Synchronization waveforms.

voltage control oscillator of the PLL circuit will adjust its speed in order to offset any phase drift between the two clock signals, thus maintaining phase lock. Signal quality, and therefore system integrity, largely depends on the quality of the reference clock. The two key performance characteristics of the reference clock signal are jitter and rise time. Jitter must be kept at an absolute minimum level.

Any excessive jitter will be induced into the PLL circuit and into the baud rate clock.

REFERENCE CLOCK RISE TIME. The reference clock rise time, the time it takes the clock signal to change from low to high, is critical. A very fast transition will eliminate the possibility of ambiguity in the input buffer and reduce

TABLE 8–26   Transmitter AC-Characteristics

Parameters	Minimum	Typical	Maximum	Symbol	Unit
Input data and test clock transition range	—	—	2.0	T_{TXCT}	bits
Input data and test clock valid time	3.0	—	—	T_{TXCV}	bits
Input data and test clock set up time	700	—	—	T_{TXS}	ps
Input data and test clock hold time	700	—	—	T_{TXH}	ps
Test clock duty cycle	35	—	65	–	%
Transmitter time delay (latency)	11-bits+1 ns	—	8-bits+2 ns	T_{TXLAT}	ns

TABLE 8–27   Receiver AC-Characteristics

Parameters	Minimum	Typical	Maximum	Symbol	Unit
Frequency lock time (from power ON)	—	—	500	f_{Lock}	μs
Synchronization bit time Kbits	—	—	2.5	B-sync	
Receiver set up time	1.5	—	—	T_{RXS}	bits
Receiver hold time	1.5	—	—	T_{RXH}	bits
Transition range	—	—	1.25	T_{RXCT}	bits
Receiver time delay (latency)	18.5-bits+2 ns	—	19.5-bits+5 ns	T_{RXLAT}	ns

TABLE 8–28 Reference Clock Requirements

Parameters	Minimum	Typical	Maximum	Symbol	Unit
Frequency range	105	—	136	f_{range}	MHz
Frequency offset	−200	—	+200	f_{offset}	ppm
Receiver baud rate clock duty cycle	35	—	65	TL_{RO}/TH_{RO}	%
Receiver baud rate clock rise/fall time	0.25	—	1.5	TR_{RO}/TF_{RO}	ns
Byte clock duty cycle	35	—	65	TL_{TC}/TH_{TC}	%
Reference clock input duty cycle	35	—	65	TL_{RF}/TH_{RF}	%
Byte clock rise and fall time	—	—	0.45	TR_{TC}/TF_{TC}	%
Reference clock rise and fall time	—	—	1.5	TR_{RF}/TF_{RF}	ns

TABLE 8–29 DC Operating Characteristics

Parameters	Minimum	Typical	Maximum	Symbol	Unit
TTL Inputs					
Input high voltage	2.0	—	5.5	V_{IH}	V
Input voltage (low)	0	—	0.8	V_{IL}	V
Input current (high)	—	50	500	I_{IH}	μA
Input current (low)	—	—	−500	I_{IL}	μA
TTL Outputs					
Output voltage (high)	2.4	—	—	V_{OH}^{-}	V
Output voltage (low)	—	—	0.5	V_{OL}	V
PECL Inputs					
Input voltage (high)	$V_{cc} - 1.1$	—	$V_{cc} - 0.7$	V_{IH}	V
Input voltage (low)	$V_{cc} - 2.0$	—	$V_{cc} - 1.5$	V_{IL}	V
Input current (high)	—	—	200	I_{IH}	μA
Input current (low)	−50	—	—	I_{IL}	μA
Input differential peak to peak swing	400	—	—	ΔV_{in}	mV
High Speed Inputs					
Differential PECL (p–p) voltage swing	0.2	—	2.6	ΔV_{in}	mV
High Speed Outputs					
Differential output (p–p) voltage swing (for 50 Ω)	1.0	—	2.2	ΔV_{out}	V
Differential output (p–p) voltage swing (for 75 Ω)	1.2	—	2.2	ΔV_{out}	V

TABLE 8–30 Absolute Maximum Values

Parameters	Minimum	Maximum	Symbol	Unit
Power supply voltage	−0.5	+ 4.0	V_{cc}	V
DC input voltage	−0.5	$V_{cc} + 0.5$	PECL inputs	V
DC input voltage	−0.5	5.5	TTL inputs	V
DC output voltage	−0.5	$V_{cc} + 0.5$	TTL outputs	V
Output current	−50	+50	PECL outputs	mA
Output current	−50	+50	TTL outputs	mA
Case temperature	−55	+125	—	°C
Storage temperature	−65	+150	—	°C

PLL jitter. Furthermore, because the external clock is used to align the internally generated clock by the PLL circuit, which is used to transfer the data from the transmitter bus to the input latch, the critical set up and hold times are subject to reference clock jitter. A small amount of jitter significantly reduces the set up and hold times. A crystal oscillator will provide the anticipated reference clock signal quality. For the optical link to perform at the anticipated level, two critical elements must comply with the design requirements at all times. That is, the reference clock signal at the transreceiver and protocol inputs must maintain phase lock and must also exhibit good quality. Another scheme for reference clock generation is to incorporate the oscillator into the protocol chip. This scheme has certain advantages and disadvantages. The main advantage is that the protocol chip is better able to meet the set up and hold time transreceiver requirements, especially when operating at 106.25 MHz (10-bit interface). Such a scheme is illustrated in Figure 8–49.

By using this method, the set up time is eased somewhat because the reference clock output buffer and the output latch of the transmit data are on track. The disadvantage of this arrangement is that the protocol chip adds jitter into the reference clock (so that it may be outside the fiber channel jitter limits). The best possible jitter results can be achieved when the external oscillator drives both the transreceiver and the protocol chip. This arrangement is illustrated in Figure 8–50.

The external 22 Ω resistors are required in order to minimize EMI reflections.

Serial Transmission (High Speed)

The differential PECL transmitter outputs require proper termination in order to maintain signal quality. These high speed (1.0625 Gb/s) 8- or 10-bit encoded signals must be able to travel at maximum distances while maintaining optimum signal quality.

Optical Fiber Termination

The maximum distance between a fiber channel transmitter and a receiver can be achieved through the optical fiber medium. The utilization of an optical fiber as the transmission medium provides maximum transmission distance and reduces the EMI effect to a minimum. For differential PECL outputs, two coupling capacitors and two pull down resistors are required. Such an arrangement is illustrated in Figure 8–51.

Assuming a short distance of approximately 5 cm between the transmitter and the receiver section on the transmitter side, two 0.01 μF capacitors and two 180 Ω resistors are required. On the receiver end, two resistors of 51.1 Ω and two 0.01 μF capacitors are needed.

Oscillations Prevention Mechanism

There is a possibility that valid data will not be detected at the receiver inputs due to a link disconnect or transmitter disabling. In this case, both inputs are internally terminated at the same bias points. When identical differential input signals are buffered, the input buffer is sensitive to noise environment with the probability that this noise will be transmitted to the receiver section. In order to prevent this from happening, the transmission lines must be properly terminated, and a small DC offset must be provided at the input of the receiver. This is accomplished through a Thevenin equivalent resistor pair, as illustrated in Figure 8–52.

An alternative oscillations prevention mechanism is illustrated in Figure 8–53.

Unused Inputs

In the event that the receiver inputs are not used, they must be properly terminated in order to avoid undesirable oscillations. The two key components needed to prevent undesirable

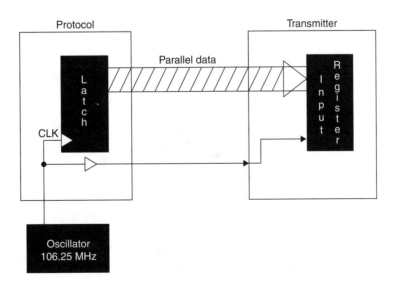

Figure 8–49. Reference clock generated by the protocol chip.

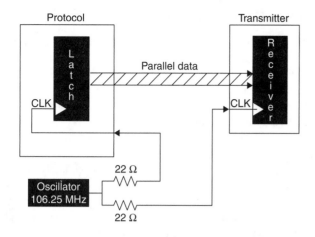

Figure 8–50. An external oscillator driving both the transreceiver and the protocol chip.

oscillations from occurring are DC offset and a low impedance noise path. A receiver termination scheme capable of satisfying these requirements is illustrated in Figure 8–54. Two external resistors of 47 kΩ each are connected in parallel to 3 kΩ internal termination resistors. The combination yields a satisfactory 50 mV DC offset voltage. Furthermore, a 0.01 μF capacitor, connected between the receiver's unused inputs, provides the low impedance path essential for noise reduction.

Power Supply Requirements

The VSC7185 transreceiver module needs a single +3.3 V (±5% power supply). A voltage conversion from 5 V to 3.3 V must be made. Two basic types of power supplies exist: the switching power supply and the linear power supply. The optical transreceiver module is powered by a linear regulator power supply because it is simpler in that its design requires a smaller number of components and exhibits higher efficiency. Another advantage for utilizing a linear regulator is that it provides complete noise isolation

between the 3.3 V output and the 5 V input. This further enhances signal quality by reducing transmitter jitter subject to power supply generated noise. A typical linear regulator ideal for the above application is the LT1117CST-3.3 developed by Linear Technology. This device provides a +3.3 regulated output voltage with a maximum current of 800 mA. A block diagram of the regulator is illustrated in Figure 8–55. Some of the linear regulator power supply's design requirements are as follows:

Optical transreceivers require that the input voltage be maintained within the ±5% range under worst case conditions. 5 V to 3.3 V voltage converters that employ linear regulators incorporate input and output capacitors of varying values, based on specific requirements.

Figure 8–55 shows that a 10μF capacitor is used at the input and another 10 μF capacitor and two 0.1 μF capacitors are used at the output to provide the required stability for the linear regulator. The output capacitors provide the bypass for a wide frequency range. The main disadvantage of linear regulators is their relatively low efficiency, which is translated into higher power dissipation. If a 5 V power supply is used, Equation (8–12) expresses the total power dissipation of the transreceiver module.

$$P_{diss} = (5.25 - 3.3) \times I_{DD} \qquad \textbf{(8–12)}$$

where P_{diss} is the total module power dissipation and I_{DD} is the total drain current. If the power dissipation from Equation (8–12) expression exceeds the recommended level, a DC to DC converter may be required. DC to DC converters may be used to convert 12 V to 3.3 V or 5 V to 3.3 V with excellent efficiency (85%–95%). An appropriate DC to DC converter, which can be used to power the VSC7185 transreceiver, is the LT1256 developed by Linear Technology. This device is capable of providing a maximum current of 1.5 A. On the down side, DC to DC converters inject noise into the 3.3 V line, require external components, and are more difficult to implement and are more expensive.

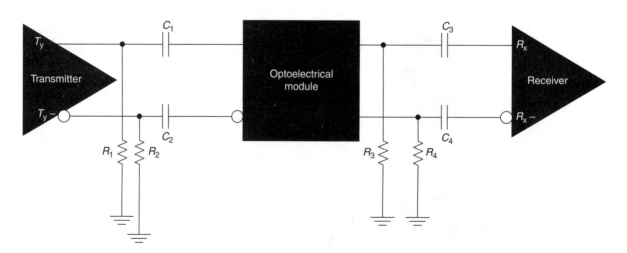

Figure 8–51. Fiber channel transreceiver termination.

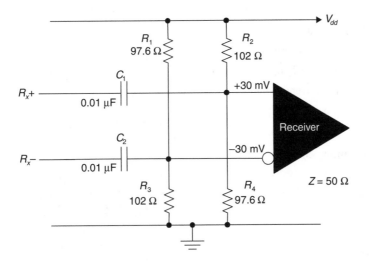

Figure 8–52. Oscillation prevention mechanism.

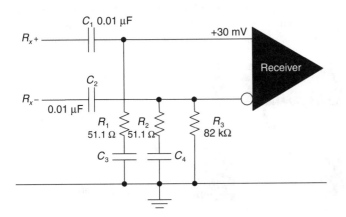

Figure 8–53. Oscillation prevention mechanism.

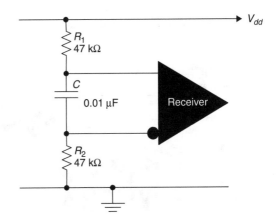

Figure 8–54. Receiver section termination scheme.

8.5 HIGH SPEED OPTICAL TRANSRECEIVERS

Optical Transreceivers Based on the Sumitomo SDG1201 Module

GENERAL DESCRIPTION. The SDG1201 2.5 Gb/s optical transreceiver, developed by Sumitomo Electric Industries, complies with the SONET OC-48 operating format. This module is designed to convert 16-bit 155.52 Mb/s parallel data into a 2.488 Gb/s serial data stream at the transmitter section, and a 2.488 Gb/s serial data stream to 16-bit 155.52 Mb/s parallel outputs at the receiver section. The SDG1201 transreceiver block diagram is illustrated in Figure 8–56. A PLL circuit is used to synthesize the 2.488 GHz clock frequency required by the transmitter and by the receiver sections. The module's main features are as follows:

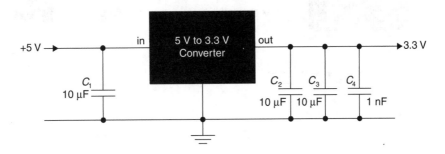

Figure 8–55. A linear regulator used as a 5 V to 3.3 V converter.

a) maximum operating speed: 2.488 Gb/s

b) complies with SONET/SDH formats

c) utilization of a single +3.3 V power supply

d) incorporation of receive loss of optical signal indicator

e) incorporation of laser diode bias current monitor

The above module is composed of the following main building blocks:

1) first in first out (FIFO) and timing control

2) parallel to serial converter

3) 2.5 GHz PLL clock synthesizer

4) optical transmitter

5) optical receiver

6) reference clock generator

7) clock and data recovery circuit

8) parallel to serial converter

Transmitter optical interface characteristics are listed in Table 8–31.

Transmitter AC operating characteristics are listed in Table 8–32.

Receiver optical interface characteristics are listed in Table 8–33.

The receiver AC operating characteristics are listed in Table 8–34.

The transreceiver maximum operating characteristics are listed in Table 8–35.

8.7 Gb/s OPTICAL TRANSRECEIVERS BASED ON THE VITESSE VSC7216 MODULE

GENERAL DESCRIPTION. The VSC7216, developed by Vitesse Corporation, is a high speed (8.7 Gb/s) fiber channel and Ethernet compliant transreceiver, intended for use in high bandwidth interconnection optical fiber systems. The module is composed of four fiber or Ethernet channels. Each channel is able to operate at a minimum rate of 784 Mb/s (8-bits at 98 MHz) and a maximum of 1088 Mb/s

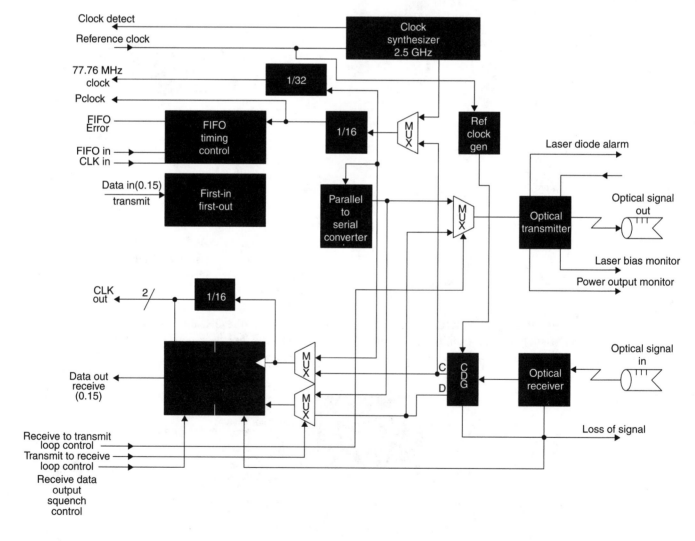

Figure 8–56. A block diagram of an optical transreceiver based on the Sumitomo SDG1201 module.

(8-bits at 136 MHz). The main building blocks of this module are as follows:

a) input registers

b) 8- or 10-bit encoders

c) parallel to serial converters

d) clock generator

e) clock and data recovery

f) serial to parallel converters

g) 8- or 10-bit decoders

h) output buffers

i) channel alignment

A block diagram of the Vitesse optical transreceiver is illustrated in Figure 8–57.

The main characteristics of the optical transreceiver are as follows:

1) maximum speed : 8.7 Gb/s

2) 4-fiber channel (4ANSI X3T11) and IEEE 802.3z Gigabit Ethernet compliant

TABLE 8–31   Transmitter Optical Interface Characteristics

Parameters	Minimum	Typical	Maximum	Symbol	Unit
Center wavelength (output)	1265	—	1360	Lc	nm
Center wavelength (output)	1500	—	1580	Lc	nm
Spectral width (rms)	0	—	4	dL	nm
Spectral width at –20 dB	—	—	1	dL	nm
Side mode suppression ratio	30	—	—	SSR	dB
Optical power output (avg)	–10	—	–3	P_0	dBm
Disabled power output	—	—	–45	$P_{0(OFF)}$	dBm
Extinction ratio	—	—	8.2	E_r	dB
Jitter (rms)	0	—	0.01	$T_{jitter(rms)}$	UI
Jitter (p–p)	0	—	0.1	$T_{jitter(p–p)}$	UI
Optical pulse mask		GR-253 compliance			

TABLE 8–32   Transmitter AC Operating Characteristics

Parameters	Minimum	Typical	Maximum	Symbol	Unit
Clock frequency in	—	155.520	—	$f_{clock(in)}$	MHz
Clock frequency tolerance	–50	—	+50	$df_{clock(in)}$	ppm
Clock frequency in jitter	—	—	–140	$Jitter_{in}$	dBc/Hz
Clock frequency duty cycle	40	—	60	DC_{in}	%
Clock frequency in rise/fall time	—	—	1.5	t_r/t_f	ns
Skew drift	5	—	5	$t_{sw.d}$	ns
Response time (ON)	—	—	3	t_{ON}	ms
Response time (OFF)	—	—	100	t_{OFF}	ms
Set up time	1.5	—	—	$t_{set\ up}$	ns
Hold time	1.5	—	—	t_{hold}	ns

TABLE 8–33   Receiver Optical Interface Characteristics

Parameters	Minimum	Typical	Maximum	Symbol	Unit
Center wavelength (input)	1265	—	1360	Lc	nm
Center wavelength (input)	1500	—	1580	Lc	nm
Optical power (input)	–18	—	0	$P_{in(opt)}$	dBm
Optical power (input)	–28	—	–9	$P_{in(opt)}$	dBm
Jitter tolerance		ITU-T G958 and GR-253 compliance			
Set up time	2.2	—	—	$t_{set\ up}$	ns
Hold time	2.0	—	—	t_{hold}	ns
Loss of optical power time response	2.3	—	100	t_{LOS}	µs

TABLE 8–34   Receiver AC Operating Characteristics

Parameters	Minimum	Typical	Maximum	Symbol	Unit
Clock frequency output	—	155.520	—	$f_{clock(out)}$	MHz
Clock frequency output tolerance	–100	—	+100	$df_{clock(out)}$	ppm
Clock frequency duty cycle	45	—	55	DC_{out}	%
Clock frequency-output rise/fall time	—	—	1.5	t_r/t_f	ns

TABLE 8–35   Transreceiver Maximum Operating Characteristics

Parameters	Minimum	Maximum	Symbol	Unit
Supply voltage (digital analog)	0	+5	V_{cc}	V
PECL input voltage	0	+3.3	$V_{cc}33$	V
Source current	0	24	I_{source}	mA
Open collector voltage	0	+6	V_{oco}	V
TTL input voltage	0	$V_{cc}33$	V_{TTL}	V
Operating ambient temperature	0	+70	T_{amb}	°C
Storage temperature	–40	+85	T_{stg}	°C

3) 8- or 10-bit encoder/decoder per channel

4) received data aligned to received clock or local reference clock

5) receive signal detect equalization

6) internal loopback function between receiver and transmitter and transmitter and receiver sections used for data verification and testing

7) CMU for baud rate clock

8) automatic lock to reference function

9) built-in self-test

10) single +3.3 /3 W power supply

The Transmitter Section

THE CLOCK SYNTHESIZER: A low at the DUAL input (see Figure 8–57) will enable the clock synthesizer to multiply the signal applied at the reference clock (REFCLK) input by either ten or twenty for a DUAL high input, thus achieving a baud rate clock of between 980 MHz and 1.36 GHz. The external 0.1 µF capacitor connected between CAP0 and CAP1 is a loop filter capacitor for the PLL circuit, and is used for clock generation. It is recommended that the capacitor exhibit a good temperature coefficient with a minimum working voltage of 5 V. The main function of the external capacitor is to minimize the effect of the power supply common mode noise that affects the CMU.

The CMU: The function of the CMU is to lock on to an external clock and to generate an internal ×10 or × 20 clock frequency. The CMU incorporates a PLL and a low noise VCO. A block diagram of the CMU is illustrated in Figure 8–58.

The main operating characteristics of the CMU are as follows:

a) Input frequency (external clock): 50 MHz to 625 MHz

b) Output frequency (clock): 100 MHz to 1.25 GHz

c) Input signal: differential PECL

d) Output: differential PECL

e) Jitter: 4 ps_{rms} or 25 $ps_{(p–p)}$

f) Lock time: 10 µs

g) Skew between outputs: < 50 ps

h) Supply voltage: single +3.3 V ± 5% /2 W

i) Operating temperature: 0° to +70°C

Applications: The clock multiplier unit can be used in optical transmission systems that require high precision, low jitter, and low skew clock signals. The input/output frequencies generated by the VSC6112 CMU are listed in Table 8–36.

CMU AC timing characteristics are listed in Table 8–37. CMU DC characteristics are listed in Table 8–38.

For single-ended applications, the ECL input DC levels are listed in Table 8–39.

The maximum current drain is 0.6 A and the maximum power dissipation of the CMU is 2 W. The absolute maximum ratings of the CMU are listed in Table 8–40.

The input reference clock (REFCLK) can either be a differential PECL or a single ended TTL. For single ended TTL operation, the TTL output is connected to the positive reference clock (REFCLKP), while the inverted reference clock (REFCLKN) is left unconnected. For PECL operation, each differ-

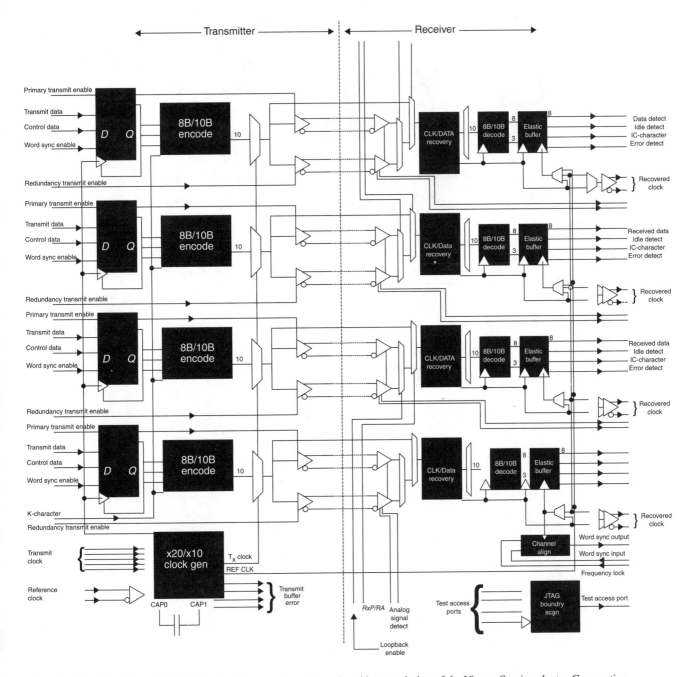

Figure 8–57. The Vitesse VSC7216 optical transreceiver. Reproduced by permission of the Vitesse Semiconductor Corporation.

ential PECL output is connected to the corresponding reference clock positive or reference clock negative inputs. An internal resistive network sets the required DC bias level to 50% of the V_{dd}.

The Data Bus: The 8-bit character is applied at each input buffer, which also incorporates two control inputs: one for control/data for channel-n (C/Dn), and the other for word sync enable for channel-n (WSEN/Dn). The C/Dn determines whether a **K-character** or normal data is transmitted while the WSEN/Dn control enables a 16-character word to be transmitted for receive channel synchronization. Both C/Dn and WSEN/Dn inputs are clocked at the positive going transition (PGT) of the reference clock (REFCLK) signal.

The 8- or 10-bit Encoder: Each of the four input channels contains an 8- or 10-bit encoder whose function is to convert the 8-byte input data to a 10-bit encoded character. Special inputs are also available at each channel, and are intended to allow special K-characters to be transmitted upon demand. The transmit data controls and the encoder response are listed in Table 8–41.

A selection of K-characters is listed in Table 8–42.

Word Synchronization: The word synchronization sequence is implemented so as to align the receiver channels through sixteen consecutive K28.5 characters. In this way the four receive data output streams are aligned with the

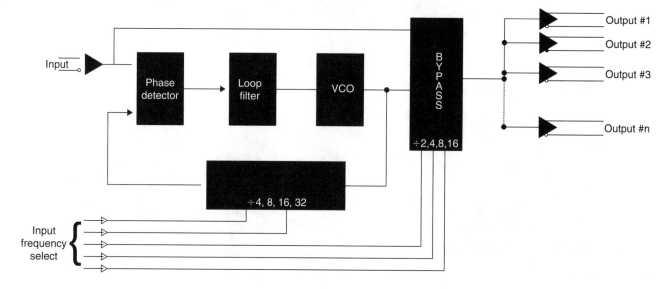

Figure 8–58. A black diagram of the CMU.

TABLE 8–36   CMU Input/Output Frequencies

f_{in}(MHz)	f_{out}(MHz)
77.5	155
155	311
311	622
622	1244

TABLE 8–37   CMU AC Timing Characteristics

Parameters	Minimum	Typical	Maximum	Symbol	Unit
Reference clock duty cycle	40	—	60	$f_{REF(DC)}$	%
Reference clock jitter	—	—	10 ps$_{(p–p)}$	$f_{REF(J(p-p))}$	ps
Output clock duty cycle	40	—	60	$f_{O.clock(DC)}$	%
Output clock jitter	—	—	25 ps$_{(p–p)}$	$f_{O.clock(J(p-p))}$	ps
Skew	–50	—	+50	t_{skew}	ps
Lock time	—	—	10	t_{lock}	µs
Output rise/fall time (20%–80%)	—	100	150	$t_{r/f}$	ps

TABLE 8–38   CMU DC Characteristics

Parameters	Minimum	Typical	Maximum	Symbol	Unit
Input voltage (diff.)	200	—	—	$V_{in(dif.)}$	mV
Input voltage (common mode)	–1.5	—	–0.5	$V_{in(CM)}$	V
Output voltage (diff.)	400	—	800	$V_{out(diff)}$	mV
Output voltage (common mode)	–1.6	—	–1.2	$V_{in(CM)}$	V

TABLE 8–39 Single Ended ECL DC Input Levels

Parameters	Minimum	Typical	Maximum	Symbol	Unit
Input voltage (high)	–1020	—	–700	V_{IH}	mV
Input voltage (low)	–2000	—	–1650	V_{IL}	mV

TABLE 8–40 Absolute Maximum Ratings

Parameters	Minimum	Maximum	Symbol	Unit
Power supply voltage	–0.5	+4.3	V_{cc}	V
Output current	—	50	I_{out}	mA
Case temperature	–55	+125	T_{case}	°C
Storage temperature	–65	+150	$T_{storage}$	°C

TABLE 8–41 Transmit Data Control Signals

WSEN/Dn	C/Dn	K-Character	Encoded 10-bit
0	0	X	Data character
0	1	0	K28.5 character
0	1	1	Special K-character
1	X	X	16-character word sync sequence

4-byte data word appearing at the input of the transmit channels.

Parallel to Serial Conversion: The 10-bit characters generated by the encoder are fed into the multiplexer circuit, which produces the serial data stream. The LSB of the 10-bit character is transmitted first. Two output ports out the primary and the redundant. Both consist of differential PECL and operate at either ×10 or ×20 of the reference clock frequency.

THE RECEIVER SECTION. Similar to the transmitter, each channel at the receiver section incorporates two differential PECL ports and a control input for selecting the required port.

The Signal Detection Process: The receiver input buffer incorporates a signal detect output used to continuously monitor the selected and nonselected inputs. These signal detect outputs go high only when the input signal amplitude is higher than 200 mV. A low at the signal detect output will indicate that a signal has been detected at the input port, with an amplitude less than 100 mV.

The Equalizer: The received signal at the buffer inputs contains a certain level of inter symbol interference (ISI), which directly contributes to an increase of the system BER. In order to minimize the effect of ISI, equalizer circuits are incorporated at each buffer and are intended to increase the high frequency component of the signal response and, consequently, to increase the SNR.

Data and Clock Recovery: At the receiver end, each of the four channels incorporates a clock recovery unit (CRU) that is designed to extract the high speed clock, and also to retime the data from the serial input data steam. In the case of data absence, the clock recovery circuit will automatically lock ON to the reference clock, thus maintaining normal operation for the rest of the receiver circuitry.

TABLE 8–42 A Selection of K-Characters

Code	Transmit Data for channel (n)	Function
K28.0	00011100	Defined by customer
K28.1	00111100	Defined by customer
K28.2	01011100	Defined by customer
K28.3	01111100	Defined by customer
K28.4	10011100	Defined by customer
K28.5	10111100	IDLE
K28.6	11011100	Defined by customer
K28.7	11111100	Testing
K23.7	11110111	Defined by customer
K27.7	11110111	Defined by customer
K29.7	11111101	Defined by customer

SERIAL-TO-PARALLEL CONVERSION: The retrieved and retimed serial data stream is converted to a 10-bit character by the serial to parallel converter. The receiver recognizes an 8-bit K-character (comma pattern) used to identify the boundary of the 10-bit character. If the data framing boundary and the comma character are misaligned, the incoming data stream will be resynchronized by the receiver to be aligned with the next comma character. Character synchronization is essential for maintaining common transmit/receive bit location.

THE 8- OR 10-BIT DECODER. The 10B/8B decoder is used to decode the 10-bit character by generating the 8-bit data byte, while a 3-bit word is used for a status report.

SUMMARY

In Chapter 8, optical transreceivers were introduced. The chapter began with an introduction of transreceivers as an integral part of optical communications systems that are intended to perform the function of data serialization/deserialization. The chapter continued with the classification of transreceivers, based on the type of the optical source used for their design, as LED or laser based transreceivers. Several modules of LED and laser transreceivers from leading manufacturers of fiber optics components such as Hewlett Packard, Lucent Technologies, Microcosm, Philips, Nortel, Siemens, Texas Instruments, and Vitesse were introduced. Block diagrams, schematics, circuit descriptions, and performance characteristics were presented for comparison, followed by detailed guidelines for the design of fiber-channel transreceivers. The chapter concluded with a description of high speed optical transreceivers developed by Sumitomo and Vitesse.

QUESTIONS

Section 8.1

1. What is the function of an optical transreceiver?
2. What are the two basic types of optical transreceivers? Explain.
3. With the assistance of a block diagram, list the main components of an optical transreceiver and briefly explain its operation.

Section 8.2

4. Explain why special NAND gates must be used as LED drivers, while standard NAND gates must not (refer to Figure 8–3).
5. What is the maximum current sourcing and current sinking of the totem pole output stage of each NAND gate of the LED driver circuit?
6. Define the role of decoupling capacitors LED driver circuits.
7. Draw a block diagram of the Microcosm MC2042-4 driver circuit, and explain its operation.
8. Name some of the most important operating characteristics of the MC2042-4 driver circuit.
9. How is the LED turn ON time improved? Explain.
10. Give the expression and explain the significance of transient current for both turn ON and turn OFF times.
11. Why is LED clamping required? Explain.
12. Give the expression and explain the necessity for pulse width adjust.
13. Explain why photodetector sensitivity must be very high, perhaps down to nanowatt levels.

14. Define *optical OSNR*.
15. Refer to Figure 8–5 and explain (in detail) its operation.
16. What are the critical elements determining the switching time of a photodiode?
17. Describe the role of a preamplifier and a postamplifier incorporated into the receiver section of a transreceiver module.
18. List the operating characteristics of Analog Device's AD8015 preamplifier.
19. Draw a block diagram of the Infinion SFH551/1 preamplifier, and briefly explain its operation.
20. List the building blocks of the MC2207 optical receiver.
21. List the building blocks and operating characteristics of the preamplifier circuit incorporated into the MC2207 optical receiver.
22. Describe the operation of the postamplifier circuit incorporated into the design of the MC2207 device.
23. What is the role of the autozero circuit?
24. Why is a level detector circuit required? Explain in detail.

Section 8.3

25. Why was it necessary to develop a large number of optical transreceivers?
26. Briefly describe the operation of the Philips TZA3005H optical transreceiver.
27. What are the functions and operating characteristics of the transmitter section of the Philips TZA3005H optical transreceiver?

28. Describe the function and operating characteristics of the clock synthesizer in the TZA3005H.

29. Describe the function and operating characteristics of the clock driver circuit as part of the TZA3005H optical transreceiver.

30. Briefly discuss the function of the receiver section as part of the TZA3005H optical transreceiver.

31. What is the function of the frame byte and boundary detection circuit?

32. How can recognition of byte boundaries be accomplished?

33. Draw a block diagram of the transmitter section of the Hewlett Packard HFCT 5200 series optical transreceiver, and briefly explain its operation.

34. Why can single ended operations increase duty cycle distortion?

35. How is duty cycle distortion usually rectified?

36. How can constant extinction ratio and duty cycle distortion be accomplished?

37. What is the function of the laser current bias monitor?

38. List the main components of of the receiver section of the Hewlett Packard HFCT 5200 series optical transreceiver.

39. What parameters determine the value of the AC coupling capacitor?

40. How can a reduction of electromagnetic interference and an enhancement of receiver sensitivity be accomplished?

41. The receiver clock frequency must exhibit zero drift. How is this accomplished?

42. Briefly describe the operation of the Lucent Technologies TB16 2.5 Gb/s optical transreceiver.

43. List the main components that compose the transmitter section of the TB16.

44. Explain the functional relationship between the various components of the transmitter's parallel to serial converter circuit.

45. What are the roles of the driver and phase detector circuits?

46. Describe (in detail) the dual function of the timing generator circuit.

47. Name the components that compose the receiver section of the TB16.

48. Briefly explain the operation of the TB16.

49. Define the roles of the frame and boundary detectors.

50. What is the function of the timing generator circuit?

51. Describe the difference between line loopback and diagnostic loopback.

52. Describe two ways of line termination of the transreceiver input required to achieve maximum power transfer.

53. Draw a block diagram of the Texas Instruments SN75FC1000 optical transreceiver and briefly describe its operation.

54. Describe the operation of the transmitter section of the SN75FC1000.

55. List the main components of the receiver section, and explain its operation.

56. What is the role of the feedback loop?

57. How does the feedback loop maintain lock?

58. Describe how 10-bit word recognition and alignment is achieved.

59. What is the K28.5 synchronization character?

60. Explain the role and function of loopback.

61. With the assistance of a block diagram, list the main components that compose the Microcosm MC4663 optical transreceiver.

62. Describe the operation of the transmitter section of the MC4663.

63. What is the role of the squelch circuit?

64. Sketch a block diagram and list the sections that compose the Vitesse VSC7185 transreceiver.

65. Briefly describe the operation of the VSC7185.

66. List the main sections that compose the receiver section of the VSC7185.

67. Explain the function of series to parallel conversion.

68. What is the comma character?

69. Describe the function of word alignment.

Section 8.4

70. List the guidelines essential in the design of a fiber channel transreceiver.

71. How is phase lock between the internal and external signals accomplished?

72. In the design of the reference clock, a crystal oscillator is used. Explain why.

73. What is the best way to achieve jitter?

74. Describe the arrangement by which maximum transmission distance and minimum EMI can be achieved.

75. With the assistance of a schematic diagram, explain the oscillation prevention mechanism.

76. Why must unused input be properly terminated?

77. Explain why linear regulator power supplies are required to power optical transreceivers.

Section 8.5

78. List the operating characteristics of the Sumitomo SDG1201 transreceiver.

79. Define the design objectives of the SDG1201.

80. List the main building blocks of the SDG1201.

81. List the building blocks of the Vitesse VSC7216 8.7 Gb/s optical transreceiver.

82. Describe the main operating characteristics of the VSC7216.

83. With the assistance of a block diagram, describe the operation of the CMU.

84. In what application is a clock multiplier unit highly desirable?

85. List the frequencies generated by the CMU.

86. What is the role of the 8- or 10-bit encoder?

87. List the binary codes for K28.0 to K28-4.

88. Why is word synchronization required?

89. Describe the signal detection process.

90. What is the role of the equalizer circuit?

9

OPTICS
OPTICAL FIBERS

OBJECTIVES

1. Introduce the fundamental laws relevant to the transmission of light through dielectric media

2. Explain the mechanisms and the contributing factors in the attenuation absorption and scattering of optical waves traveling through fibers

3. Classify optical fibers in terms of their operating characteristics and material composition

4. Describe the material composition and operating characteristics of optical fibers for **Gb.E** applications

5. Establish the need for the development of Non-Zero Dispersion Shifted Fibers (NZ-DSF) and large area (NZ-DSF)

6. Describe the fiber nonlinear processes, such as four wave mixing (FWM), self phase modulation (SPM), and cross phase modulation (XPM), and their effects in DWDM optical systems

7. Establish the need for fiber upgrading and describe the role of dispersion compensation fibers

8. Describe the various methods of fiber alignment

KEY TERMS

absorption
angle of acceptance
angle of incident
angle of refraction
attenuation
bandwidth-length product
Bragg grating method
channel bandwidth
channel bit rate
cladding

critical angle
cross-phase-modulation
cut-off wavelength
dielectric medium
differential group delay
    (DGD)
dispersion
dispersion compensation
    method
dispersion slope

dispersion wavelength
distributed modal dispersion
distribution variance
electric field
electric flux density
electromagnetic compatibility
extrinsic
fiber index profile
fictive temperature
field angular velocity

four wave mixing
graded index
group delay
group velocity
high order mode
intensity modulation
intermediate order mode
intermodal
intramodal
intrinsic

9.1 INTRODUCTION

The notion of transmitting light through a **dielectric medium** was contemplated as early as 1910. In the 1920s, experimental work began which attempted to transmit light through a silicon rod that has a **refractive index** of 1.5 and does not have **cladding.** Although the early experiments were unsuccessful as anticipated, the idea remained strong. Further studies and experimental work carried on intensively. In the 1950s, the cladding optical fiber was proposed (Figure 9–1).

Figure 9–1 illustrates the proposed **optical waveguide** composed of a core of index (n_1) surrounded by a cladding transparent material with a refractive index (n_2) that is slightly smaller than (n_1). Therefore, n_1 is greater than n_2. Controlling the refractive index of the cladding, it was possible for a large amount of optical power to be propagated through the core of the fiber.

The concept that optical waveguides can be used as **transmission medium** in communications systems was contemplated in the mid-1960s. Until that time, optical waveguides were considered impractical because of their very large optical **attenuation.** It was reported that such waveguides, operating at the 800 nm to 900 nm **wavelength window,** exhibited an average attenuation on the order of 1000 dB/km.

Researchers and manufacturers, upon recognizing the potential of optical fibers in communications applications, focused their attention on the improvement of fiber performance through material purification and sophisticated fabrication techniques, while simultaneously attempting to extend the wavelength operating window to the 1500 nm range.

The combined efforts resulted in optical fibers that operate in both short and long wavelengths, exhibiting attenuation losses on the order of 1 dB/km. Further improvements, both in material and fabrication techniques, significantly improved fiber performance to attenuation levels of 0.2 dB/km. To achieve lower attenuation, optical fibers other than those that have a silicon base were proposed and experimentally tested. Results showed that fluoride glass fibers that operate at the 2500 nm (infrared) wavelength window, exhibited attenuation levels on the order of 0.01 dB/km. For practical optical communications systems, advancements in optical fiber technology must be combined with parallel advancements in other areas such as optical sources and optical detectors that operate at compatible wavelength windows. Currently, the two wavelength windows applicable to optical fiber communications systems are the 800 nm to 900 nm short wavelength and the 1310 nm to 1550 nm long wavelength. Usable wavelength windows depend on the operating characteristics of the optical sources, detectors, and amplifiers in use.

9.2 THE RAY THEORY

Propagation of light through the core of an optical fiber is subject to the materials that compose both the core and the cladding and their refractive index difference.

Figure 9–1. An optical fiber composed of a core with refractive index (n_1) and cladding with refractive index (n_2).

Snell's Law

Snell's law states that, when an optical ray is incident with an angle (θ_1) upon the interface between two dielectric mediums with different indices ($n_1 > n_2$) (Figure 9–2), the following relationships hold true (Equations (9–1) and (9–2)). Figure 9–2 shows the **angle of refraction** to be larger than the angle of incident ($\theta_2 > \theta_1$).

$$n_1 \sin \theta_1 = n_2 \sin \theta_2 \qquad \textbf{(9–1)}$$

or

$$\frac{\sin \theta_1}{\sin \theta_2} = \frac{n_2}{n_1} \qquad \textbf{(9–2)}$$

If the **angle of incident** (θ_1) is progressively increased, and the angle of refraction (θ_2) undergoes a corresponding increase, there will be a point where θ_2 will be equal to $90°$ and will be parallel to the interface (Figure 9–3). At this point, angle θ_1 (less than $90°$) is referred to as the **critical angle** (θ_0), and is defined by the relationship expressed in Equation (9–3).

$$n_1 \sin \theta_1 = n_2 \sin \theta_2 \qquad \textbf{(9–3)}$$

or

$$\sin \theta_1 = \frac{n_2 \sin \theta_2}{n_1}$$

Because $\sin \theta_2 = \sin 90° = 1$, and $\theta_1 = \theta_0$,

then, $\sin \theta_0 = \dfrac{n_2 (1)}{n_1} = \dfrac{n_2}{n_1}$

Therefore,

$$\sin \theta_0 = \frac{n_2}{n_1} \qquad \textbf{(9–4)}$$

Equation (9–4) defines the critical angle (θ_0) as the ratio of the refractive index of the cladding to the refractive index of the core.

If the angle of incident (θ_1) is increased slightly beyond the critical angle (θ_0), the angle (θ_2) will also be increased beyond the $90°$ level, resulting in a full ray reflection toward the medium with the higher refractive index (n_1) and with an efficiency level of approximately 99.8% (Figure 9–4).

It is this principle which supported the notion that light can be propagated through a dielectric medium of refractive index (n_1), surrounded by a cladding dielectric material of refractive index (n_2) ($n_1 > n_2$) in a zigzag mode and for an incident angle $\theta_1 > \theta_0$ (Figure 9–5).

In Figure 9–5, light transmission is assumed under ideal dielectric materials for both the core and cladding.

Angle of Acceptance

The **angle of acceptance** is the maximum half conical angle incident upon the core of the optical fiber which is achieving full internal reflection. If the incident ray is propagated through the core of the fiber in a zigzag mode, the maximum

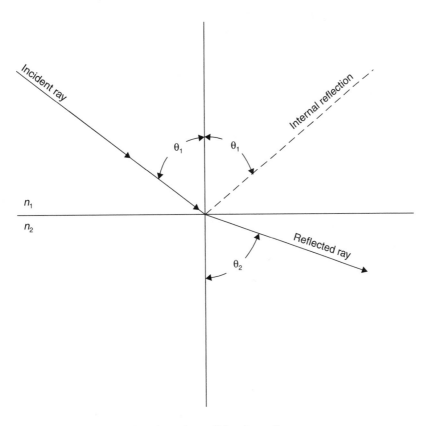

Figure 9–2. Behavior of an incident ray upon the interface of two dielectric mediums.

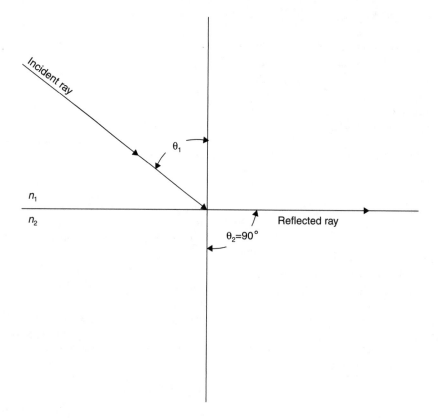

Figure 9–3. 90° refraction.

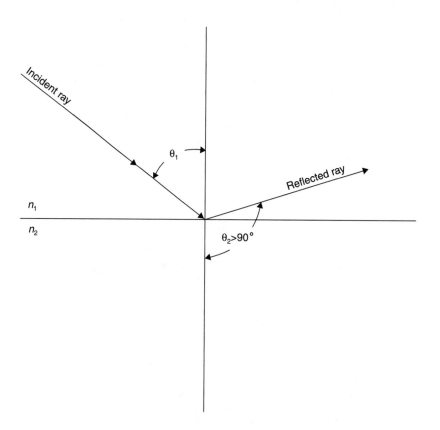

Figure 9–4. Full internal reflection.

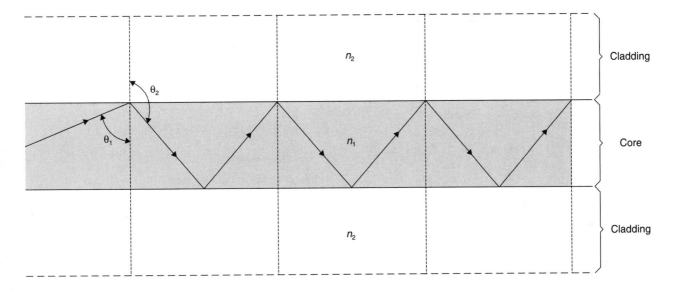

Figure 9–5. Full internal ray reflection (propagation of light through the core of the fiber).

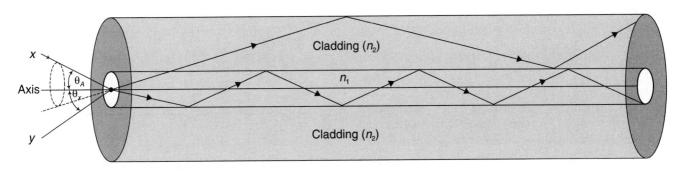

Figure 9–6. Acceptance angle θ_A $(\theta_A < \theta_X)$.

angle of incident is referred to as the angle of acceptance (θ_A) (Figure 9–6).

Numerical Aperture

The relationship between the angle of acceptance and the refractive indices of the three mediums incorporated into the fiber, those of air cladding and core, are unified by the **numerical aperture (NA)** (Figure 9–7).

Figure 9–7 illustrates the process whereby an optical ray is incident upon the core of the fiber from the air. Assuming air has a refractive index n_0, the core of the fiber has a refractive index n_1, and applying Snell's law, we have:

$$n_0 \sin\theta_1 = n_1 \sin\theta_2$$

where n_0 is the refractive index of air, θ_1 is the angle of acceptance from air to core, n_1 is the refractive index of the core, and θ_2 is the angle of refraction in the core.

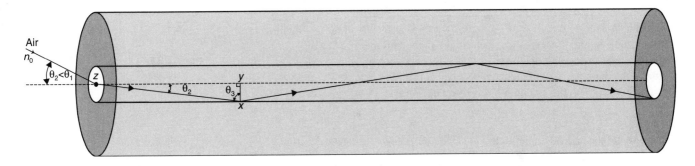

Figure 9–7. Air to fiber ray path.

From Figure 9–7, triangle *XYZ* is a right angle triangle, and the value of angle θ_3 can be established.

$$\theta_3 = 90° - \theta_2$$

Because $\theta_3 > \theta_2$, Snell's equation can be expressed as, $n_0 \sin\theta_1 = n_1 \cos\theta_3$.

Applying the trigonometric relationship,
$\sin^2\theta_3 + \cos^2\theta_3 = 1$.

Solving for $\cos^2\theta_3$, we have $\cos^2\theta_3 = 1 - \sin^2\theta_3$.

Substituting to Snell's equation, we have
$$n_0 \sin\theta_1 = n\sqrt{(1 - \sin^2\theta_3)}.$$

Equation (9–4) shows the critical angle for full internal reflection is equal to $\sin\theta_0 = \dfrac{n_2}{n_1}$.

If we also assume θ_1 to be the acceptance angle (θ_A), then Snell's equation becomes

$$n_0 \sin\theta_A = n_1\left[1 - \left(\frac{n_2}{n_1}\right)^2\right]^{1/2} = n_1\left[1 - \frac{n_2^2}{n_1^2}\right]^{1/2}$$

$$= n_1\left[\frac{n_1^2 - n_2^2}{n_1^2}\right]^{1/2} = (n_1^2 - n_2^2)^{1/2}$$

or

$$n_0 \sin\theta_A = (n_1^2 - n_2^2)^{1/2} \qquad \text{(9–5)}$$

It is evident that Equation (9–5) relates the four basic parameters: the angle of acceptance and the refractive indices of the air, core, and cladding. This very important relationship is referred to as the numerical aperture (NA) and is expressed either by Equation (9–6) or Equation (9–7).

$$NA = n_0 \sin\theta_A \qquad \text{(9–6)}$$

or

$$NA = (n_1^2 - n_2^2)^{1/2} \qquad \text{(9–7)}$$

NA is normally referred to as the fiber's core to air interface. Because $n_0 = 1$ (air)
then,

$$NA = \sin\theta_A \qquad \text{(9–8)}$$

Equation (9–8) directly relates the numerical aperture to the angle of acceptance.

Numerical aperture also relates the refractive indices of both the core and cladding and expresses the ability of the fiber to collect light. This holds true only for fibers that have core diameters equal to or larger than 8 μm.

Example 9–1
Compute the angle of acceptance, the critical angle, and the numerical aperture of an optical fiber that satisfies the minimum core diameter requirements. The fiber core index of refraction is 1.4 and the cladding index of refraction is 1.35.

Solution
i) Compute the critical angle (θ_0).

$$\sin\theta_0 = \frac{n_2}{n_1}$$

or

$$\theta_0 = \sin^{-1}\left(\frac{n_2}{n_1}\right)$$

$$= \sin^{-1}\left(\frac{1.35}{1.4}\right)$$

$$= 74.64$$

Therefore, $\theta_0 = 74.64°$.

ii) Compute the numerical aperture.

$$NA = \sqrt{(n_1^2 - n_2^2)}$$

$$= \sqrt{(1.4^2 - 1.35^2)}$$

$$= 0.37$$

Therefore, $NA = 0.37$.

iii) Compute the angle of acceptance (θ_A).

Because $NA = \sin\theta_A$, then,

$$\theta_A = \sin^{-1}(NA)$$

$$= \sin^{-1}(0.37)$$

$$= 21.71°$$

Therefore, $\theta_A = 21.71°$.

9.3 THEORY OF OPTICAL WAVE PROPAGATION

The propagation of light through optical fibers obeys the same laws as those applied to electromagnetic wave propagation defined by Maxwell's equations. An electromagnetic field is composed of an **electric field** (E) and a **magnetic field** (H) perpendicular to each other, and both perpendicular to the direction of propagation (Z). The relationship between the electric field (E), the magnetic field (H), the **magnetic flux density** (B), and the **electric flux density** (D), in an infinite transmission medium, is expressed by Equations (9–6), (9–7), (9–8), and (9–9).

$$\nabla E = -\frac{\partial B}{\partial t} \qquad \text{(9–6)}$$

$$\nabla H = \frac{\partial D}{\partial t} \qquad \text{(9–7)}$$

where, ∇ is the vector operator. Other than in free-space conditions,

$$\nabla D = 0 \qquad \text{(9–8)}$$

$$\nabla H = 0 \qquad \text{(9–9)}$$

or

$$\nabla x \nabla x E = -\mu\varepsilon\left(\frac{d^2 E}{dt^2}\right)$$

Using the operator $\nabla x \nabla x E = \nabla(\nabla E) - \nabla^2 E$, we have $\nabla(\nabla E) - \nabla^2 E = -\mu\varepsilon\left(\frac{d^2 E}{dt^2}\right)$.

Compute ∇E.

$$\nabla D = \nabla\varepsilon E = \nabla\varepsilon E + \varepsilon\nabla E = 0$$

Therefore, $\nabla E = -E\left(\frac{\nabla\varepsilon}{\varepsilon}\right)$.

From the above, $\nabla^2 E - \mu\varepsilon\left(\frac{d^2}{dt^2}\right) = -\nabla[E(\nabla\varepsilon/\varepsilon)]$.

Because $-\nabla[E(\nabla\varepsilon/\varepsilon)] \approx 0$, then, $\nabla^2 E - \mu\varepsilon\left(\frac{d^2 E}{dt^2}\right) = 0$ (homogenous wave equation)

The electric flux density (D) and magnetic flux density (B) are defined by Equations (9–10) and (9–11).

$$D = \varepsilon E \qquad (9\text{–}10)$$

$$B = \mu H \qquad (9\text{–}11)$$

where ε is the **permittivity** of the dielectric medium (F/m) and μ is the permeability of the dielectric medium (H/m). Substituting (D) and (B) into Equations (9–6) and (9–7) yields Equations (9–12) and (9–13).

$$\nabla(\nabla E) = -\mu\varepsilon\frac{\partial^2 E}{\partial t^2} \qquad (9\text{–}12)$$

$$\nabla(\nabla H) = -\mu\varepsilon\frac{\partial^2 H}{\partial t^2} \qquad (9\text{–}13)$$

The vector identity and divergence condition illustrated in Equation (9–14)

$$\nabla(\nabla Y) - \nabla^2(Y) \qquad (9\text{–}14)$$

yield the (nondispersive) wave equations (Equations (9–15) and (9–16).

$$\nabla^2 E = \mu\varepsilon\frac{\partial^2 E}{\partial t^2} \qquad (9\text{–}15)$$

$$\nabla^2 H = \mu\varepsilon\frac{\partial^2 H}{\partial t^2} \qquad (9\text{–}16)$$

where ∇^2 is the operator (Laplacian). Equations (9–10) and (9–11) hold true for cylindrical, polar coordinates as well as for rectangular Cartesian, and for all vector components that satisfy the following scalar wave (Equation (9–17)).

$$\nabla^2\Psi = \left(\frac{1}{v_p^2}\right)\left(\frac{\partial^2\Psi}{\partial t^2}\right) \qquad (9\text{–}17)$$

where Ψ is any electric (E) or magnetic (H) field component and v_P is the wave **phase velocity** in the dielectric medium expressed by Equation (9–18).

$$v_P = \frac{1}{\sqrt{\mu\varepsilon}} \qquad (9\text{–}18)$$

where $\mu = \mu_m\mu_0, \varepsilon = \varepsilon_m\varepsilon_0, \mu_m = $ **permeability** of the dielectric, $\mu_0 = $ permeability of the free space (1.26×10^{-6} H/m), $\varepsilon_m = $ permittivity of the dielectric, and $\varepsilon_0 = $ permittivity of the free space (8.85×10^{-12} F/m). The velocity of light in free space is expressed by Equation (9–19).

$$c = \frac{1}{\sqrt{\mu_o\varepsilon_0}}$$

$$= \frac{1}{\sqrt{1.26 \times 10^{-6} \times 8.85 \times 10^{-12}}}$$

$$= 3 \times 10^8 \text{ m/s} \qquad (9\text{–}19)$$

Therefore, velocity of light in vacuum is equal to 3×10^8 m/s.

The Laplacian operator (∇^2) for planar waveguides, expressed either by rectangular Cartesian coordinates (*xyz*) or circular fibers expressed by polar coordinates (*rks*), is given by Equations (9–20) and (9–21).

$$\nabla^2\Psi = \frac{\partial^2\Psi}{\partial x^2} + \frac{\partial^2\Psi}{\partial y^2} + \frac{\partial^2\Psi}{\partial z^2} \text{ (rectangular cartesian)} \qquad (9\text{–}20)$$

$$\nabla^2\Psi = \frac{\partial^2\Psi}{\partial r^2} + \frac{\partial^2\Psi}{\partial k^2} + \frac{\partial^2\Psi}{\partial s^2} \text{ (polar coordinates)} \qquad (9\text{–}21)$$

The study of the propagation of light into a fiber requires the utilization of both forms of equations. The wave equation may take a single sinusoidal waveform (uniform plane) expressed by Equation (9–22).

$$\Psi = \Psi_0\, e^{j(\omega t - Kr)} \qquad (9\text{–}22)$$

where ω is the **field angular velocity**, K is the direction of propagation vector and rate of phase change with distance ($K = \frac{2\pi}{\lambda}$ free space wave number), t is the time, r is the coordinate point of field observation, and λ is the optical wavelength in a vacuum.

Optical Modes

To fully understand the behavior of light as it propagates through an optical fiber, we must first study the behavior of light in a planar guide that is composed of a dielectric of reflective index n_1 and is confined between two cladding layers of refractive index n_2, assuming a monochromatic ray that propagates in a zigzag mode (Figure 9–8).

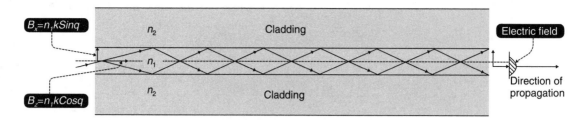

Figure 9–8. Optical wave propagation through a planar dielectric guide.

The two components of a plane wave propagation are expressed by Equations (9–23) and (9–24).

$$\beta_X = n_1 k \sin\theta \qquad (9\text{–}23)$$

$$\beta_Z = n_1 k \cos\theta \qquad (9\text{–}24)$$

where $k = \dfrac{2\pi}{\lambda}$ (free space wave number) and θ is the angle between wave propagation and direction of propagation (guide axis).

From Figure 9–8, a light ray is treated as an electromagnetic wave of a specified frequency. Therefore, the laws of electromagnetic propagation will be observed. By definition, an electromagnetic wave is composed of an electric field (E) and a magnetic field (H) perpendicular to each other, and both perpendicular to the direction of propagation. The optical wave entering the guide at an angle (θ) is defined by the wave vector and the direction of propagation.

The X vector component is fully reflected at the core-clad interface, forming a standing wave after two full reflections, with a total phase shift of $2\pi(m)$ rad. (m = integer). Through this interaction, the two plane waves converge to form an electric field with the strongest intensity at the center of the guide, while varying sinusoidally as they propagate along the guide. Therefore, light traveling through a guide is propagated in different modes specified by the angle of incident of the various rays that compose the light, and also forming standing waves by way of positive interference at the converging points.

When electromagnetic waves propagate through a guide that has the electric field perpendicular to the direction of propagation, the mode is said to be transverse electric (TE). When the magnetic field is perpendicular to the direction of propagation, the mode is said to be transverse magnetic (TM). When both the electric and the magnetic fields are perpendicular to the direction of propagation, the mode is referred to as transverse electromagnetic (TEM). In terms of modal number, they are further expressed as (TEm), (TMm), or (TEMm). From these three options, the TEm and TMm will be considered, because TEMm is rarely applicable.

Phase and Group Velocities

When a light wave propagates through an optical waveguide, specific points along the direction of propagation exist where the phase of the wave is constant. This can be understood as the wave traveling at the direction of propagation with a velocity referred to as phase velocity (v_p) (Equation (9–25)).

$$v_p = \frac{\omega}{\beta} \qquad (9\text{–}25)$$

where ω is the wave angular velocity and β is the **propagation constant**. Phase velocity is applicable to an ideal monochromatic wave traveling through an optical waveguide. In reality, a light wave is composed of discrete waves, that have slightly different frequencies, traveling through the transmission medium in a packet—not as a single ray (Figure 9–9).

Therefore, phase velocity Equation (9–25) no longer applies to the packet wave. Instead, the concept of **group velocity** is introduced (Equation (9–26)).

$$v_g = \frac{d\omega}{d\beta} \qquad (9\text{–}26)$$

where $d\omega$ is the rate of change of angular frequency, $d\beta$ is the rate of change of propagation constant, and v_g is the group velocity. A group of rays **(packet waves)** will be used to study the propagation characteristics of optical fibers.

When a packet wave is propagated through an optical waveguide of infinite length that has an index of refraction n_1 surrounded by cladding of refractive index n_2 ($n_1 > n_2$), the relationship between the core refractive index (n_1) and the operating wavelength (λ) is expressed by Equation (9–27).

$$\beta = \frac{2\pi(n_1)}{\lambda} \qquad (9\text{–}27)$$

because

$$\lambda = \frac{2\pi(c)}{\omega} \qquad (9\text{–}28)$$

Substituting Equation (9–28) into Equation (9–27) yields Equation (9–29).

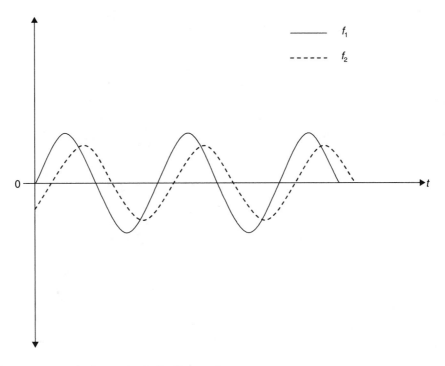

Figure 9–9. A packet optical wave composed of two very similar frequencies.

$$\beta = \frac{2\pi(n_1)}{\underset{\omega}{2\pi(c)}} = \frac{\omega(n_1)}{c}$$

therefore,

$$\beta = \frac{\omega(n_1)}{c} \qquad (9\text{–}29)$$

where ω = angular velocity of the wave, n_1 = core refractive index, and c = velocity of light in vacuum. Substituting for β in Equation (9–25) yields Equation (9–30).

$$v_p = \frac{\omega}{\dfrac{\omega(n_1)}{c}} = \frac{c}{n_1}$$

therefore,

$$v_p = \frac{c}{n_1} \qquad (9\text{–}30)$$

For a packet wave, group velocity is expressed by Equation (9–31).

$$v_g = \frac{d\lambda}{d\beta} \times \left(\frac{d\omega}{d\lambda} \right) = \frac{d}{d\lambda} \left(n_1 \frac{2\pi}{\lambda} \right)^{-1} \left(-\frac{\omega}{\lambda} \right)$$

$$v_g = -\frac{\omega}{2\pi\lambda} \left(\frac{1}{\lambda} \frac{dn_1}{d\lambda} - \frac{n_1}{\lambda^2} \right)^{-1}$$

$$v_g = \frac{c}{\left[n_1 - \lambda \left(\dfrac{dn_1}{d\lambda} \right) \right]} \qquad (9\text{–}31)$$

where $\left[n_1 - \lambda \left(\dfrac{dn_1}{d\lambda} \right) \right]$ is referred to as the guide group index, and is denoted as N_g (Equation (9–32)).

Therefore,

$$v_g = \frac{c}{N_g} \qquad (9\text{–}32)$$

9.4 OPTICAL FIBER ATTENUATION (α)

One of the most critical factors that determines optical fiber applicability in communications systems is signal attenuation. The introduction stated that fiber attenuation started at 200 dB/km and was reduced progressively to 5 dB/km. At this level, fiber attenuation was in parity with copper wire lines. However, if optical fibers were to have wider application in communications links, the signal attenuation had to be further reduced. Today, attenuation on the order of 0.2 dB/km is normal.

Optical attenuation is referred to as the progressive amplitude reduction of the light ray traveling through the fiber. This amplitude reduction or signal attenuation is expressed in terms of power loss in dB/km, and is expressed by Equation (9–33).

$$\alpha = -10 \log \frac{P_o}{P_i} \qquad \text{(9–33)}$$

where α is the **optical fiber loss** (dB), P_o is the output optical power (W), and P_i is the input optical power (W).

Example 9–2

The optical power launched at the input of a 15 km fiber is 50 mW. If the optical power measured at the output of the fiber is 8 mW, compute the fiber attenuation.

Solution

i) Compute total fiber attenuation.

$$L = -10 \log \frac{P_o}{P_i} = -10 \log \frac{8 \times 10^{-3} \, (W)}{50 \times 10^{-3} \, (W)}$$

$$= -10 \log (-0.796) = 7.96 \text{ dB}$$

Therefore, $L = 7.96$ dB/15 km.

ii) Compute fiber attenuation in dB/km.

Because the total fiber attenuation is almost 8 dB/15 km, therefore $\alpha = 0.53$ dB/km.

Example 9–3

Compute the maximum length of an optical fiber that exhibits 0.8 dB/km attenuation if the output optical power is 10 mW and the power launched at the input is 150 mW.

Solution

i) Compute total fiber attenuation.

$$L = 10 \log \frac{P_o}{P_i} = 10 \log \frac{10 \times 10^{-3}}{150 \times 10^{-3}}$$

$$= 10 \log (0.067) = 11.74 \text{ dB}$$

Therefore, $L = 11.74$ dB.

ii) Compute the maximum length of the fiber.

Because the fiber exhibits 0.8 dB/km attenuation, for a total attenuation of 11.74 dB, the fiber length is as follows:

$$l = \frac{11.74 \text{ dB}}{0.8 \text{ dB/km}} = 14.67 \text{ km}$$

Therefore, maximum optical fiber length (l) = 14.67 km.

The maximum optical power received at the output of the fiber is not only a function of the length of the fiber, but is also a function of the absorption coefficient (a) expressed by Equation (9–34). A more detailed discussion of optical fiber **absorption** follows.

$$P_o = P_i e^{-al} \qquad \text{(9–34)}$$

where P_o is the output optical power (W), P_i is the input optical power (W), a is the absorption coefficient, and l is the length of the fiber (m).

9.5 OPTICAL FIBER ABSORPTION

Optical fiber absorption is material specific and is divided into two basic categories: **intrinsic absorption** and **extrinsic absorption**.

Intrinsic Absorption

Intrinsic absorption is the result of the interaction of free electrons and the operating wavelength within the fiber material. The propagating light wave covers a wide wavelength spectrum from the ultraviolet to the infrared region. Therefore, a particular wavelength within the optical spectrum interacts differently with the atoms of the fiber material.

For example, the ultraviolet region (short wavelengths) interacts with the outer shell electrons in the atoms, while the infrared region (longer wavelengths) interacts with the lattice structure in the atoms. If the fiber is fabricated from pure silicon glass, intrinsic absorption is kept to an absolute minimum at wavelengths near the infrared region. Figure 9–10 illustrates the absorption of optical rays versus wavelength for a single mode fiber (SMF).

From Figure 9–10, it is evident that peak absorption occurs at the ultraviolet region and is extended very close to the infrared region. The absorption observed at the ultraviolet region is expressed by Equation (9–35).

$$a_{uv} = C e^{\frac{\lambda_{uv}}{\lambda}} \qquad \text{(9–35)}$$

where a_{uv} is the absorption at (uv) region, C is the constant (1.108×10^{-3} dB/km), λ_{uv} is the 4.582 μm, and λ is the wavelength in the (uv) region (160 nm–400 nm). The absorption of the operating wavelength, due to its interaction with the material lattice structure, is given by Equation (9–36).

$$a_{(lt)} = a_{(lo)} e^{\frac{\lambda_{(lt)}}{\lambda}} \qquad \text{(9–36)}$$

where, $a_{(lt)}$ is the absorption due to lattice structure (dB/km), $a_{(lo)}$ is the constant (4×10^{-11} dB/km), $\lambda_{(lt)}$ is the constant reference wavelength (48 μm), and λ is the operating wavelength.

Extrinsic Absorption

Extrinsic absorption is attributed to impurities that are unintentionally injected into the optical fiber mix during the fabrication process. The most undesirable impurities in an optical fiber mix are metal ions. The presence of metal ions

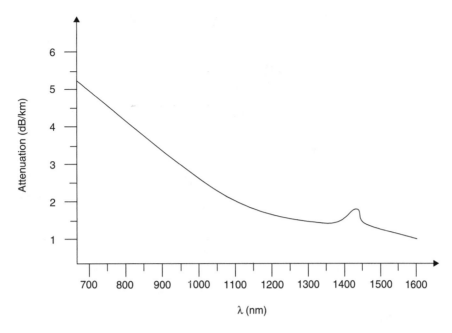

Figure 9–10. Absorption *v.* wavelength of SMF.

in the optical fiber alters the characteristic transmission properties of the fiber, resulting in unacceptable optical power losses. The metal ion concentration in the fiber mix is so critical that it must not exceed the level of one part per billion (ppb). The metallic ions, peak wavelengths, and reflected attenuations are listed in Table 9–1.

Metal contaminants such iron (Fe), chromium (Gr), and nickel (Ni) are significant contributors to optical power absorption. Figure 9–11 illustrates absorption levels in dB/km versus wavelength.

Modern fabrication techniques have all but eliminated metal ion contamination from optical fibers, as well as other contaminants such as OH^-. Hydroxyl groups (OH^-) embedded into the fiber material structures absorb optical energy at wavelengths between 2700 nm and 4200 nm, with significant overtones at the 7200 nm, 9500 nm, and 13,800 nm wavelength windows (Figure 9–12).

TABLE 9–1 Metallic Ion Concentration

Metallic Ion	Peak Wavelength (nm)	Part per Billion Attenuation (dB/km)
F_e^{3+}	400	0.15
M_n^{3+}	460	0.20
C_r^{3+}	625	1.60
N_i^{2+}	650	0.10
C^{2+}	685	0.10
C_u^{2+}	850	1.10
F_e^{2+}	110	0.68

9.6 SCATTERING

Another phenomenon that is detrimental to system performance, and that occurs during the transmission of optical waves through optical fibers, is that of **scattering.** Scattering is divided into two categories: linear scattering and nonlinear scattering.

Linear Scattering

Linear scattering occurs when optical energy is transferred from the dominant mode of operation to adjacent modes. It is proportional to the input optical power injected into the dominant mode. This optical energy spillover is instrumental in promoting intersymbol interference (ISI), which contributes to an increase in the BER and consequently to a decrease in the system performance. Linear scattering is also divided into two categories: **Mie scattering** and **Rayleigh scattering.** Both Mie and Rayleigh scattering are phenomena based on physical anomalies embedded into the optical fiber material structure as a result of imperfect fabrication techniques.

MIE SCATTERING. **Mie scattering** occurs when the size of physical anomalies within the optical fiber is larger than 1/10 of the diameter of the operating wavelength. These imperfections are the result of improper mixing of fabrication materials during the manufacturing process. Recent advancements in optical fiber manufacturing techniques have significantly contributed to improvements in areas such as wave guidance and precise control of coating, and a reduction of all the relevant manufacturing

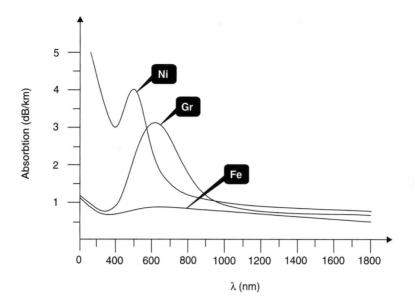

Figure 9–11. Absorption *v.* wavelength.

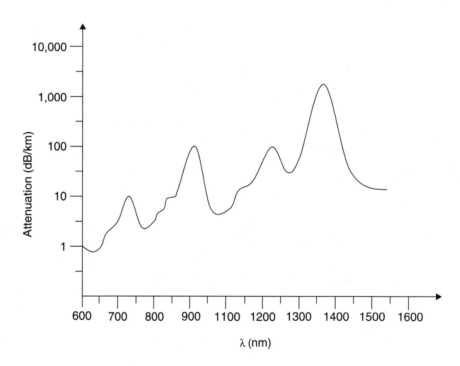

Figure 9–12. Absorption *v.* wavelength due to (OH^-).

imperfections. Today, fiber manufacturing techniques have advanced to the point where Mie scattering is almost completely eliminated.

RAYLEIGH SCATTERING. Rayleigh scattering is caused when material anomalies within the fiber material are 1/10 of the diameter (or less) of the operating wavelength. Scattering occurs when the optical wave propagating through the fiber encounters density irregularities with associated different refractive indices. Rayleigh scattering causes

a small part of the optical ray to escape from its predetermined path through the fiber, thus causing small attenuation to occur in the forward propagating optical wave. This attenuation is proportional to $(1/\lambda^4)$. Rayleigh scattering is expressed by Equation (9–37).

$$\gamma_R = \left(\frac{8\pi^3}{3\lambda^4}\right)(n^8)(\rho^2)(\beta_c)(KT_F) \qquad (9\text{–}37)$$

where γ_R = **scattering coefficient** (Rayleigh), λ = optical wavelength (m), n = refractive index of the fiber material,

ρ = **photoelastic coefficient**, β_c = **isothermal compressibility**, T_F = **fictive temperature** (thermal equilibrium), and K = Boltzmann's constant $(1.38 \times 10^{-23}$ J/K$)$.

It is evident from Equation (9–37) that Rayleigh scattering is very dependent on wavelength. Therefore, at long wavelengths, it is significantly reduced. The Rayleigh scattering coefficient is used to establish the transmission loss factor (F_R), expressed by Equation (9–38).

$$F_R = e^{-\gamma_R l} \qquad (9\text{–}38)$$

where F_R is the transmission loss factor (Rayleigh), γ_R is the Rayleigh scattering coefficient, and l is the fiber length.

Example 9–4

Compute the Rayleigh attenuation of an optical fiber fabricated with a silicon core. Use the available data. (Operating wavelengths: 630 nm, 1330 nm, and 1550 nm)

- $\beta_c = 92 \times 10^{-12}$ m^2/N
- $T_F = 1550$ K
- $n = 1.46$
- $\rho = 0.29$
- $K = 1.38 \times 10^{-23}$ J/K
- $l = 1$ km

Solution

i) Compute the Rayleigh scattering coefficient (γ_R).

$$
\begin{aligned}
\gamma_R &= \frac{8\pi^3\, n^8 \rho^2 \beta_c\, K T_F}{3\lambda^4} \\[4pt]
&= \frac{\dfrac{(248)(20.65)(0.0841)(92 \times 10^{-12})}{(1.55 \times 10^3)(1.38 \times 10^{-23})}}{3\lambda^4} \\[4pt]
&= \frac{2.825 \times 10^{-28}}{\lambda^4}
\end{aligned}
$$

For $\lambda = 630$ nm,

$$\gamma_R = \frac{2.825 \times 10^{-28}}{(630 \times 10^{-9})^4} = 1.79 \times 10^{-3}$$

Therefore, $\gamma_R = 1.79 \times 10^{-3}$.

For $\lambda = 1330$ nm,

$$\gamma_R = \frac{2.825 \times 10^{-28}}{(1330 \times 10^{-9})^4} = 0.9 \times 10^{-4}$$

Therefore, $\gamma_R = 0.9 \times 10^{-4}$.

For $\lambda = 1550$ nm,

$$\gamma_R = \frac{2.825 \times 10^{-28}}{(1550 \times 10^{-9})^4} = 1.82 \times 10^{-5}$$

Therefore, $\gamma_R = 1.82 \times 10^{-5}$.

ii) Compute the Rayleigh scattering attenuation factor (F_R).

At: 630 nm, $\gamma_R = 1.79 \times 10^{-3}$

$$F_R = e^{-\gamma_R l} = e^{-(1.79 \times 10^{-3})(1 \times 10^3)} = 0.17$$

$$F_R = 0.17$$

At: 1330 nm, $\gamma = 0.9 \times 10^{-4}$

$$F_R = e^{(-0.9 \times 10^{-4})(1 \times 10^3)} = 0.91$$

At:1550 nm, $\gamma_R = 1.82 \times 10^{-5}$

$$F_R = e^{(-1.82 \times 10^{-5})(1 \times 10^3)} = 0.98$$

$$F_R = 0.98$$

iii) Compute the Rayleigh scattering attenuation (α_R).

For: $F_R = 0.17$

$$\alpha_R = 10 \log(F^{-1}_R) = 7.7 \text{ dB/km}$$

Therefore, $\alpha_R = 7.7$ dB/km.
For: $F_R = 0.91$

$$\alpha_R = 10 \log (F^{-1}) = 0.45 \text{ dB/km}$$

Therefore, $\alpha_R = 0.45$ dB/km.
For: $F_R = 0.98$

$$\alpha_R = 10 \log (F_R^{-1}) = 0.087 \text{ dB/km}$$

Therefore, $\alpha_R = 0.087$ dB/km.

From the preceding example, it is evident that the optical fiber attenuation based on Rayleigh scattering diminishes at long wavelengths.

Nonlinear Scattering

Scattering losses in a fiber also occur due to fiber nonlinearities. That is, if the optical power at the output of a fiber does not change proportionally with the power change at the input of the fiber, the optical fiber is said to be operating in a nonlinear mode. Such behavior leads to scattering losses. Optical power loss due to nonlinear effects occurs when optical waves of a particular transmission mode travel in both forward and reverse directions, resulting in a slight shift of the centered wavelength. The level of loss and wavelength shift is a function of a critical optical power level launched into the input of the fiber. This power is referred to as threshold optical power. Two basic types of nonlinear scattering within the optical fiber exist. These are **stimulated Brillouin scattering (SBS)** and **stimulated Raman scattering (SRS)**. The two scattering mechanisms exhibit two fundamentally different characteristics. They attenuate optical signals of a particular wavelength, while shifting the operating wavelength to a point at which an optical power gain is also observed.

Although the SBS is detrimental to fiber operating performance, SRS, which contributes to wavelength shifting, will be fully utilized during the design process of erbium doped fiber amplifiers (EDFA).

STIMULATED BRILLOUIN SCATTERING. SBS occurs when the incident optical power at the input of a single mode fiber is progressively modulated along the traveling path by thermally vibrated molecules. An optical spectral display may show a carrier wave at the center and sidebands, above and below the center wavelength, which is indicative of undesirable modulation. Through the scattering process, the incident photons generate other scattering photons as well as elastic waves of acoustic frequency that shift in accordance with a shift of the acoustic wavelength. The larger frequency shift, due to Brillouin scattering, occurs in the opposite direction of propagation while little or no frequency shift is encountered in the forward direction. This indicates substantial optical power attenuation at the operating wavelength. As mentioned previously, stimulated Brillouin scattering occurs at a power level above a set threshold. This threshold optical power is expressed by Equation (9–39).

$$P_{o(Th)} = 4.4 \times 10^{-3} (\lambda^2)(d^2)(\alpha)(B_W) \qquad \textbf{(9–39)}$$

where $P_{o(Th)}$ = threshold optical power (Brillouin) (W), λ = operating wavelength (μm), d = core diameter (μm), α = fiber attenuation (dB/km), and B_W = source bandwidth (GHz). In order for stimulated Brillouin scattering to take place, the optical power launched into the fiber must be equal to or larger than the value established by Equation (9–39).

Example 9–5
Compute the SBS threshold optical power for a long single mode optical fiber with 5 μm core diameter, fiber attenuation of 0.5 dB/km, laser diode bandwidth of 1 GHz, and operating wavelength of 850 nm.

Solution

$$P_{o(Th)} = 4.4 \times 10^{-3} (\lambda^2)(d^2)(\alpha)(B_W)$$

$$= 4.4 \times 10^{-3} (0.85)^2 (5)^2 (0.5)(1)$$

$$= 4.4 \times 10^{2-3} (0.7225)(25)(0.5)(1)$$

$$= 0.034 \text{ W}$$

or

$$P_{o(Th)} \cong 40 \text{ mW}$$

Therefore, the optical power threshold before stimulated Brillouin scattering is initiated is 40 mW.

Example 9–6
Use the same parameters as Example 9–5, but with a 1330 nm operating wavelength.

Solution

$$P_{o(Th)} = 4.4 \times 10^{-3} (1.33)^2 (5)^2 (0.5)(1)$$

$$= 4.4 \times 10^{-3} (1.7689)(25)(0.5)(1)$$

$$= 97 \text{ mW}$$

Therefore, $P_{o(Th)} = 97$ mW.

Example 9–7
Use the same parameters as Example 9–5, but with an operating wavelength of 1550 nm.

Solution

$$P_{o(Th)} = 4.4 \times 10^{-3} (1.55)^2 (5)^2 (0.5)(1)$$

$$= 4.4 \times 10^{-3} (2.4)(25)(0.5)(1)$$

$$= 132 \text{ W}$$

Therefore, $P_{o(Th)} = 132$ mW.

Example 9–8
Use the same parameters as Example 9–5, but use a core diameter of 8 μm.

Solution

$$P_{o(Th)} = 4.4 \times 10^{-3} (0.85)^2 (8)^2 (0.5)(1)$$

$$= 0.101 \text{ W}$$

Therefore, $P_{o(Th)} = 101$ mW.

Example 9–9
Use the same parameters as Example 9–5, but use a core diameter of 10 μm.

Solution

$$P_{o(Th)} = 4.4 \times 10^{-3} (0.85)^2 (10)^2 (0.5)(1)$$

$$= 159 \text{ mW}$$

Therefore, $P_{o(Th)} = 159$ mW.

The preceding examples indicate that either increasing the operating wavelength, the fiber core diameter, or both can increase the stimulated Brillouin scattering threshold voltage. Therefore, a combination of both may accomplish the desired Brillouin threshold and fiber signal attenuation.

STIMULATED RAMAN SCATTERING. SRS is the result of the interaction between vibrating atoms in the crystalline lattice and the optical waves. That is, the vibrating atoms absorb some energy from the optical waves and, in conjunction with the vibration energy (characteristic of the atoms in the crystalline lattice), this energy is almost instantaneously reemitted in the form of photons. This emitted photon/atom vibration energy is a kind of light scattering, translated into wavelength shifting.

SRS contributes to signal attenuation at operating wavelengths. However, optical amplification is evident at the shifted frequencies. Therefore, the negative effect of SRS must be eliminated during the propagation of the optical wave through the fiber, while the amplification phenomenon at the shifted frequencies must be fully exploited in the design and fabrication of all optical amplifiers.

The stimulated Raman scattering threshold optical power is expressed by Equation (9–40).

$$P_{o(Th)} = 5.9 \times 10^{-2} (\lambda)(\alpha)(d^2) \qquad \textbf{(9–40)}$$

where $P_{o(Th)}$ is the SRS threshold optical power (W), λ is the operating wavelength (μm), α is the fiber attenuation (dB/km), and d is the core diameter (μm).

Example 9–10
Compute the stimulated Raman scattering threshold power in a long single mode fiber with core diameter of 5 μm, attenuation 0.4 dB/km, and operating wavelength of 850 nm.

Solution
$$P_{o(Th)} = 5.9 \times 10^{-2} (\lambda)(\alpha)(d)^2$$
$$= 5.9 \times 10^{-2} (0.85)(0.4)(5)^2$$
$$= 501 \text{ mW}$$

Therefore, $P_{o(Th)} = 501$ mW.

Example 9–11
Use the same parameters as Example 9–11, but use an operating wavelength of 1330 nm.

Solution
$$P_{o(Th)} = 5.9 \times 10^{-2} (1.33)(0.4)(5)^2$$
$$= 785 \text{ nm}$$

Therefore, $P_{o(Th)} = 785$ nm.

Example 9–12
Use the same parameters as Example 9–11, but use an operating wavelength of 1550 nm.

Solution
$$P_{o(Th)} = 5.9 \times 10^{-2} (1.55)(0.4)(5)^5$$
$$= 915 \text{ nm}$$

Therefore, $P_{o(Th)} = 915$ nm.

If the fiber core diameter is increased while the wavelength remains constant, an even higher increase in the threshold voltage will be encountered. A combination of selected operating wavelengths and core diameter can practically eliminate stimulated Raman scattering.

9.7 FIBER BEND LOSSES

Another factor that contributes significantly to optical fiber power loss is fiber bending. Optical fibers were studied under the assumption that they would operate in a straight line. This, of course, is not always possible. Therefore, the study of fibers as waveguides, under transmission conditions other than a straight line, is absolutely essential. When an optical fiber is bent, a portion of the optical energy escapes from the core to the cladding. If optical signal attenuation due to bending is to be avoided, both the wave traveling through the core and the wave traveling through the cladding must form a common wave front at the output of the fiber. The velocity of the wave traveling through the cladding must be higher than the velocity of the light in order to cover the longer distance. Because the core velocity is equal to the velocity of light (3×10^8 m/s), the velocity in the cladding must be higher than the velocity of light. This is impossible. Therefore, the wave traveling through the cladding will be lost in the form of radiation. This radiation loss is an optical power loss. Two categories of bending losses exist. One is due to **macrobending** (Figure 9–13), and the other is due to **microbending.**

Macrobending

Macrobending losses are subject to damages to the fiber during the packaging, transportation, and field installation processes. Fiber losses caused by macrobending can be expressed in terms of an attenuation coefficient (Equation (9–41)).

$$C_B = K_1 e^{-K_2 R} \qquad \textbf{(9–41)}$$

where C_B is the attenuation coefficient due to macrobending, K_1, K_2 are constants, and R is the macrobending radius.

Comparing the effect of macrobending for both single-mode and multimode fibers has shown that multimode fibers are more susceptible to macrobending than single-mode fibers are. The attenuation coefficient in Equation (9–41) becomes significant beyond a specific value of the curvature radius (R), referred to as the *critical radius*.

The critical radius of a multimode fiber is expressed by Equation (9–42).

$$R_{C_{MM}} = \frac{3\lambda n_1^2}{4\pi(n_1^2 - n_2^2)} \qquad \textbf{(9–42)}$$

where $R_{C_{MM}}$ is the critical radius, λ is the operating wavelength, n_1 is the core refractive index, and n_2 is the cladding refractive index. From Equation (9–42), it is evident that macrobending is proportional to the operating wavelength and core refractive index, and inversely proportional to the refractive index difference between core and cladding. Therefore, an increase of the operating wavelength in conjunction with an increase of the refractive index difference between the core and cladding will substantially reduce macrobending losses.

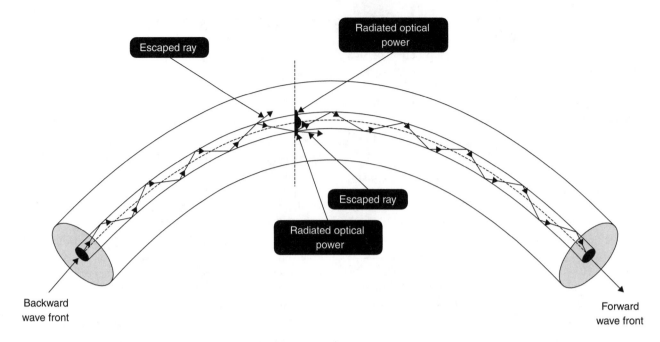

Figure 9–13. Optical fiber power loss due to macrobending.

For single-mode fibers, the critical bending radius is expressed by Equation (9–43).

$$R_{C_{SM}} = \frac{20\lambda}{\sqrt{(n_1^2 - n_2^2)}} \left(2.748 - \frac{0.996\lambda}{\lambda_C} \right)^{-3} \quad \textbf{(9–43)}$$

where $R_{C_{SM}}$ is the critical radius for single-mode fiber, λ is the operating wavelength, λ_C is the **cut off wavelength,** n_1 is the core refractive index, and n_2 is the cladding refractive index.

Microbending

Microbending of an optical fiber is defined as a slight deviation of the fiber core center axis. Microbending can be induced into the fiber during the fabrication process, installation process, or maintenance. The main factor that causes microbending is temperature. Temperature variations can cause local contractions and expansions of the core material, thus slightly altering the physical characteristics of the core, and ultimately affecting the optical fiber performance. That is, physical variations of the core can cause modal coupling, which can significantly contribute to modal dispersion in a multimode optical fiber. Modal coupling is referred to as the process whereby optical energy from one mode of transmission is transferred to other modes. Modal coupling increases signal attenuation, especially at high order modes. Microbending is affected by the optical power modal distribution, which is classified as either uniform or equilibrium mode distribution.

Uniform mode distribution is the process whereby all the available modes in the fiber are launched simultaneously with maximum sustainable optical power.

Equilibrium mode distribution is the process whereby the optical power carried by the individual modes in the fiber remain constant beyond the one kilometer fiber length and remain independent of the optical power launched at the input of the fiber. That is, any increase of the input optical power launched into a particular mode will be coupled to other modes, thus maintaining a constant modal optical power. It is evident that microbending is modal dependent and, consequently, wavelength dependent. Research has shown that optical fiber attenuation due to microbending decreases with an increase of the operating wavelength and that it is more susceptible to equilibrium modal distribution rather than to uniform modal distribution.

For single-mode fibers, attenuation due to microbending increases almost exponentially beyond the 1500 nm wavelength window, while minimum attenuation is encountered at wavelengths below the 1300 nm range. In order to minimize the microbending effect in optical fibers, certain precautions must be taken to eliminate externally imposed permanent pressures during the cabling process. At the same time, optical fibers must be fully protected from environmental changes, especially temperature variations.

9.8 CLASSIFICATION OF OPTICAL FIBERS

Optical fibers used in electronic communications are two-layer dielectric cylindrical systems capable of conveying electromagnetic waves that occupy the visible spectrum. An optical fiber consists of a dielectric cylindrical inner core, surrounded by an outer dielectric layer called cladding. Light waves are propagated through the core of the fiber in a zigzag mode, using the cladding as the reflecting medium. In order for the propagation of light to take place through the core of the fiber, the refractive index of the core's dielectric medium must be higher than that of the cladding. Other plastic layers that provide mechanical and environmental protection surround both the core and the cladding (Figure 9–14).

Materials used in the fabrication of optical fibers are SiO_2, boric oxide silica, and others. In order to achieve different refractive indices as required by the core (larger) and the cladding (smaller), different substances are chemically diffused into the basic materials through a very complicated manufacturing process. For example, to increase the refractive index of the core, germanium oxide (GeO_2) is added to the base material during the fabrication process, while fluorine is added to the cladding.

Based on their operating characteristics, optical fibers are classified into two **optical modes:** either single mode or multimode fibers (Figure 9–15). Single-mode fibers are capable of carrying only one signal of a specific wavelength. This single mode of operation has the advantage of minimizing signal distortion due to microbending, as well as minimizing overall signal attenuation and pulse expansion. Multimode fibers are capable of carrying the same optical wavelength through different paths that correspond to different arrival times at the end of the fiber. Multimode fibers can operate from a very small number of modes to a very large number. The core of a multimode fiber is substantially larger than a single-mode fiber, thus allowing for better core alignment and easier splicing. Optical fibers are also classified, according to their material characteristics, as either **step index** or **graded index** fibers (Figures 9–16 and 9–17).

Single-Mode Step Index Fibers

Step index refers to those fibers composed of a uniform cylindrical dielectric core of refractive index n_1, surrounded by a cladding material of slightly lesser refractive index n_2 ($n_1 > n_2$). *Step index* reflects the behavior of the light at the core-clad interface where an abrupt step (completely internal) reflection of the optical wave takes place. Because only one ray path is utilized in a single-mode fiber, the core diameter is substantially smaller than multimode fibers. **Single-mode step index (SM-SI)** fibers exhibit lower **intermodal** dispersion and higher operating bandwidth in comparison to multimode step index fibers. Their main disadvantages relate to the coupling difficulties with incoherent optical sources such as LEDs, small numerical aperture, and finally, critical tolerance in splicing and connections with other fibers.

The design and fabrication of advanced single mode fibers have increased during the last few years in response to high capacity demands in local area and long distance networks that operate at both the 1310 nm and the 1550 nm wavelengths. One of the most advanced single-mode fibers available in the market today is the SMF-28 developed by Corning. The SMF-28 is fabricated through outside vapor deposition (OVD). OVD is capable of achieving a very high purity synthetic fiber with excellent geometric and optical properties such as high strength and low attenuation. Test results have demonstrated that SMF-28 optical fibers are within the specifications set forth by Bellcore GR-20, ITU G-650, and IEC 60799-1 recommendations. The technical specifications of SMF-28 are illustrated in Tables 9–2, 9–3, 9–4, 9–5, and 9–6.

DISPERSION.

Zero dispersion wavelength (λ_0)

$1301.5 \text{ nm} \leq \lambda_0 \leq 1321.5 \text{ nm}$

Zero dispersion slope ($S_0 \leq 0.092 \text{ ps/nm}^2\text{km}$)

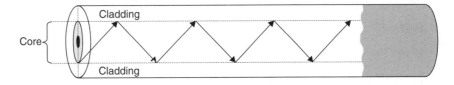

Figure 9–14. An optical fiber.

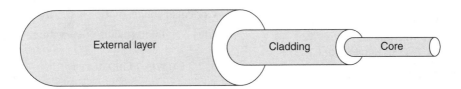

Figure 9–15. A single-mode step index fiber.

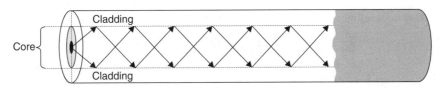

Figure 9–16. A multimode, step index fiber.

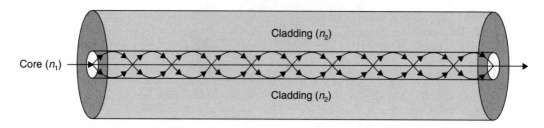

Figure 9–17. A multimode, graded index fiber.

TABLE 9–2 Standard Specifications

| Operating Wavelength (nm) | Attenuation (dB/km) | |
	Standard	Premium
1310	≤ 0.4	≤ 0.35
1550	≤ 0.25	≤ 0.3

TABLE 9–3 Attenuation _v._ Wavelength

Range (nm)	Ref (λ) nm	α_{max} (dB/km)
1285–1330	1310	≤ 0.05
1525–1575	1550	≤ 0.05

FIBER PARAMETERS.

Core diameter: 8.2 μm

Numerical aperture: 0.14 (measured at 1% power level at 1310 nm of one dimensional far field scan)

Zero dispersion wavelength (λ_0): 1312 nm

Zero dispersion slope (S_0): 0.090 ps/nm²km

Refractive index difference: 0.36%

Effective group index of refraction (N_{eff}): 1310 nm = 1.4677, 1550 nm = 1.4682

Fatigue resistance parameter (n_R): 20

Coating strip force: = Dry 0.6 lb (2.7 N), Wet 0.6 lb (2.7 N)

MECHANICAL SPECIFICATIONS.

Tensile proof stress ≥ 100 kpsi

DIMENSIONS.

Length (standard): 2.2–25.2 (km/reel)

TABLE 9–4 Attenuation _v._ Bending

Mandrel Diameter (mm)	Number of Turns	Wavelength (nm)	Induced α (dB/km)
32	1	1550	≤ 0.5
50	100	1310	≤ 0.05
50	100	1550	≤ 0.1

TABLE 9–5 Polarization Modal Dispersion (PMD)

	Value $(ps/\sqrt{km})$
PMD (link value)	≤ 0.1
Individual fiber (max)	≤ 0.2

TABLE 9–6 Environmental Specifications

	α(dB/km)	
Environmental test conditions	1310 nm	1550 nm
Temperature: −60°C to+85°C	≤ 0.05	≤ 0.05
Temperature humidity: −10°C to +85°C Up to 98% ref.	≤ 0.05	≤ 0.05
Water immersion: 23°C ± 2°C	≤ 0.05	≤ 0.05
Heat aging: 85°C ± 2°C	≤ 0.05	≤ 0.05

CLASS GEOMETRY.

Cladding diameter: 125.0 μm ±1.0 μm

Core/clad connectivity: ≤ 0.5 μm

Curl: ≥ 4.0 m (radius of curvature)

COATING GEOMETRY.

Coating diameter: 245 μm ± 5 μm

Coating cladding concentricity: < 12 μm

SPECTRAL ATTENUATION.

See Figure 9–18.

REFRACTIVE INDEX PROFILE.

See Figure 9–19.

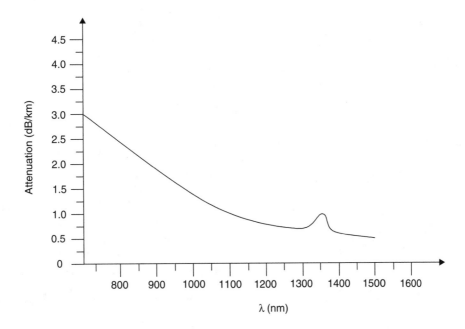

Figure 9–18. Spectral attenuation.

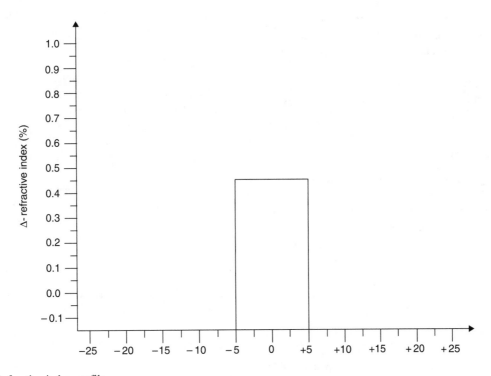

Figure 9–19. Refractive index profile.

Multimode Step Index Fibers

Multimode step index (MM-SI) fibers are very similar to single-mode step index fibers, but differ in their core diameter. MM-SI fiber core diameters are around 50 μm. Recent developments in advanced optical fiber technology has produced fibers with core diameters in excess of 80 μm. Those fibers will be discussed later in this chapter. The number of modes in a multimode optical fiber is determined by its physical and operating characteristics such as core radius, operating

wavelength expressed by the normalized frequency (V), and relative refractive index difference (Δ).

The normalized frequency (V) is a dimensionless parameter relating three important fiber variables (see Equation (9–44)). The three variables are

1) core radius (R)

2) operating wavelength (λ)

3) relative refractive index variation (Δ)

$$V = \frac{2\pi}{\lambda}(R)(NA) \qquad \textbf{(9--44)}$$

where V is the **normalized frequency,** λ is the operating wavelength, R is the core radius, and NA is the numerical aperture.

or

$$V = \frac{2\pi}{\lambda}(an_1)\sqrt{(2\Delta)} \qquad \textbf{(9--45)}$$

where $\Delta = \dfrac{n_1^2 - n_2^2}{2n_1^2}$, n_1 is the core refractive index, and n_2 is the cladding refractive index. The maximum number of modes is determined by the cut off value of the normalized frequency (V_C), expressed by Equation (9--46).

$$V_C = \frac{V^2}{2} \qquad \textbf{(9--46)}$$

Equation (9--46) indicates that no transmission mode exists below the normalized cut off frequency. From the equation, it is evident that optical power of a specific wavelength can be launched into the core of a fiber at a particular mode (θ). Under ideal conditions, this power is transmitted to the other end of the fiber without interfering with the optical power propagating through other modes within the same fiber. For ideal operation, the number of modes propagated through the fiber must be kept below the maximum determined by the normalized cut off frequency (V_C).

Graded Index Fibers

Graded index fibers are defined by a progressive decrease of the refractive index from the center of the core with radius (a) toward the cladding, while still maintaining the fundamental relationship ($n_1 > n_2$). The refractive index at any given point between the center of the core and the core cladding interface is expressed by Equation (9--47).

$$n_{(r)} = n_1\sqrt{\left[1 - 2\Delta\left(\frac{r}{a}\right)^y\right]} \qquad \textbf{(9--47)}$$

where a is the radius of the fiber core, $n_{(r)}$ is the refractive index between the center of the core and the core cladding int., r is the cylindrical coordinate with values ($r<a$), Δ is the relative refractive index, R is the core radius, and y is the core characteristic **refractive index profile:**

1) step index profile, $y = \infty$

2) **triangular profile,** $y = 1$

3) **parabolic profile,** $y = 2$

The three standard index profiles are illustrated in Figure 9--20.

Of the three index profiles, the most desirable is the parabolic index profile ($y = 2$), illustrated in Figure 9--21.

The parabolic path followed by the incident optical wave is constructed by a very large number of discrete segments dictated by the degree of refractive index graduality that is required. A perfect parabolic path is totally dependent on the fabrication process of the core. Regardless of the process, the incident ray must be fully reflected back toward the center of the core at the core-cladding interface.

Performance Characteristics

The distinctive advantage of multimode graded index fibers is that they exhibit far less modal dispersion than step index multimode fibers. Because of their unique refractive index profile, group velocities are normalized, thus providing a self-compensating dispersion mechanism.

Picture two rays within the optical wave, one traveling very close to the core and the other very close to the core-cladding interface. The ray path close to core-cladding interface is longer than the path traveled close to the center of the core. Therefore, the arrival time of the two rays will be different. This time difference is instrumental in promoting dispersion or pulse broadening. The potential problem is avoided because the refractive index closer to the center of the core is larger and, therefore, the velocity of the ray closer to the core is smaller than the velocity closer to the core-clad interface, thus compensating for the shorter traveling distance. Through this mechanism, the two rays traveling different distances at different speeds can be made to arrive at the output of the fiber at the same time, thus minimizing or completely eliminating modal dispersion. The transmission bandwidth capabilities of graded index optical fibers are larger than those of step index multimode fibers, but smaller than those of single-mode step index fibers. This multimode graded index fiber disadvantage has been almost eliminated through the introduction of larger core diameter fibers, *large effective area fibers* that have performance capabilities close to those of single-mode fibers.

Modal Number

The maximum **modal number** supported by graded index multimode optical fiber is given by Equations (9--48) and (9--49).

$$M_{max} = \left(\frac{y}{y+2}\right)(n_1 kR)^2\left(\frac{n_1^2 - n_2^2}{2n_1^2}\right) \qquad \textbf{(9--48)}$$

or

$$M_{max} = \left(\frac{y}{y+2}\right)\left(\frac{V^2}{2}\right) \qquad \textbf{(9--49)}$$

where M_{max} is the maximum number of guided modes, y is the graded index characteristic profile (parabolic $= 2$), n_1 is the core refractive index, n_2 is the cladding refractive index, and R is the core radius.

Example 9--13

Compute the maximum modal number of an optical fiber that has a core radius of 25 μm, an operating wavelength

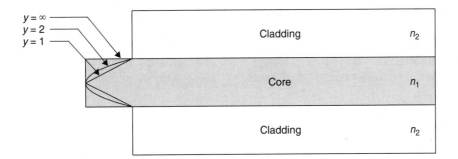

Figure 9–20. The three standard index profiles.

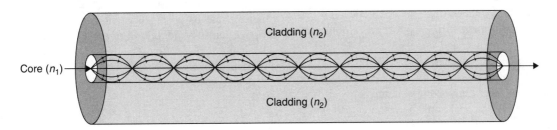

Figure 9–21. Graded index/parabolic index profile.

of 1000 nm, and NA equal to 0.2. Assume graded index characteristic profile (parabolic).

Solution

i) Compute the normalized frequency (V).

$$V = \frac{2\pi}{\lambda}(R)(\text{NA})$$

$$= \frac{2\pi}{1 \times 10^{-6}}(25 \times 10^{-6})(0.2)$$

$$= 31.4$$

ii) Compute the maximum number of modes (M_{max}).

$$M_{max} = \left(\frac{y}{y+2}\right)\left(\frac{V^2}{2}\right)$$

$$= \frac{2}{4} \times \frac{3.14^2}{2} = 27.7$$

Therefore, the maximum number of modes is equal to 27.

Cut Off Wavelength

In order for the cut off wavelength of an optical fiber to be determined, it is necessary to go back to the beginning of the chapter and review the section related to the propagation of light through an optical fiber. The light traveling through an optical fiber is an electromagnetic wave, and an electromagnetic wave is a transverse wave. It is composed of an electric and a magnetic field that are perpendicular to each other. Both are perpendicular to the direction of propagation. The analysis of the electric field component of the electromagnetic wave through the implementation of the wave equations leads

to two different conclusions. First, through the utilization of the polar component of the electric field, the precise number of the modal components can be established. This approach to establish the modal components of an optical fiber exhibits certain disadvantages because it requires the incorporation of broad approximations into the mathematical formula. The second method, utilizing the Cartesian component of the electric field, leads to the establishment of the linear polarized field (LP). This method, although it provides only an approximate representation of the field distribution, is fairly accurate and easy to calculate.

One of the most important parameters that determines the guiding mechanism of an optical fiber is its normalized frequency (V). This normalized frequency is related to the physical and operating characteristics of the optical fiber (Equation (9–50)).

$$V = \frac{2\pi}{\lambda}(R)(\text{NA}) \qquad \textbf{(9–50)}$$

where V is the normalized frequency, λ is the operating wavelength, R is the core radius, and NA is the numerical aperture.

From the preceding relationship, it is evident that an increase of the operating wavelength (λ) will decrease the normalized frequency, while an increase in the aperture or core radius will increase it. A value of the normalized frequency (V) exists below which no mode can propagate. This value is referred to as the cut off normalized frequency (V_C). For a specific value of normalized frequency, a specific number of modes will propagate. As the normalized frequency decreases, the propagating modes also decrease. A progressive decrease of (V) will reach a point at which only one mode will be able to propagate. This mode of propagation is referred to as the fundamental mode (LP_{01}). Optical fibers that operate at the fundamental mode are referred to as single-mode fibers. Beyond this value of (V), the number of modes increases. Therefore, at cut off normalized frequency, a wavelength referred to as the "cut-off–wavelength" (λ_C) exists (Equation (9–51)).

$$\lambda_{C_{TH}} = \frac{2\pi R(\text{NA})}{V_C} \qquad \textbf{(9–51)}$$

where $\lambda_{C_{TH}}$ is the cut-off wavelength (theoretical), R is the core radius, NA is the numerical aperture, and V_C is the cut off normalized frequency.

Equation (9–51) shows that the theoretical cut off wavelength is proportional to the optical fiber numerical aperture and core radius, and inversely proportional to the cut off normalized frequency (V_C). The theoretical cut off wavelength is also subject to various external stresses, such as macrobending, microbending, and optical fiber length. Therefore, in practical applications, a cut off wavelength lower than the theoretical wavelength is selected in order to compensate for the losses.

9.9 DISPERSION

Dispersion is the phenomenon whereby the modulating electric signal is broadened while traveling through the core of the fiber (Figure 9–22). Such signal chirping promotes intersymbol interference, resulting in an increase of the system BER.

The causes of pulse dispersion are several and are classified in two broad categories as **intramodal** and **intermodal.** These two categories will be examined in some detail later in this chapter.

At distance L_1 in Figure 9–22, although pulse spreading is evident, overlapping has not occurred and the optical amplifier (error free) can successfully detect the data pattern. At distance L_2, spreading has caused pulse overlap, and the optical detector cannot successfully recover the binary data, resulting in a significant increase of ISI.

Dispersion that causes **pulse chirping** is one of the fundamental elements that limits the overall system capacity by imposing limits on the maximum usable system bandwidth. The relationship between dispersion and bandwidth is given in Equation (9–52).

$$f_b \leq \frac{1}{2\tau} \qquad \textbf{(9–52)}$$

where f_b = **channel bit rate** and τ = pulse duration. Therefore, the maximum allowable pulse dispersion is equal to (2τ).

A more precise determination of the maximum channel bit rate based on dispersion is given by assuming the optical pulse at the output of the fiber to be Gaussian with standard deviation (σ). The optical power variation in reference to time (t) at the output of the fiber is expressed by Equation (9–53).

$$P_o(t) = \frac{1}{\sqrt{2\pi}} e^{-\frac{t^2}{2\sigma^2}} \qquad \textbf{(9–53)}$$

where $P_o(t)$ = optical power output, t = time, σ = **standard deviation,** and σ^2 = **distribution variance.**

Let $\dfrac{P_o(t)}{P_o(0)} = \tau$

where τ = pulse width.

Therefore, $P_o(0) = \dfrac{1}{\sqrt{2\pi}}$

or

$$\frac{P_o(t)}{P_o(0)} = \frac{\frac{1}{\sqrt{2\pi}} e^{-\frac{t^2}{2\sigma^2}}}{\frac{1}{\sqrt{2\pi}}} = e^{-\frac{t^2}{2\sigma^2}}$$

$$\frac{t^2}{2\sigma^2} = 1$$

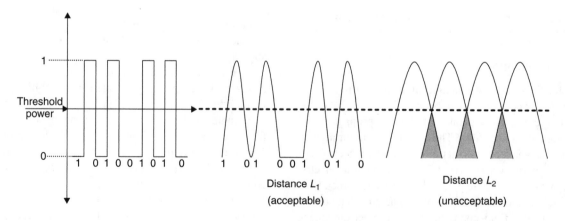

Figure 9–22. Pulse dispersion.

Solve for (t).

$$t = \sqrt{2}\sigma$$

Pulse full-width: $\tau = 2t$ or $\tau = 2\sigma\sqrt{2}$

The standard deviation is equal to pulse width (rms).

The fourier transform (FT) of $P_o(t) = \dfrac{1}{\sqrt{2\pi}} e^{-\frac{t^2}{2\sigma^2}}$ yields Equation (9–54).

$$P(\omega) = \frac{1}{\sqrt{2\pi}} e^{-\frac{\omega^2\sigma^2}{2}} \qquad (9\text{–}54)$$

At -3 dB, the optical bandwidth ($B_{W(o)}$) is defined between zero and the 3 dB point at which the required optical power is half of the input signal power.

The exponent of Equation (9–54) yields,

$$e^{\frac{\omega^2\sigma^2}{2}} = 0.5(P_o)$$

$$\frac{\omega^2\sigma^2}{2} = \ln(0.5) = 0.693$$

or

$$\frac{\omega^2\sigma^2}{2} = 0.693$$

Solve for (ω).

$$\omega = \sqrt{\frac{1.386}{\sigma^2}} = \frac{1.177}{\sigma}$$

Because $\omega = 2\pi f$, and assuming (f) the operating bandwidth (f_{B_w}), then $2\pi f_{B_w} = \dfrac{1.177}{\sigma}$

or

$$f_{B_w} = \frac{0.187}{\sigma} \text{ Hz}$$

When an optical wave is digitally modulated with an NRZ binary sequence, the maximum bit rate is equal to the available channel bandwidth (Equation (9–55)).

$$f_b = f_{B_w}$$

or

$$f_b = \frac{0.187}{\sigma} \text{ b/s} \qquad (9\text{–}55)$$

where f_b = channel bit rate (b/s) and f_{B_w} = **channel bandwidth** (Hz). In reference to pulse width,

$$\tau_P = 2\sqrt{2}\sigma$$

or

$$\sigma = \frac{\tau_P}{2\sqrt{2}}$$

Substituting this into Equation (9–55) yields

$$f_b = \frac{0.187}{\dfrac{\tau_p}{2\sqrt{2}}} = \frac{0.187 \times (2\sqrt{2})}{\tau_P} = \frac{0.529}{\tau_P}$$

Therefore, maximum bit rate is expressed by Equation (9–56).

$$f_b = \frac{0.529}{\tau_P} \qquad (9\text{–}56)$$

The usable bandwidth of a fiber employed in an optical communications link based on dispersion is different for the three basic optical fibers. The best performance is exhibited by step index single-mode fibers and the worst performance is exhibited by step index multimode fibers. Graded index multimode fibers exhibit performance levels that fall between those of step index single-mode and step index multimode fibers. The correlation between pulse chirping (pulse dispersion) and the type of fiber is illustrated in Figure 9–23.

However, the usable bandwidth of an optical fiber is also subject to its operating length. That is, bandwidth limitation is not only attributed to the dispersion properties of the fiber, but also to the distance that the pulse has to travel within the fiber, expressed in (nm/km). It is, therefore, evident that the ability of an optical fiber to convey digital information is based on its usable bandwidth, which is limited by the dispersion properties of the fiber, and its operating length. Usable bandwidth and operating length are combined to form a new optical fiber performance parameter called the **bandwidth length product** ($f_{B_w} \times L$). Because the three basic optical fiber types exhibit different dispersive properties, they also exhibit different bandwidth length products. Figure 9–23 indicates that the step index single-mode fibers have the best bandwidth length product, estimated at (100 GHz), followed by the graded index multimode fiber (1 GHz), and the step index multimode fibers, which have the worst bandwidth length performance of 20 MHz.

Example 9–14

Compute the maximum operating bandwidth (f_{B_w}), the pulse dispersion of a step index single-mode fiber that exhibits a pulse chirping (**dispersion**) of 0.25 µs and has a total fiber length of 50 km and the bandwidth length product ($f_{B_w} \times L$).

Solution

i) Use Equation (9–56) to compute the maximum operating bandwidth (f_{B_w}).

$$f_b = \frac{0.529}{\tau_P} = \frac{0.529}{0.25 \times 10^{-6}} = 2.116 \text{ Mb/s}$$

$$\therefore f_b = 2.116 \text{ Mb/s}$$

Assuming a zero ISI and an RZ input data signal, the operating bandwidth is equal to the system bit rate.

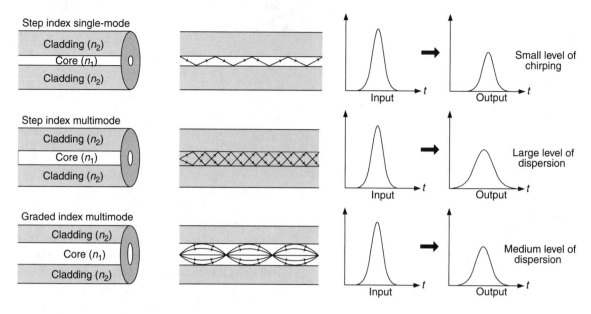

Figure 9–23. Pulse chirping for the three basic types of fiber.

Therefore,

$$f_{B_W} = f_b = 2.116 \text{ MHz}$$

$$\therefore f_{B_W} = 2.116 \text{ MHz}$$

ii) Compute the pulse chirping (dispersion). Because the total pulse dispersion for this application is 0.25 μ/50 km,

$$\text{Dispersion} = \frac{0.25 \times 10^{-6} \text{ s}}{50 \text{ km}} = 5 \times 10^{-9} \text{ s/km}$$

$$\therefore \text{Dispersion} = 5 \text{ ns/km}$$

iii) Compute the bandwidth length product $(f_{B_W} \times L)$.

$$f_{B_W} \times L = (2.116 \times 10^6)(50 \times 10^3)$$

$$= 105.8 \text{ MHz/km}$$

Example 9–15
Repeat the above example for a graded index multimode fiber that has a total dispersion of 2 μs and is 50 km long.

Solution

i) Compute the maximum operating bandwidth.

$$f_b = \frac{0.529}{2 \times 10^{-6}} \cong 0.26 \text{ MHz}$$

Applying the same criteria in Example 9–14, the system bandwidth is:

$$f_{B_W} = 0.26 \text{ MHz}$$

ii) Compute the pulse chirping (dispersion).

$$\text{Dispersion} = \frac{2 \times 10^{-6} \text{ s}}{50 \text{ km}} = 40 \text{ nm/km}$$

iii) Compute the bandwidth length product.

$$f_{B_W} \times L = 0.26 \text{ MHz} \times 50 \text{ km} = 13 \text{ MHz/km}$$

$$\therefore f_{B_W} \times L = 13 \text{ MHz/km}$$

Example 9–16
Compute the maximum operating bandwidth, the pulse dispersion of a step index multimode fiber that exhibits a total dispersion of 5 μs and has a total fiber length of 50 km. Finally, compute the bandwidth length product.

Solution

i) Compute the maximum operating bandwidth.

$$f_b = \frac{0.529}{5 \times 10^{-6}} \cong 0.106 \text{ Mb/s}$$

Applying the same criteria as per Example 9–14, the operating bandwidth is:

$$f_{B_W} = 0.106 \text{ MHz}$$

ii) Compute the pulse chirping (dispersion).

$$\text{Dispersion} = \frac{5 \times 10^{-6} \text{ s}}{50 \text{ km}} = 0.1 \text{ μs/km}$$

iii) Compute the bandwidth length product.

$$f_{B_W} \times L = 0.106 \text{ MHz} \times 50 \text{ km} = 5.3 \text{ MHz/km}$$

$$f_{B_W} \times L = 5.3 \text{ MHz/km}$$

From Examples 9–14, 9–15, and 9–16 the superior transmission capabilities of step index single-mode fibers are evident.

Intramodal Dispersion

Intramodal (chromatic) dispersion results from the inability of the optical sources to emit a single optical wavelength. Because the optical signal launched into the fiber mode is composed of more than one frequency component (very close to the main frequency), the interaction of the different spectral components of the optical signal with the dispersive and guidance properties of the fiber induce signal delays, which cause pulse broadening or chirping. Intramodal dispersion is subdivided into two basic categories: material dispersion and waveguide dispersion.

MATERIAL DISPERSION. Pulse chirping is the result of a nonlinear change of the phase velocity in reference to wavelength of a plane wave that is composed of different frequency components and is propagating through a dielectric medium. The different frequency components of the plane wave launched into the input of the fiber reflect different group velocities that directly contribute to pulse chirping at the output of the fiber. Therefore, pulse width chirping can be expressed by the total **group delay** at the output of the fiber, an inverse function of the group velocity (v_g).

Invert Equation (9–26), which indicates the group velocity, and obtain the group delay (τ) (Equation (9–57)).

$$v_g = \frac{c}{\left[n_1 - \lambda \left(\frac{dn_1}{d\lambda} \right) \right]}$$

or

$$\tau = \frac{1}{v_g} = \frac{\left[n_1 - \lambda \left(\frac{dn_1}{d\lambda} \right) \right]}{c} \qquad (9–57)$$

where τ is the group delay, based on material dispersion, v_g is the group velocity, n_1 is the core refractive index, λ is the operating wavelength, and c is the velocity of light in free space.

Pulse chirping due to material dispersion at the end of an optical fiber of length (L) is based on Equation (9–40) multiplied by the length of the fiber (Equation (9–58)).

$$\tau_{md} = \frac{\left[n_1 - \lambda \left(\frac{dn_1}{d\lambda} \right) \right]}{c} \times L \qquad (9–58)$$

where τ_{md} is the group delay based on material dispersion for the length of the fiber and L is the length of the fiber. Equation (9–58) indicates that pulse broadening due to material dispersion is also proportional to the length of the fiber. For a more detailed evaluation of pulse chirping due to material dispersion, the **source spectral line width** ($S_{sl\,(rms)}$) and mean wavelength (λ) are taken into account to form an approximate expression (Equation (9–59)).

$$d_{m(rms)} = S_{sl(rms)} \times \frac{d\tau_{md}}{d\lambda} \qquad (9–59)$$

where $d_{m(rms)}$ is the material dispersion pulse chirping, $S_{sl(rms)}$ is the source spectral line width, and τ_{md} is the group delay based on material dispersion for the length of the fiber.

Rearranging Equation (9–57) yields Equation (9–60).

$$\tau_{md} = \frac{\lambda L}{c} \left(\frac{n_1}{\lambda} - \frac{dn_1}{d\lambda} \right) \qquad (9–60)$$

The derivative of the above expression yields the following:

$$\frac{d\tau_{md}}{d\lambda} = \frac{\lambda L}{c} \left(\frac{dn_1}{d\lambda} - \frac{d^2 n}{d\lambda^2} - \frac{dn_1}{d\lambda} \right)$$

$$\therefore \frac{d\tau_{md}}{d\lambda} = \left(\frac{\lambda L}{c} \right) \left(\frac{d^2 n_1}{d\lambda^2} \right)$$

Substitute Equation (9–60) into Equation (9–59) yields Equation (9–61).

$$d_{m(rms)} = S_{sl} \times \left(\frac{L}{c} \right) \left(\frac{\lambda d^2 n_1}{d\lambda^2} \right) \qquad (9–61)$$

The term $\left[\left(\frac{1}{\lambda_c} \right) \left| \frac{\lambda^2 d^2 n_1}{d\lambda^2} \right| \right]$ is referred to as the material dispersion parameter (M_{dp}) (Equation (9–62)).

$$\therefore M_{dp} = \left(\frac{1}{\lambda_c} \right) \left| \frac{\lambda^2 d^2 n_1}{d\lambda^2} \right| \, ps/nm \cdot km \qquad (9–62)$$

The term $\left(\frac{\lambda^2 d^2 n_1}{d\lambda^2} \right)$ is referred to as the optical fiber material dispersion. Therefore, pulse chirping is expressed by Equation (9–63).

$$d_{md} = S_{sl} \times L \times M_{md} \qquad (9–63)$$

Figure 9–24 relates the material dispersion parameter (M_{md}) to wavelength.

Example 9–17

An LED source with an rms spectral linewidth of 25 nm is coupled into an optical fiber that exhibits material dispersion of 0.020. Compute the rms pulse chirping for the 850 nm operating wavelength.

Solution

i) Compute the material dispersion parameter (M_{dp}).

$$M_{dp} = \frac{1}{\lambda c} \left| \frac{\lambda^2 d^2 n_1}{d\lambda^2} \right| = \frac{1}{(870 \text{ nm})(3 \times 10^5 \text{ km})}(0.02)$$

$$= 76 \times 10^{-12} \text{ s/nm} \cdot \text{km}$$

or

$$M_{dp} = 76 \text{ ps/nm} \cdot \text{km}$$

ii) Compute the pulse chirping.

$$d_{md} = S_{lw} \times L \times M_{dp}$$

$$= 25 \text{ nm} \times 1 \text{ km} \times 76 \text{ ps/nm} \cdot \text{km}$$

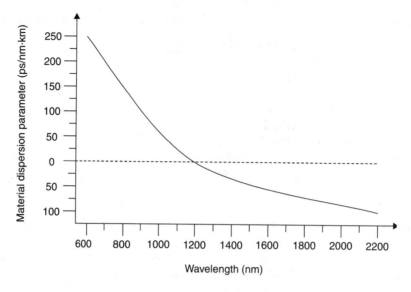

Figure 9–24. Material dispersion parameter v. wavelength. Courtesy of W. A. Gambling and D. N. Payne.

$$= 1.9 \times 10^{-9}\,\text{s}$$

Therefore, $d_{md} = 1.9$ ns/km.

Example 9–18

In Example 9–17, the source is replaced by a laser diode with a spectral line width of 2 nm while the rest of the data remains the same. Compute the rms pulse broadening and compare it to the result of the previous example.

Solution

i) Compute the material dispersion parameter (M_{dp}).

$$M_{dp} = \frac{1}{\lambda c}\left[\lambda^2 \frac{d^2 n_1}{d\lambda^2}\right] = \frac{0.02}{(870)(3 \times 10^5)} = 76 \times 10^{-12}$$

Therefore, $M_{dp} = 76$ ps/nm·km.

ii) Compute the pulse chirping (d_{md}).

$$d_{md} = S_{sl} \times L \times M_{dp}$$

$$= (2\ \text{nm}) \times (1\ \text{km}) \times (76\ \text{ps})$$

$$= 152\ \text{ps/km}$$

Therefore, $d_{md} = 0.152$ ns/km.

Laser diodes with very narrow spectral line width reduce chirping induced by material dispersion.

Waveguide Dispersion

Waveguide dispersion occurs when the ray angle of incident in relation to fiber core axis deviates from the origin in accordance with the change of the center wavelength. This slight deviation of the angle of incident reflects a change in the group velocity and leads to pulse spreading. Waveguide dispersion is mostly noticeable in single-mode fibers, while multimode operations are affected very little or not at all because they are operating at wavelengths far below the cut off wavelengths.

Intermodal Dispersion

Intermodal dispersion is noticed mainly with multimode step index fibers, much less with graded index fibers. Intermodal dispersion is the result of the different group velocities along the various modes of propagation within the optical fiber. Because these modes follow different transmission paths, they arrive at the output of the fiber at different times, which promotes undesirable pulse spreading. Step index multimode fibers are far more susceptible to intermodal spreading because of the different modes that arrive at the output at different times. Graded index multimode fibers are less susceptible to intermodal dispersion because the varying index profile of the fiber core compensates for the different group velocities generated by the propagating modes. Therefore, pulse spreading is significantly reduced or altogether eliminated.

Step index single-mode fibers are not susceptible to intermodal dispersion because only one mode of transmission is used. Furthermore, because such fibers promote the least amount of chirping, they exhibit maximum modal bandwidth utilization.

Polarization Modal Dispersion

In reference to the effect on pulse spreading or chirping, definite similarities between chromatic (intramodal) and polarization mode dispersion (PMD) exist. Like chromatic dispersion, PMD is a limiting factor in DWDM long distance optical links, especially at transmission rates beyond the 10 Gb/s mark. Because chromatic dispersion can be compensated for by the insertion of dispersion compensation modules, the effect of PMD must be evaluated and its effect on system performance must be closely monitored. Short distance optical links composed of only a single fiber induce insignificant levels of PMD, which has a minimal impact on system performance. The evaluation of PMD of optical fiber systems is based on statistical models. This is because it requires a relatively large number of fibers to induce a level of PMD capable of effecting system quality. Therefore, statistical methods of evaluation are deemed appropriate.

PMD is a function of the operating wavelength, and it is evaluated by measuring the delay time between two orthogonal polarized modes within a single-mode fiber. The measured time difference is referred to as differential group delay (DGD). Applying Maxwell's probability density function (PDF), the PMD variations can be expressed with a coefficient that expresses the instantaneous differential group delay average spread over the operating wavelength window and normalized in quadrature over the entire link distance. The quadrature or rms average is used to statistically evaluate the PMD distribution, ultimately becoming a crucial factor in determining optical link performance.

By averaging the rms over a large range of coefficients obtained from all the individual fibers that compose the entire optical link, the PMD can now be estimated. Because the PMD coefficients are randomly distributed, the total link PMD coefficients are also randomly distributed. The PMD distribution average variations are more evident with single fibers than with the total number of fibers composing the entire optical link.

The PMD coefficient of an optical cable composed of a number of equal length fibers is expressed by Equation (9–64).

$$X_M = \sqrt{\frac{\sum_i^M x_i^2}{M}} \qquad (9\text{--}64)$$

where X_M is the PMD coefficient of a number of optical fibers linked together (cable), x_i is the PMD coefficient of a single fiber (ps/nm^2km), and M is the number of equal length fibers (fiber length ≤ 10 km).

Methods of Determining PMD Link Value

The PMD upper limit can be statistically established for a combined number of fibers incorporated into a system through either the Monte Carlo numeric method, the model independent analysis method, or the gamma distribution analytic method.

These three methods can be applied to calculate the PMD probabilistic coefficient distribution. Only the Monte Carlo method will be examined in this chapter.

THE MONTE CARLO METHOD. Through the Monte Carlo method, a very large number of fibers are used to

individually measure the distributive mode dispersion (DMD) coefficient. The data obtained is then used to calculate the PMD over the entire link. From the pool of values obtained, a number of them are randomly selected (not less than ten thousand) in order to provide statistical credibility. They are then compiled and the rms is evaluated and identified as the Q probability. For an optical link composed of a relatively large number of optical fibers, the value of PMD (link value) is defined by the link PMD. The Q probability is expressed by Equation (9–65).

$$\text{Probability}(X_M > \text{PDM}) = Q \qquad (9\text{--}65)$$

If the number of fibers in the optical link is larger than a set number (M), then $X_M < Q$.

Assuming twenty fibers of equal length are incorporated into an optical link and the number of samples taken is ten thousand, the probability level Q is set at 10^{-4}. It is therefore evident that a larger number of fibers in an optical link will require a larger number of samples.

9.10 MULTIMODE FIBERS FOR Gb.E APPLICATIONS (IEEE 802.37) LAN

The revolutionary development in the area of sophisticated optical devices in the last five years has had a great impact, not only on long distance optical transmission systems in terms of span lengths and system capacities, but also on local area networks (LANs). The crucial factor that helped to propel the expansion of LANs to the Gb/s transmission range, is the standardization of the Ethernet Gb.E (IEEE-802.37) protocol. The implementation of the Gb.E standard protocol required not only sophisticated active devices but also advanced optical fibers. Until very recently, the basic optical transmitter exclusively utilized in local area networks was the LED. The operating principle of LEDs is based on spontaneous emission of radiation that has, consequently, a large carrier lifetime. This is a disadvantage that imposes limits on the system operating bandwidth, most particularly on the Gb/s range demanded by the Gb.E LAN protocol. In contrast to LEDs, laser diode operation is based on stimulation emission of radiation, reflecting a much shorter lifetime, and therefore, a much higher operating bandwidth. Another critical element that promotes the selection of lasers over LEDs in Gb.E LAN applications is spectral line width. The spectral content of a laser diode is much narrower than that of LEDs (see Figures 9–25 and 9–26).

A large spectral line width contributes to wavelength dispersion, which promotes **cross phase modulation (XPM)** and **self phase modulation (SPM)**—two critical elements that adversely affect overall optical system performance.

The third element critical to high data rate transmission is the wavelength bandwidth, which is subject to optical source fiber launch conditions. Higher bandwidths are capable of transmitting higher data rates. Laser diode launch

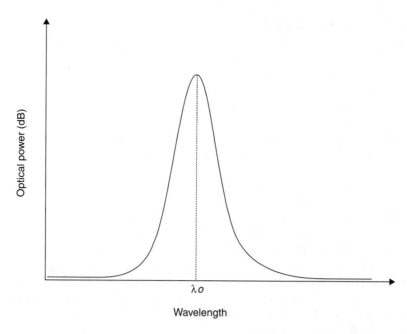

Figure 9–25. Laser diode spectral line width.

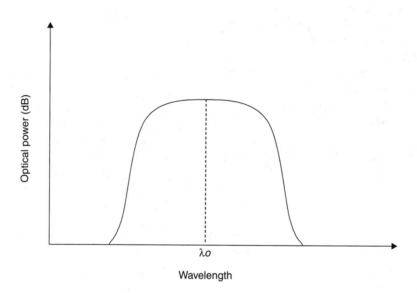

Figure 9–26. LED spectral line width.

conditions are more favorable to those of LEDs, allowing for higher operating bandwidths. The four limiting factors in an optical fiber intended for high data rate applications are attenuation, zero dispersion wavelength, dispersion slope, and modal bandwidth.

Figure 9–27 illustrates the relationship between the attenuation of a typical multimode optical fiber and its operating wavelength.

From Figure 9–27, it is evident that fiber attenuation decreases progressively with operating wavelength. Therefore, laser diodes that generate longer wavelengths than LEDs are better suited for implementation in high speed local networks. They can also enhance the overall network operating distance.

Another characteristic of optical fiber data rate limitations is chromatic absorption, which is defined by the zero **dispersion wavelength** and **dispersion slope.** Optical sources that have larger spectral line width promote high chromatic absorption, while laser diodes that have a narrow spectral line width exhibit a smaller level of chromatic absorption and are better suited for local area applications.

Fiber Index Profile

The **fiber index profile** in Figure 9–28 relates multimode optical fiber bandwidth operating characteristics to core diameter, fiber length, and launch conditions versus operating wavelength.

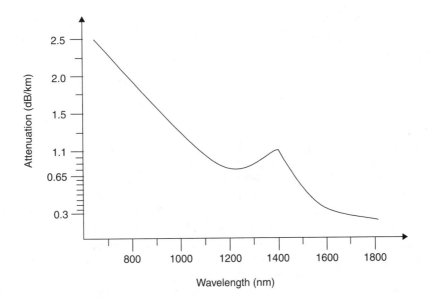

Figure 9–27. Multimode optical fiber attenuation *v.* wavelength.

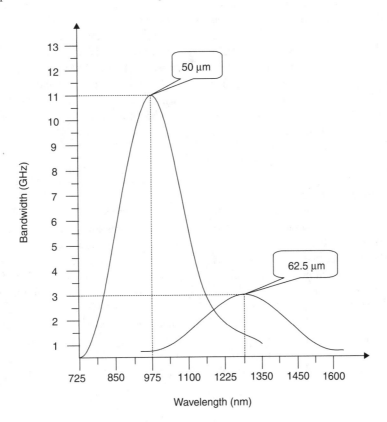

Figure 9–28. Fiber index profile.

The two basic types of multimode fibers used with LEDs in traditional local area networks are 50 μm and 62.5 μm core fibers, operating at short wavelengths. The fibers are not suited for longer wavelength operations that incorporate laser diodes, unless the basic operating characteristics such as attenuation, zero dispersion, and slope content are significantly altered in order to satisfy the high speed network requirements. This can be achieved through the manipulation of the modal bandwidth and optical source launch conditions. As mentioned previously, the basic idea behind multimode optical fibers is to allow several optical paths to be established within the core. In nongraded multimode fibers, the pulses traveling through the core tend to broaden as they travel along different path lengths. In a perfectly graded index profile fiber, a mechanism is incorporated into the fabrication process through which pulses that travel the longer paths travel faster than pulses that follow the shorter paths. The speed of the pulses within the core of the fiber is controlled by different refractive indices. It is an established fact that light travels faster through materials composed of lower refractive indices. Therefore, by controlling the material refractive index, the speed of the pulse through the core can also be controlled. The high capacity demands of the Gb.E protocol can therefore be satisfied either by interfacing VCSELs with standard multimode fibers or by reengineering high performance multimode fibers coupled with advanced designs of laser diodes. In the first case, VCSELs are capable of producing optical power in the mW range with very low threshold current and they can directly be modulated with data rates up to 2Gb/s. Furthermore, because VCSELs operate at short wavelengths approximately at 850 nm they are compatible with multimode fibers that operate at the same wavelength windows. At the receiver end, low cost silicon optical detectors that exhibit excellent responsivity and high speed can be used. The need for higher speed LANs required the establishment of new standards. Under the leadership of the Telecommunications Industry Association (TIA), various committees were established to develop the required standards. IEEE developed the Gb.E (IEEE 802.37) and the TIA FO-2.2 developed the modal fiber dependence of bandwidth. The information carrying capacity of a multimode optical fiber is given in terms of bandwidth length product (MHz·km). This parameter is indicative to the data and distance capabilities of the MMF. The simple model in Figure 9–29 illustrates the bandwidth operating characteristics of a multimode optical fiber.

In the **low order mode (LOM),** the optical rays propagate very close to the center of the fiber core. In the **intermediate order mode (IOM),** the optical rays propagate midway between the center of the core and cladding, and the **high order mode (HOM)** operates close to the border line between the core and cladding. At the output of the MMF, the optical power profile indicates a number of power distributions among the three modal areas described. It is evident that the signal at the output of the fiber is not the same shape as the input signal. This pulse spreading is referred to as pulse dissipation. Pulse dissipation is one of the most important characteristics that imposes limits on the fiber transmission capabilities.

Output signal broadening (Figure 9–29) results when all the modes of the fiber are simultaneously excited under maximum optical power. LEDs that generate a broad line width optical spectrum are capable of simultaneous modal excitation of MMFs, and are therefore unsuitable for Gb.E operation. By controlling the modal excitation, the fiber bandwidth can be enhanced. For example, if only the lower order mode is excited, the bandwidth will increase. Therefore, it is obvious that the operating bandwidth of multimode fibers is subject to modal launch. If the fiber modes are slightly out of tune, a change in the operating bandwidth can occur with differential input power variations.

The modal number and the bandwidth of an MMF are inversely proportional. That is, the higher the number of modes (optical paths) the lower the fiber operating bandwidth. The 62.5 μm fiber is capable of providing two and a half times more operating modes than the 50 μm fiber core. Based on the described characteristics, 62.5 μm multimode fibers are best suited for LANs that employ LED sources, while the 50 μm MMFs are better suited for networks that employ laser diode devices. The bandwidth length product performance characteristics of a 62.5 μm MMF that operates in the 850 nm wavelength window is approximately 160 MHz/km, while the same fiber exhibits a 500 MHz/km bandwidth length product in the 1300 nm wavelength window. Tunable mode 50 μm fibers can achieve a 500 MHz/km bandwidth length product at both the 850 nm and 1300 nm wavelength windows.

9.11 NONZERO DISPERSION SHIFTED FIBERS (NZ-DSF)

The communications industry recently underwent a phenomenal transformation that was driven by high capacity demands (mainly for the Internet) and the need for longer transmission distances. These requirements could only be satisfied through the implementation of optical fiber systems. Standard optical fibers utilized in optical communications links were able to partially satisfy the growing demands. However, it was clear that they would not be able to accommodate the projected growth in system capacity. The immediate development of a new type of fiber was essential. The optical fiber manufacturers, led by Corning, answered the call through the development of the next generation of fibers such as the nonzero

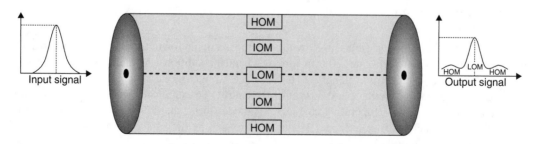

Figure 9–29. A multimode optical fiber.

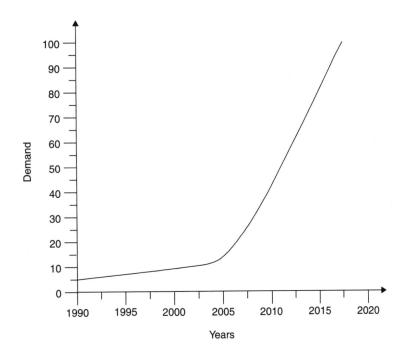

Figure 9–30. Demand for optical bandwidth.

dispersion shifted fibers (NZ-DSF), large effective area NZ-DSF, and dispersion compensation fibers. These advanced fibers, coupled with other relevant inventions in the area of optical devices such as EDFAs, add/drop multiplexers (ADOM), LiNbO$_3$ modulators, and optical filters, were instrumental in satisfying current and future capacity demands. It is estimated that the demand for bandwidth growth to facilitate these increasing data rates will be higher than 35% per annum beyond year 2000 levels. For voice, the growth is estimated to be around 10% per month (Figure 9–30).

NZ-DSFs are suitable for implementation in optical systems that operate within OC-48 (2.5 Gb/s) or OC-192 (10 Gb/s) and DWDM protocols. For longer than 300 km optical spans, even the use of NZ-DSF fibers require a small amount of dispersion compensation relative to the type of fiber used, the number of channels, and the span distances.

9.12 LARGE EFFECTIVE AREA NZ-DSF

In standard NZ-DSF fiber structures, fiber core effective area is set at 55 μm^2. However, fiber nonlinearities are inversely proportional to the core effective area, and therefore they impose limits on optical systems intended for operations within DWDM capacities.

An increase of the effective area of the fiber core will ultimately reduce nonlinearities through the reduction of the light density propagating along the core of fiber. Corning Corporation has developed an optical fiber with an effective area of 72 μm^2 as compared to 55 μm^2 standard NZ-DSF. The utilization of such a fiber in DWDM optical systems con-

tributes significantly to the increase of the distance between amplifiers, improves the optical signal to noise ratio (OSNR), and reduces the BER, thus enhancing the overall system performance. A comparison between a standard NZ-DSF and a large effective area NZ-DSF is illustrated in Figure 9–31.

From Figure 9–31, it is evident that the SBS and **four wave mixing (FWM)** have little or no effect on large effective area fibers, while a significant attenuation is observed for standard NZ-DSF.

Large effective area fibers employed in DWDM optical systems have contributed significantly to the achievement of transmission rates beyond 300 Gb/s. The fundamental difference between a standard NZ-DSF and a large effective area fiber is that large area fibers can practically eliminate four wave mixing while maintaining the levels of XPM and SPM below the 1 dB level. These two critical advantages contribute to the enhancement of span distance while operating at preset high transmission rates. For a detailed description of XPM and other system nonlinearities see Section 9.13.

An experimental optical system that utilizes the 72 μm^2 large effective area fiber, with the following operating characteristics, has been implemented by Corning (Courtesy of Corning Inc.).

number of optical channels: 32

channel capacity: 10 Gb/s

channel spacing: 100 GH/s

span distance: 90 km

number of spans: 5

number of EDFA: 5

maximum link distance: 450 km

total link capacity: 320 Gb/s

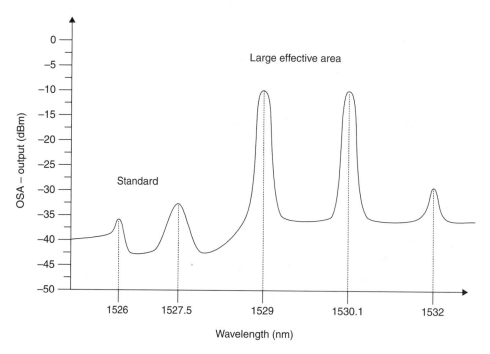

Figure 9–31. A comparison of standard and large effective area NZ-DSF fibers. Courtesy of Corning Inc.

In the early stages of optical fiber system development, fiber nonlinearities based on SBS, XPM, FWM, SRS, and SPM were negligible. In systems that employ DWDM, where more than one wavelength is propagated through a single fiber, the high data rate and higher optical power demands required to achieve longer distances directly contribute to the increase of fiber nonlinearities. Therefore, nonlinearities of DWDM optical systems are expressed by Equation (9–66).

$$NL \rightarrow \frac{n_2\,PL}{A_{eff}} \qquad (9\text{–}66)$$

where NL = nonlinearity, n_2 = nonlinear index coefficient, P = optical power at the input of the fiber, L = length of the optical link, and A_{eff} = effective cross-sectional area of the fiber core.

From Equation 9-66, it is evident that the effective cross-sectional area of the fiber core is inversely proportional to fiber nonlinearities. Therefore, an increase in the core effective cross-sectional area will substantially reduce fiber nonlinearities. Although a rather detailed explanation of the various forms of scattering is given in this chapter, a summary description of the major causes and effects is as follows.

SBS in DWDM Optical Systems

SBS is the result of light and acoustic wave interaction. It affects each of the wavelengths propagating through the fiber and is subject to the wavelength optical power level. An effective way to reduce SBS is to broaden the optical spectrum of each wavelength.

SRS in DWDM Optical Systems

SRS is subject to the total of all the optical power generated by the individual laser diodes coupled into the single fiber. SRS becomes a serious problem in systems that employ DWDM modes of operation and require relatively high levels of optical power (approximately 1 W).

For XPM and FWM, see the appropriate sections of this chapter.

Classification of Optical Fibers Based on Applications

Optical fiber communications links are designed for applications such as interbuilding, feeders, and metropolitan and long distance systems. Because the requirements for each of the applications are different, different optical fibers are also required. A short list of applications and recommended fiber usages are given in Table 9–7.

9.13 OPTICAL FIBER NONLINEAR EFFECTS

Fiber nonlinearities directly contribute to refractive index changes in relationship to input optical power variations. This relationship is expressed by Equation (9–67).

TABLE 9–7   Applications and Fiber Type Requirements

Applications	Fiber Type
Interbuilding	Multimode
Feeder	Single-mode
Metropolitan Short (60 km)	Standard single-mode
Metropolitan Long	Standard single-mode or NZ-DSF
Long distance	NZ-DSF
Submarine	NZ-DSF (negative dispersion)

$$n = n_o + \frac{n_2 P}{A_{eff}} \qquad (9\text{–}67)$$

where n is the refractive index based on nonlinearities, n_o is the ideal refractive index, n_2 is the nonlinear refractive index coefficient $(2.35 \times 10^{-2}\ \text{m}^2/\text{W})$, P is the input optical power, and A_{eff} is the effective area of the fiber core.

The nonlinear components that affect the performance characteristics of an optical fiber are divided into two basic categories.

Category I:

a) FWM

b) SPM

c) XPM

Category II:

a) SBS

b) SRS

SBS and SRS have already been discussed. Here, the three components of the first category will be dealt with in some detail. From the three impairments in Category I FWM and XPM promote cross talk, while SPM contributes to the increase of the BER. To determine the exact impact of FWM and XPM in a DWDM optical system, both must be treated as noise, while SPM must be treated as distortion.

Four-Wave Mixing

In high speed optical networks, the required optical power launched into the individual channels is much higher than the power required in conventional optical systems. This high optical power is necessary to maintain a pre-set (usually high) OSNR.

FWM is the process whereby optical power from one channel in a multi-channel system is spilled over into an adjacent channel (Figure 9–32).

Figure 9–33 represents the change when Channel 1 and Channel 4 are utilized for transmission. In this case, it is evident that Channel 1 and Channel 4 are composed of two components, the home optical power and the spillover power from the adjacent channels. Therefore, the OSNR for Channels 1 and 4 is significantly degraded.

FWM can be substantially reduced or perhaps completely eliminated through the following steps.

a) channel power reduction

b) increased dispersion

c) increased channel spacing

d) fiber photon power peak reduction (through the increase of fiber A_{eff})

The contribution of a large effective area fiber to the elimination of the FWM can be demonstrated by the following example.

Assume an 8-channel WDM optical link is implemented with two types of fibers: one with a small effective area G.655 (NZ-DSF) and the other with a large effective area fiber (G.655 large area). The optical power for each channel is set at equal levels for a fixed distance. Both links operate just below the FWM threshold optical power level. If one of the optical channels is eliminated, its optical power will be distributed among the other seven remaining channels, ultimately bringing their power levels above the FWM threshold level. This is illustrated in Figure 9–34.

From Figure 9–34, the presence of FWM in small effective area fibers is evident, while large effective area fibers have all but eliminated the effect of FWM.

Self-Phase Modulation

Single channel systems are subject to SPM. That is, when an optical pulse travels through the fiber, an increase or decrease of the light intensity occurs. These intensity variations ultimately affect the fiber refractive index, because the refractive index is subject to photon intensity. As a consequence, refractive index changes affect different parts of the optical pulse, resulting in phase nonlinearities (chirping).

Pulse broadening (chirping) results in an increase of the spectral content of the traveling light, thus making it more susceptible to dispersion levels. Therefore, it is the chromatic dispersion (an inherent element within the fiber) that causes pulse chirping and not the SPM (Figure 9–35).

SPM is therefore the result of signal intensity modulation based on refractive index nonlinearities, enhancing chromatic dispersion, and ultimately leading to chirping.

Cross-Phase Modulation

A great similarity between XPM and SPM exists. SPM is a nonlinear degradation that is normally encountered in single channel systems, and XPM is mainly encountered in multi-channel WDM systems. An increase in dispersion reduces XPM because of the smaller interaction time between pulses. To establish system impairments caused by XPM is a very difficult task, which requires complex computation models or highly sophisticated laboratory tests. Nonlinearity, such XPM, SPM, and FWM, all imposing system degradations,

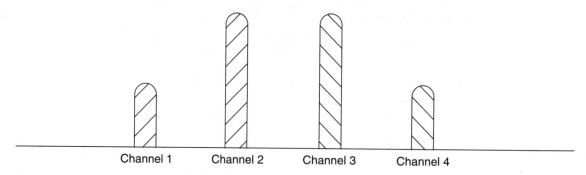

Figure 9–32. FWM in a two-channel optical transmitter.

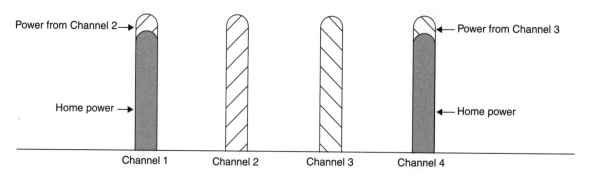

Figure 9–33. Four-channel transmission.

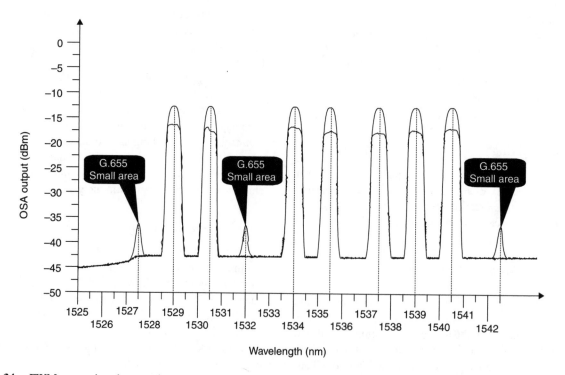

Figure 9–34. FWM comparison between large and small effective area fibers. Courtesy of Corning Inc.

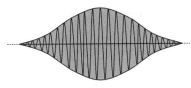
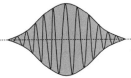
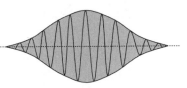

| Change of photon intensity leads to change in reflective index | Change in reflective index leads to phase change | Change in phase and dispersion results in pulse spreading |

Figure 9–35. The relationship between refractive index nonlinearities and chirping.

can be drastically reduced or altogether eliminated through the implementation of large effective area fibers.

In the case where XPM dominates the nonlinear field, link impairments are a function of the information bit pattern. In DWDM optical systems, the dominant nonlinear effects are FWM and XPM. Optical system degradations based on FWM and XPM nonlinearities can be calculated with the nonlinear Shroedinger equation (NLSE). The NLSE method of optical link impairment evaluation, based on XPM and FWM, is quite lengthy and complex. Its complexity is subject to system characteristics such as span length, channel capacity, and channel number within the DWDM system.

System nonlinearities induced by FWM can be reduced or eliminated through the utilization of NZ-DSF or standard single-mode fibers (SMF). Assuming FWM is now eliminated, the effect of XPM becomes more and more important and its level is directly related to the binary data patterns utilized in optical fiber transmission. In order for the effect of XPM to be determined, it must first be separated from the FWM nonlinear effect. It was mentioned previously that a relatively large dispersion fiber, in conjunction with relatively low channel capacity, should reduce or completely eliminate FWM. The dominant XPM nonlinear component is a function referred to as **walk off length,** and is expressed by Equation (9–68).

$$L = \frac{1}{f_b D \Delta_V} \qquad (9\text{–}68)$$

where L is the walk off length, f_b is the channel capacity, D is the fiber dispersion, and Δ_V is the channel spacing. The relationship between fiber dispersion, based only on XPM and channel capacity, can be illustrated by the following examples.

Example 9–19
The data capacity of a single channel in a DWDM optical system is 2.5 Gb/s propagating through a fiber with a 20 ps/nm·km dispersion based only on XPM impairment. Compute the channel capacity for a dispersion of 5 ps/nm·km, assuming that channel spacing and walk off length remain constant.

Solution
i) Applying Equation (9–68), solve for L.

$$L_1 = \frac{1}{f_{b_1} D_1 \Delta_V}$$

where $f_{b_1} = 2.5$ Gb/s and $D_1 = 20$ ps/nm·km.

$$L_2 = \frac{1}{f_{b_2} D_2 \Delta_V}$$

where $f_{b_2} = ?$ and $D_2 = 5$ ps/nm·km.

ii) Equalizing the walk-off distance and assuming a constant channel separation, solve for f_{b_2}.

$$f_{b_2} = \frac{D_2}{D_1}(f_{b_1}) = \frac{20 \text{ ps/nm·km}}{5 \text{ ps/nm·km}} \times 2.5 \text{ Gb/s}$$

$$= 10 \text{ Gb/s (OC–192)}$$

Therefore, $f_{b_2} = 10$ Gb/s (OC–192).

From the preceding example, it is evident that a reduction of XPM proportionally increases the channel capacity from OC-48 to OC-192.

Example 9–20
Use the same parameters as in Example 9–20, but channel separation (Δ_{V_2}) is reduced by 50% (from 100 GHz to 50 GHz).

Solution

$$f_{b_2} = \frac{\Delta_1}{\Delta_2}(f_{b_1}) = \frac{100 \text{ GHz}}{50 \text{ GHz}}(2.5 \text{ Gb/s}) = 5 \text{ Gb/s}$$

Therefore, $f_{b_2} = 5$ Gb/s.

From both examples, it is evident that dispersions based on XPM and channel separation proportionally affect the channel capacity.

System degradations reflect the degree by which the eye diagram closure is observed at the receiver end. Relevant measurements can be performed to establish the effect of XPM in an optical channel within a DWDM system. For example, if a standard data pattern is fed into the system, the eye closure can be measured at different levels of dispersion and the results can be plotted to illustrate the eye closure wavelength profile. Or, maintaining a constant dispersion level while varying the input data, a second eye v. wavelength profile can be generated. Both theoretical and experimental studies have shown that data patterns contribute significantly to system impairments when XPM is the dominant nonlinearity factor.

Effects of XPM and FWM in a DWDM Optical System

In order to assess the impact of XPM and FWM, both impairments must be treated as noise. These nonlinearities are power dependent, that is, the higher the power launched into the input of the fiber, the higher the level of XPM and FWM. Because DWDM systems require the highest possible input optical power, the impacts of both nonlinearities are critical. DWDM fiber impairments can be classified into two groups.

Group I (promotes channel crosstalk)

a) XPM

b) FWM

Group II (promotes distortion of the data pattern)

a) SPM

b) Linear dispersion

For a complete system performance evaluation, the quality (Q) parameter will be used (Corning Inc.) (Equation (9–69)).

$$Q = \frac{I(1) - I(0)}{\sqrt{\sigma^2(1)} + \sqrt{\sigma^2(0)}} \quad (9\text{–}69)$$

where Q is the link performance parameter, $I(1)$, $I(0)$ are currents ($I(1)$ is the current at logic-1 and $I(0)$ is the current at logic-0), and $\sigma^2(1)$, $\sigma^2(0)$ are total variances. In an optical system that incorporates a large number of channels and long spans, the Gaussian approximation method can be applied to establish the photocurrent variances for both XPM and FWM.

FWM. The normalized photocurrent variance σ^2 is expressed by Equation (9–70).

$$\sigma^2_{\text{FWM}} = \frac{2\left(\sum_{xyz} \mu_{xyz} P_{xyz}\right)}{P_S} \quad (9\text{–}70)$$

where σ^2 is the normalized photocurrent variance, P_S is the selected channel optical power, μ_{xyz} is the NRZ modulation factor, and P_{xyz} is the FWM power tone generated by xyz channels.

XPM. The photocurrent variance for XPM is expressed by Equation (9–71).

$$\sigma^2_{\text{XPM}} = \sum \left(\frac{1}{2\pi}\right) \int_{-\infty}^{+\infty} |X_{xyz}(\omega, L)|^2 |R(\omega)|^2 P_{yz}(\omega)\, d\omega \quad (9\text{–}71)$$

where σ^2_{xyz} is the normalized photocurrent variance, X_{xyz} is the XPM noise variance (X-the selected channel), $R(\omega)$ is the receiver transfer function, and $P_{yz}(\omega)$ is the spectral power of optical intensity for channel-y beyond the zth amplifier.

The selected channel photocurrent variance based on XPM is calculated through the integration of the power spectrum induced fluctuations. Based on previous calculations for both FWM and XPM photocurrent variances, a graphical comparison (Figure 9–36) can be made. To obtain Figure 9–36, Corning Inc. used an optical system with the following operating parameters.

number of channels: 40

channel capacity: 10 Gb/s

channel separation: 100 GHz

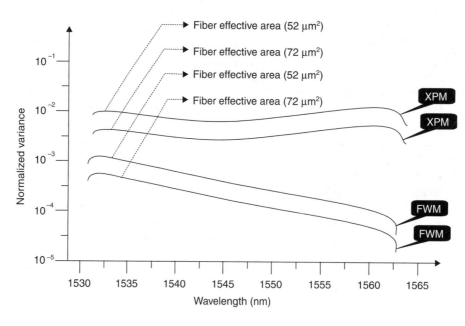

Figure 9–36. XPM and FWM noise variance for a DWDM optical system. Courtesy of Corning Inc.

operating wavelengths:
a) 1513 nm (72 μm^2), b) 1452 nm (72 μm^2), c) 1513 nm (52 μm^2), and d) 1452 nm (52 μm^2)

link distance: 400 km (5 × 80 km)

optical amplifiers: 5 EDFAs with noise figure of 5.5 dB

optical power per channel: 4 dBm

dispersion compensation incorporated at the receiver end for minimum XPM.

fiber dispersion slope: a) S_o = 0.045 ps/nm^2 · km at wavelength: λ_o = 1452 nm and b) S_o = 0.1 ps/nm^2 · km at wavelength: λ_o = 1513 nm.

From Figure 9–36 it is evident that XPM is higher than FWM. While XPM maintains a steady state level up to 1560 nm, FWM exhibits a progressive decrease at higher wavelengths. Both nonlinear impairments are also dispersion slope and fiber core cross-sectional effective area dependent. Furthermore, the total system quality factor (Q), plotted in reference to wavelength, is illustrated in Figures 9–37, 9–38, 9–39, and 9–40.

Dispersion compensation is inserted at the transmitter end to maximize channel quality (worst channel). It is evident that quality factor variations are larger at higher fiber dispersion slopes, while the overall system quality factor is a function of the operating wavelengths and fiber core cross-sectional effective area. It is also apparent that it is substantially higher when considered only with FWM.

Finally, it is safe to assume that in a DWDM optical system, the dominant impairment is the XPM. Both FWM and XPM can be substantially reduced through the implementation of large effective area fibers across the entire channel scheme. Furthermore, a decrease of the dispersion slope improves the XPM dispersion compensation effectiveness, resulting in a significant improvement in the overall link quality factor.

Large Effective Area Fiber Design

The design of a large effective area NZ-DSF is based on a triangular core and a raised index ring. The triangular ring shifts the zero dispersion wavelength (λ_o) to the 1550 nm range. The additional ring enhances the cross-sectional effective area while simultaneously achieving a relatively low bending loss.

In reference to attenuation, the large effective area NZ-DSF performs better than standard NZ-DSF fibers (Figure 9–41).

From Figure 9–41, the large effective area fibers exhibit better attenuation performance than standard NZ-DSF over the 1500 nm and 1565 nm wavelength range. The FWM and dispersion performance characteristic curve is illustrated in Figure 9–42.

Figure 9–42 illustrates the superior dispersion/FWM performance characteristics of large area fibers in comparison to standard NZ-DSF.

EXPERIMENTAL RESULTS. Corning has tested large effective area fibers in 4-channel and 8-channel WDM optical systems at the 1530 nm to 1540 nm and 1540 nm to 1560 nm wavelength windows. These wavelength ranges were selected because of the compatibility with the EDFA operating wavelength windows used in the experiments.

Experiment I

a) wavelength range: 1549.3 nm to 1560.6 nm

b) number of channels: 8

c) separation frequency: 200 GHz

d) system capacity: 80 Gb/s

e) distance between repeaters: 90 km

f) link distance: 450 km

g) number of EDFAs: 5 (5 × 90 km = 450 km)

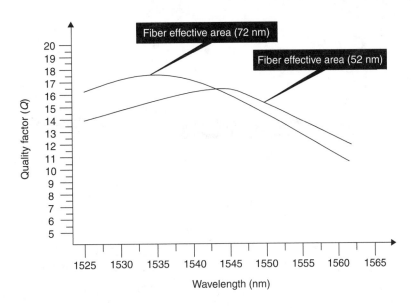

Figure 9–37. Quality factor (Q) for a single-channel system.

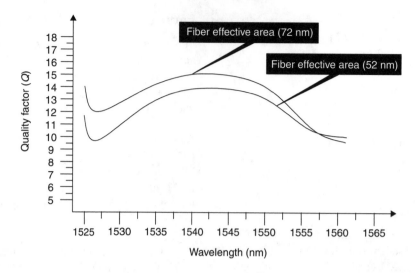

Figure 9–38. Quality factor (Q) with the FWM effect.

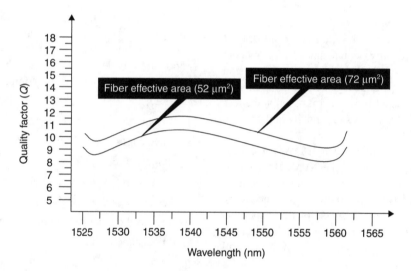

Figure 9–39. Quality factor (Q) with the XPM effect.

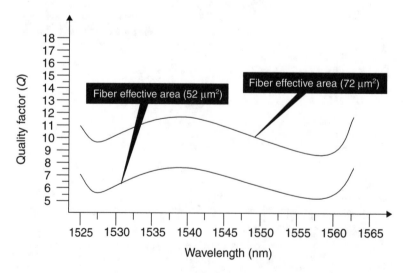

Figure 9–40. Total system quality factor (Q).

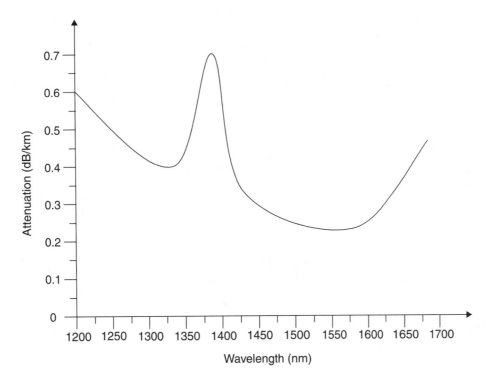

Figure 9–41. Attenuation *v.* wavelength of a large effective area optical fiber.

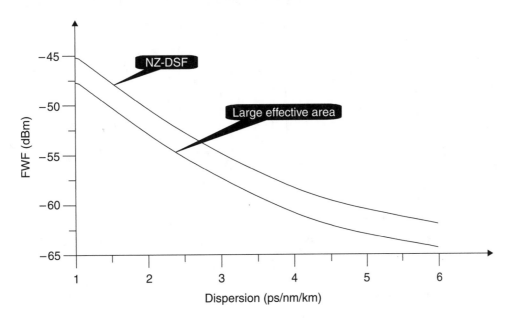

Figure 9–42. FWM *v.* dispersion performance comparison between large area and NZ-DSF.

Experiment II

a) wavelength range: 1531.9 nm to 1536.6 nm

b) number of channels: 4

c) separation frequency: 200 GHz

d) system capacity: 40 Gb/s

e) distance between repeaters: 90 km

f) link distance: 450 km

g) number of EDFAs: 5 (5 × 90 km = 450 km)

In both experiments, pre- and postdispersion compensation were used.

OBSERVATIONS. The results of both experiments using large effective area optical fibers are illustrated in Figures 9–43 and 9–44.

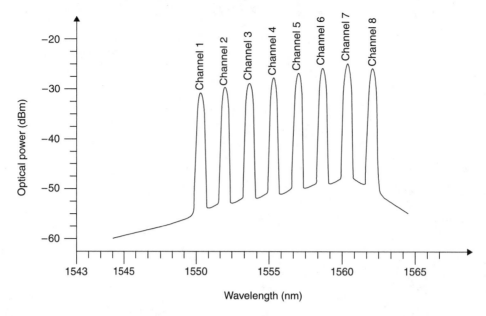

Figure 9–43. FWM in the 8-channel system in reference to channel power *v.* wavelength. Courtesy of Corning Inc.

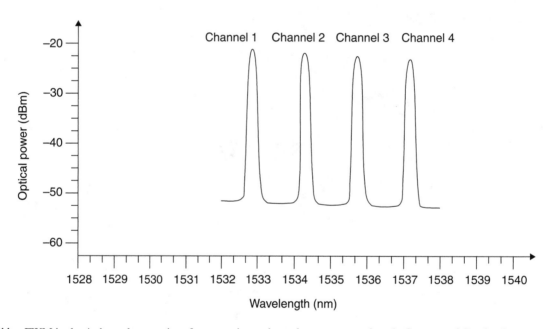

Figure 9–44. FWM in the 4-channel system in reference to input channel power *v.* wavelength. Courtesy of Corning Inc.

It is clear that both experiments show no evidence of FWM. As far as XPM and SPM are concerned, both impairments can be almost eliminated through dispersion compensation incorporated into the optical link.

FINAL OBSERVATIONS. 40 Gb/s and 80 Gb/s error free transmission capacities over a 450 km link are possible through the utilization of large effective area optical fibers.

FUTURE APPLICATIONS OF LARGE EFFECTIVE AREA FIBERS.

16-Channel Optical Link

a) operating wavelength: 1545.3 nm to 1557.4 nm

b) wavelength separation: 100 GHz

c) system capacity: 40 Gb/s (2.5 Gb/s × 16)

d) link distance: 2000 km

e) distance between repeaters: 40 km

f) number of EDFAs: 50

Large Effective Area Optical Fiber Performance Characteristics

a) high optical power handling capability

b) improved OSNR

c) longer amplifier spacing

d) maximum DWDM channel plan

e) uniform reduction of all nonlinear system degradations

f) require fewer optical amplifiers per set distance, thus highly economical

g) compatible with already installed fibers, in terms of concentricity

h) easy to splice and improved splicing performance

i) easy and economical to enhance system carrying capacity

Applicability

a) exceptionally suited to operate at 2.5 Gb/s DWDM

b) capable of operating with 32-channel 10 Gb/s DWDM system

c) expandable to 40-channel 10 Gb/s DWDM system

d) strict specifications on PMD

e) capable of transmission capacity rates beyond the 10 Gb/s range

Physical Characteristics

a) protected by Corning CPC coating system

b) CPC coating mechanically stripped with an outside diameter of 245 μm

Table 9–8 Optical Specifications

Wavelength(λ)	Attenuation
1550 nm	≤ 0.25 dB/km
1625 nm	≤ 0.25 dB/km
13183 nm	1.0 dB/km (max)

Table 9–9 Bending Induced Attenuation

Mandrel Diameter (mm)	Number of Turns	Wavelength (nm)	Attenuation (dB)
32	1	1550–1625	≤ 0.5
75	100	1550–1625	≤ 0.25

Mode field diameter

a) 9.2 μm to 10 μm at 1550 nm

Dispersion

a) 2 ps/nm·km to 6 ps/nm·km at wavelength range of 1530 nm to 1565 nm

b) 4.5 ps/nm·km to 11.2 ps/nm·km at wavelength range of 1565 nm to 1625 nm

Table 9–10 Fiber *v.* PMD

Fiber (PMD)	ps/√km
PMD link value	≤ 0.08
Maximum individual fiber	≤ 0.2

Table 9–11 Environmental Specifications

At $\lambda_0 = 1550\ nm$	Induced Attenuation (dB/km)
−60°C to +85°C	≤ 0.05
−100°C to +85°C (98% humidity)	≤ 0.05
Water immersion: (23°C)	≤ 0.05
Heat aging: 85°C (At 23°C Ref.)	≤ 0.05

PMD link value refers to the polarization modal dispersion of several fiber lengths, and is used to determine the statistical upper limit of the system performance.

Dimensions

a) standard length: 4.4 km/reel to 25 km/reel

Glass Geometry

a) cladding diameter: 125 μm ± 1μm

b) core/clad concentricity: < 0.5 μm

c) fiber curl: ≥ 4.0 m radius of curvature

d) cladding noncircularity: ≤ 1.0% (calculated as follows)

$$1 - \frac{\text{cladding} - \text{diameter(min)}}{\text{cladding} - \text{diameter(max)}} \times 100$$

Mechanical Specifications

a) proof testing: Tensile proof test ≥ 100 Kpsi

b) effective area ($A_{eff} = 72\ \mu m^2$)

c) effective group index of refraction: 1.496 at 1550 nm

d) fatigue resistance parameter: 20

Coating Strip Force

a) Dry: 0.6 lb (3.0 N)

b) Wet: 14-day room temperature, 0.6 lb (3.0 N)

Coating Geometry

a) Coating diameter: 245 μm ± 5 μm

b) Coating/Gladding concentricity: (< 12 μm)

Nonlinear effects such as SPM, XPM, and FWM are the most serious performance limiting factors in DWDM multichannel systems. Furthermore, they are capable of operating at both conventional optical bandwidth (C = band: 1530 nm–1565 nm) and longer bandwidth (L = band: 1565 nm–1625 nm). In both bands, these fibers exhibit superior transmission performance by simultaneously reducing the nonlinear effects and increasing the number of usable channels.

Intensity Modulation

In DWDM optical systems that employ standard optical fibers, the **intensity modulated** optical wavelengths alter the refractive index of the fiber core, thus promoting XPM. Through the implementation of NZ-DSFs in such systems, XPM can be estimated (Equation (9–72)).

$$\sigma_{XPM} \rightarrow \left(\frac{P}{A_{eff}} \right) \qquad \textbf{(9–72)}$$

where σ_{XPM} is the XPM variance, P is the optical input power, and A_{eff} is the fiber effective cross-sectional area. For example, a standard single-mode fiber operating at the 1550 nm wavelength has a transmission capability of 10 Gb/s for an operating distance of 60 km.

For a link span of 250 km and for the same 10 Gb/s link capacity, the allowable chromatic dispersion is set at 4 ps/nm·km. Beyond this value, dispersion compensation is absolutely essential.

9.14 ABILITY TO UPGRADE ALREADY INSTALLED STANDARD SINGLE-MODE FIBERS

Assume that a carrier currently operating with a standard single-mode fiber at the 1310 nm wavelength intends to upgrade the system, taking advantage of the latest development in the area of optical amplifiers (EDFA) operating at the 1550 nm wavelength window and WDM operating scheme. Proposed upgrading of the existing system at transmission rates of 2.5 Gb/s (OC-48) can be implemented without major difficulties. For transmission rates of 10 Gb/s (OC-192), the standard single-mode fiber exhibits dispersion levels on the order of 17 ps/nm·km. At lengths beyond 60 km and at proposed operating wavelengths of 1550 nm, this dispersion level is accumulative and very quickly adds to pulse chirping, which induces unacceptable levels of system degradations. To remedy this problem, optical fibers are fabricated with dispersions equal and opposite to link dispersions (added on), practically eliminating the dispersion problem. These types of fibers are referred to as Dispersion Compensation Fibers (DCF). The only minor disadvantage to this scheme is the additional insertion loss. It is, therefore, evident that an already installed optical link can be upgraded by taking advantage of the new technologies economically and without major structural changes.

The data and distance limiting capabilities of optical fibers are directly related to linear and nonlinear effects. Chromatic dispersion (linear), as discussed previously, can be rectified by using compensating methods. Nonlinear

effects, related to high data transmission rates and high channel transmission links, are inherent to the system and therefore cannot be removed by standard NZ-DSF. However, their overall effect can be substantially reduced by fibers with larger effective area (A_{eff}). Ideally, an optical pulse traveling through a silicon fiber consists of a single wavelength. Because the transmitted pulse has a finite width, it contains wavelength components slightly above and below the center wavelength (λ_o). Wavelengths above the center wavelength will travel faster, while wavelengths below will travel slower than the center wavelength (Figure 9–45).

Wavelength components above the center wavelength are to have positive chromatic dispersion, while wavelength components below the center wavelength are to have negative chromatic dispersion. Chromatic dispersion can be controlled by inducing components that have opposite dispersion coefficients into the light path, thus accelerating or decelerating the photon velocity through the core of the fiber. Two basic methods of chromatic dispersion compensation exist. These are the Bragg-grating method and the dispersion compensation method. At relatively low transmission speeds, up to 2.5 Gb/s, chromatic dispersion is not considered a major problem. However, at transmission speeds of 10 Gb/s, chromatic dispersion becomes a serious problem and it must be taken into account during the design process. The total link chromatic dispersion is given by Equation (9–73).

$$d_{TL} = L \times d_{AV} \qquad \textbf{(9–73)}$$

where d_{TL} is the total link chromatic dispersion (ps), L is the link distance (km), and d_{AV} is the average chromatic dispersion (ps/km).

Example 9–21
Compute the total chromatic dispersion of a 400 km optical fiber link operating at the 1550 nm wavelength window, assuming the link is using an NZ-DSF fiber with an average chromatic dispersion of 4 ps/km.

Solution

$$d_{TL} = L \times d_{AV} = 400 \text{ km} \times 4 \text{ ps} = 1600 \text{ ps/nm} \cdot \text{km}$$

Therefore, $d_{TL} = 1.6 \text{ ns/nm} \cdot \text{km}$.

A practical optical receiver that operates at 10 Gb/s is capable of tolerating a maximum chromatic dispersion of 1500 ps. In this example, because the calculated chromatic dispersion is above the maximum, a dispersion compensation module is required. Either a standard dispersion compensation module (DCM-90) with dispersion coefficient –1530 ps can be used, or the optical link must be reduced by 90 km. To compute the correct compensation, a chromatic dispersion map may be required (Figure 9–46).

Two dispersion compensation modes should be incorporated into the link: one after the transmitter, and the other before the receiver. The total dispersion coefficient is equal or close to the total link dispersion. For example, the DCM-50

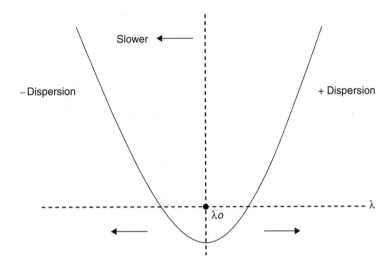

Figure 9–45. Dispersion-time delay (relative).

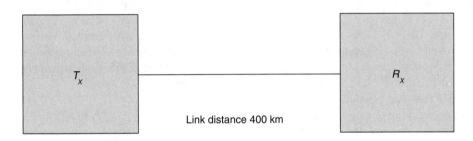

Figure 9–46. Chromatic dispersion compensation map.

with –850 ps/km dispersion compensation and the DCM-40 with –320 ps/km dispersion compensation can be used to a total negative compensation of 1170 ps. When the same link is to be implemented with the G.652 optical fiber, the following difficulties are encountered. Because the G.652 fiber exhibits a chromatic dispersion of 17 ps/nm·km for a 400 Km optical link, the total chromatic dispersion is equal to:

$$d_{TL} = L \times d_{AV} = 400 \text{ km} \times 17 \text{ ps/km} = 6800 \text{ ps/nm}$$

Therefore, $d_{TL} = 6800$ ps/nm·km.

The high chromatic dispersion exhibited by the G.652 fiber in the 1550 nm wavelength window also exhibits a tendency towards nonlinear degradations such as SPM. These disadvantages of the G.652 led to the implementation of a superior performance optical fiber: G.655.

9.15 PLASTIC OPTICAL FIBERS (POF)

In the relentless drive to optimize performance and reduce cost, some fiber manufacturers have focused their attention on the development of optical fibers that are not silicon based. The phenomenal increase in digital controls for auto-

motive and commercial aircraft has forced the industry to seek cost-effective ways to take advantage of new fiber technologies without sacrificing safety and performance.

One alternative to standard silicon fiber is the graded index plastic optical fiber (GIPOF). From the early 1990s, research has shown that GIPOFs could be an alternative to twisted pair and coaxial cables in applications such as LANs and optical sensors. GIPOFs exhibit excellent data handling capabilities, data integrity, and secure electromagnetic compatibility (EMC). These fibers also exhibit bandwidth characteristics able to accommodate data at Gb/s rates for short distances.

Plastic fibers made their appearance in the early 1960s. They were manufactured with a polystyrene (PS) core that exhibits very high transmission losses. A few years later, DuPont developed a plastic optical fiber that had a core composed of polymethylmethacrylate (PMMA). This fiber exhibited somewhat less transmission losses than previous fibers, about 500 dB/km at the 650 nm wavelength window.

In 1977, plastic fiber technology had solved some of the fundamental fabrication problems, resulting in a substantial improvement of fiber attenuation. The critical attenuation barrier below the 200 dB/km level was achieved with deuterated PMMA, operating at longer wavelength windows.

Following this, plastic optical fiber fabrication methods were revolutionized. Nippon Telephone and Telegraph (NTT), in cooperation with Mitsubishi, were able to solve some of the inherent problems encountered during the fabrication process, the most important of which is the maintenance of the core optical characteristics during thefabrication process. In the 1980s, step index plastic optical fibers (SI-POFs) measuring attenuation levels below the 170 dB/km mark made their appearance. This attenuation level was considered acceptable for short distances and low system capacity applications. Today, commercial SI-POFs with 1 mm core diameters exhibit attenuation levels of about 150 dB/km at 650 nm wavelengths, and 200 MHz–100 m operating bandwidths exist. The operating bandwidth of 200 MH–100 m for an SI-POF falls far short of the bandwidths required by today's optical links for voice, video, and data transmissions. Because of this, attention has been focused on GI-POFs capable of operating in bandwidths of 3 GHz–100 m.

GI-POFs made their appearance in the mid-1970s, with a prototype developed in Japan's Keio University. Further development has enhanced the performance of polymethylmethacrylate (PMMA) GI-POF that have attenuations as low as 160 dB/km. Currently, GI-POF fibers that operate near infrared wavelengths exhibit even lower attenuations. Various methods of fabricating plastic optical fibers exist. One of these methods, adapted by Boston Fibers, is the Interfacial Gel Polymerization (IGP) technique, which utilizes dopands that have low molecular weight in order to induce refractive index variations within the core of the fiber. Monomer dopand IGP yields compositions with gradient refractive indices as a result of diffusion coefficient differences between the high refractive index dopands and monomers.

GI-POFs hold the potential of becoming the preferable fiber for new LAN and copper wire network replacement. With material and installation costs below those of copper and glass fibers, and superior mechanical characteristics and relatively high operating bandwidths, GI-POFs are the fibers of choice.

9.16   OPTICAL FIBER GEOMETRY

Fiber geometric characteristics are critical in influencing splice losses. Through the process of new fiber installation, the fiber physical characteristics (geometry) play a crucial role in splicing and its consequent effect on system performance. In terms of economics, it is estimated that 40% of the total network cost, including material and labor, is attributed to the process of splicing.

9.17   SPLICING EFFICIENCY

The three fundamental elements that contribute directly to splice efficiency are outer cladding diameter, fiber curl, and core/clad concentricity. In reference to cladding diameter, the size of the fiber is determined by the outer cladding diameter. If two fibers are to be connected together, their physical geometry must be exact. That is, for perfect alignment, they must have the same cladding diameter. This, of course, is more important when connectors such as mating field installable and sizing ferrules are used. *Core-to-core* alignment of two optical fibers will greatly be affected if one or both cores are not concentric. Concentricity is how well the core is centered into the cladding cross section. Two perfectly concentric core fibers joined together exhibit minimal splice loss. *Fiber curl* is the degree of curvature along the entire length of the optical fiber. A relatively high degree of curvature, or fiber curl, greatly influences the splicing alignment, consequently inducing undesirable levels of splice losses. All optical fiber properties are controlled during the manufacturing process. The degree of success will ultimately determine installation costs and splicing losses. Leading optical fiber manufacturers, like Corning, have achieved fiber-cladding diameters of about 125 μm($\pm$ 1.0 μm), with core/cladding concentricity down to 0.5 μm. Curl specifications for SMFs are set at an impressive $\geq$ 4.0 m. Optical fiber splicing is an unavoidable procedure if the length of the link is to be increased beyond the length of a single fiber. For both SMF and NZ-DSFs, splicing and the measurement of splicing losses always present a formidable challenge. Fiber splicing procedures can be classified as follows:

Fiber Splicing Procedures

a) microsplicing

b) light integration and detection splicing (LID)

c) core profile alignment splicing (PAS)

Critical Parameters

a) cleave end angle

b) cleanness of the splicing

c) procedures

Equipment Required

a) passive system

b) active system

Passive system mechanically aligns the fibers with precision V-grooves. These mechanical apertures, although inexpensive, are capable of inducing higher splicing losses. Active alignment can also be classified into two different categories:

Active Alignment Categories

a) light injection and detection (LID)

b) profile alignment system (PAS)

Applying the light injection and detection methods, optical power is optimized during splicing. With the profile alignment method, the cores of the two fibers are optimally aligned. Because of the higher degree of sensitivity, the conditions prior to splicing and the actual process itself play a greater role with NZ-DSF than with SMF. Therefore, splicing procedures, cleanliness, and angle cleave play a crucial role in determining splice losses; more so in an NZ-DSF than in a standard SMF. For example, an angle cleave greater than two degrees is capable of increasing splice losses more than 100%. Laboratory experiments have shown that NZ-DSFs, spliced with both active and passive techniques, yield the following results (Table 9–12).

9.18   OPTICAL FIBER ALIGNMENT

In any communications system that includes fiber optics, both the transmitter and the receiver must be interfaced with the transmission medium. The coupling of the transmitter and receiver to the medium is associated with valuable power loss. Furthermore, several optical fiber design systems require a relatively large number of fibers to be connected together in a series configuration in order to increase the overall link distance. Three types of connections in an optical fiber link exist.

a) source-to-fiber

b) fiber-to-fiber

c) fiber-to-receiver

In all cases, coupling losses exist, even under ideal joint conditions. These losses play a significant role in the overall power budget calculations and are critical factors in determining fiber link distance. In a fiber-to-fiber coupling, although the joint is perfectly aligned, a very small fraction of the transmitted light will be reflected back toward the optical source. This reflection of the light is caused by the different refractive indices between the core of the two fibers and the refractive index of the medium between the two fibers. The reflected power is optical power loss and is expressed in the form of attenuation. The relationship that expresses the reflected optical power is expressed by the Fresnel equation (Equation 9–74).

$$P_{o(\text{refl})} = \left[\frac{n_1 - n_2}{n_1 + n_2} \right]^2 \qquad (9\text{–}74)$$

where $P_{o(refl)}$ is the reflected optical power, n_1 is the refractive index of the fiber core, and n_2 is the refractive index of the medium between the fibers at the joint. If an assumption is made that the medium between the fibers is air $(n_2 = 1)$, then Equation (9–74) becomes Equation (9–75).

TABLE 9–12   NZ-DSF Splicing Performance

NZ-DSF	Microsplicer	PAS	LID
Number of splices	90	90	90
Splice losses (dB)	0.057	0.041	0.033
Maximum losses (dB)	0.11	0.10	0.105
Standard deviation (dB)	0.026	0.022	0.024

$$P_{o(\text{refl})} = \left[\frac{n_1 - 1}{n_1 + 1} \right]^2 \qquad (9\text{–}75)$$

The optical signal attenuation (in dB) is calculated as follows in Equation (9–76).

$$L_{(\text{refl})\text{dB}} = -10 \log(1 - P_{o(\text{refl})}) \qquad (9\text{–}76)$$

where $P_{o(refl)}$ is the reflected optical power.

Example 9–22
Compute the optical attenuation induced at the joint of two fibers with both fiber core refractive indices equal to 1.35. Assume that the medium at the joint is air.

Solution
From Fresnel reflection,

$$P_{o(\text{refl})} = \left[\frac{n_1 - n_o}{n_1 + n_o} \right]^2$$

$$= \left[\frac{1.35 - 1}{1.35 + 1} \right]^2$$

$$= \left[\frac{0.35}{2.35} \right]^2 = 0.022$$

compute attenuation (dB).

$$L_{(\text{refl})\text{dB}} = -10 \log(1 - P_{o(\text{refl})})$$

$$= 10 \log(1 - 0.022) = 0.1 \text{ dB}$$

Therefore, the attenuation loss at the fiber joint is approximately 0.1 dB. For more than one joint, the Fresnel reflections are cumulative.

Example 9–23
Compute the total attenuation, based on Fresnel reflections, of an optical link composed of five optical fibers with refractive indices equal to 1.55. Assume the medium between the fibers to be air.

Solution

i) Compute the Fresnel reflective coefficient.

$$P_{o(\text{refl})} = \left[\frac{n_1 - n_o}{n_1 + n_o} \right]^2 = \left[\frac{1.55 - 1}{1.55 + 1} \right]^2$$

$$= \left[\frac{0.55}{2.55} \right]^2 = 0.04652$$

Therefore, $P_{o(\text{refl})} = 0.04652$.

ii) Compute the attenuation based on the Fresnel reflection coefficient.

$$L_{refl} = -10\log(1-P_{O(refl)})$$
$$= -10\log(1-0.04652) = 0.2 \text{ dB}$$

Therefore, $L_{refl} = 0.2$ dB

iii) Compute the total link attenuation.

$$L_{Total} = 5 \times 0.2 = 1.0 \text{ dB}$$

Therefore, the total link attenuation based on the Fresnel reflections = 1.0 dB.

Joint Losses Based on Fiber Misalignment

The combined fiber joint losses are composed of Fresnel reflection losses and joint misalignments. To determine fiber joint losses, all fibers must be compatible. That is, they must be produced by the same manufacturer, have the exact size, and have the same tolerance. Even when all the conditions are met, joint losses may still exist as a result of fiber misalignments. Fiber misalignments are classified into three categories.

Fiber Misalignments

a) lateral

b) longitudinal

c) angular

Lateral misalignment occurs when the core axis of both fibers to be joined are misaligned (Figure 9–47).

Longitudinal misalignment occurs when the two fiber cross-sectional areas, although exact, leave some space between them (Figure 9–48).

Angular misalignment occurs when the core of the fibers are not parallel to each other (Figure 9–49).

The insertion losses based on joint misalignment are related to modal power, fiber type, and fiber core effective area. The relationship between insertion loss and fiber core displacement for longitudinal effects, lateral effects, and angular misalignment is illustrated in Figures 9–50, 9–51, and 9–52.

Comparing Figures 9–50, 9–51, and 9–52 shows that the largest insertion loss is caused by lateral displacement, followed by insertion loss caused by angle misalignment, while the lowest level is due to longitudinal displacement. From Figure 9–52, the angle of displacement is a function of the fiber NA.

STEP INDEX MULTIMODE FIBER INSERTION LOSS DUE TO LATERAL MISALIGNMENT. Because both lateral and angular misalignments are the main contributors to insertion losses, it is imperative that they be discussed in some detail. It has been theoretically calculated and experimentally verified that the overall effect of lateral misalignment can be translated as an equivalent reduction in the fiber core effective area of a multimode fiber based on a uniform modal excitation. Insertion losses based on lateral misalignment are expressed by Equation (9–77).

$$L_{lat} = 10 \log C_{lat} \qquad (9\text{–}77)$$

where L_{lat} is the insertion loss based on the lateral coefficient and C_{lat} is the lateral coupling coefficient.

The lateral coupling coefficient (C_{lat}) is expressed by Equation (9–78).

$$C_{SIM\,lat} = \frac{16\left(\dfrac{n_1}{n_2}\right)^2}{\pi\left[1 + \left(\dfrac{n_1}{n_2}\right)\right]^4}$$

$$\left[2\cos^{-1}\left(\frac{y}{2a}\right) - \left(\frac{y}{a}\right) \times \left[1-\left(\frac{y}{2a}\right)^2\right]^{1/2}\right] \qquad (9\text{–}78)$$

where $C_{SIM\,lat}$ is the lateral coupling coefficient (step index multimode), n_1 is the core refractive index, n_2 is the refractive index between the fibers, y is the lateral offset of the fiber core axis, and a is the fiber core radius.

Experimental data has shown parity with the theoretical calculations based on Equations (9–77) and (9–78).

Example 9–24

Compute the insertion loss based on lateral misalignment of two step-index multimode fibers with core radius of 25 μm, lateral core offset 3 μm and core refractive index 1.5. Assume air to be the medium between fibers.

Solution

i) Compute the lateral coupling coefficient ($C_{SIM_{lat}}$).

$$C_{SIM_{lat}} = \frac{16\left(\dfrac{n_1}{n_2}\right)^2}{\pi\left[1 + \left(\dfrac{n_1}{n_2}\right)\right]^4}$$

$$\times \left[2\cos^{-1}\left(\frac{y}{2a}\right) - \left(\frac{y}{a}\right) \times \left[1-\left(\frac{y}{2a}\right)^2\right]^{1/2}\right]$$

$$= \frac{16\left(\dfrac{1.5}{1}\right)^2}{\pi\left[1 + \left(\dfrac{1.5}{1}\right)\right]^4}$$

$$\times \left[2\cos^{-1}\left(\frac{3}{2 \times 25}\right) - \left(\frac{3}{25}\right) \times \left[1-\left(\frac{3}{2 \times 25}\right)^2\right]^{1/2}\right]$$

$$\cong 0.07$$

ii) Compute the insertion losses ($L_{SIM_{lat}}$).

$$L_{SIM_{lat}} = -10 \log(1-C_{SIM_{lat}})$$

$$= -10 \log(1-0.07) = 0.3 \text{ dB}$$

Therefore, $L_{SIM_{lat}} = 0.3$ dB.

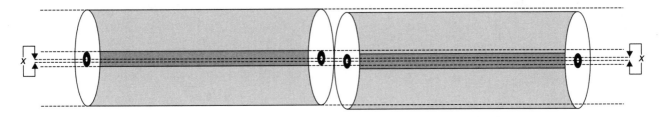

Figure 9–47. Lateral fiber misalignment.

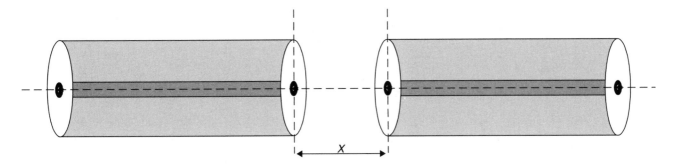

Figure 9–48. Longitudinal misalignment.

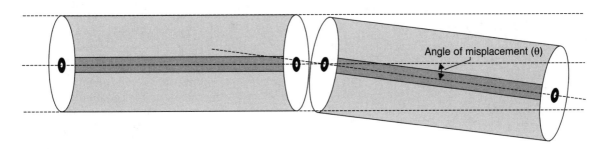

Figure 9–49. Angular misalignment.

GRADED INDEX MULTIMODE FIBER INSERTION LOSSES BASED ON LATERAL MISALIGNMENT. For graded index multimode fibers, insertion losses based on lateral misalignments are expressed by Equation (9–79).

$$C_{\text{GIM}_{\text{lat}}} = \left(\frac{2}{\pi}\right)\left(\frac{y}{a}\right)\left(\frac{\alpha + 2}{\alpha + 1}\right) \quad \textbf{(9–79)}$$

where $C_{\text{GIM}_{\text{lat}}}$ is the lateral coupling coefficient (graded index multimode), a is the fiber core radius, α is the **refractive index profile,** and y is the lateral core offset.

Equation (9–79) is valid for a lateral core offset equal to or less than 20% of the core radius, and is fully dependent on the refractive index gradient coefficient (α). The insertion loss based on lateral misalignment of a graded multimode optical fiber is expressed by Equation (9–80).

$$L_{\text{GIM}_{\text{lat}}} = -10 \log(1 - C_{\text{GIM}_{\text{lat}}}) \quad \textbf{(9–80)}$$

where $L_{\text{GIM}_{\text{lat}}}$ is the insertion loss based on lateral misalignment (graded index multimode fibers) and $C_{\text{GIM}_{\text{lat}}}$ is the lateral coupling coefficient (graded index multimode fibers).

Example 9–25

Compute the insertion loss based on lateral misalignment of two graded index multimode optical fibers with 25 μm core radius, refractive index profile ($\alpha = 2$) and lateral core offset ($y = 2$ μm).

Solution

i) Compute the lateral coupling coefficient ($C_{\text{GIM}_{\text{lat}}}$).

$$C_{\text{GIM}_{\text{lat}}} = \frac{2}{\pi}\left(\frac{y}{a}\right)\left(\frac{\alpha + 2}{\alpha + 1}\right)$$

$$= \frac{2}{\pi}\left(\frac{2}{25}\right)\left(\frac{2 + 2}{2 + 1}\right) = \frac{16}{\pi(75)} = 0.068$$

Therefore, $C_{\text{GIM}_{\text{lat}}} = 0.068$.

ii) Compute the insertion loss.

$$L_{\text{GIM}_{\text{lat}}} = -10\log(1 - C_{\text{GIM}_{\text{lat}}})$$

$$= -10\log(1 - 0.068) = 0.3 \text{ dB}$$

Therefore, $L_{\text{GIM}_{\text{lat}}} = 0.3$ dB.

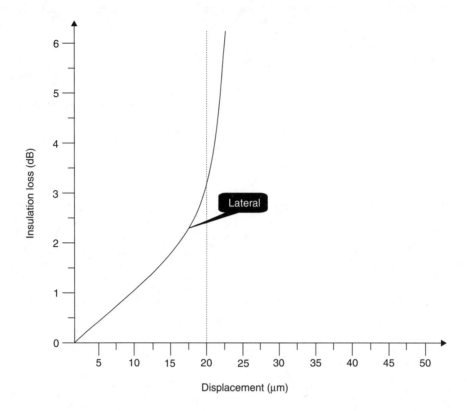

Figure 9–50. Insertion loss based on lateral core displacement.

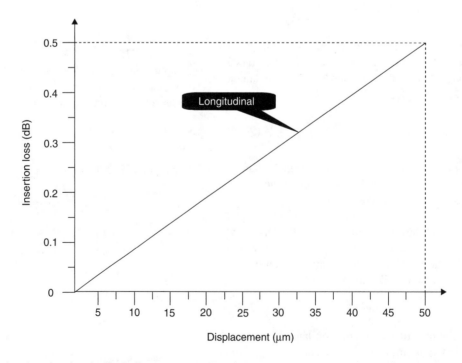

Figure 9–51. Insertion loss based on longitudinal effects.

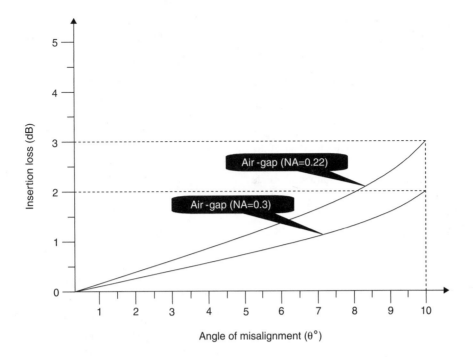

Figure 9–52. Insertion loss based on angle of misalignment.

INSERTION LOSSES OF STEP INDEX MULTIMODE FIBERS BASED ON ANGULAR MISALIGNMENT. For step index multimode fibers, the insertion loss due to angular misalignment is expressed by Equation (9–81).

$$C_{SIM_{ang}} = \frac{16\left(\dfrac{n_1}{n_2}\right)^2}{\left[1 + \left(\dfrac{n_1}{n_2}\right)^4\right]}\left[1 - \frac{n_2\theta}{\pi n(2\Delta)^{1/2}}\right] \quad (9-81)$$

where $C_{SIM_{ang}}$ is the angular coupling coefficient (step index multimode fibers), n_1 is the fiber core refractive index, n_2 is the core refractive index of the medium between fibers, θ is the angular displacement (rad), and Δ is the refractive index difference. The insertion losses of step index multimode fibers, based on angular displacement, is calculated by Equation (9–82).

$$L_{SIM_{ang}} = -10\log(1 - C_{SIM_{ang}}) \quad (9-82)$$

where $L_{SIM_{ang}}$ is the insertion loss due to angle misalignment and $C_{SIM_{ang}}$ is the angular coupling coefficient.

Example 9–26

Calculate the insertion loss at the joint of two multimode step index fibers with numerical apertures 0.3 and 0.4, respectively, core refractive index 1.5 (both), and angular misalignment 2.5°. Assume air to be the medium between fibers.

Solution

i) Compute the angular coupling coefficient ($C_{SIM_{ang}}$).

$$C_{SIM_{ang}} = \frac{16\left(\dfrac{n_1}{n_2}\right)^2}{\left[1 + \left(\dfrac{n_1}{n_2}\right)^4\right]}\left[1 - \frac{n_2\theta}{\pi n_1(2\Delta)^{1/2}}\right]$$

where $n_1(2\Delta)^{1/2} = NA$

Fiber #1: For NA $= 0.3$

$$C_{SIM_{ang}} = \frac{16\left(\dfrac{1.5}{1.5}\right)^2}{\left[1 + \left(\dfrac{1.5}{1.5}\right)^4\right]}\left[1 - \frac{1.5 \times 2.5}{180 \times 0.3}\right] = 0.930$$

Therefore $C_{SIM_{ang}} = 0.930$.

ii) Compute the insertion loss ($L_{SIM_{ang}}$).

$$L_{SIM_{ang}} = -10\log(C_{SIM_{ang}})$$
$$= -10\log(0.930) = 0.312 \text{ dB}$$

Therefore, $L_{SIM_{ang}} = 0.312$ dB.

Fiber #2: For NA $= 0.4$, repeat the previous process with the only change being the numerical aperture.

$$C_{SIM_{ang}} = 0.948$$

Therefore, $L_{SIM_{ang}} = 0.23$ dB.

The previous examples illustrate that insertion losses, based on angular misalignment, are inversely proportional to the fiber numerical aperture.

INSERTION LOSSES OF STEP INDEX SINGLE-MODE FIBERS BASED ON LATERAL MISALIGNMENT. The insertion loss at the joint of two step index single-mode fibers, due to lateral misalignment, is expressed by Equation (9–83).

$$L_{SM_{lat}} = 2.17 \left(\frac{y}{\omega} \right)^2 \qquad (9–83)$$

where, $L_{SM_{lat}}$ is the insertion loss, y is the lateral offset of the fiber core axis, and ω is the fundamental mode (LP_{01}) spot size (normalized) μm. The normalized fundamental mode spot size is expressed by Equation (9–84).

$$\omega = \frac{a(0.65 + 1.62V^{-3/2} + 2.88V^{-6})}{\sqrt{2}} \qquad (9–84)$$

where ω is the fundamental mode spot size and V is the normalized frequency of the fiber.

INSERTION LOSSES OF STEP INDEX SINGLE-MODE FIBERS DUE TO ANGULAR MISALIGNMENT. The insertion loss at the joint of two step index single-mode fibers, based on angular misalignment, is expressed by Equation (9–85).

$$L_{SM_{ins}} = 2.17 \left(\frac{\theta \omega n_1 V}{a NA} \right)^2 dB \qquad (9–85)$$

where $L_{SM_{ins}}$ is the insertion loss, θ is the angular displacement (rad), ω is the fundamental mode spot size, n is the core refractive index, V is the normalized frequency of the fiber, a is the core radius, and NA is the numerical aperture.

Example 9–27

Compute the total insertion loss at the joint of two step index single-mode fibers that have a core radius of 4 μm, a refractive index of 1.45, a numerical aperture of 0.2, a normalized frequency of 2.4, a lateral misalignment of 1 μm, and angular misalignment of 1°.

Solution

i) Compute the normalized spot size (ω).

$$\omega = \frac{a \left[0.65 + 1.62(V)^{-3/2} + 2.88(V)^{-6} \right]}{\sqrt{2}}$$

$$= \frac{4 \left[0.65 + 1.62(2.4)^{-3/2} + 2.88(2.4)^{-6} \right]}{\sqrt{2}}$$

$$= 3.12 \ \mu m$$

Therefore, $\omega = 3.12 \ \mu m$.

ii) Compute the insertion loss.

$$L_{SM_{lat}} = 2.17 \left(\frac{y}{\omega} \right)^2 = 2.17 \left(\frac{1 \ \mu m}{3.12 \ \mu m} \right)^2 = 0.48 \ dB$$

Therefore, $L_{SM_{lat}} = 0.48$ dB.

iii) Compute the insertion loss due to angular misalignment.

$$L_{SM_{ang}} = 2.17 \left(\frac{\theta \omega n_1 V}{a NA} \right)^2$$

$$= 2.17 \left(\frac{1 \times \pi / 180 \times 3.12 \times 1.45 \times 2.4}{4 \times 0.2} \right)$$

$$= 0.12 \ dB$$

Therefore, $L_{SM_{ang}} = 0.12$ dB.

iv) Compute the total insertion loss.

$$L_{SM_{Total}} = L_{SM_{lat}} + L_{SM_{ang}}$$

$$= 0.48 \ dB + 0.12 \ dB = 0.6 \ dB$$

Therefore, $L_{SM_{Total}} = 0.6$ dB.

With step index single-mode fibers, insertion losses induced both by lateral and by angular misalignments are strongly dependent on the normalized frequency (V) of the fiber. For desirable performance, fiber losses due to angular misalignment must be less than 0.3 dB, reflecting an angle of misalignment of less than 1°. From the previous discussion, it can be concluded that insertion losses should be classified into two main categories: those based on extrinsic factors such as lateral, angular, longitudinal, and Fresnel, and those based on intrinsic factors such as fiber interconnections. The calculations of the intrinsic coupling losses are expressed by Equation (9–86).

$$L_{intr} = -10 \log \left[4 \left(\frac{\omega_{02}}{\omega_{01}} + \frac{\omega_{01}}{\omega_{02}} \right)^{-2} \right] \qquad (9–86)$$

where L_{intr} is the insertion loss, ω_{01} is the spot size of the transmitting fiber, and ω_{02} is the spot size of the receiving fiber. Further to intrinsic calculations, Equation (9–86) also takes fiber core cross-sectional misalignments into account.

Example 9–28

Compute the insertion losses at the joint of two step-index single mode fibers with modal field diameters 9 μm and 8 μm, respectively. Assume no extrinsic losses present.

Solution

i) Use Equation (9–86) to compute (ω_{01}).

$$\omega_{01} = \frac{a(0.65 + 1.62V^{-1/2} + 2.88V^{-6})}{\sqrt{2}}$$

$$= \frac{4.5 \left[0.65 + 1.62(2.1)^{-1/2} + 2.88(2.1)^{-6} \right]}{\sqrt{2}}$$

$$= 5.7$$

Therefore, $\omega_{01} = 5.7$.

ii) Compute (ω_{02}). Apply the same equation as above.

Therefore, $\omega_{02} = 5.0$.

iii) Compute intrinsic losses (L_{intr}).

$$L_{intr} = -10 \log \left[4 \left(\frac{5.7}{5} + \frac{5}{5.7} \right)^{-2} \right] = 0.73 \text{ dB}$$

Therefore, $L_{intr} = 0.7 \text{ dB}$.

SUMMARY

In Chapter 9, optical fibers, the backbone of optical fiber communications systems, were discussed in length. The chapter began with the introduction of the principal laws that govern the transmission of light through optical fibers, followed by the presentation of critical topics relevant to transmission of light through fibers such as attenuation, absorption, and scattering. The chapter continued with the classification and detailed description of fibers as single-mode step index (SM-SI) and multimode step index (MM-SI) followed by an extensive presentation of the important topic of dispersion and its effect in optical signal transmission. Limitations of fiber transmission capabilities were identified, and the need for the development of advanced fibers, including non-zero dispersion shifted fibers (NZ-DSF) and large effective area NZ-DSFs were introduced. The adverse effects of four wave mixing (FWM), self phase modulation (SPM), and cross phase modulation (XPM) in DWDM optical systems was also presented. The chapter concluded with the introduction of some additional topics, such as fiber upgrading, splicing efficiency, and fiber alignment.

QUESTIONS

Section 9.1

1. State the year that transmission of light through a dielectric medium was first contemplated.
2. When was the cladding fiber introduced?
3. Why, until the middle of 1960s, were optical waveguides considered impractical?
4. Name the two wavelength windows currently applicable to optical fiber communications systems.

Section 9.2

5. What does Snell's law state? With the assistance of a diagram, explain this law.
6. Derive the equation of the critical angle.
7. Define *numerical aperture* and derive its relationship to the angle of acceptance.

Section 9.3

8. Define *electromagnetic wave* and derive the homogenous wave equation.
9. Calculate the velocity of light in vacuum.
10. Describe the concepts of transverse electric (TE) and transverse magnetic (TM) fields.
11. Define *phase velocity* and *group velocity*.
12. Derive the relationship of phase velocity in terms of core refractive index and velocity of light in a vacuum.
13. Derive the packet wave group velocity relationship.

Section 9.4

14. Define *optical attenuation*.
15. With the assistance of a formula, describe the input/output relationship of a signal traveling through an optical fiber.

Section 9.5

16. Describe *intrinsic absorption*.
17. With the assistance of a graph and a formula, show the relationship between absorption and wavelength.
18. Describe optical fiber absorption resulting from the interaction between the operating wavelength and fiber material lattice structure.
19. What is *extrinsic absorption*?
20. Name the metal contaminants that contribute to optical power absorption.
21. At what wavelength window do hydroxyl groups absorb optical energy?

Section 9.6

22. Define *linear scattering*.
23. How does linear scattering contribute to the increase of the system BER?
24. Describe *Mie scattering*.
25. Explain the difference between Mie scattering and Rayleigh scattering.
26. List all the parameters that contribute to Rayleigh scattering and their interrelationship.
27. Define *non linear scattering*.
28. When does optical power loss occur? (relevant to the nonlinear scattering effect)
29. What is *threshold optical power*?
30. What are the fundamental differences between SBS and SRS?
31. Describe, in detail, SBS.
32. With the assistance of a formula, state the optical power level beyond which SBS occurs.

33. Describe, in detail, *Raman scattering*.

34. With the assistance of a formula, state the optical power level beyond which SRS occurs, and list all the contributing parameters.

35. How is it possible to eliminate Raman scattering?

Section 9.7

36. How does fiber bending contribute to optical power loss? Explain, in detail.

37. With the assistance of a formula, describe *macrobending* and list the contributing parameters.

38. List the parameters relevant to *critical radius* for multi-mode fibers, and state their interrelationship.

39. With the assistance of a formula, describe *microbending* and list the contributing parameters.

40. List the parameters relevant to *critical radius* for single-mode fibers and state the interrelationship.

Section 9.8

41. Give a brief definition of the optical fibers used in communications systems.

42. What are the common materials used in the fabrication of optical fibers?

43. Name the substances diffused into the core in order to increase its refractive index.

44. Name the two basic classes of optical fibers.

45. Describe, in detail, the composition and operating characteristics of single-mode step index optical fibers.

46. Describe, in detail, the composition and operating characteristics of multimode step index optical fibers.

47. What is *normalized frequency* and how does it relate to multimode step index optical fibers?

48. Define *graded index fibers*.

49. Sketch the graded index/parabolic index profile of graded index optical fibers.

50. Describe, in detail, the performance characteristics of graded index optical fibers.

51. List the parameters and the relationship contributing to the establishment of the maximum modal number of graded-index fibers.

52. What is a *cut off wavelength*? Explain in detail.

53. Show the relationship between cut off wavelength, numerical aperture, fiber-core radius, and normalized frequency.

54. Explain why a practical cut off wavelength is considered instead of the theoretical.

Section 9.9

55. Define *dispersion*.

56. Why is dispersion so important in optical transmission systems?

57. Name the two types of dispersion.

58. What is the relationship between dispersion and system operating bandwidth?

59. State the relationship between pulse width and maximum system bit rate.

60. Sketch pulse chirping for step index single-mode, step index multimode, and graded index fibers.

61. Describe *intramodal dispersion*.

62. Name the two basic categories of intramodal dispersion.

63. What is *waveguide dispersion*?

64. Describe *intermodal dispersion*.

65. Define *polarization modal dispersion*.

66. Why is polarization modal dispersion mainly evaluated through statistical methods?

67. Give the expression relating the PMD coefficient and a number of equal length fibers.

68. Name the three statistical methods employed in the evaluation of PMD.

69. Describe the Monte Carlo numerical method.

Section 9.10

70. What was the determining factor which led to the significant expansion of LANs?

71. Why do LED devices impose limits on the system operating bandwidth?

72. Name the two distinct advantages of laser diodes over LED in Gb.E-LAN applications.

73. What are the four limiting factors of optical fibers intended for high bit rate transmission applications?

74. Explain the relationship between chromatic absorption and spectral linewidth in an optical source.

75. Describe, in detail, the *index profile* of an optical fiber.

76. Describe two ways through which the high capacity demands of Gb.E can be satisfied.

77. How is the information carrying capacity of a multimode fiber usually expressed?

78. Define *LOM, IOM,* and *HOM*.

79. Explain the relationship between the bandwidth and the modal number of a multimode fiber.

Section 9.11

80. Why was the development of advanced performance optical fibers necessary?

81. Name the three new types of fibers developed by Corning.

82. What are the other three devices, developed parallel to advanced fibers, which significantly contributed to the advancement of fiber optic technology?

83. Name the data protocols that NZ-DSF fibers are suitable for.

84. Can NZ-DSF fibers be used for optical spans beyond 300 km without dispersion compensation?

Section 9.12

85. Describe the relationship between fiber core effective area and system nonlinearities.

86. Name the application(s) where large effective area fibers are suitable.

87. List the possible improvements to system performance because of the utilization of large effective area fibers.

88. With the assistance of a formula, describe the relationship between fiber core cross-sectional effective area and fiber nonlinearities.

Section 9.13

89. What is the direct effect of fiber nonlinearity?

90. With the assistance of a formula, explain the relationship between fiber cross-sectional effective area, refractive index, and input optical power.

91. Name the fiber nonlinearities included in the first category.

92. Define *FWM*.

93. How can FWM be reduced or completely eliminated?

94. Define *SPM*.

95. State the effect of SPM in an optical transmission.

96. Define *XPM*.

97. Describe the effect of XPM in optical transmission.

98. Describe, in detail, the effect of both FWM and XPM in DWDM optical systems.

Section 9.14

99. How can an already existing optical fiber system that operates at 2.5 Gb/s and employs standard single mode fibers be upgraded to operate at 10 Gb/s and at optical spans beyond 60 km?

100. What is the disadvantage of implementing the upgrade in Question 99?

101. Describe the ways in which chromatic dispersion can be controlled.

Section 9.15

102. What is the incentive for the development of plastic fibers?

103. Name some areas in which plastic fibers can be successfully applied.

104. How was the critical attenuation level of less than 200 dB/km accomplished?

105. What is the operating bandwidth of a GI-POFs?

106. Why are GI-POFs considered the fibers of choice for LAN networks?

Section 9.17

107. Name the three critical elements that contribute to fiber efficiency.

108. What is *concentricity*?

109. Define *fiber curl*.

110. Name the three basic procedures and critical parameters for splicing.

Section 9.18

111. Name the types of connections employed in optical fiber links.

112. Where is the impact of connector loss most felt?

113. With the assistance of a schematic diagram, describe lateral, longitudinal, and angular misalignment.

114. Name the factor that contributes to insertion loss by both lateral and angular misalignment of a step index single-mode fiber.

PROBLEMS

1. An optical fiber is fabricated with a core refractive index of 1.65 and a cladding of 1.45. Calculate the angle of acceptance, the critical angle, and the numerical aperture.

2. An optical fiber exhibits an attenuation of 0.5 dB/km. If the optical power measured at the output of the fiber is 25 mW, calculate the optical power at the input.

3. Calculate the absorption coefficient *(a)* of an optical fiber operating with the following parameters:

  a) fiber attenuation 0.55 dB/km

  b) optical power launched at the input of the fiber 35 mW

  c) optical power measured at the output of the fiber 2.5 mW

4. Calculate the Rayleigh attenuation in an optical fiber operating in the 1420 nm wavelength and with fictive temperature of 1600°K, photoelastic coefficient of 0.23, refractive index of 1.5, isothermal compressibility of $90 \times 10^{-12}\, m^2N$, and fiber length of 2.5 km.

5. If the stimulated Brillouin threshold optical power of a long length fiber is 28 mW, the core diameter is 8 μm, the fiber attenuation is 0.25 dB/km, and the optical source operating wavelength is 1530 nm, calculate the laser diode bandwidth. If the diode operates at the 1530 nm wavelength, calculate the laser diode operating bandwidth.

6. Calculate the Raman scattering threshold power of a single mode, 8 μm core diameter fiber that exhibits an attenuation of 0.5 dB/km and operating at 1310 nm wavelength.

7. If an optical fiber with a core radius of 8 μm and a numerical aperture of 0.4 operates at a wavelength of 1310 nm, calculate its maximum modal number. Assume graded index characteristic profile (parabolic).

8. A single-mode step index optical fiber of 85 km length exhibits a pulse dispersion of 0.15 μs. Calculate:
   a) operating bandwidth
   b) dispersion/km
   c) bandwidth length product

9. Calculate and then compare the pulse broadening, or pulse chirping, generated when an LED source with an rms spectral line width of 18 nm and a laser diode with an rms spectral line width of 2.5 nm are successively connected at the input of an optical fiber that exhibits material dispersion of 0.015. Assume an operating wavelength for both sources of 900 nm.

10. An optical link is composed of ten optical fibers that were fabricated with a core refractive index of 1.44. If the medium between the fibers is air, calculate the attenuation based on Fresnel reflections.

10

OPTICAL
MODULATION

OBJECTIVES

1. Establish the necessity for external modulation in high speed optical communications systems

2. Describe the principle of operation of the Mach Zehnder interferometer

3. Describe, in detail, the principle of operation and performance characteristics of the Mach Zehnder LiNbO$_3$ optical modulator

4. List and explain all the production steps to be followed in the design of the Mach Zehnder LiNbO$_3$ optical modulator

5. Present and discuss the operating characteristics of the Lucent Technologies Mach Zehnder LiNbO$_3$ optical modulator

6. Describe, in some detail, the operation and performance characteristics of various modulator driver circuits

7. Define the need for special laser diodes for external optical modulators

8. Describe the function of operation and performance characteristics of DFB lasers and PMF fibers designed to interface with external optical modulators

9. Explain the operation and performance characteristics of electroabsorption modulated lasers (EML)

KEY TERMS

asynchronous transfer mode (ATM)
back facet
bandwidth
bias control
buffer
carrier to noise ratio (CNR)
carrier wave (CW)

CATV
chirping
dark current
diffusion
distributed feedback Bragg (DFB) laser
dynamic range
efficiency

electro-absorptive modulated laser (EML)
electro-absorptive modulator (EAM)
erbium doped fiber amplifier (EDFA)
extinction ratio
feedback

forward voltage
immunity
insertion loss
jitter
laser driver
light source
line width
Mach Zehnder interferometer

10.1 INTRODUCTION

Until recently, optical communications systems have been operating with direct modulation. That is, the input binary data directly modulates the laser diode biasing current in order to produce a time varying output **optical power.** This method of modulation has proven satisfactory for low transmission rates. However, for high speed digital (Gb/s) and **CATV** optical systems, direct modulation imposes limits to system performance because of laser diode nonlinearities and broader **spectral line width.** These two factors are very critical to high speed optical systems because they contribute to undesirable pulse dispersion and the generation of relevant distortion products. More specifically, direct modulation broadens the spectral linewidth of the laser diode by increasing the refractive index of the cavity. This refractive index increase induces large variations of the center wavelength, resulting in an enlargement of the generated spectral line width. However, a large spectral line width, coupled with fiber dispersion, increases pulse broadening (**chirping**) and ultimately imposes limits on the speed and length of the optical link. Chirping, which causes ISI and crosstalk can be dramatically reduced or altogether eliminated by a very narrow laser diode spectral line width. A narrow line width can be produced by the laser diodes operating in the **carrier wave (CW)** mode of operation. An external modulator, inducing practically zero pulse dispersion, can modulate the generated narrow line width CW optical power. Two types of external modulators exist: the **Mach Zehnder LiNbO$_3$ modulator** and the **electro-absorptive modulator.**

10.2 THE MACH ZEHNDER INTERFEROMETER

The basic **Mach Zehnder (MZ) interferometer** illustrated in Figure 10–1 is composed of four mirrors and four plates that incorporate spacers for optical path comparison, where M_1, M_2, M_3, and M_4 are mirrors and P_1, P_2, P_3, and P_4 are plates that incorporate spacers for optical path comparison.

Operation of the Mach Zehnder Interferometer

Mirror M_1 is used to divide the light beam into two equal paths. They are recombined at mirror M_4 with constructive interference. If one of the two paths is phase shifted, a destructive interference will be encountered at the beam select mirror M_4. This consideration is used in the design of a LiNbO$_3$ modulator. In a Mach Zehnder LiNbO$_3$ optical modulator, 3 dB Y optical splitters and combiners are used, while the required phase shift is induced in one of the paths by changing the waveguide (path) refractive index through a change of an applied electric field.

10.3 THE MACH ZEHNDER (LiNbO$_3$) OPTICAL MODULATOR

General Description

The cross sectional area of a Mach Zehnder optical modulator is illustrated in Figure 10–2.

The MZ LiNbO$_3$ optical modulator in Figure 10–2 is composed of two phase modulators, a 3 dB Y junction **optical splitter,** and a 3 dB Y junction combiner, and it incorporates three input ports and an output port. The CW input port connects to the output of the laser diode, the binary data is connected to the RF port, the DC biasing voltage is connected to the bias port, and the modulated optical power is collected at the output port. A narrow **line width** CW polarized optical signal, generated by the laser diode, is applied at the input of the MZ modulator. The optical signal is then divided and fed into two equal paths by the 3 dB Y junction splitter. If a voltage $V(t)$ is applied at the RF input port, an interaction between the optical signal and the RF electrical signal will alter the phase of the optical signal. Through this process, two sets of optical signals are generated: in-phase and out-of-phase signals. At the Y junction combiner, the in-phase signals are coupled into the optical fiber, while the out-of-phase signals, transformed into a higher order mode, are dissipated into the substrate of the modulator. The transfer function of the modulator is expressed by Equation (10–1).

$$I(t) = aI(\theta)\cos^2\left(\frac{V_{(t)}\pi}{2\,V}\right) \qquad \textbf{(10–1)}$$

Where $I(t)$ is the intensity of the output modulated signal, a is the **insertion loss** of the modulator, $I(\theta)$ is the laser diode intensity at the input of the modulator, $V_{(t)}$ is the voltage applied at the RF port, and $V_{(t)}\pi$ is the **modulator**

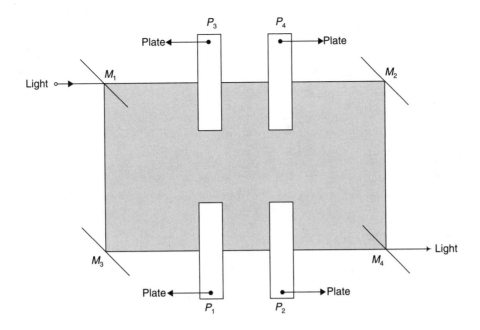

Figure 10–1. A Mach Zehnder interferometer.

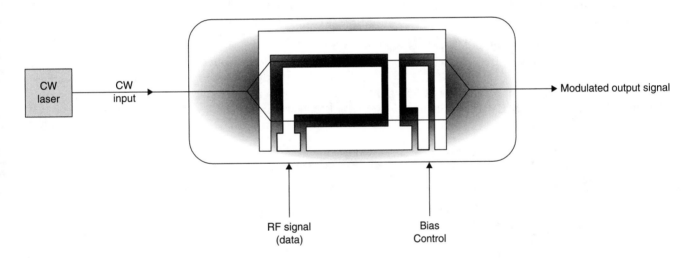

Figure 10–2. A cross section of a Mach Zehnder LiNbO₃ optical modulator.

driver voltage required to achieve 180° phase shift. A typical MZ LiNbO₃ modulator transfer function curve is illustrated in Figure 10–3.

Operation of the MZ LiNbO₃ Optical Modulator

The generated optical CW signal for a DFB laser diode that has a very narrow line width is fed into the MZ LiNbO₃ modulator. The modulator transimpedance is a function of the $V_{(t)}$ signal applied at the RF input. Therefore, a binary data stream applied at the RF port of the modulator can alter the optical CW signal. The benefits of the utilization of an external modulator in optical fiber communications links are:

a) higher optical output power coupled into the optical fiber
b) very small pulse broadening (chirping)
c) longer link distance
d) higher transmission rates
e) fewer EDFAs required
f) no **jitter**
g) electronic noise rejection
h) >20 dB **dynamic range**
i) **immunity** to environmental changes
j) supports wide band transmission
k) cost **efficiency**

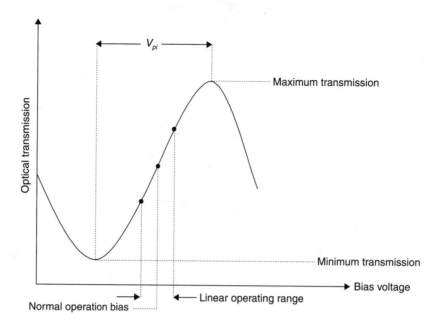

Figure 10–3. The transfer function curve for a Mach Zehnder LiNbO₃ modulator.

Driving Elements of an MZ LiNbO₃ Optical Modulator

A more detailed circuit diagram of an MZ LiNbO₃ optical modulator is illustrated in Figure 10–4.

THE LIGHT SOURCE. For a satisfactory CNR, a **distributed feedback Bragg (DFB)** CW laser diode may be used

for the light source. It must have a very narrow line width and a **relative intensity noise (RIN)** of larger than 150 dB/Hz.

THE LASER DRIVER. Through an internal feedback mechanism, the **laser driver** circuit maintains the laser threshold current at a fixed level, a necessary condition for constant laser output signal intensity.

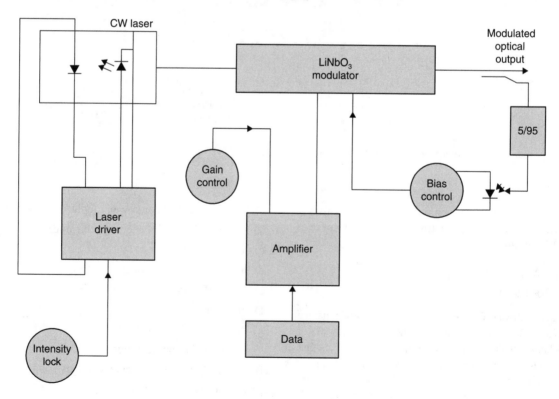

Figure 10–4. Driving circuitry of an MZ optical modulator.

DC Biasing. The DC biasing voltage is applied at the bias port of the modulator. Quadrature bias can be obtained under a DC bias voltage of ± 7 V.

Modulator Driver. Because LiNbO$_3$ modulators respond to voltage rather than current, a voltage amplifier is used as the modulator driver. In order to achieve 100% modulation, the pulse voltage must be equal to V_{pi} (see Figure 10–3).

Applications

Mach Zehnder LiNbO$_3$ optical modulators, in conjunction with **EDFAs,** are the ideal devices for enhancing the performance of optical systems such as:

a) **asynchronous transfer mode (ATM)**

b) **SONET/SDH** optical add/drop multiplexers (OADMs)

c) **Soliton** systems

d) testing and measuring systems

e) long distance communications systems

f) undersea systems

g) terrestrial communications links < 100 km

h) CATV

i) RF down links from satellites

Device Characteristics. Mach Zehnder optical modulators display the following characteristics:

a) high **extinction ratio**

b) excellent stability

c) extremely low chirp

d) EDFA compatibility

Typical modulator operating characteristics are listed in Table 10–1, Table 10–2, and Table 10–3.

TABLE 10–1 Modulator Operating Characteristics

Parameters	Minimum	Typical	Maximum	Units
Wavelength	1525	—	1575	nm
Insertion loss	2.5	3.8	5.0	dB
Chirp	−0.2	—	0.2	
Extinction ratio	20	—	—	dB

TABLE 10–2 RF Input

Parameters	Minimum	Typical	Maximum	Units
Return loss	—	—	−8	dB
Rise/fall time	—	—	120	ps
Input impedance	27	29	32	Ω
Amplitude ripple	—	—	1	dB

TABLE 10–3 Bias Input

Parameters	Minimum	Maximum	Units
Voltage range (DC)	−7	+7	V
$V\pi$ (DC)	4	8	V

10.4 THE MZ LiNbO$_3$ DESIGN PROCESS

The design process of an MZ LiNbO$_3$ optical modulator is standard across the industry. Nevertheless, the following procedures are based on an AT&T design process illustrated by the flow chart in Figure 10–5.

The AT&T fabrication process of an MZ optical modulator employs the z cut crystal orientation illustrated in Figure 10–6. The vertical orientation of the z axis is the smallest of the three dimensions (xyz) in the waveguide. Proportionally, it requires the smallest electric field in order to interact with the optical signal.

LiNbO$_3$ crystals are highly polarization sensitive as a result of crystal asymetricity. It is therefore important that the appropriate polarization mode be launched into the modulator. If the **transverse electric (TE)** mode were to be launched into the modulator, a much stronger electric field and, consequently, a larger $V\pi$ would have been required to interact with the optical signal. In order to achieve the required **transverse magnetic (TM)** polarization, a **polarization maintenance fiber (PMF)** is used between the laser diode output and the modulator optical CW input signal.

Diffusion

The LiNbO$_3$ substrate is now ready for the **diffusion** process. To generate the optical waveguides, Ti (titanium) is deposited over the area of the substrate that is marked as the optical waveguide, then heated to a temperature of 1000°C in order for the diffusion process to be completed. The required diffractive index difference between the titanium and the LiNbO$_3$ substrate is achieved through control of the amount of titanium used in the diffusion process, temperature, and the length of the baking process.

Buffer (SiO$_2$)

The **buffer** layer has two functions. One is to increase the modulator impedance from approximately 23 Ω without the buffer to 43 Ω under critical layer thickness control; 43 Ω is much closer to 50 Ω, matching the impedance required for most applications. The other function is to maintain almost zero attenuation.

Electric Conducts

There are certain considerations that must be taken into account during the electrode deposition process. One of the

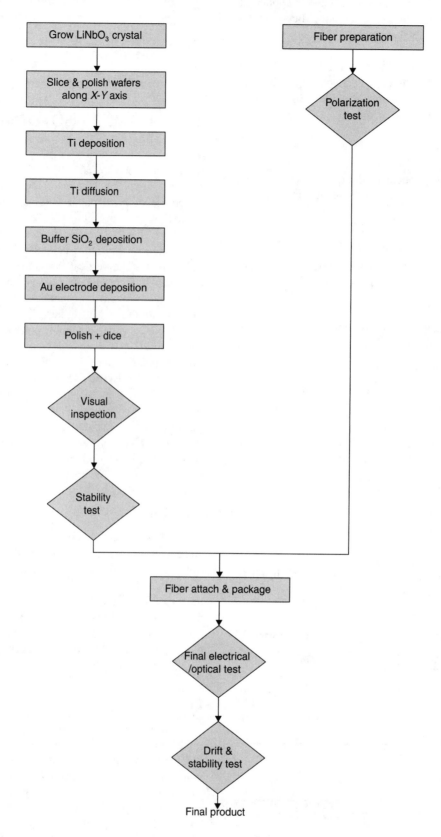

Figure 10–5. MZ LiNbO$_3$ design flow chart.

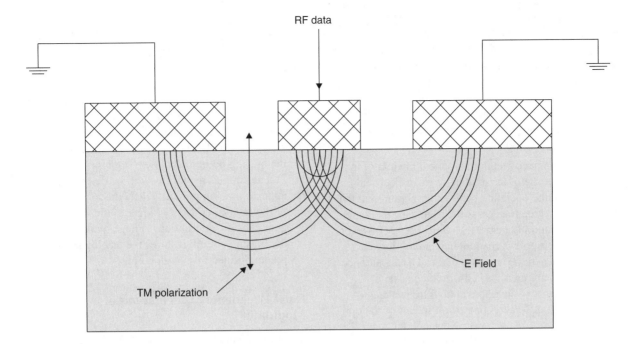

Figure 10–6. *z*-cut of a LiNbO$_3$ crystal for single drive modulator.

most critical considerations is the length of the electrodes. Long length electrodes are preferable, because they enhance the much needed electric/optic field interaction and reduce the otherwise required high bias voltage ($V\pi$).

However, long length electrodes induce phase mismatch between the optical and electrical signals. Therefore, the lengths of the electrodes must be selected to maintain a balance between the advantages and disadvantages. Typical electrode lengths vary between 1 cm and 4 cm. Furthermore, the position of the electrodes, in reference to the position of the waveguides within the fabrication structure, varies between an *x* cut and a *z* cut LiNbO$_3$ crystal. The fundamental requirement for both cuts is a perfect coupling efficiency between the electric and the optical fields. To achieve an almost perfect match, it requires the direction of the electric field and the polarization of the optical field must be aligned to the direction of the *z* axis. Therefore, the electrodes in a *z* cut arrangement must be deposited directly above the optical waveguides. Likewise, in an *x* crystal cut, the electrodes must be deposited off the waveguide, thus achieving parallel (TE) propagation.

LiNbO$_3$ Modulator Design

LiNbO$_3$ modulators must be optimally designed to operate at the 1300 nm or 1550 nm wavelengths. If a modulator designed to operate at the 1550 nm wavelength is coupled to a 1310 nm laser source, a double mode will appear at the modulator waveguide and will promote an inferior **extinction ratio** and higher loss. Likewise, if a modulator designed to operate at the 1310 nm wavelength is coupled to a 1550 nm laser source, it will also promote inferior extinction ratio and higher loss.

Mach Zehnder LiNbO$_3$ optical modulators incorporate at least three input ports into their design as well as an optical power input port, a DC bias input port that also accommodates the RF signal, and an output port from which the modulated optical signal is collected. Modern designs incorporate a separate port for the RF signal and a separate port for the bias DC voltage.

INPUT SPECIFICATIONS.

a) *Optical input port:* The CW optical power generated by the laser diode at either the 1310 nm or the 1550 nm wavelength range must be polarized in order to minimize the RIN and to maximize the SNR. Optical signal polarization from the laser diode to the input of the modulator can be accomplished with a properly spliced PMF or through an appropriate polarization controller.

b) *The DC bias port:* At the bias input port, DC voltage is required in order to set the operating point (inflection point) of the modulator to the position at which a small variation of the signal at the RF port will generate a maximum and linear change of the output optical signal. It is evident that a very stable DC bias voltage is essential in order to maintain the modulator at its maximum performance. However, environmental, mechanical, and other factors can induce drift of the inflection point and degrade the modulator's performance. Recent modulator designs have rectified this problem by incorporating appropriate DC bias **feedback** loops capable of completely eliminating the DC bias drift.

c) *Extinction ratio:* The extinction ratio of an MZ LiNbO$_3$ modulator is the ratio of the optical signal at the output of the modulator generated by a binary-1, to the optical power generated by a binary-0. This ratio must be kept at a maximum. Titanium diffused waveguides promote some polarization mixing between the TE and the TM modes. Because optical modulators operate in a TM mode, a polarizer to prevent any optical power from escaping into the TE mode will be required. The incorporation of the polarizer will substantially improve the modulator extinction ratio.

d) *Bandwidth:* The bandwidth of an optical modulator is the optical response of the modulator to an electrical signal. Ideally, the operating bandwidth measured at the −3 dB point must be flat across the frequency band and must roll off smoothly below the −3 dB point. However, in reality, some ripple is encountered at both low and high frequencies of the response curve. The appearance of ripple at low frequencies is due to the package acoustic response, while ripple at high frequencies is the result of the different propagation speeds of the optical and electrical signals traveling through the waveguide, thus inducing undesirable phase shifts. Also detrimental to the normal operation of the modulator is electrical back reflection. Back reflections originate at the input ports of the modulator and are caused by impedance mismatch. They are capable of inducing undesirable signal distortions. In order to eliminate the impact of back reflections, they must be kept 15 dB below the modulator operating bandwidth through proper coupling of the incoming data at the RF port (impedance matching).

e) *Voltage swing:* **Swing voltage** ($V\pi$) is defined as the voltage required to shift the phase of the optical signal in one of the modulator waveguides by $180°$. The phase difference between the two branches generates a minimum optical power at the output of the modulator through the process of destructive interference. $V\pi$ is a function of the frequency of the input RF signal. That is, if the frequency of the input signal increases, so does the $V\pi$. This is a result of the increased signal attenuation and the walk off phenomenon for both electrical and optical signals propagating through the corresponding electrodes. Therefore, it is imperative that trade-offs must be considered between short electrodes that enhance modulator operating **bandwidth (B_W)** and long electrodes that require lower $V\pi$. Optical modulators, such as the AT&T 2122, require $V\pi$ between 3.8 V and 4.0 V for electrode lengths of 4 cm, operating at bandwidths of 4 GHz and at the 1550 nm wavelength. For data rates of 2.5 Gb/s, $V\pi$ is set between 5.8 V and 5.9 V.

f) *Automatic bias control:* To eliminate DC biasing drifts of the modulator, an external feedback loop is required to provide the automatic DC bias control (Figure 10–7). The feedback circuit of the automatic bias control mechanism is driven by a relatively low frequency (1 KHz). A small level of the optical power is obtained from a 90/10 optical splitter, which is fed into the input of a low bandwidth PIN photodetector diode capable of demodulating only the 1 KHz frequency. The 1 KHz output signal from the photodetector diode is fed into a multiplier circuit and through a low pass filter into the input of the bias circuit. At the proper biasing point, where the maximum optical output occurs (inflection point), the signal of the PIN diode output will measure 1 KHz. If the biasing point drifts away from the inflection point, 2nd harmonic frequencies will be detected and will appear at the output of the PIN diode. The combination of the multiplier and LPF detects these harmonics and generates the appropriate voltage, which, when fed into the biasing circuit, restores the biasing level to the predetermined inflection point.

Pulse Broadening Calculations of an MZ LiNbO$_3$ Modulator

The fundamental operating concept of a dual MZ LiNbO$_3$ modulator is illustrated in Figure 10–8.

The optical power (E_o) generated by a very narrow line width laser diode is fed into the optical input port of the modulator. The input optical power is split by a 3 dB Y junction splitter and phase shifted at both arms of the modulator by externally applied voltages $V_1(t)$ and $V_2(t)$. The Y junction combiner recombines the phase shifted optical fields of both arms and the resultant output optical signal is represented by Equation (10–2).

$$E = E_o \cos\frac{(\phi_2 - \phi_1)}{2} \; e^{j\frac{(\phi_2 - \phi_1)}{2}} \qquad \textbf{(10–2)}$$

where E is the output optical field, E_o is the input optical field, ϕ_1 is the phase change in waveguide 1, and ϕ_2 is the phase change in waveguide 2.

The modulator pulse broadening (chirping) is the ratio of the phase to amplitude modulation and is expressed by Equation (10–3).

$$\alpha = 2E^2 \frac{\left(\dfrac{d\phi}{dt}\right)}{\left(\dfrac{dE^2}{dt}\right)} \qquad \textbf{(10–3)}$$

where α is the pulse broadening (chirp), ϕ is the average phase change $\left(\dfrac{\phi_2 + \phi_1}{2}\right)$, and E^2 is the **optical field intensity,**

or

$$\alpha = \left[\frac{\dfrac{d\phi_1}{dt} + \dfrac{d\phi_2}{dt}}{\dfrac{d\phi_1}{dt} - \dfrac{d\phi_2}{dt}}\right] \cot\left(\frac{\phi_2 - \phi_1}{2}\right) \qquad \textbf{(10–4)}$$

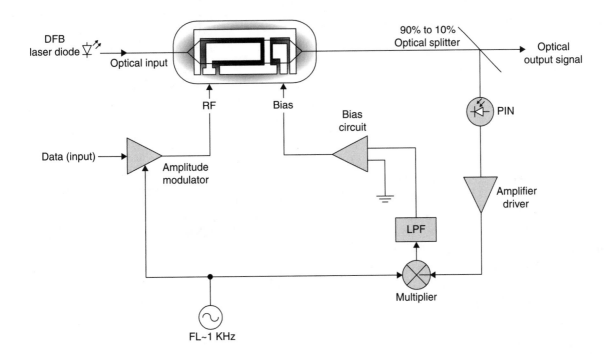

Figure 10–7. An automatic bias control circuit.

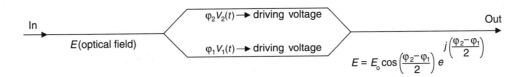

Figure 10–8. MZ-LiNbO₃ modulator concept.

The phase change induced at each arm of the modulator is a function of the length of each waveguide that composes the arms and the corresponding index of refraction (Equation (10–5)).

$$\phi = \left(\frac{2\pi}{\lambda_o} \right) nL \qquad (10\text{–}5)$$

where λ_o is the wavelength in vacuum, n is the **refractive index,** and L is the length of the waveguide (length of the arm). Substituting Equation 10–5 into Equation 10–4 yields Equation 10–6, which expresses the modulator pulse broadening (chirp), in relation to refractive index change.

$$\alpha = \left[\frac{\dfrac{dn_1}{dt} + \dfrac{dn_2}{dt}}{\dfrac{dn_1}{dt} - \dfrac{dn_2}{dt}} \right] \cot^2 \left[\left(\frac{\pi L}{\lambda_o} \right) (n_2 - n_1) \right] \qquad (10\text{–}6)$$

The refractive index in each arm of the modulator is a complex one. It has two components: one is the refractive index (n_o) of the material that composes the waveguide, and the other is the refractive index change (Δn) subject to the voltage change at the RF input port of the modulator. (Δn) is expressed by Equation (10–7).

$$\Delta n(t) = \Gamma \left(\frac{n_o^3}{2} \right) r \left(\frac{V_j(t)}{d} \right) \qquad (10\text{–}7)$$

where Γ is the overlap integral between electric and optical fields, r is the coefficient of the electro-optic tensor, and d is the electrode separation.

Equation (10–7) can be simplified to represent a combined index change in reference to voltage change (Equation (10–8)).

$$\Delta n_x(t) = \eta_x V_x(t) \qquad (10\text{–}8)$$

where η_x is the refractive index change/voltage and V_x is the applied voltage. If a voltage (V_x) is applied to one of the arms of the modulator, Equations (10–9) and (10–10) express the combine voltage at each arm.

$$V_1(t) = V_1 \sin \omega t \qquad (10\text{–}9)$$

$$V_2(t) = V_2 \sin \omega t + V_x \qquad (10\text{–}10)$$

The refractive index in waveguide 2 can be expressed in terms of the driving voltage (Equation (10–11)).

$$n_2 = n_o + \Delta n_2 = \eta_2 V_2(t) \qquad (10\text{–}11)$$

Substituting Equation (10–10) into Equation (10–11) yields Equation (10–12).

$$n_2 = n_o + \eta_2 V_2 \sin(\omega t + \eta_2 V_x) \quad \textbf{(10–12)}$$

The derivative of Equation (10–12) yields Equation (10–13).

$$\frac{dn_2}{dt} = \eta_2 V_2 \omega \cos \omega t \quad \textbf{(10–13)}$$

The refractive index in waveguide 1 is expressed in the same way as that of waveguide 2 (Equation (10–14)).

$$\frac{dn_1}{dt} = \eta_1 V_1 \omega \cos \omega t \quad \textbf{(10–14)}$$

Substituting Equations (10–13) and (10–14) into Equation (10–6) yields Equation (10–15).

$$\alpha = \left[\frac{\eta V_1 + \eta_2 V_2}{\eta_1 V_1 - \eta_2 V_2} \right]$$

$$\times \cot\left[\left(\frac{\pi L}{\lambda_o} \right)(\eta_2 V_2) \sin \omega t + \eta_2 V_x - \eta_1 V_1 \omega t \right] \quad \textbf{(10–15)}$$

Because the modulator is to operate at the half power point, the intensity is also at half value (Equation (10–16)).

$$I = I_o \cos^2 \frac{(\phi_2 - \phi_1)}{2} \quad \textbf{(10–16)}$$

where I is the optical intensity at each arm and I_o is the optical intensity at the input of the modulator. Assuming $\phi_1 = \phi_2$, Equation (10–16) yields Equation (10–17).

$$I = \frac{I_o}{2} \quad \textbf{(10–17)}$$

From Equation (10–17), it is evident that the optical intensity in each arm is one-half of the input intensity. At the normal biasing point (inflection) from Equation (10–6), we have

$$\cot^2\left[\left(\frac{\pi L}{\lambda_o} \right) \eta_2 V_x \right] \cong 1$$

Therefore, modulator pulse broadening (chirping) is expressed by Equation (10–18).

$$\alpha = \frac{\eta_1 V_1 + \eta_2 V_2}{\eta_1 V_1 - \eta_2 V_2} \quad \textbf{(10–18)}$$

If both waveguides are the same, then the refractive index/per volt rate of change is the same for each, then the modulator chirp is a direct function of the applied voltage Equation (10–19).

$$\alpha = \left(\frac{V_1 + V_2}{V_1 - V_2} \right) \quad \textbf{(10–19)}$$

Because the peak to peak takes a maximum equal to $V\pi$, Equation (10–20) holds true.

$$\alpha = \frac{V_1 + V_2}{V_1 - V_2} = \frac{V_1 + (V_1 - V\pi)}{V\pi} = \frac{2V_1 - V\pi}{V\pi}$$

$$= \frac{2V_1}{V\pi} - \frac{V\pi}{V\pi} = \frac{2V_1}{V\pi} - 1$$

Therefore,

$$\alpha = \frac{2V_1}{V\pi} - 1 \quad \textbf{(10–20)}$$

Example 10–1

Compute the required biasing voltage at waveguide 2, if the chirp of an MZ LiNbO$_3$ optical modulator is to be maintained at a level of 0.25.

Solution

i) Use Equation 10–20, and solve for V_1.

$$V_1 = \frac{\alpha V\pi + V\pi}{2}$$

ii) *Compute* (V_1).

$$V_1 = \frac{0.25 \, V\pi + V\pi}{2} = \frac{1.25 \, V\pi}{2} = 0.625 \, V\pi$$

$$\therefore V_1 = 0.625 \, V\pi$$

iii) *Compute* (V_2).

Because $V_2 = V_1 - V\pi$,

then, $V_2 = 0.625 \, V\pi - V\pi = -0.375 \, V\pi$

$$\therefore V_2 = -0.375 \, V\pi$$

Example 10–2

Determine the biasing range of a Mach Zehnder LiNbO$_3$ optical modulator with an allowable pulse broadening range of between 0.3 and 0.8.

Solution

i) For $\alpha = 0.3$, compute V_1 and V_2.

$$\alpha = \frac{2V_1}{V\pi} - 1$$

or

$$0.3 = \frac{2V_1}{V\pi} - 1$$

Solving for V_1,

$$V_1 = 0.65 \, V\pi$$

Because $V_2 = V_1 - V\pi$, then,

$$V_2 = -0.35$$

For $\alpha = 0.8$, repeating the preceding process, we have

$$V_1 = 0.9 \, V\pi$$

$$V_2 = -0.1 \, V\pi$$

Therefore, the biasing range is: $0.65 \, V\pi \leq V_1 \leq 0.9 \, V\pi$

$$-0.1 \, V\pi \leq V_2 \leq -0.35 \, V\pi$$

10.5 MZ LiNbO₃ OPTICAL MODULATOR OPERATING REQUIREMENTS

The following operating requirements are based on the Lucent Technologies Mach Zehnder LiNbO₃ optical modulator. The biasing arrangement of such a modulator is illustrated in Figure 10–9.

The devices in Figure 10–9 may differ slightly in order to satisfy different system requirements, however, the overall operating characteristics are very similar. MZ LiNbO₃ optical modulators, designed with either three or four electrodes, require a highly polarized optical signal coupled into the input

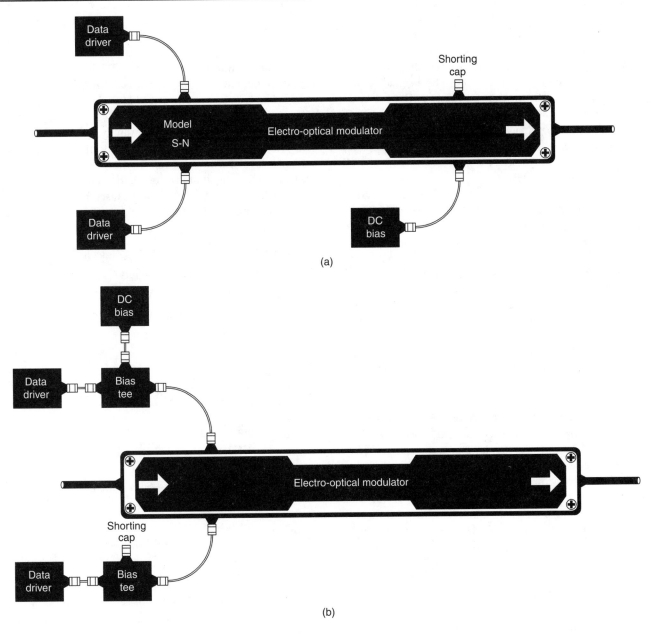

Figure 10–9. a) Biasing arrangement of an MZ LiNbO₃ optical modulator that uses separate electrodes for RF and DC biasing (three-input port). b) Biasing arrangement of an MZ LiNbO₃ optical modulator that uses the same electrode for both RF and DC biasing (two-input port).

optical port. This very narrow spectral line width optical signal is usually generated by a DFB laser diode. Such devices also require an RF signal applied at the RF input port, and a DC biasing voltage applied at the biasing input port for a three-input device, or applied to the same input port with the RF signal for a two-input device.

Operating Requirements

More detailed signal requirements for a three-input MZ LiNbO$_3$ optical modulator follow.

OPTICAL INPUT SIGNAL REQUIREMENTS. The optical signal required at the input of an MZ LiNbO$_3$ modulator is generated by a DFB laser diode, which generates a narrow CW at the 1550 nm wavelength that has a very low RIN ratio. The very low RIN is essential if a correspondingly high SNR is to be maintained, especially when the device is used to transmit analog signals such as CATV channels. Because the modulator operates in a TM mode, the optical signal applied at the input port must be highly polarized. Optical beam polarization can be achieved either by incorporating a PMF between the laser diode and the input of the modulator, or by inserting a polarization controller. Polarization controllers accept a standard single-mode signal and rotate it to all possible polarization states.

ELECTRICAL INPUT REQUIREMENTS. All MZ LiNbO$_3$ modulators incorporate either one or two electrical input ports, based on their intended applications. For example, if the modulator is designed to operate at high frequency applications above 8 GHz, two ports are normally required, one for the RF input and the other for the DC biasing input. For applications below 8 GHz, a single electrical input port is used, facilitating both the RF and the DC biasing signals.

Electrical port input impedance: In a single or dual port modulator, impedance matching is absolutely essential for maximum optical power transfer. In almost all applications, the required input impedance is 50 Ω. However, because the electrical port input impedance is a function of the transmission line geometry and composition of the material used for the waveguide, a compromise has been reached. The acceptable value of the input impedance is set at 43 Ω. To achieve impedance matching, the input transmission line is internally terminated by a 43 Ω thin film resistor. For a single electrode input, a capacitor (in series with the internally terminated 43 Ω resistor) is required in order to block the DC biasing voltage applied at the input.

10.6 MODULATOR DRIVERS

For digital applications, an appropriate modulator drive circuit is essential. Figure 10–10 illustrates a block diagram of a modulator driver circuit. A differential amplifier, used as the modulator driver, is capable of sourcing 100 mA of cur-

rent for 2.5 Gb/s applications, and 300 mA of current for 10 Gb/s applications. However, because the modulator is a voltage-driven device, a current-to-voltage (I/V) conversion circuit is also required. An AC-coupling capacitor is used between the output of the driver circuit and the bias tee in order to compensate for the different ground requirements of the driver circuit and the modulator (capacitive ground). The bias tee applies the DC bias voltage to the common electrical input of the modulator.

Modulator Driver Operating Characteristics

As mentioned previously, driving circuits are required to provide the necessary voltages for the proper operation of an MZ LiNbO$_3$ optical modulator. The operating characteristics of such a circuit are based on the Lucent Technologies LG1626DXC modulator driver circuit (Figure 10–11).

The circuit in Figure 10–11 is a GaAs IC circuit that was designed to provide the required voltage levels for an optical modulator used in high speed NRZ digital applications. The circuit is composed of four GaAs FET cascaded stages and NiCr precision resistors. The device uses a single negative (–5.2 V) power supply, and the output is capable of driving 50 Ω loads with an input capability of 100 kΩ levels. The amplifier is designed to control the optical output signal of the modulator, and also to compensate for the modulator DC offsets caused by power supply and temperature drifts. Modulator driver operating characteristics are as follows:

a) adjustable output voltage: 3 V$_{p-p}$ max at 50 Ω
b) differential or single ended inputs
c) single power supply: (–5.2 V)
d) rise/fall time: approximately 90 ps
e) enabling control

Pin Descriptions

Pin	Description	Symbol
1, 3, 4, 9, 13, 14, 15	Ground	GND
2	Data input	V_{in}
5	Complementary data input	$\overline{V_{in}}$
6	Complementary threshold control input	$\overline{V_{TH}}$
7	–2.5 V band gap reference	BG2P5
8	Modulation enable	MOD-E
11	Complementary mark density output	$\overline{MK}$
12	Mark density output	MK
16	AC couple output (50 Ω) to modulator	V_{out}
17	Modulator DC offset output	$V_{out-DC_{OFFSET}}$
18	Modulator DC offset control input	V_{DC-ADJ}
19	V_{ss2} supply (–5.2 V) for output prebias	V_{ss2}
20	V_{ss3} supply (–5.2 V) for output modulation	V_{ss3}

Continued

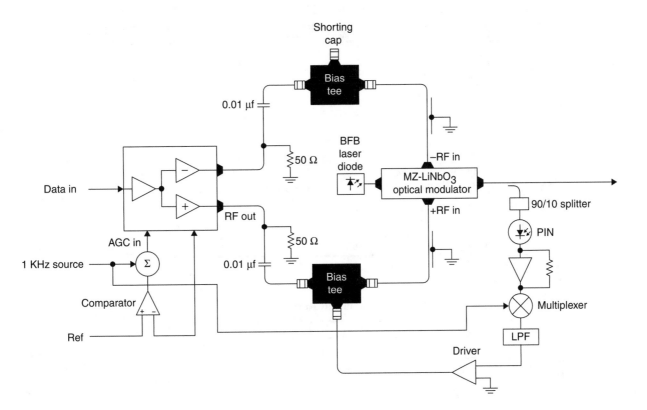

Figure 10–10. A block diagram of a modulator driver.

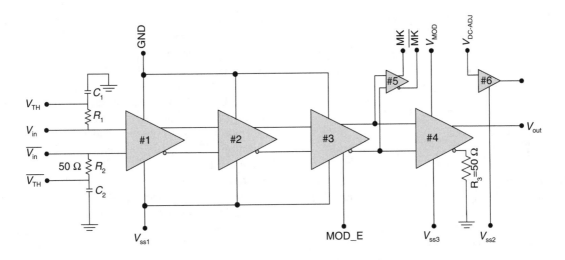

Figure 10–11. Lucent Technologies LG1626DXC modulator driver circuit. Courtesy Agere Systems (Lucent Technologies).

Pin Descriptions, *(Continued)*

Pin	Description	Symbol
21	Output modulation control input	V_{MOD}
22, 23	V_{ss1} supply (–5.2 V)	V_{ss1}
24	**Threshold control**	V_{TH}

Applications:

Modulator drivers can be used as:

a) optical transmitters

b) SONET/SDH

c) SONET/SDH testing equipment

TABLE 10–4 Electrical Characteristics (25°C)

Parameters	Minimum	Typical	Maximum	Units	Symbol
Data input voltage (p–p)	0.3	0.6	1.0	V	V_{in}
Voltage control (mod. output)	–5.5	—	–4	V	V_{mod}
Modulated output voltage (max)	2.7	—	3	V	$V_{out(max)}$
Modulated output voltage (min)	0	—	0.2	V	$V_{out(min)}$
Rise/fall time (output)	—	90	—	ps	t_r, t_f
Power supply voltage	–5.5	–5.2	–4.9	V	$V_{ss1}, V_{ss2}, V_{ss3}$
Power supply current	0.1	0.14	0.18	A	I_{ss}
Mark density	—	–0.5	—	V	MK
DC offset (max)	1.2	—	1.5	V	$V_{out(DC-offset)max}$
DC offset (min)	0	—	0.1	V	$V_{out(DC-offset)min}$
Offset voltage control	–5.5	—	–3	V	$V_{DC-Adjust}$

TABLE 10–5 Absolute Maximum Ratings (25°C)

Parameters	Minimum	Maximum	Units	Symbol
Supply voltage	—	5.7	V	V_S
Input voltage	GND	V_S	V	V_I
Power dissipation	—	1	W	P_D
Operating temperature	0	100	°C	T_C
Storage temp. range	–40	125	°C	T_{stor}

The electrical characteristics of the modulator driver circuit are given in Table 10–4.

The driver circuit absolute maximum ratings are given in Table 10–5.

10.7 DFB LASER DIODE WITH PMFs FOR EXTERNAL MODULATORS

Lucent Technologies, a leader in optical device design and manufacturing, has developed optical modules that incorporate DFB laser diodes and PMFs that are ready to be connected directly to MZ LiNbO$_3$ optical modulators. The D254xPxLaser 2000 isolated DFB laser module with PMF is one such device. It is specifically designed to operate in conjunction with an MZ LiNbO$_3$ optical modulator, without a polarization controller. It was mentioned previously that optical signal polarization is essential because it allows the laser diodes to operate in a CW mode with a very narrow line width. This arrangement is very critical to system performance, because it contributes to significant reductions or elimination of pulse broadening (chirping). Chirping is a detriment to optical systems that are intended for applications such as high speed digital transmission, CATV, and undersea optical lines. The laser diode incorporated into the module is usually an MQW structure that exhibits optical power output above the 50 mW range. A

schematic diagram of such an MQW-DFB laser diode is illustrated in Figure 10–12.

One of the basic characteristics of this family of laser diodes is the incorporation of an optical isolator within their structures. Such an isolator reduces the reflection of the light back to the diode at about –30 dB. Furthermore, the incorporation of a thermoelectric cooler (TEC) guarantees normal laser operation independent of temperature variations. That is, the temperature inside the module is maintained at a constant 25°C under environmental temperature variations of between –40°C and +65°C. The module also includes an InGaAs PIN photodetector used as a **back facet** monitor. The function of the photodetector is to monitor the optical emission at the back of the laser facet, and to control the optical power launched into the polarization maintenance fiber by using a control circuit.

The electrical characteristics of the above module are shown in Table 10–6.

The optical characteristics of the module are shown in Table 10–7.

The absolute maximum ratings of the module are shown in Table 10–8.

SUMMARY OF DESIGN CHARACTERISTICS

a) high performance MQW DFB laser diode

b) 14 pin butterfly package

c) operating temperature range: $-40°C$ to $+65°C$

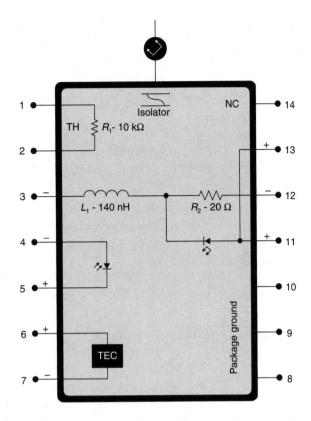

Figure 10–12. Schematic of an MQW DFB laser diode. The pin description of the device is in the table below. Courtesy Agere Systems (Lucent Technologies).

MQW DFB Laser Diode Pin Descriptions

Pin-number	Description
1	Thermistor
2	Thermistor
3	Laser cathode DC bias (−)
4	Back facet monitor control anode (−)
5	Back facet monitor control cathode (+)
6	Thermoelectric cooler (+)
7	Thermoelectric cooler (−)
8	Case ground
9	Case ground
10	Case ground
11	Laser anode (+)
12	RF laser input cathode (−)
13	Laser anode (+)
14	No connect

d) InGaAs PIN photodetector for back facet monitor control

e) integrated optical isolator

f) low **threshold current**

g) PMF

h) directly coupled to MZ LiNbO$_3$ optical modulator without the need for separate polarization controller

Applications

10 Gb/s MZ LiNbO$_3$ Modulator. An MZ LiNbO$_3$ optical modulator, designed to operate at 10 Gb/s data rates, is illustrated in Figure 10–13.

The device in Figure 10–13 and its associate biasing circuitry are designed to operate as a single-mode, externally modulated circuit at long wavelengths. The CW optical

TABLE 10–6 Electrical Characteristics (25°C)

Parameters	Minimum	Typical	Maximum	Units	Symbol
Forward voltage	—	1.3	2.0	V	V_{LF}
Threshold current	—	20	50	mA	I_{TH}
Current above threshold	—	350	400	mA	—
Slope efficiency	0.1	0.14	—	mW/mA	η
Dark current	—	0.01	0.1	μA	I_D
Input impedance	—	25	—	Ω	Z_{in}
TEC current	—	—	1.5	A	I_{TEC}
TEC voltage	—	—	3.5	V	V_{TEC}

TABLE 10–7 Optical Characteristics

Parameters	Minimum	Typical	Maximum	Units	Symbol
Peak optical power	40	—	—	mW	P_{peak}
Center wavelength	1540	—	1560	nm	λ_C
Line width	—	1	3	MHz	$\Delta\lambda_C$
Side mode suppression	35	—	—	dB	SMSR
Optical isolation	30	—	—	dB	
Optical polarization extinction ratio	20	—	—	dB	
RIN	—	–165	–155	dB/Hz	RIN

TABLE 10–8 Absolute Maximum Ratings

Parameters	Minimum	Maximum	Units	Symbol
Laser reverse voltage	—	2	V	V_{RLmax}
DC forward current	—	500	mA	I_{FLmax}
Operating case temperature	–40	65	°C	t_C
Photodiode reverse voltage	—	10	V	V_{RPDmax}
Photodiode forward current	—	2	mA	I_{FPmax}

signal generated by a DFB laser diode is converted into a time varying optical signal by the external modulator. Some of the characteristic features of such a modulator are as follows:

a) operating wavelength: 1550 nm

b) bandwidth: 15 GHz

c) operating temperature: 0°C – 70°C

d) 43 Ω (minimizing reflections)

The typical electrical and optical characteristics of the modulator are listed in Table 10–9.

The modulator absolute maximum ratings are listed in Table 10–10.

10 Gb/s MZ LiNbO₃ external modulators can be used in applications such as,

a) telecommunications

b) **SONET**/SDH, OC-48/STM-16, OC-192/STM-64

c) long and ultra-long distance

d) undersea

e) CATV

f) analog video

g) digital video

Engineers at Lucent Technologies are currently engaged in the development of a 40 Gb/s MZ LiNbO₃ optical modulator.

10.8 ELECTRO-ABSORPTION MODULATED LASERS

Three optical modulation schemes for modulated lasers exist: the directly modulated laser diode, the externally modulated laser diode, and the **electro-absorptive modulated laser (EML)** diode.

The development of the EML allowed longer transmission distances with a minimum degree of pulse broadening (chirping). Because directly modulated laser diodes exhibited large amounts of chirping, they were considered insufficient for long distance optical transmissions. On the other hand, EML modules that exhibit chirp levels of about 0.2 A° at 2.5 Gb/s bit rates over distances in excess of 400 km were considered ideal for long haul applications.

An EML module incorporates a CW laser diode and an integrated modulator in a single chip. This module is designed to complement the external MZ LiNbO₃ modulator, offering certain advantage such as ease of implementation and reasonable cost. The device is intended for long

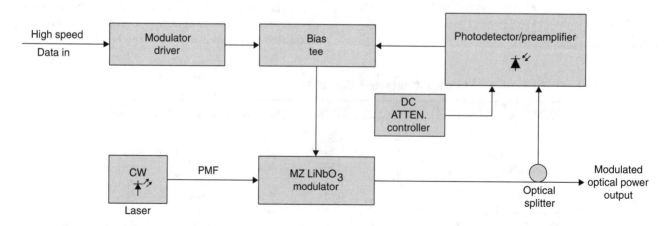

Figure 10–13. A 10 Gb/s, three-port MZ LiNbO₃ modulator.

TABLE 10–9   Electrical and Operating Characteristics

Parameters	Minimum	Typical	Maximum	Units
Operating wavelength	1525	1550	1565	nm
Insertion loss	3	4	6	dB
Bandwidth	8	10	15	GHz
Extinction ratio (at DC)	20	27	—	dB
Extinction ratio (at RF)	—	12	—	dB
$V\pi_{DC}$	2.5	3.0	4.5	V
$V\pi_{RF}$	3.5	4.0	5.5	V
Return loss (optical)	—	—	−35	dB
Return loss (electrical)	—	−15	−6	dB
Over the operating bandwidth				
Electrode impedance	—	43	—	Ω

TABLE 10–10   Absolute Maximum Ratings

Parameters	Minimum	Maximum	Symbol	Units
Optical input power	—	30	P_{in}	mW
Operating temperature	0	70	t_{op}	°C
Storage temperature	−40	85	t_{stg}	°C
Voltage (RF in)	—	20	V_{RF}	V
Voltage (RF in)	−20	20	$V_{DC(RF)}$	V

distance optical transmissions applications (600 km), and can be used for 2.5 Gb/s or 10 Gb/s bit rates with the addition of an integrated driver circuit. EML modules are designed to operate at the 1550 nm wavelength window and at discrete wavelengths across the ITU-T frequency grid.

One critical element that must be taken into consideration when an EML module is employed in an optical transmitter is the required laser diode forward biasing current. This biasing current ultimately determines the optical power output generated by the laser. Lucent Technologies' E2500 series EML module requires forward current levels based on applications between 50 mA and 100 mA. Below or above the current limits, the device will not perform the way it was intended to. For example, if the biasing current is less than 50 mA, the laser diode exhibits a substantial increase in the spectral line width. This line width increase is due to a reduction of the laser relaxation oscillation frequency. If, on the other hand, the biasing current is above the 50 mA level but below 100 mA, the device demonstrates improved performance in areas such as optical power output and noise. However, there is a slight decrease of the extinction ratio. Therefore, trade-offs between required optical power and noise levels (RIN) on the one hand and extinction ratio on the other, must be considered. The EML module performance can be further enhanced through modulator applied voltage optimization. That is, in order to reduce chirp, a small voltage level representing the 1-bit is required. However, a small ON state voltage contributes to a reduction of the optical output power. Again, a compromise must be reached between op-

tical power requirements and chirp during the design process.

Biasing Characteristics of EML Modules

The transfer function of an EML modulator is illustrated in Figure 10–14.

Explanation

CASE 1.   Assume the electric signal applied at the input of the modulator to be free of noise and distortion and with the ON state set at zero volts. The generated output optical signal is identical to the input electrical signal (Figure 10–15).

CASE 2.   Assuming that the electrical signal at the ON state is somewhat distorted and noisy, it can be intentionally reduced to some negative value (Figure 10–16). This voltage reduction is intended to reduce chirp. Because the input signal swings symmetrically around the midpoint of the transfer function (almost linear), the output optical signal will be identical to the input electrical signal, with noise and distortion also present.

CASE 3: IMPEDANCE OPTIMIZATION.   The degradation of the input electrical signal applied at the input of the modulator is mainly the result of impedance mismatch between the driver circuit output port and the modulator input port. The driver circuits are designed to interface with 50 Ω loads at all times. However, the modulator input impedance varies in

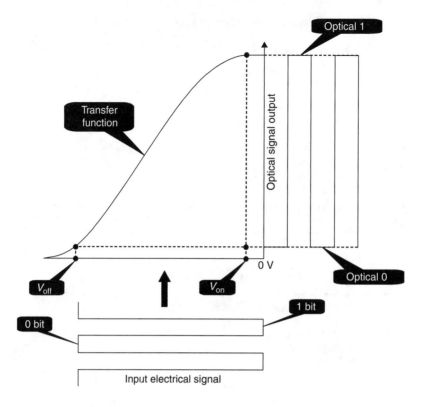

Figure 10–14. EML modulator transfer function.

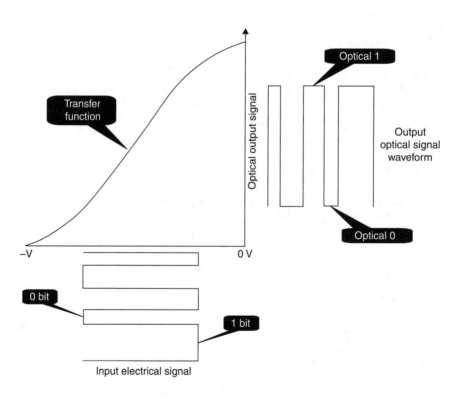

Figure 10–15. Modulated output optical signal *v.* ideal input electrical signal.

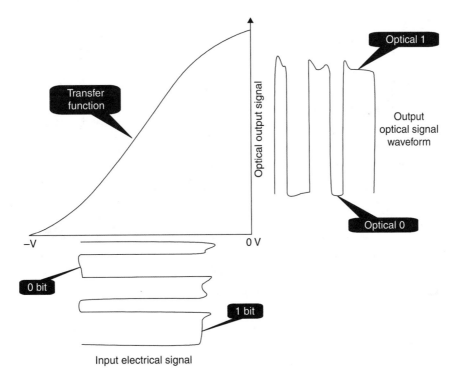

Figure 10–16. Modulated output optical signal *v.* non-ideal input electrical signal.

accordance with the applied input electrical signal between the ON state and OFF state. This difference induces impedance mismatch between the driver and the modulator circuits, which results in a substantial deterioration of the input electrical signal. Sophisticated designs incorporate passive components, such as inductors and capacitors, for impedance optimization between ON and OFF input signal swings. Through impedance matching, input signal distortion is substantially reduced.

Another futuristic design incorporates a dynamic modification of the modulator transfer function by reducing *dL/dV* at the ON state. Such a reduction will practically eliminate the noise and distortion at the output optical signal (Figure 10–17).

EML performance characteristics, such as optical power output, chirp, extinction ratio, and input signal distortion can be substantially improved by controlling the laser diode bias current and impedance matching, and modifying the transfer function. An EML module incorporating an integral driver is illustrated in Figure 10–18.

The module in Figure 10–18 is composed of an electro-absorptive modulator, a **thermistor**, an integral driver, a CW laser diode, a monitor photodiode, and a thermoelectric cooler.

Operation

a) integral driver IC: controls the function of the modulator

b) *The modulator offset:* controls the bias voltage applied to the modulator at the ON state and OFF level

c) *Peak current monitor:* intended to perform the peak current monitor function

d) *Duty cycle monitor:* intended to perform the duty cycle monitor function

e) *Duty cycle adjust:* through this input, the pulse width can be adjusted to the required level, normally 50%. It can also adjust (slightly increase) eye crossing, thus achieving maximum performance.

f) *Thermistor/laser/case:* facilitates the much-needed common ground between the thermistor, the laser, and the case

g) *RF-input:* set for a 50 Ω input impedance with a required input voltage of between 0.5 V and 1.0 V

A block diagram of a driver IC is illustrated in Figure 10–19.

Applications

A 10 Gb/s EML Module. E2560/80 10 Gb/s EML modules are designed and fabricated by Lucent Technologies. The E2560 does not incorporate a driver IC, while the E2580 does. The module is designed for TDM and DWDM optical schemes, and incorporates a CW laser diode and an electro-absorptive modulator. It can achieve transmission distances of between 40 km and 80 km before amplification. If coupled to a number of optical amplifiers (EDFAs), the transmission distance can be increased to perhaps 1000 km. The device also includes a rear facet monitor photodiode, a thermoelectric cooler, a thermistor, and an optical isolator. The module com-

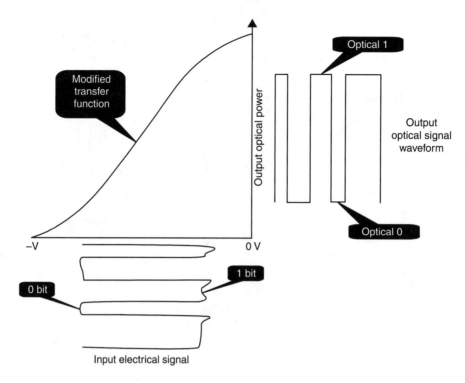

Figure 10–17. Modulator imperfect input electrical signal *v.* perfect output optical signal.

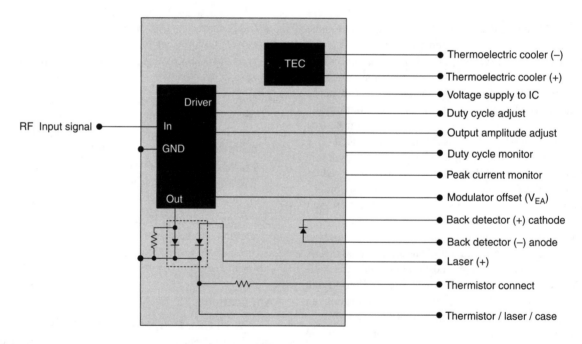

Figure 10–18. EML module with internal driver. Courtesy Agere Systems (Lucent Technologies).

plies with the ITU-T frequency grid for DWDM operating schemes (10 Gb/s per channel). The device also exhibits 100 GHz channel spacing and excellent wavelength stability. The electrical and optical operating characteristics are listed in Table 10–11.

The electrical/optical operating characteristics of the individual circuits within the module are listed in Tables 10–12a, b, and c.

The EML module absolute maximum ratings are listed in Table 10–13.

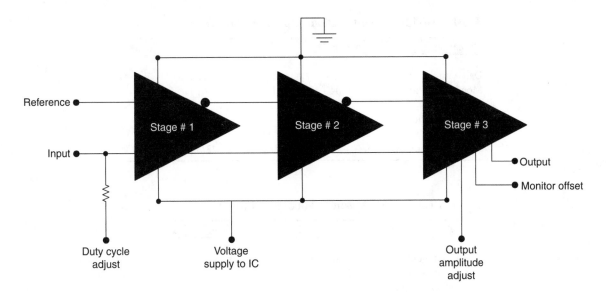

Figure 10–19. A block diagram of a three-stage driver.

TABLE 10–11 Electrical/Optical Operating Characteristics of the EML-Module

Parameters	Minimum	Maximum	Symbol	Units
Forward voltage	—	2.2	V_F	V
Threshold current	5	35	I_{TH}	mA
Operating current	50	100	I_{OP}	mA
Threshold power	—	80	P_{TH}	μW
Fiber output power (peak)	1	—	P_{PK}	dBm
Wavelength (peak)	1530	1563	λ_o	nm
Side mode suppression ratio	30	—	SMSR	dB
Dispersion penalty at BER $= 10^{-10}$	—	2	DP	dB

TABLE 10–12a Modulator Driver

Parameters	Minimum	Maximum	Symbol	Units
Extinction ratio	11	—	ER_{RF}	dB
Return loss (0 GHz–6 GHz)	10	—	S_{11}	dB
Return loss (6 GHz–8 GHz)	7	—	S_{11}	dB
Return loss (8 GHz–10 GHz)	5	—	S_{11}	dB
Bandwidth (-3dB)	11	—	B_W	GHz
Input voltage (pk to pk)				
AC coupled	0.5	1.0	V_{in}	V
Rise/fall time	—	40	t_r/t_f	ps

TABLE 10–12b Thermistor

Parameters	Minimum	Maximum	Symbol	Units
Resistance	9.5	10.5	R_{therm}	kΩ
Thermistor current	10	100	I_{TC}	μA
Thermoelectric cooler current	—	1.1	I_{TEC}	A
Thermoelectric cooler voltage	—	2.6	V_{TEC}	V
Thermoelectric cooler power	—	2.9	P_{TEC}	W
Thermoelectric cooler capacity	55	—	ΔT	°C

TABLE 10–12c   Monitor Diode/Optical Isolator/Package

Parameters	Minimum	Maximum	Symbol	Units
Dark current	—	0.1	I_d	μA
Monitor current	0.04	1.1	I_M	mA
Capacitance	—	25	C	pF
Optical isolation	30	—	—	dB
Wavelength/ case temperature	—	0.5	$d\lambda/dT$	pm/°C

TABLE 10–13   Absolute Maximum Ratings

Parameters	Maximum	Units
Laser diode forward current (CW)	150	mA
Laser diode reverse voltage (CW)	2	V
Optical output power (CW)	10	mW
Modulator reverse voltage	5	V
Modulator **forward voltage**	1	V
Monitor diode reverse voltage	10	V
Monitor diode forward voltage	1	V
Operating temperature	−10 to + 85	°C
Storage temperature	−40 to + 85	°C

SUMMARY

In Chapter 10, the need for external modulation in high speed optical communications systems was identified. The chapter began with the description of the principles and operating characteristics of the Mach Zehnder interferometer, leading to the development of the Mach Zehnder LiNbO$_3$ optical modulator. A substantial part of the chapter was devoted to the detailed description of the design steps leading to the full development of a Mach Zehnder LiNbO$_3$ optical modulator.

The chapter continued with the introduction of Lucent Technologies Mach Zehnder LiNbO$_3$ optical modulator, followed by a brief description of various modulation driver circuits. Because external modulators require special DFB lasers and polarization maintenance fibers, the principles and operating characteristics of such devices was briefly discussed. The chapter concluded with a description of the principles of operation and performance characteristics of electro-absorption modulated lasers (EML).

QUESTIONS

Section 10.1

1. Explain the need for external modulation in optical communications systems.

2. How does direct modulation broaden the spectral line width of the laser diode?

3. Describe the ways by which chirping can be reduced or altogether eliminated.

4. Name the two types of external optical modulators.

Section 10.2

5. With the assistance of a schematic diagram, describe the operation of a Mach Zehnder interferometer.

Section 10.3

6. Draw a cross sectional area of the MZ LiNbO$_3$ optical modulator and explain its operation.

7. List all the parameters incorporated into the modulator transfer function and elaborate on their interrelationships.

8. What does the modulator transfer function curve represent? Explain.

9. List all the benefits that arise from the utilization of external modulators in optical communications systems.

10. Draw a block diagram of a modulator driving circuit and briefly explain its operation.

11. Name a few of the optical systems and networks that external modulators are considered absolutely essential for.

Section 10.4

12. Draw the AT&T flow chart for designing MZ LiNbO$_3$ modulators and describe, in detail, the design process.

13. Why is a PMF required between the modulator and the laser diode?

14. What is the role of titanium in the modulator diffusion process?

15. Explain the function of the SiO$_2$ buffer in the fabrication of optical modulators.

16. What will happen if a modulator designed to operate in the 1550 nm wavelength window is coupled to a 1310 nm laser source?

17. How many ports does an optical modulator require?

18. Explain the role of each port.

19. Define *extinction ratio*.

20. Why must this ratio be kept to a minimum?

21. How can the extinction ratio of an optical modulator be further reduced?

22. Define *bandwidth* of an optical modulator.

23. What causes the appearance of ripple at both low and high frequencies?

24. Describe back reflections.

25. Name the source or sources of back reflections.

26. What are the negative effects of back reflections in optical modulators?

27. Describe, in detail, the function of voltage swing (V_π).

28. What is the role of the external feedback loop?

29. Sketch an automatic bias control circuit and explain, in detail, its operation.

30. Use the appropriate mathematical formulas and describe the pulse broadening process of an MZ modulator.

Section 10.5

31. Describe the characteristics of the optical signals required at the input of an external modulator.

32. Why are DFB lasers used to operate with MZ modulators?

33. How is optical beam polarization achieved in external modulators?

34. When do MZ LiNbO$_3$ modulators require two electric ports?

35. Why is electric port input impedance matching essential? Name this impedance value.

36. Describe the reasons why a single electrode input requires a capacitor in series with the internal resistor of 43 Ω.

Section 10.6

37. Draw a block diagram of a modulator driver circuit and briefly explain its operation.

38. For the driver module in Question 37, explain why a current to voltage circuit is required.

39. What is the function of the AC coupling capacitor?

40. What is the function of the bias tee?

41. Draw and briefly explain the operation of the LG1626DXC driver circuit.

Section 10.7

42. Name the distinct advantages of using the Lucent Technologies D254xPx laser module.

43. Why is optical signal polarization critical? Explain.

44. Explain the necessity of incorporating an MQM laser into the D254xPx.

45. What is the function of a thermoelectric cooler?

46. Describe the function of the InGaAs PIN photodetector.

47. List the general performance characteristics of MZ optical modulators.

48. Draw a block diagram of a 10 Gb/s MZ LiNbO$_3$ modulator, list its operating characteristics, and briefly explain its operation.

49. Name some of the applications that the modulator in Question 48 is suitable for.

Section 10.8

50. Name the three types of electro-absorption modulators.

51. What are the main building blocks of electro-absorption modulators?

52. List the advantages and disadvantages of electro-absorption modulators compared to MZ LiNbO$_3$ modulators.

53. Describe the critical element to be taken into account when an electro-absorption modulator is employed in an optical transmitter.

54. Name the three critical operating characteristics of an electroabsorption modulator, and explain their interrelationships.

55. Draw and interpret the EAM modulator transfer function.

56. Draw and interpret the EAM modulator output/input signal power relationship.

57. What is the main cause of degradation of the optical signal at the input of a modulator?

58. How can noise and distortion at the output of an EAM be altogether eliminated?

59. Describe the method through which the chirp extinction ratio and signal distortion of an EAM can be substantially improved.

60. Draw a block diagram of an EAM that incorporates an internal driver, and explain its operation.

61. What is the design objective of the Lucent Technologies E2560/80ELM module?

62. List, and explain, some of the most important optical and electrical operation characteristics of the E2560/80ELM module.

11
MULTIPLEXING

OBJECTIVES

1. Establish the role of multiplexing in optical communications systems

2. Present the various modulation techniques

3. Describe the differences and similarities between FDM, TDM, CDMA, and WDM

4. Establish the need for advanced multiplexing techniques required for high speed optical communications systems

5. Describe, in detail, the function and operating characteristics of dense wavelength division multiplexing (DWDM)

6. Describe the principle of operation and performance characteristics of add/drop multiplexing

7. Define *demultiplexing*

8. Explain the function of demultiplexing in optical communications systems

9. List the main building blocks of an optical demultiplexer module and explain the function of each block

10. Describe the principles of operation and performance characteristics of ultra high speed demultiplexers

KEY TERMS

add/drop multiplexer
ALOHA
AND gate
array waveguide grading (AWG)
ATM
attenuation
attenuation coefficient

bandwidth length product
bandwidth polarization dependence
bipolar code sequence
CCITT
CEPT
channel baseband
channel power

channel separation
channel spacing
channel wavelength accuracy
charge pump current
chirp
chromatic dispersion
clock jitter

clock synthesizer
code division multiplex access (CDMA)
code division multiplexing (CDM)
coherent
coherent optical signal processing

11.1   INTRODUCTION

Multiplexing is the process whereby several optical signals are combined and then transmitted through a single fiber. The continuous drive for higher system capacities, flexibility, and simplicity necessitated the implementation of multiplexing in optical fiber links. Multiplexing, as applied to optical fiber systems, underwent a transformation from electrical, to partially optical, to fully optical. Initially, the three well known and widely applied multiplexing techniques in traditional digital communications systems such as **frequency division multiplexing (FDM), time division multiplexing (TDM),** and **code division multiplexing (CDM)** were utilized before the development of fully optical networks. However, beyond these three methods of multiplexing and before the development of complete optical networks, the **space division multiplexing (SDM)** technique was developed and implemented with limited success.

The basic design philosophy of an SDM system is to transmit only one signal through a single fiber, then to combine a number of fibers into a single cable. Although this method of multiplexing increases the overall link capacity, additional component requirements and extra provisions for optimum channel isolation rendered the technique uneco-

nomical and impractical. In the following sections, the three electrical multiplexing techniques (FDM, TDM, and CDM) will be discussed briefly, while the latest of all optical methods of multiplexing (**wavelength division multiplexing (WDM)** and **dense wavelength division multiplexing (DWDM)**) will be fully explored.

11.2   FREQUENCY DIVISION MULTIPLEXING

An FDM system incorporates several **voice channels.** Each voice channel must have a bandwidth of 4 KHz. These voice channels are combined to form a single FDM signal at the output of the multiplexer. An LPF of 4 KHz bandwidth is required to confine the voice signal bandwidth to 4 KHz, thus eliminating any frequency components outside the 4 KHz limit.

The **subcarrier** oscillator generates the subcarrier and the pilot frequencies required for modulation. Each modulator modulates the appropriate subcarrier with the corresponding voice channel. The output band-pass filter makes sure that each channel occupies the preassigned frequency band. This band-pass filter contributes significantly to an efficient and accurate demultiplexing process.

The pilot or **reference frequency** is required to ensure that the demultiplexing subcarrier oscillators generate the exact same frequencies as the multiplexing subcarrier oscillators. Any frequency difference will result in signal distortion. A 4 KHz frequency **guard band** is assigned for each voice channel, in order to avoid channel overlap.

CCITT Groups

THE STANDARD GROUP. In its G.232 recommendation, the **International Telegraph and Telephone Consultative Committee (CCITT),** a branch of the processor of the **international telecommunication union (ITU-T),** defined a **standard group** as twelve voice channels that occupy the frequency range between 60 and 108 KHz (each voice channel has a bandwidth of 4 KHz). The usual bandwidth of a voice channel is between 300 and 3400 Hz, rounded off to 4 KHz (Figure 11–1).

THE STANDARD SUPER GROUP. The **standard super group** consists of five standard groups, totaling sixty voice channels, and occupies the frequency range between 312 KHz and 552 KHz. The five subcarrier frequencies required for the formation of a standard super group are 420 KHz, 468 KHz, 516 KHz, 564 KHz, and 612 KHz (Figure 11–2).

THE STANDARD MASTER GROUP. The **standard master group** FDM scheme incorporates five super groups, totaling three hundred voice channels, and occupies the frequency range between 812 KHz and 2.044 MHz (Figure 11–3).

STANDARD SUPER MASTER GROUP. The **standard super master group** is composed of three super groups and occupies the frequency range between 8.516 MHz and 12.388 MHz (Figure 11–4). The three subcarrier frequencies for the standard super master group are 10.56 MHz, 11.88 MHz, and 13.2 MHz. Generation of the above frequencies is accomplished by utilizing the harmonic components of standard 4 KHz, 12 KHz, and 124 KHz oscillator frequencies. The harmonics of a 4 KHz oscillator cover the standard group range between 64 KHz and 108 KHz. The harmonics of a 12 KHz oscillator are utilized to cover the frequency range between 420 KHz and 612 KHz (super group). The harmonics of a 124 KHz oscillator are used to cover the range between 1052 KHz and 2044 KHz (master group).

In an FDM system, the presence of pilot frequencies is absolutely necessary for two reasons. One is to regulate the system level and maintain level fluctuations within the ±0.5 dB range, and the other is to maintain a constant frequency generation for the entire spectrum of frequencies required by all standard FDM groups. None of these frequencies must deviate more than ±2 Hz from their preassigned range. In all FDM systems, a master frequency source is used to generate all the required FDM frequency components. The receiver section of the system must be in absolute synchronization with the transmitter. This is

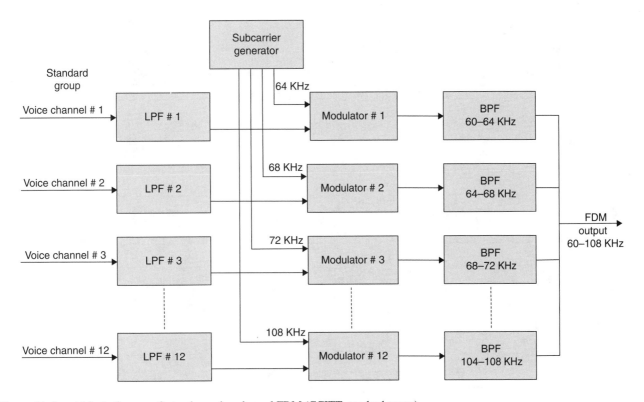

Figure 11–1. A block diagram of a twelve-voice-channel FDM (CCITT standard group).

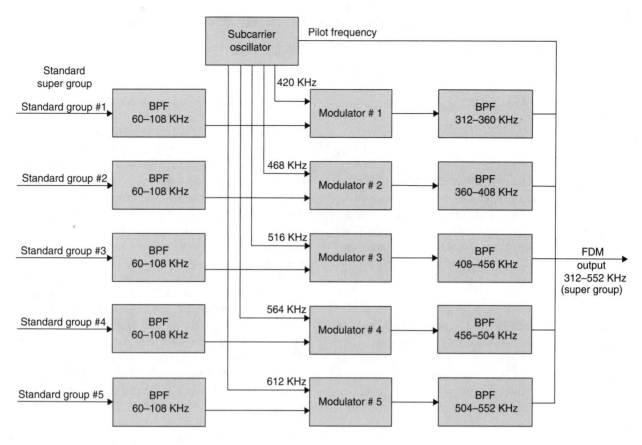

Figure 11–2. A block diagram of a sixty-voice-channel FDM (CCITT standard supergroup).

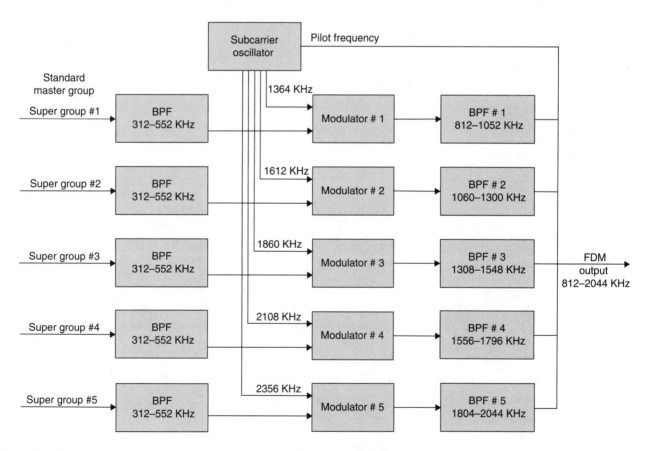

Figure 11–3. A block diagram of a 300-voice-channel FDM (CCITT standard master group).

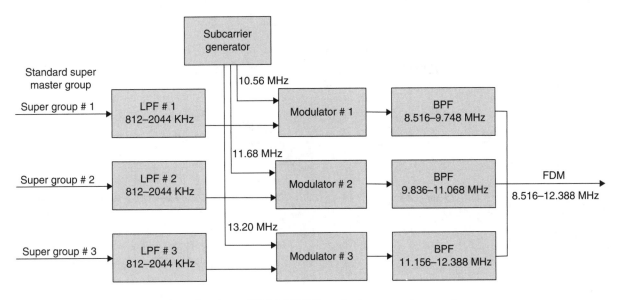

Figure 11–4. A block diagram of a 900-voice-channel FDM (CCITT standard super master group).

accomplished through the pilot frequency. For digital transmissions (TDM), synchronization is achieved through coherent detection.

11.3 TIME DIVISION MULTIPLEXING OF PCM SIGNALS

In a large communications system, several voice signals are combined and then transmitted through a single channel. The processes of digital signal combining is referred to as digital multiplexing, or more specifically, time division multiplexing (TDM) (Figure 11–5).

Figure 11–5 illustrates the TDM process. Here, three voice channels are sampled in a logical sequence and at a rate satisfying Nyquist's sampling relationship. The combined amplitude-modulation signal is shown at the output. A more detailed logic circuit that implements the TDM process is shown in Figure 11–6.

Figure 11–6 illustrates a 4-channel **PCM-TDM** logic circuit. In this circuit, each **AND gate** will be enabled only when all its inputs go to logic-high. The control logic provides the required timing sequence, which enables each AND gate in accordance to the timing sequence.

In 1957, the CCITT was established in order to set standards and make recommendations that relate to matters of international communications to the International Telecommunications Union (ITU-T), the coordinating body of all member countries. The international consultative committees are divided into study groups. Each study group reports its findings to an assembly, which generates the final recommendations. These recommendations are accepted by all member countries and have established the basis for the present operations and the future development of global communications. Recommendations G.732 and G.733 are referred to as primary PCM systems.

The North American (BELL) 24-Channel T1 System (G.733)

This system accommodates 24 voice channels. It is time division multiplexed and has a sampling rate of 8000 samples per second per channel and 8-bit word for coding.

The T1 system generates a primary bit rate equal to 1.544 Mb/s, as follows:

24 voice channels × 8-bits/word = 192 bits

Add one bit for framing: 192 + 1 = 193

The frame or synchronization bit is necessary in order for the receiver to recognize which bit corresponds to which of the 24 channels. Therefore, 193 bits × 8000 samples per second equals 1.544 Mb/s.

Nyquist's Theorem

In 1933, Henry Nyquist developed a mathematical relationship between the analog input signal and the sampling frequency as follows:

$$f_s = 2B_W$$

where f_s is the clock frequency (Hz) and B_W is the bandwidth of the sampled signal (Hz). This relationship, known as the **Nyquist theorem,** determines the minimum required clock frequency necessary for sampling a continuous time varying waveform without the possibility of distortion at the signal reproduction stage.

Time frame

The first bit for each of the 24 channels, plus the synchronization bits, constitute the **time frame** of the scheme. Time frame is calculated as follows:

Each voice channel has a bandwidth of 3.3 KHz, which is rounded off to 4 KHz. In order to satisfy Nyquist requirement, the voice channel is sampled with a sampling

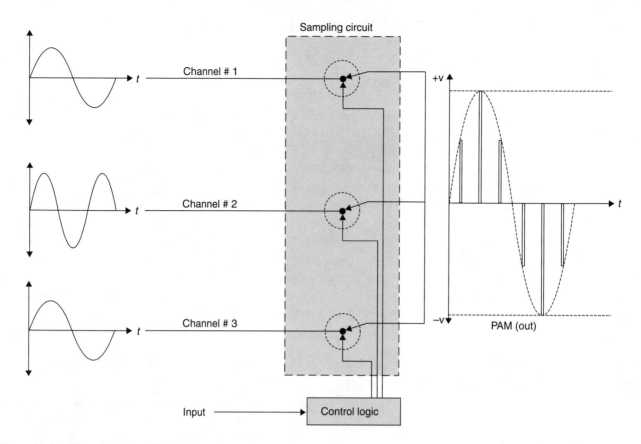

Figure 11–5. Time division multiplexing.

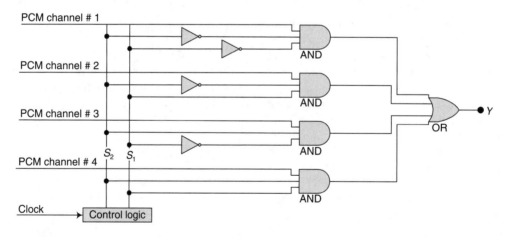

Figure 11–6. A simplified logic diagram of a TDM transmitter circuit.

rate (f_s) equal to 4 KHz $\times$ 2 = 8000 samples per second (Equation (11–1))

$$\text{time frame } t(s) \;=\; \frac{1}{8000} = 125 \ \mu s \qquad \textbf{(11–1)}$$

Therefore, the T1 system is composed of a total of 8000 frames that each occupy a time slot of 125 μs. Each time slot accommodates 193 bits of information. The pulse duration of each bit is 125 μs, divided by 193, equals 0.647668393 μs. From this figure, the system bit rate can now be established (Equation (11–2)).

$$\text{T1 (24-channel) bit rate} \;=\; \frac{1}{\text{bit}-\text{time}} \qquad \textbf{(11–2)}$$

$$\text{T1 (24-channel) bit rate} \;=\; \frac{1}{\dfrac{\text{frame}-\text{time}(s)}{193 \text{ bits}}}$$

$$=\; \frac{1}{\dfrac{125 \times 10^{-6}\text{s}}{193 \text{ bits}}} = 0.647668$$

$$\therefore \text{T1 bit rate} = 1.544 \text{ Mb/s}$$

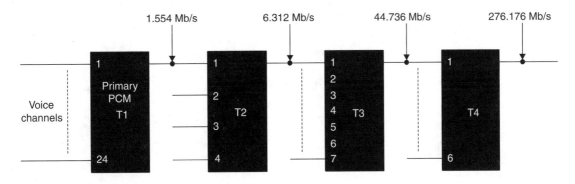

Figure 11–7. The North American digital hierarchy.

T1 systems are further combined to generate higher order of digital multiplexing hierarchies such as T2, T3, and T4 schemes (Figure 11–7).

The T2 scheme accommodates 96 voice channels and has a corresponding bit stream of 6.312 Mb/s. The T3 accommodates 672 voice channels and has a corresponding bit rate of 44.736 Mb/s. Finally, the T4, which has a total of 4032 voice channels, generates a bit stream of 274.176 Mb/s.

The second level digital hierarchy has a total bit stream of 6.312 Mb/s. If the four primary levels divide this second level, a 1.578 Mb/s is indicated instead of 1.544 Mb/s. The additional bits are required for framing, alignment, and control.

The Japanese Digital Hierarchy

Japanese and North American digital hierarchies are the same for T1 and T2, but differ for higher groups (Figure 11–8). The third level Japanese hierarchy accommodates five T2 groups instead of the seven for North America, and has a bit stream equal to 32.064 Mb/s. The fourth digital hierarchy accommodates three T1 groups instead of the six for North America, and has bit streams of 97.729 Mb/s. Furthermore, a fifth level is added, which generates a total bit stream of 397.2 Mb/s and has a voice channel capacity of 5760 channels, which is more than the 4232 voice channels for the North American system.

The European 30-Channel System CEPT (G732)

The European system accommodates 30 voice channels through TDM. It uses 8000 samples per second per channel **(sampling rate),** and 8-bit word for coding (Figure 11–9).

The European system generates a primary rate of 2.048 Mb/s, as follows:

30 voice channels $\times$ 8-bit quantization = 240 bits

Add 8 bits for signaling = 248 bits

Add 8 bits for framing = 256 bits

256 bits per frame $\times$ 8000 frames = 2.048 Mb/s

The frame period is equal to 125µs divided by 30 voice channels, plus two for signaling and framing, which gives 3.90625 µs (frame period) or 125µs, divided by 256 bits per frame, which generates a 488 ns bit period. Inverting 1/488 ns is equal to 2.048 Mb/s.

The clock used in the multiplexing process is referred to as the **writing clock,** and the clock used in the demultiplexing process is referred to as the **reading clock.** Both reading and writing clocks must be absolutely synchronized in order to avoid the possibility of error. Under absolutely ideal synchronization conditions, the ratio of data stream to synchronization bits is set as forty-five to one. In the event that the reading clock is slightly out of phase with the writing clock, an imbalance will occur between the writing and the reading rates. The **justification process** rectifies this rate of imbalance.

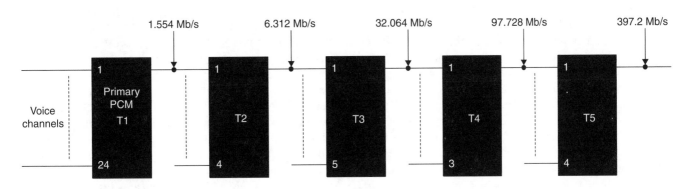

Figure 11–8. The Japanese digital hierarchy.

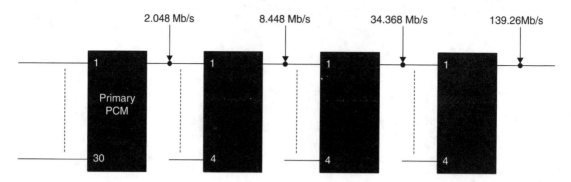

| 2.048 Mb/s | 8.448 Mb/s | 34.368 Mb/s | 139.26Mb/s |

Figure 11–9. The European digital hierarchy.

Justification is the process whereby a time slot is added to or subtracted from the frame in order to maintain the balance between writing and reading bits. Because two clocks are used in the multiplexing/demultiplexing process, three possibilities will exist. When the writing clock is slightly faster than the reading clock, the reading clock is unable to completely follow the writing clock, and one bit must be subtracted from the frame. Because of the reduction of a bit from the writing frame, the process is called **negative justification.** On the other hand, if the reading clock is faster than the writing clock, a bit must be added to the writing clock. Because the addition of a bit to the writing frame is required, the process is referred to as **positive justification.** If no difference exists between writing and reading clocks, no justification bits are required. The writing and reading clock drifts described above are referred to as multiplex-jitter.

11.4   CODE DIVISION MULTIPLEX ACCESS

Code division multiplex access (CDMA) techniques are used mainly for LANs. The basic objective in such an optical system is to utilize the available channel bandwidth for the highest possible data rate transmission, and to make the bandwidth accessible to the largest possible number of users.

In traditional communications systems that employ **time domain CDMA,** every bit of the input binary data sequence is encoded and the resulting waveform, which represents the destination address of the binary data, is broadened by a specific factor. Although this method can be effectively utilized at relatively high data rates, it exhibits certain limitations at higher data rates due to the performance of the electronic components employed in the system. To overcome the deficiencies, optical CDMA systems that have different encoding techniques have been proposed. Very large bandwidth expansion, required by the spread spectrum code division multiple access, can be accomplished through a combination of **incoherent optical signal processing** and optical fiber channel

transmission. The required code generation and correlation are accomplished through incoherent optical signal processing, while the optical channel provides the necessary wide bandwidth.

CDMA owes its development to the rapid growth of mobile and satellite communications systems. The utilization of code division multiple access in LANs allows for the implementation of schemes such as **ALOHA** and carrier sense multiple access with collision detection **(CSMA/CD).** The efficient utilization of the optical channel is based mainly on the fact that network users can access the channel in an asynchronous mode. That is, a large number of users can simultaneously access the same channel through the asynchronous mode of transmission without having to wait until the channel is idle. This is a definite advantage, considering that the alterative to TDMA, in which each user is assigned only a portion of the available channel. Therefore, the addition of new users can easily be implemented through a CDMA scheme in very dense, high rate LAN networks. The drive to accommodate an ever increasing number of users reflects a proportional increase of the system expansion capabilities, which translates into larger channel bandwidths and very narrow pulse width. Optical fibers, that exhibit a bandwidth length product (BL) of about 1 THz/km and attenuation of about 0.2 dB/km, are ideal for CDMA optical LAN applications. Performance of CDMA optical LANs is subject to electronic component limitations used to convert signals from electrical to optical and from optical to electrical. This significant performance limitation, encountered in such networks, can be remedied through the design of an all optical signal processing mechanism inserted at the points in the link where they are required most. In such schemes, optical encoders and decoders can be utilized in conjunction with single-mode optical sources and optical modulators. An all optical signal processing system eliminates the need for wide bandwidth photodetector diodes, and significantly reduces the probability of data congestion subject to optical sequence generation and correlation. Optical signal processing techniques in CDMA LAN systems are classified as either **coherent** or **incoherent.**

Coherent Optical Processing

In a **coherent optical signal processing** scheme, the optical pulses are converted to phase coherent code sequences by the optical encoder, then transmitted through a single fiber to the input of the optical decoder, where the opposite process takes place. This process allows for the use of bipolar instead of **unipolar code sequences. Bipolar code sequences** correlate to those used in mobile and satellite systems that utilize spread spectrum technology. Coherent optical decoders exhibit significant interference reduction from other code sequences, while allowing for better user access. The decoder interference reduction capability is subject to coherent pulse sequence characteristics such as amplitudes, polarization states, and phase shifts.

The critical factor for an absolute coherent correlation between the encoder and the decoder in a CDMA optical system is the time correlation between the laser diode and the encoder and decoder modules. Experimental work has shown that utilizing a laser diode that emits at the 1550 nm wavelength with a **full wave at half maximum (FWHM)** of 2 nm requires a matching delay of 798 µm or better for a single-mode fiber. Furthermore, because alignment of the individual pulses, in reference to vector and phase polarization within the pulse sequence, is absolutely essential, the utilization of polarization-maintaining fibers is necessary.

Incoherent Optical Processing

Incoherent optical processing is based on light intensity variations: light is either in the ON state or OFF state. This method of optical signal processing is much simpler than the coherent method. Because it is based on optical intensity variations of the decoder module, the environmental interferences encountered with coherent signal processing are eliminated. Furthermore, incoherent optical signal processing can only process signals that have a small code weight-to-length sequence ratio, which is required to maintain a minimum level of interference. Because it is very difficult to achieve interference reduction at acceptable levels with unipolar signals, the use of such a scheme will ultimately limit the number of simultaneous users. Based on the difficulties with incoherent optical signal processing, research has concentrated its effort on the development of a **synchronous CDMA** all-optical system for LANs.

However, either coherent and incoherent optical CDMA systems may be implemented, depending on data requirements and related technical and economic considerations. For example, coherent processing schemes are highly desirable for digital video transmission, but for voice and data transmission, the less difficult and less expensive incoherent processing scheme can be effectively implemented. Moreover, the multitude of data and traffic requirements can also be implemented through the utilization of both TDMA and synchronous code division multiple access (S/CDMA) schemes.

Synchronous-CDMA with Modified Prime Sequence

S/CDMA utilizes unipolar data (1,0). The 1-bit is represented by a code sequence waveform that represents the bit address. The length (N) of the sequence is subject to the bit period (T), which is subdivided into smaller units referred to as chips, whose number is determined by the required code sequence. Therefore, the number of chips is given by the ratio of the bit period (T) to the chip pulse duration (τ) (Equation (11–3)).

$$\text{number of chips} = \frac{T}{\tau} \qquad \textbf{(11–3)}$$

where T is the bit period(s) and τ is the chip period(s). At the decoder input, the received optical sequence is correlated with the decoder address. If the received signal has arrived at the desired address, autocorrelation will occur (high peak). On the other hand, if the received signal has arrived at the wrong address, cross-correlation will occur. It is therefore necessary that each receiver employs a strong autocorrelation function in order to differentiate the desired signals from undesired signals.

Performance: Synchronous CDMA v. Standard CDMA

In optical CDMA systems that utilize the **primary sequence code (PSC),** the number of **pseudo-orthogonal code sequences (POCS)** is equal to the prime number (P) used in primary sequence code. This number is also indicative of the maximum number of subscribers effectively utilizing the synchronous-CDMA optical system simultaneously and virtually error free. The maximum number of subscribers for such a system is expressed by Equation (11–4).

$$K = P - 1 \qquad \textbf{(11–4)}$$

where K is the maximum number of subscribers and P is the prime number used in the sequence. In the case where the number of subscribers becomes equal to or larger than P ($K \geq P$), the receiver will be unable to differentiate between autocorrelation and cross-correlation peaks, which increases the probability of error.

ERROR PROBABILITY ($P_{(e)}$). In the case where $K > P$, the **error probability** is expressed by Equation (11–5).

$$P_{(e)} = \frac{P}{2K^2}\left(1 - \frac{P^P}{P^2}\right) + \sum_{P+1}^{K}\frac{K}{2^K} \qquad \textbf{(11–5)}$$

where $P_{(e)}$ is the error probability, K is the number of users, and P is the prime number used in the sequence. For synchronous CDMA, the error probability is expressed by Equation (11–6).

$$P_{(e)} = \frac{1}{\sqrt{2\pi}}\int_{-\infty}^{\infty} e^{-y/2}\,dy\left(\frac{-P}{\sqrt{1.16(K-1)}}\right) \qquad \textbf{(11–6)}$$

where

$$\frac{1}{\sqrt{2\pi}} \int\limits_{-\infty}^{\infty} e^{-y/2} \, dy$$

is the cumulative distribution function, K = number of users, and P = prime number used in the sequence.

An error probability comparison of S/CDMA and CDMA is illustrated in Figure 11–10.

Figure 11–10 illustrates that S/CDMA schemes exhibit better overall error probability than CDMA schemes. Maintaining a fixed error probability of $P_{(e)} = 10^{-9}$ for both S/CDMA and CDMA schemes, research has shown the following results (Table 11–1).

The S/CDMA Optical System

A block diagram of an S/CDMA optical system is illustrated in Figure 11–11.

TABLE 11–1   Multiplexing Schemes $v.$ Number of Subscribers for $P_{(e)} = 10^{-9}$

	S/CDMA	CDMA
For $P = 31$	961 subscribers	31 subscribers
	31 simultaneous users	23 simultaneous users

OPERATION OF AN S/CDMA OPTICAL SYSTEM. S/CDMA links are designed to process 10 Mb/s of binary data. The 10 Mb/s input data is fed into the pulse generator, which produces a 2 ns pulse every time the 1-bit appears at its input. The 2 ns electric pulse is used to drive the laser diode, which generates a 2 ns optical pulse at the 1300 nm wavelength. The 2 ns optical pulse generated by the laser diode is equal to the desired chip width. The optical pulse is then fed into the input of the optical encoder designed to generate the required address sequence of the corresponding receiver. In the encoder, a 1×7 splitter is used to split the

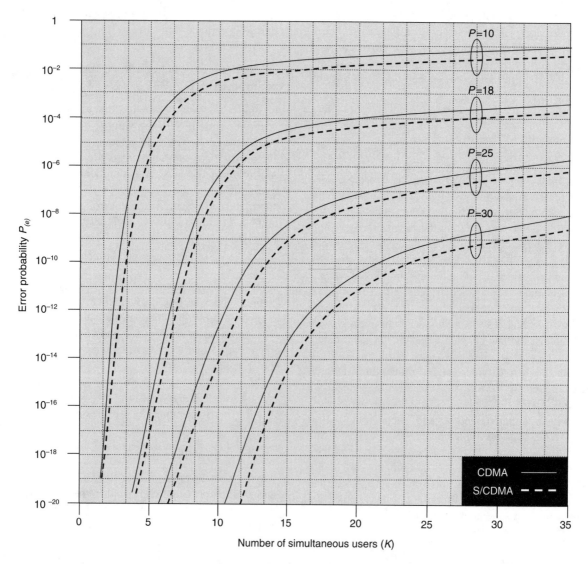

Figure 11–10.   Error probability comparison of S/CDMA and CDMA.

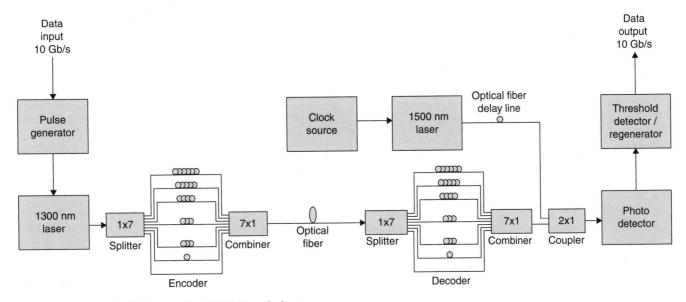

Figure 11–11. A block diagram of an S/CDMA optical system.

optical pulse, then delay it through seven-fiber delay lines designed to reflect only the 1 s at the corresponding receiver address. A 7 × 1 optical combiner is used to combine the delayed pulses and then form the desired code sequence. Furthermore, a directional coupler is used to process the formed optical code sequence through the fiber to the input of the photodetector diode in order for the signal to be converted back to electrical. The optical decoder is designed to recognize the exact address sequence of the transmitting encoder. The required absolute synchronization between the transmitter and the receiver is accomplished through the 2 nm pulse, triggering a 1500 nm laser diode. The 2 nm optical pulse generated by the laser diode is fed into an optical delay line, through a 2 × 1 directional coupler, and is transmitted to the receiver end as the clock signal. The clock signal must be correlated to the receiver peak autocorrelation function. A threshold detector is used to retrieve the autocorrelation peaks from the clock signal.

CDMA Optical Encoders and Decoders

A block diagram of an optical encoder is illustrated in Figure 11–12. It is composed of an optical power splitter, a number of parallel optical fiber delay lines equal to the prime number (P) used to generate the code sequence, and a power combiner.

OPERATION OF CDMA ENCODERS AND DE-CODERS. The optical pulse, generated by the laser diode, that represents the 1-bit is split into a number of pulses by the optical power splitter. Each pulse is then delayed by a fiber delay line controlled by an address selector. The delayed pulses are then recombined by the optical combiner to form the CDMA signal. If more than one channel is required, the same process is applied and the output of each of the channels is then fed into the multiplexer circuit to form the combined CDMA code sequence.

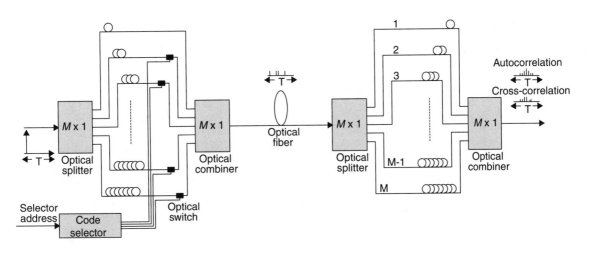

Figure 11–12. A block diagram of an optical encoder.

2^n Encoders and Decoders. Parallel/serial architecture is used for code sequence generation, selection, and correlation. 2^n encoders/decoders, consist of optical switches, a set of M address codes and an $M \times 1$ optical combiner. Each address code is configured to generate its own distinct code sequence (same for the decoder) and is composed of 2×2 passive couplers and fiber delays. At the input of the address code, the optical pulse is divided into two pulses: one delayed and the other nondelayed. The two pulses are fed into the input of the next 2×2 coupler, and then split again to 4 pulses. This process is repeated for a number of couplers. The total number of generated pulses is equal to 2^n, where n is the number of required couplers. The final code sequence is referred to as a 2^n code sequence. For example, if the prime number, $P=4$, is used, the number of directional couplers required is $n = 2$ ($2^n=2^2$) and the number of chips for stage-one is equal to four and for stage two is equal to eight. Therefore, the code word generated is 1000100010001.

Signaling Requirements for TDMA and S/CDMA Schemes

The fundamental problem associated with the implementation of dual schemes, such as TDMA and S/CDMA (which incorporate different data rates), is synchronization. That is, each bit stream for video, voice, and data requires a different clock signal. It has been proposed that synchronization problems may be eliminated if only one clock signal is used with a frequency equal to a multiple of the bit rate of the channel with the highest bit rate in the system. If the bit rate of the video signal is $1/T_{\text{video}}$, where T_{video} is the pulse duration of the video signal, then the rate must be a multiple of the voice bit rate (Equation (11–7)).

$$\frac{1}{T_{\text{video}}} = M \left(\frac{1}{T_{\text{voice}}} \right) \tag{11--7}$$

where M is an integer. Similarly, the voice channel bit rate must be a multiple of the data bit rate (Equation (11–8)).

$$\frac{1}{T_{\text{voice}}} = N \left(\frac{1}{T_{\text{data}}} \right) \tag{11--8}$$

where N is an integer. Finally, the system clock frequency required is expressed by Equation (11–9).

$$\frac{1}{T_{\text{clock}}} = Q \left(\frac{1}{T_{\text{highest}}} \right) \tag{11--9}$$

where Q is an integer. Based on the above, it is possible to multiplex a number of video, voice, and data signals to a common transmission frame that is composed of a number of slots (the same number of service channels incorporated into the system). The number of slots in the frame is determined by the clock pulse width, and is expressed by Equation (11–10).

$$N_{\text{slots}} = \frac{1}{\tau} \tag{11--10}$$

where, τ is the clock pulse width. A number of slots within the frame (see Figure 11–13) are assigned to video channels (wide band) that utilize the TDMA mode of transmission, while the remaining slots are assigned to voice and data channels that utilize the S/CDMA mode of transmission.

Each time slot in the low band can accommodate more than one simultaneous user, the number of which depends on the code sequence length used. Furthermore, the number of chips within the code sequence is subject to the ratio of the signal bit length (low band) to the frame length. The data and voice signals will have different chip numbers. Therefore, each slot in the lower band will carry one chip for each user. At the receiver end, the autocorrelation peak will occur when the entire code sequence has been received at the correct destination address. Because autocorrelation peaks for different users occur at different time slots, it is possible for each receiver to be synchronized for achieving autocorrelation at its home code sequence address.

A Comparison of S/CDMA and TDMA

An oversimplified comparison between TDMA and S/CDMA reveals that TDMA is a special case of S/CDMA. In a TDMA scheme, the 1-bit is represented by a coded waveform composed of a number of time slots specific to the bit destination address, while the 0-bit is not coded. In the generated TDMA code sequence, only a single pulse located at the appropriate time slot represents the destination address. Therefore, a number of destination addresses can be accommodated in a TDMA code sequence, allowing for a number of users to simultaneously transmit through the same code sequence. At the receiver end, synchronous reception is accomplished by synchronizing the receiver to the subscriber destination address. Comparing the two schemes, TDMA is superior to S/CDMA for applications such as high density traffic or transmission systems that require bit rates to the Mb/s levels (digital video transmission). However, TDMA exhibits certain deficiencies in applications where the channel is not fully utilized. S/CDMA is better suited for low density applications such as voice and data. Therefore, it has been suggested that for optical networks that must simultaneously process low density and high density traffic, both TDMA and S/CDMA schemes should implemented for maximum system efficiency.

11.5 WAVELENGTH-DIVISION-MULTIPLEXING

During the last five years, long distance transmission of information has moved very rapidly from microwave line-of-sight and satellite transmission links to optical fiber transmission. For the next five years, the industry estimates an annual growth of twenty percent.

The benefits arising from the utilization of optical networks to business customers, TV networks, and utility customers that have already started to replace coaxial

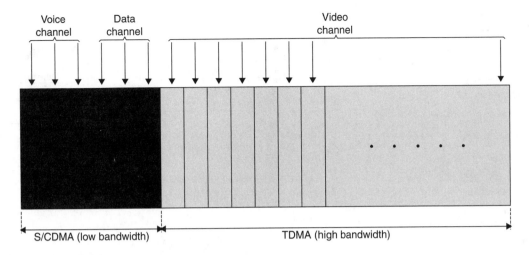

Figure 11–13. TDMA-S/CDMA transmission frame.

cables with optical fibers is evident. Because of the Deregulation Bill of 1996, fierce competition has developed between regional and long distance telephone providers to upgrade their existing lines with optical fibers or to install all new optical networks.

The ultimate goal of an optical fiber link is to optimize the data transmission capabilities and to allow for future expansions that are economical and lack major operating disruptions. Increases of data transmission rates can be achieved in two ways. One is through TDM and the other is through DWDM. Through TDM the overall link capacity can be increased by increasing the data rate of a single wavelength transmitted through a single fiber. For example, if an optical fiber link designed to operate with a maximum capacity of 1.25 Gb/s is to be upgraded to 2.5 Gb/s, all the terminal equipment must be replaced by new equipment that is capable of handling the new required transmission rates of 2.5 Gb/s.

The quest for finding new ways to increase the optical fiber system capacity led to WDM. Through WDM, multiple optical carriers of different wavelengths utilize the same optical fiber. With such an operating scheme, a dramatic increase of the system capacity can be achieved. For example, two or more laser diodes that transmit at different (but very close) wavelengths can be coupled into the same fiber, which results in a significant increase of the system capacity without the need for the installation of additional optical fibers. Figure 11–14 illustrates a block di-

agram of a simplified conventional optical fiber system that is composed of a single wavelength, while Figure 11–15 illustrates an optical fiber system that is composed of five optical wavelengths, which are processed through a WDM mode of operation.

Comparing Figures 11–14 and 11–15, the following can be observed: The system in Figure 11–14 is capable of transmitting data rates of 1.25 Gb/s on a single optical fiber. If the system is designed for long distance transmission, it will require several repeaters that convert optical signals to electrical, amplify them, and then convert them to optical signals for retransmission to the next repeater. These repeaters are essential if the system is to maintain a satisfactory SNR or BER. The system in Figure 11–15 is capable of transmitting data rates of 5 Gb/s at the same distance through a single fiber, using only one in-line **erbium doped fiber amplifier (EDFA)**.

The development of the EDFA pushed the WDM concept to the next level, DWDM.

The ever-increasing demand for higher system capacity requires that all new optical networks incorporate provisions for future system upgrades into the design. Future upgrades must be economical and must be implemented with no service interruptions. New services, such as video on demand, home shopping, video games, home banking and the Internet, have an expansion rate of one hundred percent per month. The demand for system capacities can only be satisfied by sophisticated optical networks.

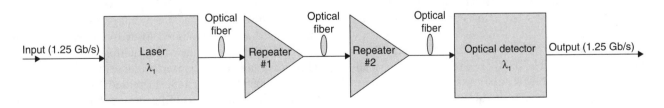

Figure 11–14. A conventional single wavelength optical fiber system.

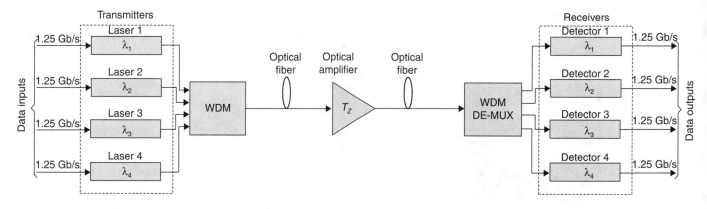

Figure 11–15. A 4-channel WDM optical fiber system.

Since its inception in early 1990, and its progressive implementation, WDM has revolutionized the entire communications industry by substantially increasing the distance between repeaters, allowing for flexible wavelength management and a dramatic increase of link capacity. In the early 1990s, the highest carrying capacity of a fiber optics line was 1.25 Gb/s, utilizing TDM. This transmission rate was considered totally inefficient for the emerging demands. This deficiency left carriers with few choices. They could increase the bit rate of the existing system to a level far above the established 1.25 Gb/s maximum, and by doing so, stretch the performance capabilities of the TDM transport system to its limits with no possibility for future link upgrades. Or, they could install more fibers, an uneconomical solution particularly for small service providers. In 1991, a 4-fiber bidirectional ring that accommodates 2.5 Gb/s (OC-48) was contemplated. While the data rate was twice the previous system's maximum capacity, it could not even partially satisfy the projected exponential growth in demand for capacity. In 1995, the concept of 4-channel WDM was developed, followed by 8-, 16-, and 40-channel systems. It was evident that the ever increasing system capacity demands and longer transmission distance requirements, coupled with economic considerations, could only be satisfied through WDM, eventually expanded to DWDM. The implementation of DWDM not only increased the number of wavelengths transmitted through a single fiber, but also facilitated the wavelength add/drop mechanism.

In communications links that employ DWDM, the optical wavelengths are very closely spaced in a single fiber with a separation distance of about 0.8 nm. This concept, coupled with advanced performance optical amplifiers (EDFAs), was instrumental in revolutionizing not only long distance optical systems but also metropolitan and local networks.

These two innovative concepts allowed networks to switch from an inflexible TDM mode of operation to an all-optical system, capable of transmission rates of up to 100 Gb/s through a single fiber (**SONET/SDH, ATM,** and **PDH** protocols). These optical systems proved economical

and efficient. Experimental DWDM optical systems have achieved transmission rates above the 1 Tb/s range. Optical fiber transport systems that employ DWDM are identified in terms of the total number of wavelengths they can carry and the maximum distance between repeaters. The latest optical transport system developed by Lucent Technologies, which incorporates eighty optical channels in a DWDM mode of operation, is capable of transmitting a maximum of 800 Gb/s at a distance of 640 km. A comparison of single channel TDM and multichannel DWDM optical systems, in terms of data transmission capabilities, is illustrated in Figure 11–16.

Figure 11–16 illustrates optical systems that employ the TDM mode of operation and exhibit limited data transmission capabilities, while systems that employ the DWDM mode of operation have data transmission capabilities limited only by technological innovations. DWDM systems eliminate the complex conversion from optical to electrical and from electrical to optical layers. Today, optical networks designed to perform functions such as cross-connect, wavelength add/drop, and signal restoration are all optical. The implementation of an all-optical network was made possible through advancements in areas such as optical amplifiers (EDFAs), laser diodes, optical multiplexers/demultiplexers, fibers, and optical filters.

Although optical systems that employ DWDM for long distance transmissions have demonstrated extraordinary performance in terms of data capacity, flexibility, expandability, and cost, their application in metropolitan and local area networks can produce the same benefits.

The function of WDM is performed by multiplexer/demultiplexer circuits. A wavelength division multiplexer combines several closely spaced optical wavelengths. The main device that performs WDM is the combiner circuit. Initially, the fundamental disadvantage of combiner circuits was that they exhibited **insertion losses** that were proportional to the number of the multiplexed wavelengths. Therefore, unacceptable insertion losses were encountered as the number of wavelengths increased. Recently, innovative multiplexer and optical filter designs have dramatically improved overall circuit performance. A block diagram

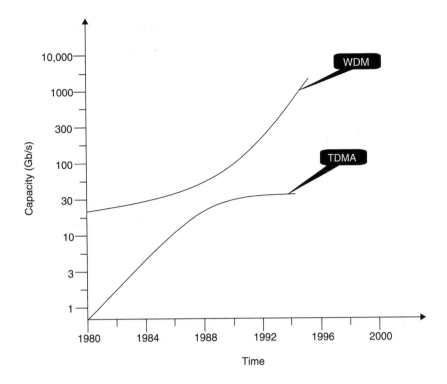

Figure 11–16. A comparison of TDM and DWDM data transmission capabilities.

of a 16-channel 155 Mb/s multiplexer circuit is illustrated in Figure 11–17.

Any increase in link capacity must be implemented through the replacement of the transmitter/receiver and the repeater modules with units designed to handle new transmission data requirements. The optical link in Figure 11–17 utilizes more than one wavelength, thus allowing for substantial increases in system capacity. Another advantage of the system is the incorporation of EDFAs into the design. These amplifiers do not require conversion from optical to electrical and back to optical, thus eliminating unnecessary signal losses and additional costs. The optical link in Figure 11–15 can easily be upgraded by replacing one or more transmitter/receiver sets with others designed to operate at data rates set forth by con-

sumer demands. A comparison between TDM, WDM, and DWDM optical systems, in terms of the type and number of wavelengths, used is illustrated in Figure 11–18.

The degree of difficulty in the design and implementation of optical links that employ the TDM mode of operation and links that employ the DWDM mode of operation is illustrated by Figure 11–19.

Figure 11–19 illustrates that optical links that employ the TDM mode of operation exhibit limitations at transmission rates beyond the 10 Gb/s mark, while links that employ the DWDM mode of operation encounter significant design and operating difficulties at transmission rates beyond 100 Gb/s.

The potential data transmission capabilities of optical fiber systems are far greater than today's demands. While the

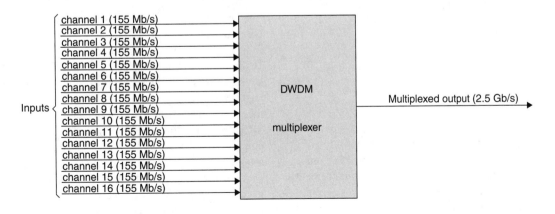

Figure 11–17. A 16-channel optical multiplexer.

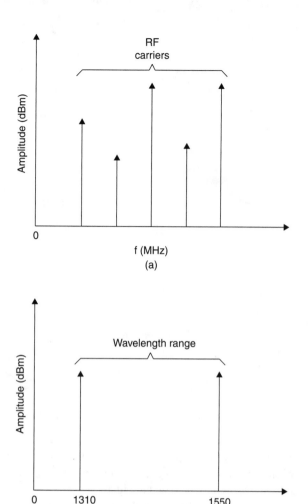

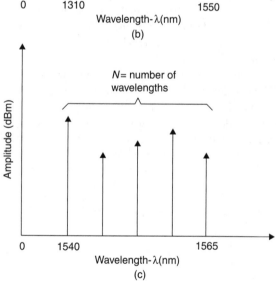

Figure 11–18. a) TDM, b) WDM, c) DWDM.

have evolved from operating data rates as low as 140 Mb/s up to 2.5 Gb/s, and lately up to 10 Gb/s.

Currently, researchers are engaged in the development of optical devices that are capable of operating at 20 Gb/s and 160 Gb/s data rates. At these operating rates, the design and fabrication of such optical devices become difficult and uneconomical. The other critical element of an optical fiber link is the achievable maximum distance between repeaters in a conventional optical system, or the distance between two optical EDFAs in a DWDM system. The two components that adversely affect path distance are fiber losses and pulse dispersions. In a DWDM system, the incorporation of EDFAs has practically eliminated the fiber attenuation problem, but pulse dispersion remains a serious concern, especially at very high data transmission rates. Pulse dispersion can be reduced, but not completely eliminated, through the utilization of very narrow line width DFB laser diodes that operate at wavelength windows that induce minimum dispersion. Directly modulated DFB laser diodes that operate at 10 Gb/s exhibit a line width of 1.2 nm at –20 dB, but MQW DFB lasers exhibit a 0.2 nm line width modulated by the same 10 Gb/s data rate. Pulse dispersion of an optical wavelength traveling through a fiber is proportional to the square of the data rate. Therefore, an increase of the transmitted data rate will substantially decrease the optical length. In order to increase the optical length, the installation of additional repeaters or **dispersion** shifted fibers will be required.

Traditional optical communications links operate at either 1310 nm or 1550 nm wavelengths. Some of these links utilize both the 1310 nm and the 1550 nm wavelengths in order to increase their data transmission capabilities. These links are referred to as the interband optical system. DWDM, however, utilizes only one optical window within the same fiber, which accommodates a number of different optical wavelengths with a 0.8 nm separation. The implementation of DWDM would not have been possible without the development of EDFAs. Because these amplifiers operate close to the 1550 nm wavelength range, they are compatible with optical fibers that also operate in the same 1550 nm wavelength window.

Induced Crosstalk Impairment in WDM Systems

The negative impact of fiber **nonlinearities** in optical communications systems has already been discussed. Link impairment, based on nonlinearities induced by **stimulated Brillouin scattering (SBS),** XPM, SPM, and FWM, can be substantially reduced or altogether eliminated through sophisticated design techniques. However, the performance of the optical link is still subject to fiber nonlinearities induced by SRS. These nonlinearities are critical in optical systems that employ WDM and DWDM schemes, because such systems require relatively higher optical power per-channel in order to operate. The impact of SRS on a WDM optical system is translated as an excessive optical power spill-over from one channel to the next, causing **crosstalk** to increase to unacceptable levels and ultimately affecting the system BER.

peak traffic demand in the U.S. at present is approximately 0.4 Tb/s, the optical range between 1540 nm and 1565 nm is capable of handling data rates of about 4 Tb/s. Enhancement of optical fiber data transmission performance was recently achieved through substantial innovations in the design of optical transmitters and optical receivers. Optical devices

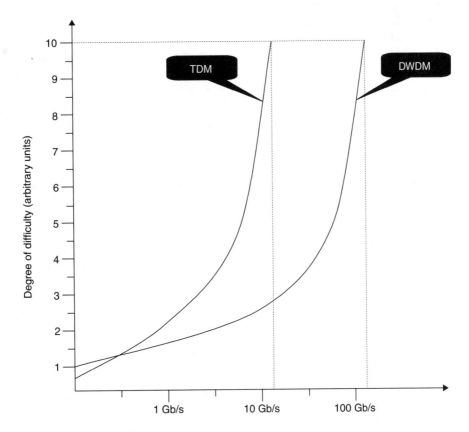

Figure 11–19. A comparison of TDM and DWDM.

In order to thoroughly examine the impact of SRS in a WDM optical system, several assumptions must be made. Assume a WDM optical system with (N) number of channels, a constant separation frequency between channels (Δf), and with the induced optical power (P_n) equally distributed among all the channels, thus maintaining a linear SRS profile. By definition, the optical power in such a system will be transferred from the lowest channel to the highest channel through the process of SRS, in a manner proportional to the channel frequency, provided the total channel bandwidth does not exceed the 13.2 THz level. In such a system, the optical power spillover can be determined by Equation (11–11).

$$\frac{dP_n(L)}{dL} + \alpha P_n(L) + \left(\frac{q\Delta f}{2A}\right)P_n(L)\sum_{m=1}^{N}(m-n)\,P_m(L)=0$$

$$(11\text{–}11)$$

where P_n is the optical channel power, L is the **propagation distance**, α is the **attenuation coefficient** of fiber, q is the slope of the SRS gain profile $\left(\dfrac{dq}{df}\right)$, Δf is the **channel spacing**, P_m is the **modal optical power,** and A is the fiber effective cross-sectional area.

Solving Equation (11–3), the SRS worst case scenario for a WDM optical system can be established.

$$P_n(L) = \frac{P_n\,J_o\,e^{-\alpha L}\exp[\,GJ_o\,(n-1)\,L_e]}{\left[\sum_{m=1}^{n}P_{m_o}{}^{GJ_o(m-1)L_e}\right]} \quad (11\text{–}12)$$

where J_o is the total input power of the system $(J_o = \Sigma\,P_{m_o})$,

$G = \dfrac{q\Delta f}{2A}$, $L_e = \dfrac{1-e^{-\alpha L}}{\alpha}$ (effective length), and P_{n_o} is the input power of the nth channel.

Based on the assumption made above that all channels are at equal power, $P_{n_o} = P_o$, the channel power is expressed by Equation (11–13).

$$P_n = NP_o e^{-\alpha L}\exp\left[\left(\frac{GJ_o L_e}{2}\right)(2n-N-1)\right]$$
$$\left[\frac{\sinh\left(\dfrac{GJ_o L_e}{2}\right)}{\sinh\left(\dfrac{NGJ_o L_e}{2}\right)}\right] \quad (11\text{–}13)$$

The maximum crosstalk, based on the SRS, is expressed by Equation (11–14).

$$CT_{(L)} = \frac{[P_o\,e^{-\alpha L}-P_1(L)]}{P_o\,e^{-\alpha L}} \quad (11\text{–}14)$$

where $CT(L)$ is the crosstalk fractional power based on SRS in the lowest channel and P_1 is the optical power of channel one. Assuming a negligible interchannel cross talk, the crosstalk based only on SRS is expressed by Equation (11–15).

$$CT(L) = \frac{N(N-1)(P_o GL_e)}{2} \qquad \textbf{(11–15)}$$

Example 11–1

Compute and graph the crosstalk in reference to the number of channels of a WDM optical system that operates with the following parameters:

- operating wavelength: $\lambda = 1550$ nm
- fiber effective cross-sectional area: $S = 50 \ \mu m^2$
- channel spacing: 120 GHz
- fiber attenuation: 0.047/km or $\alpha_{dB} = -10 \log(1 - 0.047) = 0.21$
- Raman gain slope: $q = 4.9 \times 10^{-18}$ m/W·GHz
- number of system channels: 20, 40, 60, 80, 100
- optical power: 3.5 mW

Solution

a) Compute (G).

$$G = \frac{q \Delta f}{2A} = \frac{4.9 \times 10^{-18} \times 100}{2 \times 50 \times 10^{-6}} = 4.9 \times 10^{-3}$$

Therefore, $G = 4.9 \times 10^{-6}$.

b) Compute $CT(L)$.

$$CT(L) = \frac{N(N-1)(P_o GL_e)}{2}$$

$$= \frac{20(20-1)(3.5 \times 10^{-3})(4.9 \times 10^{-6})(10 \times 10^3)}{2} = 0.03$$

Therefore, $CT(L) = 0.03$.

Applying the same relationships as in Example 11–1, the corresponding crosstalk (CD) figures for different numbers of channels and effective distances are listed in the following table.

Number of Channels	Effective Distance (L_e) km		
	Crosstalk 10	20	40
20	0.03	0.065	0.130
40	0.13	0.270	0.540
60	0.30	0.600	1.200
80	0.54	1.100	2.200
100	0.85	1.700	3.400

Optical Link Redundancy

Point to point optical communications systems are susceptible to a complete outage. To remedy the deficiency, modern design approaches favor the traffic rerouting method (Figure 11–20).

In a DWDM optical system under normal operation, transmission data is processed through λ_1 and λ_2. In the event that the ring is interrupted at wavelength λ_1, the data stream automatically switches to wavelength λ_3. The λ_3 wavelength uses the same path as λ_1 but it travels in the opposite direction. Therefore, customers that are located beyond the brake point are still able to access the data stream through λ_3. Likewise, if λ_2 is interrupted, the data will be automatically switched to λ_4, which travels on the same path as λ_2 but in a counterclockwise direction. It is therefore evident that access ring optical networks provide the necessary back-up mechanism for complete and uninterrupted link operation (**optical link redundancy**).

Component Requirements for WDM System Implementation

The implementation of DWDM in an optical communications link requires special optical devices and optical modules such as low chirp integrated laser/modulator modules, EDFAs, **fiber grating add/drop multiplexers, optical circulators,** and **phase arrays.**

Laser Diode Characteristics for DWDM Applications

In low rate optical fiber links, LEDs or DBF laser diodes are traditionally employed. The input data stream amplitude modulates the optical source by switching the diode driver circuit ON or OFF. This method of modulation induces line width broadening (**chirp**) of about 0.2 nm to 1.2 nm.

Laser diodes are the only sources employed in DWDM optical systems. The basic operating characteristics of a laser diode are:

a) optical power

b) wavelength

c) line width

d) chirp

In the majority of applications, WDM networks employ DBF single-mode laser devices. Laser diodes are divided into two basic categories: directly modulated and externally modulated.

Directly modulated laser diodes generate an output wavelength that has an amplitude that changes in accordance with the change of the laser DC biasing current. Because DWDM optical systems are designed to handle data transmission rates beyond 2.5 Gb/s, the line width (chirp) phenomenon becomes a very serious problem. It imposes limits on both the transmission system capacity (usually at the 2.5 Gb/s rate) and the link span (at 150 km). For distances beyond the 150 km mark and for system capacities larger than 2.5 Gb/s, externally modulated laser diodes are almost always required. These devices are more expensive to fabricate and, because they are very complex, exhibit higher losses.

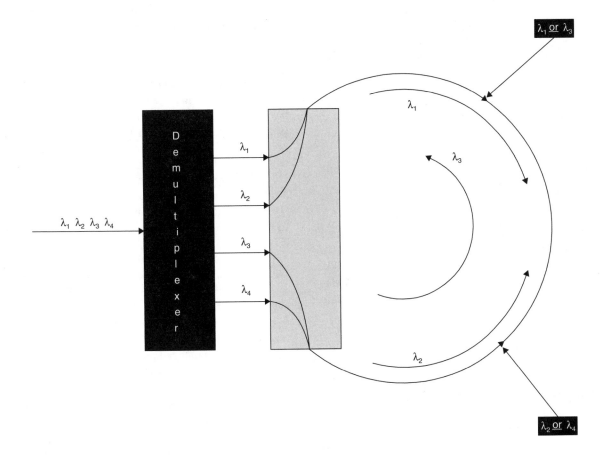

Figure 11–20. Traffic rerouting.

The constant amplitude optical wavelength generated by the laser diode is coupled into port 1 of an external modulator, while port 2 of the external modulator is connected to the desired channel data stream. At the output port of the external modulator, the amplitude of the single optical wavelength generated by the laser diode will change in accordance with the input signal bit rate. The most commonly used external modulator is a lithium niobate **Mach Zehnder modulator.**

A brief description of the operating principle of the Mach Zehnder modulator is as follows (Figure 11–21).

A Brief Description of Mach Zehnder Modulators

If the output of the laser diode is coupled into one of the two ports of the modulator, the modulator will divide the input wavelength into two parallel lines, which converge to form a

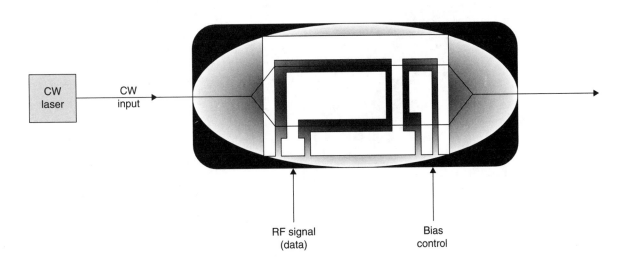

Figure 11–21. A block diagram of a Mach Zehnder modulator.

single line at the output port of the modulator (identical to the input wavelength). If a signal is applied at the second port of the modulator, with an electric field strong enough to invert the phase of one of the split lines, the combined output signal will be completely suppressed. In essence, the optical signal at the output of the modulator will be switched ON or OFF in accordance with the input data rate. This type of modulation is the equivalent of the **on-off keying (OOK)** of a conventional digital communications system. With the utilization of the external modulator, the laser diode biasing current is maintained at a constant and, therefore, pulse broadening is practically eliminated, leading to higher data transmission rates and longer transmission spans.

The utilization of the external modulator also requires additional electronic driving circuitry, which results in the induction of undesirable insertion losses into the optical link. Integrated optical devices that are composed of laser diodes and modulator circuits and are fabricated into the same microchip can rectify this problem (Figure 11–22).

The fundamental requirement of the laser diode is to generate a center optical wavelength, that is consistent with the channel wavelength, without any drift throughout the operating life of the link. Future laser diodes will be tunable to the desired wavelength through a multi-electrode mechanism that controls the driver current of the diode and facilitates electro-absorption control of the modulator circuit. The fine wavelength tuning of a DBF laser diode employed in a DWDM optical system can be accomplished by altering the temperature of the laser/modulator circuit. The performance characteristics and long term behavior of a DFB laser diode are established through precise measurements of the center wavelength, wavelength drift in reference to time, side-mode spacing, side-mode suppression, optical signal amplitude, and wavelength chirp (under direct modulation conditions).

A Brief Description of EDFAs

A detailed description of EDFAs is dealt with in Chapter 5. Here, a brief review of the basic performance characteristics will be given, as they apply to DWDM optical systems. It is universally accepted that DWDM optical systems would not have been possible without the development of EDFAs (Figure 11–23). EDFAs can be used as preamplifiers and postamplifiers, and are capable of increasing the optical path length up to 300 km.

The performance characteristics of EDFAs are established by the device amplification capabilities (gain flatness and gain tilt). The amplification or gain factor of an EDFA device is a function of the operating wavelength (Figure 11–24).

Gain flatness is defined as the gain difference between successive wavelengths, while gain tilt is defined as the steepness of the slope of the gain. From Figure 11–24, it is evident that EDFAs operating at more than one wavelength in a DWDM optical system exhibit different gain factors. For short optical links, where very few EDFAs are required, gain flatness is very small and presents no significant problem to the link performance. For long distance transmissions where a relatively large number of EDFAs is required, gain flatness is a crucial factor, which must be compensated. **Gain flatness compensation** can be accomplished through pre-emphasis. That is, if an input optical signal of a precise wavelength is very small in amplitude, it must be amplified to the level of the next wavelength. Or, if an input signal of a specific wavelength is very large, it is attenuated to the level of the next signal. Another method of controlling the gain flatness is to employ filter gratings at both the input and output sections of the EDFA.

All the wavelengths processed through an EDFA must maintain an amplitude level below SBS, or the entire link will be rendered nonfunctional.

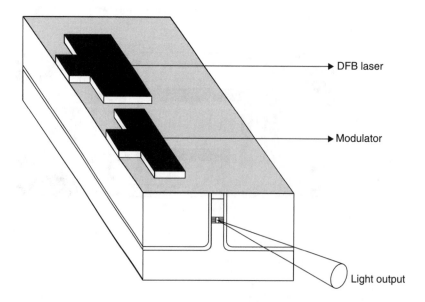

Figure 11–22. A DBF modulator integrated optical module.

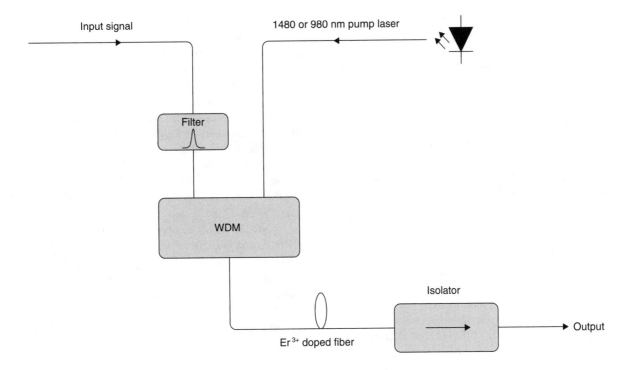

Figure 11–23. A block diagram of an EDFA.

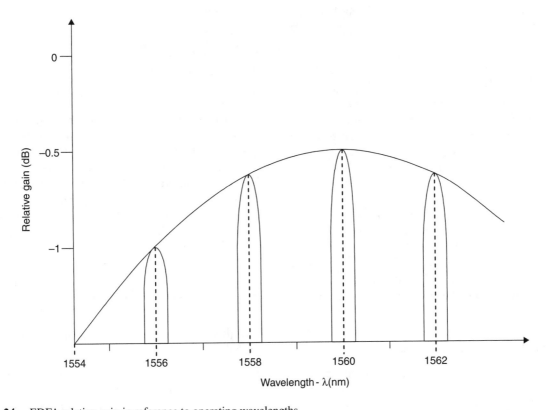

Figure 11–24. EDFA relative gain in reference to operating wavelengths.

Stimulated-Brillouin-Scattering

SBS is a serious problem often encountered in optical fiber transmission systems. SBS scattering is capable of inducing nonlinearities in the optical fiber, and consequently increasing the fiber attenuation to undesirable levels. This attenuation is the result of a wave generated by the SBS and traveling in a direction opposite to the normal. The interaction of the backward traveling SBS signal and the forward traveling normal signal results in a substantial attenuation of the forward traveling normal signal.

In order for SBS to occur, two conditions must be present: The power of the optical signal at the output of the EDFA amplifier must be higher than a set level, and the wavelength line width generated by the laser diode cannot be very narrow.

Gain Competition of EDFAs

For the proper operation of a DWDM optical system, the maintenance of strict laser diode wavelength line width specifications is of paramount importance. This eliminates one of the two causes of SBS. As for the first condition, that of the amplitude level, all signals within the DWDM scheme, must have equal amplitudes and must also be kept below the SBS level. The following scenario can trigger Brillouin scattering, resulting in a complete system failure.

Let us suppose that the EDFA employed in a DWDM optical system simultaneously amplifies five optical wavelengths that have amplitudes that are kept just below SBS levels. If one of the five wavelengths generated by a laser diode degrades or completely collapses, the amplitudes of the remaining four wavelengths will increase, because the output optical power level of the EDFA remains constant (characteristic of the EDFA). If this amplitude increase is higher than the SBS set limit, stimulated Brillouin scattering will occur, with the possibility of complete system failure.

Optical Detector Requirements

The principal function of an optical detector is to receive the multiwavelength signal from the fiber and convert it back into the original electrical signal. An optical detector is one of the major components, as are transimpedance amplifiers, AGC postamplifiers, a clock and data recovery circuit, and a demultiplexer circuit. All the modules constitute the receiver section of an optical system.

Optical receivers are designed to operate in both the 1310 nm and 1550 nm wavelength range. The combination of optical detectors and optical amplifiers utilized into the receiver design induces imbalanced noise levels. This is because the detected noise is higher at 1-bit than at 0-bit. Therefore, a provision must be made at the decision point to offset noise imbalance.

Optical Fiber Requirements

The physical and optical characteristics of optical fibers, as well as their applicability in optical links, are dealt with in Chapter 9. Although the incorporation of EDFAs in a DWDM optical system allows for a practically unlimited link distance, the spacing between amplifiers is subject to fiber dispersion limitations. From the two major types of dispersion, chromatic and polarization modes, chromatic dispersion of single-mode fibers can be compensated for through the use of fibers with negative dispersion coefficients or fiber grating. Polarization mode dispersion can affect long distance optical links

that operate at very high data rates. At very high data rates, pulse chirping increases substantially, reflecting a proportional increase in the channel BER and a decrease of the optical system performance.

Optical fibers are divided into two basic types: single-mode and multimode.

Single-mode fibers that have a core diameter of about 10 μm are capable of transmitting at the lowest bound mode of a particular wavelength, and are normally used for high data rates/long distance transmissions. These types of fibers are subdivided into three categories: nondispersion shifted (NDSF), nonzero dispersion shifted (NZ-DSF), and dispersion shifted (DSF).

NDSFs are used in optical links that operate at both the 1310 nm and the 150 nm wavelengths. When operating at the 1310 nm wavelength range, they exhibit a zero dispersion shift, while the maximum dispersion shift at the 1550 nm range is approximately 18 ps/nm-km. NZ-DSFs operating at the 1550 nm wavelength range exhibit a 2 ps/nm-km dispersion shift. DSFs are special fibers that operate at the 1550 nm range.

Overall, the performance of an optical fiber is determined by the following parameters:

a) **attenuation**

b) dispersion

c) nonlinearity

d) **connectivity**

e) **splicing**

Attenuation limits the link distance. In today's advanced optical systems, fiber attenuation incorporated into the system design is on the order of 0.25 dB/km. Optical fiber dispersion imposes limits on the link span between repeaters, and, because it is data transmission dependent, it limits the channel capacity beyond the 10 Gb/s range. Nonlinearity, connectivity, and splicing contribute to an increase in the fiber loss, signal distortion, and noise.

11.6 FUTURE OPTICAL DEVICES FOR DWDM SCHEMES

Tunable LEDs and Photodetector Diodes for WDM Applications

Two significant contributors to the enhancement of the performance of an optical link in terms of maximum capacity and distance are device speed and maximum utilization of the optical fiber bandwidth capability. An optical device capable of satisfying DWDM requirements is the **tunable extended vertical cavity (TEVC).** Researchers at Boston University have demonstrated that such diodes are ideal for WDM optical schemes because they can be tuned over a relatively wide wavelength range with an almost flat response and with a very small level of interchannel crosstalk. **Vertical cavity (VC)** tunable diodes

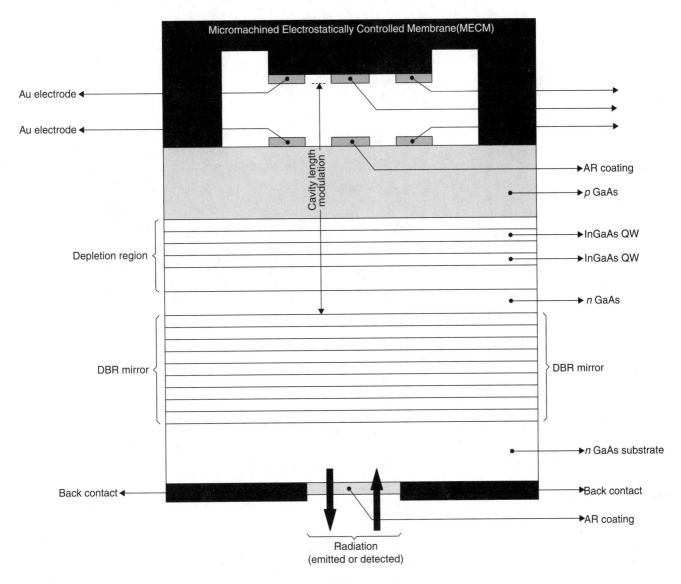

Figure 11–25. TEVC diode cross-sectional area structure.

are very difficult to produce because of the fabrication complexity involved during the crystal growth. To overcome these difficulties, researchers are working toward the development of TEVC diodes that have tunable capabilities over the 100 nm range. They are relatively inexpensive and have a high yield. Such a diode structure is illustrated in Figure 11–25.

The device in Figure 11–25 is a tunable extended vertical LED structure based on InAlGaAs compounds. The design philosophy of such a device is based on a conventional VC structure that has the top mirror embedded under a micromachined membrane. The device is also identified by the two QW active regions that are separated by λ/4 and confined by GaAs layers. Au is used for the upper mirror and an AlGaAs distributed Bragg reflector is used for the bottom mirror. Such a device is capable of providing constant performance during cavity tuning. Furthermore, the diode bandwidth can be further enhanced by electron-hole transit time equalization between the contacts. The almost flat response across the entire operat-

ing wavelength is attributed to the antireflection coating between the Au mirror and the top surface. The high efficiency is attributed to the coating of the GaAs substrate, which promotes a higher degree of radiation coupling. The spectral response of a TEVC diode is illustrated in Figure 11–26.

11.7 DWDM DEMULTIPLEXING

DWDM networks can be used in point-to-point topology and are capable of carrying, on average, four different optical wavelengths in the 1550 nm range. These optical links are able to transmit a much higher volume of binary data without exceeding the fiber dispersion limits so critical in the 1550 nm wavelength range. Point-to-point optical links that incorporate wavelength routing are the results of highly sophisticated and complexly designed optical networks capable of routinely transmitting in the terabit range. One of the

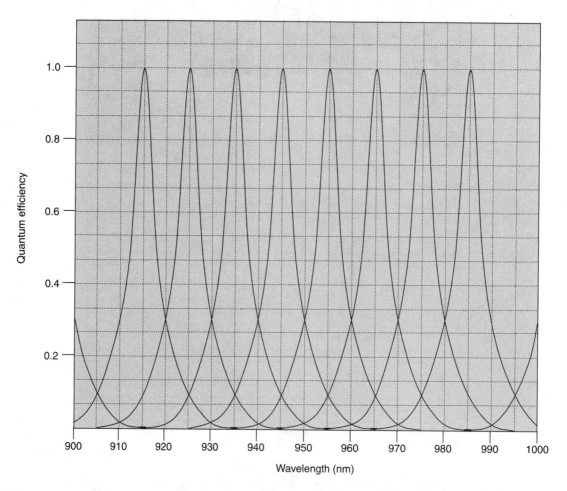

Figure 11–26. TEVC spectral response for 8-channel WDM applications.

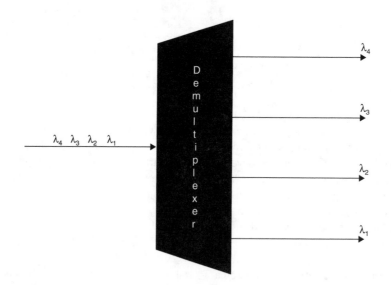

Figure 11–27. A block diagram of a DWDM demultiplexer.

fundamental components in a DWDM optical system is the demultiplexer circuit in Figure 11–27. The basic function of a demultiplexer circuit in a DWDM optical system is to separate the multiwavelength optical signal applied at the input into discrete wavelengths at the output.

The demultiplexing process in a DWDM mode of operation can be performed by phase arrays, fiber gratings, and optical circulators.

The center wavelength of the demultiplexer must be the same as the channel wavelength. This requirement is most

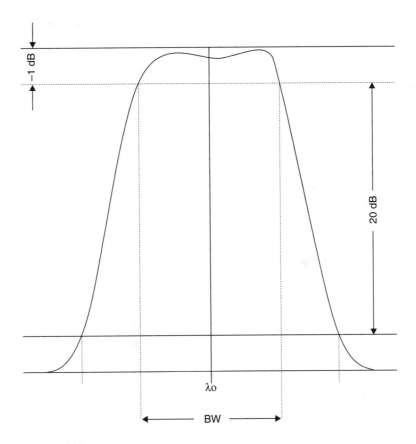

Figure 11–28. Demultiplexer bandwidth.

critical in DWDM systems because of the very narrow spacing between channels. The **channel baseband** is measured between 0 dB and 1 dB, while the complete wavelength bandwidth is measured between 0 dB and 20 dB (Figure 11–28). If the attenuation levels of optical wavelengths, processed through a DWDM demultiplexer, are equal to or higher than 1 dB, they are considered unacceptable.

For demultiplexers that employ fiber grating technology, the falling slope of the wavelength response must be known in order to determine the **filter ripple,** a parameter indicative of fiber grating quality. Demultiplexers that employ phase arrays exhibit a higher degree of polarization sensitivity, which may lead to a shift of the center wavelength or filter bandwidth drift. Optical demultiplexers employed in DWDM induce insertion losses in optical systems. These losses are the sum of the coupling losses between the optical fiber and the demultiplexer, and other losses such as waveguide and splicing losses.

For fiber grating demultiplexers, the coupling losses are very small, but they can be significant for phase array modules. The **link power budget** must compensate for insertion losses through the addition of an extra EDFA. Future optical demultiplexers will incorporate EDFAs into their design to compensate for insertion losses. Finally, a crucial parameter that determines system performance is crosstalk. Crosstalk is the amount of optical power coupled into a channel by the two adjacent channels, that causes sys-

tem performance degradations by significantly increasing the system BER. Crosstalk is attributed to the broadband characteristics of either the APD or PIN photodetector diodes. Standard demultiplexers employed in conventional optical networks exhibit crosstalk levels of about −25 dB, and demultiplexers employed in DWDM optical systems require crosstalk levels of about −40 to −50 dB.

Demultiplexing Through Multilayer Interference Filters

The process of optical demultiplexing through multilayer interference filters is illustrated in Figure 11–29. Multilayer interference filters are color coded rectangles, transparent only to a single optical wavelength, while reflecting the rest of the incident multiwavelength optical beam. If the optical demultiplexer is composed of a number of rectangular filters such as light blue coded, gray coded, or black coded, and a multiwavelength beam is incident at a specific angle upon the light blue filter, then the light blue wavelength will be processed through while the rest of the beam will be fully reflected and incident upon the next gray coded filter. Similarly, this filter will pass only the gray wavelength, while the rest of the beam will be fully reflected and incident to the next filter.

From Figure 11–29, it is evident that a complete separation of a multiwavelength beam into individual

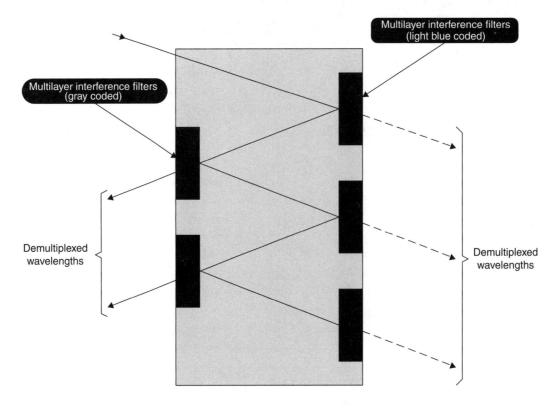

Figure 11–29. Multilayer interference filters.

wavelengths can be accomplished. Thus, by employing multilayer interference filters, optical demultiplexing of a DWDM signal can be successfully performed.

Demultiplexer Parameters

The following operating characteristics must be known and controlled if the demultiplexer is to perform within the design parameters.

NOMINAL WAVELENGTH. The **nominal wavelength** of a demultiplexer output is the wavelength which the output was designed to carry, and is one of the system wavelengths. The spacing between two consecutive nominal wavelengths must also be in compliance with ITU-T recommendations specified at 100 GHz (Figure 11–30).

PEAK WAVELENGTH. The **peak wavelength** of a DWDM demultiplexer output is the wavelength that appears at a particular output and experiences the lowest possible attenuation (insertion loss) (Figure 11–31).

MEAN CENTER WAVELENGTH. The **mean center wavelength** is defined by the mean value of the two wavelengths that occupy the band edges of a specific demultiplexer output signal (Figure 11–32).

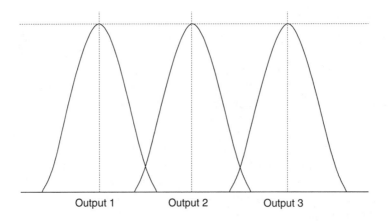

Figure 11–30. Nominal output demultiplexer wavelengths.

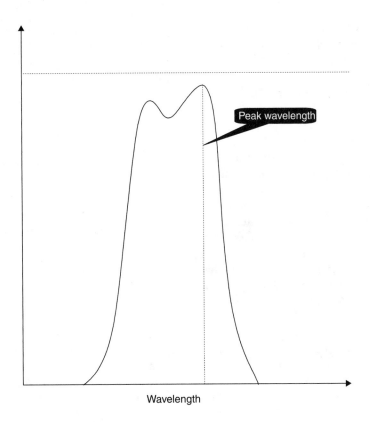

Figure 11–31. DWDM demultiplexer peak output wavelength.

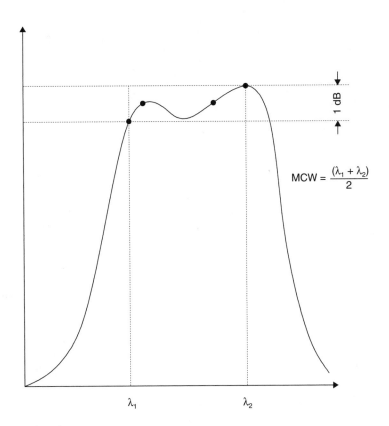

Figure 11–32. Mean center wavelength.

WAVELENGTH BANDWIDTH. The bandwidth (B_W) of an optical wavelength (**wavelength bandwidth**) in a DWDM system is the filter width at the 1 dB or 1.5 dB point below peak (Figure 11–33).

For highly complex optical systems, the 1 dB or 1.5 dB value is unacceptable and must be further reduced. The comparison of the bandwidth at 1 dB and the bandwidth at 20 dB defines the slope steepness of the filter. Ideally, the two bandwidths must be the same in order to reduce or completely eliminate the otherwise anticipated adjacent channel interference.

In a real optical system design, the pass-band of the filter, the channel spacing, and the laser diode center wavelength must be taken into consideration. For example, if the filter pass-band is significantly wider than the channel spacing, the transmitter design criteria may be somewhat relaxed. That is, the laser diode may exhibit a higher level of wavelength drift, providing that the drift does not exceed the established bandwidth.

INSERTION LOSSES. Insertion loss of a DWDM demultiplexer circuit is the optical power difference between the input of the demultiplexer and a specific output which is measured at the nominal wavelength of the channel under test (Figure 11–34).

Demultiplexer insertion losses include input connector losses and demultiplexer losses. If the connector losses are known, then the demultiplexer losses can easily be established.

CROSS TALK. **Demultiplexer crosstalk** is the level of optical power spilled over from one channel to adjacent channels (Figure 11–35).

The level of crosstalk can be minimized through sophisticated optical filtering. The steeper the filter slope for both the adjacent channels, the less the crosstalk levels. An excessive optical power spillover or crosstalk will increase the channel BER, thus contributing significantly to system performance degradation.

RETURN LOSS. **Return loss** is the ratio of the reflected optical power to the incident power. Return loss is measured at the input of a DWDM demultiplexer and is expressed in dB (Figure 11–36).

In a single mode TDM optical system, return loss is highly undesirable. In a DWDM optical system that employs add/drop demultiplexers with multigrating fibers, complete elimination of return loss is essential for selected wavelengths.

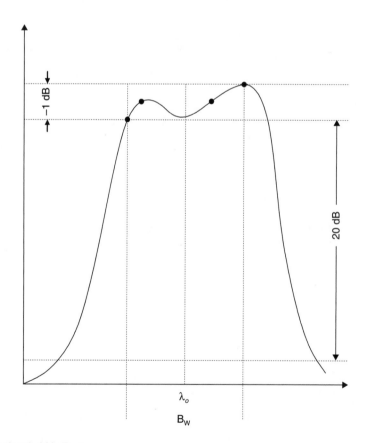

Figure 11–33. Demultiplexer bandwidth (B_W).

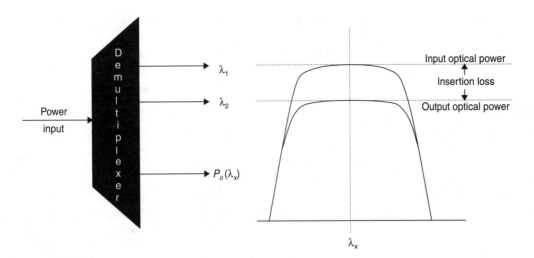

Figure 11–34. Demultiplexer insertion losses.

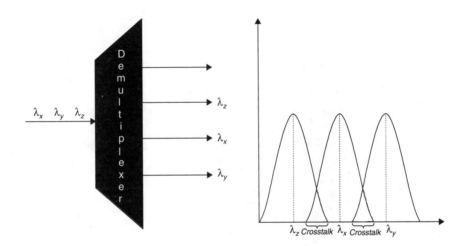

Figure 11–35. Demultiplexer crosstalk.

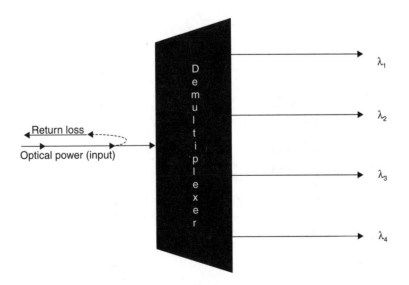

Figure 11–36. DWDM demultiplexer return loss.

POLARIZATION DEPENDENT LOSSES. **Polarization dependent** loss is the peak-to-peak output power variation of a specific demultiplexer output channel when the input of the demultiplexer is fully polarized (all polarization states) (Figure 11–37).

Polarization dependent loss is fiber dependent. Fiber gratings exhibit very small polarization dependency, while integrated optics exhibit high polarization losses. In such devices, the symmetry of the device plays a crucial role in the reduction of the polarization dependent loss. Highly symmetrical devices exhibit smaller polarization dependent losses than asymmetrical optical wavequides.

Polarization dependent losses contribute directly to an increase of crosstalk and insertion losses.

CENTER WAVELENGTH POLARIZATION DEPENDENCE. The all-state input signal polarization of a DWDM demultiplexer circuit also affects the center wavelength of the output channel by promoting a peak-to-peak deviation of the center wavelength (Figure 11–38).

Demultiplexer devices that employ fiber gratings exhibit a very small peak-to-peak center wavelength deviation due to the input fiber polarization, while phase array demultiplexers are more susceptible to input signal polarization. Center wavelength drift due to input signal polarization results in the reduction of the demultiplexer 1 dB channel output bandwidth.

BANDWIDTH POLARIZATION DEPENDENCY. Because input signal polarization has an effect on the center wavelength, it will have the same effect on the channel bandwidth (Figure 11–39).

Bandwidth polarization dependence is the difference between the maximum bandwidth and the minimum bandwidth of an output demultiplexer channel. Bandwidth polarization dependence has the same effect on system performance as center wavelength drift. That is, it causes a deviation from the 1 dB channel bandwidth.

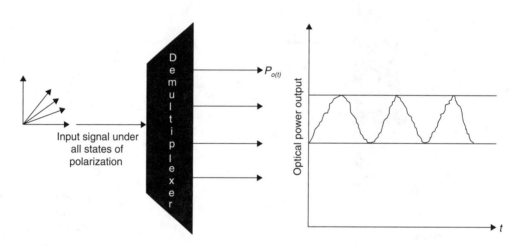

Figure 11–37. Polarization dependent losses.

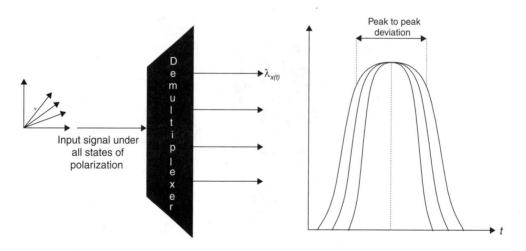

Figure 11–38. Center wavelength polarization dependence.

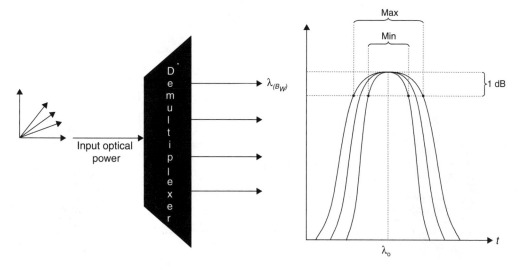

CHAPTER 11

Figure 11–39. Bandwidth polarization dependence.

11.8 AWG MULTIPLEXERS/ DEMULTIPLEXERS FOR DWDM SYSTEMS

One of the challenging problems facing providers of optical fiber systems is upgrading. Because the demand for higher data rates is ever increasing, system providers must respond to the challenge by either upgrading the already existing optical networks or replacing them with new networks. An effective way of providing future network upgrading (easily *and* economically) is through **array waveguide grating (AWG)** multiplexers/demultiplexers.

Corado Dragone of Bell Labs invented the waveguide routing method of optical demultiplexing. To honor this invention, the device is referred to as the Dragone router (Figure 11–40).

The operation of the Dragone router is based on the principle of optical interference. The demultiplexer in Figure 11–40 is composed of an array of curved channel waveqides that have an equal path length difference among adjacent channels connected to a set of input and output planar free-space slab waveguides.

The input slab diffracts the entering multiwavelength beam, and the scattered light entering the array is phase shifted by a different degree in relationship to the adjacent waveguides, thus generating interference at the output slab. This process generates maximum interferences at different wavelengths and at different output locations. In essence, it demultiplexes the incoming multiwavelength beam into individual wavelengths.

Performance Characteristics

The Dragone router exhibits the following characteristics.

a) excellent crosstalk

b) high thermal stability

c) small insertion loss

d) excellent long term reliability

e) independence of transmission protocol, signal format, and line speed

Furthermore, when AWG multiplexers/demultiplexers are applied in DWDM systems, they dramatically increase the system capacity and also provide for inexpensive future upgrades. The transmission architecture of such an optical system and a detailed block diagram is illustrated in Figure 11–41.

The typical performance characteristics of an AWG DWDM module are as follows:

Number of channels: 8, 16, 32, or 40

Operating wavelength: 1310 or 1550 (nm)

Channel spacing: 200 (8-channel), 100 (16-channel), 100 (32-channel), 100 (40-channel) (GHz)

Bandwidth (3 dB): 80 (8-channel), 40 (16-channel), 40 (32-channel), 40 (40-channel) (GHz)

Insertion loss: 5 dB to 7 dB

Return loss: better than 35 dB

Polarization dependent loss: 0.5 dB

Directivity: better than 35 dB

Temperature stability: $< \pm 0.5$ dB

Temperature sensitivity: $< \pm 0.01$ nm

AWG Temperature Control

For best performance, AWG routers require an absolute temperature control on the order of $\pm 0.2°$C. Temperature control mechanisms can either be designed and built by the router user or purchased. The temperature control module is composed of two components. The **temperature stabilization substrate (TSS)** is used as a temperature source that provides the required internal temperature for the AWG device. The resistance

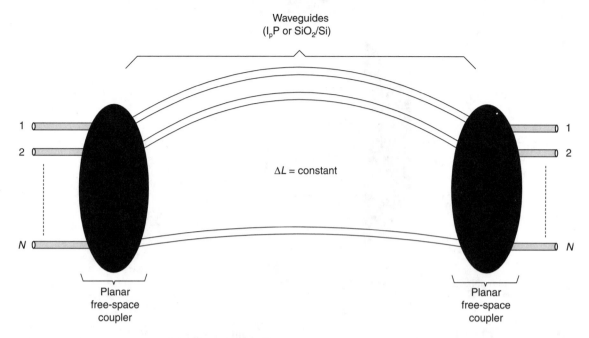

Figure 11–40. Optical demultiplexing by waveguide routing (Dragone router).

temperature detection (RTD) is used to provide the required feedback mechanism. The TSS is used to maintain AWG internal temperature control, and incorporates two leads that are American gauge number 26, Teflon coated wire. (Teflon is produced by duPont.) The RTD incorporates two **resistance temperature detectors (RTD):** RTD-1 and RTD-2.

RTD-1 is used to monitor the internal temperature of the AWG device, and RTD-2 provides the required negative feedback mechanism for maintaining the constant temperature. Furthermore, RTD-1 and RTD-2 complement each other in case of failure, thus providing the necessary redundancy mechanism. TSS characteristics are listed in Table 11–2.

Resistance temperature detector characteristics are as follows:

Number of leads: 3

Calibration: DIN specification (43760) for class A detector (platinum wire: 100 Ω/0°C)

Thermal resistance coefficient: 3.85×10^{-5} Ω/°C

There are other methods of temperature control. Such methods are:

a) **The proportional integral and derivative controller (PID):** This controller incorporates an integral autotune function and a maximum cycle time of 1 s.

b) **The mercury displacement relay (MDR):** This temperature controller is very reliable and has the longest operational lifetime.

c) **The MOSFET solid state relay (SSR):** This relatively new temperature control device is not so reliable. It may cause thermal runaway by failing in a closed-conduct state.

The general temperature control system specifications are listed in Table 11–3.

11.9 40-CHANNEL AWG MULTIPLEXER/DEMULTIPLEXER

A 40-channel AWG multiplexer/demultiplexer is designed and fabricated by Lucent Technologies. It is to be used in high density DWDM optical systems. The device exhibits very stable characteristics over all operating wavelengths. The design is based on different path length fibers with a diffraction grating composite function. The device complies with ITU-T channel spacing requirements, while exhibiting excellent operating characteristics.

Device characteristics

a) high channel capacity

b) precision interchannel spacing

c) low insertion loss

d) low crosstalk

e) high stability

f) high reliability

Device applications

a) DWDM

b) wavelength selection routing

c) DWDM **optical add/drop multiplexing (OADM)**

d) **Optical signal processing (OSP)**

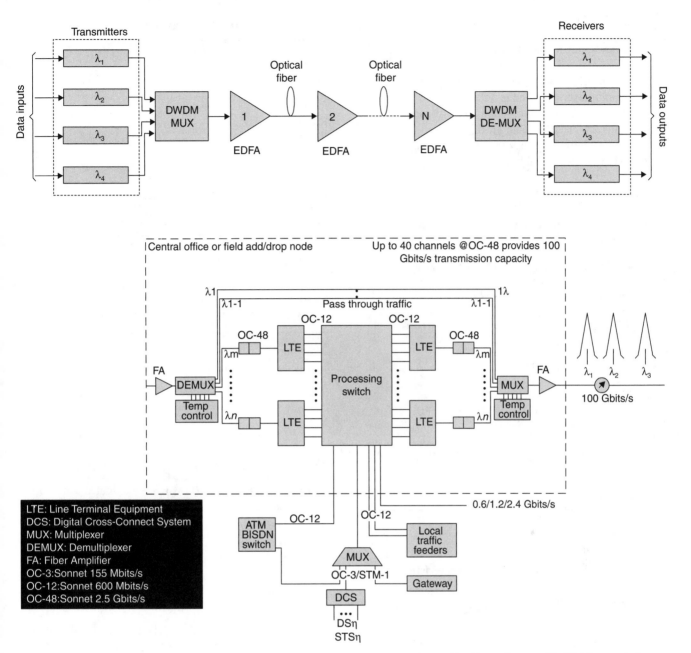

Figure 11–41. a) Simplified block diagram of a DWDM transmission architecture. b) Detailed block diagram of DWDM transmission architecture.

TABLE 11–2 Temperature Stabilization Substrate Characteristics

Parameters	Specifications
Resistance	2.78 Ω
Leads	American wire gauge #26 (Teflon coated)
Applied voltage (max)	5 Vdc
Power output (max)	5 W
Power consumption (max)	3.5 W

TABLE 11–3 Temperature Control System Specifications

Temperature-Control-System	Specifications
Resistance temperature detector	DIN specifications 43760 (class-A)
PID controller	Integral autotune function
Cycle time	≤ 1 s
Discrete power control	SSR, MDR

Electrical-Optical Characteristics

The electrical-optical characteristics of a 40-channel AWG multiplexer are listed in Table 11–4.

The electrical-optical characteristics of a 40-channel AWG demultiplexer are listed in Table 11–5.

The absolute maximum ratings of a 40-channel AWG Mux/Demux are listed in Table 11–6.

The operating frequency range of a 40-channel AWG multiplexer is shown in Table 11–7.

TABLE 11–4   Electrical-Optical Characteristics (Multiplexer)

Parameters	Minimum	Maximum	Symbol	Unit
Channel spacing	100	100	Δf	GH
Channel wavelength accuracy	–0.1	0.1	WL_{OFF}	nm
Insertion loss	0.0	7.5	L	dB
Insertion loss variations	0.0	2.0	ΔL	dB
Return loss	40.0	—	L_{RE}	dB
Polarization dependent loss	0.0	0.6	PDL	dB
1 dB bandwidth	0.35	—	$\Delta\lambda_{1dB}$	nm
3 dB bandwidth	0.6	—	$\Delta\lambda_{3dB}$	nm
Temperature stability	–0.5	0.5	T_{STAB}	dB
Power supply requirement	5.0	5.0	—	V_{DC}
Power consumption	6.0	6.0	—	—

TABLE 11–5   Electrical-Optical Characteristics (Demultiplexer)

Parameters	Minimum	Maximum	Symbol	Unit
Channel spacing	100	100	Δf	GH
Channel wavelength accuracy	–0.1	0.1	WL_{OFF}	nm
Insertion loss	0.0	7.0	L	dB
Insertion loss variations	0.0	2.0	ΔL	dB
Return loss	40.0	—	L_{RE}	dB
Polarization dependent loss	0.0	0.6	PDL	dB
1 dB bandwidth	0.2	—	$\Delta\lambda_{1dB}$	nm
3 dB bandwidth	0.4	—	$\Delta\lambda_{3dB}$	nm
Temperature stability	–0.5	0.5	T_{STAB}	dB
Adjacent crosstalk	22	—	AX	dB
Nonadjacent crosstalk	30	—	NX	dB
Total crosstalk	20	—	TX	db
Power supply requirement	5.0	5.0	—	V_{DC}
Power consumption	6.0	6.0	—	W

TABLE 11–6   Absolute Maximum Ratings

Parameters	Minimum	Maximum	Symbol	Unit
Storage temperature	–40	65	T_{stg}	°C
Operating case temperature	0.0	65	T_C	°C
Relative humidity	—	85	H_R	%
Heater current	—	2.0	I_{HEATER}	A
Heater voltage	—	5.0	V_{HEATER}	V
Heater power dissipation	—	5.0	P_{HEATER}	W
RTD resistance	100	132.81	R_{RTD}	Ω
RTD thermal coefficient	—	3.85×10^{-5}	—	$\Omega/°C$

TABLE 11–7   40-Channel AWG Multiplexer/Demultiplexer Operating Frequencies

Frequency Range	Channel Spacing
191.90 THz to 195.80 THz	100 GHz
191.95 THz to 195.85 THz	100 GHz

AWG Multiplexer/Demultiplexer Vocabulary

a) ***channel center (fixed)*** **(λ_C):** The calibrated center of the channel band.

b) ***channel wavelength accuracy*** **(WL_{OFF}):** The sum of three components: a) temperature sensitivity, b) polarization dependent wavelength, and c) channel wavelength offset.

c) ***channel separation*** **(ΔF):** The separation between fixed channel centers.

d) ***insertion loss (optical)*** **(L):** The loss level at the point of the measured center frequency (λ_c).

e) ***insertion loss variations:*** The difference between the maximum and the minimum of losses across the channel.

f) ***1 dB bandwidth*** **($\Delta\lambda_{1dB}$):** The difference between the upper and the lower wavelengths (λ_A and λ_B), where the filter response passes the level 1 dB down from the peak (Figure 11–42). The region of the 1 dB bandwidth is expressed as: $\Delta\lambda_{1dB} = \lambda_A - \lambda_B$.

g) ***3 dB bandwidth*** **($\Delta\lambda_{3dB}$):** The upper and the lower wavelength (λ_A and λ_B) where the filter response first passes through the –3 dB level from the peak ($\Delta\lambda_{3db} = \lambda_A - \lambda_B$).

h) ***measured center wavelength*** **(λ_C):** The measured center frequency defined as the 3 dB point $\left(\lambda_C = \dfrac{\lambda_A + \lambda_B}{2}\right)$.

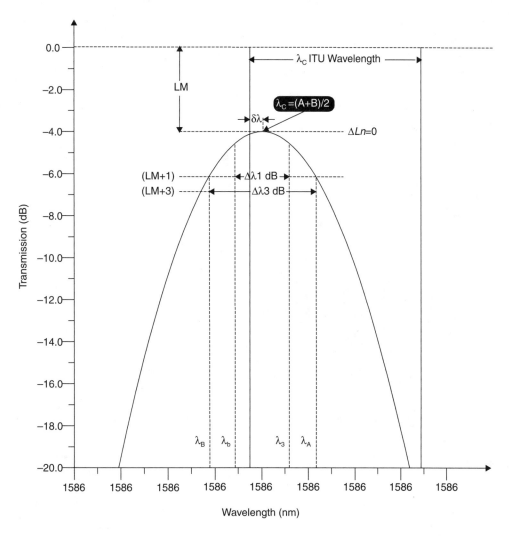

Figure 11–42. Optical filter pass-band.

i) **crosstalk:** Is the result of worst leakage optical power suppression at ±15 GHz bandwidth, above and below adjacent channel centers (Figure 11–43).

j) **polarization mode dispersion (PMD):** The polarization that arises from differences in the speed of two polarization modes of light in a fiber (vertical and horizontal).

k) **chromatic dispersion (CD):** Is the result of the interaction between the light velocity in the fiber and the operating wavelength.

l) **return loss (R):** Is the ratio of the reflected back optical power to the source power.

m) **polarization dependent loss (PDL):** Is the insertion loss variation over all polarization states. It is also the peak intensity difference between the TE and the TM modes.

n) **thermal stability (T_{STAB}):** Is a measure of operating insertion loss variation for each port over the full range of local ambient operating temperatures while the device is thermally controlled.

11.10 ADD/DROP OPTICAL MULTIPLEXING/DEMULTIPLEXING (OADM)

In a modern all-optic communications system, it has become increasingly necessary to serve not only the customers located at the two end points, but also the customer needs located between the end points. In traditional optical links, a single wavelength add/drop can be accomplished by demultiplexing the entire data stream. The only requirement in such a scheme is the addition of an EDFA and another stage that performs the function of conversion from optical to electrical and back to optical signals (Figure 11–44).

For an optical link that employs the DWDM mode of operation, the add/drop process can be accomplished without great difficulty (Figure 11–45). The incoming complex optical signal (multiwavelength) is processed through an optical demultiplexer module. At the output of the demultiplexer, some of the incoming wavelengths will be collected by the terminal equipment for local processing,

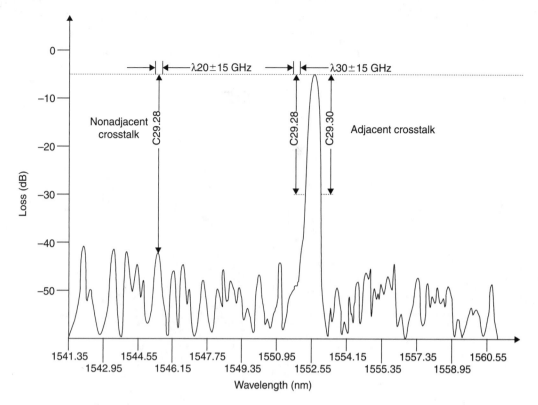

Figure 11–43. Adjacent and nonadjacent crosstalk.

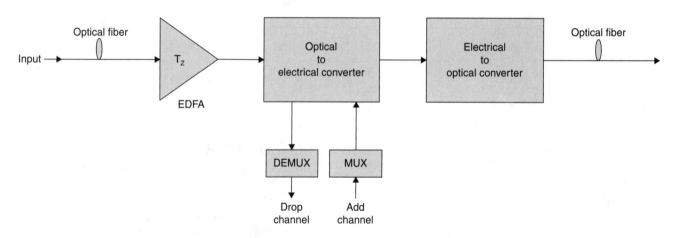

Figure 11–44. A traditional optical system that employs a single wavelength add/drop mechanism.

while the remaining wavelengths will be recombined with additional wavelengths from the local traffic and processed through the multiplexer circuit to form a complete multiwavelength output. If higher capacity upgrading is required at the add/drop points, only the end equipment will need to be replaced to facilitate add/drop point new data requirements.

A disadvantage often encountered with the implementation of add/drop multiplexers in optical communications systems is that they induce optical signal narrowing and possible signal distortion due to the multifiltering requirement at each add/drop stage.

11.11 PROGRAMMABLE ADD/DROP MULTIPLEXERS/DEMULTIPLEXERS

In general, optical add/drop multiplexers must exhibit the following performance characteristics.

a) must be able to combine all the assigned wavelengths without optical power loss or signal distortion

b) must be able to add/drop one or more wavelengths in any order

c) the add/drop process must not interfere with the normal operation of the remaining wavelengths

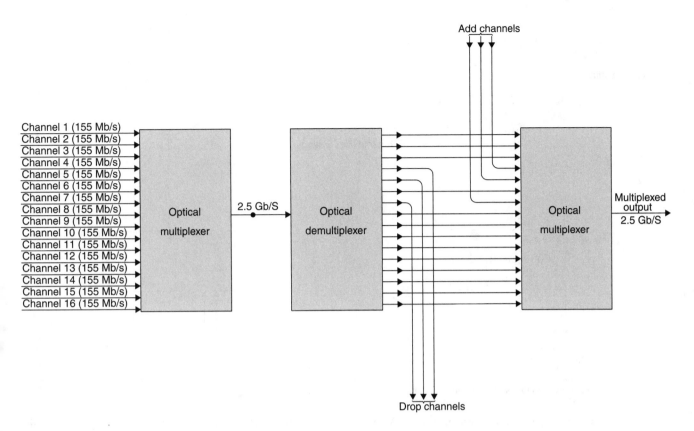

Channel 1 (155 Mb/s)
Channel 2 (155 Mb/s)
Channel 3 (155 Mb/s)
Channel 4 (155 Mb/s)
Channel 5 (155 Mb/s)
Channel 6 (155 Mb/s)
Channel 7 (155 Mb/s)
Channel 8 (155 Mb/s)
Channel 9 (155 Mb/s)
Channel 10 (155 Mb/s)
Channel 11 (155 Mb/s)
Channel 12 (155 Mb/s)
Channel 13 (155 Mb/s)
Channel 14 (155 Mb/s)
Channel 15 (155 Mb/s)
Channel 16 (155 Mb/s)

Figure 11–45. Optical add/drop multiplexing

d) the system must exhibit a high OSNR.

The add/drop multiplexer in Figure 11–45 is unable to satisfy all the stated performance objectives. An improved version of this multiplexer is the **programmable add/drop multiplexer** developed by Lucent Technologies (Figure 11–46).

Operation of Programmable Add/Drop Multiplexers/Demultiplexers

The programmable add/drop multiplexer in Figure 11–46 is composed of two 3-port circulators, a demultiplexer, a multiplexer, and tunable fiber gratings.

A multiwavelength beam is applied at port A of circulator 1, and exits at port B. Port B is connected to a number of tunable fiber gratings. The fiber gratings can exist in either a tuned or untuned state. In the untuned state, they are capable of passing through all the wavelengths processed from port A to port B by circulator 1. When in the tuned state, the wavelengths corresponding to the tuned fiber gratings will be fully reflected back at circulator 1 and exit at port C. Port C of circulator 1 is the input to the demultiplexer circuit. The reflected wavelengths will be demultiplexed and therefore appear at the corresponding outputs of the demultiplexer.

The original optical beam, minus the drop wavelengths, enters port A of circulator 2. The output of the multiplexer circuit is connected to port C of the same circulator, thus inducing the added wavelengths into circulator 2 with the combined wavelength appearing at the output port B of the same circulator. No wavelength reflections occur from circulator 2 to circulator 1 because the filter gratings are tuned only at the desired drop wavelengths.

Summary of DWDM Performance Characteristics

The fundamental objective of an optical communications system is to transmit the highest possible data rate for the longest distance and at the lowest possible cost, while maintaining system performance as set forth by the SNR or BER levels.

To achieve maximum system performance, the system designer must select the laser diode operating wavelengths that are optimally compatible with the operating optical fiber wavelength range, and also with the operating wavelengths of the EDFAs employed in the system. Because the laser diode operating wavelengths are very close to each other, wavelength drift must be kept to an absolute minimum in order to avoid channel crossover. Channel crossover induces intersymbol interference into the system, ultimately leading to an increase of the BER and a degradation of system performance.

Another significant element that must be taken into account is the optical power generated by each laser diode. A constant laser optical output power is essential if a constant SNR is to be maintained for that wavelength or for the entire optical system.

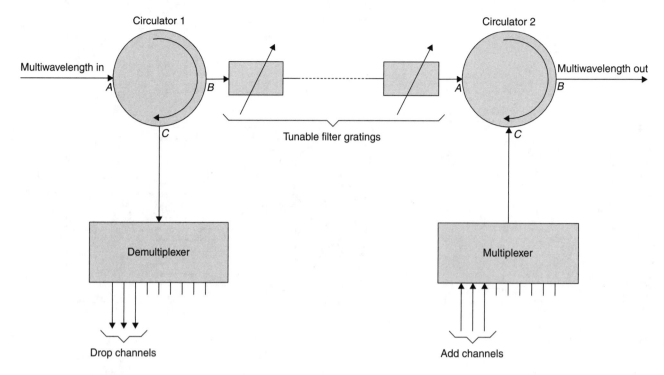

Figure 11–46. A programmable add/drop multiplexer.

In summary, the operating performance of a DWDM optical system is determined by the following parameters:

a) carrier wavelength

b) optical power of each laser diode

c) channel separation

d) drift

e) flatness

f) SNR or BER

11.12 2.5 Gb/s 8:1 MULTIPLEXER CIRCUIT AND LASER DIODE DRIVER

The system diagram of a 2.5 Gb/s 8:1 multiplexer and laser diode driver is illustrated in Figure 11–47. This circuit is based on the Nortel AC30 multiplexer and laser diode driver circuit.

Operation of the 2.5 Gb/s 8:1 Multiplexer

The multiplexer in Figure 11–47 can be used for the SONET/SDH protocol at 2.488 Gb/s data rates. It incorporates the 8:1 multiplexer circuit, the clock driver, and the output stage. The function of the multiplexer is to combine eight 311.04 Mb/s data inputs into one 2.488 Gb/s output. The multiplexer is also capable of detecting parity errors by comparing each input data scheme (311.04 Mb/s) with the parity signal. If a parity error is detected, the circuit will activate (active high) a parity alarm. The circuit also incorporates a mechanism for an internal VCO or an external 2.488 Gb/s clock selection.

The clock driver circuit provides the control of the multiplexer by generating a 311 MHz signal required by the processor. A block diagram of multiplexer/driver is illustrated in Figure 11–48.

THE CONTROL CIRCUIT. The **loop bandwidth** and the **damping coefficient** for the control circuit are set by the external biasing components connected at filter (P) and filter (N), while the internal VCO center frequency is set by the passive components connected with the internal components of the VCO circuit.

OUTPUT STAGE. The output stage of the multiplexer/driver circuit is a high power current switch capable of directly modulating a laser source. The pulse width of the differential signal can be adjusted through the pulse width adjust input, while the output level can be constantly monitored by the monitor output. A pin description of the device is shown in Table 11–8.

The top view of the AC30 multiplexer device is illustrated in Figure 11–49.

DESIGN CONSIDERATIONS. The multiplexer/driver device must be properly biased in order to achieve maximum performance. Full biasing arrangements are illustrated in Figure 11–50 (see page 410).

PLL BANDWIDTH. The external biasing component arrangement is illustrated in Figure 11–51 (see page 411).

PLL EXTERNAL PASSIVE COMPONENT CALCULATIONS. The (PLL) bandwidth of a multiplexer/ driver circuit

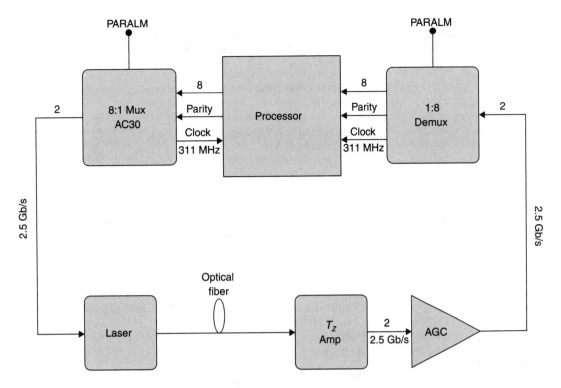

Figure 11–47. A 2.5 Gb/s 8:1 multiplexer and laser diode system diagram (based on the Nortel AC30). Reproduced with the permission of Nortel Networks.

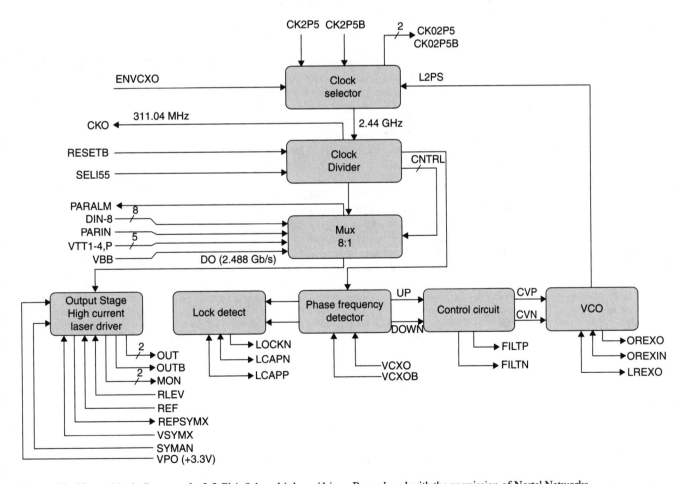

Figure 11–48. A block diagram of a 2.5 Gb/s 8:1 multiplexer/driver. Reproduced with the permission of Nortel Networks.

TABLE 11–8   PIN Description

PIN	Name	Description
6	DIN1.	Input data, 311 Mb/s NRZ (MSB)
8	DIN2.	Input data, 311 Mb/s NRZ
10	DIN3.	Input data, 311 Mb/s NRZ
12	DIN4.	Input data, 311 Mb/s NRZ
14	DIN5.	Input data, 311 Mb/s NRZ
16	DIN6.	Input data, 311 Mb/s NRZ
18	DIN7.	Input data, 311 Mb/s NRZ
20	DIN8.	Input data, 311 Mb/s NRZ (LSB)
22	PARIN	Parity input, 311 Mb/s
7	VTT1	Termination voltage for DIN1, DIN2 (–2 V)
11	VTT2	Termination voltage for DIN3, DIN4 (–2 V)
15	VTT3	Termination voltage for DIN5, DIN6 (–2 V)
19	VTT4.	Termination voltage for DIN7, DIN8 (–2 V)
23	VTTP	Termination voltage for PARIN, RESETB, SYMXEN, SEL155
36	RESETB	Counter reset, ELC input terminated 1 K to VTT (active-low)
43	SYMXEN	Symmetry adjust Enable, ECL-input terminated 1 K to VTT
13	I/O	–1.3 V threshold for DIN1-8, PARIN, RESETB, SYMXEN, ENVCXO
93	CKO	Low speed (311 MHz) output clock
48	CKO2P5	2.488 GHz output clock
49	CKO2P5B	2.488 GHz output clock. Inverted
24	PARALM	Parity alarm output, active high
29	LOCKN	Out of lock indicator, active high
88	CK2P5	2.488 GHz input clock, AC-coupled, internally terminated (ECL)
87	CK2P5B	2.488 GHz inverted input clock, AC-coupled, internally terminated to VTT39
4	VCXO	38.88/155.52 MHz differential reference clock input
3	VCXOB	Clock input
48	VTT39	Termination voltage (–2 V) for VCXO, ENVCXO, CK2P5, CK2P5B
97	ENVCXO	Clock select control
40	SEL115	Reference clock frequency select input: High (GND)-155.52 MHz. Low (VTT)-38.88 MHz
53	VSYMX	Input offset voltage for output pulse symmetry control, 10 K to REFSYMX
52	REFSYMX	Output reference voltage for VYSYMX
79	ORENIX	Internal VCO bias input
77	OREXO	Internal VCO bias output
78	LREXO	Internal VCO bias output
73	FILTP	Analog connections for VCO loop filter passive components
74	FILTN	Analog connections for VCO loop filter passive components
28	LCAPP	Capacitor connections for lock detect circuit
27	LCAPN	Capacitor connections for lock detect circuit
54	RLEV	Laser driver low current knee control
72	REF	Laser driver reference input : RIN>50 K
71	MON	Laser driver output level monitor
60, 61, 62, 63	OUT	Non-inverted data output 2.488 Gb/s
64, 65, 66, 67	OUTB	Inverted data output 2.488 Gb/s
59, 60, 62, 63, 64, 65, 67, 68	VPO	Positive power supply for laser driver current switch (+3.3 V)
31, 32, 39, 40, 41, 42	VEE	Negative power supply (−5.2 V)
81, 84, 85	VEEA	Negative analog power supply (−5.2 V)
5, 21, 30, 55, 56, 57, 58, 59, 68, 69, 70, 91, 96	VEE	Negative power supply, high speed. It requires 1 nF capacitor very close to the pins.
2, 9, 17, 33, 35, 37, 38, 44, 45, 46, 47, 86, 90, 95, 99	GND	Power supply, 0 V, GND
80, 82, 83	GNDA	Power supply, 0 V, GND analog
92, 94	EGND	Separate GND for CKO, ECL output driver
1, 25, 26, 50, 51, 75, 76, 100	FL	Fused corner leads connected to die pad
89	TDIODE	On chip diode connection for die temperature monitoring

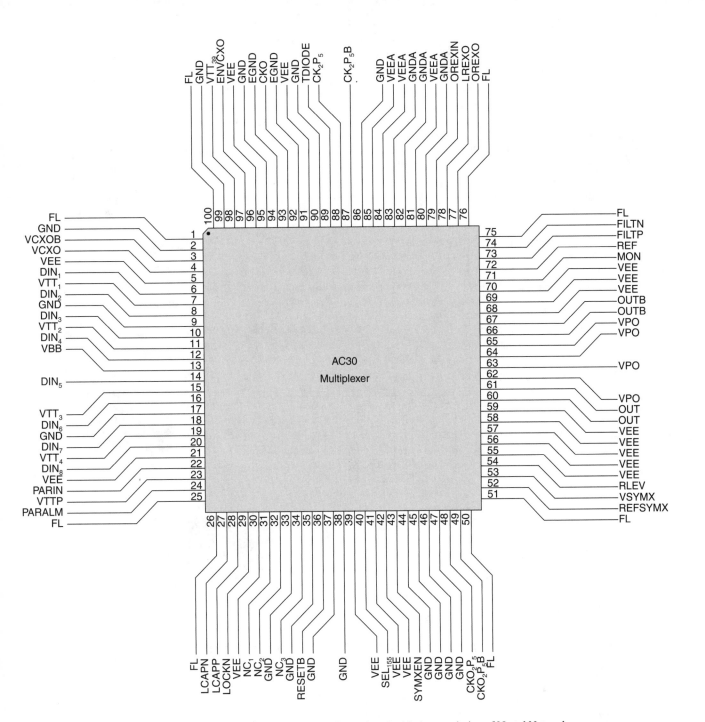

Figure 11–49. Top view of the Nortel AC30 multiplexer device. Reproduced with the permission of Nortel Networks.

is a function of the external passive R, and C components, and is expressed by Equation (11–16).

$$f_{-3dB} = \frac{K_o V_o}{(2\pi)^2 N} \quad (11\text{–}16)$$

where f_{-3dB} is the bandwidth, K_o is the VCO gain, $V_o = (I_p \times R) \leq 0.7$ V, and $1/N$ is the data transmission density.

Calculate K_o: $K_o = (2\pi)(625 \text{ MHz/V})$

Calculate V_o: $V_o = I_p \times R$

For: $I_p = 0.5$ mA and $R = 500\ \Omega$ (assumed), $V_o = 0.5$ mA $\times$ $500\ \Omega = 0.25$ V

Calculate f_{-3dB}:

$$f_{-3dB} = \frac{K_o V_o}{(2\pi)^2 N} = \frac{(2\pi)(625 \text{ MHz})(0.25 \text{ V})}{(2\pi)^2 (64)}$$
$$= 388 \text{ KHz}$$

Therefore, $f_{-3dB} = 388$ KHz.

Calculate C: The value of the capacitor (C) of the PLL circuit is expressed by Equation (11–17).

$$C = \frac{4 K_o I_P \zeta^2}{2\pi N w_{-3dB}^2} \quad (11\text{–}17)$$

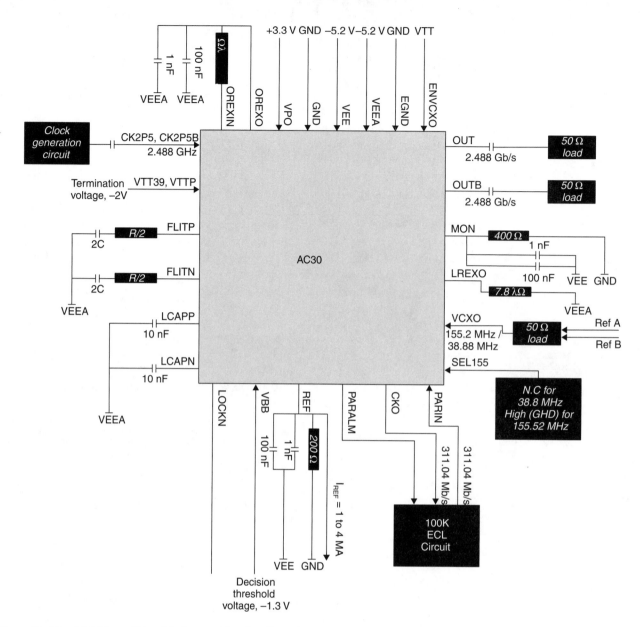

Figure 11–50. Multiplexer/driver biasing consideration. Reproduced with the permission of Nortel Networks.

where C is the capacitance of the PLL, K_o is the **VCO gain,** I_p is the **charge pump current,** $1/N$ is the data transmission density, w_{-3dB} is the loop bandwidth at –3 dB, and ζ is the damping coefficient.

Note: For SONET/SDH jitter specifications at 0.1 dB maximum, the dumping coefficient is set at $\zeta \geq 3$.

Example 11–2

Calculate the capacitor value of the PLL section in a multiplexer/ laser driver circuit based on the following parameters:

- $K_o = (2\pi)(625 \text{ MHz/V})$
- $I_p = 0.6 \text{ mA}$
- $N = 64$
- $w = 2.44 \text{ MHz}$

- $\zeta = 5$
- $V_{\text{out}} = 5 \text{ V}$

Solution

$$C = \frac{4K_o\, I_P \zeta^2}{2\pi N w^2_{-3dB}} = \frac{(4)\,(2\pi)\,(625 \text{ MHz/5})(0.6 \text{ mA})(25)}{(2\pi)(64)(2.44 \text{ MHz})^2}$$

$$\cong 20 \times 10^{-9}$$

Therefore, $C = 20 \text{ nF}$.

Example 11–3

Calculate the value of the damping coefficient (ζ) of the PLL circuit and determine whether the calculated value

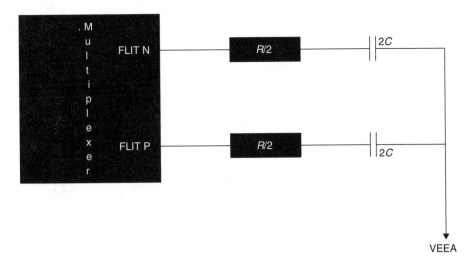

Figure 11–51. External passive component arrangement for setting the PLL bandwidth.

satisfies SONET/SDH specifications. Use the following values:

- $K_o = (2\pi)(625 \text{ MHz/V})$
- $V_o = 4 \text{ V}$
- $w_{-3dB} = 2.44 \text{ MHz}$
- $I_p = 0.35 \text{ mA}$
- $N = 64$
- $C = 22 \text{ nF}$

Solution

Use Equation (11–17) to solve for ζ.

$$\zeta = \sqrt{\frac{(2\pi)(64)(2.44 \text{ MHz})^2 (22 \times 10^{-9})}{(4)(2\pi)(625 \text{ MHz})(0.25)(0.35 \text{ mA})}} \cong 6$$

Therefore, $\zeta \cong 6$.

For an assumed standard capacitance value of 22 nF, the damping coefficient is higher than the recommended value of $\zeta \geq 3$ ($\zeta = 6$).

11.13   2.5 Gb/s 16:1 MULTIPLEXER

Operation

The following 16:1 data multiplexer and clock synthesizer is based on the Lucent Technologies TTRN012G5 (Figure 11–52).

The module accepts one 155.52 MHz clock and sixteen different PECL data inputs and produces a CML of 2.5 Gb/s output in addition to 2.5 GHz clock signal. This eliminates the need for an external clock. It also incorporates a parity check for the sixteen input lines, a second 2.5 Gb/s data output, and a PLL configurable bandwidth adjustable to customer specifications. The main function of the multiplexer is to combine sixteen 155 Mb/s data inputs and to generate a single bit

stream of 2.5 Gb/s at the output, in compliance with the OC-48/STM-16 (Bellcore, ITU-T) format and also to synthesize the required clock frequency. The input 155 Mb/s per channel parallel data is first checked for parity verification and the multiplexed output with a reference clock is used to synthesize the 2.5 GHz frequency required to retrieve the serial output data. For easier interface, output data stream polarity can also be accessed in an inverted format.

The Clock Synthesizer

A PLL circuit is used to synthesize the 2.5 GHz clock frequency. The input parallel data is clocked by a 155 MHz frequency derived from the synthesized 2.5 GHz clock frequency. For jitter of less than 0.1 dB, a loop filter with an acceptable damping coefficient ($\zeta > 3$) must be connected between LF(P)/VC(P) and LF(N)/VC(N) (Figure 11–53).

For a 2 MHz loop bandwidth, the values of the passive components of the filter are:

$C_1 = 0.1 \text{ }\mu\text{F}$
$C_2 = C_3 = 10 \text{ pF}$
$R_1 = 680 \text{ }\Omega$

When a valid SONET/SDH stream of $2^{23} - 1$ PRBS data stream is applied, the clock synthesizer will seek phase/frequency lock. The time required for phase/frequency lock is subject to loop bandwidth. For a loop bandwidth of 2 MHz, the phase/frequency lock will be obtained in 5 ms.

Loss of Lock

The lock-detect function compares the phases of the internally generated 155 MHz clock with the 155 MHz input clock. If the phase difference between the two signals is very close to zero, the clock detect signal "LOCK-LOSS" is at the logic high state. If the phase difference between the two clock signals is changing at a rate higher than the filter cut-off frequency, the LOCK-LOSS signal goes to logic low, indicating the **loss of lock** state.

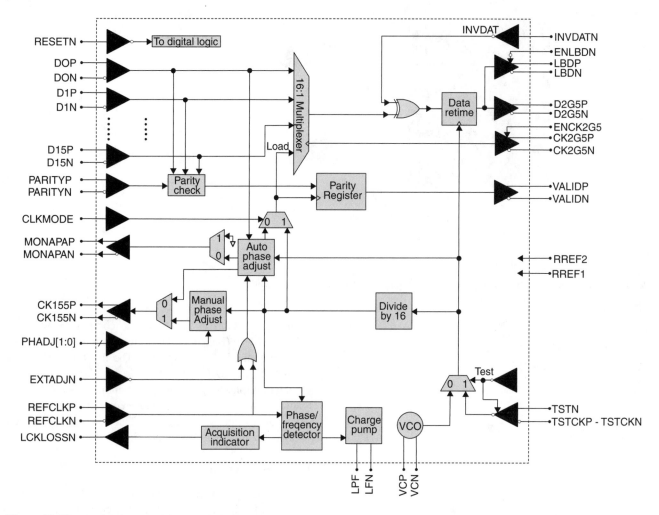

Figure 11–52 A 16:1 data multiplexer and clock synthesizer (based on Lucent Technologies TTRN012G5 device).

Clock Jitter

Typical **clock jitter** performance of an internally generated 2.5GHz synthesizer clock signal is shown in Table 11–10.

The above data is obtained with a band-pass filter that has bandwidth of 20 KHz–2 MHz. The jitter frequency response is illustrated in Figure 11–54.

CML-Output Architecture

The CML output of the multiplexer is composed of a current control mechanism and an amplifier circuit (Figure 11–55). The arrangement provides output current control, and ultimately control of the output voltage swing by controlling the value of the outside terminating resistors. These resistors are connected between the V_{cc} and the CML outputs, in par-

allel with two internal 100 Ω pull-up resistors, and are used to provide the DC current path with AC coupled loads.

With a typical output voltage swing of 0.4 V, the CML output contributes significantly to the reduction of noise transients, EMI interference, and crosstalk. At the same time, it reduces the power dissipation of the external resistors by 50% in comparison to ECL output architecture. Furthermore, CML outputs allow for greater flexibility when customized designs are required. The reference resistor R_{REF1} controls the CML output driver current. This resistor value, in conjunction with R_{REF2}, is capable of controlling the output signal voltage swing to the level required by the particular application. Let us assume an output voltage swing of 0.5 V is required.

Calculate the value of R_{REF1}. For a fixed value of R_{REF2} equal to 1.5 kΩ, the output current is expressed by Equation (11–20).

$$I_o = \frac{(1.21)(18)}{R_{REF1}} \qquad (11\text{–}20)$$

Calculate the load impedance (R_L). The CML output load impedance (R_L) is calculated as follows in Equation (11–21).

TABLE 11–10 Clock Jitter Performance

Parameters	Typical	Maximum	Units
Jitter$_{(p-p)}$	0.02	0.09	UI$_{(p-p)}$
Jitter$_{(rms)}$	0.002	0.009	UI$_{(rms)}$

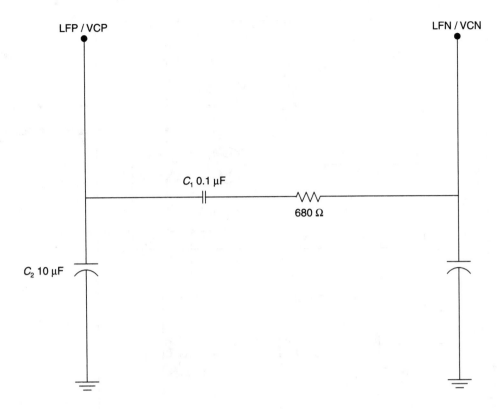

Figure 11–53. A PLL loop filter.

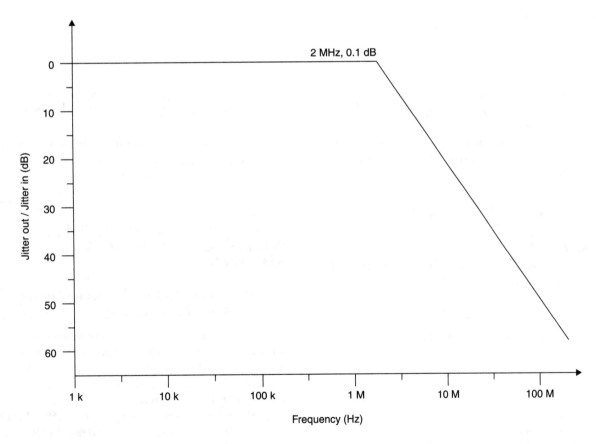

Figure 11–54. Jitter frequency response.

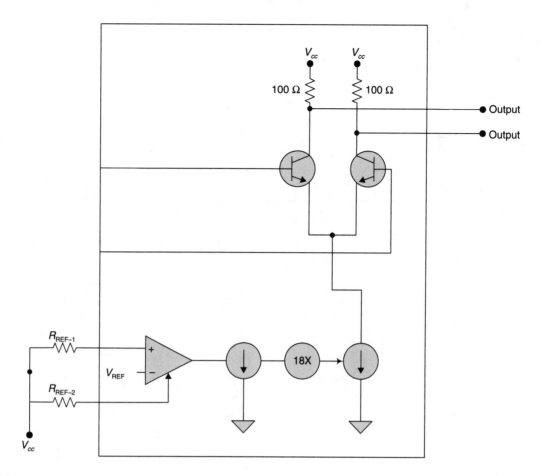

Figure 11–55. CML output stage

$$R_L = \frac{(R_{\text{pull up}})(R_o)}{R_{\text{pull up}} + R_o} \qquad \textbf{(11–21)}$$

where R_L is the load resistance, $R_{\text{pull up}}$ is the pull up resistance (100 Ω), and R_o is the external resistance (50 Ω).

or

$$R_L = \frac{(100)\,(50)}{100 + 50} = 33.3\ \Omega$$

Therefore, $R_L \cong 33\ \Omega$.

Calculate output current (I_o). The output current value is calculated as follows:

$$V_o = I_o \times R_L$$

Because $V_o = 0.5$ V and $R_L = 33\ \Omega$, then

$$I_o = \frac{0.5\ \text{V}}{33\ \Omega} = 15\ \text{mA}$$

Therefore, $I_o = 15$ mA.

Calculate R_{REF1}.

From Equation (11–20), solve for R_{REF1}.

$$R_{\text{REF1}} = \frac{(1.21)(18)}{I_o} = \frac{21.78\ \text{V}}{15\ \text{mA}}\ 1.45\ \text{k}\Omega$$

For an output voltage swing of 0.7 V, and repeating the previous process, R_{REF1} must be equal to 1.2 kΩ.

Therefore, $R_{\text{REF1}} = 1.2$ kΩ.

11.14 2.5 Gb/s 1:8 DEMULTIPLEXER WITH CLOCK AND DATA RECOVERY

Operation of the 2.5 Gb/s 1:8 Demultiplexer with Clock and Data Recovery

The primary function of a 1:8 demultiplexer circuit is to demultiplex a high speed serial input data (2.448 Gb/s) into eight (311.04 Mb/s) parallel streams of data. The block diagram of a 1:8 demultiplexer clock and data recovery circuit is illustrated in Figure 11–56.

The demultiplexer circuit in Figure 11–56, which operates at 2.448 Gb/s, is used for SONET/SDH protocols. It is composed of a 1:8 demultiplexer and a clock and data recovery

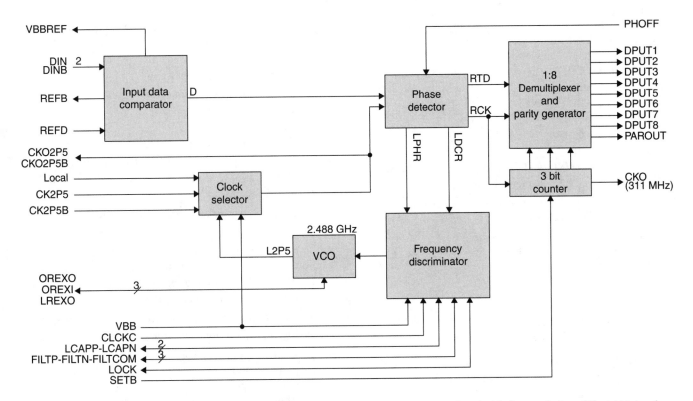

Figure 11–56. A block diagram of 1:8 demultiplexer clock and data recovery circuit. Reproduced with the permission of Nortel Networks.

circuit. The differential data applied at the input is recovered by the data recovery and then applied at the input of the phase detector circuit. The phase detector compares the recovered data stream with the 2.448 GHz clock frequency generated by the clock selector circuit. However, the main function of the phase detector is to generate the required control signal for the frequency discriminator, and also to provide frequency acquisition control. As soon as the frequency acquisition is completed, the discriminator circuit is disabled through an internal lock detection circuit. The demultiplexer circuit generates eight 311.04 Mb/s data signals, a 311.04 MHz clock signal, and a parity bit for each byte of parallel data outputs. For clock recovery, the circuit incorporates an internal PLL circuit. Both the PLL and the VCO circuits require external biasing components for proper operation.

PLL LOOP BANDWIDTH. The elements that control the PLL bandwidth are

a) the **charge pump current** ($I_p = 0.25$ mA) nominal value

b) VCO gain $K_o = (2\pi)(300 \text{ MHz/V})$ nominal value

c) loop filter: R, C elements

For a satisfactory dynamic range, the VCO control voltage is limited to ≤ 1.2 V, while the loop voltage is limited to ≤ 0.75 V.

RESISTOR VALUE (R). The maximum value of the filter resistance is calculated as follows.

Because $V_o = I_p \times R \leq 0.75$ V, then

$$R = \frac{0.75 \text{ V}}{I_P} = \frac{0.75 \text{ V}}{0.25 \text{ mA}} = 3 \text{ k}\Omega$$

Therefore, $R = 3$ kΩ.

LOOP BANDWIDTH. The elements that control the loop bandwidth are the VCO gain (K_o), the voltage drop across loop resistor (V_o), and data transmission density ($1/N$). At –3 dB, the loop bandwidth is given by the expression in Equation (11–22).

$$w_{-3\text{dB}} = \frac{K_o V_o}{2\pi N} \times 1.75 \qquad \textbf{(11–22)}$$

For $K_o = (2\pi)(300 \text{ MHz/V})$ and assuming $R = 200 \ \Omega$ and $N = 4$ then,

$$V_o = I_P \times R = 0.25 \text{ mA} \times 200 \ \Omega = 0.05 \text{ V}$$

Substituting V_o into Equation (11–22) yields

$$w_{-3\text{dB}} = \frac{(2\pi)(300 \text{ MHz/V})(0.05)}{(2\pi)(4)} \times 1.75 \cong 6.6 \times 10^6 \text{ rad}$$

Therefore,

$$f_{-3\text{dB}} = \frac{w_{-3\text{dB}}}{2\pi} = \frac{6.6 \times 10^6}{2\pi} = 1.05 \text{ MHz}$$

$$\therefore f_{-3\text{db}} = 1.05 \text{ MHz}$$

By appropriately selecting the resistor value, the PLL bandwidth can be adjusted to the required levels.

DAMPING COEFFICIENT (ζ). As with the multiplexer circuit, the damping coefficient (ζ) must take values larger than three in order to satisfy SONET/SDH jitter specifications set at 0.1 dB maximum.

Therefore, $\zeta \geq 3$.

CAPACITOR VALUE (C). For a $\zeta \geq 3$, the capacitor value is determined by Equation (11–23).

$$C = \frac{4 K_o I_P \zeta^2}{2\pi N w_{-3dB}^2} \qquad (11\text{–}23)$$

where $K_o = (2\pi)$, $I_P = 0.25$ mA, $\zeta = 5$ (assumed), $N = 4$, and $w_{-3dB} = 6.6 \times 10^6$ Rad,

or

$$C = \frac{(2\pi)(300 \times 10^6 \,/V)(0.25 \times 10^{-3})(5^2)}{(2\pi)(4)(6.6 \times 10^6)^2}$$

$$\cong 11 \times 10^{-9} \, F$$

Therefore, $C = 11$ nF.

If a standard capacitor value of 22 nF is to be used, then the damping efficiency (ζ) must be recalculated based on the selected capacitor value. The recalculation of the damping coefficiency is absolutely essential and must be at least ($\zeta = 3$) in order to comply with the system's preset requirements.

Therefore,

$$\zeta = \sqrt{\frac{(2\pi)N w_{-3dB} \, C}{4 K_o \, I_P}}$$

$$= \sqrt{\frac{(2\pi)(4)(6.6 \times 10^6)^2 \, (22 \times 10^{-9})}{4 \times (2\pi)(300 \times 10^6)(0.25 \times 10^{-3})}} \cong 3.8$$

The value of (ζ) equal to 3.8 satisfies SONET/DSH protocol requirements for a jitter level of 0.1 dB.

Application Considerations

a) The input circuit must be DC coupled to the AGC post amplifier

b) The DIN and DINB inputs can either be AC or DC coupled (AC coupled is recommended for maximum performance)

c) If only one input is used (single-ended), the unused input must be connected to the ground via a 50 Ω resistor

d) The external clock CK2P5 must be AC coupled

e) If an external clock is used, the control signals must be connected to VTT (-2 V) or to a valid low-level ECL

f) If the control signals are connected to ground or to a valid high-level ECL, the 2.448 GHz internal clock is selected

g) For the PLL operation, two coupling capacitors and two external resistors are required in order to control the VCO center frequency

h) DOUT-1 to DOUT-8, PARALM, and CKO are 100 kΩ signals and must be terminated at VTT (-2 V) through a 50 Ω resistor at the receiver end

Power Supply Considerations

a) The negative power supply (-5.2 V) is intended for digital CML gates, while the VEE supply is intended for analog blocks

b) Clock and data recovery circuits should be connected to VEEA

c) To eliminate the noise generated on the analog power supply, which ultimately leads to circuit performance degradations, a second order LC filter with a maximum corner frequency of 10 KHz must be used

d) The fused leads (FL) must be connected to the negative analog supply for noise minimization

e) Decoupling capacitors must be connected to GRDA (analog); ground connections are divided into three categories: a) power ground (GRD-PWR), for ECL output pads, b) analog ground (GRDA), for analog circuits, and c) digital ground (GRDD), for CML blocks and pads.

The three ground levels must be connected to the main ground plane for the elimination of noise propagation from one area to the other.

It is important to mention that the highest noise level is generated in the ECL output pads (GND-PWR). For an effective reduction of the high frequency noise, the VCO decoupling capacitors must be connected to the analog negative supply very close to the device's pin. Furthermore, in order to avoid common mode noise converted into differential noise, loop filter capacitors must be connected to VEEA. In addition, loop filter resistors must be connected very close to the device's pins.

11.15   2.5 Gb/s 1:16 DEMULTIPLEXER

Operation of the 2.5 Gb/s 1:16 Demultiplexer

The function of a 2.5 Gb/s 1:16 demultiplexer is to convert the 2.5 Gb/s data stream into a 16-bit 155 Mb/s parallel stream. The block diagram of a 2.5 Gb/s to 16-bit parallel stream device is illustrated in Figure 11–57.

The demultiplexer circuit can be used with an external clock and a data recovery circuit. The incoming 2.5 Gb/s data stream is compared to the positive going pulse transition (PGT) of the 2.5 GHz clock frequency. The demultiplexer select mechanism provides the required 155 MHz (falling edge) clock, thus dividing (by 16) the 2.5 GHz clock frequency in order to obtain the 155 MHz 16-bit parallel

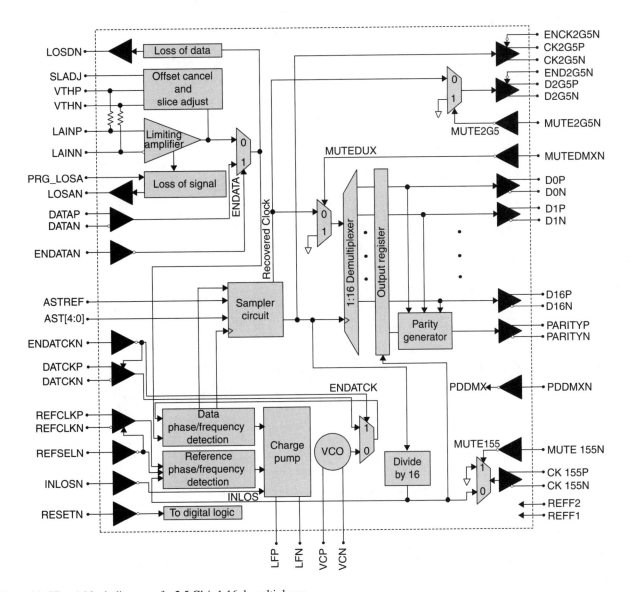

Figure 11–57. A block diagram of a 2.5 Gb/s 1:16 demultiplexer.

output data. The input signals, both data and clock, are fed into the demultiplexer through the clock and data recovery modules illustrated in Figure 11–58.

Because the 2.5 GHz data and clock signals at the output of the clock and data recovery circuits are changing at the falling edge, the 1:16 demultiplexer detects these input signals at the rising edge. In a high speed (2.5 Gb/s) format, the demultiplexer inputs are differential input pairs (CML) (Figure 11–59).

DEMULTIPLEXER DIFFERENTIAL INPUT/OUTPUT STAGES. The differential inputs of the demultiplexer are internally terminated by a 100 Ω resistor, and require a differential input voltage of between 60 mV and 350 mV supplied by the output stage of the clock and data recovery modules.

11.16   ULTRA HIGH SPEED DEMULTIPLEXERS (40 Gb/s–100 Gb/s)

The ever-increasing demand for high speed (above 10 Gb/s) LANs and long haul optical systems necessitates the design and implementation of very sophisticated multiplexer/demultiplexer modules. Electronic switching for high speed systems has been proven insufficient because of semiconductor device speed limitations. These limitations generated the incentive for the development of exotic multiplexing/demultiplexing devices. In the quest to achieve speeds well beyond the 10 Gb/s range, researchers have developed an all-optical scheme for the design of a very high speed demultiplexer circuit by utilizing semiconductor laser and single arm interferometer (SAI) geometry.

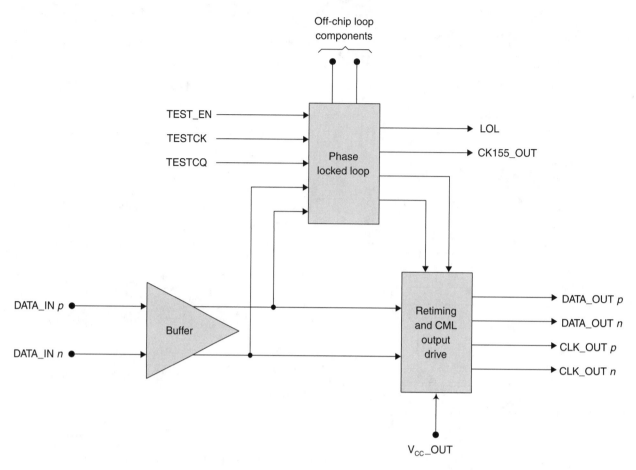

Figure 11–58. A block diagram of a Demultiplexer clock and data recovery arrangement.

Based on the geometry of a SAI, the transmitted data is independent of the long lived refractive index nonlinearities. Through the combination of **nonlinear loop optical mirrors (NLOM), semiconductor laser amplifiers (SLA),** and SAIs, researchers were able to achieve transmission speeds of 160 Gb/s. Further experimentation, based on counter propagating pulses in a single arm interferometer, has shown that, because the transmitted data is immune to long lived refractive index nonlinearities, the need of optical filters is no longer necessary, provided that the employed devices are utilized in a cascaded mode. The fundamental difference between the NOLM and the **ultra fast nonlinear interferometer** method (UNI) is the fact that UNI schemes do not require polarization sensitive elements in order to reject the control signals. Furthermore, with a 2 ps maximum walk off time range and linear geometry, the UNI method allows for device integration. The UNI setup is illustrated in Figure 11–60.

Operation of a UNI Demultiplexer

The operation of the UNI demultiplexer scheme is based on cooperative control geometry. The input signal is fed into the polarization sensitive isolator (PSI), then divided into two orthogonal polarized signals and delayed by 12.5 ps through a

BRF of 7.5 m in length. The control pulse is combined with the orthogonal signals through a 50/50 optical combiner. The control pulse is now orthogonal with the delayed optical signal. Both signals are then recombined through another 7.5 m long birefringent fiber and a process of inverse polarization takes place through the second polarization sensitive isolator. An optical filter is used to reject the control signal, while the input signal is detected at the output. A polarization controller biases the interferometer appropriately in order to provide the required signal components' phase delays. A more detailed experimental setup for UNI is illustrated in Figure 11–61. A 20 Gb/s or 40 Gb/s RF signal drives the mode locked extended cavity laser (ML-ECL) diode, operating at the 1564.7 nm wavelength and 8 ps pulse line width. A 10 Gb/s pseudo random binary stream (PRBS) modulates the optical pulse through an electro-optic modulator.

A passive multiplexer is designed to generate the 20 Gb/s optical PRBS signal with a length of $2^{31}-1$. The UNI output is fed into the receiver module composed of an EDFA, and a PIN photodetector diode. The receiver output is then fed into either a bit error rate tester (BERT), a high speed oscilloscope, or both. Measurements show that the performance of the 10 Gb/s or 20 Gb/s demultiplexer induces a 3 dB power penalty for a BER of 10^{-9}, and a 3.3 dB power penalty for a 40 Gb/s demultiplexer.

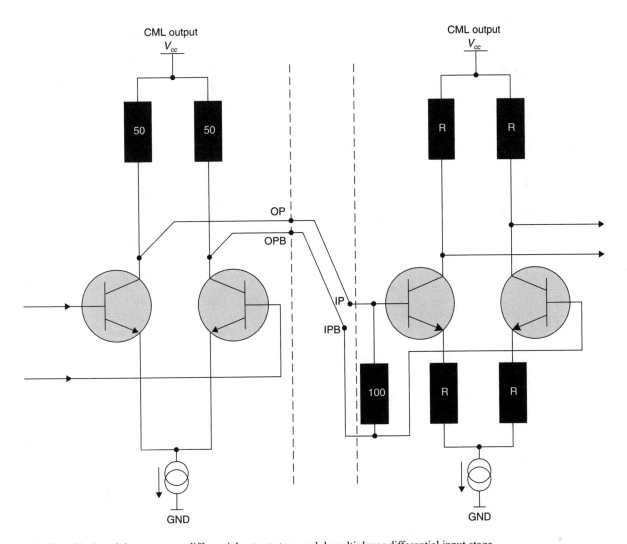

Figure 11–59. Clock and data recovery differential output stage and demultiplexer differential input stage.

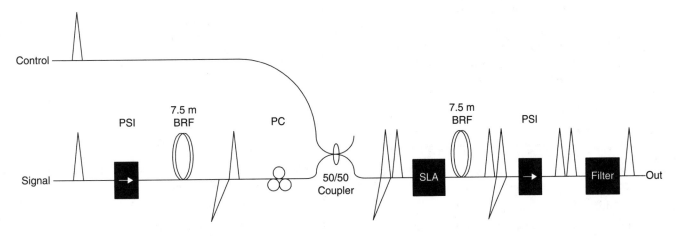

Figure 11–60. UNI demultiplexer setup.

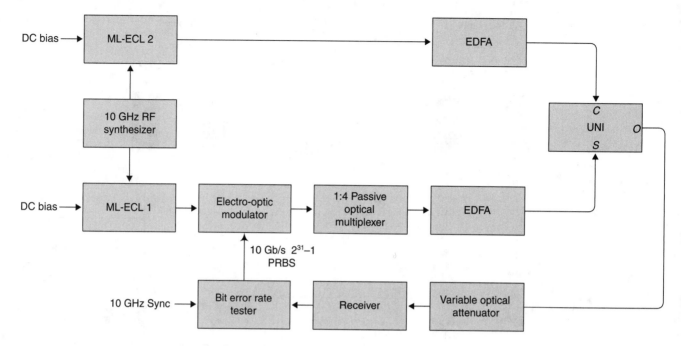

Figure 11–61. A 20 Gb/s or 40 Gb/s UNI experimental scheme.

SUMMARY

In Chapter 11 the important role of multiplexing in optical communications systems was presented. The chapter began the introduction of various multiplexing techniques such as FDM, TDM, CDMA, and WDM. The chapter continued with the introduction of DWDM and add/drop multiplexers, essential elements in high density optical communications systems. Furthermore, the concept of optical demultiplexing and its operating characteristics was also discussed in some length. The chapter concluded with a brief discussion of the composition and operating characteristics of ultra high speed optical demultiplexers.

QUESTIONS

Section 11.1

1. Define the process of multiplexing.
2. Name the three traditional multiplexing techniques applied to communications systems.
3. What was the interim multiplexing step prior to the development of complete optical communications systems?
4. Describe the basic design philosophy of SDM.

Section 11.2

5. Define *FDM*.
6. What is the voice signal bandwidth of an FDM system?
7. Describe the role of the subcarrier oscillator incorporated into an FDM scheme.
8. What does CCITT stands for?
9. With the assistance of a block diagram, describe the operation of an FDM standard group.

10. List the subcarrier oscillator frequencies of a standard master group.
11. How many voice channels does an FDM standard super master group carry?
12. Draw a block diagram of an FDM standard master group and describe its operation.
13. Describe the role of the pilot frequency in an FDM scheme and list its important operating parameters.

Section 11.3

14. Define *TDM*.
15. With the assistance of a block diagram, describe the operation of a TDM scheme.
16. Derive the bit-rate for the North American T1 system.
17. What does the Nyquist theorem state?
18. What is the length of the time frame of the T1 system?

19. Draw a block diagram of the North American digital hierarchy and indicate the bit-rates of each level.

20. Draw a block diagram of the Japanese digital hierarchy and indicate the bit-rates of each level.

21. What is the maximum voice signal transmission capacity of the North American T1 system, and how does it compare with its Japanese equivalent?

22. Draw a block diagram of the European CEPT digital hierarchy and indicate the bit-rates of each level.

23. Compare the Bell North American T1 system with the European CEPT and its Japanese equivalent systems.

24. Describe the process of *justification*.

25. What is meant by *positive justification* and what is meant by *negative justification?*

26. Define multiplex *jitter*.

Section 11.4

27. Describe, in detail, the concept of *CDMA*.

28. What triggered the development of CDMA?

29. Explain why fiber optics systems are ideal for the utilization of CDMA schemes.

30. List the classifications of the optical signal processing techniques in CDMA LAN systems.

31. Describe *coherent optical processing*.

32. Describe *incoherent optical processing*.

33. Why must optical receivers exhibit a strong autocorrelation function?

34. Compare the performance of a standard CDMA and an S/CDMA scheme.

35. Derive the error probability function of an S/CDMA scheme.

36. Draw a block diagram of an S/CDMA scheme and describe its operation.

37. Draw a block diagram of an optical encoder employed in an S/CDMA scheme and describe its operation in detail.

38. Describe the signaling requirements for TDMA and S/CDMA multiplexing schemes.

39. List the differences and similarities between TDMA and S/CDMA multiplexing schemes.

Section 11.5

40. Define *WDM*.

41. Name the driving forces behind the development of WDM.

42. What are the major design goals of an optical communications system?

43. Describe *DWDM*.

44. Name the two innovative concepts that led to the development of all-optical networks.

45. Draw a block diagram of a 4-channel WDM optical system and explain its operation in detail.

46. With the assistance of a graph, compare TDM and DWDM transmission systems.

47. What are the two components that adversely affect the optical transmission distance?

48. Name one method of optical link redundancy.

49. List the main components required in a WDM optical system.

50. Briefly describe the principle operation of a Mach Zehnder modulator.

51. Briefly describe the principle operation of an EDFA.

Section 11.6

52. What are the two significant contributors to the enhancement of the performance of an optical link?

53. Name the type of LEDs considered ideal for DWDM applications.

54. With the assistance of a schematic diagram, describe the operation of a tunable extended vertical cavity diode.

Section 11.7

55. Define the concept of dense wavelength division demultiplexing.

56. How is the demultiplexing process performed in a DWDM system?

57. Compare fiber grading and phase array demultiplexers in terms of their insertion loss into the optical power budget calculations.

58. Compare the crosstalk levels between standard and DWDM demultiplexers.

59. Describe the process of optical demultiplexing through a multiplayer interference filter.

60. List the operating characteristics that must be controlled to achieve maximum demultiplexer performance.

61. Define the terms *mean center wavelength, return loss, polarization dependent loss*, and *bandwidth polarization dependency*.

Section 11.8

62. What is the most effective way of future optical upgrading?

63. Describe the operation of the *Dragone router*.

64. List the performance characteristics of AWG.

65. With the assistance of a block diagram, describe the design philosophy of DWDM transmission architecture.

66. Explain why temperature control in AWG devices is essential.

Section 11.10

67. Draw a block diagram of an add/drop multiplexer and briefly explain its operation.

68. Why is add/drop multiplexing essential in all-optical fiber optics systems?

69. How can upgrading be accomplished in a high capacity system?

70. Define *PMD*.

71. Define *chromatic dispersion*.

Section 11.11

72. List the performance characteristics of an add/drop multiplexer.

73. Explain why programmable add/drop multiplexers were necessary.

74. Draw a block diagram of a programmable add/drop multiplexer and explain its operation.

Section 11.12

75. Draw a block diagram of the Nortel AC30 2.5 Gb/s multiplexer and laser driver and briefly explain its operation.

Section 11.13

76. Draw a block diagram of the Lucent Technologies TTRN 012G5 demultiplexer and clock synthesizer and explain its operation.

Section 11.14

77. Explain the reasons for the development of ultra high speed demultiplexers.

78. Draw a block diagram of an ultra fast nonlinear interferometer demultiplexer and explain its operation.

12

OPTICAL SYSTEMS

OBJECTIVES

1. Explain the concept of optical fiber link systems
2. Describe, in detail, the rules and procedures for optical power budget analysis
3. Establish the parameters and specifications necessary for complete optical fiber link design
4. Explain the concepts of dispersion and wave polarization and their effects on fiber optics transmission systems
5. Present a complete design example of an analog optical system
6. Describe the rules and procedures for the design of undersea optical systems
7. Present several experimental results in the area of Tb/s data transmission
8. Report on the experimental results in ultra high speed single channel OTDM transmission
9. Describe the concept of soliton
10. Describe in detail a 160 Gb/s soliton transoceanic system

KEY TERMS

acoustic effect
add/drop multiplexer
adjacent channel interference
amplified spontaneous
 emission (ASE)
amplitude modulated
 vestigial sideband
 (AM-VSB)
amplitude shift keying (ASK)
ANSI

beat length
bit error rate (BER)
carrier-to-noise ratio (CNR)
chromatic dispersion
chromatic second order
 (CSO) dispersion
chromatic third order
 (CTO) dispersion
clock recovery
composite second order (CSO)

composite third order (CTO)
conversion efficiency
cross phase modulation (XPM)
data recovery
deemphasis
degree of freedom (DOF)
dispersion coefficient
dispersion compensation fiber
 (DCF)
dispersion management

dispersion mapping
dispersion penalty
dispersion shifted fibers (DSF)
dispersion slope
 compensition (DSC)
dynamic range
effective gain
effective noise figure
fiber distributed data
 interface (FDDI)

12.1 INTRODUCTION

Optical fiber links are designed to connect point *A* with point *B* through a light signal. The components required for the establishment of a very basic optical communications link are the transmitter, the receiver, and the optical fiber. The function of the transmitter is to convert electrical signals into optical signals, which are then coupled into the optical fiber. The function of the optical fiber is to guide the optical signals to the input of the receiver module, then to detect these optical signals and convert them into electrical signals that are identical to the transmitted signals. The level of optical power launched into the fiber is subject to the fiber size. Light can be guided through an optical fiber in either a **single-mode (SM)** or **multimode (MM)** format. Single-mode optical transmission is suitable for long distance transmission, and multimode is suited for short distance transmission, to a maximum link distance of 5 km. The difference between the optical power launched into the fiber and the power required at the input of the receiver to generate a usable electrical signal is referred to as the link power budget or power margin. Several system parameters affect the optical power budget or **power margin;** perhaps the most important is **chromatic dispersion** and **modal dispersion.** Fiber attenuation caused by chromatic and modal dispersion is commonly referred to as a **dispersion penalty,** and is expressed in dB. Power margin is also referred to as the optical power at the input of the receiver above a set minimum that is defined by the receiver sensitivity. Excessive power margin can be used to either extend the overall link distance or to allow for the incorporation of additional splicing of directional couplers into the system. On the other hand, low power margin can contribute directly to a higher system **bit error rate (BER).**

Optical fiber links can be very simple and capable of transmitting a few Mb of data. They can also be very complex (all photonic) and capable of transmitting data in the terra bit range over several hundred kilometers without repeaters. For point-to-point optical fiber links, several physical layers, such as the **ANSI**x3T9.5 (FFDI) that has a maximum capacity of 100 Mb/s, can be used in parallel to increase the system capacity to desirable levels. Since 1995, the design and implementation of optical communications links underwent an extensive and radical evolution. Until then optical links, designed for either short distance or long haul, incorporated electrooptical conversion and electronic switching. The maximum system capacity was limited by the speed of the electronic devices incorporated into the system. Since then, the exponential growth of data demands has necessitated the design and implementation of completely optical (photonic) systems. The development of optical modulators, **add/drop multiplexers, wavelength conversion,** and regeneration allowed for the development of completely optical systems with capacities in the terra bit range and link spans several hundred kilometers apart. In this chapter, the design process of completely photonic systems, from the very simple to more complex, will be presented. The phenomenal data transmission capability of optical fiber systems has already been emphasized. This transmission capability is mainly due to the very large operating bandwidth available to optical fibers. Although this bandwidth is very large, only a fraction of it is utilized because of various factors within the transmitting fiber, such as chromatic dispersion **(group velocity dispersion (GVD))**, and **amplified spontaneous emission (ASE)**, limit the full exploitation of the optical fiber bandwidth. However, recent advances in fiber and optical device technology have been able to minimize the GVD and ASE noise that affect the optical link transmission capacity through appropriate dispersion management techniques. Optical fibers can be classified in terms of their optical length as short distance, medium distance, long distance, and very long distance, or in terms of their multiplexing scheme, as single channel **time division multiplexing (TDM)** or WDM. DWDM is an extension of the WDM scheme. Short length optical links are those links whose maximum length does not exceed 500 km, medium length optical links fall between 500 km and 2000 km, long length links fall between 2000 km and 5000 km and very long length links are those between 5000 km and 10,000 km. Transoceanic optical links fall into the last category (5000 km to 10,000 km). In terms of multiplexing, TDM was the multiplexing of choice until recently. With TDM, transmission rates of 10 Gb/s were possible. Currently, various research groups have concentrated their efforts in improving the TDM transmission capabilities beyond the 10 Gb/s to 20 Gb/s range and even 40 Gb/s with some success. The drive to enhance the transmission capability of TDM schemes is based on economic considerations. That is, for optical fiber links already installed, it makes economic sense to increase the transmission capacity instead of replacing them. However, for newly installed systems, the multiplexing scheme of choice is WDM. Through this method, low capacity optical channels are utilized and transmitted simultaneously, forming a very high capacity optical system. Another important advantage of the WDM system is transparency. That is, low bit rate optical channels can be dropped off or added onto the optical layer without the need of conversion from optical to electrical and electrical to optical signals. Regardless of the length of the optical link or the multiplexing scheme used (TDM or WDM) all optical links are composed of an optical transmitter, a transmission fiber, and an optical receiver.

Modern links have replaced their optical/electrical and electrical/optical in-line repeaters with EDFAs. Although optical amplifiers are capable of compensating for signal losses that occur during transmission through the optical fiber, they are themselves sources of ASE noise and are unable to compensate for the signal distortion accumulated through the transmission fiber. The majority of modern optical systems designed for digital transmission are **intensity modulated direct detection (IM-DD)** systems utilizing either **non-return to zero (NRZ)** or **return to zero (RZ)** data formats. A block diagram of an intensity modulation direct detection optical link is illustrated in Figure 12–1. The optical link in Figure 12–1 is composed of an optical transmitter, a number of fibers of length (L), and a number of optical repeaters composed of optical amplifiers, filters, and dispersion compensation modules. Data transmission limitations for systems such as these are the result of degradations induced by all the components incorporated into the optical link. However, different components induce different types and levels of degradation. For example, in low transmission rate systems (below 20 Gb/s), the main impairment contributors are the optical amplifiers and the transmission fiber. In high rate transmission systems, all the main components (including the transmitter and the receiver modules) contribute significantly to system limitations. The transmitter section is composed mainly of a laser diode that is directly modulated by the input electrical signal or a laser diode and an external modulator. For a detailed description of optical transmitters, see Chapter 7.

Direct modulation laser diodes induce unwanted pulse chirping, while optical transmitters that use external modulators do not. Modern optical systems employ external LiNbO$_3$ Mach Zehnder modulators. For best performance, any pulse broadening or chirping must not exceed 20% of the pulse width of the transmitted signal (Equation (12–1)).

$$20\% = \left(\frac{t}{T_{\text{FWHM}}}\right)^2 \qquad \textbf{(12–1)}$$

where t is the time and T_{FWHM} is the **full width at half maximum (FWHM)** of the transmitted pulse.

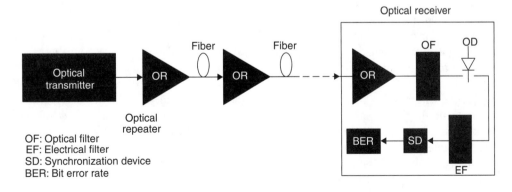

Figure 12–1.   A block diagram of an IM-DD optical link.

The modulation signal can either be an NRZ or RZ format. For NRZ, the optical signal is amplitude modulated with a resultant shape of a risen cosine function (T_{rise}) expressed by Equation (12–2).

$$T_{rise} = \frac{T}{4} \qquad (12\text{--}2)$$

or

$$T_{rise} = \frac{1}{4R}$$

where T is the inverse of the transmission rate and R is the transmission bit rate. For the RZ data format, the resultant optical signal is either a time dependent super Gaussian pulse expressed by Equation (12–3),

$$S(t) = Ae^{-0.5\left(\frac{t}{T_s}\right)^2} \qquad (12\text{--}3)$$

or it can also be a hyperbolic secant expressed by Equation (12–4).

$$S(t) = \frac{A}{\left(e^{\frac{-t}{T_s}} + e^{\frac{t}{T_s}}\right)} \qquad (12\text{--}4)$$

where A is the amplitude and $T_S = \frac{T_{FWHM}}{1.763}$. Under certain conditions, a hyperbolic RZ optically modulated signal will be able to compensate, avoid, or fix certain dispersion effects that will be discussed later in the chapter. This is related to **soliton.** A detailed discussion of soliton transmission will be given later in the chapter. The optically modulated signal at the transmitter output is launched into the core of the optical fiber, which allows the signal to propagate through it in a zig-zag mode. In the early days of fiber development, propagation of light through optical fibers was subject to substantial attenuation. However, the latest manufacturing techniques have reduced these losses to levels as low as 0.4 dB/km at the 1300 nm wavelength window and 0.2 dB/km at the 1500 nm wavelength window. Although advanced fabrication techniques have reduced signal attenuation, there are other parameters that set limits on the propagation of signals through optical fibers. These parameters are GVD, **polarization modal dispersion (PMD),** fiber nonlinearities, and **third order chromatic dispersion.** GVD is one of the main factors that induces pulse broadening. However, pulse broadening can be remedied with the introduction of **dispersion shifted fibers (DSF)** that operate at the 1550 nm wavelength window that have an effective length of 1000 km. For longer optical distances, chromatic dispersion is still a problem that seriously affects link transmission capabilities. On the other hand, PMD is a nonlinear effect capable of inducing pulse chirping and is based on the fiber's dual polarization nature. Because optical fibers have two polarization modes, each mode induces a separate chromatic dispersion

that ultimately leads to pulse broadening. It is, therefore, evident that pulse broadening depends on the **state of polarization (SOP)** of the input signal. PMD can also be caused by factors such as **waveguide ellipticity,** fiber stresses, and fiber bending. These factors can induce variations to group velocities. The study of PMD requires the incorporation of an important parameter, that of the *characteristic length* (L_c). Characteristic length is defined as the fiber length at which fifty percent of the optical power launched into a polarization state is transferred to an **orthogonal modal state.** PMD can be expressed by its mean value ($\delta\tau \sqrt{z}$). This value holds true if the characteristic length is much smaller than that of the optical fiber length. That is, $L_c < z$. The PMD mean value for modern fibers is on the order of 0.1 ps/km^{-1}. Although this may be considered insignificant for short, low capacity links, it can be very problematic for high capacity long distance links. The optical system performance can also be impaired by polarization effects such as **polarization dependent gain (PDG)** induced by inline optical amplifiers, **polarization dependent loss (PDL),** and **polarization hole burning (PHB).** Furthermore, other nonlinearities such as the **Kerr effect,** Brillouin scattering, and Raman scattering evident in single-mode fibers can induce associated noise at the input of the optical receiver and consequently reduce the system's OSNR. The effect of Raman scattering is mainly felt in optical systems that employ WDM schemes, which cause optical energy to be transferred from one optical channel to the adjacent channels, resulting in **intersymbol interference (ISI)** or crosstalk. The impact of Raman scattering on ISI can be marginalized if the condition expressed by Equation (12–5) is satisfied.

$$P_{total}B_{total}L_E < (9\text{mWTHzkm}) \qquad (12\text{--}5)$$

where P_{total} is the total optical power, B_{total} is the total optical bandwidth, $L_E = \frac{z}{L_{eff}}$, and $L_{eff} = \frac{1 - e^{-az}}{\alpha}$ or $L_{eff} \approx \frac{1}{\alpha}$, z is the link length, and α is the fiber attenuation. The third nonlinearity, the Kerr effect, is caused by the interaction of the field intensity with the fiber refractive index. Such an interaction causes signal broadening through the generation of additional frequencies. These frequencies induce an undesirable signal modulation referred to as **self phase modulation (SPM).** Although the Kerr effect contributes to SPM, it can also be used to compensate for system degradations induced by chromatic dispersion in single channel systems. However, the negative effect of Kerr nonlinearities is more evident in WDM and DWDM schemes. In these schemes, Kerr nonlinearities induce other phenomena such as **cross-phase-modulation (XPM)** and FWM. The generation of new frequencies is subject to the following condition expressed by Equation (12–6).

$$4\pi^2\beta_2\Delta f^2 L_i \ll 1 \qquad (12\text{--}6)$$

where β_2 is the chromatic dispersion, Δf is the frequency variations, and L_i is the fiber length harboring frequency interaction (fiber length where chromatic dispersion is constant). The propagation of signals through an optical fiber in the presence of GVD, PMD, fiber losses, and Kerr nonlinearities can be expressed by the nonlinear Schröendiger equation (Equation (12–7)).

$$\frac{\partial U}{\partial \zeta} + \frac{i}{2} \text{sign}(\beta_2)\frac{\partial^2 U}{\partial \zeta^2} + \frac{1}{2}\frac{L_D}{L_\alpha}U$$
$$= i\frac{L_D}{L_{NL}}|U|^2 U + \frac{1}{6}\frac{L_D}{L'_D}\frac{\partial^3 U}{\partial \tau^3} \tag{12–7}$$

where $U(z,t) = \dfrac{A(z,t)}{\sqrt{P_o}}, \tau = \dfrac{t}{T_o}, \zeta = \dfrac{z|\beta_2|}{T_o^2}, L_D = \dfrac{T_o^2}{|\beta_2|}$,

$L_{NL} = \dfrac{1}{\gamma.P_o}, L'_D = \dfrac{T_o^3}{|\beta_3|}, L_a = \dfrac{1}{\alpha}, P_o = $ peak signal power,

L_D = dispersion length, L_{NL} = nonlinear length, L'_D = third order dispersion length, L_α = attenuation length, A = electric field, β_2 = second order chromatic dispersion, β_3 = **chromatic third order dispersion,** and T_o = arbitrary constant (i.e., FWHM duration or pulse rise time).

In order to provide the necessary signal amplification at specific length intervals along the link, optical amplifiers are introduced. These amplifiers may be either EDFAs or SOAs. EDFAs and SOAs exhibit unique operating characteristics. For example, an SOA exhibits faster gain saturation than an EDFA. EDFAs also exhibit constant gain across the entire operating bandwidth. Although EDFAs provide signal amplification, they also add on noise in the form of ASEs. ASE noise is Gaussian across the operating bandwidth with an average power expressed by Equation (12–8).

$$\text{ASE}_{\text{noise}} = Fhv(G-1)B_A \tag{12–8}$$

where F is the incomplete polarization inversion, h is the Planck's constant, v is the carrier frequency, G is the gain, and B_A is the amplifier bandwidth. Commercial devices exhibit bandwidths of about 30 nm to 40 nm, while fluoride doped EDFAs exhibit bandwidths of about 160 nm, operating at the 1550 nm wavelength window. The above bandwidths are valid if only one amplifier is in use. If more than one optical amplifier is incorporated into the link, the effective bandwidth is then expressed by Equation (12–9).

$$B_{A(\text{eff})} = \frac{B_A}{N^{1/2}} \tag{12–9}$$

where $B_{A(\text{eff})}$ = effective bandwidth, B_A = single amplifier bandwidth, and N = number of amplifiers used in the optical link. The noise spectral components are independent Gaussian variables with variance (σ^2), expressed by Equation (12–10).

$$\sigma^2 = \frac{Fhv(G-1)\Delta v}{2} \tag{12–10}$$

where Δv is the bandwidth discrete Fourier component. The SOA gain is expressed by Equation (12–11).

$$\frac{dG}{dt} = \frac{G_o - G}{\tau_c} - \frac{P_{\text{in}}(t)}{E_{\text{sat}}}(e^G - 1) \tag{12–11}$$

where P_{in} is the amplifier input signal power, τ_c is the spontaneous carrier life time, G_o is the ideal gain required to compensate fiber losses ($G_o = e^{\alpha L_A}$), G is the **effective gain,** E_{sat} is the saturation energy ($\tau_c P_{\text{sat}}$), P_{sat} is the **saturation power,** α is the fiber loss, and L is the fiber length. Furthermore, the SOA ASE noise is expressed by Equation (12–12).

$$\text{ASE}_{\text{noise}} = Fhv[G(t) - 1] \tag{12–12}$$

In order to minimize the effect of SE and ASE, optical fiber links, especially those that incorporate dispersion management techniques, use **optical band pass filters (OBPF).** The transfer function of a Fabry-Perot (FP) filter is expressed by Equation (12–13).

$$H(V) = \frac{1 - R}{1 - R \, exp\left[\dfrac{i(V - V_o)}{\Delta V}\right]} \tag{12–13}$$

where ΔV is the spacing between adjacent resonant frequencies, R is the **mirror reflectivity (interferometer),** and $1 - R$ is the interferometer loss. For good filter quality, $1 - R$ must satisfy the condition $1 - R < 1$. The filter bandwidth is expressed by Equation (12–14).

$$BW = \frac{1 - R}{\pi\sqrt{R}}(\Delta V) \tag{12–14}$$

where B is the filter bandwidth. At the receiver end, the function of the optical receiver is to convert the optical power detected at the input to an electrical current by the PIN photodetector. A PLL is incorporated for timing recovery, while a threshold decision circuit is used to generate the transmitted binary stream. The advantage of such a coherent system is indicated by a substantial increase in the receiver SNR.

However, with the incorporation of EDFAs in the preamplifier stage, the optical levels can be substantially increased. An additional benefit arising from the use of an EDFA in the preamplifier stage is the reduction of the ASE noise at the input of the receiver through the incorporation of an OBPF. This decrease of the noise level also reflects an improvement of the receiver SNR. Another method of increasing receiver sensitivity is **forward error correction (FEC).** The FEC scheme requires that the transmitted data be specially encoded at the transmitter level, while the receiver is programmed to recognize any changes in the received data and perform the appropriate corrections. When an

IM-DD receiver is used, such as that in Figure 12–1, an electrical filter must be placed between the output of the photodiode and the receiver front end. Finding the correlation of the input and output signals achieves the required synchronization between the transmitted signal and the received signal. Knowledge of the signal delay (transmit received) determines the sampling instant. The electrical filter bandwidth inserted between the PIN and the receiver front end is subject to transmitted data format. For example, an **NRZ** data format requires a filter bandwidth between $4R$ and $10R$, subject to chromatic dispersion levels and the dispersion management scheme employed. An RZ data format requires a filter bandwidth of $4R$ (R is the transmission bit rate). The optical receiver overall performance in the presence of ASE and Kerr nonlinearities can be evaluated by the level of BER and **Q-factor.** Equation (12–15) expresses the relationship between these two parameters.

$$\text{BER} = \frac{1}{2} \, erfc\left(\frac{Q}{\sqrt{2}}\right) \qquad \textbf{(12–15)}$$

where Q-factor is the SNR at the input of the decision circuit, expressed in terms of voltage or current. Although Equation (12–15) is fairly accurate for an NRZ signal pattern, it does not fully satisfy the receiver performance criteria for an RZ signal pattern. For a complete receiver performance evaluation in which an RZ data format is used, jitter must also be taken into consideration. For a satisfactory optical receiver performance evaluation detecting RZ data format, the error probability $P_{(e)}$ must not be higher than 10^{-9} at a corresponding Q-factor of six and jitter standard deviation of no more than 6% of the bit pulse width. The maximum bit rate ($R_{\max}$) that an optical link is capable of transmitting is primarily subject to second order chromatic dispersion (β_2) and transmission distance (z), assuming that no pulse chirping is induced by the optical source. The relationship between the second order chromatic dispersion (β_2), transmission distance (z), and maximum system bit rate ($R_{\max}$) is expressed by Equation (12–16).

$$R_{\max} = \left(\frac{0.1}{\beta_2 z}\right)^{1/2} \qquad \textbf{(12–16)}$$

where $R_{\max}$ is the maximum system bit rate, β_2 is the second order GVD, and z is the optical link distance. As mentioned, chromatic dispersion is the major factor that contributes to pulse broadening, the degree of which is a function of the dispersion coefficient (β_2). For the proper study of the link performance, two levels of chromatic dispersion are required, a low GVD and a high GVD. A low chromatic dispersion link can be defined by the relationship in Equation (12–17).

$$R \, \bar{\beta} \, z < 1 \qquad \textbf{(12–17)}$$

where $\bar{\beta}$ is the average chromatic dispersion coefficient.

If the chromatic dispersion varies along the fiber link, the shape of the transmitted bit also varies proportionally to chromatic dispersion variations. However, if the accumulated GVD at the end of the link is very small, its effect on the transmission data is also very small. Although the shape of the transmitted data is not generally affected by a small chromatic dispersion, it is affected by other fiber nonlinearities and system noise. If the accumulated chromatic dispersion at the end of the optical link is at a level which causes substantial pulse broadening, then the relationship between system bit rate average second order chromatic dispersion and link distance is as follows in Equation (12–18).

$$R \, \bar{\beta} \, z > 1 \qquad \textbf{(12–18)}$$

In some cases, the chromatic dispersion effect on pulse chirping can be substantially reduced, or perhaps altogether eliminated, by the Kerr nonlinear effect. In WDM schemes, the effect of FWM can be detrimental to system performance. In optical links with local high GVD, dispersion compensation is essential. The reduction of chromatic dispersion can be accomplished through the process of dispersion management. Normally, this process involves the incorporation of a compensation fiber of length L_{comp} with a dispersion coefficient equal and opposite to the accumulated chromatic dispersion through the transmission fiber. The graphical representation of this dispersion compensation scheme has a sawtooth distribution form that indicates a periodic compensation of the **chromatic second order (CSO) dispersion** β_2 at a distance L_L along the transmission fiber, utilizing another fiber of length L_{comp} with negative dispersion coefficient β'_2. Based on this dispersion management scheme, the average dispersion coefficient $\bar{\beta}$ can be expressed by Equation (12–19).

$$\bar{\beta} = \frac{L_L \beta_2 + L_{\text{comp}} \beta'_2}{L_{\text{comp}} + L_L} \qquad \textbf{(12–19)}$$

where $\bar{\beta}_2$ = Average **dispersion coefficient,** L_L = fiber length at the end of which dispersion compensation is required, L_{comp} = dispersion compensation fiber length, β'_2 = dispersion coefficient of the dispersion compensation fiber, and β_2 = dispersion coefficient of the transmission fiber.

The described method of **dispersion management** can also be applied to already existing optical links. In such a system, the compensation fiber (L_{comp}) can be induced at the same location as that of the optical amplifier. Another method of dispersion management is to incorporate a specially designed **dispersion compensation fiber (DCF).** High chromatic dispersion severely impedes the system bandwidth-length product and must therefore be properly managed. Other dispersion management techniques can be implemented at either the transmitter module or receiver module for relatively short optical links, (100 km to 200 km). For long length high bit rate systems, the recommended dispersion management scheme is that of **phase conjugation** or **spectral inversion.** With this dispersion management scheme, Kerr nonlinearities can also be phase conjugated, contributing to chromatic dispersion compensation. A more detailed examination of spectral inversion, or phase-conjugation, is presented later in this chapter. Kerr nonlinearities can fully compensate for pulse distortion, which is the result of chromatic dispersion under certain conditions such as when the GVD must be locally high and variable across the fiber length, and the propagating pulse must have a sech-hyperbolic shape (soliton), with a peak optical power P_{peak} expressed by Equation (12–20).

$$P_{peak} = \frac{|\beta_2|}{\gamma \tau_s^2} \qquad (12\text{–}20)$$

where P_{peak} = soliton peak pulse power, β_2 = dispersion coefficient of the transmission fiber, γ = Kerr nonlinearity coefficient, and $\tau_s = \dfrac{T_{FWHM}}{1.763}$. Soliton pulses can theoretically propagate at infinite distances without experiencing any pulse distortion. However, distortion does occur and is subject to fiber losses. If fiber losses are periodically compensated, for example, through incorporation into the line of EDFAs, then the transmission fiber may be considered as one that has zero power losses. In this case, the soliton RZ peak pulse power P_{peak} is modified as follows in Equation (12–21).

$$P_{peak} = \frac{\alpha L_A\, G|\beta_2|}{(G-1)\gamma \tau_s^2} \qquad (12\text{–}21)$$

where $\dfrac{\alpha L_A G}{G-1}$ expresses power fluctuations along the fiber length, L_A is the optical amplifier fiber length, and α is the fiber loss. Soliton transmission through an amplified optical system exhibits a level of instability caused by a decrease of the optical pulse power during transmission. A decrease of the soliton power promotes broadening of the pulse width. However, if soliton properties are to be maintained, the **soliton period** (z_o) must have a set value expressed by Equation (12–22).

$$z_o = \left(\frac{\pi}{2}\right)\left(\frac{\tau_s^2}{\beta_2}\right) \qquad (12\text{–}22)$$

where z_o = soliton period, $\tau_s = \dfrac{T_{FWHM}}{1.763}$, and β_2 = dispersion coefficient of the transmission fiber. If soliton propagation in a lossless optical transmission link (optically amplified system) is to be achieved with zero pulse chirping, it must satisfy the theoretical condition $z_o > L_A$. However, in practical applications, this condition can be modified as follows in Equation (12–23).

$$z_o \geq 1.25\, L_A \qquad (12\text{–}23)$$

Equation (12–23) must always be satisfied if pulse chirping is to be avoided. In a case in which Equation (12–23) is not satisfied, soliton pulse broadening will occur as a result of **power loss.** The power loss is converted into a wide spectrum radiative wave with frequencies outside the soliton spectral components. These frequencies travel through the fiber at different group velocities. The Kerr nonlinearities promote interaction between the soliton pulses and the parasitic frequencies that lead to excessive jitter. Therefore, it is important that a minimum relationship must exist between amplifier fiber length (L_A) and soliton period (z_o) (Equation (12–23)). Soliton transmission is also affected by soliton interaction and the time jitter generated by frequency components of the soliton pulses and the ASE of the optical amplifiers incorporated into the link. This phenomenon is the **Gordon-Haus effect.** An increase of the time jitter beyond a predetermined level promotes an increase of the system error probability ($P_{(e)}$). If the optically amplified soliton transmission link does not employ soliton inline control, then the jitter variance (σ_{jitter}) is expressed by Equation (12–24).

$$\sigma_{jitter} = \left(\frac{h\nu\alpha\gamma|\beta_2|FL^3}{9\tau_s}\right)^{1/2}\left[\frac{(G-1)^2}{G[\ln(G)]^2}\right]^{1/2} \qquad (12\text{–}24)$$

The error probability ($P_{(e)}$) can be maintained at levels below 10^{-9} if the following condition is satisfied (Equation (12–25)).

$$\sigma_{jitter}\, R < 0.06 \qquad (12\text{–}25)$$

where σ_{jitter} is the Gordon-Haus jitter variance and R is the transmission bit rate. Another critical requirement in soliton transmission is the maintenance of a relatively high $SNR_{electrical}$ at the input of the optical receiver, that is, a high signal power and low noise power at the receiver input. Equation (12–26) expresses the electrical $SNR_{electrical}$.

$$SNR_{elect} = \frac{\overline{P}}{N_a\, h\nu F\,(G-1)\, B_e} \qquad (12\text{–}26)$$

where N_a is the number of optical amplifiers incorporated into the optical link, h is Planck's constant, ν is the carrier frequency, F is the receiver noise figure, $\overline{P}$ is the average received optical power at bit intervals, and B_e is the electrical bandwidth. System performance can also be expressed in terms of the Q-factor. The relationship between the Q-factor and $SNR_{electrical}$ is expressed by Equation (12–27).

$$Q = \sqrt{SNR_{electrical}} \qquad (12\text{–}27)$$

A Q-factor larger than six ($Q > 6$) is considered a good performance indicator. In this introductory discussion of soliton propagation, it is assumed for simplicity that the propagation occurs in a single mode. However, this assumption is not entirely correct. Soliton propagation that occurs in a dual polarization state medium is scalar and, as a consequence, the required peak power (P_{peak}) is slightly higher than that expressed by Equation (12–21). The required soliton peak power (P_{peak}), assuming a dual polarization mode of propagation, is expressed by Equation (12–28).

$$P_{peak(A)} = 1.125 \left[\frac{\alpha L_A \, G |\beta_2|}{(G-1)\gamma \tau_s^2} \right] \qquad \textbf{(12–28)}$$

A typical relationship between system bit rate (R) and pulse time duration (τ_s) for a soliton is illustrated in Figure 12–2.

The bit rate and time duration as a function of Gordon Haus generated jitter is illustrated in Figure 12–3.

12.2 OPTICAL POWER BUDGET ANALYSIS

Link Power Margin

As mentioned in the introduction of this chapter, power margin refers to the optical power at the input of the receiver above a set minimum. The set minimum is subject to receiver sensitivity. Traditionally, optical power budget analysis is performed through the application of the linear worst case analysis, while for a more accurate result the statistical method is preferable.

THE LINEAR WORST CASE METHOD. For power budget calculations based on the first order approximation or linear worst case scenario, the following assumptions are made:

a) transmitter optical power: minimum

b) link power losses: maximum

c) receiver sensitivity: minimum

The optical power margin, based on the above assumptions, is expressed by Equation (12–29).

$$P_{M(dB)} = P_{t(dB)} - P_{r(dB)} - P_{L(dB)} \qquad \textbf{(12–29)}$$

where P_M is the power margin (dB), P_t is the transmitter power output (min) (dB), P_r is the receiver sensitivity (min) (dB), and P_L is the link optical power losses (max) (dB). When multimode fibers are incorporated into the system, an extra 0.5 dB must be added to the losses, and 1 dB must be added to the clock and data recovery module. Link losses may also include chromatic and modal dispersion losses, the level of which is subject to the type of fiber used and its operating wavelength, fiber attenuation, and connector and splice losses. The linear worst case method is a conservative method of optical power budget calculations, because it is based on a worst case performance of all the components incorporated into the system design. However, because this is not always the case, a more accurate method for power margin calculations is required. The method, which provides a more reliable and accurate method of power margin evaluation, is the *statistical method*. The system designer can employ the statistical method of power margin evaluation when a cost effective and highly efficient optical link is to be implemented.

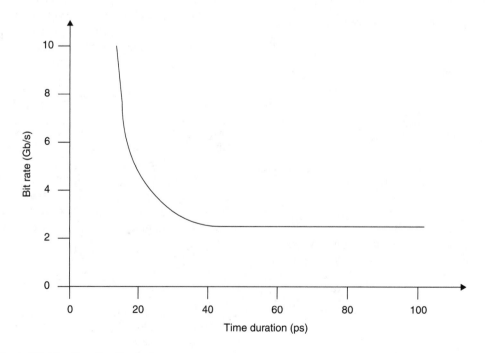

Figure 12–2. Bit rate (Gb/s) *v.* time duration (ps).

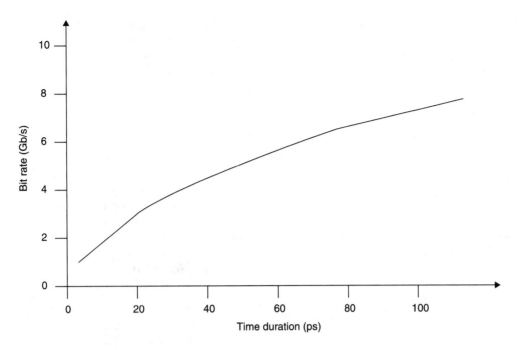

Figure 12–3. Bit rate (Gb/s) *v.* time duration (ps) relevant to Gordon Haus effect.

Example 12–1

Calculate the power margin of an optical communications link that has a transmitter power of 10 µW and receiver sensitivity of 1 µW.

Solution

$$P_{M \, (\mathrm{dB})} \; = \; P_{t \, (\mathrm{dB})} - P_{r_{\min} (\mathrm{dB})}$$

where $P_{t(\mathrm{dBm})}$ is the transmitter power and $P_{r\min(\mathrm{dBm})}$ is the **receiver sensitivity.**

Transmitter power $(P_{t \, (\mathrm{dBm})}) \; = \; 10 \log \left(\dfrac{P_t}{1 \; \mathrm{mW}} \right)$

$$= \; 10 \log \left(\frac{10 \times 10^{-6} \; \mathrm{W}}{1 \times 10^{-3} \; \mathrm{W}} \right) = -20 \; \mathrm{dBm}$$

Therefore, $P_{t \, (\mathrm{dBm})} \; = \; -20 \; \mathrm{dBm}$.

Receiver sensitivity $(\mathrm{P}_{t(\mathrm{dBm})})$

$$= \; 10 \log \left(\frac{1 \times 10^{-6} \; \mathrm{W}}{1 \times 10^{-3} \; \mathrm{W}} \right) = -30 \; \mathrm{dBm}$$

Therefore, $P_{t \, (\mathrm{dBm})} \; = \; -30 \; \mathrm{dBm}$.

Substituting for values:

$$P_{M \, (\mathrm{dB})} \; = \; P_{t \, (\mathrm{dB})} - P_{r_{\min} (\mathrm{dB})}$$
$$= \; -20 \; \mathrm{dBm} - (-30 \; \mathrm{dBm})$$
$$= \; -20 \; + \; 30$$

$$= \; 10 \; \mathrm{dB}$$

Therefore, power margin $(P_M) = 10 \; \mathrm{dB}$.

Example 12–2

Compute the receiver sensitivity incorporated into an optical link that has a required power margin of 15 dB and available transmitter power of 25 µW.

Solution

$$P_{M \, (\mathrm{dB})} \; = \; P_{t(\mathrm{dB})} - P_{r_{\min}(\mathrm{dB})}$$

Solve for $P_{r_{\min} \, (\mathrm{dB})}$.

$$P_{r_{\min} \, (\mathrm{dB})} \; = \; P_{t(\mathrm{dBm})} - P_{M \, (\mathrm{dBm})}$$

Transmitter power $(P_{t \, (\mathrm{dBm})}) = 10 \log \left(\dfrac{P_t}{1 \; \mathrm{mW}} \right)$

$$= \; 10 \log \left(\frac{25 \times 10^{-6} \; \mathrm{W}}{1 \times 10^{-3} \; \mathrm{W}} \right) = -16 \; \mathrm{dBm}$$

Substituting for values:

$$P_{r_{\min} \, (\mathrm{dB})} \; = \; P_{t(\mathrm{dBm})} - P_{M \, (\mathrm{dBm})}$$
$$= \; -16 \; \mathrm{dBm} - 15 \; \mathrm{dB}$$
$$= \; -31 \; \mathrm{dBm}$$

Therefore, $P_{r_{\min} \, (\mathrm{dB})} \; = \; -31 \; \mathrm{dBm}$. The minimum optical receiver sensitivity must be –31 dBm in order to comply with the link requirements.

TABLE 12–1 Optical Power in Reference to Fiber Diameter

Fiber (μm)	Ratio	dB
50/50	1	0
50/62.5	2.19	3.4
50/85	3.76	5.7
50/100	5.8	7.6
50/200	32	15.1
62.5/50	0.46	–3.4
62.5/62.5	1	0.0
62.5/85	1.72	2.3
62.5/100	2.65	4.2
62.5/200	14.63	11.7
85/50	0.27	–5.7
85/62.5	0.58	–2.3
85/85	1	0.0
85/100	1.54	1.9
85/200	8.52	9.3
100/50	0.17	–7.6
100/62.5	0.32	–4.2
100/85	0.65	–1.9
100/100	1	0.0
100/200	5.52	7.4
200/50	0.03	–15.1
200/62.5	0.07	–11.7
200/85	0.12	–9.3
200/100	0.18	–7.4
200/200	1	0.0

TABLE 12–2 Losses and Numerical Aperture of Various Fiber Types

Size	Loss/km	NA
200/230	7 dB	0.27
100/140	5 dB	0.29
62.5/125	4.5 dB	0.28
50/125	4 dB	0.20

Transmitter Power Launch into the Optical Fiber Cable

The optical power launched into the fiber cable is subject to core cross-sectional area and **numerical aperture (NA).** A small core cross-sectional area and a small NA reduce the amount of optical power launched into the fiber cable. The coupled optical power from an LED transmitter into the various core diameter fibers is obtained from Table 12–1.

Example 12–3

An LED transmitter can couple –18 dBm of optical power into a 50/125 micron fiber. The same transmitter can couple –10 dBm of optical power into a 100/140 micron fiber. Calculate the level of additional power launched into the 100/140 fiber.

Solution
$$-18 \text{ dBm} - (-10 \text{ dBm}) = -8 \text{ dB}$$

Therefore, there will be 6.3 times more optical power launched into 100/125 fibers than in 50/125 micron fibers.

The attenuation and numerical aperture of various fiber cables is obtained from Table 12–2.

The connection of both ends of the fiber cable to either the transmitter or the receiver is accomplished through spe-cial connectors. The two basic types of connectors are the Amphelon SMA-905 and SSA-906, and the AT&T ST®. The ST® is a more reliable connector because it incorporates an alignment slot.

Receiver Sensitivity

Sensitivity defines the absolute minimum optical power level required at the input of the optical receiver for reliable signal detection. Receiver sensitivity is measured either in μW or dBm, and is a function of the optical link speed. The higher the system capacity, the lower the receiver sensitivity.

Link Loss Power Budget

The link loss power budget represents all the power losses in the link. For effective operation, the link power loss budget must be smaller than the link power budget or the link power budget must be higher than the link power loss budget by a level referred to as the safety margin. These three components are interrelated as follows:

link power budget = link loss budget + safety margin

Link losses are subject to optical fiber attenuation, splicing losses, and connector losses. Splicing losses are also subject to the degree of sophistication used for splicing, and can vary from as low as 0.15 dB up to 3 dB.

Example 12–4

Design an optical link with 10 dB of optical power launched into a 50/125 micron fiber. Calculate the link power budget and the required transmitter power.

Solution
link power budget = transmitter power
$$-\text{receiver sensitivity} + \text{safety margin}$$

Assuming a safety margin of 3 dB and receiver sensitivity of –30 dBm (1 μW), the transmitter power is then calculated.

$$P_t = P_{LB} + P_{r_{min}}$$
$$= 10 \text{ dB} + 3 \text{ dB} + (-30 \text{ dBm}) = -17 \text{ dB}$$

Therefore,
$$P_t = -17 \text{ dBm}$$

---

or

$$P_t = 20 \ \mu W$$

The required optical transmitter power is −17 dBm or 20 μW.

Example 12–5

An optical fiber link designed to operate at a maximum bit rate of 50 Mb/s employs an LED transmitter and a PIN photodetector diode as the receiver. The link requires a 6 dB power budget and a 3 dB safety margin. The receiver section draws 1.5 μA of current and the LED draws 100 mA. Calculate the PIN diode required operating power and the total power budget.

Solution

Select a PIN photodiode. Assume PIN sensitivity 0.5 A/W.

i) Calculate the PIN diode required operating power. Receiver current is equal to 1.5 μA,

    or

$$1.5 \times 10^{-6} = 0.5 \ A/W(X_{watts})$$

Solve for (X_{watts}).

$$(X_{watts}) = \frac{1.5 \times 10^{-6} \ A}{0.5 \ A/W} = 3 \times 10^{-6} \ W$$

Therefore, the electrical power required by the PIN diode is 3 μW or −25.2 dBm.

ii) Calculate the total power budget.

total power budget = total loss budget

    + safety margin

total loss budget = loss due to optical fiber cable

    + loss due to patch panels

    = 5 dB (specified)

Assuming a safety margin of 3 dB,

total power budget = total loss budget

    + safety margin

    = 5 dB + 3 dB = 8 dB

Because the total power budget = transmitter power − receiver sensitivity, solving for transmitter power yields the following:

transmitter power = total power budget

    + receiver sensitivity

    = 8 dB + (−25.2 dBm)

    = −17.2 dBm

Therefore, transmitter power required is −17.2 dBm.

12.3   10 Mb/s OPTICAL LINK DESIGNED FOR INDUSTRIAL APPLICATIONS

A 10 Mb/s optical link can be designed to facilitate various industrial applications such as series data interfaces used in robotic applications, printing machines, and industrial assemblies. These applications, which require fast data transfer from machine to machine can best be implemented through the utilization of optical fiber links. Another important advantage of optical fiber links in industrial applications is their immunity from stray electromagnetic fields commonly encountered in industrial settings.

Component Characteristics

THE OPTICAL TRANSMITTER.   The preferable choice for optical transmitters used in industrial applications is the HP-HFBR-1528. The transmitter incorporates a high quantum efficiency LED device based on the AlInGaP structure, and is capable of delivering −3 dBm of optical power into a 1 mm POF with 60 mA of drive current. The transmitter, which operates at the 650 nm wavelength window takes full advantage of the POF attenuation properties, because such a fiber exhibits a minimum attenuation at the same wavelength window of 650 nm. The combination of the LED based optical transmitter and the POF can achieve link distances of about 100 m, while the same transmitter coupled into **hard clad silica fibers (HCSF)** can achieve distances of 500 m. Transmitter optical power output as a function of the drive current is illustrated in Figure 12–4.

THE OPTICAL RECEIVER.   The optical receiver normally selected for industrial applications is the HP-HFBR-2528. This receiver can operate at maximum data rates of 10 Mb/s (NRZ) and is capable of providing both CMOS and TTL compatible outputs. The receiver optical sensitivity is set at −21 dBm when interfaced with a 1 mm POF, or at −23 dBm when interfaced with 200 mm HCSF. The receiver also incorporates a PWD correction circuit that limits the pulse width distortion to a maximum of ±30 nm. This relatively low PWD is due mainly to the receiver's equal t_{PLH} and t_{PHL}. Equal rise and fall times contribute to the maintenance of a constant LED drive current for various fiber types and link distances—an ideal prerequisite for applications such as gate drive and arbitrary duty-cycle links. A block diagram of the receiver is illustrated in Figure 12–5.

OPTICAL CABLES.   In terms of their applications, optical cables (the fiber and the connector) are divided into two categories: those applicable to long haul optical communications and local area networks and those applicable to short distance industrial links. For long haul optical communications and LANs, optical fibers with large bandwidths and very low attenuation are required. For example, **fiber distributed data interface (FDDI)** and Ethernet optical systems

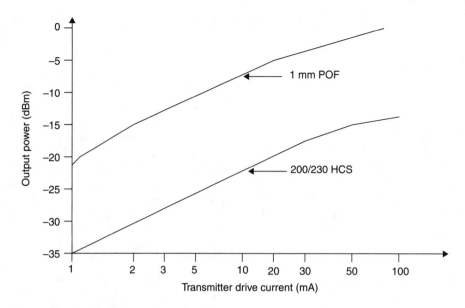

Figure 12–4. Transmitter optical power output *v.* drive current.

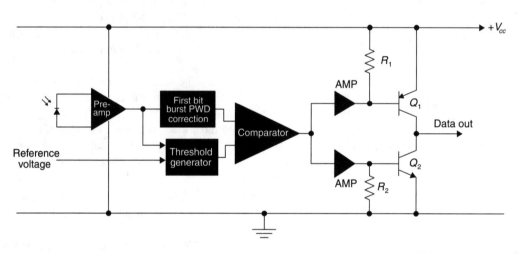

Figure 12–5. A block diagram of an industrial LED-based optical receiver.

employ the 62.5/125 μm small core glass fibers that require high precision connectors to keep coupling losses to a minimum. In industrial applications, 1 mm POF or 200 μm hard clad silica fiber (HCSF) is recommended. These fibers exhibit a high tolerance to industrial environment and require less expensive connectors; 2.2 mm jackets surround both types of fibers, providing thermal and mechanical protection, while at the same time increasing the cable strength—a requirement for industrial applications.

Furthermore, POFs are ideal for industrial applications because of their large core diameter (980 μm to 1100 μm) and high NA. Both the large core and NA are well matched to the large optical ports, thus allowing more optical power to be coupled into the fiber. This optical power can be as high as 0 dBm. High coupling power, inexpensive terminators, and relative simplicity in assembling make POFs the ideal choice for industrial applications.

The combination of LED transmitters that operate at the 650 nm wavelength window is compatible to POFs that exhibit a minimum attenuation (0.2 dB/km) at the same wavelength window. The graphical representation of POF attenuation and the operating wavelength is illustrated in Figure 12–6. The utilization of step index (200 μm) core HCSFs in the design of optical links intended for industrial applications relaxes the optical transmitter and receiver specifications considerably. The selection of such fibers allows the utilization of low cost optical transmitters and receivers. HCSFs exhibit 8 dB/km attenuation in the 650 nm wavelength window. A graphical representation of attenuation and wavelength of an HCSF is illustrated in Figure 12–7.

A comparison between the physical properties and operating characteristics of POFs and HCSFs is presented in Table 12–3.

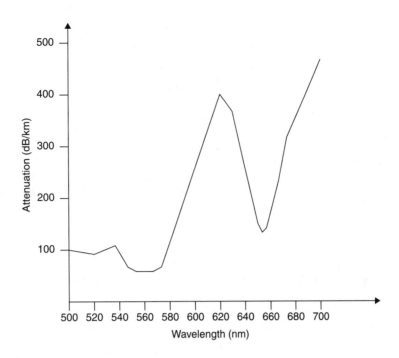

Figure 12–6. POF attenuation *v.* wavelength.

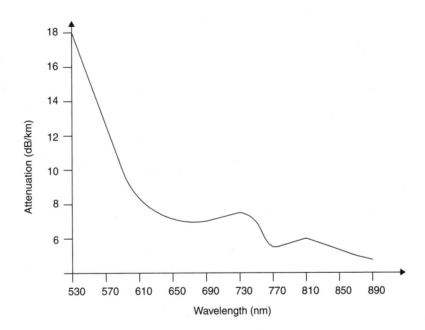

Figure 12–7. HCSF attenuation *v.* wavelength.

TABLE 12–3 Physical Properties and Operating Characteristics of POF and HCSF Fibers

Parameters	POF	HCSF	Units
Attenuation (at 660 nm)	200	6	dB/km
Tension (60 min)	50	100	N
Tension (10 year)	1	25	N
Numerical aperture (NA)	0.47	0.37	—
Bend radius	25	10	mm
Installation temperature	−20 to +70	−20 to +85	°C

12.4 OPTICAL FIBER LINK DESIGN

The proposed optical link intended for industrial applications must satisfy the following operating characteristics:

a) system bit rate: $f_b = 10$ Mb/s

b) link length: $L = 60$ m (max) with POF

c) $L = 500$ m (max) with HCSF

d) allowable power margin: 3 dB

Link Considerations

The basic components of this optical link are the LED transmitter, the fiber, and the optical receiver. The LED transmitter is modulated by an electrical signal and the modulated optical power is then coupled into the fiber. The fiber conveys the optical power into the input of the optical detector, which converts the optical power into an electrical signal identical to the transmitter electrical signal. The three fundamental parameters that determine system functionality are:

a) optical power coupled into the fiber

b) optical fiber attenuation

c) optical receiver sensitivity

Normally, every optical link is length and bandwidth limited. That is, for system bit rates beyond 100 Mb/s and at a specified link length, signal distortion will occur because of fiber dispersion. However, in industrial applications, link lengths do not exceed 100 m and system rates do not exceed 10 Mb/s. Because fiber dispersion is critical at link lengths longer than 100 m for POFs and 1 km for HCSFs, neither the length nor the system bit rate are critical to the design of such systems.

Optical Power Budget (P_{budget})

The link power budget is defined as the difference between the transmitter optical power and the receiver optical sensitivity (Equation (12–30)).

$$P_{budget} = P_{t(min)} - P_{r(min)} \qquad \textbf{(12–30)}$$

where P_{budget} is the link power budget (W), $P_{t(min)}$ is the minimum optical transmitter power (dBm), and $P_{r(min)}$ is the optical receiver sensitivity (dBm).

Link Length (L_{max})

The maximum length of the optical link (L_{max}) is determined by the power budget, optical fiber attenuation, insertion losses, and optical **power margin** (Equation (12–31)).

$$L_{max} = \frac{P_{budget} - P_{IL} - P_{OM}}{\alpha_{max}} \qquad \textbf{(12–31)}$$

where L_{max} is the maximum link length (m), P_{budget} is the optical power budget (W), P_{IL} is the insertion losses (dB), P_{OM} is the optical power margin (dB), and α_{max} is the fiber attenuation (dB/km). A graphical representation of the optical power levels at different points along the link is illustrated in Figure 12–8.

Optical Receiver Dynamic Range

Optical receiver **dynamic range** is the difference between receiver sensitivity and overdrive. Knowledge of the receiver dynamic range will determine the minimum and maximum optical link lengths. A system that operates beyond the set receiver dynamic range will induce excessive PWD.

Determining Minimum Link Distance

The minimum link distance is expressed by Equation (12–32).

$$L_{(min)} = \frac{P_{t(max)} - P_{r(max)}}{\alpha_{(min)}} \qquad \textbf{(12–32)}$$

where, $L_{(min)}$ is the minimum link distance (m), $P_{t(max)}$ is the maximum transmitter optical power (dBm), $P_{r(max)}$ is the receiver optical sensitivity (dBm), and $\alpha_{(min)}$ is the optical fiber minimum attenuation (dB/km). The relationship between PWD and receiver input optical power is illustrated in Figure 12–9.

Determining Coupling Power Losses (P_{L_C})

In an optical link, power loss occurs when power is transferred from one fiber to another that has a different core diameter and NA. Equation (12–33) expresses coupling losses.

$$P_{L_C} = 20 \log\left(\frac{d_1}{d_2}\right) + 20 \log\left(\frac{NA_1}{NA_2}\right) \qquad \textbf{(12–33)}$$

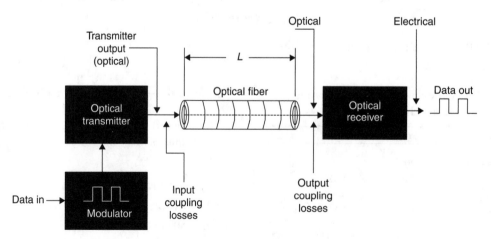

Figure 12–8. Optical power levels at various points along the link.

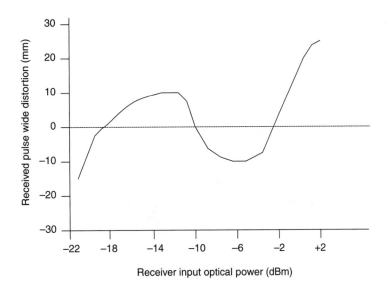

Figure 12–9. PWD *v.* receiver input optical power.

where P_{L_c} is the coupling power losses (dB), d_1 is the diameter of fiber 1 (m), NA_1 is the numerical aperture of fiber 1, d_2 is the diameter of fiber 2 (m), and NA_2 is the numerical aperture of fiber 2, assuming optical power is coupled from fiber 1 to fiber 2 and $d_1 > d_2$. If optical power is conveyed from fiber 1 to fiber 2 and $d_1 < d_2$, then coupling losses are equal to zero.

Temperature Drift

The maximum length and overall link performance are subject to transmitter optical power output drifts. Trans-

mitter optical power outputs at different temperatures can be obtained from graphs. One such graph is illustrated in Figure 12–10.

Transmitter optical output power as a function of the junction temperature is expressed by Equation (12–34).

$$P_{t(T)} = \left[P_{t(25°C)} - \left(\frac{\Delta P_t}{\Delta T} \right)(T-25) \right] \quad \textbf{(12–34)}$$

where $P_{t(T)}$ = transmitter optical power at a desirable temperature, $P_{t(25)}$ = transmitter optical power at room temperature, and $\dfrac{\Delta P_t}{\Delta T}$ = coefficient (dB/°C). The mechanism

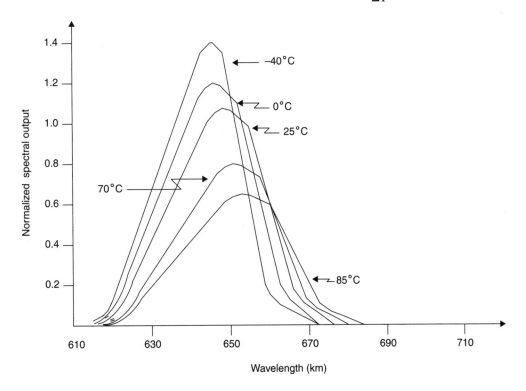

Figure 12–10. Transmitter optical power output as a function of operating temperatures.

through which the diverse effect of the junction temperature increases, mainly affecting the transmitter optical power, can be properly managed by controlling the LED forward bias voltage (V_F). Equation (12–35) expresses the LED forward voltage in relationship to operating temperature.

$$V_{F(T)} = V_{F(25)} - \left[\left(\frac{\Delta V_F}{\Delta T} \right)(T-25) \right] \quad \textbf{(12–35)}$$

where $V_{F(T)}$ = forward voltage at specified operating temperature (V), $V_{F(25)}$ = forward voltage at room temperature (V), and $\dfrac{\Delta V_F}{\Delta T}$ = coefficient ($\Delta V/\Delta °C$).

Connector Losses

Connector losses for the HP-HFBR45X5 POF 1 mm fiber are:

a) minimum: 0.7 dB

b) typical: 1.5 dB

c) maximum: 2.8 dB

LED Driver Circuits

The LED transmitter in this application can either be driven in a series or parallel configuration. A series driving circuit is illustrated in Figure 12–11.

From Figure 12–11, the LED is in series configuration with the collector of the driver transistor and a current limiting resistor R_2 connected to V_{cc}. The function of R_2 is to set the LED drive current, and the function of R_1 is to provide a discharge path at the turn OFF state of the LED. The value of R_1 is expressed by Equation (12–36).

$$R_1 = \frac{V_{cc} - V_{CE} - V_F}{I_F} \quad \textbf{(12–36)}$$

where V_{cc} = power supply voltage (V), V_{CE} = collector to emitter voltage (V), V_F = LED forward voltage (V), and I_F = forward current (A). R_1 provides a low impedance path that allows the LED to discharge very rapidly, which decreases the optical fall time substantially. For best PWD performance, the value of R_1 is set at 2 kΩ. The function of C_1 and C_2 is to reduce the power supply noise during the ON and OFF switching times. The relationship between rise/fall times and PWD as a function of resistor R_1 is illustrated in Figure 12–12.

The significant deficiency of the series LED driver circuit is based on the fact that the LED forward current is temperature dependent. The LED forward current is also subject to power supply variations, current limiting resistor tolerance, and V_{CE} variations. However, HFBR 1528 LED transmitters that exhibit a negative temperature coefficient maintain a relatively constant LED forward bias current. A parallel LED driving circuit is illustrated in Figure 12–13.

The design of the parallel LED driving circuit in Figure 12–13 is based on both simplicity and performance. The circuit is composed of a PNP transistor in parallel with the LED, which can be interfaced with either CMOS or TTL gates. The speed of the driver circuit is a function of the transistor nonsaturation mode of operation, low LED impedance at the OFF state, and the prebias of the emitter base junction. The only deficiency in this circuit is the increase of power supply current consumption, due to the constant current flow through the bias resistor (R). However, the constant current drain minimizes the power supply generated noise and therefore relaxes the optical receiver sensitivity strict requirements somewhat.

The Statistical Method of Power Margin Evaluation

For a more accurate and reliable evaluation of the total link losses, a statistical model is used. The statistical model

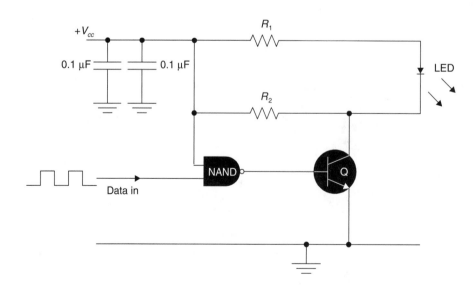

Figure 12–11. A series LED driving circuit.

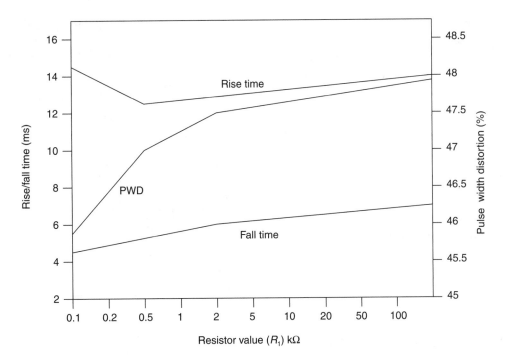

Figure 12–12. R_1 v. PWD (For $I_F = 60$ mA, $T_A = 25°C$, 1 mm HCSF)

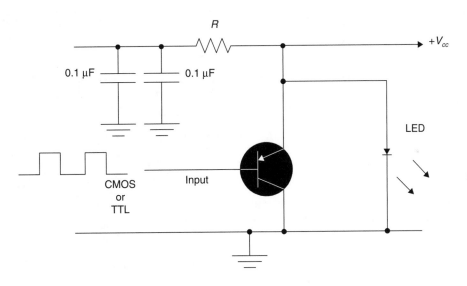

Figure 12–13. A parallel LED driving circuit.

employed to evaluate link losses incorporates terms relating to higher order mode losses and dispersion. Finally, the total link loss is compared to the difference between the mean transmitter power and receiver sensitivity. The terms incorporated in the statistical model are expressed by Equation (12–37) (mean).

$$\mu M = \mu t - \mu r - (\mu cL + \mu coNco$$
$$+ \mu sNs + \mu D + \mu H + \mu CR) \quad \textbf{(12–37)}$$

where μM is the mean link margin, μt is the mean transmitter power, μr is the mean receiver sensitivity, μc is the mean cable loss, L is the system length, μco is the mean

connector loss, Nco is the number of connectors, μs is the mean splice loss, Ns is the number of splices, μD is the mean dispersion penalty, μH is the mean higher order mode loss, and μCR is the mean **clock recovery** and **data recovery** module loss.

The terms incorporated into the statistical model are expressed by Equation (12–38) (sigma).

$$\sigma M = [\sigma t^2 + \sigma r^2 + \sigma c^2\, LL_S + \sigma co^2\, Nco +$$
$$\sigma s^2\, Ns + \sigma D^2 \sigma H^2 + \sigma CR^2\,]^{\frac{1}{2}} \quad \textbf{(12–38)}$$

where σM = sigma link margin, σt = sigma transmitter power, σr = sigma receiver sensitivity, σc = sigma cable loss, L_S =

TABLE 12–4 Statistical Mean and Sigma Values

Parameters	Symbol	Value	Unit
Mean link margin	μM	Output	dB
Mean higher order mode loss	μH	0.5[*]	dB
Mean CRM loss	μCR	1.5[**]	dB
Sigma link margin	σM	Output	dB
Sigma higher order mode loss	σH	0.0[*]	dB
Sigma CRM loss	σCR	0.25	dB

[*] 0.0 dB for single mode fiber systems
[**] 1.5 dB for *AT&T* T7032 (50 Mb/s) CRM or 0.5 dB for *AT&T* T7035 (50 Mb/s-250 Mb/s) CRM (Clock Recovery Module).

average section loss, σco = sigma connector loss, Nco = number of connectors, σs = sigma splice loss, Ns = number of splices, σD = sigma dispersion penalty, σH = sigma higher order mode loss, and σCR = sigma clock recovery module loss.

The statistical mean and sigma values of the above parameters are listed in Table 12–4.

The building and outside plant for the AT&T standard fiber cable parameters are listed in Table 12–5.

The statistical values for AT&T optical receivers and transmitters, as applied to various optical systems, are listed in Table 12–6.

The statistical values for AT&T connector losses are listed in Table 12–7.

The statistical values for AT&T splice losses are listed in Table 12–8.

12.5 THE DISPERSION EFFECT

In an optical communications link, the most important parameter, capable of limiting the link distance, is the pulse dispersion through the optical fiber. The two types of dispersion are modal and chromatic. The optical source and the type of fiber used in the link will determine whether the dispersion is modal, chromatic, or both. For example, if a laser source is used with a single–mode (SM) fiber, the dispersion effect is chromatic. With the use of a multimode (MM) fiber, the effect is modal. On the other hand, if an LED source is used with a multimode fiber, the dispersion effects are both modal and chromatic, but when the same LED is used with a single-mode fiber, the effect is only chromatic. In a multimode fiber, each mode travels different paths; therefore, all the different modes will arrive at different times, resulting in a broadening of the optical signal. If the optical pulse broadening exceeds certain well defined limits, pulse overlapping will occur, which results in optical power transfer at the leading or lagging edge of one pulse to the next. An increase in the system BER will occur. In modern system designs, this disruptive dispersion effect has been rectified to a great extent by the

TABLE 12–5 AT&T Standard Fiber Cable Parameters

(Building) Mean (μ) Values			
Type of Fiber Cable	Operating Wavelength	Loss (dB/km)	Dispersion (ps/nm-km)
8/125 µm	1300 nm	0.4	2.8 (max)
8/125 µm	1500 nm	0.3	18 (max)

(Building) Sigma (σ) Values			
Type of Fiber Cable	Operating Wavelength	Loss (dB/km)	Dispersion (ps/nm-km)
8/125 µm	1300 nm	0.0	2.8 (max)
8/125 µm	1500 nm	0.0	18 (max)

(Outside Plant) Mean (μ) Values			
Type of Fiber Cable	Operating Wavelength	Loss (dB/km)	Dispersion (ps/nm-km)
8/125 µm	1300 nm	0.35/0.4	2.8 (max)
8/125 µm	1500 nm	0.23/0.3	18 (max)

(Outside Plant) Sigma (σ) Values			
Type of Fiber Cable	Operating Wavelength	Loss (dB/km)	Dispersion (ps/nm-km)
8/125 µm	1300 nm	0.0	2.8 (max)
8/125 µm	1500 nm	0.0	18 (max)

(Building) Mean (μ) Values			
Type of Fiber Cable	Operating Wavelength	Loss (dB/km)	Bandwidth (MHz-km)
62.5/125 µm	850 nm	3.4	200 (min)
62.5/125 µm	1300 nm	1.0	500 (min)

(Building) Sigma (σ) Values			
Type of Fiber Cable	Operating Wavelength	Loss (dB/km)	Bandwidth (MHz-km)
62.5/125 µm	850 nm	0.0	200 (min)
62.5/125 µm	1300 nm	0.0	500 (min)

(Outside Plant) Mean (μ) Values			
Type of Fiber Cable	Operating Wavelength	Loss (dB/km)	Bandwidth (MHz-km)
62.5/125 µm	850 nm	3.4	200 (min)
62.5/125 µm	1300 nm	1.0	500 (min)

(Outside Plant) Sigma (σ) Values			
Type of Fiber Cable	Operating Wavelength	Loss (dB/km)	Bandwidth (MHz-km)
62.5/125 µm	850 nm	0.0	200 (min)
62.5/125 µm	1300 nm	0.0	500 (min)

incorporation of graded index fibers into the system (for a detailed description, see Chapter 10). Although chromatic dispersion generates the same pulse broadening effect, its cause is different. The index of diffraction of the optical fiber is wavelength dependent. Therefore, the different wavelengths, generated by broad wavelength spectrum devices (such as LEDs), will arrive at the end of the fiber at different times, which results in an increase of ISI and a consequent BER system degradation. Chromatic dispersion is therefore dominant in systems that employ LEDs,

TABLE 12–6 Statistical Value for AT&T Optical Transreceivers

Mean- μ				
Device Type	Code	Application	Power Output	Receiver Sensitivity BER=10^{-9}
SONET-Rx	1310C	OC-3Rx	—	–38.1 dBm
SONET-Rx	1310MC	OC-12Rx	—	–32.0 dBm
SONET-Rx	1310PC	OC-12Rx	—	–32.1 dBm
Fiber-ch.	1318A	1.5 Gb/sRx	—	–27.4 dBm
SONET-Rx	1320C	OC-12 with CR	—	–31.5 dBm
Laser-Tx	1227P	OC-3 (1300 nm)	–11 dBm	—
Laser-Tx	1227J	OC-3 (1300 nm)	–4.9 dBm	—
Laser-Tx	1227AF	OC-12 (1300 nm)	0.1 dBm	—

Sigma-σ				
Device Type	Code	Application	Power Output	Receiver Sensitivity BER=10^{-9}
SONET-Rx	1310C	OC-3Rx	—	0.48 dB
SONET-Rx	1310MC	OC-12Rx	—	0.28 dB
SONET-Rx	1310PC	OC-12Rx	—	0.98 dB
Fiber-ch.	1318A	1.5 Gb/sRx	—	0.58 dB
SONET-Rx	1320C	OC-12 with CR	—	0.3 dB
Laser-Tx	1227P	OC-3 (1300 nm)	0.15 dB	—
Laser-Tx	1227J	OC-3 (1300 nm)	0.50 dB	—
Laser-Tx	1227AF	OC-12 (1300 nm)	0.15 dB	—

TABLE 12–7 Statistical Value for AT&T Optical Transreceivers

Mean-μ				
Connector Type	50/125 μm	62.5/125 μm	8/125 μm	Units
ST® II	0.50	0.40	0.35	dB
SC	0.30	0.25	0.12	dB
FDDI-MIC	—	0.50	—	dB
ESCON	—	0.30	0.40	dB
FC-APC (Not AT&T)	—	—	0.09	dB

Sigma-σ				
Connector Type	50/125 μm	62.5/125 μm	8/125 μm	Units
ST® II	0.20	0.20	0.20	dB
SC	0.20	0.15	0.08	dB
FDDI-MIC	—	0.25	—	dB
ESCON	—	0.20	0.30	dB
FC-APC	—	—	0.05	dB

TABLE 12–8 Statistical Value for AT&T Splice Losses

Mean-μ				
Splice Type	50/125 μm	62.5/125 μm	8/125 μm	Units
Rotary	0.25	0.25	0.20	dB
CSL	0.15	0.15	0.15	dB

Sigma-σ				
Splice Type	50/125 μm	62.5/125 μm	8/125 μm	Units
Rotary	0.10	0.10	0.10	dB
CSL	0.10	0.10	0.10	dB

total dispersion. This extra optical power is referred to as the dispersion penalty. It is a penalty because the extra optical power is directly subtracted from the link power budget. For a statistical random method evaluation of the total link budget, the mean dispersion penalty is established in conjunction with the **standard deviation**. The statistical random method is mostly applicable to systems that employ multimode fibers, but in systems that employ single-mode fibers, the worst case dispersion method can be applied. Because dispersion effects result from the combined use of the optical source and the fiber, and because two types of sources (LEDs and lasers) exist, and two types of fibers (single mode and multimode) exist, there are four possibilities for examining dispersion effects. These are as follows:

a) LED + single-mode fiber

b) LED + multimode fiber

c) laser + single-mode fiber

d) laser + multimode fiber

From these four possibilities, only three are applicable. The LED plus the single-mode fiber is not realistic because the low level optical power generated by the LED is insufficient for coupling into the small core size single-mode fiber that is employed for long distance transmission. LEDs are mainly used with multimode fibers for short distance transmissions. Before we proceed with a detailed discussion of optical system designs, a brief review of chromatic dispersion and its effect in long haul optical systems is essential.

Chromatic Dispersion Compensation

As mentioned previously, the distance between spans and system bit rate for long haul optical links that utilize laser diodes and single mode fibers is affected by chromatic dispersion. Until the middle 1990s, long distance optical communications links operating at 2.5 Gb/s transmission rates could achieve a maximum link span of 40 km before signal regeneration, reshaping, and retiming were required. By the mid-1990s, increasing demand for higher system bit rates and longer spans was met through the utilization of EDFAs and add/drop multiplexers. The development of EDFAs and their implementation eliminated the

while modal dispersion is dominant with systems that employ laser diodes. For long distance optical links that utilize single-mode fibers, the modal dispersion effect is completely eliminated, but chromatic dispersion remains a major concern. Increasing the optical power at the input of the receiver module can compensate for the negative effect of chromatic and modal dispersion in an optical link design, reflecting a proportional increase of the OSNR. It is therefore the principal task of the designer to determine the exact level of optical power required to compensate for the need for signal regeneration (electrical-to-optical and optical-to-electrical signal conversion). However, in optical systems that employ EDFA and standard SM fibers, chromatic dispersion is accumulative and requires some form of compensation. If existing optical links are to be upgraded by incorporating EDFAs, dispersion compensation is also required. This is because the existing systems employ zero dispersion fibers that operate close to the 1300 nm wavelength window, while EDFAs exhibit optical gain in the 1530 nm to 1560 nm wavelength window. Therefore, zero dispersion fibers operating in EDFA wavelengths will induce substantial levels of dispersion. Therefore, optical systems that employ multiwavelength transmission formats require dispersion compensation. As mentioned previously, optical systems that operate with bit rates close to 1 Tb/s also require dispersion compensation. Although in some cases zero dispersion fibers are used to compensate for system nonlinearities, dispersion compensation is also essential in those systems.

Chromatic Dispersion Overview

The linearly polarized electric field component of an optical signal transmitted through an SM fiber is expressed by Equation (12–39).

$$\overrightarrow{E}(x,y,z,t) = \frac{\hat{n}}{2} 3 E(x,y,z,t) e^{j\,3\beta(\omega)z - \omega t 4} + Cc4 \qquad \textbf{(12–39)}$$

where $\overrightarrow{E}$ = electric field component, $\hat{n}$ = unit vector, $E(x,y,z,t)$ = complex scalar varying along the direction of propagation (z), in time (t), $\beta(\omega)$ = propagation constant, ω = angular velocity, and Cc = Complex conjugate.

The pulse propagation, assuming zero fiber nonlinearities, is expressed by Equation (12–40).

$$\beta(\omega) = \sum_{n=0}^{\infty} \frac{1}{n} \beta_n (\omega - \omega)^n \qquad \textbf{(12–40)}$$

where $\beta_n = \left. \dfrac{\partial \beta}{\partial \omega^n} \right|_{\omega_0}.$

The propagating pulse group velocity is expressed by Equation (12–41).

$$v_g = \frac{1}{\beta_1} \qquad \textbf{(12–41)}$$

The phase velocity is expressed by Equation (12–42).

$$v_p = \frac{\omega_o}{\beta_o} \qquad \textbf{(12–42)}$$

Chromatic dispersion is synonymous with group velocity variations in a fiber and is a function of optical frequency variations. The result of group velocity variations (or chromatic dispersion) is pulse chirping, which is a detrimental effect on system performance relevant to distance

span and system capacity. Pulses propagating through a fiber contain a spectrum of wavelengths (shorter and longer). Shorter wavelengths will travel faster than longer wavelengths. As a result, the optical pulse detected at the input of the receiver will be chirped, maybe to the extent that it might increase the system BER. Chromatic dispersion expresses the level by which an optical pulse traveling through a fiber suffers broadening, measured in ps/nm-km. More specifically, chromatic dispersion is the measure of pulse broadening (in ps) of an optical pulse with bandwidth of 1 nm traveling through an optical fiber of 1 km in length. Equation (12–43) expresses chromatic dispersion:

$$D = \frac{d}{d\lambda} \left(\frac{1}{v_g} \right) \qquad \textbf{(12–43)}$$

where D is the optical pulse dispersion (ps/nm), λ is the operating wavelength (nm), and v_g is the group velocity. Chromatic dispersion can also be expressed by Equation (12–44).

$$D = \frac{d^2\beta}{d\lambda} \left(\frac{1}{d\omega} \right) \qquad \textbf{(12–44)}$$

where β = propagation constant and ω = angular velocity. A very close approximation of Equation (12–44) is Equation (12–45).

$$D = \frac{2\pi c}{\lambda^2} \qquad \textbf{(12–45)}$$

where λ = wavelength and c = velocity of light. Because chromatic dispersion is wavelength dependent, step index single-mode fibers (SMF) with zero chromatic dispersion at the 1310 nm wavelength window will exhibit 17 ps/nm-km dispersion at the 1550 nm wavelength window. WDM optical systems that operate at long wavelengths and do not employ EDFAs exhibit a significant level of chromatic dispersion. Therefore, in such systems, dispersion compensation may be required for the different wavelengths within the WDM scheme. As mentioned previously, chromatic dispersion (pulse chirping) imposes a limit in the optical link span length. This maximum distance before amplification is expressed by Equation (12–46).

$$L_{\text{max}} = \frac{1}{(f_b)\,(\Delta\lambda)\,(D)} \qquad \textbf{(12–46)}$$

where L_{max} = maximum link distance before amplification (km), f_b = system bit rate (Mb/s), $\Delta\lambda$ = bandwidth (nm), and D = dispersion (ps/nm-km).

Example 12–6

An optical system operates at the 1550 nm wavelength window with a system bit rate of 1.25 Gb/s. The system utilizes a step index single-mode fiber (SI-SMF) with 0.5 nm bandwidth and dispersion of 15 ps/nm-km. Compute the maximum link distance before amplification.

Solution

Data:

a) $f_b = 1.25$ Gb/s

b) $D = 17$ ps/nm-km

c) $\Delta\lambda = 0.5$ nm

$$L_{max} = \frac{1}{(f_b)(\Delta\lambda)(D)}$$

$$= \frac{1}{(1.25 \times 10^9)(0.5 \times 10^{-9})(17 \times 10^{-12}/1 \times 10^{-9})\,\text{km}}$$

$$= \frac{1 \times 10^3}{(10.625)}\,\text{km}$$

$$= 94\,\text{km}$$

Therefore, maximum link span $(L_{max}) = 94$ km.

Example 12–7

An optical link that employs a WDM scheme is designed to operate at 2.5 Gb/s. Compute the maximum allowable chromatic dispersion (D) for a fiber bandwidth $(\Delta\lambda)$ of 0.4 nm and 50 km of link distance (L_{max}) before amplification.

Solution

Data:

a) $f_b = 2.5$ Gb/s

b) $L_{max} = 50$ km

c) $\Delta\lambda = 0.4$ nm

$$L_{max} = \frac{1}{(f_b)\,(\Delta\lambda)\,(D)}$$

Solve for D:

$$D = \frac{1}{(f_b)(\Delta\lambda)L_{max}}$$

$$= \frac{1}{(2.5 \times 10^9)(0.4 \times 10^{-9})(50\,\text{km})}$$

$$= 20\,\text{ps/nm-km}$$

Therefore, maximum allowable chromatic dispersion $(D) = 20$ ps/nm-km.

Example 12–8

Calculate the maximum bit rate (f_b) of an optical link that operates at the 1550 nm wavelength window and employs a standard SI-SMF that has a 0.5 nm bandwidth and a span length of 60 km.

Solution

Data:

a) $L_{max} = 60$ km

b) $\Delta\lambda = 0.5$ nm

c) $D = 17$ ps/nm-km (for a standard SI-SMF operating at 1550 nm)

$$L_{max} = \frac{1}{(f_b)(\Delta\lambda)(D)}$$

Solve for f_b:

$$f_b = \frac{1}{(\Delta\lambda)(D)(L_{max})}$$

$$= \frac{1}{(0.5 \times 10^{-9})(17 \times 10^{-12}/1 \times 10^{-9}\,\text{km})(60\,\text{km})}$$

$$= 1.96\,\text{Gb/s}$$

Therefore, the maximum system bit rate $= 1.96$ Gb/s.

For optical links that require higher bit rates and longer link spans, EDFAs and MZ external modulators must be used in DWDM schemes. With the utilization of external modulators, the system bit rate can reach 83% of the optical bandwidth. In such an optical system, the span distance is expressed by Equation (12–47).

$$L_{max} = \frac{6.1 \times 10^3}{f_b^2}\,\text{km} \qquad (12\text{–}47)$$

where f_b is in Gb/s.

Example 12–9

Calculate the maximum link span for the following system rates:

- 4-channel, 2.5 Gb/s per channel optical link
- 8-channel, 2.5 Gb/s per channel optical link
- 16-channel, 2.5 Gb/s per channel optical link

Solution

i)
$$L_{max} = \frac{6.1 \times 10^3}{f_b^2}\,\text{km}$$

$$= \frac{6.1 \times 10^3}{(4 \times 2.5)^2\,\text{Gb/s}}$$

$$= 61\,\text{km}$$

Therefore, $L_{max} = 61$ km.

ii)
$$L_{max} = \frac{6.1 \times 10^3}{f_b^2}\,\text{km}$$

$$= \frac{6.1 \times 10^3}{(8 \times 2.5)^2\,\text{Gb/s}}$$

$$= 15.25\,\text{km}$$

Therefore, $L_{max} = 15.25$ km.

iii)
$$L_{max} = \frac{6.1 \times 10^3}{f_b^2} \text{ km}$$

$$= \frac{6.1 \times 10^3}{(16 \times 2.5)^2 \text{ Gb/s}}$$

$$= 3.8 \text{ km}$$

Therefore, $L_{max} = 3.8$ km.

From Example 12–9, it is evident that an increase of the system bit rate decreases the link span substantially. A graphical representation of system bit rate (f_b) and the corresponding span length (before amplification) for both standard and DSFs is illustrated in Figure 12–14.

The graphical representation of the optical link span length and chromatic dispersion penalty is illustrated in Figure 12–15.

Demands for longer distances and higher system bit rates necessitated the exploration of different design methods. In order to achieve longer span distances and higher bit rates, two factors must be considered:

a) Either the operating wavelength window must be 1310 nm, at which standard fibers exhibit zero dispersion, or

b) The operating wavelength window must be 1550 nm, in conjunction with DSFs.

The phenomenal potential of EDFAs and low attenuation optical fibers both operating at the 1550 nm wavelength window generated the incentive for a completely optical link that uses the 1550 nm wavelength range. DSFs are ca-

pable of shifting the zero chromatic dispersion (also known as waveguide dispersion) to the 1550 nm range with an almost constant chromatic dispersion slope ($dD/d\lambda$) of 0.08 ps/nm²-km. Although optical system designs may be successful in terms of managing chromatic dispersion, they are still sensitive to FWM in WDM and DWDM schemes. Therefore, instead of dispersion shifted fibers, nondispersion shifted fibers that use ± 2 ps/nm-km to ± 4 ps/nm-km chromatic dispersion may be used. The accumulated chromatic dispersion through the optical fiber must be compensated for at the transmitter or receiver end by an appropriate compensation scheme, or through the optical fiber. Upgrading preexisting optical links requires modifications suited for standard fibers, while new systems may use standard DSFs or NZ–DSFs. Recently, several dispersion compensation techniques have been developed, some of which are briefly described as follows.

Dispersion Compensation

In order to implement a dispersion compensation scheme, it is necessary to calculate the total chromatic dispersion accumulated through the fiber. These calculations can be demonstrated by the following example.

Example 12–10
Calculate the chromatic dispersion induced by a 200 km fiber employed in an optical link that operates at the 1550 nm wavelength window, exhibits chromatic dispersion of 17 ps/nm-km, and has a 10 Gb/s transmission rate.

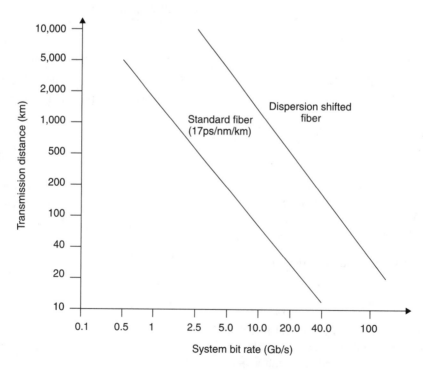

Figure 12–14. Maximum span length *v.* system bit rate.

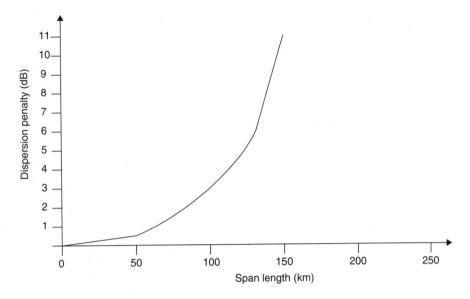

Figure 12–15. Maximum span length *v.* chromatic dispersion penalty.

Solution

A standard fiber that operates at the 1550 nm wavelength and at a 10 Gb/s bit rate exhibits a chromatic dispersion of 17 ps/nm-km.

Therefore, $(200 \text{ km})(17 \text{ ps/nm-km}) = 3.4 \times 10^3 \text{ps/nm}$.

Assuming the selected modulation scheme requires a bandwidth of 0.1 nm, the total pulse broadening is:

$$(0.1 \text{ nm})(3.4 \times 10^3 \text{ ps/nm}) = 0.34 \times 10^3 \text{ ps}$$

Therefore, total chromatic dispersion (pulse chirping) $(D) = 0.34 \times 10^3$ ps.

The ratio of pulse dispersion to the available bandwidth is:

$$\frac{0.34 \text{ ps}}{0.1 \text{ nm}} = 3 \text{ ps/nm}$$

This implies that at the end of the fiber length (single span), the traveled pulse will occupy three time slots. It is obvious that dispersion compensation is absolutely essential. Chromatic dispersion can be compensated at the transmitter, the receiver, or the fiber stage.

Dispersion Compensation at the Transmitter Stage

Dispersion compensation at the transmitter stage is achieved by prechirping the optical pulse at the transmitter stage. That is, the leading edge of the optical pulse is spread at a higher than average wavelength, while the lagging edge is spread at a shorter than average wavelength. The longer wavelength of the leading edge will travel through the time slot with no possibility of overlapping with the following optical pulse, thus achieving zero ISI. Prechirping the optical signal at the transmitter stage can be accomplished in a number of ways. Some of the most important are as follows.

ADDING PHASE MODULATION. In this method, an MZ modulator is used to generate the required prechirping of the optical pulse by inducing a modest level of chromatic dispersion. The two unequal length arms of the MZ modulator allow the optical pulse to arrive at different times at the output, resulting in a slight broadening of the combined pulse. The intended pulse chirping is subject to the arm difference of the MZ modulator.

FREQUENCY MODULATING (FM) THE LASER DIODE. The laser diode is driven by an FM signal, resulting in a broadening of the optical signal applied at the input of the MZ modulator. If such a dispersion compensation scheme is applied to an optical link that operates with a 10 Gb/s capacity at a span length of 100 km, then only a 2 dB dispersion penalty is required.

DISPERSION SUPPORTED TRANSMISSION. This pulse chirping technique is the most successful. With this method, the transmitter module generates an FM signal relevant to the link span. The FM signal is then converted to an AM signal by the fiber dispersion. The optical receiver easily detects such a signal. In this method of dispersion compensation, link spans of 200 km and system bit rates of 10 Gb/s have been achieved with a 6 dB receiver sensitivity penalty. The major disadvantages of this dispersion compensation method are:

a) laser drive current must be relevant to the link span

b) laser diode must exhibit excellent FM bandwidth response

c) optical receiver must be capable of detecting a three level signal

A combination of the three techniques achieves levels of dispersion compensation, which allows for link spans of up to 350 km with 10 Gb/s transmission rates.

Dispersion Compensation Through the Optical Fiber

The transmitter based dispersion compensation technique examined previously involves electrical processing, while dispersion compensation achieved through the fiber length is based on optical signal processing. The optical signal generated by the transmitter module is coupled into the fiber. As mentioned, the fiber induced dispersion substantially broadens the optical pulse at the end of the fiber. However, if a negative dispersion component that has optical characteristics exactly opposite of those of the fiber is induced into the fiber, it will restore the transmitted optical pulse to its original shape. The negative dispersion compensation module can also be installed at the beginning of the fiber span, or every 100 km. The main disadvantage of this scheme is that it induces a small level of signal attenuation. However, if incorporated into the EDFA module, the optical gain provided by the EDFA will compensate for the losses. Chromatic dispersion compensation techniques induced into the fiber can be achieved through:

a) interferometers

b) negative dispersion fibers (NDF)

c) phase conjugation

INTERFEROMETERS. The Gires-Tournois interferometer dispersion compensation scheme allows the various spectral components of the optical signal to travel different wavelength dependent paths at different speeds, thus reconstructing the optical signal at the end of the fiber into its original shape. Dispersion compensation through the interferometer method can be successfully applied to WDM schemes and can provide dispersion compensation to a maximum of 2000 ps/nm. Experimental results have shown that such a dispersion compensation scheme employed in an optical link that operates at 10 Gb/s with a span of 160 km required a dispersion penalty of 0.2 dB.

The silica on silicon planar circuit dispersion compensation technique involves a number of MZ modulators that operate in a cascading mode. The light at the input of the modulator is divided into two beams, which travel the two unequal length arms of the MZ modulator and are then reconstructed at the output with a phase delay proportional to the total arm length difference. The short wavelength spectral components travel through the longer arm of the modulator, while the longer wavelength components travel the shorter arm. A number of modulators in cascade can provide the required phase delay. A silica on silicon interferometer is capable of achieving dispersion compensation to a maximum of 836 pA/nm. This level can satisfy the dispersion compensation requirements of an optical link that operates at 10 Gb/s with a 50 km span. Dispersion compensation techniques involving MZ modulators exhibit a narrow bandwidth and are polarization dependent. On the positive side, this scheme provides wide tunable capabilities and is suited for WDM schemes that provide simultaneous channel dispersion compensation.

A very promising dispersion compensation technique is the chirp fiber Bragg grating technique. Through this technique, the refractive index of the fiber changes progressively along its length. If the wavelength of the light entering the fiber is twice the grating period, it will be fully reflected. In dispersion compensation based on fiber Bragg grating schemes, the grating period is progressively reduced (linearly) along the length of the device so that the short wavelength spectral components of the optical pulse will be reflected at a point farther into the device than that for longer wavelength spectral components. Thus, the short wavelength spectral components are delayed. A fiber Bragg grating dispersion compensation device of 50 cm length can satisfy the dispersion compensation requirements of a 10 Gb/s, 300 km optical link. For WDM schemes, fiber Bragg grating devices in a series configuration can provide up to 8000 ps/nm dispersion compensation, satisfying the dispersion compensation requirements of an optical link operating at 10 Gb/s with a maximum distance of 650 km and dispersion penalty of 1.4 dB.

NEGATIVE DISPERSION FIBER. The most commonly used method of dispersion compensation is implemented through a **negative dispersion fiber (NDF)** commonly known as a dispersion compensation fiber (DCF). For a more detailed discussion, see Chapter 12. Dispersion compensation fibers composed of a single core and cladding may exhibit negative dispersion of about -100 ps/nm with a corresponding attenuation of 0.35 db/km. Spans of 1000 km have been achieved in optical systems designed for 16-channel, 10 Gb/s per channel WDM schemes. Other applications that employ DCF include 8-channel WDM optical systems with 20 Gb/s per channel, which achieve spans of 232 km with a dispersion slope of 0.02 ps/nm^2-km.

Spectral Inversion

The interferometer and dispersion compensation fiber methods are subject to the optical link span, while the spectral inversion method requires only one dispersion compensation component, and is independent of the optical link span. The spectral inversion compensation technique works as follows. The span of the optical link is theoretically divided into two parts. The transmitted optical signal is intentionally dispersed in the first part of the fiber with the short wavelength spectral components of the pulse leading the long wavelength spectral components. At an appropriate point within the fiber (close to the midpoint), the dispersed optical signal is inverted (phase conjugation) with the long wavelength spectral component leading the short wavelength.

Optimum Fiber Length of Spectral Inversion.
The most noticeable disadvantage of a spectral inversion scheme that uses FWM is the increase of the effective noise figure (NF_{eff}). This undesirable effect of spectral inversion can be managed with the insertion of OBPFs at the signal channel, at the pump channel, or at both. Experimental work has confirmed that substantial improvement of the NF_{eff} of the spectral inversion scheme can be achieved through the implementation of OBPFs. Spectral inversion, implemented for dispersion compensation, is characterized by the **conversion efficiency** (η) defining the level of the optical signal attenuation caused by spectral inversion. If the combined effect of ASE, RIN, and SE are taken into consideration, then the combined effect can be expressed as a degradation of the OSNR. The overall noise performance of the spectral inverter, based on FWM, can be defined by its NF_{eff}.

Polarization Sensitive Spectral Inverter.
The principal noise components in a polarization sensitive optical spectral inversion scheme are:

a) ASE generated by the EDFAs

b) SE

c) **relative intensity noise (RIN)** generated by the pump channel

Both SE and ASE noise are broadband signals, but RIN is a narrowband signal in comparison to spectral inversion bandwidth. A block diagram of a polarization sensitive spectral inversion scheme is illustrated in Figure 12–16.

Figure 12–16 is composed of a number of optical amplifiers which are identical in terms of gain (G) and noise figure (NF), DSFs, a polarization sensitive spectral inverter, and three OBPFs. The spectral inverter module is composed of an optical laser pump, followed by an optical amplifier of a specified G and NF, used to elevate the optical power generated by the pump to the required level. The OBPF inserted at the output of the spectral inversion module is used to reduce ASE generated by the pump channel. The OBPF at the output of the spectral inversion fiber is used to extract the spectral inverted signal from the FWM, which is then processed through the inversion fiber. The function of the optical amplifier (OA) at the output is to elevate the level of the optical signal to that of the input. Of the three OBPFs incorporated into the dispersion compensation scheme, BPF1 and BPF2 are used for noise reduction, and BPF3 is used for the extrapolation of the spectrally inverted input signal. The identification of the various parameters in the system is as follows:

a) NF = noise figure

b) G_S = gain

c) P_{in} = optical input power

d) Φ_P = pump noise power spectral density

e) P_S = signal channel power

f) Φ_S = signal noise power spectral density

g) Φ_C = channel noise power spectral density

h) ω_P = pump wave frequency

i) ω_s = input signal wave frequency

j) ω_C = spectral inverted signal wave frequency

It is important to note that ASE, RIN, and SE are considered to have the same polarization as the input signal. The performance indicator of the spectral inverter is the NF_{eff}, defined as the ratio of the SNR at the input of the spectral

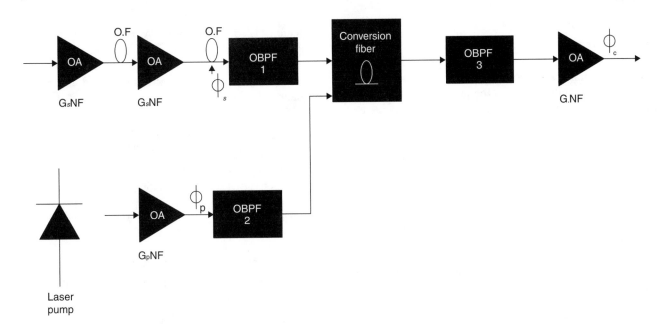

Figure 12–16. A block diagram of a polarization sensitive spectral inversion scheme.

inverter fiber to the SNR at the output of the same fiber. This ratio is expressed by Equation (12–48).

$$NF_{eff} = \frac{SNR_{input}}{SNR_{output}} \quad \textbf{(12–48)}$$

where NF_{eff} = effective noise figure of the spectral inverter, SNR_{input} = SNR at the input of the spectral inverter, and SNR_{output} = SNR ratio at the output of the spectral inverter. Another parameter that determines spectral inverter performance is conversion efficiency (η), expressed by Equation (12–49).

$$\eta = \frac{P_{in}}{P_{out}} \quad \textbf{(12–49)}$$

where η is the **spectral inverter efficiency,** P_{in} is the optical power at the input of the spectral inverter, and P_{out} is the optical power at the output of the spectral inverter.

A relationship between η and G can be established. Because G is defined as the ratio of the output power to the input power (P_{out}/P_{in}), and η is defined as the ratio of the input power to output power, then the gain-efficiency relationship is expressed by Equation (12–50).

$$G = \frac{1}{\eta} \quad \textbf{(12–50)}$$

Because the noise signal at the input of the spectral inversion fiber is undistinguishable from the input signal, noise spectral inversion also takes place. The theoretical limit of the effective noise figure (NF_{eff}) is set by the nondegenerate FWM that cannot be filtered out. A more detailed expression of η is given by Equation (12–51).

$$\eta = (P_P\gamma)^2 \left| \frac{12e^{i\left(\Delta\beta - 2\gamma\bar{P}_P\right)L2\alpha L}}{i\left(\Delta\beta - 2\gamma\bar{P}_P\right) - \alpha} \right|^2 \exp(-\alpha L) \quad \textbf{(12–51)}$$

where γ is the **nonlinear coefficient,** L is the conversion fiber length, $\Delta\beta$ is the linear phase mismatch, α is the fiber loss per unit length, $\bar{P}_P$ is the pump average wave power/conversion fiber length, and P_P is the pump optical power at the input of the spectral inversion. Equation (12–52) expresses the noise at the spectral inversion induced by the pump channel.

$$\Phi_{pump} = \left(\Phi_{SP} + \frac{NF_P\omega}{2}\right)\left(\frac{P_P}{P'_P}\right) \quad \textbf{(12–52)}$$

where Φ_{pump} is the combined noise power spectral density generated by the pump laser, Φ_{SP} is the power spectral density of spontaneous emission noise, NF_P is the pump optical amplifier noise figure, ω_P is the pump wave frequency, P_P is the pump power at the input of the spectral inversion, and P'_P is the optical power generated by the pump laser.

The division by two of the second term of Equation (12–52) is indicative of two polarization states. The ASE

noise signal density induced by the signal channel into the spectral inversion fiber is expressed by Equation (12–53).

$$\Phi_{signal} = \frac{NF_a\omega_s N}{2} \quad \textbf{(12–53)}$$

where Φ_{signal} is the ASE noise signal density induced by the source, NF_a is the amplifier noise figure, ω_s is the input signal wave frequency, and N is the number of inline amplifiers (before the spectral inversion). Equation (12–54) expresses the noise power spectral density of the RIN converted by the nondegenerate FWM.

$$\Phi_{ch(RIN)} = (4\eta P_{in} RIN)$$

$$\left| \frac{1 - \exp[-\alpha L - i\gamma(\bar{P}_P - \bar{P}_{in})L + i\Delta\beta L]}{i\Delta\beta - i\gamma(\bar{P}_P - \bar{P}_{in}) - \alpha} \right|^2 \over \left| \frac{1 - \exp(-\alpha L + i\Delta\beta L - i\gamma 2\bar{P}_P L)}{i\Delta\beta - i\gamma 2\bar{P}_P - \alpha} \right|^2 \quad \textbf{(12–54)}$$

where $\Phi_{ch(RIN)}$ is the inversion channel noise power spectral density of the RIN converted by FWM, η is the conversion efficiency, α is the fiber loss, L is the fiber length, γ is the nonlinear coefficient, $\Delta\beta$ is the linear phase mismatch, RIN is the relative intensity noise, $\bar{P}_P$ is the pump average wave power/conversion fiber length, and $\bar{P}_{in}$ is the optical power at the input of the spectral inverter.

If both Φ_{ch} and Φ_{signal} are known, the NF_{eff} of the spectral inverter can be estimated (Equation (12–55)).

$$NF_{eff} = \frac{\Phi_{ch}}{\Phi_{signal}} \quad \textbf{(12–55)}$$

Equation (12–55) is valid for $N \neq 0$. If no inline optical amplifiers are incorporated, then the shot noise must be taken into account. In such a case, the effective noise figure expressed by Equation (12–55) is modified as follows in Equation (12–56):

$$NF_{eff} = \frac{\Phi_{ch}}{h\omega} \quad \textbf{(12–56)}$$

where h is Planck's constant (6.62×10^{-34} J/s). A team of researchers has investigated the effect of filters on system performance. The dispersion compensation incorporating spectral inversion was investigated under the following operating parameters:

a) optical amplifier gain: $G = 11.5$ dB

b) optical power at the input of the conversion fiber: $P_{in} = -5$ dBm

c) inline amplifier(s) **noise figure:** $NF = 6$ dB

d) signal ASE noise density: $\Phi_{signal} = 3 \times 10^{-19}$ W/Hz

e) Optical power at the output of the OA (pump): $P_P = 5$ dBm

f) Pump ASE noise density: $\Phi_{\text{pump}} = 3 \times 10^{-19}$ W/Hz

g) Pump OA noise figure: $NF_{\text{pump}} = 3$ dB

h) RIN: -150 dB/Hz

i) Pump optical power output: $P'_P = 3$ dBm

j) Kerr nonlinearity coefficient: $\gamma = 2.05$ km^{-1} W^{-1}

k) Conversion fiber length: $L = 20$ km

l) Conversion fiber loss: $\alpha = 0.23$ dB/km

m) Linear phase mismatch of the polarization sensitive phase conjugated FWM wave: $\Delta\beta = 0$

Case 1: Performance Evaluation of a System Without Filters

Assuming that an input noise is a band limited white noise, then Equation (12–57) can express the noise power spectral density in the spectral inverted channel.

$$\Phi_{\text{ch}} = \left(\Phi_{\text{pump}} + \Phi_{\text{signal}}\right)\left[\eta + \exp(-\alpha L)\right]\left(\frac{1}{\eta}\right)$$

$$+ \; 4\eta P_{\text{signal}}\left(\text{RIN}\right)\left(\frac{1}{\eta}\right) + \Phi_{\text{OA}} \qquad \textbf{(12–57)}$$

where Φ_{ch} is the total noise power spectral density in the inverted signal channel, $(\Phi_{\text{pump}} + \Phi_{\text{signal}})\exp(-\alpha L)$ is the input noise to inverted channel by both signal and pump noise power spectral density attenuated by the fiber loss, $(\Phi_{\text{pump}} + \Phi_{\text{signal}})\eta$ is the noise power spectral density converted to spectral inverted channel, and $4\eta P_{\text{signal}}(\text{RIN})(\zeta)$ $\left(\dfrac{1}{\eta}\right)$ is the noise converted by FWM.

The spectral inverter **effective noise figure** (NF_{eff}) is then expressed by Equation (12–58).

$$NF_{\text{eff}} = \left(\frac{\Phi_{\text{pump}}}{\Phi_{\text{signal}}} + 1\right)\left(\frac{e^{-\alpha L}}{\eta} + 1\right)$$

$$+ \frac{4P_{\text{in}}\text{RIN}\zeta + \Phi_{\text{OA}}}{\Phi_{\text{signal}}} \qquad \textbf{(12–58)}$$

where $\Phi_{\text{OA}} = NFh\omega$ (noise power spectral density of the optical amplifier).

$$\zeta = \frac{\left|\dfrac{1-\exp[-\alpha L-\gamma(\bar{P}_P-\bar{P}_S)\,L + i\Delta\beta L\,]^2}{i\Delta\beta-i\gamma(\bar{P}_P-\bar{P}_S)-\alpha}\right|}{\left|\dfrac{1-\exp(-\alpha L + i\Delta\beta L-i\gamma 2\bar{P}_P L\,)^2}{i\Delta\beta-i\gamma 2\bar{P}_P-\alpha}\right|} = \;\geq 1$$

The relationship between the length of the conversion fiber and the corresponding effective noise figure (NF_{eff}) of the spectral inverter is illustrated in Figure 12–17.

From Figure 12–17, it is evident that an exponential increase of the effective noise figure occurs with a decrease of the conversion fiber length. At the 20 km conversion fiber length, 5 dB of effective noise figure can be obtained. The graph also indicates that a 1 dB improvement in the effective noise figure will require doubling the conversion fiber length to 40 km. Therefore, the optimum effective noise figure of the spectral inversion can be obtained at a 20 km conversion fiber length, assuming that a relatively large number of optical amplifiers (≈ 50) are incorporated into the system. The relationship between the RIN and effective noise figure for the above dispersion compensation scheme is illustrated in Figure 12–18.

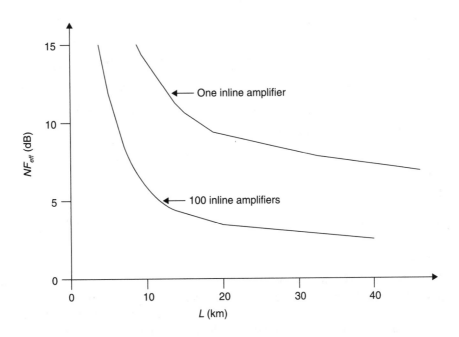

Figure 12–17. Effective noise figure (NF_{eff}) v. spectral conversion fiber length (L).

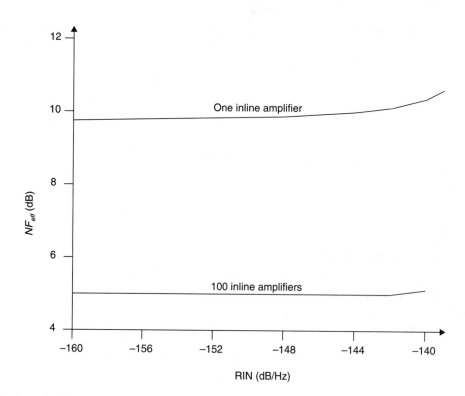

Figure 12–18. Effective noise figure (NF_{eff}) *v.* RIN.

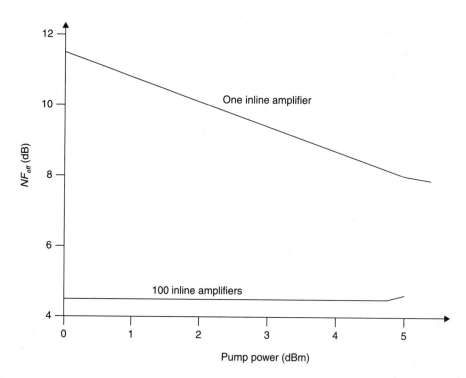

Figure 12–19. Effective noise figure NF_{eff} *v.* pump power.

Figure 12–18 shows that the RIN effect on NF_{eff} is minimal. Pump laser power has a larger effect on the effective noise figure (Figure 12–19).

Figure 12–19 shows that with only one optical amplifier incorporated into the system, a pump power increase from 0 dBm to 5 dBm will correspondingly decrease the effective noise figure from 11.5 dB to 8 dB; a considerable improvement of the effective noise figure. However, if the system incorporates more than ten inline optical amplifiers, pump power variations between 0 dBm and 5 dBm will

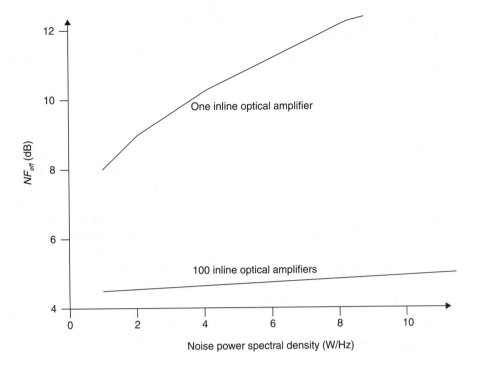

Figure 12–20. NF_{eff} v. noise power spectral density (Φ_{OA}).

produce a maximum effective noise figure reduction of less than 1 dB. The relationship between effective noise figure and noise power spectral density of the optical amplifier is illustrated in Figure 12–20.

Figure 12–20 shows a rather substantial increase of NF_{eff} in relation to a linear increase of Φ_{OA} when only one optical amplifier is inserted into the system. As the number of optical amplifiers progressively increases, the NF_{eff} rate of change is substantially reduced. If the number of inline optical amplifiers

is increased to fifty, the NF_{eff} remains constant across the 1×10^{-19} W/Hz $- 1 \times 10^{-18}$ W/Hz range of Φ_{OA}.

Case 2: Performance Evaluation of a System with Only One Filter

The incorporation of an optical bandpass filter at the output of the spectral conversion fiber is illustrated in Figure 12–21. The insertion of the bandpass filter at the output of

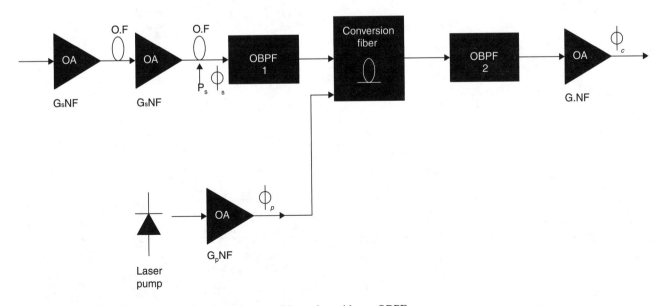

Figure 12–21. Dispersion compensation through spectral inversion with one OBPF.

the Nth optical amplifier substantially improves the performance of the spectral inversion scheme and the overall optical link performance. Through this filter, the accumulated amplified stimulation emission (ASE) noise of the optical amplifiers is completely eliminated, allowing only the input signal to be spectrally converted with the pump channel spontaneous emission noise. Therefore, the inverted noise spectral density in the conversion channel is expressed by Equation (12–59).

$$\Phi_{ch} = \left[\frac{\Phi_{pump} \cdot e^{-\alpha L} + (\Phi_{pump} + \Phi_{signal})\eta}{\eta} \right]$$
$$+ 4\eta P_{in}(RIN)\zeta\left(\frac{1}{\eta}\right) + \Phi_{OA}$$
$$= \frac{\Phi_{pump} \cdot e^{-\alpha L}}{\eta} + (\Phi_{signal} + \Phi_{pump})$$
$$+ 4\eta P_{in}(RIN)\zeta\left(\frac{1}{\eta}\right) + \Phi_{OA} \qquad \textbf{(12–59)}$$

The spectral inverter NF_{eff} is expressed by Equation (12–60).

$$NF_{eff} = 1 + \frac{\Phi_{Pump}}{\Phi_{signal}}\left(\frac{e^{-\alpha L}}{\eta} + 1\right)$$
$$+ \left(\frac{4P_{in}RIN\zeta + \Phi_{OA}}{\Phi_{signal}}\right) \qquad \textbf{(12–60)}$$

PERFORMANCE EVALUATION. The NF_{eff} of the spectral inverter in relation to conversion fiber length (L), where only one OBPF is inserted at the output of the Nth optical amplifier, is illustrated in Figure 12–22.

A substantial improvement of the NF_{eff} occurs with the incorporation of the OBPF. It is also evident that 20 km of conversion fiber is the ideal length. The NF_{eff} in relation to RIN, pump optical power, and spontaneous emission noise power spectral density is illustrated in Figures 12–23, 12–24, and 12–25.

Case 3: Performance Evaluation of a System with Only One Filter

In this case, an optical bandpass filter is inserted at the output of the laser pump channel. The purpose of inserting the narrow bandwidth optical filter is to eliminate the SE noise and ASE noise generated by the laser pump. The only broadband noise present is that generated by the signal channel. This noise is spectrally inverted along with the input signal by the conversion fiber. Equation (12–61) expresses the noise power spectral density at the output of the conversion fiber.

$$\Phi_{ch} = \Phi_{signal}(\eta + e^{-\alpha L})\left(\frac{1}{\eta}\right)$$
$$+ 4\eta P_S(RIN)\zeta\left(\frac{1}{\eta}\right) + \Phi_{OA} \qquad \textbf{(12–61)}$$

Equation (12–62) expresses the spectral inversion effective noise figure.

$$NF_{eff} = 1 + \frac{e^{-\alpha L}}{\eta} + \frac{4P_S(RIN)\zeta + \Phi_{OA}}{\Phi_{signal}} \qquad \textbf{(12–62)}$$

PERFORMANCE EVALUATION. The NF_{eff} of the aforementioned spectral inversion dispersion compensation scheme in relation to conversion fiber length (L), RIN, and input signal power (P_{in}), is illustrated in Figures 12–26, 12–27, and 12–28.

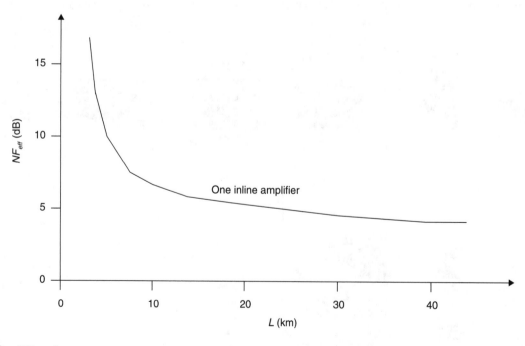

Figure 12–22. NF_{eff} v. L.

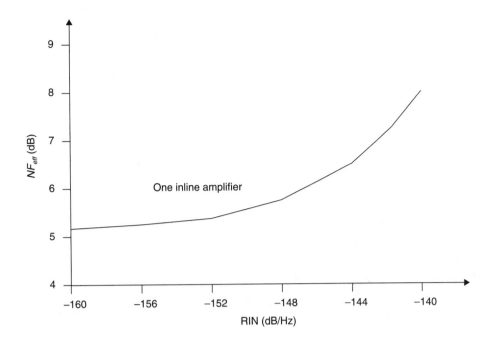

Figure 12–23. NF_{eff} v. RIN.

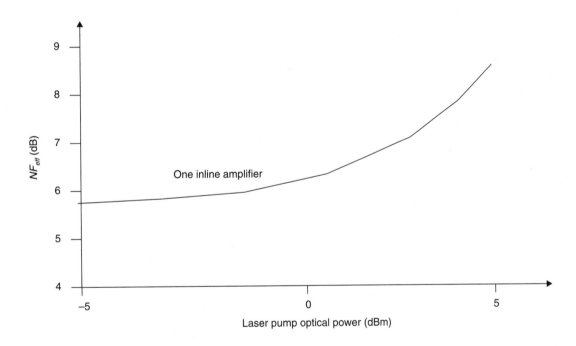

Figure 12–24. NF_{eff} v. laser pump optical power (P_P).

The benefit of placing the OBPF at the output of the pump channel is a reduction of the spectral inverter effective noise figure. An NF_{eff} reduction from 10 dB to 5 dB can be achieved through a corresponding increase of the conversion fiber from 3 km to 20 km. Therefore, to maintain a constant 5 dB NF_{eff}, the optimum length of the conversion fiber is 20 km. However, because pump ASE noise and SE noise is filtered out by the narrowband OBPF, only the signal channel RIN will be spectrally converted along with the input signal.

Case 4: Performance Evaluation of a System with Two OBPFs

In this case, two OBPFs are inserted at both the signal channel (at the output of the Nth optical amplifier) and the pump channel outputs. Through this arrangement the benefits obtained from both Case 2 and Case 3 are compiled, resulting in a substantial decrease of the NF_{eff}. The dispersion compensation scheme that employs spectral inversion with two OBPFs is illustrated in Figure 12–29.

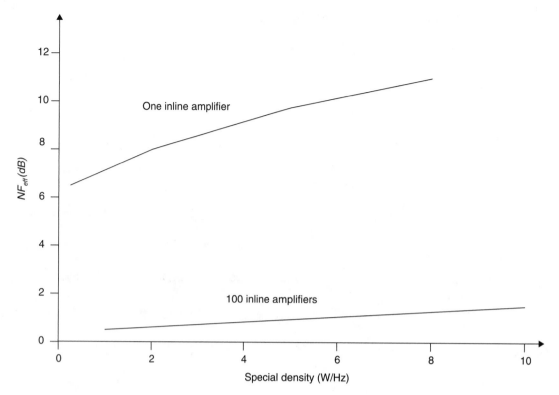

Figure 12–25. NF_{eff} v. SE spectral density (Φ_{OA}).

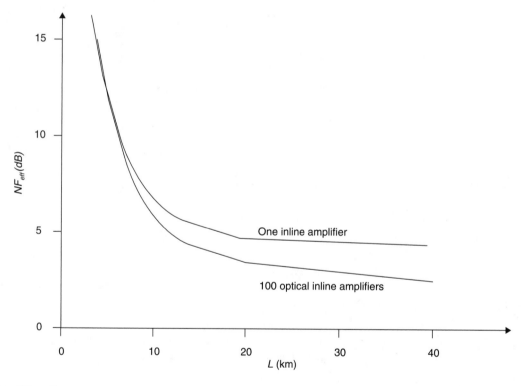

Figure 12–26. NF_{eff} v. L.

The incorporation of OBPFs in both the signal channel and the pump channel eliminates the SE and ASE generated by the laser pump. The noise power spectral density of the conversion channel (Φ_{ch}) is expressed by Equation (12–63).

$$\Phi_{ch} = \Phi_{signal} + 4\eta P_S (\text{RIN})\zeta\left(\frac{1}{\eta}\right) + \Phi_{OA} \quad \textbf{(12–63)}$$

Equation (12–64) expresses the NF_{eff}.

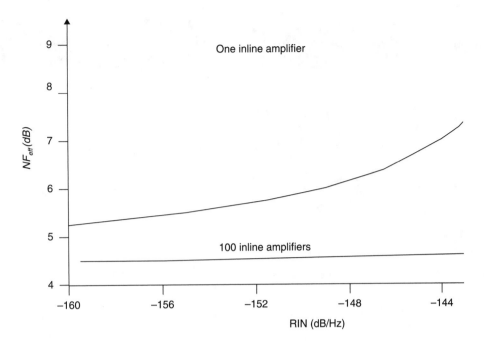

Figure 12–27. NF_{eff} v. RIN.

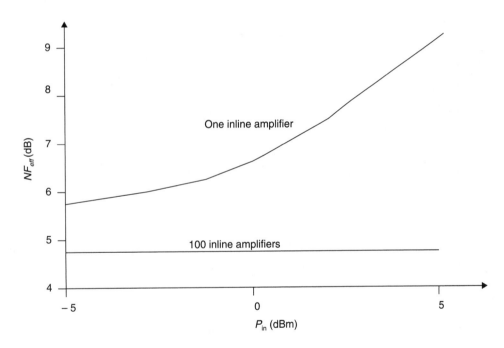

Figure 12–28. NF_{eff} v. P_P.

$$NF_{eff} = 1 + \left[\frac{4P_s (\text{RIN}) \zeta + \Phi_{OA}}{\Phi_{signal}} \right] \quad \textbf{(12–64)}$$

It was anticipated that the incorporation of both OBPFs would improve the spectral inversion performance. In fact, the NF_{eff} on this scheme was substantially improved, compared to the other cases. The relationship between the NF_{eff} and L is illustrated in Figure 12–30.

From Figure 12–30, it is evident that an NF_{eff} of 0.5 dB can be maintained with an L of between 3 km and 40 km and with the addition of fifty optical amplifiers. The

NF_{eff} as related to RIN and P_{in} is illustrated in Figures 12–31 and 12–32.

From Figures 12–31 and 12–32, it is evident that a longer conversion fiber is required in order to maintain low NF_{eff}. However, if the **SBS** on the pump wave is to be marginalized, the conversion fiber must be relatively short. Furthermore, in order to minimize the effect of **SPM** and XPM in the FWM (a direct contributor to pulse chirping of the spectrally inverted signal), a short conversion fiber is also required. The insertion of the narrowband OBPF at the output of the pump channel eliminates the effect of

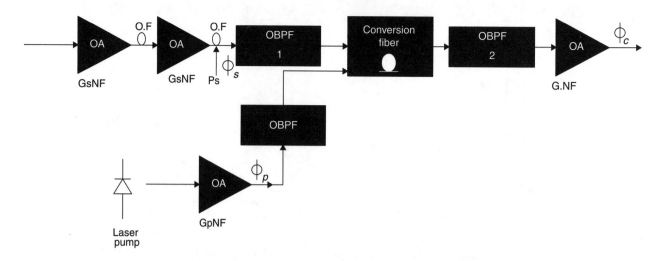

Figure 12–29.   Dispersion compensation scheme that employs two OBPFs.

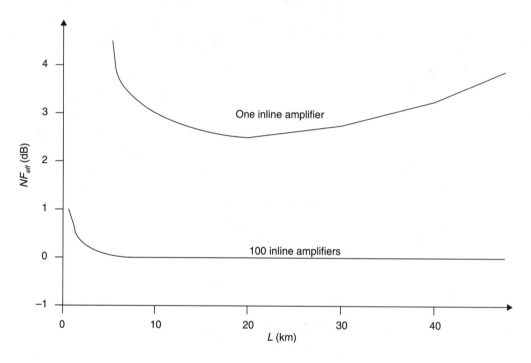

Figure 12–30. NF_{eff} v. L.

SBS, and pulse chirping can be reduced with the reduction of the input signal power. Shorter conversion fiber is also required to offset the reduction of the conversion efficiency caused by zero dispersion wavelength fluctuations in the conversion fiber. With proper management of all the fibers in the system, the anomaly (shorter v. longer conversion fibers) can be rectified. Therefore, the optimum conversion fiber length must be selected so as to satisfy all the requirements, the most important of which is the spectral inverter NF_{eff}.

Noise in a Polarization Insensitive Optical Spectral Inversion

A block diagram of a polarization insensitive optical spectral inverter is illustrated in Figure 12–33. This spectral inversion scheme uses a DSF that makes use of nondegenerate FWM and two orthogonally polarized laser pumps.

The optical power generated by the two pumps must be identical in order for the RIN to be identical as well. The optical power generated by each laser pump is fed into the corresponding optical amplifiers, which exhibit an identical power gain. The outputs from each optical amplifier are fed into a polarization beam splitter (PBS) that is capable of producing a constant optical pump power (P_P) and noise power spectral density (Φ_{pump}) at its output. Because the ASE and SE are wideband noise signals, the pump eliminates their effect on the spectral inversion channel and on the signal channel filters. The implementation of such a spectral inverter is polarization dependent, and the two generated polarization states increase the complexity of

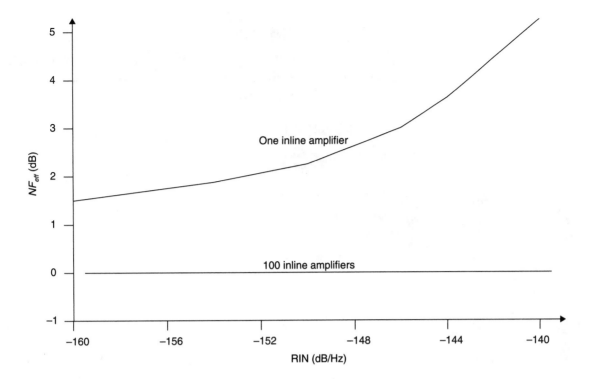

Figure 12–31. *NF$_{eff}$ v.* RIN.

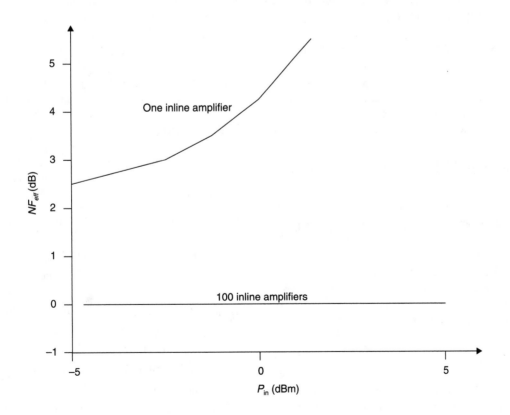

Figure 12–32. *NF$_{eff}$ v. P$_{in}$.*

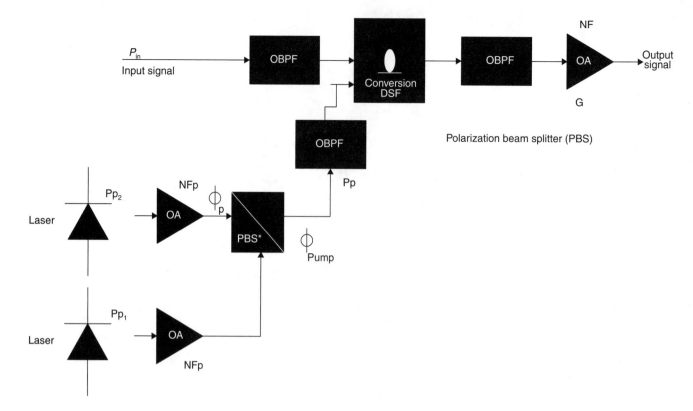

Figure 12–33. A polarization insensitive spectral inverter.

the spectral analysis. The NF_{eff} for this spectral inversion scheme is expressed by Equation (12–65).

$$NF_{eff} = \frac{\sigma_c^2}{\sigma_s^2} \qquad (12\text{–}65)$$

where σ_c^2 is the spectral inverted output beating noise variance and σ_s^2 is the input beating noise variance. Equation (12–66) expresses the power spectral density, which is equal for both pump channels.

$$\Phi_{pump} = \frac{P_P}{P_{P_1}} \left(\frac{NF_P\, h\omega}{2} + \Phi_{OA} \right) \qquad (12\text{–}66)$$

where Φ_{pump} = noise power spectral density, P_P = optical pump power at the input of the spectral inverter, P_{P_1} = optical power generated by laser pump 1 equal to laser pump 2, NF_P = noise figure of the pump optical amplifier, ω = signal wave frequency, h = Planck's constant (6.62×10^{-34} J/s), and Φ_{OA} = noise power spectral density of the optical amplifier. Likewise, Equation (12–67) expresses the ASE power spectral density from the signal channel for one polarization state.

$$\Phi_{signal} = \left(\frac{\Phi_{signal(V)} + \Phi_{signal(H)}}{2} \right) \qquad (12\text{–}67)$$

Because $\Phi_{signal(V)} = \Phi_{signal(H)}$, then $\Phi_{signal} = \Phi_{signal(V)} = \Phi_{signal(H)}$, where $\Phi_{signal\,(V)}$ is the noise power spectral density at *vertical* polarization and $\Phi_{signal\,(H)}$ is the noise power spectral density

at *horizontal* polarization. The power spectral density of the inverted RIN is expressed by Equation (12–68).

$$\Phi_{ch(RIN)} = \eta P_S (RIN)\zeta \qquad (12\text{–}68)$$

where $\zeta \approx 1$. Comparing $\Phi_{ch(RIN)}$ for both polarization sensitive and polarization insensitive optical spectral inversion, it is evident that noise spectral density of the polarization insensitive inversion scheme is smaller than that of the polarization sensitive scheme. This can be attributed to the fact that the linear phase mismatch of FWM is twice as large in the polarization sensitive scheme. In order to evaluate the spectral inversion scheme, the following system parameters were used:

a) optical power from both laser pumps ($P_{P_1} = P_{P_2}$): 12 dBm

b) signal channel amplifier gain (G_{signal}): 11.5 dB

c) optical power at the input of the spectral inverter (P_P): 15 dB

d) noise figure of the OA at the channel output (NF_{OA}): 6 dB

e) optical power at the signal channel (P_S): –5 dBm

f) spectral inversion fiber length (L): 20 km

g) fiber attenuation (α): 0.23 dB/km

h) group velocity dispersion of conversion fiber (GVD): 0.07 ps/nm²-km

i) kerr nonlinearity coefficient (γ): 2.05 km^{-1} W^{-1}

j) relative intensity noise (RIN): –150 dB/Hz

k) pump frequency separation $\left(\dfrac{\Omega}{2\pi}\right)$: 500 GHz

Case 1. Performance Evaluation of a System Without Filters

A polarization insensitive spectral inversion scheme without the incorporation of optical filters is illustrated in Figure 12–34.

The noise components of the spectral inverted signal channel in the $x \rightarrow$ direction are expressed by Equation (12–69).

$$n_{\text{ch}(x)} = (n_{P_1} + n_{s(x)} \cos\theta + n_{s(v)} \sin\theta)\exp\left(\frac{-\alpha L}{2}\right)(G_{\text{ch}})^{1/2}$$

$$+ \ (G_{\text{ch}}\, n_{xy})^{1/2}\,(n_{P_2} + n_{s(y)} \cos\theta + n_{s(x)} \sin\theta)$$

$$+ \ n_{\text{ch(RIN)}}\,(G_{\text{ch}})^{1/2} + n_{OA(x)} \qquad\qquad \textbf{(12–69)}$$

where $n_{\text{ch}(x)}$ is the total noise components of the spectral inverted signal channel in the $x \rightarrow$ direction, $(n_{P_1} + n_{s(x)} \cos\theta + n_{s(v)} \sin\theta)\exp\left(\frac{-\alpha L}{2}\right)(G_{\text{ch}})^{1/2}$ is the combined noise of the input signal channel and pump channel into the spectral inversion channel in the $x \rightarrow$ direction. This noise component is attenuated by the conversion fiber. $(G_{\text{ch}}\, n_{xy})^{1/2}\,(n_{P_2}\, n_{s\,(v)} \cos\theta + n_{s(H)} \sin\theta)$ is the noise of the input signal channel and pump channel into the spectral inversion in the $y \rightarrow$ direction, $n_{\text{ch(RIN)}}$; $G_{\text{ch}}^{1/2}$ is the spectral inversion of the RIN, $n_{OA(x)}$ is the noise generated by the optical amplifier connected at the input of the spectral inversion fiber, and θ is the polarization orientation of the two pump waves, signal waves, inverted waves, and the corresponding noise signals.

The noise components of the spectral inversion signal channel in the $x \rightarrow$ direction are expressed by Equation (12–70).

$$n_{\text{ch}(y)} = (n_{P_2} + n_{s(y)} \cos\theta + n_{s(x)} \sin\theta)\exp\left(\frac{-\alpha L}{2}\right)(G_{\text{ch}})^{1/2}$$

$$+ \ (G_{\text{ch}}\, n_{xy})^{1/2}\,(n_{P_2} + n_{s(x)} \cos\theta + n_{s(y)} \sin\theta)$$

$$+ \ n_{\text{ch(RIN)}}\,(G_{\text{ch}})^{1/2} + n_{OA(y)} \qquad\qquad \textbf{(12–70)}$$

Both Equations (12–34) and (12–35) incorporate the same noise components in the two polarization states. Assuming that both noise fields are statistically independent with zero means, the **mean-square variance** of the above relationships can be expressed by Equations (12–71) and (12–72).

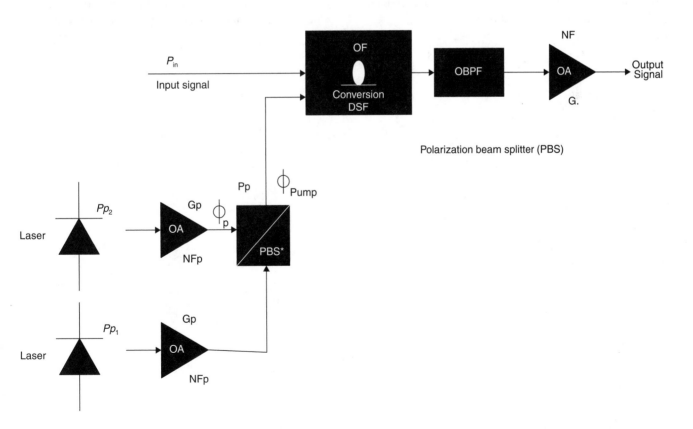

Figure 12–34. Polarization insensitive spectral inversion without OBPFs.

$x \rightarrow$ Direction:

$$\langle |n_{ch(x)}|^2 \rangle = [(\Phi_{pump} + \Phi_{signal})\exp(-\alpha L)/\eta + (\Phi_{pump}$$
$$+ \Phi_{signal}) + \Phi_{channel.RIN(x)}/\eta + \Phi_{OA}] B_o \quad \textbf{(12–71)}$$

$$\langle |n_{ch(y)}|^2 \rangle = [(\Phi_{pump} + \Phi_{signal})\exp(-\alpha L)/\eta + (\Phi_{pump}$$
$$+ \Phi_{signal}) + \Phi_{channel.RIN(x)}/\eta + \Phi_{OA}] B_o \quad \textbf{(12–72)}$$

where $|n_{ch(x)}|^2$ = mean square variance (noise channel) in the $x \rightarrow$ direction, $|n_{ch(y)}|^2$ = mean square variance (noise channel) in the $y \rightarrow$ direction, Φ_{pump} = pump noise power spectral density, Φ_{signal} = signal noise power spectral density, η = spectral inversion channel efficiency, Φ_{OA} = optical amplifier noise power spectral density, and B_o = filter bandwidth.

The inverted signal channel beat noise variance is expressed by Equation (12–73).

$$\sigma_{ch}^2 = P_S [< |n_{ch(x)}|^2 > \sin^2 \theta + < |n_{ch(y)}|^2 > \cos^2 \theta] \textbf{(12–73)}$$

Substituting Equations (12–71) and (12–72) into (12–73) yields Equation (12–74).

$$\sigma_{ch}^2 = P_S [(\Phi_{pump}\Phi_{signal})\exp(-\alpha L)/\eta + \Phi_{pump}$$
$$+ \Phi_{channel} + \Phi_{ch(RIN)} \left(\frac{1}{\eta} \right) + \Phi_{OA}] B_o \quad \textbf{(12–74)}$$

Finally, based on Equation (12–74) the NF_{eff} of the polarization insensitive spectral inversion without the incorporation of OBPFs is expressed by Equation (12–75).

$$NF_{eff} = \left[\frac{\Phi_{pump}}{\Phi_{signal}} + 1 \right] \left[\frac{\exp(-\alpha L)}{\eta} + 1 \right]$$
$$+ \left[\frac{P_S RIN\zeta + \Phi_{OA}}{\Phi_{signal}} \right] \quad \textbf{(12–75)}$$

Comparing Equation (12–40) with Equation (12–23) shows that the only difference between the polarization sensitive and polarization insensitive spectral inversion schemes is the RIN. That is, in the polarization sensitive scheme, the RIN is four times higher than that in the polarization insensitive spectral inversion scheme, which indicates a better effective noise figure performance in the latter case.

Performance Evaluation

The NF_{eff} of the polarization insensitive spectral inversion in relation to L is illustrated in Figure 12–35.

Figure 12–35 illustrates that an NF_{eff} of almost 9 dB can be maintained with a 20 km conversion fiber and the addition of 100 optical amplifiers. Furthermore, the NF_{eff} of this scheme is practically unaffected by the RIN maintained at 8.5 dB across the –160 dB/Hz to –140 dB/Hz range. The relationship between NF_{eff} and RIN is illustrated in Figure 12–36.

The effect of the pump power output also indicates a very small change in the NF_{eff} of between 0 dBm and 5 dBm. The relationship between NF_{eff} and pump output optical power is illustrated in Figure 12–37.

This change is less than 0.5 dB, while the NF_{eff} change due to signal channel input power is zero. The relationship

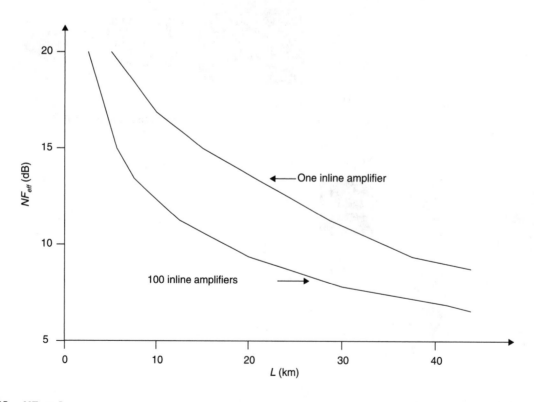

Figure 12–35. NF_{eff} v. L.

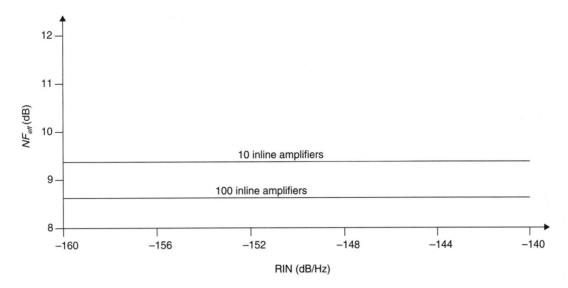

Figure 12–36. NF_{eff} v. *RIN*.

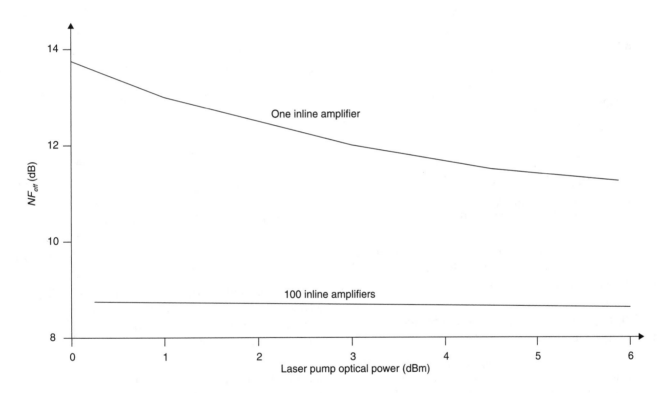

Figure 12–37. NF_{eff} v. laser pump optical power (P_P).

between NF_{eff} and input signal optical power is illustrated in Figure 12–38.

Case 2. Performance Evaluation of a System with Only One Filter

If an OBPF is inserted into the signal channel and a comparison between the NF_{eff} of a polarization sensitive and polarization insensitive spectral inversion is made, then a difference between the two schemes is evident in the $4P_{in}RIN$ term as follows:

$$NF_{eff} = 1 + \frac{\Phi_{pump}}{\Phi_{signal}}\left(\frac{e^{-\alpha L}}{\eta} + 1\right)$$
$$+ \left(\frac{P_{in}RIN\zeta + \Phi_{OA}}{\Phi_{signal}}\right)\text{(polarization insensitive)}$$

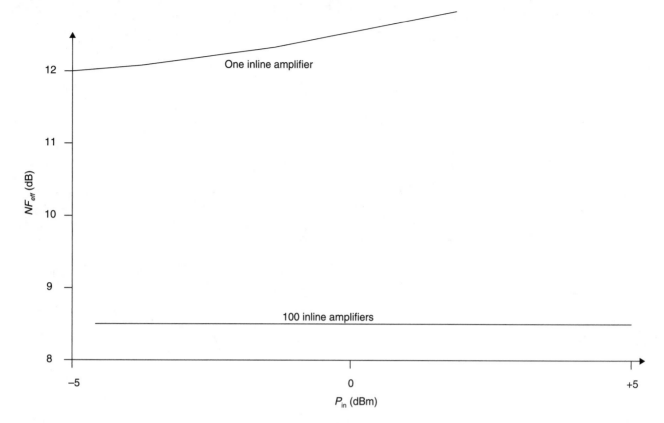

Figure 12–38. NF_{eff} v. P_{in}.

$$NF_{eff} = 1 + \frac{\Phi_{pump}}{\Phi_{signal}}\left(\frac{e^{-\alpha L}}{\eta} + 1\right)$$
$$+ \left(\frac{4P_{in}RIN\zeta + \Phi_{OA}}{\Phi_{signal}}\right)(\text{polarization sensitive})$$

From the above, it is evident that the NF_{eff} of the polarization sensitive spectral inversion scheme is higher than that of the polarization insensitive scheme.

Case 3. Performance Evaluation of a System with Only One Filter

If an OBPF is inserted into the pump channel, it will eliminate the ASE noise and SE noise for both laser pumps. The only noise that is spectrally inverted is the signal channel noise. Therefore, the noise components in the inverted channel for both x and y polarization are expressed by Equations (12–76) and (12–77).

$$n_{ch(x)} = (n_{s(x)}\cos\theta + n_{s(y)}\sin\theta)\exp\left(\frac{-\alpha L}{2}\right)(G_{ch})^{1/2}$$
$$+ (G_{ch}\eta_{(xy)})^{1/2}(n_{s(y)}\cos\theta + n_{s(x)}\sin\theta) \quad \textbf{(12–76)}$$
$$+ (G_{ch})^{1/2}n_{ch(RIN)} + n_{OA(x)}$$

$$n_{ch(y)} = (n_{s(y)}\cos\theta + n_{s(x)}\sin\theta)\exp\left(\frac{-\alpha L}{2}\right)(G_{ch})^{1/2}$$

$$+ (G_{ch}\eta_{(yx)})^{1/2}(n_{s(x)}\cos\theta + n_{s(y)}\sin\theta)$$
$$+ (G_{ch})^{1/2}n_{ch(RIN)_y} + n_{OA_y} \quad \textbf{(12–77)}$$

Equation (12–78), in this case, expresses the NF_{eff}.

$$NF_{eff} = 1 + \frac{e^{-\alpha L}}{\eta} + \frac{P_sRIN\zeta + \Phi_{OA}}{\Phi_{signal}} \quad \textbf{(12–78)}$$

There is very little difference between the NF_{eff} of polarization sensitive and polarization insensitive spectral inversion schemes. However, the spectral inversion efficiency, in this case, is somewhat lower, reflecting a lower NF_{eff}.

Case 4. Performance Evaluation of a System with Two OBPFs

Finally, in the polarization insensitive spectral inversion scheme, two OBPFs are incorporated: one at the pump channel and the other in the signal channel. The system is then evaluated in terms of NF_{eff}. Equations (12–79) and (12–80) express the noise components for vertical and horizontal modes of polarization in the spectral inverting channel.

$$n_{ch(x)} = (G_{ch}\eta_{(xy)})^{1/2}(n_{s(y)}\cos\theta + n_{s(x)}\sin\theta)$$
$$+ (G_{ch})^{1/2}n_{c(RIN)_x} + n_{OA_{(x)}} \quad \textbf{(12–79)}$$

$$n_{ch(xy)} = (G_{ch}\eta_{(yx)})^{1/2} (n_{s(x)} \cos \theta + n_{s(y)} \sin \theta)$$
$$+ (G_{ch})^{1/2} n_{c(RIN)_y} + n_{OA_{(y)}} \qquad \textbf{(12–80)}$$

Equation (12–81) expresses the NF_{eff} for this case.

$$NF_{eff} = 1 + \frac{P_S RIN\zeta + \Phi_{OA}}{\Phi_{signal}} \qquad \textbf{(12–81)}$$

Comparison shows that polarization sensitive schemes and polarization insensitive schemes are the same. A comparison of the effective noise-figure (NF_{eff}) to conversion fiber length (L) for all four cases is illustrated in Figure 12–39.

From Figure 12–39, it is evident that the incorporation of both OBPFs in a polarization insensitive spectral inversion with no optical amplifiers involved can achieve an NF_{eff} of 16 dB, by comparison to 17 dB and 24 dB, respectively, for the other two cases.

12.6   20 Gb/s AND 40 Gb/s EXPERIMENTAL SYSTEMS THAT EMPLOY SPECTRAL INVERSION

The bandwidth–length product (BL) and the system transmission bit rate of any optical communications system are limited by the chromatic dispersion and fiber nonlinearities. The utilization of multichannel WDM schemes and dispersion compensation fibers in optical systems has dramatically increased link distance and transmission capacity. However, demands for higher channel bit rates beyond the 2.5 Gb/s, up to 20 Gb/s and 40 Gb/s, have proven to be viable experi-

mentally (especially of ultra high speed systems). Optical systems that implement such dense channel capacities are limited by the induced GVD and SPM. Therefore, if such systems are to be successfully deployed, GVD and SPM dispersion compensation are essential. Various electrical and optical dispersion compensation techniques such as the use of two mode fibers, prechirping of the transmitted signal, and the use of the equalizer fibers. Although these dispersion compensation techniques can deal effectively with GVD, they are unable to compensate for SPM. An effective way of compensating for SPM is through soliton transmission (see soliton optical transmission, in this chapter). However, both GVD and SPM can be effectively compensated for through the implementation of **optical phase conjugation (OPC)**. Through this dispersion compensation method, the induced GVD pulse distortion of the transmission fiber (1st fiber) is compensated for by OPC. After this, the signal is then transmitted by the second fiber that exhibits the same GVD. Here, the same process takes place. One of the most effective OPC dispersion compensation methods is one that exploits the parametric FWM concept. Experimental results have shown that very fast optical transmission systems can be achieved by implementing the OPC dispersion compensation method. It has also been proven that the FWM technique is effective for both positive and negative GVD, and is independent of the adopted modulation format. Furthermore, this dispersion compensation technique can be applied with equal efficiency in both standard SMF and DSF. However, the implementation of OPC for SPM induced dispersion compensation is somewhat limited because of the ampli-

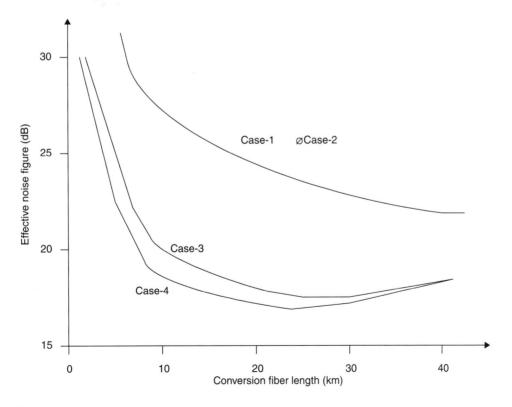

Figure 12–39. NF_{eff} v. L.

fier gain and the intensity variations induced by fiber losses. Therefore, dispersion compensation by OPC will be most effective at relatively small fiber intensity variations. GVD and SPM can effectively be compensated for through proper fiber designs. The effectiveness of optical phase conjugation is most apparent in WDM optical systems. OPC technique is capable of providing dispersion compensation for each channel within the WDM scheme, regardless of the level of dispersion identified by its dispersion slope.

GVD and SPM Dispersion Compensation by OPC

The method of GVD and SPM dispersion compensation by OPC is illustrated in Figure 12–40.

The fiber properties employed in optical systems are as follows:

a) length (L)

b) fiber loss (α)

c) dispersion (D)

d) optical nonlinearity coefficient (γ)

The input signal of a specified optical power is transmitted through fiber 1 of length (L_1) and dispersion (D_1). The OPC converts the input signal into phase conjugate (spectral inversion), in relation to pump frequency. The phase conjugate signal is then transmitted to fiber 2, of length (L_2) and dispersion (D_2). If fiber 2 exhibits the same dispersion as fiber 1, then complete dispersion compensation can be achieved. Therefore, Equation (12–82) expresses the first condition for dispersion compensation through optical phase conjugation.

$$L_1 D_1 = L_2 D_2 \qquad (12\text{–}82)$$

where L_1 = length of fiber 1, L_2 = length of fiber 2, D_1 = dispersion of fiber 1, and D_2 = dispersion of fiber 2. Likewise, equal SPM induced phase shift (in both fibers) is the second requirement for achieving complete dispersion compensation through OPC. Equation (12–83) expresses this second requirement.

$$L_1 \bar{P}_j \gamma_1 = L_2 \bar{P}_j \gamma_2 \qquad (12\text{–}83)$$

where (phase average power) $\bar{P}_j = \dfrac{1}{L_j} \int P_j(z)\, dz$ and $j = 1,2,3$ ... (integer). Therefore, for small intensity variations $\overline{\overline{P}}_j = 0$, Equation (12–82) is sufficient. Furthermore, both Equations (12–82) and (12–83) indicate that neither channel bit rate nor modulation formats are subject to OPC. However, SPM induced dispersion compensation, via OPC, is subject to fiber nonlinearities and amplifier gain. Effective dispersion compensation can be achieved for a small fiber phase shift. Small fiber phase shift is generated when the amplifier spacing along the link (l) is much shorter than the fiber nonlinear length. Complete GVD and SPM dispersion compensation can be achieved by OPC, if the ratios of dispersion-to-fiber nonlinearity at two points (x_1 and x_2) below and above the OPC (0) location are the same. Equation (12–84) expresses this ratio.

$$\frac{D_1(-x_1)}{\gamma_1(-x_1)\, P_1(-x_1)} = \frac{D(x_2)}{\gamma_2(x_2)\, P_2(x_2)} \qquad (12\text{–}84)$$

where D_1 = dispersion (fiber 1), D_2 = dispersion (fiber 2), P_1 = optical power launched into fiber 1, P_2 = optical power launched into fiber 2, γ_1 = optical nonlinearity coefficient: fiber 1, γ_2 = optical nonlinearity coefficient: fiber 2, $-x_1$ = location before OPC, and x_2 = location after OPC. Because the nonlinearities are the same, the denominator of Equation (12–84) can be expressed as follows in Equation (12–85).

$$\int_0^{x_2} \gamma_2(x)\, P_2(x)\, dx = -\int_0^{-x_1} \gamma_1(x)\, P_1(x)\, dx \qquad (12\text{–}85)$$

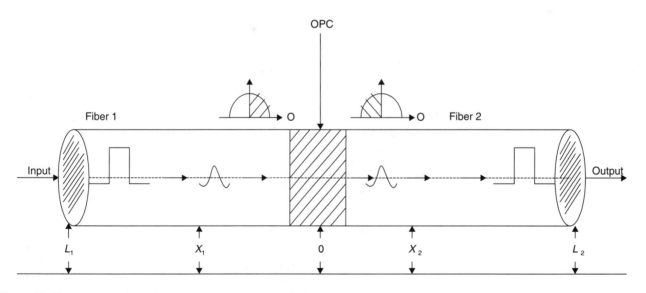

Figure 12–40. GVD and SPM dispersion compensation via OPC.

Equation (12–86) expresses the numerator.

$$\int_0^{x_2} D_2(x)\, dx = -\int^{-x_1} D_1(x)\, dx \qquad \textbf{(12–86)}$$

Both Equations (12–84) and (12–85) indicate that the optical wave at the output of the second fiber is phase conjugate of the optical wave at the input of the first fiber. That is, any phase shift induced in the optical wave by GVD and SPM at any $(-x_1)$ location can be fully compensated for by the phase distortion encountered at the (x_2) location along the length of the second fiber. Therefore, the necessary condition for complete dispersion compensation for both GVD and SPM through the application of OPC is that expressed by Equation (12–84). Designing optical fibers that have special operating characteristics to satisfy Equation (12–84) will allow full dispersion compensation. The equalization of the dispersion and nonlinearity ratio at appropriate positions along both fibers will provide full dispersion compensation, thus allowing for a substantial increase in the system BL products. However, other factors such as GVD slope, OSNR, and polarization modal dispersion set the limit for dispersion compensation efficiency.

Optical Spectral Inversion or Phase Conjugation (OPC) Schemes

Optical phase conjugation schemes can be implemented in three ways.

The OPC can be placed in the middle of equal length fibers (Figure 12–41).

The OPC can also be placed at the output of the optical transmitter. This is referred to as a *precompensation scheme*. Through this method, illustrated in Figure 12–42, a fiber with large dispersion is used at the output of the optical transmitter, while a fiber with low dispersion is used as the transmission line. The benefits gained from the precompensation scheme are several. EDFAs, in addition to dispersion shifted fibers, can provide effective dispersion compensation by fine tuning the OPC operating wavelength to the peak gain wavelength of the EDFA.

OPC precompensation also provides better stability than the other two compensation schemes. Furthermore, it can be implemented in WDM systems. Each channel in the WDM system is individually compensated for in terms of GVD and SPM, regardless of the GVD slope difference of each channel.

The OPC can be placed at the input of the optical receiver. This is referred to as a *postcompensation* scheme. Postcompensation schemes are more difficult to implement because of the randomized state of polarization of the optical signal. However, postcompensation can effectively be implemented in a completely optical signal regeneration that uses pulse shaping. Therefore, dispersion compensation requirements of optical communications links can be satisfied by both precompensation and postcompensation OPC methods. Both compensation schemes can be implemented through the utilization of dispersion compensation fibers that have relatively large dispersion and small mode-field diameter. That is, DCFs with dispersion decreasing characteristics as a function of optical intensity changes can be used with OPC for a successful dispersion compensation for both GVD and SPM induced dispersion.

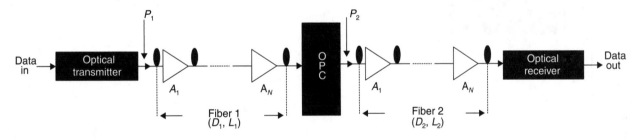

Figure 12–41. Dispersion compensation with OPC placed in the middle of two equal fibers.

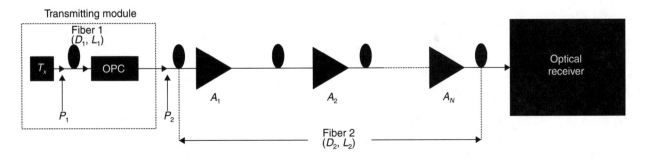

Figure 12–42. OPC precompensation scheme.

Experimental Optical Links That Employ Phase Inversion

In an attempt to increase bit rate and transmission distance of optical communications links, various research groups have been engaged in experimental designs that involve phase inversion or OPC techniques for dispersion compensation.

MIDPOINT OPC EXPERIMENTAL LINK. An optical link utilizing the midpoint OPC scheme was designed with the following parameters.

a) input data format: 16-bit NRZ signal

b) receiver bandwidth: $0.6 \times$ data rate

c) dispersion compensation: OPC with infinite bandwidth

d) EDFA: included

e) noise figure (NF): 5 dB

f) bandwidth: 24 nm (FWHM)

g) random noise: statistical evaluation by the **Monte-Carlo method**

h) location of OPC: midpoint

i) applied data at the input: 20 Gb/s

j) used fiber:

SMF: two of equal length

fiber loss: $\alpha_1 = \alpha_2 = 0.18$ dB/km

dispersion: $D_1 = D_2 = 18$ ps/nm/km

dispersion slope: 0.08 ps/nm^2/km

nonlinear coefficient: $\gamma = 1.5$ W^{-1} km^{-1}

k) input optical power to each fiber: $P_1 = P_2 = 2.5$ dBm

l) EDFA spacing: 50 km

The performance of the experimental optical link that utilizes OPC for dispersion compensation is expressed by the eye-opening penalty in relation to transmission length (Figure 12–43).

Figure 12–43 indicates that the eye-opening penalty increases almost linearly up to 4000 km. Beyond that, an exponential penalty increase is encountered. However, if the desired eye-opening penalty is to be maintained below 1 dB, then the maximum achievable link distance with midpoint OPC scheme is 2000 km. This can be considered an impressive performance. A second optical system design that employs DSF operates with the following parameters:

a) input data: 20 Gb/s

b) fiber: DSF (2)

c) fiber loss: $\alpha = 0.2$ dB/km

d) dispersion: $D = -1.0$ ps/nm/km

e) dispersion slope: 0.08 ps/nm^2/km

f) nonlinearity coefficient: $\gamma = 1.5$ W^{-1} km^{-1}

g) input optical signal power: $P_1 = P_2 = 3.5$ dBm

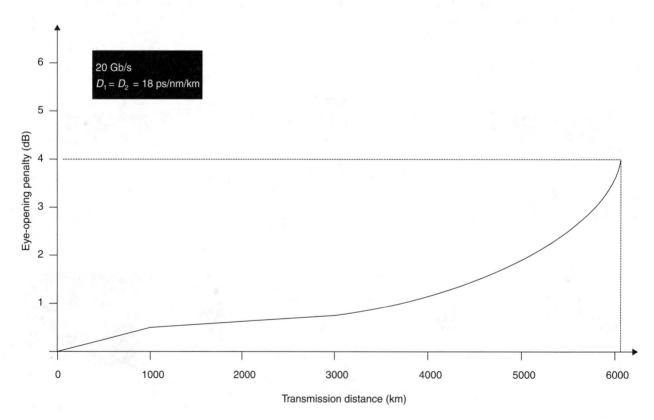

Figure 12–43. Eye-opening penalty *v.* transmission length.

A graph that depicts the design performance, expressed as the eye-opening penalty, in relation to transmission length is illustrated in Figure 12–44.

From Figure 12–44, it is evident that employing DSFs that have a midpoint OPC scheme can achieve transmission distances of up to 8000 km, while maintaining the eye-opening penalty below 1 dB.

PRECOMPENSATION OPC EXPERIMENTAL LINKS. A precompensation OPC scheme can be applied to an optical system that operates with the following parameters:

a) input data: 20 Gb/s

b) fiber 1: DD-DCF (DD-dispersion decreasing). Several lengths of fiber 1 can be used in conjunction with EDFAs in order to increase the transmission distance.

   fiber loss: 0.35 dB/km

   dispersion slope: 0.16 ps/nm^2/km

   nonlinearity: $\gamma_1 = 15.6$ W^{-1} km^{-1}

c) fiber 2: DSF (–1.0 ps/nm/km)

   nonlinearity: 2.6 W^{-1} km^{-1}

d) optical power: 3.5 dBm at the input of each fiber

The total length between the transmitter output and the OPC determines the length of each fiber. The DD-DCF consists of 5 DCFs that each have lengths equal to $L_{DD} \div 5$. The function of the DD-DCF is to compensate for each $L_2 \div N$ of fiber 2. The team of researchers working on this experiment calculated that 1000 km of transmission distance will require a dispersion decreasing fiber (DD) of 10 km. This 10 km fiber consists of 5 DD-DCFs with a Δz_1 of 2 km. Similarly, a 1500 km optical length will require a 15 km DD also composed of 5 DD-DCFs with a Δz_1 of 3 km. The power launched into each of the DD-DCFs was calculated at 13.3 dBm for the 10 km fibers and 14.2 dBm for the 15 km fibers. The performance characteristics of this experimental system, as expressed by the relationship between eye opening penalty and transmission length for both 20 Gb/s and 40 Gb/s, is illustrated in Figure 12–45.

From Figure 12–45, it is evident that in order to maintain the eye-opening penalty below 1 dB, the maximum achievable distance for 40 Gb/s is approximately 3000 km, while for 20 Gb/s the maximum transmission distance is 5000 km.

12.7   APPLICATION 1

The Impact of Chromatic Dispersion in Multi-Channel Digital Microwave Systems

In order to determine the effect of chromatic dispersion in multi-channel microwave systems transmitted over single-mode fibers, various groups have performed computer simulation studies. Assume that a number of digitally modulated

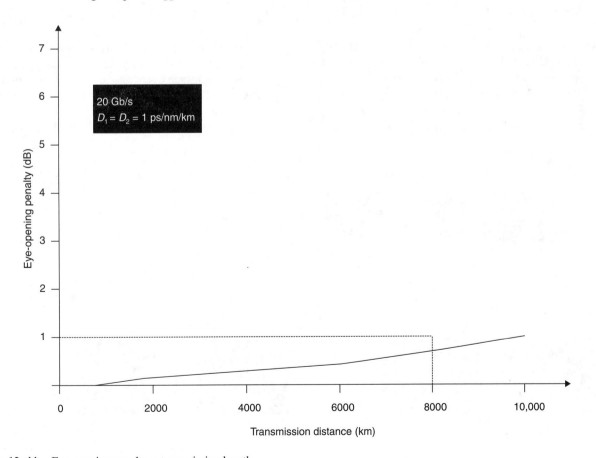

Figure 12–44.   Eye-opening penalty *v.* transmission length.

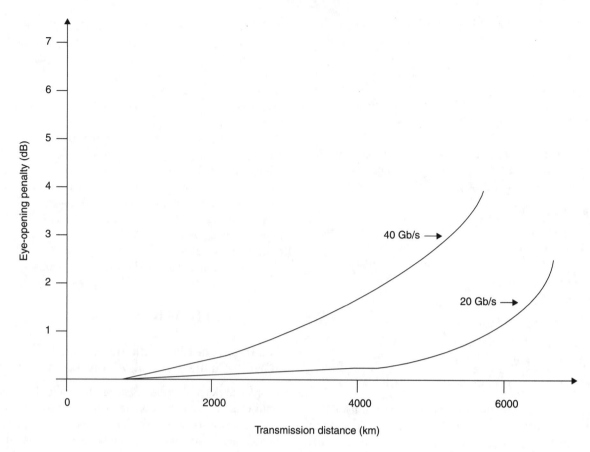

Figure 12–45. Eye-opening penalty *v.* transmission length.

(PSK or QAM modulation schemes) channels operate in the upper twenty GHz frequencies are to be distributed to corresponding transmission facilities through an optical link that employs a standard SMF and a narrow line width laser that operate at the 1550 nm wavelength window. The block diagram of such a system is illustrated in Figure 12–46.

The intent of the system is to distribute high density, digitally modulated broadband signals to individual transmission points. Although such a scheme can employ laser diodes that operate at 1310 nm wavelengths, where fiber dispersion is minimal, the 1550 nm wavelength window is preferable because it allows for longer optical lengths and therefore

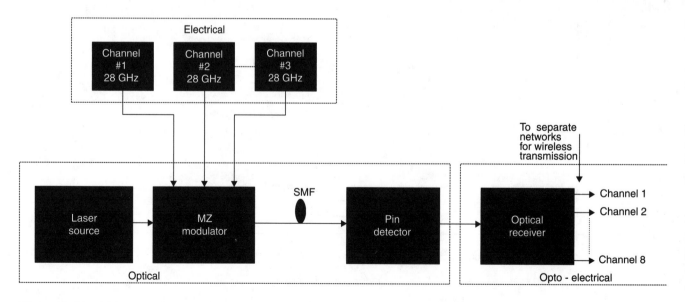

Figure 12–46. Digital multi-channel microwave transmission over optical links.

longer distribution distances. This wavelength is also the operating wavelength of EDFA, essential for long length optical systems. However, optical fibers that operate in the 1550 nm range induce chromatic dispersion that produces a negative effect, which must be taken into consideration. Chromatic dispersion induces intermodulation distortion into the system which leads to signal degradation. Furthermore, the optical span distance is directly affected by chromatic dispersion and also by EDFA and photodetector noise. The proposed system employs an external MZ optical modulator that operates linearly over the intended microwave bandwidth, a narrow line width ($\delta\lambda$) laser diode, and a standard SMF with a dispersion parameter of 18 ps/nm-km. The selection of the above optical source contributes very little to microwave carrier signal degradation. At the receiver end, each subcarrier exhibits small levels of power variation. These variations are attributed to fiber chromatic dispersion, which, in turn, is subject to the fiber length. Equation (12–87) expresses the optical power detected by the receiver in such a system.

$$P_r = \cos^2 \left[\frac{\pi D L \, (\lambda f)^2}{c} \right] \qquad \textbf{(12–87)}$$

where P_r = received optical power, D = dispersion, L = fiber length, λ = operating wavelength, f = laser diode operating wavelength, and c = velocity of light. Computer simulations have shown that at 8.85 km fiber length intervals, the individual subcarriers are not dispersion limited. The optical level of the multichannel spectrum is affected by changes in the fiber length and channel frequency because each frequency corresponds to a different fiber periodic length. Channel ISI is also caused by chromatic dispersion. Such interference can be remedied by an increase of the sampling rate to $[f_s = 64f_c \, (64 \times 28 \text{ GHz})]$ and by a channel separation of 54.68 MHz twice the selected resolution of 27.34 MHz. This channel separation generates a 984.24 MHz of broadband spectrum for an 18-channel system. The **carrier-to-noise ratio (CNR)** was observed to be 41.5 dB for the fiber distance between 49.8 km and 56.2 km.

System Specifications

The system specifications for an 18-channel, 28 GHz per channel transmission over an SMF used in the computer simulation were set as follows:

a) optical fiber: 0.25 dB/km attenuation

b) operating wavelength: 1550 nm

c) *EDFA*

operating wavelength: 1550 nm

gain: 10 dB

noise figure (NF): 10 dB

optical power output: 10 dBm

d) receiver CNR per channel: 37 dB

e) maximum fiber length: 80 km

f) typical transmitter power: 0 dBm

It is, therefore, possible for such a system to distribute a relatively large number of microwave signals over optical fibers that utilize narrow line width laser diodes that operate at the EDFA wavelength window (1550 nm) and at lengths in the 80 km range. Fiber lengths are limited by system bandwidth, operating wavelength, and fiber dispersion.

12.8 APPLICATION 2

The Effect of Chromatic Dispersion on a 2.5 Gb/s, 250 km Direct Modulation Optical Link

The fundamental difficulty encountered in high bit rate direct modulation optical systems transmitted over **nondispersive shifted single-mode fibers (NDS-SMF)** is the substantial increase of fiber chromatic dispersion. Fiber chromatic dispersion causes spreading of the optical spectrum and leads to an increase of ISI, ultimately becoming a limiting factor to the system length. In direct modulation schemes, the negative effects of laser diode phase modulation characteristics can be overcome by filtering the digitally modulated signals. Digital modulation techniques, appropriate for the above applications, are **amplitude shift keying (ASK), frequency shift keying (FSK),** and **phase shift keying (PSK).** If the optical system is to be implemented for low transmission rates (622 Mb/s to 2.5 Gb/s), the laser diode modulation frequency response exhibits nonuniformity due to temperature variations. However, if a spectral filtering method is employed, the performance characteristics of a directly modulated optical system that operates at Gb/s transmission rates and uses NDS-SM fibers will be enhanced. The proposed filtering scheme that utilizes the intensity modulation concept (IM) with a high extinction ratio is capable of significantly decreasing ISI induced by the laser nonuniform modulation response. A block diagram of such an optical system is illustrated in Figure 12–47.

The main components of the system are:

a) distributed feedback Bragg laser diode (DFB-LD)

b) spectral filter

c) optical amplifier

d) nondispersive shifted single-mode fiber (NDS-SMF)

e) inline optical amplifier

f) optical receiver

At the transmitter end, the selected laser diode may exhibit an extinction ratio of 10 dB and a spectral line width of 0.27 nm at –20 dB from the optical pick. A Fabry-Perot optical filter can be used with a 12 GHz bandwidth. Although the filter may induce insertion losses as high as 3 dB, the optical amplifier of the next stage offsets the losses. Simulation results of the above experimental system indicated the following.

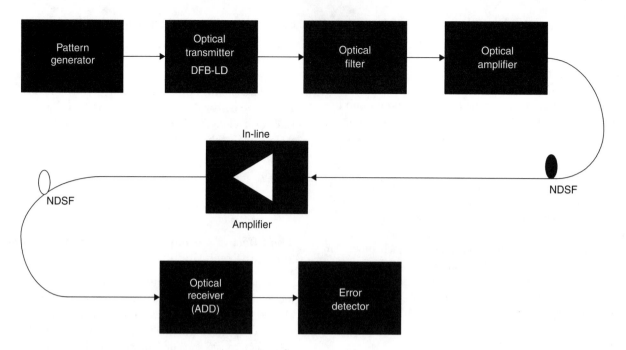

Figure 12–47. A direct modulation Gb/s optical system that uses NDS-SM fibers.

The optical sensitivity at the input of the filter was measured at −34.5 dB and −35 dB at the filter output. The system BER was measured at 10^{-10} with an optical power of 10 dBm coupled into the fiber. With the transmitter output connected to the fiber through the optical amplifier, the experiment required a power penalty of 2.2 dB at a BER of 10^{-10}. However, placing the spectral filter between the transmitter and the optical amplifier reduces the power penalty to 0.5 dB. The utilization of the in-line optical amplifier, with a gain of 22 dB, can increase the optical length from 100 km to 250 km while maintaining the same system performance. The power penalty, based on experimental results, was found to be 2.5 dB without the spectral filter and 1.4 dB with the filter. The experiment employed the following components:

a) laser diode line width enhancement factor (3)

b) input data: pseudorandom 2^7-1 of length

c) modulation current required to generate a 10 dB extinction ratio

d) optical fiber with 17 ps/nm-km dispersion

e) optical receiver bandwidth: $0.6\,f_b$ (f_b–system bit rate)

The theoretical (simulated) and experimental results that relate to optical distances and power penalties are illustrated in Figure 12–48.

From Figure 12–48, it is evident that the power penalty increases with increase of the optical length. It is evident that it is significantly smaller with the insertion of the spectral filter. The proposed optical system can offset the laser nonuniform frequency modulation response, and therefore it can accommodate transmission bit rates from 622 Mb/s to 10 Gb/s. Furthermore, optimization of the

extinction ratio and pulse broadening penalties can maximize optical lengths.

12.9 APPLICATION 3

High Speed Optical Transmission That Utilizes NDS-SMF

Digital multichannel optical systems can be designed to transmit bit rates of about 40 Gb/s per channel through NDS-SMF that operate at the 1550 nm wavelength window. NDS-SM fibers that operate at the 1550 nm exhibit dispersion levels of about 16 ps/nm-km. This wavelength window is also the operating wavelength of EDFAs, a necessary component in such a system. A block diagram of the proposed system is illustrated in Figure 12–49.

The optical system components are as follows:

a) number of channels: N

b) input data per channel: 40 Gb/s

c) WDM ($\lambda_1 \ldots \lambda_N$)

d) fiber: single mode (several lengths)

e) DCF module incorporated into each length

f) EDFAs: one for each fiber length

g) fiber spacing: 80 km

The system can be implemented by employing the spectral inversion technique (see the relevant section of this chapter) instead of DCF. In such an optical system, performance analysis for both single channel and multichannel formats must be performed. Performance analyses are required in

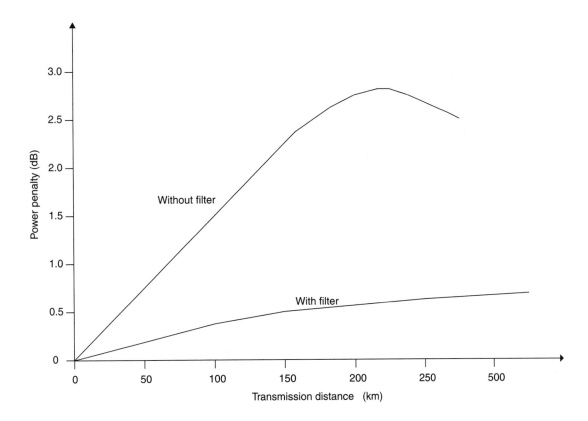

Figure 12–48. Optical distance *v.* power penalty.

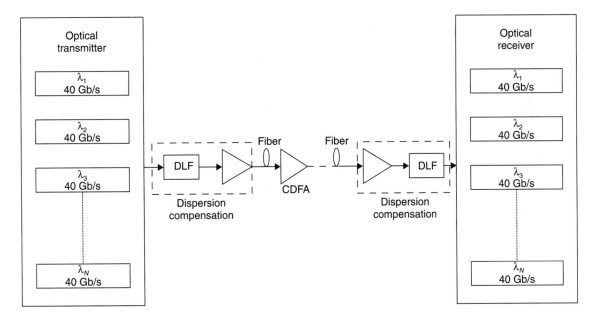

Figure 12–49. High speed optical system that uses NDS-SM fibers.

order to establish the interaction between chromatic dispersion and SPM within the optical fiber. The study of a 40 Gb/s transmission rate over a 150 km single mode nondispersion compensation fiber was performed without the use of EDFAs. Fiber dispersion compensation can be applied after the optical transmitter (precompensation) or before the optical receiver (postcompensation). However, to prevent interaction within the DCF module, the optical power coupled into the fiber should be kept at a minimum. With the pseudorandom binary sequence was selected with a length of $2^7 - 1$. The optical power penalty, as related to the average power coupled into the fiber, is illustrated in Figure 12–50.

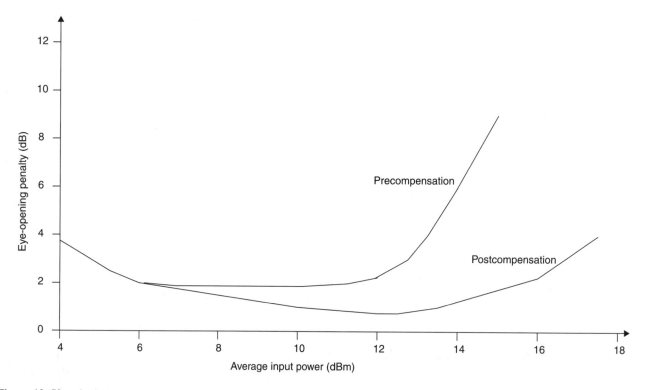

Figure 12–50. Optical power at the input of the SMF *v.* power penalty for both precompensation and postcompensation schemes.

Figure 12–50 shows that at a low input optical power, the power penalty is substantial because of the small OSNR. The power penalty is also substantial at high input power levels because of SPM. For a postcompensation scheme, the power penalty shows an exponential increase at 16 dBm, in comparison to the precompensation scheme, which exhibits an exponential increase at 13 dBm. Precompensation *v.* postcompensation are as follows:

Precompensation

Strong power fluctuations due to SPM dependency.

Post-compensation

Small power fluctuations and smaller dependency on SPM. The dispersion slope $\left(\dfrac{\Delta D}{\Delta \lambda}\right)$ induces limitations on RZ transmission signals above 80 Gb/s and at optical lengths over 160 km. For higher system bit rates, the dispersion compensation modules must be inserted very accurately into the optical fiber at a precise point. If the optical link is to incorporate several fiber lengths, the precise placement of the dispersion compensation module is critical for either 10 Gb/s or 40 Gb/s RZ data rates. It has been suggested that alternate placement of precompensation and postcompensation yields the best results; 40 Gb/s RZ transmission rates over 1000 km, with 32 km spans have been reported. In a 4-channel, 40 Gb/s per channel optical system transmitting over a 100 km span (SMF with 17.5 ps/nm-km), the fiber dispersion can be completely eliminated by the insertion of both precompensation and postcompensation modules. The performance characteristics of a 4-channel, 40 Gb/s optical system are illustrated in Figure 12–51.

Pulse chirping for the four channels was measured as 8 ps, 4.5 ps, and 6 ps, compared to residual dispersion. In an optical link, the PMD imposes a limit to the system bit rate. (For more details see the relevant section in this chapter.) For system lengths of 500 km, with bit rates of 40 Gb/s, 80 Gb/s, and 16 Gb/s NRZ signals, the corresponding polarization modal dispersions are 0.025 ps/km$^{1/2}$, 0.0063 ps/km$^{1/2}$, and 0.0016 ps/km$^{1/2}$. If RZ signals are used, the polarization dispersion is even larger because of the higher bandwidth of the RZ signal. Therefore in such systems, PMD is significant and must be compensated for.

12.10 APPLICATION 4

Highs Speed Optical Transmission Utilizing Large Effective Area Fibers

Optical links that employ DWDM can dramatically increase the link capacity. However, this increase of system bit rate is achieved at the expense of fiber nonlinearity induced power penalty. Traditionally, DWDM optical links employ NZ-DSF to reduce the impact of FWM on system performance. An alternative solution for reducing the impact of optical nonlinearities on system capacity is the incorporation of large effective area fibers (LEAF), developed by Corning, into the system. The incorporation of LEAFs into DWDM optical systems substantially increases the optical length by

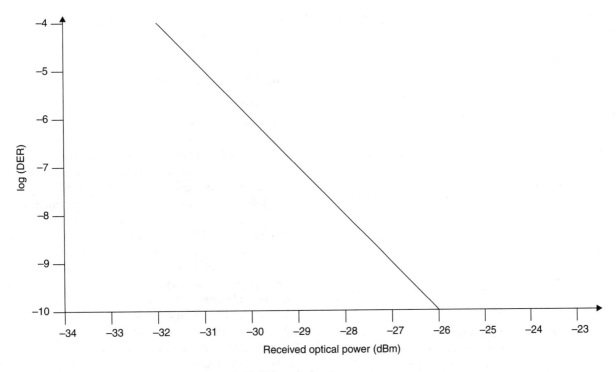

Figure 12–51. BER *v.* optical power for a 4-channel, 40 Gb/s optical system.

allowing higher optical powers to be launched into the fiber. At the same time, this decreases or completely eliminates the need for dispersion compensation. SPM and XPM can also be reduced below the 1 dB level (for an 8-channel, 10 Gb/s per channel optical system), while FWM can be completely eliminated. A 32-channel, 10 Gb/s per channel DWDM optical system that employs large effective area fibers is illustrated in Figure 12–52.

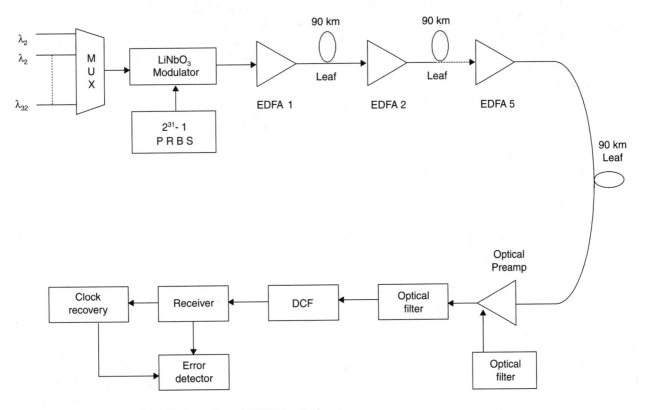

Figure 12–52. A 32-channel, 10 Gb/s per channel DWDM optical system.

The performance of the proposed link was evaluated based on the following parameters:

a) optical sources: 32 DFB lasers ($\lambda_1 = 1532.68$ nm or 195.6 THz, $\lambda_{32} = 1557.36$ nm or 192.5 THz)

b) multiplexing the laser diodes: fiber couplers

c) input data pattern: PRBS with $2^{31} - 1$ length

d) channel spacing: 100 GHz

e) EDFAs

f) modulator: LiNbO$_3$ MZ

g) optical fibers: large effective area 5 spans, 90 km per span, to a maximum 450 km

   fiber effective area: 72–78 μm^2

   fiber bandwidth: 8 nm (1506 nm to 1514 nm)

   fiber dispersion slope: 0.1 ps/nm^2/km

h) optical power coupled into each fiber span: 19 dBm

i) span loss: 24 dB (assumed)

The selection of the individual channel to be measured was accomplished through a tunable narrow band grating filter (FWHM = 0.3 nm), while precompensation and postcompensation were equally applied to all channels. The system exhibited an optical Q of 8.9 dB at a BER of 4.5×10^{-15} across the entire wavelength spectrum. The measurements indicated a power penalty of 0.9 dB (average) with 1.3 dB of channel 12. The OSNR in relation to operating wavelength is illustrated in Figure 12–53.

Figure 12–53 shows an average OSNR of 25 dB. It was also observed that with a reduction of the optical power coupled into the fiber from 19 dBm to 17 dBm with a constant OSNR, the Q was improved by 0.5 dB. Therefore, it was evident that the implementation of large effective area fibers substantially reduced the effect of nonlinearities. The impact of FWM on the adjacent channels was measured at –30 dB in relation to channel power. The relationship between FWM and channel number is illustrated in Figure 12–54.

12.11   THE WAVE POLARIZATION EFFECT IN OPTICAL SYSTEMS

Until the mid-1990s, the relatively short distances and low transmission capacities of optical links were unaffected by wave polarization, because the optical receivers were detecting power levels determined by their optical sensitivities, regardless of the wave polarization. The ever-increasing demand for higher system bit rates and for longer spans between amplification was satisfied by the development of EDFAs, add/drop multiplexers, and narrow line width ($\delta\lambda$) laser diodes. The implementation of such technologies, span distances, and transmission rates pushed the component and system operating parameters close to their limits. Although wave polarization effects were not a factor for short distance and relatively low transmission rate optical systems, they became important and had to be taken into account for high

speed, long span systems that employ EDFA and SMFs. In such systems, PDL and PMD are cumulative and must therefore be taken into consideration.

Fiber Geometry and Wave Polarization

An SMF is a cylindrical optical waveguide, and the propagation of the light through it is subject to the physical properties of the waveguide. In a fiber with ideal physical properties, the two modes of light behave as one. In practice, SMFs exhibit some degree of physical imperfection, induced during the manufacturing process or imposed upon them by external forces during transportation and installation. In this case, the two orthogonally polarized modes are no longer identical, and therefore exhibit different group (v_g) and phase (v_p) velocities, which ultimately degrade system performance. (For more details, see Chapter 10.) In long distance optical systems, the fiber's physical imperfections are not uniform and may change in magnitude along the length of the fiber. However, in order to better understand optical wave polarization, a relatively short fiber with uniform physical imperfections will be considered. Along this short fiber length and under these assumptions, there will be two propagation constants (β_1 and β_2), one for each mode. Equation (12–88) expresses the difference between the propagation constants.

$$\beta_2 - \beta_1 = \frac{\omega(n_2 - n_1)}{c} \qquad \textbf{(12–88)}$$

where β_1 is the propagation constant for mode one, β_2 is the propagation constant for mode two, ω is the angular velocity, c is the velocity of light, n_1 is the refractive index for the fast mode, and n_2 is the refractive index for the slower mode: ($n_2 > n_1$). Equation (12–89) expresses the refractive index difference between the slow and the fast modes.

$$\Delta n = n_2 - n_1 \qquad \textbf{(12–89)}$$

Substituting Equation (12–88) into Equation (12–89) yields Equation (12–90).

$$\beta_2 - \beta_1 = \frac{\omega(\Delta n)}{c} \qquad \textbf{(12–90)}$$

For fibers employed in long haul optical systems, typical values for ($\Delta n_{2,1}$) fall between 10^{-5} and 10^{-7}. These values are very small compared to the refractive index difference (3×10^{-3}) between the core and cladding of the same fiber. An imperfect fiber might exhibit several states of polarization, subject to the polarization state of the optical wave at the input of the fiber and to the types of imperfections in the fiber.

Beat Length (L_B)

If a linear polarized optical wave is launched into an imperfect fiber at 45°, a phase difference will gradually develop between the two orthogonal components: from a linear state to a cyclic state and back to linear after a number of cyclic states. The fiber length that is required in or-

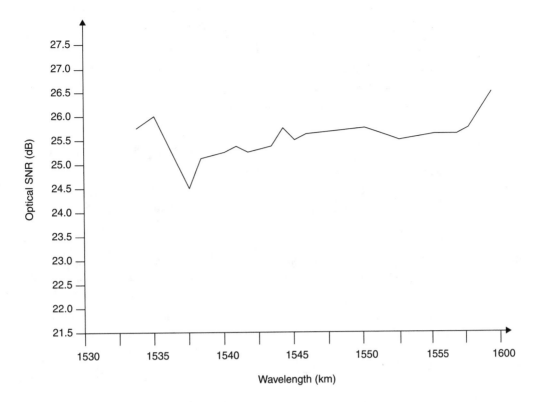

Figure 12–53. Optical SNR *v.* wavelength.

der for the linear polarized optical wave to go through the cyclic states and back to linear is referred to as the beat length (L_B). Equation (12–91) expresses the **beat length.**

$$L_B = \frac{\lambda}{\Delta n} \qquad (12\text{–}91)$$

The two polarization modes generated by the fiber's physical anomalies reflect a differential group velocity. This is a major factor that contributes to pulse chirping and, as a consequence, is a limit of the fiber bandwidth. The graphical representation of the differential group velocity is illustrated in Figure 12–55.

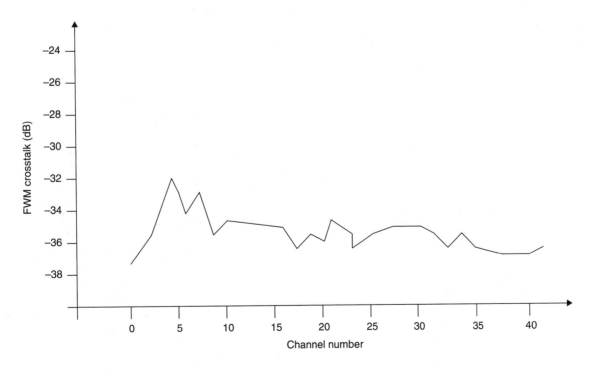

Figure 12–54. FWM *v.* channel number.

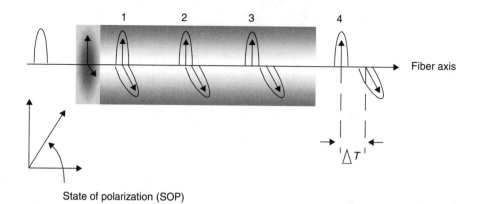

Figure 12–55. Differential group velocity.

The differential group velocity is expressed as the group delay time per unit fiber length and is derived as follows:

From Equation (12–90), $\beta_2 - \beta_1 = \dfrac{\omega(\Delta n)}{c}$, the frequency derivative yields Equation (12–92).

$$\frac{d(\beta_2 - \beta_1)}{d\omega} = \frac{\Delta n}{c} - \frac{\omega d \Delta n}{c d \omega} \qquad \textbf{(12–92)}$$

The above expression is referred to as the group delay time per unit length $\left(\dfrac{\Delta \tau}{L}\right)$ or PMD (Equation (12–93)).

Therefore,

$$\text{PMD} = \frac{\Delta \tau}{L} \text{ (ps/nm)} \qquad \textbf{(12–93)}$$

As previously mentioned, PMD is valid only for short length optical fibers because physical anomalies are assumed to be ever present (intrinsic). This, of course, is not true for long fiber lengths. Equation (12–94) expresses the PMD long length fibers.

$$\text{PMD} = \frac{\Delta \tau}{(L)^{1/2}} \text{ ps/km}^{1/2} \qquad \textbf{(12–94)}$$

PMD, expressed in time domain as $\left(\dfrac{\Delta \tau}{L}\right)$, can also be expressed in frequency domain (Equation (12–95)).

$$\Delta \omega = \frac{2\pi}{\Delta \tau} \qquad \textbf{(12–95)}$$

PMD values for various fibers employed in long haul optical systems and for different physical anomalies are listed in Table 12–9.

Since the 1970s, attempts have been made to reduce PMD in optical fiber through various techniques. These techniques involve PMD reduction within and outside the fiber. The technique used within the fiber attempts to reduce the intrinsic PMD, while external methods attempt to reduce PMD either through equalizers incorporated into the optical receiver module or through automatic polarization control on both the receiver and the transmitter stages.

TABLE 12–9 Operating Wavelength $\lambda = 1550$ nm

Fiber Physical Imperfections	Δn	PMD (ps/km)
Core ellipticity (1%)	5.0×10^{-8}	1.50
Twist (one turn per meter)	2.3×10^{-7}	0.00
Lateral stress (1 g-wt/cm)	7.0×10^{-8}	0.23
Radius bend (10 cm)	5.0×10^{-8}	0.17

PMD in Digital Optical Systems

Although PMD may have little or no effect on very short optical fiber links, it becomes a very serious problem for long spans. As mentioned previously, the development of EDFA and their employment in optical systems enabled designers to achieve lengths of about 10,000 km. With such lengths, very small fiber imperfections will induce significant levels of PMD. However, small anomalies along the fiber length do not add linearly. That is, small fiber imperfections in one span may be added to or subtracted from the next span. Therefore, PMD in long fibers is random. Two models for determining polarization modal dispersion in long length fibers exist. These are:

a) coupling power model (low coherence)

b) principal state model (high coherence)

THE LOW COHERENCE COUPLED POWER MODEL. Imagine a 2-mode optical waveguide. If an optical pulse is coupled into one of the two modes, it will travel this mode until it encounters a small anomaly (perturbation). This physical imperfection, although very small, will cause some of the power of the optical pulse to be coupled into the second mode. Farther down the fiber length, the traveling pulses encounter another perturbation, and the two become four, and so on. Because the pulses travel different modes that have different modal lengths, they will arrive at the output of the fiber at different times. This process is illustrated in Figure 12–56.

The pulse shape at the end of the fiber can be determined by adding all the individual small pulses in time (incoherent approach). Through this approach, it is evident

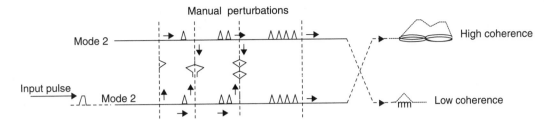

Figure 12–56. Pulse traveling a two mode, nonideal optical fiber.

that the output pulse is Gaussian. The output pulse broadening is expressed by Equation (12–96).

$$\sigma = \left(\frac{1}{2\Delta v}\right)(Ll_c)^{1/2} \quad \text{(12–96)}$$

where σ is the pulse chirping (rms), l_c is the coupling length, L is the length of the fiber, and Δv is the differential group velocity. l_c is the length at which the average power in the orthogonal polarization mode does not exceed 13% of the power in the starting mode at the input of the fiber.

Equation (12–96) indicates that pulse chirping is proportional to the square root of the fiber and coupling lengths and is inversely proportional to the differential group velocity. Although the power coupling mode scheme was developed in order to determine PMD in multimode systems that employ low coherence LED transmitters, it can also be used in some limited applications that employ SMFs. If a high coherent optical source is used, the output pulse will be coherent, with a shape determined by the phase of all the small pulses at the fiber output. Under these circumstances, the average pulse shape can be predicted. However, the model cannot predict the actual pulse shape caused by changes in the source wavelength or polarization states.

THE PRINCIPAL STATES MODEL. The effects of PMD in long span optical systems can be better understood through the high coherence model, adopting the principal states of polarization. In applying the above model, two assumptions must be made. First, the optical source coherent time must be larger than the PMD pulse shift. Second, the fiber loss must be polarization independent. In essence, in a long haul optical system, the pulse delay of the traveled pulse must be much smaller than the pulse period. Based on the above assumptions, Equation (12–97) can express the optical fiber transmission properties.

$$T(\omega) = e^{a(\omega)}\begin{bmatrix} u_1(\omega) & u_2(\omega) \\ -u'_2(\omega) & u'_1(\omega) \end{bmatrix} \quad \text{(12–97)}$$

where $T(\omega)$ is the frequency dependent transmission matrix, $a(\omega)$, $u_1(\omega)$, $u_2(\omega)$ are complex quantities, and $u_1^2 + u_2^2 = 1$. At the output of the fiber, the pulse electric field vector $E_2(\omega)$ is expressed by Equation (12–98).

$$E_2(\omega) = e^{a(\omega)}\begin{bmatrix} u_1(\omega) & u_2(\omega) \\ -u'_2(\omega) & u'_1(\omega) \end{bmatrix}[E_1(\omega)] \quad \text{(12–98)}$$

where $E_1(\omega)$ is the input pulse electric field vector. The second part of Equation (12–98) indicates that an orthogonal pair of input principal states of polarization exists. Under this principle, if a pulse at the input of the fiber is aligned to one of the principal states, it will appear at the output of the fiber, and all its spectral components will have the same polarization state. That is, the input pulse traveled through the fiber has suffered only a phase shift, while its shape has remained the same.

The adoption of the principal states model implies that any input signal can be expressed by two vector components aligned to the principal states. Therefore, the electric field vector (time varying) of the output pulse can be expressed by Equation (12–99).

$$E_2(t) = c_+\hat{e}_+ E_1(t + \tau_+) + c_-\hat{e}_- E_1(t + \tau_-) \quad \text{(12–99)}$$

where $E_1(t)$ is the input pulse electric field vector (time varying), $\hat{e}_+$, $\hat{e}_-$ are the unit vectors indicative to the output component polarization states (output principal states), and c_+, c_- are the complex coefficients. If the energy of the input optical pulse is split into two principal states, Equation (12–100) can express the arrival time at the output of the fiber.

$$\Delta\tau = \tau_+ - \tau_- \quad \text{(12–100)}$$

$\Delta\tau$ represents the pulse chirping at the output of the fiber, subject to differential time delay and complex coefficients c_+ and c_-. It has also been observed that in long span fibers, dispersion is a function of ambient temperature variations.

The Impact of PMD in Digital Systems

On numerous occasions in this text, it has been stated that pulse broadening (chirping) limits the maximum span and transmission rate of a long haul optical link. One of the elements that directly contribute to pulse spreading is the fiber PMD. Pulse broadening (chirping) decreases performance by increasing system BER. Because PMD is time dependent, the induced system degradations are also time dependent. For example, variations in operating temperature can induce corresponding changes in the system BER. Figure 12–57 illustrates such changes.

System performance variations require optical power compensation. Such optical power compensation is referred to as *power penalty*. The **power penalty** of a digital NRZ optical fiber system is expressed by Equation 12–101.

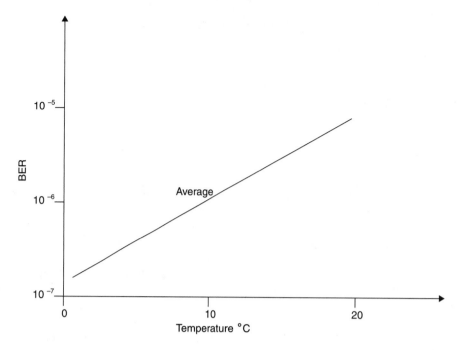

Figure 12–57. BER *v.* temperature.

$$\varepsilon_{(dB)} = \frac{A(\Delta\tau^2)(\gamma)(1-\gamma)}{T^2} \qquad \textbf{(12–101)}$$

where ε is the power penalty, T is the FWHM, A is the parameter related to pulse shape and optical receiver filter, and γ is the splitting ratio of the input pulse power ($\gamma < 1$).

$$T = \frac{1}{f_b}$$

where f_b is the system bit rate. Substituting T into Equation (12–101) yields Equation (12–102).

$$\varepsilon_{(dB)} = A(\Delta\tau^2)(\gamma)(1-\gamma)(f_b^2) \qquad \textbf{(12–102)}$$

The relationship between the A parameter and pulse shape is listed in Table 12–10.

From Equation (12–29), it is evident that for $\gamma = 0$, power penalty is also zero, indicating no pulse chirping, because all the optical power coupled into the fiber is one modal component. However, power penalty is also a function of the square of the differential time delay ($\Delta\tau^2$) and the square of the system bit rate (f_b). Under normal circumstances, power penalties relevant to PMD should not exceed the 1 dB mark. Optical power penalties are probabilistic and are expressed by Equation (12–103).

$$P_{(\varepsilon)} = \eta e^{-\eta\varepsilon} \qquad \textbf{(12–103)}$$

where $P_{(\varepsilon)}$ is the power penalty probability density function, $\eta = \dfrac{16\,T^2}{A\pi\langle\Delta\tau\rangle^2}$, $\langle\Delta\tau\rangle$, (average differential delay), and ε (probability of $\varepsilon > 1$ dB) $= e^{-\left(\frac{16\,T^2}{A*\langle D\Delta\rangle^2}\right)}$. The average differential

TABLE 12–10 Relationship Between A-Parameter and Pulse Shape

A-Parameter	Pulse Shape
12	Square
15	25% (t_r/t_f)
22	Raised cosine
24	Triangular
25	Gaussian

delay time of the principal states must not exceed 14% of the pulse period. Equation (12–104) expresses the relationship between optical span (L), system bit rate (f_b), and PMD.

$$f_b^2 L = \frac{0.02}{(PMD)^2} \qquad \textbf{(12–104)}$$

where f_b is the optical system bit rate, PMD is the polarization modal dispersion (ps/km$^{1/2}$), and L is the optical span. A graphical representation of the relationship between PMD and ($f_b^2 L$) is illustrated in Figure (12–58).

12.12 LED TRANSMITTERS COUPLED INTO MULTIMODE FIBERS

Dispersion Penalty Calculations

The mean dispersion penalty (μD), induced by an LED transmitter, coupled into a multimode fiber, and expressed in dB is subject to system bit rate and LED bandwidth and is expressed by Equation (12–105). Equation (12–105) is applied only for short wavelengths in the 850 nm range,

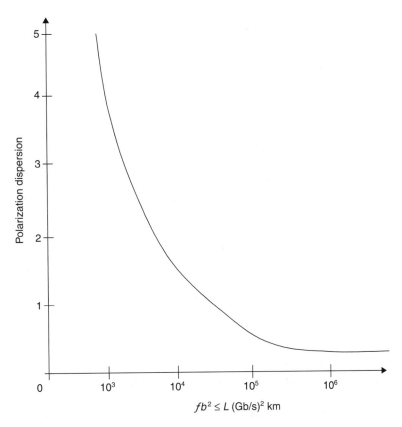

Figure 12–58. PMD *v. $f_{b^2} L$*.

while for long wavelengths in the 1300 nm operating window, Equation (12–106) is applicable.

$$\mu D_{(dB)} = 0.88 \left(\frac{f_b}{B_{W_{total}}} \right)^2 \text{ (short wavelength)} \quad \textbf{(12–105)}$$

where μD is the mean dispersion penalty, f_b is the system bit rate (Mb/s), and $B_{W_{total}}$ is the total LED exit bandwidth (MHz).

$$\mu D_{(dB)} = 1.2 \left(\frac{f_b}{B_{W_{total}}} \right)^2 \text{ (long wavelength)} \quad \textbf{(12–106)}$$

Calculation of the total LED exit bandwidth ($B_{W_{total}}$) is subject to both modal and chromatic dispersion effects. Equation (12–107) expresses the statistical sum of both dispersion effects on bandwidth.

$$B_{W_{total}} = \frac{(B_{W_{chromatic}})(B_{W_{laser}})}{\left[(B_{W_{chromatic}})^2 (B_{W_{laser}})^2 \right]^{1/2}} \quad \textbf{(12–107)}$$

Bandwidth, which is the result of modal dispersion, is equated to laser bandwidth ($B_{W_{laser}}$) because a laser source coupled into a multimode fiber induces only modal dispersion due to its narrow beam width. For multimode fibers, the spooled fiber bandwidth (B_{W_F}) is determined (measured) with a laser diode. The laser bandwidth ($B_{W_{total}}$) is proportional to the spooled bandwidth (B_{W_F}), the total system length (L), and the average section length (L_S), and is expressed by Equation (12–108).

$$B_{W_{total}} \propto \kappa(B_{W_F}) L_S^{-\beta} L^{-\gamma} \quad \textbf{(12–108)}$$

where κ is the reduction factor < 1, β is the co-catenation scaling factor <1 for average section length, and γ is the co-catenation scaling factor <1 for average system length. The bandwidth resulting from chromatic dispersion is proportional to the total length of the link and is inversely proportional to the spectral width and minimum dispersion wavelength of the used fiber (λ_{min}) Equation (12–109).

$$B_{W_{chromatic}} \propto \frac{L^{-0.7}}{(\text{FWHM})^{1/2} |\lambda_C - \lambda_{min}|} \quad \textbf{(12–109)}$$

where $B_{W_{chromatic}}$ is the chromatic dispersion bandwidth, L is the total system length, and FWHM is the full wave at half maximum. Both dispersion bandwidths (the modal and chromatic) are combined into the LED bandwidth (total exit). Typical LED spectral profiles and AT&T fiber specs are used to calculate the LED exit bandwidth.

For short wavelengths (850 nm to 870 nm) in 50/125 μm fibers, the LED bandwidth is expressed by Equation (12–110).

$$\frac{1}{B_{W_{total}}} = \frac{L^{1.96}}{(92.3)^2} + \frac{L^{1.8}}{(0.77 \, B_{W_F})^2} \quad \textbf{(12–110)}$$

For short wavelengths (850 nm to 870 nm) in 62.5/125 μm fibers, the LED bandwidth is expressed by Equation (12–111).

$$\frac{1}{B_{W_{total}}} = \frac{L^{1.8}}{(65.3)^2} \qquad \textbf{(12–111)}$$

For long wavelengths (1300 nm), and 50/125 µm fibers, with 0.23 NA, the LED bandwidth is expressed by Equation (12–112).

$$\frac{1}{B_{W_{total}}} = \frac{L^{1.3}}{(688)^2} + \frac{L^{1.4} L_S^{0.5}}{(0.71\, B_{W_F})^2} \qquad \textbf{(12–112)}$$

For long wavelengths (1300 nm), in 62.5/125 µm fibers with 0.275 NA, the LED wavelength is expressed by Equation (12–113).

$$\frac{1}{B_{W_{total}}} = \frac{L_S^{0.5} L^{1.4}}{(0.71\, B_{W_F})^2} + \frac{L^{1.38}}{(400)^2} \qquad \textbf{(12–113)}$$

12.13 LASER TRANSMITTERS COUPLED INTO MULTIMODE FIBERS

The utilization of laser diodes as optical transmitters coupled to multimode fibers has constantly gained prominence in optical fiber link designs, especially in datacom applications. Because laser sources are used as optical transmitters, the main dispersion effect is modal. The mean dispersion penalty is expressed by Equation (12–114).

$$\mu D_{dB} = 0.85\left(\frac{f_b}{B_{W_{total}}}\right)^2 \qquad \textbf{(12–114)}$$

where $B_{W_{total}}$ is the modified laser bandwidth ($B_{W_{laser}}$) and $B_{W_{laser}}$ is $F\kappa\,(B_{W_F})\,L^{-\gamma}\,LS$ or $S^{-\beta}$. For both 50/125 µm and 62.5/125 µm fibers, the $B_{W_{laser}}$ is expressed by Equation (12–115).

$$B_{W_{laser}} = 0.71\,(B_{W_F})\,L^{-0.7}\,L_S^{-0.25} \qquad \textbf{(12–115)}$$

Combining Equations (12–114) and (12–115), the mean dispersion penalty can be calculated.

Example 12–11

Calculate the dispersion penalty (μD) imposed by an optical link of 1.5 km maximum length incorporating a laser diode, as the optical source, coupled into a multimode fiber. The fiber bandwidth (B_{W_F}) is 600 MHz/km and only three splices are to be used. The required system bit rate is set at 1 Gb/s.

Solution
Data:

1) $L = 1.5$ km

2) $L_S = 0.5$ km (1.5 km ÷ 3)

3) $B_{W_F} = 600$ MHz

4) $f_b = 1$ Gb/s

$$\mu D_{dB} = 0.85\left(\frac{f_b}{B_{W_{total}}}\right)^2$$

i) Compute laser bandwidth ($B_{W_{laser}}$).
Applying Equation (12–13) yields

$$B_{W_{laser}} = 0.71(B_{W_F})\,L^{-0.7}\,L_S^{-0.25}$$

$$= 0.71(600) \times 1.5^{-0.7} \times 0.5^{-0.25}$$

$$= 380 \text{ MHz}$$

Therefore, $B_{W_{laser}} = 380$ MHz.

ii) Compute the mean dispersion penalty (μD).

$$\omega D_{dB} = 0.85\left(\frac{f_b}{B_{W_{total}}}\right)^2$$

$$= 0.85\left(\frac{1 \times 10^3 \text{ Mb/s}}{380 \text{ MHz}}\right)^2$$

$$= 5.9$$

Therefore, mean dispersion penalty (μD) = 5.9 dB.

12.14 LASER SOURCES COUPLED INTO SINGLE-MODE FIBER

When laser diodes are used in conjunction with single-mode fibers in optical links, the dispersive effect is purely chromatic for short distance optical links while PMD is present in long distance high bit rate and analog systems. For short and medium length distances, optical links that employ laser transmitters, an optimal wavelength (λ_0) that exhibits zero dispersion exists at the 1310 nm and 1550 nm windows. DSFs also exhibit a zero dispersion wavelength (λ_0) at the 1550 nm wavelength window. Theoretically, chromatic dispersion can only be eliminated if a laser source that transmitts at the appropriate wavelength with zero line width ($\delta\lambda$) is used. However, in practical systems, nonideal components are used, which leads to chromatic dispersion. Dispersion penalty calculations for medium and long haul optical links are based on a worst-case approach. For SONET and Bellcore short, medium, and long distance links, dispersion penalties of 1 dB and 2 dB are allowed to compensate for optical fiber dispersion and laser diode nonzero line width. The maximum system dispersion, laser spectrum line width, and maximum system bit-rate are expressed by Equation (12–116).

$$\varepsilon = 10^{-6}(f_b)(D_{max})(\delta\lambda) \qquad \textbf{(12–116)}$$

where ε is the time-slot width fraction, f_b is the system bit rate (Mb/s), D_{max} is the maximum fiber dispersion (ps/nm), and $\delta\lambda$ is the laser diode spectral line width (nm). For **single longitudinal mode (SLM)** laser diodes, the spectral line width

$(\delta\lambda)$ is the Gaussian spectral width at the -20 dB point and is expressed by Equation (12–117).

$$\delta\lambda = \Delta\lambda 20/6.07 \qquad (12\text{–}117)$$

For **multilongitudinal mode (MLM)** laser diodes, $\delta\lambda$ is the rms spectral width. The values of ε for both single longitudinal mode and multilongitudinal mode lasers and for either 1 dB or 2 dB dispersion penalties are listed in Table 12–11. They are based on the Bellcore TR-NWT-000253 document.

The figures from Table 12–11 are translated as follows: An optical link that employs SLM laser diode, pulse broadening must not exceed 30.6% if a 1 dB dispersion penalty is to be maintained. Likewise, for 2 dB dispersion penalty, pulse broadening must not exceed 49.1%. The figures for MLM lasers are more rigid. This is because MLM lasers also exhibit mode-partition noise that directly contributes to an increase of the system ISI. For long distance OC-18 and OC-24 transmission formats, the allowable dispersion penalty is 2 dB. For OC-48 transmission rates, a slightly higher dispersion penalty may be required.

Example 12–12

Calculate the maximum length of an optical link incorporating an MLM laser diode as the transmitter, operating at the 1310 nm wavelength window, and coupled into an SMF that exhibits a dispersion of 3.0 ps/nm-km. The system transmission rate is 622 Mb/s and the laser diode spectral line width is 4 nm. Assume a maximum dispersion penalty of 2 dB.

Solution

i) *Determine ε.*
   From Table 12–11. For an MLM laser and 2 dB dispersion penalty, $\varepsilon = 0.182$.

ii) *Determine D_{max}.*
   From Equation (12–116), solve for D_{max}.

$$\varepsilon = 10^{-6}(f_b)(D_{max})(\delta\lambda)$$

or

$$D_{max} = \frac{\varepsilon}{1 \times 10^{-6}(f_b)(\delta\lambda)}$$

$$= \frac{0.182}{1 \times 10^{-6}(622)(4)}$$

$$= 73.1 \text{ ps/nm}$$

Therefore, $D_{max} = 73.1$ ps/nm.

iii) *Calculate the link distance.*
   The maximum link distance can be calculated by dividing the D_{max} by the optical fiber dispersion (3.0 ps/nm-km).

or

$$L_{max} = \frac{D_{max}}{3.0 \text{ ps/nm-km}} = \frac{73.1 \text{ ps/nm}}{3.0 \text{ ps/nm-km}} = 24.3 \text{ km}$$

Therefore, maximum link distance (L_{max}) $= 24.3$ km.

TABLE 12–11   Laser Types and Dispersion Penalties in Reference to ε

Laser Diode	Dispersion Penalty (dB)	
SLM	1.0	0.306
	2.0	0.491
MLM	1.0	0.115
	2.0	0.182

Example 12–13

Calculate the maximum length of an optical link that incorporates an MLM laser diode as the transmitter, operating at the 1310 nm wavelength window, and coupled into an SMF that exhibits a dispersion of 3.0 ps/nm-km. The system transmission rate is 622 Mb/s and the laser diode spectral line width is 4 nm. Assume a maximum dispersion penalty of 1 dB.

Solution

i) *Determine ε.*
   From Table 12–11. For an MLM laser and 1 dB dispersion penalty, $\varepsilon = 0.115$.

ii) *Determine D_{max}.*
   From Equation (12–116), solve for D_{max}.

$$\varepsilon = 10^{-6}(f_b)(D_{max})(\delta\lambda)$$

or

$$D_{max} = \frac{\varepsilon}{1 \times 10^{-6}(f_b)(\delta\lambda)}$$

$$= \frac{0.115}{1 \times 10^{-6}(622)(4)}$$

$$= 46.2 \text{ ps/nm}$$

Therefore, $D_{max} = 46.2$ ps/nm.

iii) *Calculate the link distance.*
   The maximum link distance can be calculated by dividing the D_{max} by the optical fiber dispersion (3.0 ps/nm-km).

$$L_{max} = \frac{D_{max}}{3.0 \text{ ps/nm-km}} = \frac{46.2 \text{ ps/nm}}{3.0 \text{ ps/nm-km}} = 15.4 \text{ km}$$

Therefore, maximum link distance (L_{max}) $= 15.4$ km.

It is evident that a decrease of the dispersion penalty from 2 dB to 1 dB substantially reduces the maximum link distance from 23.3 km to 15.4 km.

Example 12–14

The following example is relevant to SONET/OC-3 applications that employ single-mode fibers. Given the following data, calculate maximum dispersion (D_{max}), mean link margin (μM), and sigma link margin (σM). Data:

- transmitter: 1237 J (laser)
- receiver: 1310 C (SONET)
- link length (L): 50 km
- system bit rate (f_b): 156 Mb/s
- optical fiber: 8.3/125 single-mode operating in the 1300 nm range
- optical connectors: ST-II lightguide (7)
- splices: rotary (4)
- spectral width (rms): 4 nm (max)
- outside plant cable
- clock recovery module
- dispersion penalty (μD): 2 dB

Solution

From Table 12–5

a) mean cable loss: (μc)=0.4 dB

b) sigma cable loss: (σc)=0.0 dB

c) dispersion (max): D_{max}=2.8 dB

From Table 12–6

a) mean transmitter power: (μt)=–4.9 dBm

b) sigma transmitter power: (σt)=0.5 dBm

c) mean receiver sensitivity: (μr)=–38.1 dBm

d) sigma receiver sensitivity: (σr)=0.48 dBm

From Tables 12–7 and 12–8

a) mean connector loss: (μco)=0.35 dB

b) sigma connector loss: (σco)=0.20 dB

From Tables 12–9 and 12–10

a) mean splice loss: (μs)=0.2 dB

b) sigma splice loss: (σs)=0.1 dB

Optical links that utilize single-mode fibers can be either attenuation limited or dispersion limited.

Case 1: Dispersion Limiting

From Equation (12–116), solve for $D_{max.}$

$$D_{max} = \frac{\varepsilon}{1 \times 10^{-6}\,(f_b)(\delta\lambda)}$$

where ε (percent of time slot width) = 0.182 for a dispersion penalty of 2 dB, f_b (system bit rate) = 156 Mb/s, and $\delta\lambda$ (spectral width)=4 nm.

$$D_{max} = \frac{\varepsilon}{1 \times 10^{-6}\,(f_b)(\delta\lambda)}$$

$$= \frac{0.182}{1 \times 10^{-6}\,(156)\,(4)}$$

$$= 292 \text{ ps/nm}$$

Therefore, D_{max} = 292 ps/nm.

The maximum dispersion for an SMF is 2.8 ps/nm-km (see Table 12–4). For a total fiber dispersion of 292 ps/nm, the maximum link distance (L_{max}) is calculated as follows:

$$L_{max} = \frac{D_{max}}{2.8 \text{ ps/nm-km}} = \frac{292 \text{ ps/nm}}{2.8 \text{ ps/nm-km}} = 104 \text{ km}$$

Therefore, maximum link distance (L_{max}) = 104 km.

Because the operating link distance is 50 km and the maximum link distance is 104 km (based on accumulated dispersion), the optical system is not dispersion limiting.

Case 2: Attenuation Limiting

From Equation (12–37), calculate the mean link margin (μM).

$$\mu M = \mu t - \mu r - (\mu cL + \mu coNco + \mu sNs$$
$$+ \mu D + \mu H + \mu CR)$$

$$= -4.9 - (-38.1) - [(0.4)\,50 + (0.35)7$$
$$+ (0.2)4 + 2 + 0 + 0.5]$$

$$= -4.9 + 38.1 - (20 + 2.45 + 0.8 + 2 + 0 + 0.5)$$

$$= -4.9 + 38.1 - 25.75$$

$$= 7.45 \text{ dB}$$

Therefore, μM = 7.45 dB.

From Equation (12–38), calculate the sigma link margin (σM).

$$\sigma M = [\sigma t^2 + \sigma r^2 + \sigma c^2\, LL_S + \sigma co^2\, Nco$$
$$+ \sigma s^2\, Ns + \sigma D^2 \sigma H^2 + \sigma CR^2\,]^{1/2}$$

$$= [(0.5)^2 + (0.48)^2 + (0)^2\,(50)\,(10)$$
$$+ (0.2)^2\,(7) + (0.1)^2\,(4) + (0.25)^2\,]^{1/2}$$

$$= 0.928 \text{ dB}$$

Therefore, σM = 0.928 dB.

The Effect of Attenuation on System Performance

The effect of attenuation on system performance is established at the 2% point of a normal distribution, expressed by Equation (12–118).

$$\mu M - 2\sigma M \qquad \text{(12–118)}$$

Substituting Equation (12–118) for values yields

$$\mu M - 2\sigma M = 7.45 - 2(0.928)$$

$$= 5.6 \text{ dB}$$

Therefore, $\mu M - 2\sigma M$ = 5.6 dB.

Allowing 1 dB of tracking error for the laser diode temperature variations, the power margin is reduced to 4.6 dB. In this design, the power margin is well above the 2 dB dispersion penalty. Therefore the link is not attenuation limited.

Example 12–15
Digital video link design (OC–12). Data:

- transmitter: 1227 AF
- receiver: 1320 C
- link length: $L = 25$ km
- system bit rate: 622 Mb/s
- optical fiber: 8.3/125 μm (operating window 1310 nm)
- connectors: SC (2)
- splices: CSL (4), $L_s = 6.25$ km
- spectral width: 4 nm (max)
- clock recovery included in the optical receiver
- dispersion penalty: $\mu D = 1$ dB

Solution
Information obtained from tables is as follows:

Tables 12–3 and 12–4

a) $\sigma D = 0$ dB

b) $\sigma H = 0$ dB

c) $\sigma CR = 0$ dB

d) $\mu H = 0$ dB

e) $\mu CR = 0$ dB

Table 12–4

a) $\sigma c = 0$ dB

b) $\mu c = 0.4$ dB

c) $D_{max} = 2.8$ ps/nm-km

Table 12–5

a) $\mu t = 0.1$ dBm

b) $\mu r = -31.5$ dBm

c) $\sigma t = 0.15$ dBm

d) $\sigma r = 0.3$ dBm

e) $\mu co = 0.12$ dB

f) $\mu s = 0.15$ dB

g) $\sigma co = 0.08$ dB

h) $\sigma s = 0.1$ dB

Case 1: Dispersion Limiting

For the optical link that incorporates a laser transmitter and a single-mode fiber, calculate the maximum allowable dispersion for a 1 dB dispersion penalty.

i) Compute maximum dispersion (D_{max}). From Equation (12–16), solve for D_{max}.

$$D_{max} = \frac{\varepsilon}{1 \times 10^{-6}\,(f_b)\,(\delta\lambda)}$$

where $\varepsilon = 0.115$ (Bellcore TR-NWT000253 for MLM lasers), $f_b = 622$ Mb/s, and $\delta\lambda = 4$ ps/nm. Substitute for values:

$$D_{max} = \frac{\varepsilon}{1 \times 10^{-6}\,(f_b)\,(\delta\lambda)}$$

$$= \frac{0.115}{1 \times 10^{-6}\,(622)\,(4)}$$

$$= 46.22 \text{ ps/nm}$$

Therefore, $D_{max} = 46.22$ ps/nm.

Because the fiber dispersion is 2.8 ps/nm-km and the system D_{max} is 46.22 ps/nm, the maximum link distance (L_{max}) is calculated as follows:

$$L_{max} = \frac{D_{max}}{\delta\lambda}$$

$$= \frac{46.22 \text{ ps/nm}}{4 \text{ ps/nm-km}}$$

$$= 16.5 \text{ km}$$

Therefore, maximum link distance (L_{max}) = 16.5 km.

Because the required link distance is 25 km and the maximum link distance based on 1 dB dispersion penalty is 16.5 km, the design is dispersion limited. To achieve the required distance of 25 km, the dispersion penalty must be increased to 2 dB, allowing for a maximum distance of 26 km ($\varepsilon = 0.182$).

ii) Calculate the 2% point of a nominal distribution ($\mu M - 2\sigma M$).

First, compute the mean link margin (μM).

$$\mu M = \mu t - \mu r - (\mu cL + \mu coNco$$
$$+ \mu sNs + \mu D + \mu H + \mu CR)$$
$$= 0.1 - (-31.5) - [(0.41)25 + (0.12)2$$
$$+ (0.15)\,4 + 1.0 + 0 + 0]$$
$$= 19.4 \text{ dB}$$

Therefore, $\mu M = 19.4$ dB.

Second, compute the sigma link margin (σM).

$$\sigma M = [\sigma t^2 + \sigma r^2 + \sigma c^2\,LL_s + \sigma co^2\,Nco$$
$$+ \sigma s^2\,Ns + \sigma D^2\sigma H^2 + \sigma CR^2]^{\frac{1}{2}}$$
$$= [(0.15)^2 + (0.3)^2 + (0.0)^2\,(60)(12)$$
$$+ (0.08)^2\,(2) + (0.1)^2\,(4)]^{1/2}$$
$$= 0.41 \text{ dB}$$

Therefore, $\sigma M = 0.41$ dB.

Substitute μM and σM into $\mu M - 2\sigma M$.

$$\mu M - 2\sigma M = 19.4 - 2(0.41)$$
$$= 18.58 \text{ dB}$$

Therefore, $\mu M - 2\sigma M = 18.58$ dB.

Subtracting a 1 dB tracking error for the laser diode, the system power margin is equal to 17.58 dB. Therefore, the optical system is not attenuation limiting.

Example 12–16

Design an optical link to be used as a high speed computer interface. The link operating parameters are as follows:

- transmitter: 1238 A
- receiver: 1318 A
- link distance: 5 km
- system bit rate: 1.062 Gb/s
- fiber type: 8.3/125 μm (single-mode, 1310 wavelength window)
- connectors: SC (8)
- splices: $4(L_s = 1 \text{ km})$
- spectral width: 6 nm (max)
- inside building cable
- clock recovery module: yes
- dispersion penalty: 1 dB

Solution

From tables, the following values are obtained:

Tables 12–3 and 12–4

a) $\sigma D = 0.0$ dB

b) $\sigma H = 0.0$ dB

c) $\sigma CR = 0.25$ dB

d) $\mu H = 0.0$ dB

e) $\mu CR = 1.0$ dB

Table 12–4

a) $\sigma c = 0.0$ dB

b) $\mu c = 0.4$ dB

c) $D_{max} = 2.8$ ps/nm-km

Table 12–5

a) $\mu t = -8$ dBm

b) $\mu r = -28.7$ dBm

c) $\sigma t = 0.6$ dBm

d) $\sigma r = 0.74$ dBm

e) $\mu co = 0.12$ dB

f) $\mu s = 0.2$ dB

g) $\sigma co = 0.08$ dB

h) $\sigma s = 0.1$ dB

i) Calculate maximum dispersion (D_{max}).

From Equation (12–116), solve for D_{max}.

$$D_{max} = \frac{\varepsilon}{1 \times 10^{-6}(f_b)(\delta\lambda)}$$
$$= \frac{0.115}{1 \times 10^{-6}(1062)(6)}$$
$$= 18 \text{ ps/nm}$$

Therefore, $D_{max} = 18$ ps/nm.

ii) Calculate maximum distance (L_{max}).

$$L_{max} = \frac{D_{max}}{\delta\lambda}$$
$$= \frac{18 \text{ ps/nm}}{2.8 \text{ ps/nm-km}}$$
$$= 6.42 \text{ km}$$

Therefore, $L_{max} = 6.42$ km.

Because the link distance is 5 km and D_{max} is 6.42 km, the link is not dispersion limited.

iii) Calculate μM.

$$\mu M = \mu t - \mu r - (\mu c L + \mu co Nco + \mu s Ns + \mu D + \mu H + \mu CR)$$
$$= -8 + 28.7 - [(0.4)5 + (0.12)8 + (0.2)4 + 1 + 0 + 1]$$
$$= 14.9 \text{ dB}$$

Therefore, $\mu M = 14.9$ dB.

iv) Calculate (σM).

$$\sigma M = [\sigma t^2 + \sigma r^2 + \sigma c^2 LL_s + \sigma co^2 Nco + \sigma s^2 Ns + \sigma D^2 \sigma H^2 + \sigma CR^2]^{\frac{1}{2}}$$
$$= [(0.6)^2 + (0.75)^2 + (0.0)^2(5)(1) + (0.08)^2 + (0.1)^2(4) + (0.25)^2]^{1/2}$$
$$= 1.03 \text{ dB}$$

Therefore, $\sigma M = 1.03$ dB.

v) Calculate the 2% point of nominal distribution.

Substitute μM and σM into $\sigma M - 2\sigma M$:

$$\mu M - 2\sigma M = 14.9 - 2(1.03)$$
$$= 12.84 \text{ dB}$$

Therefore, $\mu M - 2\sigma M = 12.84$ dB.

Allowing 1 dB for tracking error, the 2% point of normal distribution is 11.84 dB. Therefore, the optical link is not attenuation limited.

12.15 ANALOG OPTICAL SYSTEMS

Video Transmission Over Optical Fibers

Optical fiber technology has had a profound effect on multichannel video transmission. Advancements in fiber optics enabled cable providers to substantially increase the number of video channels and to enhance the quality of transmission. Furthermore, it was instrumental in the development of digital video transmission. In the late 1970s, attempts were made to transmit video channels over optical fiber links with limited success. The difficulties encountered at that time were the laser diode's inability to provide the optical power required by the system, and the high RIN. These important laser diode deficiencies were all but eliminated with the development of DFB lasers. Semiconductor lasers exhibit unacceptable levels of nonlinearities. However, the elimination of nonlinearities was achieved in the mid-1980s, resulting in a proliferation of optical systems in video multichannel transmission. Initially, analog optical links operated at the 1310 nm wavelength window and employed direct modulation schemes. Such systems were capable of transmitting up to 100 video channels at a maximum distance of 50 km. With the advent of LiNbO$_3$ modulators and EDFAs, the operating wavelength window was shifted to 1550 nm, resulting in highly sophisticated and extremely efficient optical systems.

The primary function of an analog optical link is to accept a number of FDM video channels (composite electrical signals) at the input, to directly or indirectly modulate a laser diode, and to convey the modulated optical signal to a desirable distance through an optical fiber. At the final destination, the optical signal is detected by the receiver module and converted back to an electrical signal that is ready for final distribution. Modern links are designed to carry several video channels (analog or digital), as well as voice and data services. A block diagram of an analog optical fiber link is illustrated in Figure 12–59.

The intensity of the RF signal, composed of a number of FDM analog channels, directly or indirectly modulates a laser diode. The intensity modulated optical signal is coupled into the fiber and amplified at specific fiber lengths by a number of EDFAs. At the receiver end, the optical power conveyed by the fiber is detected and the composite RF signal is generated to its original form. The main components of an analog multichannel optical system are:

a) transmitter

b) optical fiber

c) optical amplifier(s)

d) receiver

In **amplitude modulated vestigial sideband (AM-VSB)** signals that incorporate a number of video channels, the typical performance parameters are as follows.

a) minimum CNR: CNR $\geq$ 52 dB

b) composite second order distortion: CSO $\leq$ –60 dB

c) composite triple beat: CTB $\leq$ –65 dB

d) optical modulation depth reflecting the above values: OMD = 3.5% (per video channel)

In an 80-channel AM-VSB optical system, the required values at the receiver end set forth by the **National Association of Broadcasters** (NAB) are CNR: 46 dB, composite

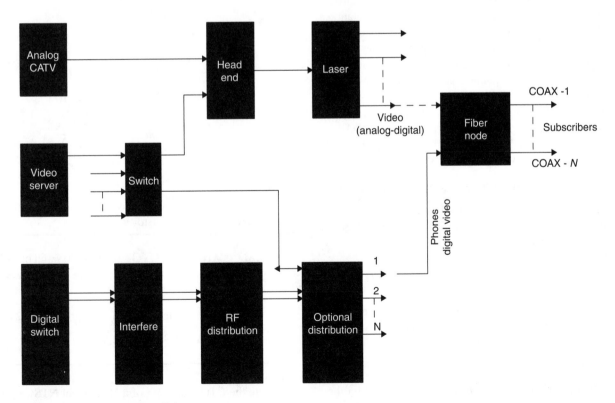

Figure 12–59. Analog optical fiber link.

second order distortion (CSO) and composite third order (CTO) distortion: –53 dB.

AM-VSB Signals Over Passive Optical Networks

Analog multichannel signals can be transmitted simultaneously with digital voice and data signals through **passive optical networks (PON)**. In such systems, the CNR at the optical receiver output must be maintained at a 46 dB minimum, while both **composite second order (CSO)** and **composite third order (CTO)** distortions should maintain a maximum level of –56 dB per channel. These performance characteristics can be achieved with an **optical amplitude modulation (OAM)** depth of 5% per channel. As mentioned earlier, the main components that constitute a multichannel analog optical system are the transmitter, the fiber, and the receiver modules. Following is a brief description of the module performance characteristics, which meet system requirements.

The Transmitter of an Analog Multichannel Optical System

Transmitters that operate in analog optical systems must exhibit performance characteristics such as high optical power output, extreme linearity, and very low noise. Furthermore, in relation to the modulation scheme, they can be classified as direct modulation or indirect (external) modulation transmitters. Both transmitter types can operate at either the 1310 nm or 1550 nm wavelength windows, with distinct operating characteristics relevant to the specific operating wavelength window. For example, direct modulation transmitters are simpler in design, fully exploit laser diode linear current intensity transfer characteristics, and are less expensive in their implementation. On the other hand, indirect or external modulation systems operate with higher optical powers and fully exploit the optical amplification characteristics of EDFAs. The most critical element in the design of optical transmitters employed in analog multichannel systems is the high level of electrical-to-optical linearity required by the AM-VSB signals. A high degree of linearity is essential in order to maintain the CSO and CTO distortion products within the required limits. Equation (12–119) expresses the optical power in the form of a Taylor series.

$$P \propto K\left(1 + x + ax^2 + bx^3 + \ldots + nx^{n-1}\right) \quad \text{(12–119)}$$

where P is the optical power, K is the bias current or voltage, and x is the modulation signal. Equation (12–120) expresses the modulation signal $x(t)$.

$$x(t) = \sum_{i}^{N} m_i(t)\cos\left[2\pi f_i t + \phi_i\right] \quad \text{(12–120)}$$

where $x(t)$ is the modulation signal, $m_i(t)$ is the normalized modulation signal for i-channel, N is the number of channels, f_i is the subcarrier frequency, and ϕ_i is the phase of the i-channel.

CALCULATE THE CSO DISTORTION. Based on the assumption that the optical power is expressed as the Taylor series expansion of the electrical signal, the CSO distortion is expressed by Equation (12–121).

$$CSO_{dB} = 10 \log\lfloor N_{CSO}(am)^2\rfloor \quad \text{(12–121)}$$

where N_{CSO} = second order product, a = coefficient related to NTSC scheme, and m = modulation depth.

Example 12–17

Calculate the CSO distortion for a 50-channel optical system that has a modulation depth of 5% and a coefficient (a) of 3.6×10^{-3} (for a 50-channel NTSC format).

Solution

$$
\begin{aligned}
CSO_{dB} &= 10 \log\lfloor N_{CSO}(am)^2\rfloor \\
&= 10 \log\lfloor 50 \times (3.6 \times 10^{-3} \times 0.05)^2\rfloor \\
&= 10 \log(1.62 \times 10^{-6}) \\
&= -57.9 \text{ dB}
\end{aligned}
$$

Therefore, CSO $= -57.9$ dB.

Example 12–18

Calculate the required AM modulation depth (m) required to maintain a CSO distortion of –59.8 dB. Assume an NTSC 50-channel video signal.

Solution
From Equation (12–121) solve for m.

$$
\begin{aligned}
-59.8 &= 10 \log(y) \\
y &= Inv \log(-5.98) \\
y &= 1.047 \times 10^{-6}
\end{aligned}
$$

Where, $y = \lfloor N_{CSO}(am)^2\rfloor$

$$50(3.6 \times 10^{-3})\, m^2 = 1.047 \times 10^{-6}$$

$$
\begin{aligned}
m^2 &= \frac{1.047 \times 10^{-6}}{6.48 \times 10^{-4}} \\
&= 1.6157 \times 10^{-3}
\end{aligned}
$$

Therefore, $m = 4\%$.

It is evident that a decrease of the AM modulation depth decreases the CSO distortion.

CALCULATE THE CTO DISTORTION. The CTO distortion component of an NTSC scheme is expressed by Equation (12–122).

$$CTO_{dB} = 10 \log\lfloor N_{CTO}(1.5\, bm)^2\rfloor \quad \text{(12–122)}$$

where $b = 1.07 \times 10^{-2}$ (for a 50-channel NTSC).

Example 12–19

Compute the CTO distortion of a 50-channel NTSC system with 5% modulation depth and b coefficient of 1.07×10^{-2}.

Solution

$$CTO_{dB} = 10 \log \lfloor N_{CTO} \, (1.5 \, bm)^2 \rfloor$$
$$= 10 \log \lfloor 50 \, (1.5 \times 1.07 \times 10^{-2} \times 0.05)^2 \rfloor$$
$$= 10 \log (6.44 \times 10^{-7})$$
$$= -61.9 \text{ dB}$$

Therefore, CTO = −61.9 dB.

Example 12–20

Calculate the CSO and CTO distortions for an 80-channel NTSC format that operates with 5% AM modulation depth and 3% modulation depth. Compare the two sets of distortions in relation to modulation depths. Assume the following coefficients:

- $a = 2.43 \times 10^{-3}$
- $b = 4.65 \times 10^{-3}$

Solution

i) Compute the CSO for $m = 5\%$.

$$CSO_{dB} = 10 \log \lfloor N_{CSO} \, (am)^2 \rfloor$$
$$= 10 \log \lfloor 80 \, (2.4 \times 10^{-3} \times 0.05)^2 \rfloor$$
$$= 10 \log (1.152 \times 10^{-6})$$
$$= -59.3 \text{ dB}$$

Therefore, CSO = −59.3 dB.

ii) Compute the CTO for $m = 5\%$.

$$CTO_{dB} = 10 \log \lfloor N_{CTO} \, (1.5 \, bm)^2 \rfloor$$
$$= 10 \log \lfloor 80 \, (1.5 \times 4.65 \times 10^{-3} \times 0.05)^2 \rfloor$$
$$= 10 \log \lfloor 9.73 \times 10^{-6} \rfloor$$
$$= -50.1 \text{ dB}$$

Therefore, CTO = −50.1 dB.

iii) Compute the CSO for $m = 53\%$.

$$CSO_{dB} = 10 \log \lfloor N_{CSO} \, (am)^2 \rfloor$$
$$= 10 \log \lfloor 80 \, (2.4 \times 10^{-3} \times 0.03)^2 \rfloor$$
$$= 10 \log (4.147 \times 10^{-7})$$
$$= -63.82$$

Therefore, CSO = −63.82 dB.

iv) Compute the CTO for $m = 3\%$.

$$CTO_{dB} = 10 \log \lfloor N_{CTO} \, (1.5 bm)^2 \rfloor$$
$$= 10 \log \lfloor 80 \, (1.5 \times 4.65 \times 10^{-3} \times 0.03)^2 \rfloor$$
$$= 10 \log \lfloor 3.5 \times 10^{-6} \rfloor$$
$$= -54.5$$

Therefore, CTO = −54.5 dB.

From Example 12–20, it is evident that a decrease in the AM modulation depth also decreases both CSO and CTO distor-

tions. It is, therefore, up to the system designer to establish the AM modulation depth reflecting acceptable levels of CSO and CTO distortions. However, an increase of modulation depth also increases the channel CNR. Assuming an increase of the modulation depth from 5% to 6%, the corresponding increase of the CNR is as follows:

$$CNR = 20 \log \left(\frac{6}{5} \right)$$
$$= 20 \log (1.2)$$
$$= 1.58 \text{ dB}$$

Therefore, CNR = 1.58 dB (improvement).

The implementation of an external modulator requires very stringent linear characteristics between the input electrical signal and the output optical signal. The utilization of an external LiNbO$_3$ MZ modulator imposes a problem, because it exhibits sinusoidal transfer characteristics. However, the CSO distortion in such an application can be eliminated through proper biasing of the modulator, while the CTO distortion can also be eliminated through the application of pre-distortion linearization techniques. Another concern, related to the indirect or external modulation scheme, is the baseband composite signal distortion. Baseband distortion (chirping) occurs when the laser diode is driven below the threshold level by the modulating electrical signal. In such an occurrence, Equation (12–123) expresses the distortion (chirping)

$$D_{(chirp)} = m \left(\frac{N_{ch}}{2} \right)^{1/2} \qquad \textbf{(12–123)}$$

where D_{chirp} is the distortion (chirp), m is the AM modulation depth, and N_{ch} is the number of channels. A graphical representation of distortion by chirping and root-mean-square modulation depth (m_{rms}) is illustrated in Figure 12–60.

Figure 12–60 shows that a rapid increase of the CSO and CTD distortion occurs between 0.2 and 0.4 rms modulation depths. Therefore, when an optical transmitter that is employed in an analog multichannel CATV system and the CSO and CTD distortions are above a set maximum, a corresponding reduction of the rms modulation depth will be required. The second, but equally important, component in the design of optical receivers incorporated in analog multichannel video transmission is noise. The noise induced into the system by the optical transmitter is classified as either RIN or **phase noise.** Noise intensity is a fundamental component that influences the system CNR. Equation (12–124) expresses the relationship between CNR and RIN.

$$CNR_{(RIN)} = \frac{m^2}{2(RIN) \, B_e} \qquad \textbf{(12–124)}$$

where m is the modulation depth, RIN is the relative intensity noise, and B_e is the channel bandwidth.

Example 12–21

Calculate the CNR for an AM modulation depth of 4% and RIN of −150 dB/Hz.

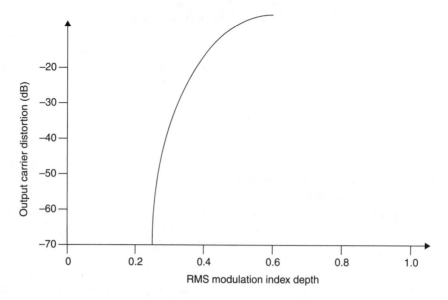

Figure 12–60. Chirp based distortion *v*. RMS modulation depth (m_{rms}).

Solution

$$\text{CNR}_{\text{RIN}} = 10 \log\left[\frac{m^2}{2(B_{ch})(\text{RIN})}\right]$$

$$= 10 \log\left[\frac{0.04^2}{2(4 \times 10^6)(1 \times 10^{-15})}\right]$$

$$= 10 \log(2 \times 10^5)$$

$$= 53 \text{ dB}$$

Therefore, CNR = 53 dB.

Example 12–22
Calculate the RIN for an AM modulation depth of 3% and CNR of 50 dB.

Solution
From Equation (12–124), solve for RIN.

$$\text{CNR}_{\text{RIN}_{dB}} = 10 \log\left[\frac{m^2}{2(B_{ch})(\text{RIN})}\right]$$

or

$$5.0 = Invlog\left[\frac{m^2}{2(B_{ch})(\text{RIN})}\right]$$

$$\frac{m^2}{2(B_{ch})(\text{RIN})} = 1 \times 10^5$$

$$\text{RIN} = \frac{0.03^2}{2(4 \times 10^6)(1 \times 10^6)}$$

$$= 1.125 \times 10^{-16}$$

Therefore, RIN = −159.5 dB/Hz.

Another important element, which must be taken into consideration when designing optical transmitters employed in multichannel analog systems, is transmitter non-linearity related to spectral distortion. The spectrum of the optical signal at the output of the transmitter is influenced by three factors:

1) laser efficiency

2) alpha parameter (α)—line width enhancement factor

3) phase noise

The α-factor relates the laser and optical modulator parameters within the transmitter module. For a direct modulation transmission scheme, the spectrum of the output signal is chirped by the laser and modulator non-linear characteristics, while for external modulation transmission schemes, the spectrum of the output signal is very narrow. Equation (12–125) expresses the relationship that incorporates various spectral components of the optical signal.

$$E(t) = A(1 + x(t))^{1/2}\exp\left[\, if_o(t) \right.$$
$$\left. + i\phi(t) + i(2\pi)\beta i_b \int\limits^{\infty} x(\tau) + i\frac{\alpha}{2}x(t)\right] \text{ (12–125)}$$

where f_o is the carrier frequency (optical), ϕ_n is the phase noise, β is the transmitter efficiency (MHz/mA), i_b is the bias current level above threshold, α is the line width enhancement factor, and i is the bias current at threshold.

Direct Modulation

The spectral time variations at the output of the transmitter of a multichannel analog optical system are Gaussian. Equation (12–126) expresses the standard deviation (σ).

$$\sigma = m\beta i_b \left(\frac{N_{ch}}{2}\right)^{1/2} \qquad \textbf{(12–126)}$$

where σ is the standard deviation, m is the modulation depth, β is the transmitter efficiency, and N_{ch} is the channels. The relationship between standard deviation (σ) and **half-wave at half-maximum (HWHM)** is expressed by Equation (12–127).

$$\text{HWHM} = \sigma(2\ln 2)^{1/2} \qquad \textbf{(12–127)}$$

or

$$\text{HWHM} = 1.1774(\sigma)$$

Laser diodes utilized in the design of optical transmitters are either DFB lasers or QW lasers. The spectral line width of each type of laser can be calculated as follows:

FOR A DFB LASER

$$\beta = 200\text{ MHz} - 600\text{ MHz}$$
$$m = 5\%$$
$$i_b = 20\text{ mA}$$

Calculate $\sigma_{(200\text{ MHz})}$.

$$\sigma = m\beta i_b \left(\frac{N_{ch}}{2}\right)^{1/2}$$
$$= (0.05)(200 \times 10^6)(20)\left(\frac{50}{2}\right)^{1/2}$$
$$= 1 \times 10^9$$

Therefore, $\sigma = 1$ GHz.
Calculate $\text{HWHM}_{(200\text{ MHz})}$.

$$\text{HWHM} = 1.1774\,(\sigma)$$
$$= 1.1774\,(1 \times 10^9)$$
$$= 1.1774\text{ GHz}$$

Therefore, $\text{HWHM}_{(200\text{ MHz})} = 1.1774$ GHz.
Calculate $\sigma_{(600\text{ MHz})}$.

$$\sigma = m\beta i_b \left(\frac{N_{ch}}{2}\right)^{1/2}$$
$$= (0.05)(600 \times 10^6)(20)\left(\frac{50}{2}\right)^{1/2}$$
$$= 3 \times 10^9$$

Therefore, $\sigma_{(600\text{ MHz})} = 3$ GHz.
Calculate $\text{HWHM}_{(600\text{ MHz})}$.

$$\text{HWHM}_{(600\text{ MHz})} = 1.1774\,(\sigma_{(600\text{ MHz})})$$
$$= 1.1774(3\text{ GHz})$$
$$= 3.53\text{ GHz}$$

or

$$\text{HWHM}_{(600\text{ MHz})} = 3.53\text{ GHz}$$

$$\text{HWHM} = \text{HWHM}_{(200\text{ MHz})} \text{ to } \text{HWHM}_{(600\text{ MHz})}$$

Therefore, HWHM = –1.1774 GHz to 3.53 GHz.

FOR A QW LASER

$$\beta = (5\text{ MHz} - 200\text{ MHz}).$$

Calculate $\sigma_{(5\text{ MHz})}$.

$$\sigma = m\beta i_b \left(\frac{N_{ch}}{2}\right)^{1/2}$$
$$= (0.05)(5 \times 10^6)(20)\left(\frac{50}{2}\right)^{1/2}$$
$$= 25 \times 10^6$$

Therefore, $\sigma = 25$ MHz.
Calculate $\text{HWHM}_{(5\text{ MHz})}$.

$$\text{HWHM} = 1.1774\,(\sigma)$$
$$= 1.1774(25 \times 10^6)$$
$$= 29.4\text{ MHz}$$

Therefore, $\text{HWHM}_{(5\text{ MHz})} = 29.4$ MHz.
Calculate $\sigma_{(200\text{ MHz})}$.

$$\sigma = m\beta i_b \left(\frac{N_{ch}}{2}\right)^{1/2}$$
$$= (0.05)(200 \times 10^6)(20)\left(\frac{50}{2}\right)^{1/2}$$
$$= 1 \times 10^9$$

Therefore, $\sigma = 1$ GHz.
Calculate $\text{HWHM}_{(200\text{ MHz})}$.

$$\text{HWHM} = 1.1774\,(\sigma)$$
$$= 1.1774(1 \times 10^9)$$
$$= 1.1774\text{ GHz}$$

Therefore, $\text{HWHM}_{(200\text{ MHz})} = 1.1774$ GHz.

$$\text{HWHM} = \text{HWHM}_{(5\text{ MHz})} \text{ to } \text{HWHM}_{(200\text{ MHz})}$$

Therefore, HWHM = 29.4 MHz to 1.1774 GHz.

From the proceding, it is evident that the spectral line width of a QW laser is much narrower than that of a DFB laser.

Indirect (External) Modulation

The spectral line width of externally modulated optical transmitters is a function of the laser line width, chirp, and modulation intensity (Equation (12–128)).

$$\sigma = \sqrt{\sum_i^{N_{ch}} \frac{1}{2}\left(\frac{m\alpha f_i}{2}\right)^2} \qquad \textbf{(12–128)}$$

where σ = standard deviation, α = line width enhancement factor, and f_i = sideband frequency.

THE FIBER IN ANALOG MULTICHANNEL OPTICAL SYSTEMS. Fiber induced linear and nonlinear effects also distort analog multichannel AM-VSB video signals that are transmitted over optical links. Analog multichannel optical systems usually operate with the following parameters:

a) fiber effective area: 8 µm2

b) attenuation: 0.35 dB/km (for λ_o = 1310 nm)

c) attenuation: 0.22 dB/km (for λ_o = 1550 nm)

d) dispersion: 0 ps/nm-km (at λ_o = 1310 nm)

e) dispersion: 17 ps/nm-km (at λ_o = 1550 nm)

As mentioned previously, video signals traveling through an optical fiber are subject to linear effects such as dispersion and multipath interference, while nonlinear effects such as SPM and SBS significantly affect the analog multichannel AM-VSB signals. SRS has no effect on such systems.

Dispersion

In several places throughout this text, the nature and effect of fiber dispersion in optical systems has been examined. Here, the dispersion of video signals that travel through a 20 km (a typical value for such systems) optical fiber will be discussed in some detail. Dispersion is classified as either chromatic or modal, and is more evident in optical systems where chirping in the transmitter module is excessive. Fiber dispersion coupled with excessive transmitter chirping leads to system performance degradation through an increase of ISI. The effect of the ratio between chromatic dispersion and intermodulation distortion is expressed by CSO distortion. Equation (12–129) expresses that relationship.

$$
\begin{aligned}
\mathrm{CSO_{dB}} = 10 \log \Bigg\{ & \left[m \left(-\frac{\lambda^2 D}{2\pi c} \right) z\Omega_d \right]^2 \\
& \left[\gamma^2 N_{\mathrm{CSO}} + \left(\frac{-\lambda^2 Dz\Omega}{8} \right)^2 \sum_i^{N_{ch}} (\Omega_i \Omega_j)^2 \right] \Bigg\}
\end{aligned}
$$ (12–129)

where z is the length of the fiber, γ is the $2\pi\beta_{FM} (I_b - I_{th})$, β_{FM} is the FM efficiency (MHz/mA), D is the dispersion coefficient (ps/nm-km), Ω_d is the intermodulation distortion frequency (rad/s), I_{th} is the threshold current, and I_b is the above the threshold current.

Because the CSO distortion product is dependent on the intermodulation distortion frequency, the highest CSO is encountered at the higher channels. Equation (12–130) expresses the composite second order distortion at the higher video channel.

$$
\begin{aligned}
\mathrm{CSO_{dB}} = 10 \log \Bigg\{ & N_{\mathrm{CSO}} \left[m \left(\frac{-\lambda^2 D}{2\pi c} \right) z\Omega_d \right]^2 \\
& \left[\gamma^2 + \left(\frac{-\lambda^2 Dz\Omega_d}{64\pi c} \right)^2 \right] \Bigg\}
\end{aligned}
$$ (12–130)

For a 50-channel AM-VSB analog system that requires a CSO $\leq$ –59 dB$_{\mathrm{ch}}$, the fiber dispersion-length product must

be equal to or less than 120 ps/nm. If the system is designed to operate at the 1550 nm wavelength window, at which a non-dispersion shifted single-mode fiber (NDS-SMF) exhibits a dispersion of 17 ps/nm-km, the maximum achievable length is seven kilometers (120 ps/nm-km ÷ 17 ps/nm), operating with an MQW laser diode. If a DFB laser is employed, dispersion compensation techniques are required, through electronic means or through the utilization of DCFs. The CTO distortion is also defined by the highest channel. Equation (12–131) expresses the CTO distortion.

$$
\begin{aligned}
\mathrm{CTO_{db}} = 20 \log \Bigg[& m \left(\frac{-\lambda^2 D}{2\pi c} z\Omega_d \right)^2 \\
& (N_{\mathrm{CTO}})^{1/2} \left(0.75\gamma^2 + \frac{\Omega_d^2}{48} \right) \Bigg]
\end{aligned}
$$ (12–131)

Comparing both CSO and CTO, in terms of their corresponding maximum values of –59 dB$_{\mathrm{ch}}$ and –1117 dB$_{\mathrm{ch}}$, it is evident that the CSO distortion is the dominant component in analog multichannel AM-VSB systems. In such applications, fiber PMD must also be kept to a minimum, or an increase noise and signal distortion may be observed. The CSO distortion that results from fiber PMD is expressed by Equation (12–132).

$$
\begin{aligned}
\mathrm{CSO_{dB}} = 10 \log \Bigg[& N_{\mathrm{CSO}} (\gamma m)^2 \left(\frac{\pi^2 \Omega_d^2 \langle \Delta_\tau \rangle^4}{2^8} \right. \\
& \left. + \frac{\pi \Delta T^2 \langle \Delta_\tau \rangle^2}{48} \right) \Bigg]
\end{aligned}
$$ (12–132)

where $\Delta\tau$ = PMD, ΔT = polarization dispersion loss, γ = chirp, m = modulation depth, N_{CSO} = channels, and Ω_d = frequency (rad/sec) intermodulation distortion.

Nonlinear effects are optical power dependent. For low optical power levels, nonlinearities are practically negligible. However, for relatively high optical power levels such as those required by the multichannel AM-VSB video signals, distortion based on nonlinearities is substantial and must be taken into account. The system parameter most affected by nonlinearities such as SPM and SBS is the power-length product.

The Effect of SPM in Analog Optical Fiber Systems

SPM is the result of the refractive index nonlinearities of the optical fiber. An intensity modulated signal traveling through a fiber with a nonlinear index of refraction will induce undesirable phase modulation. If a phase modulated signal travels through a dispersive fiber, it will be subject to an additional level of intensity modulation, ultimately translated into output signal distortion. This kind of distortion is expressed as CSO distortion. Equation (12–133) expresses the CSO distortion as a function of SPM.

$$\text{CSO}_{\text{dB}} = 20 \log \left\{ (N_{\text{CSO}})^{1/2} \, m \left(\frac{-\lambda^2 D}{2\pi c} \Omega_d^2 \right) \right.$$

$$\left. \left(\frac{\pi n P}{\lambda A_{\text{eff}}} \right) \left(\frac{1}{\alpha} \right) \left[L - \frac{1}{\alpha}(1 - e^{-\alpha L}) \right] \right\} \quad \textbf{(12–133)}$$

where P is the optical power launched into the fiber, A_{eff} is the mode effective area, α is the loss coefficient, and n is the refractive index.

The CTO distortion that results from SPM has two limits: For $L > \dfrac{1}{\alpha}$ (fiber length greater than effective length), Equation (12–134) expresses the CTO.

$$\text{CTO}_{\text{dB}} \leq \left[(N_{\text{CTO}})^{1/2} (m\Omega_d)^2 \left(\frac{-\lambda^2 D}{2\pi c} \right)^2 \right.$$

$$\left. \left(\frac{\pi n P}{6\lambda A_{\text{eff}}} L \right)^2 (\alpha^{-2}) \right] \quad \textbf{(12–134)}$$

For $\alpha = 0$ (fiber length equal to effective length), Equation (12–135) expresses the CTO.

$$\text{CTO}_{\text{dB}} \leq \left[(N_{\text{CTO}})^{1/2} (m\Omega_d)^2 \left(\frac{-\lambda^2 D}{2\pi c} \right)^2 \right.$$

$$\left. \left(\frac{\pi n P}{6\lambda A_{\text{eff}}} L \right)^2 \left(\frac{L^2}{3} \right) \right] \quad \textbf{(12–135)}$$

SBS in Analog Optical Systems

Perhaps the most important nonlinear effect of an analog optical system is SBS. For a more detailed examination of SBS, see Chapter 10. Here, the effect of SBS on analog optical systems will be briefly examined. SBS causes optical power to be reflected back into the fiber. This results in an increase in the noise intensity at the forward direction of the optical wave. This increase of the noise intensity at the forward direction occurs at a power level referred to as the Brillouin threshold. Therefore, in order to eliminate the effect of SBS, the power level launched into the fiber must be kept below the Brillouin threshold. Of course, this leads to a decrease of the system CNR. Equation (12–136) expresses the critical Brillouin threshold optical power level.

$$P_{\text{SBS}} = 21 \left[\frac{A_{\text{eff}} \alpha}{g_B \, p(1 - e^{-\alpha L})} \right] \quad \textbf{(12–136)}$$

where $P_{\text{SBS}} = $ SBS threshold optical power, $A_{\text{eff}} = $ fiber mode effective area, $\alpha = $ loss coefficient, $g_B = $ Brillouin effective gain coefficient (5×10^{-11} m/W), and $L = $ fiber length.

The Brillouin effective coefficient (5×10^{-11} m/W) holds true only for a narrow signal spectrum (narrow line width laser source). For signal spectra that are wider than the Brillouin bandwidth, applicable to direct modulation systems, SBS has no effect on the optical system.

THE RECEIVER IN ANALOG MULTICHANNEL OPTICAL SYSTEMS. For a detailed examination of optical receivers, consult Chapter 8. Here, the function and operating characteristics of optical receivers employed in multichannel AM-VSB analog systems will be examined. The function of an optical receiver is to convert optical power to an electrical signal. Perhaps the most important device employed in the design of an optical receiver is the photodetector diode. The photodetector current generated by the PIN diode is coupled into the preamplifier through a matching network. Equation (12–137) expresses the photocurrent generated by the PIN diode in relation to received optical power.

$$I_{pd} = \frac{P_{\text{in}} \eta e}{hv} \quad \textbf{(12–137)}$$

where η is the quantum efficiency (%), hv is the photon energy, P_{in} is the input optical power (W), $e = 1.6 \times 10^{-19}$ C, and I_{pd} is the PIN photocurrent.

Example 12–23

Compute the required optical power at the input of an optical receiver given the following operating parameters:

- $I_{pd} = 0.15$ mA at the 1310 nm operating wavelength
- $\eta = 75\%$

Solution

$$P_{\text{in}} = \frac{hv(I_{pd})}{\eta e}$$

$$= \frac{(1.55 \times 10^{-19})(0.15 \times 10^{-3})}{0.75 \times 1.6 \times 10^{-19}}$$

$$= 0.137591 \text{ mW}$$

Therefore, $P_{\text{in}} \cong 0.137591$ mW.

Figure 12–61 is a block diagram of such a receiver. The main concern in designing optical receivers is the proper management of the RIN generated by the preamplifier stage and the shot noise generated by the photodetector diode. Both PIN shot noise and RIN contribute directly to the receiver performance degradation through a decrease of the system CNR. Some of the important receiver parameters are as follows:

This particular optical receiver is composed of a PIN photodetector diode, a matching network, and a push pull amplifier. The PIN photodetector diode converts optical power into electric current while the matching network matches the impedance of photodetector diode to the push–pull preamplifier. The average gain of the preamplifier required in such applications is approximately 21.5 dB.

RESPONSIVITY. **Responsivity** is defined as the ratio of the output voltage (V) to the modulated input optical power (W). Equation (12–138) expresses this ratio.

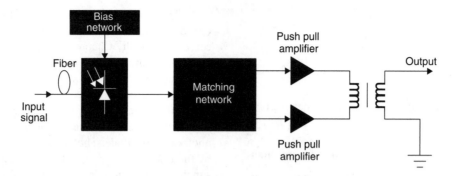

Figure 12–61. A block diagram of an optical receiver employed in CATV systems.

$$\text{responsivity} = \frac{\text{output voltage (V)}}{\text{modulated input optical power (W)}}$$
$$(12–138)$$

Optical receiver responsivity can also be expressed in *A/W* (Equation (12–139)) or *W/W* (Equation (12–140)).

$$\text{responsivity (A/W)} = \frac{\text{responsivity (V/W)}}{\text{load impedance } (\Omega)} \quad (12–139)$$

$$\text{responsivity (W/W)} = \frac{\text{responsivity (V/W)}^2}{\text{load impedance } (\Omega)} \quad (12–140)$$

The load impedance of such systems is 75 Ω.

FREQUENCY RESPONSE FLATNESS. The frequency response flatness of an optical receiver is defined by the maximum deviation from an ideal flat frequency response over a defined spectrum. Flatness can be calculated by using either the cable curve method or the straight line method.

As mentioned, frequency response flatness is determined in relation to an ideal response curve. Equation (12–141) expresses that ideal responsivity, using the cable curve method.

$$R_{\text{ideal}} = K + C \left(\frac{f_x}{f_1}\right)^{1/2} \quad (12–141)$$

where R_{ideal} is the ideal responsivity, K is the constant, C is the cable constant, f_x is the desired frequency, and f_1 is the start frequency. Equation (12–142) expresses the cable constant.

$$C = \frac{R_n - R_1}{\left(\frac{f_n}{f_1} - 1\right)^{1/2}} \quad (12–142)$$

where C is the cable constant (start), R_1 is the measured response at start frequency, R_n is the measured response at stop frequency, f_1 is the start frequency, and f_n is the stop frequency.

The value of constant (K) is selected to reflect an equal deviation ($\pm$) from the ideal response curve. The cable constant (C) is also varied by $\pm$ 0.001 dB in order

for the measured curve to be compatible with the ideal curve.

The straight line method is the most commonly used method of measuring frequency response flatness. Through this method, the optical receiver responsivity is measured at a frequency range of between 40 MHz and 870 MHz. Any deviation from the interpolation line is interpreted as flatness, measured among peaks and valleys within the defined bandwidth. For example, a peak of 0.5 dB and a valley of −0.1 dB correspond to a frequency response flatness of 0.6 dB.

THE FREQUENCY RESPONSE SLOPE. The frequency response slope is defined as the difference between the ideal gain at the start frequency f_1 and the final gain at frequency f_n. Again, the frequency response slope can be calculated by using either the cable method or the straight line method.

INPUT RETURN LOSS. The input return loss of an optical receiver is determined by the photodiode optical back reflection measured at the output of the fiber. Equation (12–143) expresses the optical back reflection.

$$\text{optical back reflection} = 10 \log\left(\frac{P_{\text{refl}}}{P_{\text{in}}}\right) \quad (12–143)$$

where P_{refl} is the power reflected back to the fiber optical power and P_{in} is the input optical power.

Example 12–24
Calculate the percentage of optical power reflected back into the fiber for a –40 dB optical back reflection.

Solution

$$-40 \text{ dB} = 10 \log\left(\frac{P_{\text{refl}}}{P_{\text{in}}}\right)$$

$$\left(\frac{P_{\text{refl}}}{P_{\text{in}}}\right) = inv\log(-4)$$

$$\frac{P_{\text{refl}}}{P_{\text{in}}} = 1 \times 10^{-4}$$

$$P_{refl} = 0.01\% \ (P_{in})$$

Therefore, 0.01% of the input optical power (P_{in}) will be reflected back into the fiber.

OUTPUT RETURN LOSS. The optical receiver output return loss is expressed in dB by the S_{22} parameter. S_{22} is the 20 log of the reflection coefficient, indicative of the matching properties between the 75 Ω characteristic impedance and the output impedance of the optical receiver.

TOTAL CURRENT CONSUMPTION. Total current (I_{total}) consumption is the current drained from the power supply by an optical receiver at a specified DC voltage.

PHOTODIODE BIAS CURRENT. Under a specified DC supply voltage, photodiode bias current is the current drained from the power supply by the photodiode bias circuit.

EQUIVALENT INPUT NOISE. The equivalent input noise power of an optical receiver is composed of three components: laser diode noise, photodiode shot noise, and optical receiver thermal noise. Equation (12–144) expresses the optical receiver equivalent input noise.

$$P_{noise} = \lfloor (\text{RIN}.I_o^2) + (2eI_o) + I_n^2 \rfloor (BR_d) \quad \textbf{(12–144)}$$

where P_{noise} is the noise power measured with a spectrum analyzer (W), RIN is the relative intensity noise of the laser (dB/Hz), I_o is the photodetector DC current, $e = 1.6 \times 10^{-19}$ C, I_n is the equivalent noise current of the receiver ($A/\sqrt{Hz}$), B is the resolution bandwidth of the spectrum analyzer (Hz), and R_d is the optical CATV amplifier resistivity (Ω).

The receiver equivalent input noise (I_n) can be calculated from Equation (12–144) based on the following assumptions:

a) RIN: 160 dB/Hz (for DFB lasers)

b) The responsivity of the device under test (DUT) is assumed to be constant over the used optical input span

c) The spectral analyzer noise floor is smaller than the optical receiver noise. Equation (12–145) expresses the theoretical calculation of the optical receiver output noise.

$$Y(I_o) = [\ (\text{RIN}I_o) + 2e\ ] (BR_d) \quad \textbf{(12–145)}$$

A graph that reflects $Y(I_o)$ output for different values of I_o can be constructed. Equation (12–146) expresses the receiver equivalent input noise I_n.

$$I_n = \left(\frac{2eP_{sa}(0)}{Y(0)} \right)^{1/2} \quad \textbf{(12–146)}$$

where P_{sa} is the measured noise power with a spectrum analyzer at the absence of optical input power, $Y(0)$ can be obtained from graph, and I_n is the relative equivalent noise current.

Second Order and Third Order Distortion

Second order (SO) distortion of a CATV signal is the difference between the peak level of the RF signal and the peak level of the signal generated by the second order modulation product ($f_1 \pm f_2$). Likewise, the third order (TO) distortion product of a CATV signal is the difference between the peak level of the RF signal and the peak level generated by the third order modulation product ($f_1 + f_2 - f_3$).

Optical Receiver Output Voltage

The amplitude modulated optical signal (AM-VSB) is converted to an RF signal by the optical receiver. The receiver output voltage is the product of receiver responsivity input optical power and the index of modulation. Equation (12–147) expresses the output voltage of such a receiver.

output voltage (V) = receiver responsivity × optical input power × modulation index, or

$$V_{out(peak)} = R \times P_{in} \times m \quad \textbf{(12–147)}$$

where $V_{out(peak)}$ is the optical receiver output voltage (mV) at 75 Ω load, R is the responsivity (V/W), P_{in} is the optical input power (mW), and m is the index of modulation (%).

Example 12–25
Calculate the output voltage of an optical receiver employed in an analog multichannel (NTSC) system that operates with an optical input power (P_{in}) of 1.5 mW, a receiver responsivity (R) 800 V/W, and a 4% index of modulation (m).

Solution

$$V_{out(peak)} = R \times P_{in} \times m$$
$$= (800 \text{ V/W})(1.5 \times 10^{-3} \text{ W}) (0.04)$$
$$= 48 \text{ mV}$$

Therefore, $V_{out(peak)} = 48$ mV.

The average output voltage is:

$$V_{avg} = \frac{V_{out(peak)}}{\sqrt{2}}$$
$$= \frac{48 \text{ mV}}{\sqrt{2}}$$
$$= 33.9 \text{ mV}$$

Therefore, $V_{avg} = 33.9$ mV.

In dB,

$$V_{avg(dB_{mV})} = 20 \log (V_{avg})$$
$$= 20 \log (33.9 \times 10^{-3})$$
$$= -29.4 \text{ dB}_{mV}$$

Therefore, $V_{avg} = -29.4 \text{ dB}_{mV}$.

Example 12–26

Determine the optical receiver responsivity required for a $20 \text{ dB}_{\text{mV}}$ output voltage, a modulation depth of 3.5%, and an input optical power of 1.2 mW.

Solution

$$V_{\text{out(peak)}} = (R)(P_{\text{in}})(m)$$

$$20 \text{ dB}_{\text{mV}} = (R)(1.2 \times 10^{-3} \text{ W})(0.035)$$

$$10 = 4.2 \times 10^{-5} R$$

$$R = \frac{10\sqrt{2}}{4.2 \times 10^{-5}}$$

$$= \frac{14.1}{4.2 \times 10^{-2}} \text{ V/W}$$

Therefore, $R = 336$ V/W.

Example 12–27

Calculate the modulation depth (m) required by an optical receiver that operates with an 800 V/W responsivity, an input optical power of 1 mW, and a $28 \text{ dB}_{\text{mV}}$ output voltage.

Solution

$$V_{\text{out(peak)}} = (R)(P_{\text{in}})(m)$$

$$m = \frac{V_{\text{out(peak)}}}{(R)(P_{\text{in}})}$$

$$= \frac{35.5 \times 10^{-3} \text{ V}}{800 \text{ V/W}(1 \times 10^{-3} \text{ W})}$$

$$= 4.4\%$$

Therefore, $m = 4.4\%$.

The relationship between the input optical power and the output voltage at a specified receiver responsivity and modulation depth must be linear, with a ratio of 1 to 2. That is, a 1 dB increase of optical power at the receiver input will generate a 2 dB voltage increase at the receiver output. This relationship is graphically illustrated in Figure 12–62.

CNR of Optical Receivers

The **CNR** of optical receivers that operate in analog multi-channel systems is influenced by the RIN of the transmitter, the optical fiber, and the photodiode shot noise. Equation (12–148) expresses the relationship between the CNR and the various parameters.

$$\text{CNR} = \frac{\lfloor 0.5\,(mI_{pd})^2 \rfloor}{(2e\,I_{pd} + \text{RIN}I_{pd}^2 + n_{th}^2)(B_W)} \qquad \textbf{(12–148)}$$

where CNR = carrier to noise ratio (dB), I_{pd} = photodiode current (A), m = modulation depth (%), RIN = relative intensity noise (dB/Hz), n_{th} = equivalent input noise of the optical receiver ($A/\sqrt{\text{Hz}}$), and B_W is the noise bandwidth per channel (4 MHz for NTSC format).

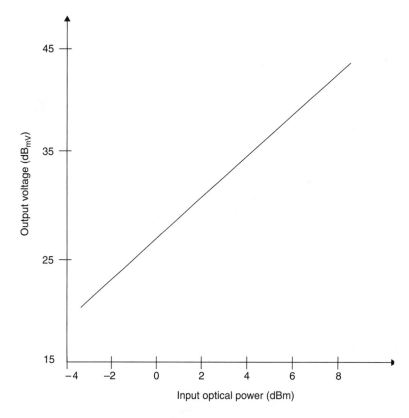

Figure 12–62. Output voltage (dB_{mV}) *v.* input optical power (dBm).

Example 12–28

Calculate the CNR of an optical receiver that operates in an analog multichannel AM-VSB optical system with the following parameters:

- $I_{pd} = 1.2$ mA
- $m = 4\%$
- RIN = -160 dB/Hz
- $e = 1.6 \times 10^{-19}$
- $n_{th} = 8$ pA/Hz (normal range: 5 pA–12 pA)
- $B_W = 4$ MHz

Solution

$$
\begin{aligned}
\text{CNR} &= \frac{[0.5\,(0.04)^2\,(1.2 \times 10^{-3})^2]}{[2(1.6 \times 10^{-19})(1.2 \times 10^{-3})} \\
&\quad + \frac{1}{(8 \times 10^{-12})^2\,]\,(4 \times 10^6)1} \\
&= \frac{1.152 \times 10^{-9}}{(3.84 \times 10^{-22} + 0.64 \times 10^{-22})\,(4 \times 10^6)} \\
&= \frac{1.152 \times 10^{-9}}{1.792^{-15}}
\end{aligned}
$$

$= 6.4 \times 10^5$

In dB, CNR = 58 dB.

A graphical representation of CNR, in relation to optical input power, is illustrated in Figure 12–63.

A graphical representation of equivalent input noise in relation to operating frequency is illustrated in Figure 12–64.

The Optical Receiver CSO and CTO Distortions

The CSO distortion of an optical receiver that operates in an analog AM-VSB optical link is defined as the difference between the amplitude of the carrier signal and the amplitude of signal generated by the CSO distortion component. The ratio of the two components is 1 to 2. That is, 1 dB in the carrier signal reflects a 2 dB increase in the amplitude of the CSO generated signal. Furthermore, because the ratio of the carrier signal and the input optical power is also 1 to 2, an increase of 1 dB in the input optical power will generate a 4 dB increase in the V_{CSO}. A graphical representation of the relationship between carrier levels and CSO distortion levels is illustrated in Figure 12–65.

Similar to CSO distortion, the CTO distortion is defined as the difference between the amplitude of the RF

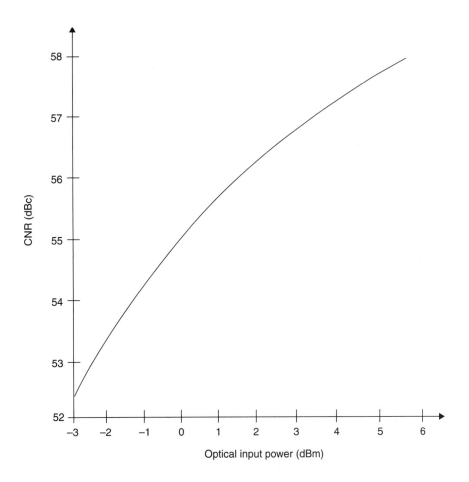

Figure 12–63. CNR *v.* optical input power.

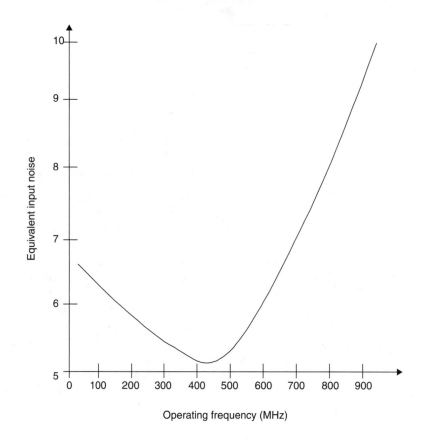

Figure 12–64. Equivalent input noise *v.* operating frequency.

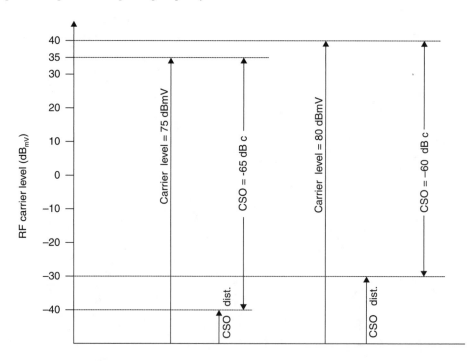

Figure 12–65. RF carrier levels *v.* CSO distortion levels.

carrier signal and the amplitude of the signal generated by the CTO distortion component. The ratio of the signals is 1 to 3. That is, a 1 dB increase of the carrier signal will generate a 3 dB increase in the CTO distortion signal. Based on the same ratio as that of the CSO, an optical increase of 1 dB at the input of the receiver will generate a 6 dB increase of the CTO distortion. A graphical representation of the relationship between carrier signal levels and CTO distortion levels is illustrated in Figure 12–66.

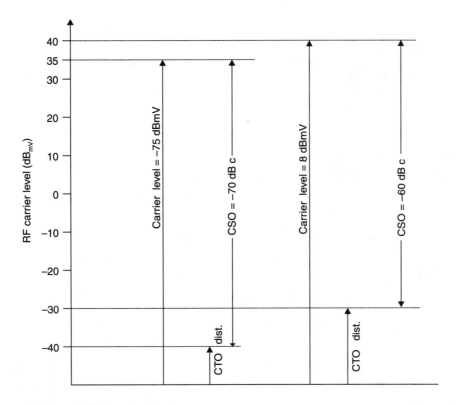

Figure 12–66. RF carrier levels *v.* CTO distortion levels.

12.16 HYBRID MULTICHANNEL ANALOG AND DIGITAL OPTICAL SYSTEMS

Research has demonstrated the viability of combined AM-VSB and M-level **quadrature amplitude modulation (QAM)** multichannel video signals for transmission over optical links to distances over 100 km utilizing NDS-SM fibers. A block diagram of such an optical link is illustrated in Figure 12–67.

The main components incorporated into the system are:

a) optical transmitter: Composed of a DFB laser and an EDFA, both operating at the 1550 nm wavelength window

b) external modulator: MZ

c) inline optical amplifiers: EDFA operating at the same 1550 nm wavelength window

d) fiber: NDS-SM

e) optical receiver

The deployment of such hybrid video optical systems is industry driven. That is, economic incentives dictated an investigation into the possibility of jointly transmitting analog and digital video signals over optical links at relatively low cost. In mid-1996, various research groups conducted experiments and were able to demonstrate that such complex optical systems can successfully be implemented. One particular group was able to transmit a combined 79-AM-VSB video channel and 4-QAM (digital)

channel signal over a 120 km optical link. If analog and digital video signals are to be simultaneously transmitted through the same optical transmitter, attention must be given to the maintenance of the required CNR at levels dictated by video channel specifications. This is because analog video channels are more susceptible to noise and nonlinear distortion. In such systems, maintaining the appropriate CNR levels will necessitate a 10 dB optical power margin. However, a 10 dB power margin imposes a 50 km maximum on the optical span. For longer optical lengths, inline EDFAs, operating at the 1550 nm wavelength window, will be required. An additional problem encountered during the experimentation with the optical links was the degradation of the link performance due to the interaction between analog and digital video channels. In studying the full impact of the interaction between analog and digital video channels and the viability of such an optical link as a whole, one experimental optical link incorporated an optical transmitter that was composed of a DFB laser diode and an EDFA, both operating at the 1550 nm wavelength window. The transmitter also incorporated a mechanism for linearization and SBS suppression. The composite (analog and digital) signal was fed into one of the two ports of the external MZ modulator incorporated into the optical transmitter. The transmitter amplitude modulated optical output power was launched into the NDS-SM fiber. A number of EDFAs were incorporated in order to achieve the desired optical length. At the receiver end, a broadband optical receiver was used to detect the optical power, which was amplitude modulated by the

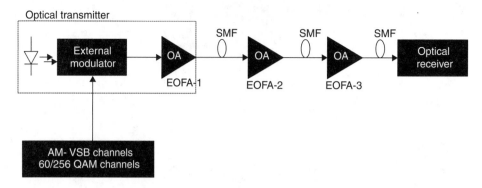

Figure 12–67.   A hybrid multichannel analog and digital video optical link.

composite (analog and digital) signal. After detection, the composite signal was separated through a filtering mechanism and the digital signal was demultiplexed to individual channels.

System Performance Evaluation Based on Analog Signals

The optical power at the input of the first EDFA was set at a level required by the channel CNR. Based on assumed component and system parameters, the CNR is expressed by Equation (12–149).

$$
CNR_{dB} = 10 \log \left[\cfrac{m^2}{\left(\cfrac{2hv\,(NF)}{P_{in}} \right) + RIN + \left(\cfrac{hv\,(NF)}{P_{in}} \right)^2 }{\cfrac{1}{(B_{opt}) + \left(\cfrac{2e}{R.P_r} \right) + \cfrac{i_{sn}^2}{(RP_r)^2}\left(\cfrac{1}{2B_{el}} \right)}} \right]
$$

$$(12\text{–}149)$$

where m is the index of modulation (per channel), RIN is the relative intensity noise, NF is the optical receiver noise figure, i_{sn}^2 is the spectral noise power density of the optical receiver, R is the receiver responsivity, B_{opt} is the optical bandwidth of the received (composite) signal, B_{el} is the electrical bandwidth of each video channel, P_{in} is the input power of the EDFA, and P_r is the received optical power at the receiver input; $\frac{2hv\,(NF)}{P_i}$ calculates signal spontaneous beat noise caused by the EDFA and $\left[\frac{2hv(NF)}{P_i} \right]^2$ calculates spontaneous spontaneous beat noise, also caused by the EDFA.

Experimental results have indicated that before the inclusion of EDFAs, the CNR was mainly influenced by the laser transmitter RIN and the loading of the RF signal. After the inclusion of EDFAs, the CNR is subject to signal spontaneous beat noise, which progressively increases with corresponding increases of the optical power at the input of the EDFA. However, when the EDFA has reached the saturation point, the signal spontaneous beat noise no longer affects the

CNR. Factors that set the limit for the CNR are the optical receiver thermal and shot noise and the received optical power. The CNR, in relation to the index of modulation, is illustrated in Figure 12–68.

With EDFA driven to saturation, an increase of the system CNR has been observed. Furthermore, the incorporation of EDFAs into the link does not affect either the CSO or the CTO distortions. This implies that chirping due to an external modulator does not affect the transmitter performance. The laser diode operating point must be selected so a maximum CNR can be achieved while maintaining CSO and CTO at low levels. Experimental results have shown that a CNR > 49 dB can be achieved with a 3.2% index of modulation while both CSO and CTO distortions are maintained at less than –63 dB per channel. The figures apply to an optical system with a maximum length of 120 km and an optical power budget of 35 dB.

System Performance Evaluation Based on Digital Signals

System performance based on digital signals was evaluated in the presence of analog signals, with or without the inclusion of EDFAs. The performance of digitally modulated signals can be established by measuring the BER in reference to the optical modulation intensity (OMI) of the QAM signal. Figure 12–69 illustrates the relationship between BER versus OMI without inline EDFAs. Figure 12–70 illustrates the relationship between BER and OMI with the inclusion of EDFAs. System measurements indicated that with the incorporation of inline optical amplifiers, a power penalty of 2 dB would be required as a result of the beat noise added by EDFAs. Furthermore, it was observed that the 256 QAM modulation scheme requires about a 300% increase of the **optical modulation intensity (OMI)** in order to maintain system performance at a BER of 10^{-9}. However, QAM signals induce certain levels of nonlinear distortion and chirping, thus decreasing the SNR and increasing the system BER. The relationship between the BER and SNR is illustrated in Figure 12–71.

For both 64 QAM and 256 QAM signals, system performance can be measured with the OMI kept constant

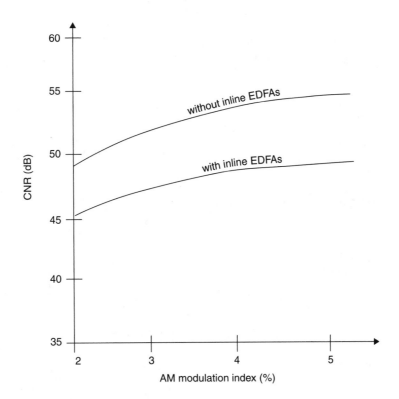

Figure 12–68. CNR *v.* index of modulation.

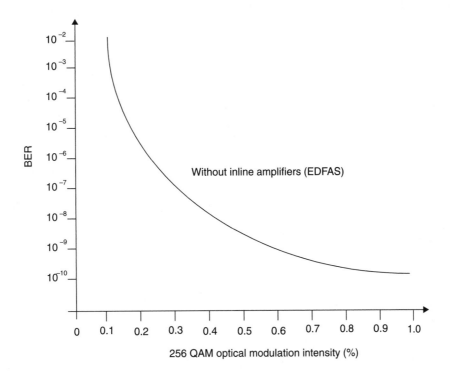

Figure 12–69. BER *v.* OMI (without EDFAs).

and at a very low level. This is achieved by progressively increasing the white noise with a simultaneous increase of the SNR, while maintaining a constant system BER. For the 64 QAM modulation scheme, it was observed that the impact of the analog channels on the digital performance was zero, while a BER constant level started to emerge at an SNR of 37.7 dB. Such system degradations can be remedied by increasing the QAM signal levels. Although this may improve the system BER, it may also decrease the quality of the analog signals through an undesirable

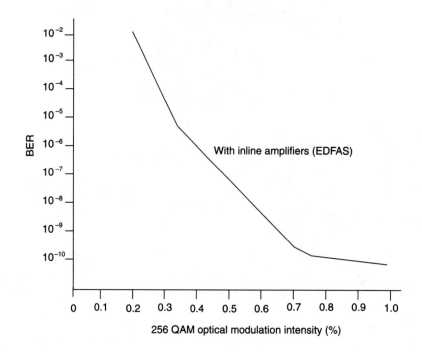

Figure 12–70. BER *v.* OMI (with EDFAs).

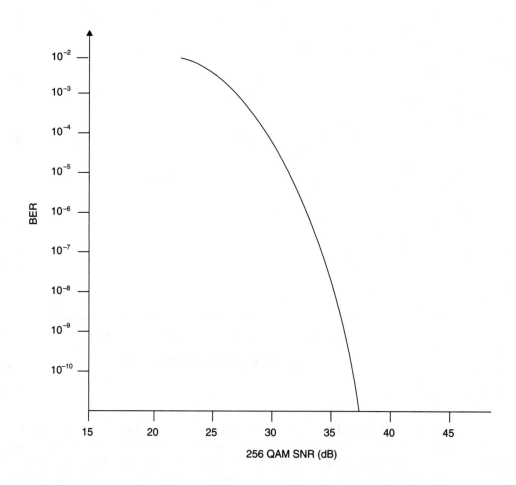

Figure 12–71. BER *v.* SNR.

increase of CSO and CTO distortions. These increases in distortion will decrease the CNR of the analog signals.

12.17. UNDERSEA OPTICAL SYSTEMS

Introduction

Attempts to connect the world's continents through an undersea cable go back as far as 1866, with the installation of the first transcontinental telegraph cable between North America and Europe. It was intended to provide telegraph services between the two continents. Almost a century later, in 1956, the first transatlantic telephone system was installed between North America and Europe. It was capable of providing only 48 voice channels of questionable quality. Since that time, the demand for high capacity systems that facilitate voice, video, and data signals has increased dramatically. Technological innovations, such as sophisticated repeater designs and high quality cables, contributed to a substantial increase of system capacity. By the early 1980s, the number of voice channels had been increased to ten thousand. However, voice, video, and data transmission via cable based systems are bandwidth limited. The newly emerging fields of digital and optical technologies were instrumental in the development and deployment of the transoceanic systems for both the Atlantic and the Pacific oceans in the late 1980s. These optical systems were able to carry 280 Mb/s per fiber incorporating electrooptical repeaters while utilizing the 1310 nm wavelength window. Progress in optical device technology allowed the system to operate at the low loss 1550 nm wavelength window, with a corresponding increase of transmission capacity, to the level of 2.5 Gb/s per fiber, and with a simultaneous increase of the fiber span. Although the demands for further capacity increases were ever present, the systems were unable to accommodate these demands as a result of limitations imposed by the speed of the electronic devices prior to the development of EDFAs.

Transmission Formats for Long Distance Systems

The two transmission formats adapted for long distance optical transmission are the NRZ, in which the pulse occupies the entire time slot, and the soliton, where the pulse occupies only one-fifth of the time slot. Both formats exhibit distinct performance characteristics. For example, NRZ signals require simpler transmitter and amplifier designs, are more tolerant to wavelength stability, and can operate at speeds of up to 10 Gb/s. Soliton signals are less tolerant to wavelength stability and require more complex transmitter and amplifier designs. On the positive side, they can operate at speeds of up to 20 Gb/s. NRZ and soliton formats are illustrated in Figure 12–72.

Furthermore, NRZ signals are easier to generate, process, and detect while soliton pulses require an additional optical pulse source and jitter control. With soliton signals, a higher channel capacity can be achieved than with NRZ signals. However, this disadvantage of the NRZ format in reference to soliton can be remedied by increasing the channel density of the WDM system. Based on system demands, the NRZ format can be used with higher density WDM systems, or the soliton format can be used with smaller channel density, thus maintaining the same system capacity. The incorporation of an EDFA into long distance optical transmission eliminates the need for coherent systems because phase control is difficult to be maintained in amplitude shift keying (ASK) modulation schemes.

60 Gb/s TDM WDM Oceanic Optical System

One of the first oceanic systems intended for commercial applications was deployed in the Okinawa–Kagoshima route, which traveled over 900 km. This system was the first to employ EDFAs for undersea repeaters. The incorporation of EDFAs made the transmission of high speed rates with a relatively simple system configuration possible. The two main advantages of the system are the capability of transmitting at either 2.5 Gb/s or 10 Gb/s and the ability to upgrade through WDM, high speed TDM, or soliton. A block diagram of the system is illustrated in Figure 12–73.

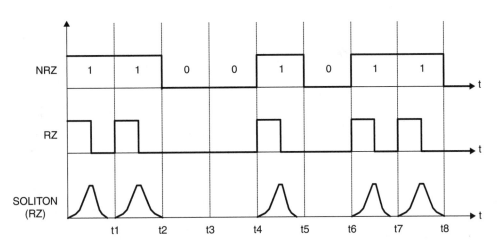

Figure 12–72. NRZ and soliton transmission formats.

System operating characteristics

a) transmission capacity: 60 Gb/s

b) transmission length: 900 km

c) optical amplifiers: EDFA (12)

d) optical repeaters: Composed of EDFA, 1480 nm laser pump, and WDM coupler

e) optical filters: BPF—one for each output port required to confine optical power within the 10 nm bandwidth

f) optical time domain: Including reflectometry (OTDR) loop back, included at each amplifier output for signal monitoring

g) receiver noise figure (NF): 6 dB (0.2 dB standard deviation)

h) polarization dispersion loss (PDL): 0.06 dB

i) polarization modal dispersion (PMD): 0.09 ps (standard deviation 0.04 ps)

j) cable has: 6-line pairs (total 12 fibers), and 90 km of DS fibers with 0.21 dB/km fiber loss, 0.072 ps/$\sqrt{km}$ PMD and 0.019 ps/$\sqrt{km}$ standard deviation. The cable is covered with a trisected steel pipe and polyethylene sheaths. Armored cable is used in shallow waters where the possibility of damage is higher.

k) A control mechanism is put in place to detect and control the optical power output. This control mechanism is composed of an optical coupler and a photodiode.

To achieve zero dispersion at the output, each fiber has normal dispersion at the input where the optical power is high and an irregular dispersion at the output where the optical power is low. Zero dispersion is achieved at the 1552 nm wavelength window. The two main concerns for such a system are excess intensity noise and waveform distortion. Excess intensity noise can be reduced by allowing for maximum fiber dispersion at signal wavelength while at the same time complying with zero dispersion wavelength.

System Performance

The system performance was measured in terms of PMD in relation to distance (Figure 12–74) and zero-dispersion wavelength in relation to distance (Figure 12–75).

Figure 12–74 shows that a PMD of less than 10 ps can be achieved for a maximum distance of 10,000 km. This PMD level is insignificant for the performance of such a system, even when operating with 10 Gb/s transmission rates.

From Figure 12–75, it is evident that a very small ZDW deviation, of about 0.13 nm, does occur, which is quite satisfactory. The required amplifier bandwidth was achieved by incorporating optical filters that have distinctly different free-space range (FSR) at the amplifier outputs. The relationship between the optical bandwidth and transmission distance is illustrated in Figure 12–76.

Figure 12–76 shows the measured center wavelength range to be within 0.2 nm. The SNR (electrical) was measured for both 2.5 Gb/s and 10 Gb/s NRZ transmission rates. To improve the SNR, scrambling was induced through a LiNbO$_3$ MZ optical modulator. The relationship between SNR and transmission distance for both 2.5 Gb/s and 10 Gb/s NRZ signals is illustrated in Figure 12–77.

Figure 12–77 shows that the higher the transmission rate, the lower the available SNR, with or without polarization scrambling which is mainly used to minimize the EDFA polarization hole burning (PHB). It is also evident that a higher polarization scrambling rate improves SNR. (A further discussion of polarization scrambling is presented later in this chapter.) When the polarization scrambling rate was increased to 20 GHz, twice the highest transmission rate, it was observed that the SNR improvement was almost the same as that of the 2.5 Gb/s transmission rate. This can be attributed to PMD and simultaneous phase modulation induced by polarization scrambling. Higher order distortion also has a negative effect on system performance and is translated into waveform distortion. Therefore, a reduction of the higher order distortion will substantially improve waveform distortion.

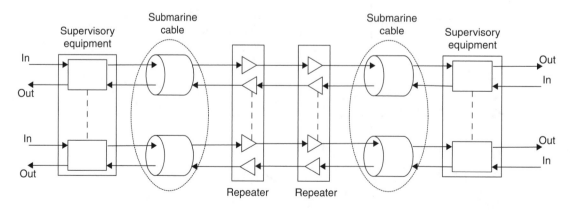

Figure 12–73. A block diagram of a 60 Gb/s, 900 km oceanic optical system.

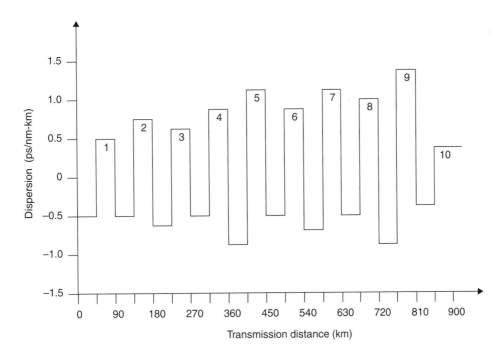

Figure 12–74. PMD *v.* transmission distance.

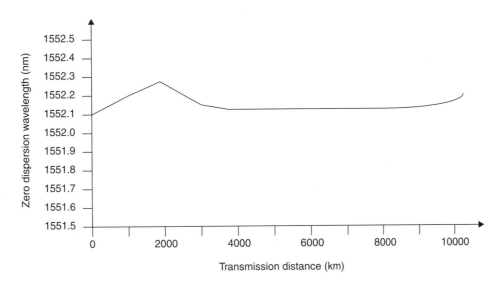

Figure 12–75. Zero dispersion wavelength (ZDW) *v.* transmission distance.

The North America to Europe Undersea Optical System

In 1996, a transatlantic ring optical network was established, connecting North America and Europe. This optical system was designed to accommodate 600,000 simultaneous voice channels and another 600,000 voice channels as standby. The network incorporated the following major components:

a) optical fibers: TAT-12 (5900 km)

TAT-13 (6300 km)

2 interconnect cables

b) optical amplifiers: EDFAs spaced at 45 km spans, a total of 132 for a required distance of 5900 km and 140 required for a distance of 6300 km

The system is based on ring architecture, with the capability of mutual service restoration in the event of failure. That is, if a failure in the cable does occur, the data will be rerouted around the ring, thus bypassing the fault point. The transatlantic system became operational in 1996 and was the first undersea link to incorporate EDFAs. The major operating components of the transatlantic system are meant to be effective for at least twenty-five years. To compensate for the aging of the components and *Q*-factor fluctuations, a substantial optical power

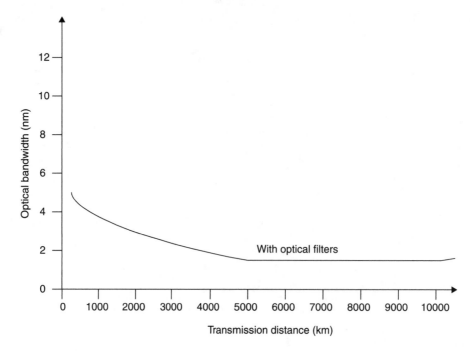

Figure 12–76. Optical bandwidth *v.* transmission distance.

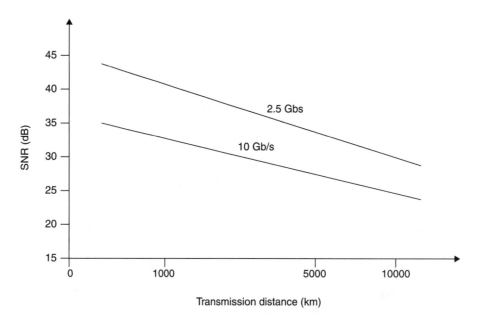

Figure 12–77. SNR *v.* transmission distance for both 2.5 Gb/s and 10 Gb/s NRZ signals.

margin was established within the system. As with any optical fiber link, the parameters that guide the design philosophy are as follows:

a) optical power launched into the fiber

b) repeater spacing

c) dispersion management

d) cost

During the design process, parameter trade-offs are essential for system reliability and performance. Performance is mea-

sured by the system BER and *Q*-factor. For this transatlantic optical ring, the BER is set at 10^{-10} and the *Q*-factor at 16.1 dB. A summary of the operating parameters of a 5 Gb/s per fiber undersea optical link are listed in Table 12–12.

A block diagram of a typical undersea optical cable system is illustrated in Figure 12–78.

One of the most important elements in the design of undersea optical links is the incorporation of a performance monitoring mechanism. Such a mechanism is accomplished through high loop back paths between EDFAs. A –45 dB signal identified by a specific delay time is allowed to travel in

TABLE 12–12   Operating Characteristics *v.* Power Penalty for a 5 Gb/s Transatlantic Optical System

Parameters	Power Penalty (dB)
Q-factor (ideal)	30.5
Beginning of life impairment	–3.5
Fluctuations and interactive impairments	–8.5
Worst case Q-factor	21.8
Aging and repairs	–0.7
Minimum end of life performance	17.1
Q-factor for BER = 10^{-10}	16.1

the opposite direction. At the land terminal, this small signal correlates with the outgoing 5 Gb/s signal, and the gain of the loop back path is established by means of a set modulation depth. If the identified modulation depth increases by 100%, then the system is out of service.

FIBER CABLES FOR THE TRANSATLANTIC PATH.

a) fiber (DS-SM): TAT12/13 four chains of EDFAs grouped as two bidirectional pairs with each fiber carrying 5 Gb/s NRZ signals

b) amplifier spacing: 45 km

c) repeater pressure tolerance: 800 atmospheres

d) repeater current: 0.9 A (DC)

e) fiber compensation for dispersion accumulation: For every 500 km, a few hundred meter SM fibers are used, operating at the 1310 nm wavelength window

EDFAs.

a) gain: 9.5 dB

b) noise figure (NF): 5 dB

c) output optical power: +3 dBm

MAXIMUM SYSTEM CAPACITY.

a) Four STM-16: Each STM-16 is capable of carrying 30,240 voice channels

b) Total traffic capacity without compression: 120,960 (4 × 30,240) voice channels. However, using a compression ratio of 5:1, the maximum number of simultaneous conversations is raised to 604,800 with the same number applied to the protection lines.

Other Transoceanic Systems

Similar to the transatlantic optical system, other systems are as follows:

a) The transpacific cable: TPC-3 and TPC-4

b) The transpacific cable-5 network: TPC-5CN

c) Asian-Pacific cable network: APCN

d) Fiber optics link around the globe: FLAG

e) Japan information highway: JIH

The trend for capacity demands on the transpacific optical system is illustrated in Figure 12–79.

The operating parameters for the transpacific ocean optical transmission cable are listed in Table 12–13.

A block diagram of a typical repeater used in an undersea optical system is illustrated in Figure 12–80.

The major problems encountered in undersea optical systems are:

a) chromatic dispersion of the transmission fiber

b) nonlinearities of the transmission fiber

c) optical noise accumulation

d) polarization dependent loss (PDL)

e) polarization hole burning (PHB)

f) polarization modal dispersion (PMD)

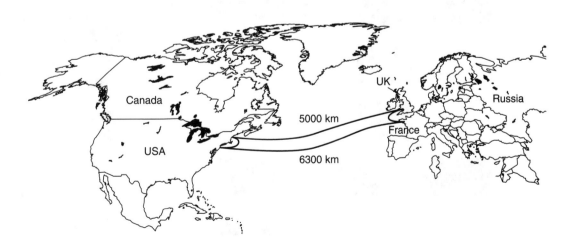

Figure 12–78.   A block diagram of an undersea optical cable.

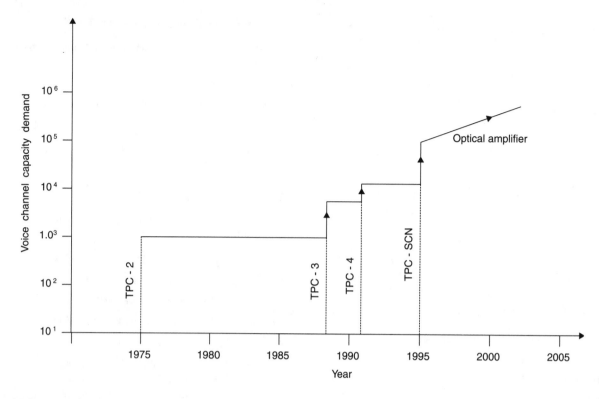

Figure 12–79. Capacity demand trends for the transpacific optical cable (TPC).

TABLE 12–13   TPC-3 and TPC-4 system parameters

Parameters	TPC-3	TPC-4
Operating wavelength	1310 nm	1550 nm
Bit rate	280 Mb/s	560 Mb/s
Modulation	IM	IM
Repeater spacing	50 km	120 km
Optical source	F.P LD	DFB
Optical detector	Ge APD	InGaAsAl
Power supply current	1.6 A (DC)	1.6 A (DC)

The operating parameters of a transoceanic optical system employing EDFAs are listed in Table 12–14.

The operating parameters of the Japanese information highway (JIH) undersea optical system are listed in Table 12–15.

Experimental Transoceanic Optical Systems

220 Gb/s – 9500 KM.   An experimental transoceanic optical system designed to carry 220 Gb/s to a distance of 9500 km through a recirculating loop is illustrated in Figure 12–81.

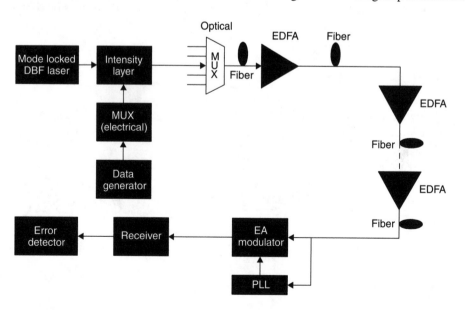

Figure 12–80.   A block diagram of an undersea optical repeater.

TABLE 12–14 Operating Parameters

Parameters	
Bit rate (per fiber)	5 Gb/s
Operating wavelength	1558.5 nm
Input signal level	–20 dBm to 0 dBm
Repeater gain	7 dB to 20 dB
Noise figure	5 dB to 6 dB
Optical amplifier	EDFA
Pump laser	InGaAsP/InP
Pump wavelength	1475 nm

TABLE 12–15 Operating Parameters for the JIH Optical System (WDM)

Parameters	
Bit rate (total capacity)	100 Gb/s (WDM)
	4×2.5 Gb/s $\times 10$
Wavelength	1530 nm to 1570 nm
Repeater noise figure (NF)	5 dB to 6 dB
Optical amplifier	EDFAs
Pump laser	InGaAsP/InP
Pump wavelength	1475 nm

The transmission performance of the 220 Gb/s, 9500 km experimental optical system expressed by the Q-factor, in relationship to the operating wavelength, while a BER $= 10^{-10}$ is maintained constant, is illustrated in Figure 12–82.

20 Gb/s , 8100 km Optical Soliton Transmission. A block diagram of an experimental 20 Gb/s, 8100 km optical soliton transmission is illustrated in Figure 12–83.

The performance of the above experimental system, measured as the Q-factor in relation to distance, is illustrated in Figure 12–84.

The EDFA in Undersea Optical Systems

EDFAs are ideal devices for use in undersea optical links. The optical amplifiers incorporated into the repeater modules provide excellent gain with very low noise. Because the operating wavelength window of these devices is in the 1550 nm range, and because they are fiber based, they can easily be coupled to the transmission fiber with relatively small signal loss. Furthermore, these devices exhibit excellent reliability over a 25-year life span. More specifically, EDFAs incorporated in the repeater chain are designed to perform the following functions:

a) provide optical amplification to offset cable induced attenuation

b) control the optical power output level

c) provide the required signal bandwidth

d) minimize pulse chirping caused by chromatic dispersion and nonlinearities

e) provide control of accumulated noise

It is important that the optical power launched into the fiber be maintained at a constant. In order to achieve this,

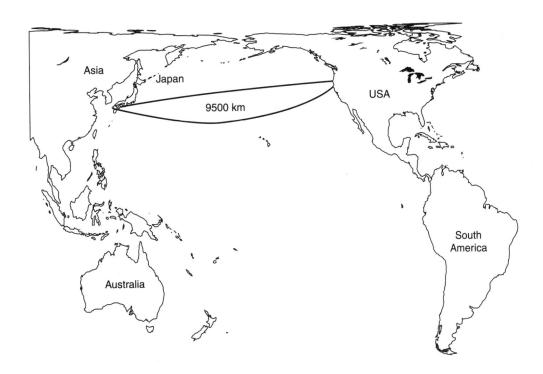

Figure 12–81. A 220 Gb/s, 9500 km experimental transoceanic optical system.

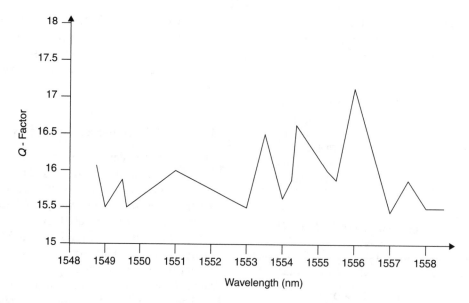

Figure 12–82. Performance characteristics: Q-factor $v. \lambda_o$.

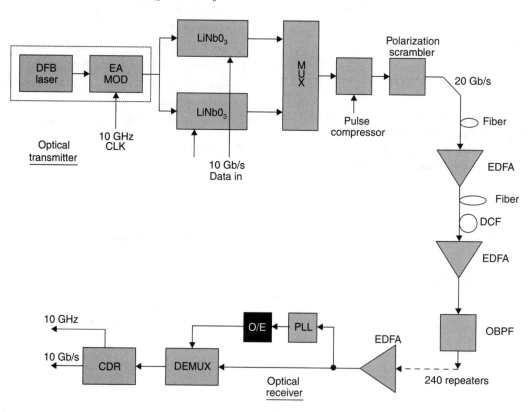

Figure 12–83. A block diagram of a 20 Gb/s, 8100 km optical soliton transmission.

the gain of the optical amplifier is varied, inversely proportional to the optical power at the input of the amplifier, which is subject to the signal attenuation of the previous fiber span. That is, if the optical power at the input of the amplifier decreases, the amplifier gain increases, thus maintaining a constant optical power output. This kind of negative feedback mechanism is achieved by appropriately setting the operating point of the optical amplifier.

EDFAs employed in undersea repeater modules require gain compensation because in a long amplifier chain their normal operating bandwidth of 35 nm is substantially reduced. Therefore, gain equalizers are required to maintain a constant relationship between gain and bandwidth. The effective bandwidth of the amplifier chain in undersea applications that do not have gain equalizers is on the order of 3.5 nm, while those that have equalizers have amplifier chain

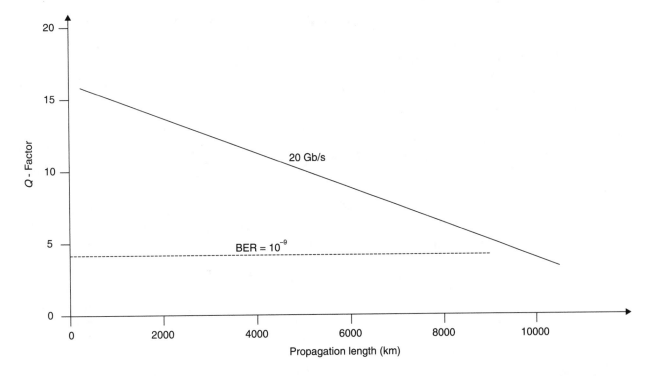

Figure 12–84. Performance characteristics: Q-factor v. distance.

bandwidth in excess of 10 nm. Optical amplifier gain equalization can be accomplished by using fiber grating filters. Fiber grating filters exhibit the following characteristics:

a) low back reflection

b) temperature insensitivity

c) low polarization dependency

d) low polarization mode dispersion

A serious problem that arises from the use of EDFAs in an undersea repeater chain is the accumulated noise generated by the optical amplifiers. This noise can affect the system performance by reducing the SNR. Equation (12–150) expresses the optical amplifier accumulated noise power.

$$P_n = 2n_f hv\,(G-1)\,B_{opt} \qquad \textbf{(12–150)}$$

where P_n=noise power, n_f=excess noise factor, hv=photon energy, G=amplifier gain, and $B_{opt.}$=optical bandwidth. The accumulated noise spectral density of such a system depends on fiber loss and repeater gain.

Example 12–29

Determine the advantage of an undersea system with a repeater spacing of 40 km over that with a repeater spacing of 200 km, in terms of the accumulated noise density. The system employs an SM optical fiber with 0.2 dB/km attenuation.

Solution

i) Compute total fiber span attenuation for the 40 km span scheme.

$$(0.2\ \text{dB/km})(40\ \text{km})\ =\ 8\ \text{dB}$$

ii) Compute total fiber span attenuation for the 200 km span scheme.

$$(0.2\ \text{dB/km})(100\ \text{km})\ =\ 20\ \text{dB}$$

The 40 km fiber span will require an EDFA with 8 dB gain, while the 200 km fiber span will require EDFAs with total gain of 40 dB. Because the accumulated noise spectral density is gain dependent, therefore:

$$20\ \text{dB}-8\ \text{dB}\ =\ 12\ \text{dB}$$

or, $Invlog(1.2) = 16$.

It is evident from Example 12–29 that the accumulated noise density of the longer fiber span will be sixteen times higher than that of the shorter fiber span. Therefore, long transmission systems will require shorter repeater spans, in order to maintain the accumulated noise spectral density at the end of the fiber at acceptable levels, without degrading the system SNR. Equation (12–151) expresses the SNR at the output of a chain of optical amplifiers.

$$\text{SNR}_{\text{dB}} = 10 \log \left[\frac{P_l}{(GhvB_{opt}N)\,(NF)} \right] \quad \textbf{(12–151)}$$

where P_l = optical power launched into the fiber, G = amplifier gain, hv = photon energy, $B_{opt.}$ = optical bandwidth, N = number of amplifiers, and NF = amplifier noise figure.

Example 12–30

Calculate the SNR of a chain repeater composed of 10 EDFAs, each with 10 dB gain, 5 dB of NF, and 5 GHz of optical bandwidth. The power launched into the fiber is 2.75 mW.

Solution

Use Equation (12–151) and solve for SNR_{dB}.

$$\text{SNR}_{\text{dB}} = 10 \log \left[\frac{2.75 \times 10^{-3}}{(10 \times 1.6 \times 10^{-19} \times 5 \times 10^9)3.16} \right]$$

$$= 10 \log(1.1 \times 10^5)$$

$$= 50 \text{ dB}$$

Therefore, SNR = 50 dB.

Example 12–31

Calculate the required optical power launched into the fiber if 40 dB of SNR is to be maintained at the input of the repeater. The repeater module is composed of 8 EDFAs that exhibit 8 dB gain, 6 dB of noise figure, and 4 GHz of effective bandwidth.

Solution

From Equation (12–151), solve for P_l.

$$10 \times 10^3 = \frac{P_l}{6.3 \times 1.6^{-19} \times 4 \times 10^9 \times 8 \times 3.98}$$

$$P_l = (10 \times 10^3)(6.3)(1.6 \times 10^{-19})(4 \times 10^9\,(8)(3.98)$$

$$P_l = 1.3 \times 10^{-3} \text{ W}$$

Therefore, 1.3 mW of optical power is required.

The Impact of Dispersion on System Performance

In several places in the text, the impact of fiber dispersion, nonlinearities, and noise was discussed. Here, a brief review of the impact of the phenomena on oceanic optical transmission systems will be presented. SMFs employed in undersea optical systems exhibit a small degree of nonlinearities, caused by the fiber diffractive index dependency on the intensity of the propagated optical signal. Such fiber nonlinearities are instrumental for the transfer of optical power from one channel to the adjacent channels, as well as for mixing optical signal power with noise power. The fiber index nonlinearities, in terms of optical power coupled into the fiber, fiber refractive index, nonlinear coefficient, and fiber effective area are expressed by Equation (12–152).

$$n = n_o + \frac{PN}{A_{eff}} \quad \textbf{(12–152)}$$

where n=refractive index nonlinearity, n_o=fiber refractive index linear point, P=optical power launched into the fiber, N = nonlinear coefficient (2.6×10^{-6} cm^2 / W), and A_{eff}=fiber effective area.

The fiber refractive index nonlinearities promote nonlinear effects such as FWM, XPM, and SPM. In relation to the available system bandwidth, it can best be obtained at the wavelength at which the fiber exhibits zero dispersion in the absence of nonlinearities. However, systems that operate at the zero dispersion wavelength encounter the problem of interaction between the signal and optical amplifier noise. If the wavelength of the noise signal is very close to the signal wavelength, an interaction may occur between the two signals. Furthermore, in a WDM optical system, the adjacent channels that operate at slightly different wavelengths can overlap; that is, the optical power can be transferred from one channel to the adjacent channels due to chromatic dispersion. This problem can be remedied at the design level, by appropriately managing the refractive index nonlinearities and chromatic dispersion through a process referred to as **dispersion mapping.** If dispersion mapping is optimally adjusted, noise and signal mixing and data distortion due to nonlinearities can be substantially reduced. The dispersion mapping process is illustrated in Figure 12–85.

From Figure 12–85, the accumulated dispersion is compensated for by inserting a negative dispersion coefficient fiber (based on accumulated dispersion level) at specified intervals along the transmission path. At the end of the transmission distance, the accumulated dispersion is almost zero.

The Impact of Polarization on System Performance

In any long distance optical fiber link, zero variations of the SNR are indicative of system performance. Polarization effects can cause SNR fluctuations. Optical fibers are nonideal, that is, the smallest physical anomaly along the fiber can cause the transmitted optical signal to switch modal states, and therefore cause PMD. Furthermore, modal variations can interact with the EDFA polarization loss dependent components, resulting in a reduction of the optical SNR. SNR fluctuations cannot be completely eliminated when using nonpolarization maintenance optical fibers. However, SNR fluctuations can be minimized, or altogether eliminated, through a reduction of the PMD, through a reduction of the PDL and an increase of the beginning of life optical power margin. Another critical element that influences system performance is PHB. A relationship between PHB and ASE exists. That is, PHB increases the accumulated ASE noise level in the orthogonal axis more quickly than in the parallel axis, resulting in a reduction of the optical SNR. Experimental results have shown that the effect of PHB on the signal quality is very small when only one EDFA is in use. However, in oceanic optical systems in

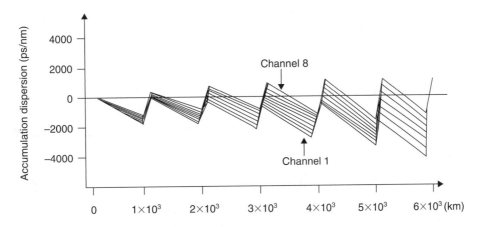

Figure 12–85. Dispersion mapping.

which several EDFAs are incorporated into the repeater modules, the effect of PHB is quite significant. The effect of PHB can be remedied through polarization scrambling (polarization modulation). If the signal is scrambled at faster rates than that at which the EDFA is capable of responding, the effect of PHB is practically eliminated. Although polarization scrambling can improve PHB, it can also generate additional problems. For example, jitter may be observed as a result of the interaction between PMD and polarization scrambling. Either low speed scrambling or high speed scrambling in relation to clock frequency, which is subject to transmission bit rate, can implement polarization scrambling. At low speed scrambling (lower than the system bit rate), the effect of PHB is significantly reduced. However, at the same time an undesirable amplitude modulation is introduced into the system through the PDL. At high speed scrambling (higher than the system bit rate), the AM level due to PDL is substantially reduced, while at the same time it impacts the system operating bandwidth negatively. Experimental results have shown that the optimum polarization scrambling rate is equal to the transmission rate.

System Margin

In long haul optical communications systems (oceanic systems), system performance is measured in terms of BER, assuming digital transmission. Any optical link must maintain a specified maximum BER at the end of life. For this reason, an optical power margin must be incorporated into the system at the beginning of life. System power margin is defined as the difference between the SNR required to maintain a preset BER (in this case equal to 10^{-10}) and the SNR measured at the receiver input. It is evident that $SNR_{rec} > SNR_{(BER)}$. As mentioned, the system BER is influenced by the following impairments:

a) fiber chromatic dispersion

b) polarization modal dispersion

c) optical noise

d) fiber nonlinearities

e) fiber polarization effects

f) amplifier component deviations

BER is also degraded by pattern dependent effects (ISI) and is affected by random noise. Random noise accumulation promotes fluctuations in the received data, the level of which can be interpreted by the closure of the eye diagram. The most important source of optical noise is that of ASE that results from the incorporation of EDFAs into the system. ISI is also the result of the impact of chromatic dispersion, fiber nonlinearities, variations in the transmitter and receiver operating characteristics, and polarization dispersion. The received digital signals at the receiver input may fall into the following four categories:

1) low noise and low distortion

2) low noise and high distortion

3) high noise and low distortion

4) high noise and high distortion

The form of the above signals is illustrated in Figure 12–86. Equation (12–153) expresses BER.

$$BER_{(V)} = \frac{1}{2}\left[erfc\left(\frac{|\mu_0 - V|}{\sigma_1} \right) + erfc\left(\frac{|V - \mu_0|}{\sigma_0} \right) \right] \quad \textbf{(12–153)}$$

where $erfc = \dfrac{1}{\sqrt{2\pi}} \displaystyle\int_y^{\infty} e^{-\alpha/2}\, d\alpha$, $\sigma_{1,0}$ is the standard deviation (voltage at the mark and space in the eye diagram), and $\mu_{1,0}$ is the means. Equation (12–154) expresses the nonideal Q-factor as a function of means ($\mu_{1,0}$) and standard deviation ($\sigma_{1,0}$).

$$Q = \frac{|\mu_1 - \mu_0|}{\sigma_1 - \sigma_0} \quad \textbf{(12–154)}$$

In dB, $Q_{dB} = 20 \log(Q)$

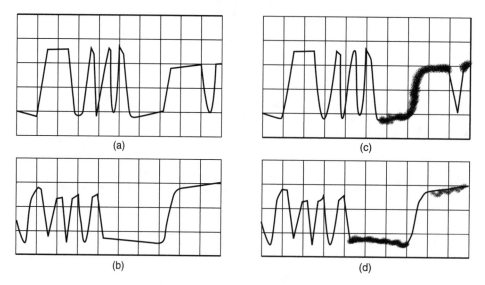

Figure 12–86. Possible forms of digital signals at the input of the optical receiver.

The ideal Q-factor, in reference to optical *SNR*, optical bandwidth, and electrical bandwidth, is expressed by Equation (12–155).

$$Q_{dB} = 20 \log \left[\frac{2SNR_o \left(\dfrac{B_o}{B_e} \right)^{1/2}}{1 + (1 + 4SNR_o)^{1/2}} \right] \quad (12–155)$$

where Q = Q-factor (dB), SNR_o = optical SNR, B_o = optical bandwidth, and B_e = electrical bandwidth.

Experimental System 1: 100 Gb/s Oceanic Optical Systems

As stated on numerous occasions, the ever expanding demand for voice, video, and data channel requirements necessitated ongoing research and development in undersea optical systems beyond the current 5 Gb/s channel capacity. Recent developments in optical devices such as optical amplifiers (EDFAs), add/drop multiplexers (OADM), and optical cross connect (OXC) allowed for new and sophisticated oceanic systems to be designed and experimentally evaluated for future applications. It is estimated that optical systems with data carrying capacities of between 100 Gb/s and 1 Tb/s, utilizing DWDM network schemes, will be designed. However, high density oceanic optical systems like any other long haul optical system are limited by:

1) SNR

2) adjacent channel interference

3) EDFA gain bandwidth limitations

4) repeater optical power limitations

SNR. The two major factors that define the SNR at the input of an undersea repeater are the optical power detected at that input and the accumulated ASE at the same input from this and from all the other optical amplifiers in the chain. Therefore, the progressive increase in the ASE noise decreases the SNR. However, if the ASE in each EDFA is decreased, the end SNR will be proportionally increased. The main source of ASE in an EDFA is the laser pump that operates at the 1470 nm wavelength window. If the operating wavelength of the laser pump is reduced to 980 nm, the ASE noise will be substantially reduced. EDFA laser pumps are designed to operate at the 1470 nm wavelength window for reliability. Therefore, fabricating high reliability laser pumps in the 980 nm region will enhance the performance of undersea optical systems.

ADJACENT CHANNEL INTERFERENCE DUE TO FIBER NONLINEARITIES. To increase system bit rates, an increase in the number of channels within the WDM scheme (unless soliton transmission is used) is required. An increase in the number of channels will also increase the probability that fiber nonlinearities will induce impairments into the system. Although the main source of impairments into the optical system is chromatic dispersion, nonlinear effects such as FWM, SPM, and SBS can be marginalized by controlling the optical power launched into the fiber and by incorporating large effective area fibers ($80 \, \mu m^2 - 100 \, \mu m^2$). The Q-factor of an undersea optical system as a function of the power launched into the fiber for different fiber effective areas is illustrated in Figure 12–87.

EDFA GAIN BANDWIDTH LIMITATIONS. Early designs of undersea repeaters utilized EDFAs that exhibited an inherent autofiltering effect. This natural impairment was used to suppress gain saturation caused by ASE. The negative result of autofiltering is a reduction of repeater

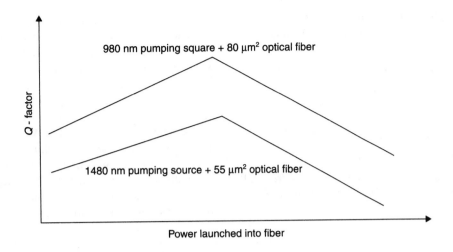

Figure 12–87. *Q*-factor *v.* optical power launched into the fiber.

gain bandwidth. Therefore, EDFAs employed in WDM oceanic optical systems require bandwidth expansion. Such bandwidth expansion can be accomplished through gain equalization. However, the optical bandwidth that is attainable after gain equalization is subject to Al concentration within the EDFA. Higher Al concentration allows for higher bandwidths. The relationship between EDFA bandwidth and Al concentration is illustrated in Figure 12–88.

REPEATER OPTICAL POWER OUTPUT LIMITATIONS. An increase in the number of channels in a DWDM optical system reflects an increase of the optical power launched into the fiber with a corresponding increase in the pump power. The assumption is that system reliability will be maintained at a constant.

Experimental System 2: 110 Gb/s Transoceanic System

An experimental optical system intended for oceanic applications was designed with the following specifications:

a) transmission capacity: 110 Gb/s

b) multiplexing scheme: WDM

c) number of channels: 22

d) data rate per channel: 5.3 Gb/s

e) transmission distance: 9500 km

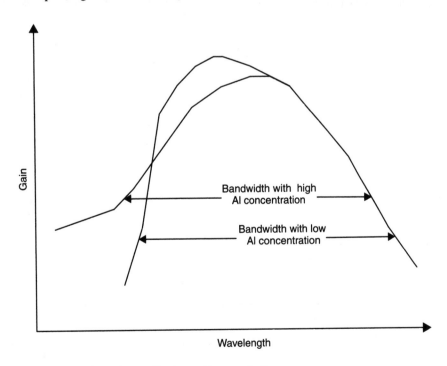

Figure 12–88. Attainable bandwidth after gain equalization *v.* Al concentration.

f) optical repeaters (19): each including an EDFA, a WDM optical coupler, an optical isolator, and a 980 nm laser diode pump

g) repeater power output: 8 dBm

h) gain equalizer: F.P Etalon tunable filters

i) **preemphasis:** included

j) fibers: 16 dispersion shifted and 2 NDS-SM

k) fiber span: 40 km

l) receiver input: ultra narrow band optical filters are used for channel selection, and band-pass filters are used for signal confinement

The measured performance characteristics of the proposed oceanic optical system are illustrated in Figure 12–89.

It was observed that higher wavelength channels show some degree of pulse broadening, which can be attributed to fiber nonlinearities. In order to establish the presence or absence of FWM, a small number of channels were removed from the system. There was no evidence of the presence of FWM at their spectrum location. A measure of the *Q*-factor, in relation to wavelength, is illustrated in Figure 12–90. A small observed variation in the *Q*-factor for different channels can be attributed to SNR variations for these channels. SNR and *Q*-factor variations will be offset through deemphasis at the transmitter module.

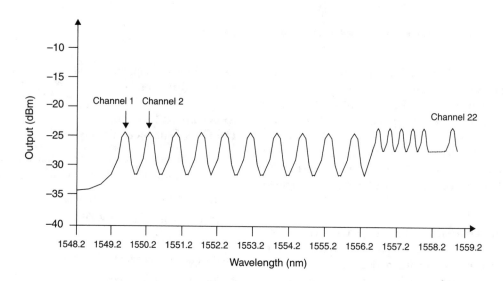

Figure 12–89. Signal spectral of 22 WDM channels *v.* wavelength.

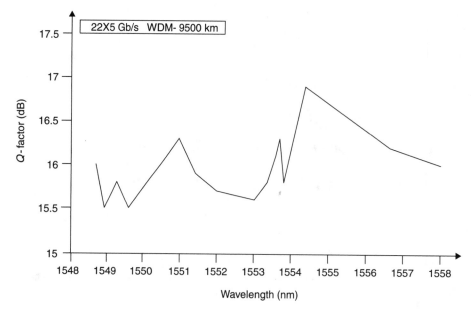

Figure 12–90. *Q*-factor *v.* wavelength.

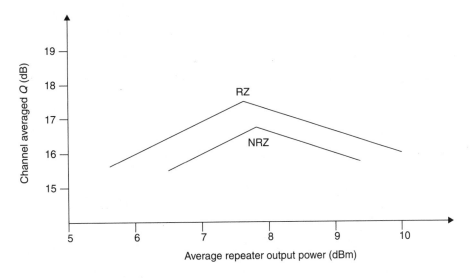

Figure 12–91. Channel average Q-factor v. average repeater optical power output.

Experimental System 3: 127 Gb/s Transoceanic Optical System

The operating characteristics of a 24-channel, 127 Gb/s WDM experimental transoceanic optical system are as follows:

a) transmission capacity: 127 Gb/s

b) multiplexing scheme: WDM

c) number of channels: 24

d) bit rate per channel: 5.3 Gb/s

e) transmission distance: 8000 km (approximately)

f) channel spacing: 0.5 nm

g) dispersion accumulation: −3870 ps/nm channel 1 to +3870 ps/nm channel 24 (no pre-DCF)

h) fiber dispersion slope: 0.086 ps/nm^2/km

i) dispersion accumulation: +1000 ps/nm channel 1 to −1000 ps/nm channel 24 (with pre-DCF)

The inclusion of a predispersion compensation fiber (pre-DCF) was instrumental in reducing the chromatic dispersion difference between the first and the last channels to the 5740 ps/nm level and in reducing waveform distortion due to SPM and chromatic dispersion variations among channels. The pre-DCF method also maintains a constant SNR in contrast to the post-DCF method, which experiences a reduction of SNR due to the power loss induced by the post-DCF.

Data Format. The repeater span in transoceanic optical systems can be enhanced through the increase of repeater optical output. However, an increase of the repeater power output, coupled with limited laser diode pump power, increases fiber nonlinearities such as FWM and XPM. These nonlinearities are capable of inducing cochannel interference, thus reducing system performance. These potential problems can be avoided by utilizing the RZ data format. The Q-factor that expresses channel performance for both RZ and NRZ data formats, in relation to repeater optical power output, is illustrated in Figure 12–91.

From Figure 12–91, it is evident that the Q-factor is higher by about 1 dB with RZ data format than with NRZ. Furthermore, the opening of the eye diagram reveals the advantage of using RZ instead of NRZ data formats, as well as the effects of XPM on jitter and FWM on noise. RZ formats exhibit smaller nonlinear channel interaction and therefore smaller ISI. Data obtained from these experimental systems also indicates a reduction in the laser diode pump power requirement with RZ formats.

Experimental System 4: 170 Gb/s Transoceanic Optical System

This experimental transoceanic optical system incorporates a 32-channel WDM scheme (5.3 Gb/s per channel) capable of transmitting 169.6 Gb/s over a distance of almost 10,000 km. The viability of the proposed optical system was based on the development of new technologies such as wideband EDFAs, sophisticated gain equalization schemes, and chromatic dispersion management techniques applied at the transmitter end. For this experimental link, RZ data format was employed. Inherently, such dense wavelength optical systems exhibit susceptibility to GVD slope and also to nonlinearities such as XPM, SPM, and FWM. Therefore, overcoming the difficulties will be essential for the successful field implementation of the experimental optical system.

Optical Gain Bandwidth Enhancement. Transoceanic optical systems require broadband EDFAs. Broadband expansion can be obtained through either gain equalization or high aluminum concentration.

EDFA gain equalization: The gain bandwidth profile of an EDFA is illustrated in Figure 12–92. From Figure 12–92,

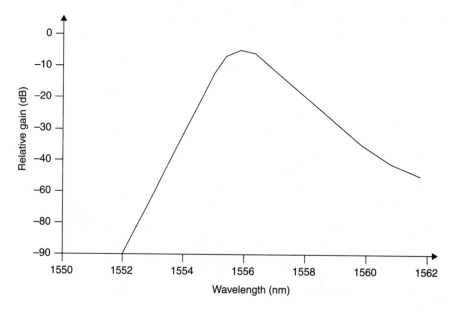

Figure 12–92. EDFA gain bandwidth profile.

it is evident that the gain bandwidth profile for such an optical amplifier is asymmetric. Experimental results have shown that a symmetric (flat) gain bandwidth profile for EDFAs incorporated into the repeater modules in transoceanic optical systems is essential for achieving maximum optical lengths.

Therefore, in order to improve the amplifier gain bandwidth profile symmetry, the following scheme was proposed. The proposed scheme incorporates two optical filters with distinctly different free spectral ranges (FSRs) as gain equalizers. The first filter, that has an FSR of 25 nm, reflects a 6 dB attenuation at the 1559.8 nm wavelength. The second filter, that has an FSR of 2 nm, reflects a 2 dB attenuation at the 1562.3 nm wavelength. Based on this scheme, the first equalizer broadens the amplifier bandwidth while the second compensates for the loss due to the first equalizer. The combined effect of both filters is a significant improvement of the amplifier gain bandwidth profile (Figure 12–93).

EDFA Bandwidth Enhancement: Long haul WDM transoceanic systems require high power and large bandwidth EDFAs. Broad bandwidth in such amplifiers can be achieved through the optimization of the amplifier operating point and the appropriate selection of aluminum dopand levels (high concentration of aluminum). The operating wavelength of the laser pump is at the 1480 nm window.

Dispersion Compensation: The techniques for establishing dispersion levels and compensation schemes to be applied here are the same as in the previous experimental systems. The operating characteristics of the experimental transoceanic optical system are as follows:

a) system bit rate: 169.6 Gb/s

b) multiplexing: WDM

c) number of channels: 32

d) bit rate per channel: 5.3 Gb/s

e) operating wavelength: 1545 nm to 1560.5 nm

f) transmission distance: 10,000 km (approximately)

g) repeater span: 50 km

h) fiber: DSF (zero dispersion at 1579 nm)

i) fiber loss: 0.21 dB/km

j) dispersion slope: 0.089 ps/nm^2/km

k) repeater optical output power: 10.5 dBm

l) receiver noise figure (NF): 4.7 dB

12.18   Tb/s EXPERIMENTAL OPTICAL SYSTEMS

The demand for high bit rate optical systems has increased almost exponentially since the middle 1980s. The implementation of high speed optical systems was mainly driven by a competition among various service providers and advancements in both optical and electronic device technology such as EDFAs, optical add/drop multiplexers (OADM), optical cross connect (OXC), high speed FETs, and high energy mobility transistors (HEMT). Economic considerations dictate that high capacity optical systems must be created by upgrading existing optical links and by creating new systems that allow for upgrades. In the proposed optical links, the multiplexing formats can be selected, based on the following system specifications:

a) **single channel electronic time division multiplexing (ETDM)**

b) **single channel optical time division multiplexing (OTDM)**

c) multichannel wavelength division multiplexing (WDM)

d) WDM-OTDM

e) WDM and polarization multiplexing

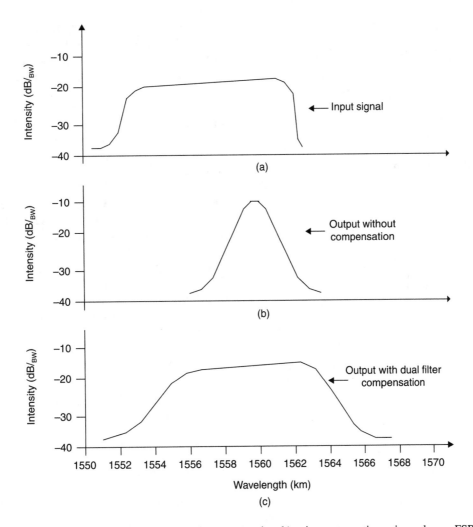

Figure 12–93. EDFA gain bandwidth profile. a) without gain compensation, b) gain compensation using only one FSR Filter, c) gain compensation using two FSR filters (full gain compensation).

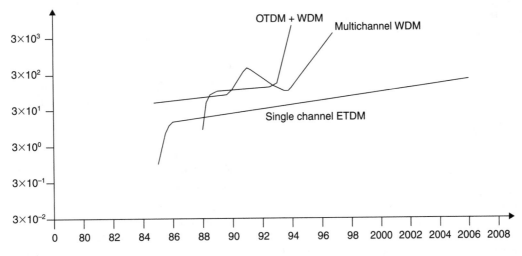

Figure 12–94. Capacity demand projections.

Capacity demand projections for all of the previously mentioned schemes, and for the time period between 1980 and 2010, are illustrated in Figure 12–94. Basic obstacles to high speed optical transmission are the incompatibility between chromatic dispersion and fiber nonlinearities. Chromatic dispersion is also a limiting factor for transmission length. The relationship between transmission length, system bit rate, and fiber dispersion is expressed by Equation (12–156).

$$f_b^2 DL \leq 10^5 \qquad \text{(12–156)}$$

where f_b is the system bit rate (Gb/s), D is the fiber dispersion (ps/nm-km), and L is the fiber length (km).

Example 12–32

Compute the transmission length of a nondispersion compensated single-mode fiber (non-DCF) operating in an optical system that has a 10 Gb/s capacity and 1550 nm operating wavelength (fiber dispersion: 17ps/nm-km).

Solution
From Equation (12–156), solve for L.

$$L_{km} = \frac{1 \times 10^5}{Df_b^2}$$

$$= \frac{1 \times 10^5}{(17)(10)^2}$$

$$= 58.8 \text{ km}$$

Therefore, $L = 58.8$ km.

Repeat the above example for a 2.5 Gb/s transmission capacity.

$$L_{km} = \frac{1 \times 10^5}{Df_b^2}$$

$$= 940 \text{ km}$$

Therefore, $L = 940$ km.

From Example 12–32, it is evident that with a decrease in the transmission rate from 10 Gb/s to 2.5 Gb/s, the link distance dramatically increases.

Example 12–33

Compute the maximum bit rate (f_b) that can be transmitted through an 80 km non-DCF operating at the 1550 nm wavelength window (fiber dispersion: 17 ps/nm-km).

Solution
From Equation (12–156), solve for f_b.

$$f_b = \sqrt{\frac{1 \times 10^5}{(80)(17)}}$$

$$= 8.6 \text{ Mb/s}$$

Therefore, $f_b = 8.6$ Mb/s.

The problem generated from the use of NDSFs was partially alleviated through the introduction of DSFs. The index profile of a DSF is designed so that the end chromatic dispersion is equal to zero at the 1550 nm wavelength window. However, zero chromatic dispersion is a perfect condition for the generation of four photonic mixing (FPM). Until the middle 1990s, system designers had two choices. They had to design either high speed single channel systems that utilize low dispersion fibers, or low capacity WDM systems that utilize high dispersion fibers. In the late 1990s, new techniques for dealing with these problems were developed. Although high speed transmission is limited by the accumulated chromatic dispersion at the end of the fiber, the effect of nonlinearities can be suppressed with the utilization of local dispersion along the optical fiber. Therefore, if optical fibers are selected that have a local dispersion that is slightly higher than a set minimum and a link dispersion that is maintained at very low levels, it is possible to transmit high bit rates within a WDM transmission scheme. The process through which chromatic dispersion is controlled to a minimum level is referred to as dispersion management. In 1993, dispersion management was experimentally demonstrated. For this experiment, an 8-channel, 10 Gb/s per channel WDM scheme was implemented. For dispersion management, the experiment utilized both NDF with –2 ps/nm-km (long length), and NDS-SM fibers (17 ps/nm-km). This, and numerous other dispersion management techniques, was successfully implemented and enhanced the system capacity to terra-bit levels (Tb/s). Prominent research groups all over the globe (Lucent Technologies, NEC, Fujitsu, and NTT) have been engaged in experimental work related to optical systems that transmit in the terra bit range.

Experiment 1: The 1 Tb/s AT&T Bell Labs Experiment

AT&T Bell Labs conducted the following high bit rate experiment, which utilized WDM and a polarization multiplexing scheme. The experimental setup is illustrated in Figure 12–95.

The main components incorporated into the above experiment are as follows:

a) laser diodes: 25 (24 EC lasers and 1 DFB)

b) star coupler: 8×8

c) waveguide grating routers (WGR): 2

d) operating wavelengths: 1542 nm to 1561.2 nm

e) polarization beam splitters (PBS): 3

f) polarization controllers (PC): 30

g) LiNbO$_3$ MZ modulators: 2 (zero chirp)

OPERATION. The laser diode outputs are fed individually into the PC, which provide independent polarization control to each laser. The outputs of the first eight polarization controllers are multiplexed by the 8×8 star coupler, while the rest of the PC outputs are multiplexed by two WGRs. The combined outputs of the star coupler and the WGRs are also multiplexed by a 3:1 optical multiplexer, the output of which is fed, after amplification, into a PBS. The function of the PBS is to provide polarization alignment. After amplification, the optical signal is fed into a 2×2 coupler, then each optical beam is fed into separate beam expanders, which independently block the two polarizations. The outputs of the polarization controllers are then fed into the optical port of the two LiNbO$_3$ MZ modulators while, at the electrical port, 20 Gb/s NRZ signals are

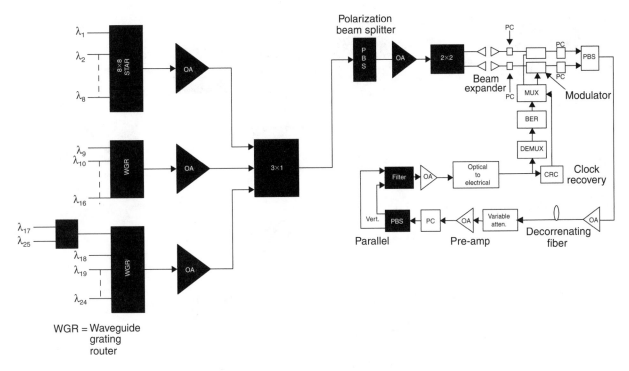

Figure 12–95. 1 Tb/s WDM polarization multiplexing experiment.

applied. The modulator outputs are fed again into two polarization controllers. The independently polarized 50 channels (2 × 25), each modulated by the 20 Gb/s bit rate, are capable of generating a total system bit rate of 1 Tb/s. The composite optical signal is then transmitted for 55 km. At the receiver end, a variable attenuator is used to maintain the constant optical power output required at the input of the preamplifier. Another polarization controller provides further polarization control, while a polarization beam splitter is used as a 1:2 demultiplexer. The 2 × 25 wavelengths are demultiplexed through grading optical filters and then converted to electrical signals by the corresponding PIN photodiodes. The electrical signals generated by the photodetector diodes are used for clock and data recovery (20 Gb/s per channel). The 20 Gb/s signal is further demultiplexed to 2 × 10 Gb/s data rates. The experiment indicated the following:

a) a relatively low PDL, with no effect on system performance

b) the signal copolarization was well suppressed

c) overall system performance was maintained at anticipated levels

Experiment 2: 1 Tb/s OTDM-WDM NTT Experiment

A 10-channel, 100 Gb/s per channel OTDM-WDM optical system was tested by Nippon Telephone and Telegraph (NTT) labs. The experimental setup is illustrated in Figure 12–96.

The main building blocks of the experiment are as follows:

a) optical source: supercontinuum (SC) broadband, composed of a mode lock erbium doped fiber ring laser (ML-EDFRL) that operates at the 1527 nm (3.5 ps) wavelength window, an EDFA 1.5 W_{Pk}, and a Supercontinuum fiber (3 km).

b) EDFA: To amplify the 10-channel, 100 Gb/s per channel signals with 5 dBm optical power output

c) 40 km DSF operating at the 1561.2 nm wavelength window

d) OBPF

e) FWM demultiplexer generating a 10 Gb/s channel bit rate

f) prescaled PLL clock recovery

g) variable attenuator

h) optical receiver composed of an EDFA and a PIN photodetector diode.

OPERATION. The supercontinuum optical source composed of an ML-EDFRL, an EDFA producing 1.5 W of optical power at the output, and a 3 km supercontinuum fiber operates at the 1527 nm wavelength window. The output of the supercontinuum source is fed into an AWG, which is used as an optical filter to separate the ten optical channels (400 GHz of spacing). The output of the AWG filter is fed into the optical port of the LiNbO$_3$ MZ modulator while, at the electrical port, a $2^{15} - 1$ pulse pattern generator is connected, providing the required 10 Gb/s input signal. The

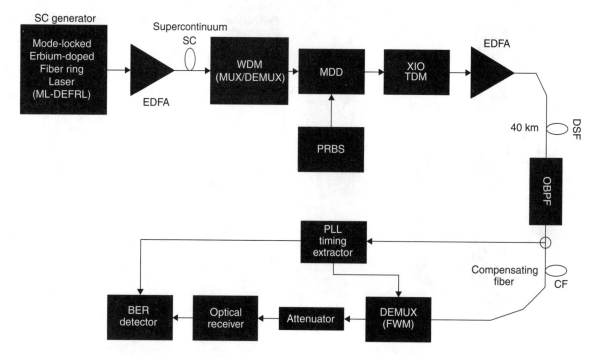

Figure 12–96. 1 Tb/s OTDM-WDM optical system (NTT experiment).

optically modulated signal is fed into the input of a planar optical circuit that is designed to perform the function of ×10 OTDM. This generates a 100 Gb/s 10-channel output. The following EDFA is used to amplify the composite signal to the required level before it is launched into the 40 km DSF, operating at a wavelength of 1561.3 nm. In the experiment, dispersion management is achieved through the use of various lengths of non-DSFs. At the end of the 40 km fiber, an OBPF is incorporated to maintain spectral confinement. The optically modulated 100 Gb/s signal is fed into the inputs of both the FWM stage used as an optical demultiplexer, and into the PLL clock recovery (CR) circuit. The demultiplexed 10 Gb/s signal is applied at the input of the optical receiver, which is composed of an EDFA, a PIN photodetector diode, and a variable attenuator. The function of the attenuator is to maintain a fixed optical power level at the input of the EDFA. The electrical signal generated by the optical amplifier is further amplified to the level required by the data recovery circuit. The results of the experiment, in terms of measured optical SNR and BER were satisfactory. However, a variation of the baseline sensitivity, attributed to FWM demultiplexer induced losses, was noticed.

Experiment 3: 1.1 Tb/s WDM Experimental System (Fujitsu Labs)

A 55-channel, 20 Gb/s per channel WDM experimental system was tested by Fujitsu labs. The main components of the experimental optical link are as follows:

a) Laser: external cavity (EC) lasers, 46 DFB lasers

b) operating wavelengths: 1531.7 nm to 1564 nm

c) channel spacing: 0.6 nm

d) polarization control (PC): once for each laser

e) LiNbO$_3$ MZ modulator: one (with chirp parameter $\alpha = -1$)

f) $2^{23} - 1$ PRBS (20 Gb/s): modulation NRZ signal

g) postamplifier: EDFA

h) fibers: three 50 km SMF (unshifted fibers)

i) repeaters: 2-composed of an EDFA, a DCF with −103 ps/nm-km dispersion, and dispersion slope −0.18 ps/nm^2-km

j) optical receiver: composed of the following: a preamplifier (DCF + EDFA), a PIN photodetector, a BPF, an amplifier (electronic), a multiplexer, and a clock and data recovery

OPERATION. The laser output is fed into the polarization control (PC) stage and the combined signal from all the polarization control outputs is fed into the optical port of the LiNbO$_3$ MZ modulator. At the electrical port of the modulator, a 2^{23} −1 long PRSB word 20 Gb/s NRZ signal is applied. The modulated optical signal from the MZ modulator is further amplified by an EDFA to the level required for launching into the fiber. A 50 km unshifted fiber (USF) with 15.2 ps/nm-km dispersion is used, followed by the repeater stage composed of an EDFA and a DCF with a negative dispersion of −103 ps/nm-km. The negative DCF is used to compensate for the accumulated dispersion caused by the 50 km fiber span.

At the end of the third 50 km fiber span, the output signal is detected by the PIN photodetector diode. The generated electrical signal is then amplified, demultiplexed, and a clock and data recovery circuit is used to extract both the clock and the data signals for further processing. In order to maintain the required SNR at the input of the receiver, a 10 dB preemphasis was necessitated. The experiment also indicated a 3 dB receiver sensitivity variation for all 55 channels.

12.19 ULTRA HIGH SPEED SINGLE CHANNEL OTDM EXPERIMENTAL SYSTEMS

Economic considerations dictate the full utilization of already existing TDM transport optical systems. Upgrading such a system to transmission rates of 40 Gb/s and beyond cannot be achieved with optoelectronic devices already in existence. A new generation of devices that employ novel design techniques and fabrication processes is required. Single channel optical TDM 40 Gb/s systems can be implemented through two methods.

a) optical TDM (4×10 Gb/s RZ signals)

b) electrical TDM (4×10 Gb/s NRZ signals)

For the implementation of such an optical system, the required building blocks are as follows:

a) optical sources: For 10 GHz or 40 GHz RZ signals

b) multiplexers: Optical TDM/electrical TDM (4×10 Gb/s)

c) modulators: LiNbO$_3$ MZ

d) dispersion management: DCFs

e) demultiplexers: Optical TDM/electrical TDM (40 Gb/s to 10 Gb/s)

f) clock recovery (CR): 40 GHz /10 GHz

g) data recovery: 4×10 Gb/s signals

Requirements for Optical Sources That Operate in 4×10 Gb/s OTDM Systems

One of the most important characteristics of optical sources that operate in 40 Gb/s single channel OTDM systems is the extinction ratio. The generated pulses of the optical source must have a very high extinction ratio in order to avoid channel overlapping, and to minimize interference between adjacent channels during the multiplexing process. The extinction ratio required in such systems is set at –30 dB. The second performance characteristic of the above optical sources is temporal stability.

GAIN-SWITCHED LASER SOURCES. Experimental results have shown that InGaAsP MQW laser structures can successfully be implemented in 4×10 Gb/s OTDM systems. DFB lasers fabricated for these applications performed impressively in terms of differential gain, bandwidth enhancement factor, modal confinement, and damp-

ing factor. The critical selection of the laser parameters was necessitated by the low jitter, narrow bandwidth, and high extinction ratio requirements.

Experimental results indicated a 1 dB reduction of receiver sensitivity when operating at 40 Gb/s rather than at 10 Gb/s; 1 dB of power penalty was attributed to the use of the 4×1 multiplexer. The overall performance of the switched DFB laser employed in a 40 Gb/s optical system was satisfactory. The highly desirable narrow width and high extinction ratio pulse generation can be accomplished through the implementation of two integrated electroabsorption modulators in tandem. Such an integrated device exhibited an extinction ratio of –55 dB, while the signal to interference noise ratio is maintained at a constant low level for delay time pulses longer than 30% of the pulse period. Because EAMs used in tandem induce very small levels of multiplexing penalty, they can be considered 4×10 Gb/s multiplexers. A summary of the component requirements for 10 Gb/s and 40 Gb/s optical systems is as follows:

a) 10 GHz signal source for 4×10 Gb/s OTDM: gain switching laser, modulators in tandem

b) 40 GHz sources for 40 Gb/s ETDM: 40 GHz integrated laser modulator (ILM), 10 GHz/40 GHz monolithic mode locked laser, injection tracked laser

c) Modulators: 40 Gb/s ETDM, 40 Gb/s electroabsorption

d) Multiplexers: 40 Gb/s ETDM, InP HBT multiplexer, GaAsP HEMT

e) Multiplexers: 40 Gb/s/80 Gb/s OTDM, SOA MZ interferometers

f) 4×10 Gb/s add/drop multiplexers

40 GHz LASER SOURCES FOR 40 Gb/s OPTICAL SYSTEMS. A 40 Gb/s ETDM optical system can either use RZ or NRZ data formats. The utilization of the RZ format allows optical demultiplexing from 40 Gb/s to 10 Gb/s, followed by low bit rate electronic detection. However, both RZ and NRZ signal formats require encoding at the driver stage. When the NRZ transmission format is implemented, a CW source prior to the modulation stage is required. When the RZ transmission format is implemented, the modulator requires a 40 GHz pulse source. For this application, extinction ratio constraints are somewhat relaxed. That is, a 15 dB extinction ratio will be quite satisfactory. The 20 dB difference between the extinction ratio required for EDTM (–15 dB) and for OTDM (–35 dB) is attributed to additional optical noise induced by the optical multiplexer.

INTEGRATED LASER MODULATOR. Modulator operating characteristics applicable to ETDM optical systems that operate at 40 Gb/s transmission rates are:

a) wide bandwidth

b) sufficient optical power

c) specified extinction ratio

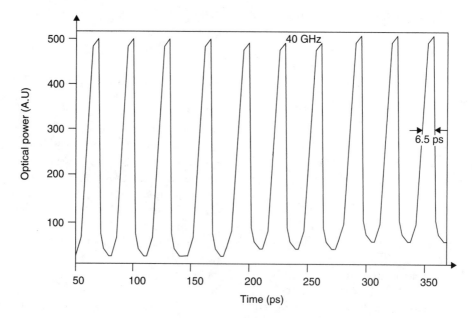

Figure 12–97. Spectrum of a 40 GHz optical pulse.

A group of researchers working on this project at the University of Bristol were able to generate a 40 GHz pulse with 3 dBm of optical power at 60 mA laser current and extinction ratio of 11.5 dB. The optical spectrum of the 40 GHz pulse is illustrated in Figure 12–97.

10 Gb/s/40 Gb/s MONOLITHIC MODE LOCKED LASER.
Special structure mode locked (ML) lasers have been fabricated to generate required 40 GHz pulses. Mode locked structures have been fabricated with 10 dBm of modulation power (electrical). Figure 12–98 illustrates the pulse profile and Figure 12–99 illustrates the optical pulse spectrum of an ML laser diode.

Figure 12–98 shows a pulse width of 6.5 ps while the FWHM of Figure 12–99 is about 0.65 nm.

40 Gb/s INJECTION TRACKED LASER OSCILLATOR.
Injection tracked lasers can obtain high power output oscillations at 40 GHz. An injection tracked laser oscillator is illustrated in Figure 12–100.

The scheme incorporates two DFB laser diodes and an optical isolator. Laser 1 is modulated with a weak clock signal, thus generating weak sidebands of about –40 dB at the output. The output signal from laser 1 is applied to laser 2 through the optical isolator. In this way, laser 2 is locked into one of the sidebands that generate two modes that

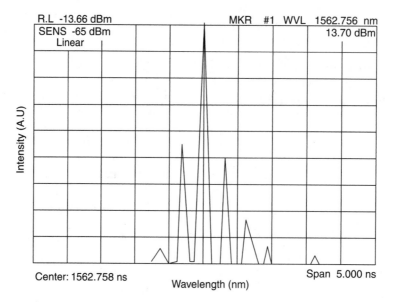

Figure 12–98. Monolithic mode locked laser diode profile.

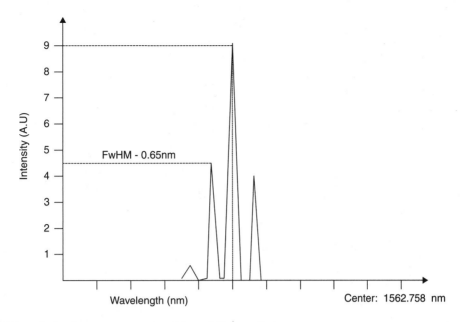

Figure 12–99. 40 GHz optical pulse spectrum generated by an ML laser diode.

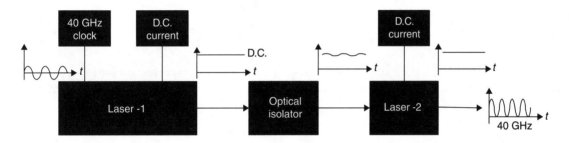

Figure 12–100. Injection tracked laser oscillator.

have amplitudes that reflect a maximum extinction ratio at its output. The electrical spectrum of the generated signal is illustrated in Figure 12–101.

The pulse signal in Figure 12–101 exhibits a very narrow line width at 40 GHz.

40 Gb/s ETDM Modulators. 40 Gb/s ETDM systems require wide bandwidth, high speed integrated circuits. A 10 Gb/s data stream is multiplexed by a 4:1 multiplexer, then amplified by a high gain amplifier, which uses the output to drive the modulator stage. To achieve the required high gain of the amplifier circuit, high energy mobility transistor (HEMT) technology is used. In order to achieve maximum performance, impedance matching between the output of the amplifier driver circuit and the electroabsorption modulator (EAM) is essential. EAMs employed in 4 × 10 Gb/s optical systems must exhibit bandwidths larger than 35 GHz. The fabrication of the modulator is based on MQW technology. The achieved EAM bandwidth with this structure was larger than 35 GHz (Figure 12–102).

The bandwidth measured at the –6 dB point is 37 GHz, with an almost constant response up to 36 GHz. This is compatible with electronic driver and multiplexer circuits.

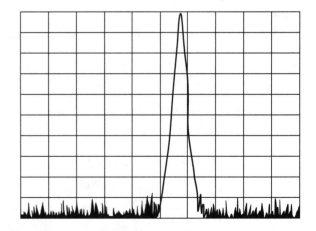

Figure 12–101. 40 GHz pulse spectrum (electrical).

THE 40 Gb/s Driver Circuit. The driver stage must be designed as a high gain, large output amplifier. The fabrication of such a circuit is based on GaAs HEMT 0.1 μm gate technology and double distributed architecture. Experimental driver circuit designs have exhibited a 19 dB voltage gain at the 40 GHz bandwidth and a 4.2 V peak-

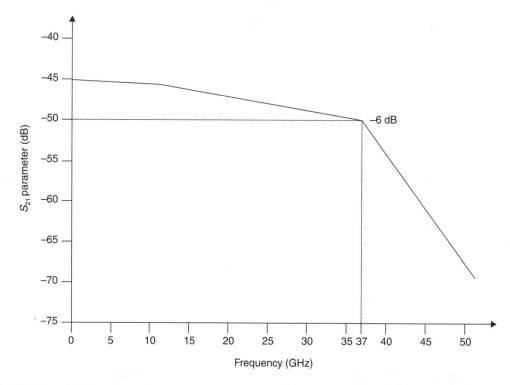

Figure 12–102. EAM bandwidth.

to-peak output voltage developed across a 50 Ω load. For satisfactory impedance matching of the driver output and the electroabsorption modulator input, a matching circuit is essential.

THE 40 Gb/s 2:1 MULTIPLEXER. The design of a 40 Gb/s multiplexer circuit is based on InGaAs/InP technology. The experimental device, developed by the same group of researchers, incorporates switch buffers (at the input current) required to perform single ended to differential transformation within the microchip. It also incorporates a selector gate and an output buffer. The multiplexer exhibits a current gain of 40 kA/cm² and an operating bandwidth of 56 GHz. In order to minimize the impact of multiplexer reflection to distortion and jitter, the clock and data inputs are terminated at 50 Ω loads,

while a 100 Ω output resistor is used as a device to offset the jitter produced by the internal current switch. The performance characteristics measured by the multiplexer are as follows:

a) operating bit rate: 40 Gb/s

b) output voltage: 500 mV$_{p-p}$

c) switching current: 15 mA

d) power consumption: 1.2 W

The overall performance of the experimental multiplexer, which operates at 40 Gb/s transmission rates expressed by the eye diagram, is illustrated in Figure 12–103.

40 Gb/s CLOCK RECOVERY CIRCUIT. Clock recovery circuits are used for multiplexer synchronization as well as for extracting clock signals. Clock signals of 10 GHz and

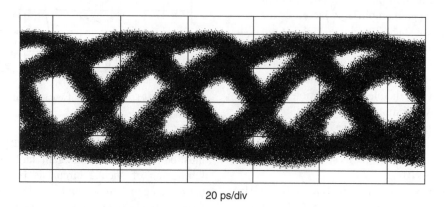

20 ps/div

Figure 12–103. 40 Gb/s 2:1 multiplexer eye diagram.

20 GHz must be extracted from the incoming 40 Gb/s TDM signal. The Bristol University research team developed two methods of optical clock recovery and one method of electrical clock recovery for both 10 GHz and 20 GHz signals. The first method uses a self pulsating laser (SPL) diode as the key device. It incorporates a DFB laser structure, utilizing its dispersive Q-switching mechanism to generate self pulsations. The device is composed of three sections: a DFB section (130 µm long and acting as the dispersive reflector), a 300 µm long DFB section intended to provide the required laser gain, and another 300 µm long DFB section integrated between the first and second and performing a phase tuning function (Figure 12–104). Self pulsating is initiated by the DC current at both the reflector and the phase sections of the device while the required pulsating frequency can be adjusted by the DC current at the laser section. The device bandwidth range was measured at between 4 GHz and 21 GHz. It is evident that such devices can be used for clock recovery if an optical signal capable of adjusting the self pulsating frequency to the transmitting data rate, regardless of the polarization of the data stream, is applied at the input.

The values for the required DC current levels for the reflector, phase, and laser are 20 mA, 20 mA, and 200 mA, respectively. Performance evaluation of the optical clock recovery circuit was performed by applying a $2^{31}-1$ PRBS signal at the input port. The operating wavelength window was set at 1550 nm and the optical power at 3 mW. The 10 GHz self pulsating oscillator was adjusted by appropriately setting the DC current at the reflector and phase sections. The 20 GHz clock recovery is illustrated in Figure 12–105.

The 10 GHz or 20 GHz clock recovery modes of operation can be obtained by an appropriate adjustment of the DC current levels. Finally, measurements have shown an impressive jitter performance. For example, the 10 GHz mode of operation exhibited 2 ps of jitter, while the 20 GHz mode exhibited 2.5 ps of jitter. Both figures are satisfactory.

Another design approach for optical clock recovery is through an SOA phase detector. The operating principle of optical clock recovery through SOA phase detection is based on the PLL operating principle. The three building blocks of a PLL circuit are the VCO, the PD, and the LPF. Phase/frequency synchronization of the VCO is obtained by detecting the phase difference of the input/output signal by the phase detector. A Michelson interferometer (MI) that exploits differential XPM provides the required phase detection. The experiment is illustrated in Figure 12–106.

Operation: The 40 Gb/s optical signal is applied at the input of the MI, while the control signal to the interferometer is provided by the DFB laser. The gain switching action of the DFB laser is triggered by the output signal of the sideband modulator. The 10 GHz signal generated by the VCO is applied at the input of the single sideband modulator, while a 2 MHz offset generator provides the modulating signal.

Two frequency doublers double the 2 MHz signal, and the generated 8 MHz signal is used to demodulate the control signal from the input signal. The control signal at the output of the WMC is applied to the control input of the VCO through the LPF. Therefore it is possible that an SOA based

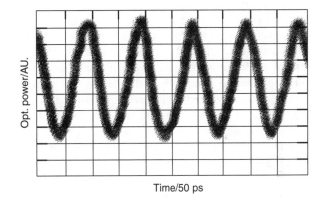

Time/50 ps

Figure 12–105.   10 GHz clock recovery.

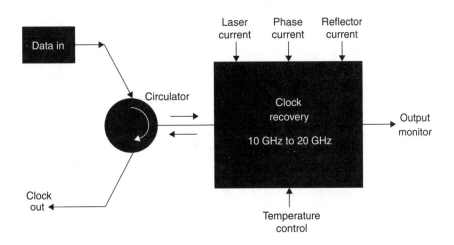

Figure 12–104.   Clock recovery based on an SPL.

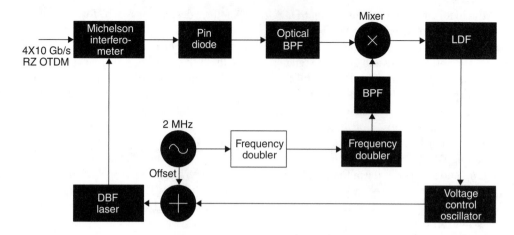

Figure 12–106. Optical clock recovery by the SOA phase detection method.

(MI) technology, and an offset modulator can be used as the phase detector components of a PLL circuit employed in an optical clock recovery scheme.

40 Gb/s OPTICAL DEMULTIPLEXING. Optical demultiplexing is required to extrapolate 10 Gb/s from the 40 Gb/s data stream. High bit rate demultiplexing has been performed through the utilization of either an MZ optical modulator or an add/drop multiplexer. A block diagram of an MZ demultiplexing scheme is illustrated in Figure 12–107.

Operation: The 40 Gb/s OTDM signal is applied at port 1 of the demultiplexer circuit. Differential delayed short control pulses are injected into the demultiplexer, which generates an adjustable switching window. The phase difference generated by the different arm lengths within the demultiplexer allows the 10 Gb/s channel to appear at port 4, while the remaining 30 Gb/s bit stream appears at port 3. Measurements conducted for both 40 Gb/s to 10 Gb/s and 80 Gb/s to 10 Gb/s demultiplexing processes, in terms of their corresponding BER levels and various input optical power levels, are illustrated in Figure 12–108.

From Figure 12–108, it is evident that the 40 Gb/s/ 10 Gb/s demultiplexer requires, on average, 4 dB less power than the 80 Gb/s/10 Gb/s demultiplexer in order to maintain the same BER.

DISPERSION MANAGEMENT. The accumulated chromatic dispersion by the entire transmission length can be compensated for through the implementation of mid-span spectral inversion. The experimental setup for this dispersion management technique is illustrated in Figure 12–109.

The experimental setup consists of the following:

a) EDFA: 3

b) OBPF: 4

c) SOA: 1

SYSTEM REQUIREMENTS. Phase conjugation for the 40 Gb/s system applied at the SOA stage is based on FWM. For phase conjugation, the required OSNR must be larger than 1.5 dB/nm of bandwidth, while the optical power output must be larger than –18 dBm. These conditions are necessary in order to limit the add-on ASE and also to lower distortion levels induced by the data pattern modulation gain of the SOA. The required high OSNR (larger than 1.5 dB/nm) and low SOA distortion can be achieved by a pump optical power higher than 10 dB. Such an optical power is necessary in order to drive the SOA into saturation. Furthermore, the required high conjugate optical power (FWM increased efficiency) is accomplished through an increase of the device length. Measurements obtained for an FWM of 1 mm and 2 mm in length are illustrated in Figure 12–110.

From Figure 12–110, it is evident that a 2 mm length provides a better conjugate output power. However, an improvement in FWM efficiency can also be obtained by optimizing the SOA active region. Active region optimization results in a reduction of "density of state" and a reduc-

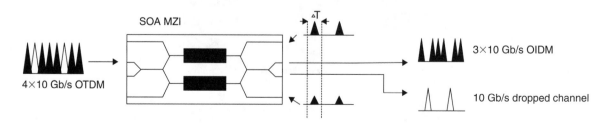

Figure 12–107. MZ interferometer demultiplexer.

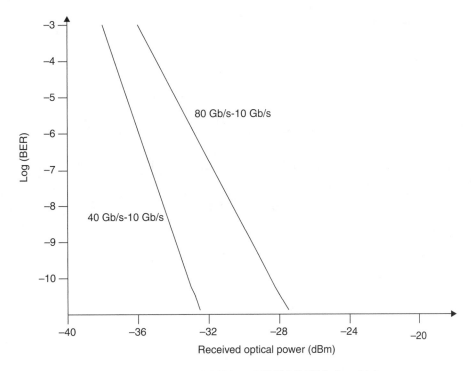

Figure 12–108. BER *v.* input optical power for both 40Gb/s/10Gb/s and 80Gb/s/10Gb/s demultiplexers.

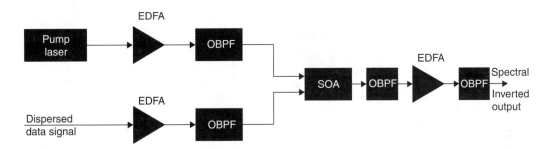

Figure 12–109. Dispersion management for a 40Gb/s optical system.

tion of the amplifier NF. An additional element that contributes to FWM efficiency increase is the increase of the line width enhancement factor to a point at which the operating wavelength is very close to the SOA band gap wavelength.

40Gb/s/10Gb/s Add/Drop Multiplexer. The method used for 40 Gb/s10 Gb/s demultiplexing can also be used for the add/drop multiplexing process. The experimental setup of an optical add/drop multiplexer is illustrated in Figure 12–111.

The 4 × 10 Gb/s optical TDM signal is applied at port 1 of the add/drop multiplexer, while the 10 Gb/s optical signal is applied at port 2. The differentially delayed short control pulses are applied at the control ports. Through this method, the resultant OTDM signal (4×10 Gb/s) is obtained from port 3, while the dropped channel (10 Gb/s) appears at port 4. The performance characteristics of the OAD multiplexer are illustrated in Figure 12–112.

12.20 SOLITON TRANSMISSION

Introduction

In September of 1844, John Scott Russell (a Scottish engineer) filed his *Report on Waves* with the British association for the advancement of science. An extract of this report reads as follows:

"I was observing the motion of a boat which was rapidly drawn along a narrow channel by a pair of horses, when the boat suddenly stopped—not so the mass of the water in the channel which it had put in motion; it accumulated round the prow of the vessel in a state of violent agitation, then suddenly leaving it behind, rolled forward with great velocity, assuming the form of a large **solitary** elevation, a rounded, smooth and well defined hip of water, which continued its course along the channel apparently without change of form or diminution of speed."

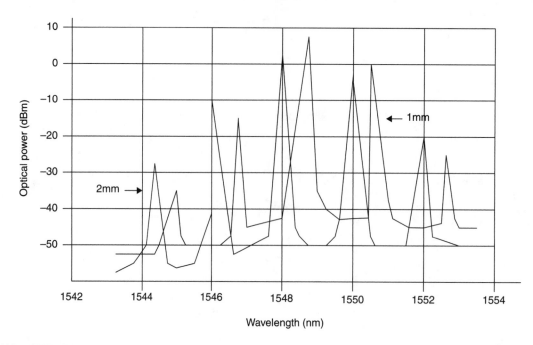

Figure 12–110. FWM for 1 mm and 2 mm semiconductor optical amplifiers.

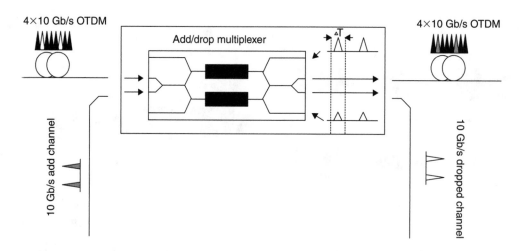

Figure 12–111. 40 Gb/s add/drop multiplexer.

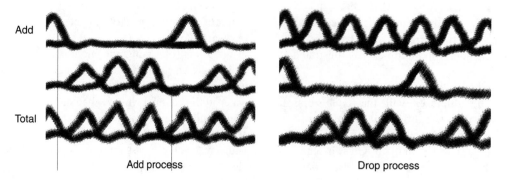

Figure 12–112. An eye diagram of a 40 Gb/s/10 Gb/s OAD multiplexer.

He continued:

"Such in the month of August 1834 was my first chance interview with the singular and beautiful phenomenon which I have called **wave of translation**."

Russell continued his experimentation on wave translation and formulated an experiment that related the wave speed in relation to its amplitude and channel depth. He also established that higher waves travel faster than lower waves. That is, wave speed is proportional to its height. However, several leading scientists (Russell's contemporaries) disputed the notion that waves of permanency could exist. In the mid-1870s, Reyleigh and Boussinesq verified Russell's theory of wave permanency. In 1895, Korteweg and deVries developed an equation that expressed the wave profile, relating the wave amplitude changes in space and time. This equation is referred to as the KdV equation, in honor of the inventors. The KdV equation, which expresses wave permanency, was largely ignored for over sixty years. In 1955, Fermi, Pasta, and Ulam, while engaged in the study of finite heat conductivity in solids, were able to conclude that heat can flow through solids with the temperature at the output equal to that of the input, thus implying an infinite thermal conductivity of the solid. This is the equivalent of wave permanency. However, in subsequent experimentation, Fermi, Pasta and Ulam were faced with the following paradox. They were trying to prove that the initial thermal energy could be distributed throughout the system's modes. Instead, they found that the thermal energy flowed back and forth through all the modes and could not be regenerated to its initial state. In 1967, Kruskal and Zabusky of Princeton University further investigated this unexpected outcome. Their results were the same as those of Korteweg and deVries. Kruskal's and Zabusky's computer simulation results indicated that, if the initial condition is a sine wave, it disintegrates into a train of pulses identical to Russell's wave translation, and is capable of being reconstructed at the output to form a waveform identical to that of the input. Furthermore, Kruskal and Zabusky observed that if two pulses of the pulse train collide, they emerge after the interaction without any distortion. Because their behavior is similar to particles rather than to waves, they were named solitons. If the soliton concept was to be fully explained, the Korteweg/deVries equation that related wave temporal changes to spectral changes must be solved. Through the KdV equation, one must be able to establish the shape of the wave at any instant of time and to project the wave profile at the next instant of time or at any future instant of time. Kruskal and Zabusky successfully solved the KdV equation by relating it to the Schroedinger nonlinear equation. The soliton concept has several applications, the most prominent of which is in optical fiber communications. If a more detailed study of soliton transmission is to be performed, a basic knowledge of the wave transmission properties through the fiber are essential. For a detailed description, see Chapter 10. If a monochromatic optical wave with angular velocity (ω) is launched into a fiber, the wavelength of the optical wave (λ) is a function of the fiber refractive index and the fiber waveguide properties and its phase $\phi(z,t)$ is expressed by Equation (12–157).

$$\phi(z, t) = \frac{2\pi}{\lambda_f} z - \omega t \qquad (12\text{–}157)$$

where λ_f is the optical wavelength in the fiber, z is the distance along the fiber, and ω is the angular velocity. The key to optical wave propagation in a fiber is the relationship between the frequency components, the wavelength, and the velocity of light in a vacuum. However, the wavelength of a plane wave in an isotropic transparent medium is $n\omega/c$, where n is the refractive index at frequency ω. For a single-mode fiber (cladding-core), the above relationship holds true, assuming that n has a value between the cladding and the core, depending on transverse mode shape. Equation (12–158) expresses the wave phase change in relation to time.

$$\frac{d\phi(z, t)}{dt} = \frac{n\omega(v)}{c} - \omega \qquad (12\text{–}158)$$

where v is the velocity $\left(\dfrac{dz}{dt}\right)$. For a constant phase, the required velocity is derived as follows in Equation (12–159).

$$v = \frac{\omega}{\dfrac{n\omega}{c}}$$
$$= \frac{\omega c}{\omega n}$$
$$= \frac{c}{n}$$

Therefore,

$$v = \frac{c}{n} \qquad (12\text{–}159)$$

Any point of a constant phase wave traveling along the fiber is referred to as phase velocity. Assuming that a second wave, with an angular velocity that is (ω_2) marginally different from the first wave (ω_1), is transmitted through the same fiber, the interaction of the two optical waves will generate a third modulated wave. The amplitude of this wave will exhibit a peak (add) at the point where both principal waves are in phase, and a valley (subtract) where the principal waves have their maximum phase difference. However, the rate of change of the phase difference between the two waves must be equal to zero for a nonmaterial dispersive case (Equation (12–160).

$$\frac{n\omega_1}{c}(v) - \omega_1 = \frac{n\omega_2}{c} - \omega_2 \qquad (12\text{–}160)$$

Equation (12–161) expresses the required velocity.

$$[v_g(\omega)]^{-1} = \frac{d\left(\dfrac{\omega}{c}\right)}{d\omega} \qquad \textbf{(12–161)}$$

where $v_g(\omega)$ is the group velocity. This can be applied to more than two frequency components. A group of frequencies will exhibit a peak if all the frequencies are in phase at the same point in time along the fiber. The modulated signal will propagate along the fiber with a constant group velocity. However, if the average group velocity changes with frequency (dispersion), it will result in pulse broadening (chirping) because the various frequency components tend to separate in time.

The Unit of Soliton

The soliton FWHM (Δt) is determined by Equation (12–162).

$$\Delta t = 2\lfloor \cosh^{-1}(\sqrt{2}) \rfloor \qquad \textbf{(12–162)}$$

$$= 1.7627\ldots$$

As mentioned in the introduction, the soliton pulse width can be expressed by (τ_s) as follows in Equation (12–163).

$$\tau_s = \frac{T_{FWHM}}{\Delta t}$$

$$= \frac{T_{FWHM}}{1.7627}$$

Therefore,

$$\tau_s = \frac{T_{FWHM}}{1.763} \qquad \textbf{(12–163)}$$

The unit distance (z_s) is determined by Equation (12–164).

$$z_s = \frac{1}{(1.763)^2}\left[\frac{2\pi c\, T_{FWHM}^2}{\lambda^2 D}\right] \qquad \textbf{(12–164)}$$

where z_s is the soliton characteristic length (km), D is the dispersion constant (ps/nm/km), c is the velocity of light in vacuum (3×10^5 km/s), and λ is the operating wavelength (nm),

or

$$z_s = \frac{1}{(1.763)^2}\frac{(2\pi)}{(1557)^2}\frac{T_{FWHM}^2}{D}$$

$$\approx 0.25\frac{T_{FWHM}^2}{D}\ \text{(km)}$$

Example 12–34

Compute the soliton characteristic length (z_s) for a dispersion of 0.2 ps/nm/km and an FWHM of 18 ps.

Solution

$$z_s = 0.25\frac{T_{FWHM}^2}{D}\ \text{(km)}$$

$$= 0.25\frac{(18)^2}{0.2}$$

$$= 400\ \text{km}$$

Therefore, $z_s = 400$ km.

Example 12–35

Determine the maximum dispersion (D) allowable for a soliton transmission with an FWHM of 20 ps, a soliton characteristic length 600 km, and an operating wavelength 1550 nm.

Solution

From Equation (12–164), solve for D.

$$D = \frac{1}{(1.763)^2\left[\dfrac{2\pi(3 \times 10^5) \times 20^2}{1550^2 \times 600}\right]}$$

$$= 0.17\ \text{ps/nm/km}$$

Therefore, $D = 0.17$ ps/nm/km.

Example 12–36

Calculate the soliton pulse width required in a soliton transmission with a soliton characteristic length of 550 km, a maximum dispersion coefficient of 0.25 ps/nm/km, and an operating wavelength 1557 nm.

Solution

From Equation (12–164), solve for (T_{FWHM}).

$$T_{FWHM} = \left[\frac{(1.763)^2\,(\lambda^2)(D)(z_s)}{2\pi c}\right]^{1/2}$$

$$= 23\ \text{ps}$$

Therefore, $T_{FWHM} = 23$ ps.

The power of soliton: The unit of soliton power is defined by the soliton peak power and is expressed by Equation (12–165).

$$P_s = \left(\frac{A_{eff}}{2\pi n_2}\right)\left(\frac{\lambda}{z_s}\right) \qquad \textbf{(12–165)}$$

In Equation (12–165), substituting for z_s (12–164) yields Equation (12–166).

$$P_s = \left(\frac{1.763}{2\pi}\right)^2\left[\frac{A_{eff}\,(\lambda)^3\,(D)}{n_2 c\,(T_{FWHM})^2}\right] \qquad \textbf{(12–166)}$$

where A_{eff} is the fiber core effective area (μm^2) and n_2 is the refractive index average over all polarization states. The value for silica glass fibers is 2.6×10^{-16} cm^2/W.

Example 12–37

Calculate the soliton peak pulse power (P_s) for the following parameters:

- $A_{eff} = 55 \ \mu m^2$
- $\lambda = 1557$ nm
- $n_2 = 2.6 \times 10^{-16} \ cm^2/W$
- $D = 0.20$ ps/nm/km
- $T_{FWHM} = 30$ ps

Solution

i) Calculate z_s.

$$z_s = \frac{1}{(1.763)^2} \left[\frac{2\pi c T_{FWHM}^2}{\lambda^2 D} \right]$$

$$= \frac{1}{(1.763)^2} \left[\frac{2\pi (3 \times 10^5)(30)^2}{(1557)^2 (0.2)} \right]$$

$$\approx 1120 \ km$$

Therefore, $z_s = 1120$ km.

ii) Calculate P_s.

$$P_s = \frac{A_{eff} \lambda}{2\pi n_2 z_s}$$

$$= \frac{(55 \times 10^{-12})(1557 \times 10^{-9})}{2\pi (2.6 \times 10^{-20})(z_s) \times 10^3}$$

$$= \frac{0.5244 \ W}{z_{s \ (km)}}$$

or

$$P_s = \frac{5244 \ mW}{z_{s(km)}}$$

Because $z_s = 1120$ km, then

$$P_s = \frac{5244 \ mW}{1120}$$

$$= 4.68 \ mW$$

Therefore, $P_s = 4.68$ mW.

In soliton transmission, the Kerr effect cancels out the accumulated chromatic dispersion of the fiber. In other words, the dispersion and nonlinear terms of the nonlinear Schroedinger equation are canceled out. The first order effects of the dispersive and nonlinear terms are phase shifted (complementary to each other). For the nonlinear term, the phase shift ($d\phi(t)$) is expressed by Equation (12–167).

$$d\phi(t) = |u(t)|^2 \ dz \qquad \textbf{(12–167)}$$

For the dispersive effect, the phase shift ($d\phi(t)$) is derived as follows:

$$\frac{\partial u}{\partial z} = if(z,t)u$$

The phase change is:

$$d\phi(t) = f(0,t) \ dz$$

The reduced NLS equation can be written as follows:

$$\frac{\partial u}{\partial z} = \left(\frac{i}{2u} \frac{\partial^2 u}{\partial t^2} \right) u$$

Therefore, the phase shift resulting from the dispersive and nonlinear terms are expressed as follows in Equations (12–168) and (12–169).

$$d\phi = \left(\frac{1}{2u} \frac{\partial^2 u}{\partial t^2} \right) dz$$

where $u = sech(t)$.
Therefore,

$$d\phi_{nonlinear} = sech^2(t) \ dz \qquad \textbf{(12–168)}$$

and

$$d\phi_{dispersive} = \left[\frac{1}{2} - sech^2(t) \right] dz \qquad \textbf{(12–169)}$$

The differential sum of the constant phase through integration yields a phase shift ($\frac{z}{2}$) common for the entire pulse. Finally, Equation (12–170) expresses the soliton equation.

$$u(z,t) = sech(t) \ e^{i(z/2)} \qquad \textbf{(12–170)}$$

From Equation (12–170), it is evident that the soliton pulse is not dispersive in either the frequency or temporal domain. A graphical representation of the nonlinear and dispersive phase shifts of a soliton pulse is illustrated in Figure 12–113.

Path Averaging Soliton

The optical power loss induced by fiber attenuation along the transmission path is compensated for through the insertion of optical amplifiers. These amplifiers are spaced at specified lengths along the fiber, referred to as amplifier lengths (L_A), and the distances are calculated for maximum system efficiency. The dispersion power profiles of the fiber lengths between two optical amplifiers are not the same. This continuous change of the optical power and variation in dispersion also promotes differential phase shift variations for both the nonlinear ($d\phi_{nonlinear}$) and dispersive ($d\phi_{dispersive}$) terms. These two terms do not cancel each other out. However, soliton pulse transmission characteristics can be maintained if the relationship (Equation (12–171)) between z_s and L_A is maintained and if the path average values of both dispersion (D) and intensity (I) are maintained equally for all optical spans.

$$z_s \gg L_A \qquad \textbf{(12–171)}$$

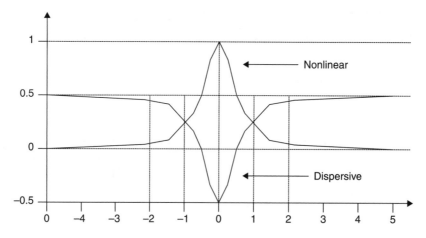

Figure 12–113. Nonlinear and dispersive phase shifts of a soliton pulse.

where z_s is the soliton characteristic length and L_A is the optical amplifier length.

The key element for maintaining soliton is that the spectral content of the pulse must be maintained at a constant. In optical systems where a large number of optical amplifiers are used, soliton pulse characteristics can also be maintained by tapering the dispersion parameter (D_z). This can be accomplished by allowing the dispersion characteristic curve to follow the intensity characteristic curve.

The Gordon-Haus Effect

The soliton central frequency can also be affected by ASE induced by optical amplifiers. ASE induced noise produces random frequency variations (σ_{jitter}), which can be translated into time variations at the pulse output. This phenomenon is the Gordon-Haus effect. Soliton transmission is susceptible to the Gordon-Haus effect, which results in a reduction of the overall system transmission capability. The soliton pulses are affected by two sources. One is the fluctuation of the pulse energy of the soliton, and the other is the pulse arrival time at the end of the fiber. The induced ASE noise from the ED-FAs into the transmission affect soliton critical components such as pulse energy, central frequency, mean time, and phase in a random mode. However, the most critical parameters affected by the EDFA generated soliton jitter are energy fluctuation and central frequency deviation. It is imperative that the jitter generated by the Gordon-Haus effect and its implications in the soliton transmission be fully understood. The mean path average noise energy generated by an optical amplifier noise field is expressed as $\left(\frac{1}{2}\right)\bar{P}(v)$. Based on the sampling theorem, a detected optical field has 2 m **degree of freedom (DOF)** and a soliton pulse will occupy one. Consequently, the noise optical field will shift the central frequency of the soliton. The effective noise field component can be expressed as follows in Equation (12–172).

$$\delta u = iau \tanh(t) \qquad (12\text{–}172)$$

The noise field expressed by Equation (12–172) can shift the soliton central frequency by a level expressed in Equation (12–173).

$$\delta\Omega = \frac{2a}{3} \qquad (12\text{–}173)$$

where a is the Gaussian random variable. Equation (12–174) expresses the random frequency shift variance.

$$\langle\delta\Omega\rangle_A^2 = \left(\frac{1}{3}\right)\bar{P}(v) \qquad (12\text{–}174)$$

where v is the soliton frequency and $\bar{P}_A$ is the mean path noise energy generated by the Gordon-Haus effect. Equation (12–175) expresses the resulting soliton pulse time shift as a result of Equation (12–174).

$$\delta t = -\sum \delta\Omega_N z_N \qquad (12\text{–}175)$$

where z_N is the distance between the N amplifier and the end. The time variance of Equation (12–175) is expressed as follows in Equation (12–176).

$$\langle\delta t^2\rangle = \langle\delta\Omega_N^2\rangle \sum_{A(\text{total})} z_N^2$$

Because

$$\langle\delta\Omega\rangle_A^2 = \left(\frac{1}{3}\right)\bar{P}(v),$$

then

$$\langle\delta t^2\rangle = \left(\frac{1}{3}\right)\bar{P}(v)\left(\frac{Z}{3L_A}\right) \qquad (12\text{–}176)$$

where Z is the total system length, L_A is the optical amplifier length $\left(L_A = \dfrac{\ln G}{\alpha}\right)$, G is the optical amplifier gain, and α is the fiber loss.

Substituting $\left(L_A = \dfrac{\ln G}{\alpha}\right)$ into Equation (12–176) yields Equation (12–177).

$$\langle \delta t^2 \rangle = \left(\frac{1}{9}\right) \alpha n_{sp} F(G) \, hvz^3 \qquad \textbf{(12–177)}$$

The time variance $\langle \delta t^2 \rangle$ can also be expressed as standard deviation σ^2 expressed by Equation (12–178).

$$\sigma^2_{G-H} = 3.6 \times 10^3 \, n_{sp} \, F(G) \left(\frac{\alpha}{A_{eff}}\right) \left(\frac{D}{T_{FWHM}}\right)(Z^3) \qquad \textbf{(12–178)}$$

where σ_{G-H} is the standard deviation (ps), α is the fiber loss (km^{-1}), A_{eff} is the fiber core cross section area (μm^2), D is the group delay distortion (ps/nm/km), T_{FWHM} is the soliton FWHM (ps), and Z is the total system length (mm).

Example 12–38

Compute the standard deviation (σ) of a soliton pulse transmitted under the following parameters:

- total system length: $Z = 10$ mm
- soliton FWHM: $T_{FWHM} = 22$ ps
- fiber chromatic dispersion (GVD): $D = 0.5$ ps/nm/km
- fiber effective area: $A_{eff} = 55 \, \mu$m^2
- fiber attenuation: $\alpha = 0.05$ km^{-1}
- spontaneous emission noise: $n_{sp} = 1.5$
- amplifier noise figure: $F = 2$

Solution

$$\sigma^2_{G-H} = 3.6 \times 10^3 \, n_{sp} \, F(G) \left(\frac{\alpha}{A_{eff}}\right) \left(\frac{D}{T_{FWHM}}\right)(Z^3)$$

$$= 3.6 \times 10^3 \, (1.5)(2.0) \left(\frac{0.05}{55}\right)\left(\frac{0.5}{22}\right)(10)^3$$

$$= 15 \text{ ps}$$

Therefore, $\sigma = 15$ ps.

The impact of the Gordon-Haus effect on soliton transmission system BER is as follows. A bit error will occur when a pulse that represents binary 1 is received outside the optical receiver detection window (w) as a result of the Gordon-Haus effect. The system BER can be expressed as a function of the Q-factor given by the approximate relationship in Equation (12–179).

$$\text{BER} = \left[2\pi(Q^2 + 2)\right]^{-1/2} e\left(-\frac{Q^2}{2}\right) \qquad \textbf{(12–179)}$$

where Q is the quality factor ($Q \geqslant 3$).

Example 12–39

Calculate the system BER of a soliton transmission for the following Q-factors.

- Q-factor equal to 4
- Q-factor equal to 6

Solution

i) For Q equal to 4,

$$\text{BER} = \left[2\pi(Q^2 + 2)\right]^{-1/2} e\left(-\frac{Q^2}{2}\right)$$

$$= \left[2\pi(4^2 + 2)\right]^{-1/2} e\left(-\frac{4^2}{2}\right)$$

$$= 3 \times 10^{-5}$$

Therefore, BER $= 3 \times 10^{-5}$.

ii) For Q equal to 6,

$$\text{BER} = \left[2\pi(Q^2 + 2)\right]^{-1/2} e\left(-\frac{Q^2}{2}\right)$$

$$= \left[2\pi(6^2 + 2)\right]^{-1/2} e\left(-\frac{6^2}{2}\right)$$

$$= 9.1 \times 10^{-10}$$

Therefore, BER $= 9.1 \times 10^{-10}$.

It is evident from Example 12–39 that an increase of the Q-factor from four to six dramatically improves the system BER. The relationship between standard deviation (σ_{G-H}), optical receiver operating window ($2w$), and quality factor (Q) is as follows in Equation (12–180).

$$Q = \frac{w}{\sigma_{G-H}} \qquad \textbf{(12–180)}$$

where Q is the quality factor, w is the optical receiver operating window (ps), and σ_{G-H} is the standard deviation (Gordon-Haus effect) (ps). Assuming Q equal to six (a desirable value) then,

$$w = 6\sigma_{G-H}$$

or

$$2w = 12\sigma_{G-H}$$

For σ_{G-H} equal to 15 ps, the operating receiver window is:

$$2w = 12(15)$$

$$= 180 \text{ ps}$$

or,

$$w = 90 \text{ ps}$$

Therefore, an optical receiver operating window of 90 ps will be required in order to maintain a BER of 9.1×10^{-10} and a standard deviation of 15 ps, based on the Gordon-Haus effect.

The Acoustic Effect

From the preceding examples, it was determined that jitter induced by the Gordon-Haus effect is a limiting factor in soliton transmission. It was also evident that the Gordon-Haus effect is independent of the system bit rate. However, there is another effect that also induces a certain level of jitter on the soliton transmission for long distance, high bit rate systems. In such systems, while soliton pulses may be well separated and not affected by nonlinear interaction, they may alter each other's frequency components, thus promoting different pulse arrival times. This is the result of the interaction of an acoustic wave that affects the fiber refractive index and soliton pulses. For more on acoustic waves, see Chapter 1. Soliton pulses under the influence of an acoustic wave demonstrate that the rate of change of the inverse group velocity is a function of the local slope of the induced refractive index change. That is, dv_g^{-1}/dZ, or $\delta v_g^{-1} \propto Z$, which can be translated as a displacement in time as $\delta t \propto Z^2$. The resultant standard deviation (σ) of the acoustic wave is expressed by Equation (12–181).

$$\sigma_{\text{acoustic}} = 8.6 \left(\frac{D^2}{\tau_s} \right) \left(\frac{Z^2}{2} \right) (f_b - 0.99^{1/2}) \quad \textbf{(12–181)}$$

where σ_{acoustic} = standard deviation based on acoustic effect (ps), D = dispersion coefficient (ps/nm/km), τ_s = soliton pulse width (ps), f_b = system transmission rate Gb/s, and Z = system total length (Mm).

Example 12–40

Compute the standard deviation resulting from the acoustic wave effect of a soliton transmission system operating under the following parameters:

- $D = 0.5$ ps/nm/km
- $\tau_s = 22$ ps
- $f_b = 10$ Gb/s
- $Z = 1$ Mm and 10 Mm

Compute and compare the standard deviations for both 1 Mm and 10 Mm transmission distances, and then compare them to the standard deviation induced by the Gordon-Haus effect.

Solution

i) Solve for Z equal to 1,000 km (1 Mm).

$$\sigma_{\text{acoustic}} = 8.6 \left(\frac{D^2}{\tau_s} \right) \left(\frac{Z^2}{2} \right) (f_b - 0.99^{1/2})$$

$$= 8.6 \left(\frac{0.5^2}{22} \right) \left(\frac{1^2}{2} \right) (10 - 0.99)^{1/2}$$

$$= 0.15 \text{ ps}$$

Therefore, $\sigma_{\text{acoustic}} = 0.15$ ps.

ii) Solve for Z equal to 10,000 km (10 Mm).

$$\sigma_{\text{acoustic}} = 8.6 \left(\frac{D^2}{\tau_s} \right) \left(\frac{Z^2}{2} \right) (f_b - 0.99^{1/2})$$

$$= 8.6 \left(\frac{0.5^2}{22} \right) \left(\frac{10^2}{2} \right) (10 - 0.99)^{1/2}$$

$$= 15 \text{ ps}$$

Therefore, $\sigma_{\text{acoustic}} = 15$ ps.

From Example 12–40, it is evident that for relatively short transmission distances, the jitter generated by the acoustic effect is very small. At long (transoceanic) distances, the jitter is compatible to that generated by the Gordon-Haus effect.

The soliton pulse energy is a factor that determines timing and energy errors. That is, high soliton pulse energy will decrease the energy error while simultaneously increasing the error induced by the Gordon-Haus effect. Therefore, the soliton pulse energy must be set at a level at which the combined errors are at a minimum. Because the soliton pulse energy is proportional to soliton pulse width, measured at FWHM and the dispersion coefficient ratio (τ_s/D), it is therefore possible to maintain the required soliton pulse energy by controlling this ratio. However, the value of the pulse width must not exceed 25% of the pulse period if soliton pulse interaction is to be avoided. Another way to determine soliton pulse energy is through established graphs that represent BER as a function of τ_s/D and energy levels. Such a graph is illustrated in Figure 12–114. This figure illustrates the relationship between timing and energy related errors and soliton pulse energy levels and τ_s/D. For small soliton pulse energy levels, the BER is quite high. As the soliton pulse energy increases, the BER decreases exponentially. However, for soliton pulse energy of a specific level, the timing error level begins to increase significantly. It is also evident that the timing errors are bit rate dependent. Therefore, a medium level of soliton pulse energy must be established to satisfy both the energy and timing error requirements. Although the graph seems to indicate a transmission capacity limit of 5 Gb/s, employing filter techniques such as those of a guiding filter which allows much higher transmission capacities.

Soliton Collision

Soliton collision is the process in which a soliton pulse from one channel in a WDM optical scheme overtakes another pulse of an adjacent channel within the scheme. Here, a pulse interaction takes place. The time during which overlapping of the two soliton pulses is encountered is referred to as *soliton collision*. If a soliton pulse of wavelength λ_1 in channel 1 is ahead of a soliton pulse of wavelength λ_2 in channel 2, and assuming $\lambda_1 > \lambda_2$, (Figure 12–115), the pulse with the shortest wavelength (λ_2) will pass the pulse with the longest wavelength at a distance referred as the *collision length* ($L_{\text{collision}}$). Collision length is an important parameter, expressed in terms of soliton pulse width (τ_s), wavelength difference ($\Delta\lambda$), and dispersion coefficient (D) (Equation (12–182)).

$$L_{\text{collision}} = \frac{2\tau_s}{(\Delta\lambda D)} \quad \textbf{(12–182)}$$

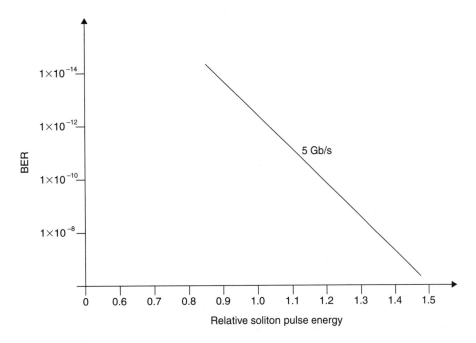

Figure 12–114. Soliton pulse energy *v*. BER.

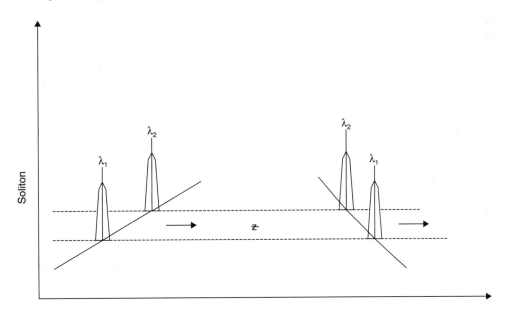

Figure 12–115. Collision representation of two soliton pulses.

where $L_{\text{collision}}$ is the collision length (km), τ_s is the soliton pulse width (ps), $\Delta\lambda = \lambda_1 - \lambda_2$, and D is the dispersion coefficient (ps/nm/km).

Example 12–41

Calculate the collision length ($L_{\text{collision}}$) of two soliton pulses in a WDM optical scheme traveling through a fiber that has a dispersion coefficient of 0.45 ps/nm/km, a wavelength difference of 0.5 nm, and a 22 ps soliton pulse width.

Solution

$$L_{\text{collision}} = \frac{2\tau_s}{(\Delta\lambda D)}$$

$$= \frac{2(22)}{(0.5)(0.45)}$$

$$= 195.5 \text{ km}$$

Therefore, $L_{\text{collision}} = 195.5$ km.

Soliton collision is the result of the implementation of the nonlinear term of the NLS equation. In single channel transmission, the interaction of the soliton pulses due to the nonlinear term of the NLS equation generates SPM. Likewise, in a WDM optical scheme, the interaction of the soliton pulses of one channel with the soliton pulses of another channel promotes refractive index nonlinear changes. The resultant XPM induces slight variations on the pulse group velocity (v_g). Through this, new frequencies such as $\omega_A = 2\omega_1 - \omega_2$ and $\omega_B = 2\omega_2 - \omega_1$ are produced. In a different chapter in this text, the new frequency components were referred to as FWM, which results in optical power exchange between channels. As a consequence, the overall system performance is decreased. However, the negative results from the collision of two soliton pulses are temporary, and both emerge unchanged from such a collision. This advantage is fundamental to the implementation of soliton transmission in WDM optical systems.

SOLITON COLLISION IN AN IDEAL FIBER. An ideal optical fiber is one that has constant dispersion along its entire length and that exhibits zero losses. The general form of a soliton pulse is expressed by Equation (12–183).

$$u = A\,\text{sech}\left[A(t + \Omega z)\right] e^{i(\phi - \Omega t)} \qquad (12\text{–}183)$$

where $\phi = (A^2 - \Omega^2)\left(\dfrac{z}{2}\right)$, z is the frequency shift $\left(\dfrac{1}{\tau}\right)$, and A is the amplitude of the soliton pulse. Based on Equation (12–183), the equations for two colliding soliton can be expressed by Equations (12–184) and (12–185).

$$u_1 = \text{sech}(t + \Omega_1 z)e^{-i\Omega_1 t}\, e^{i(1 - \Omega_1^2)(z/2)} \qquad (12\text{–}184)$$

$$u_2 = \text{sech}(t + \Omega_2 z)e^{-i\Omega_2 t}\, e^{i(1 - \Omega_2^2)(z/2)} \qquad (12\text{–}185)$$

Assuming that both soliton pulses are copolarized, then the above become Equation (12–186).

$$\frac{\partial u_1}{\partial z} = i\frac{1}{2}\frac{\partial^2 u_1}{\partial t^2} + i|u_1|^2 u_1 + 2i|u_2|^2 u_1 \qquad (12\text{–}186)$$

where $i\dfrac{1}{2}\dfrac{\partial^2 u_1}{\partial t^2}$ and $i|u_1|^2 u_1$ are the NLS for soliton pulse –1 and $2i|u_2|^2 u_1$ represents XPM. This term is considered zero except for pulse overlapping. The generated transient phase shift is expressed by Equation (12–187).

$$\frac{d\phi_1 (z,t)}{dz} = 2|u_2 (z,t)|^2 \qquad (12\text{–}187)$$

The rate of frequency shift ($\partial\omega_1/\partial z$) is expressed by Equation (12–188).

$$\frac{\partial\omega_1}{\partial z} = \frac{\partial}{\partial z}\frac{\partial\phi_1}{\partial t}$$

$$= \frac{\partial}{\partial t}\frac{\partial\phi_1}{\partial z}$$

$$= \frac{\partial}{\partial t}\left(2|u_2|^2\right) \qquad (12\text{–}188)$$

or

$$\frac{\partial\Omega_1}{\partial z} = \int_{-\infty}^{\infty} \text{sech}^2 (x)\, dx \int_{-\infty}^{\infty} |u_1|^2 \frac{\partial}{\partial t}|u_2|^2\, dt$$

Because $\displaystyle\int_{-\infty}^{\infty} \text{sech}^2 (x)\, dx = 2$, then

$$\frac{\partial\Omega_1}{\partial z} = 2\int_{-\infty}^{\infty} |u_1|^2 \frac{\partial}{\partial t}|u_2|^2\, dt \qquad (12\text{–}189)$$

Substituting Equations (12–184) and (12–185) into Equation (12–189) yields Equation (12–190).

Therefore,

$$\qquad (12\text{–}190)$$
$$\frac{\partial\Omega_1}{\partial z} = \int_{-\infty}^{\infty} \text{sech}^2 (t + \Omega_1 z)\frac{\partial}{\partial t}\text{sech}^2 (t + \Omega_2 z)\, dt$$

Equation (12–190) represents the rate of change of the inverse group velocity (v_g^{-1}) in terms of distance (z).

For $\Omega \gg 1$, then, $\dfrac{\delta\Omega}{\Omega} \ll 1$.

Replacing Ω_1 and Ω_2 with Ω in Equation (12–190), yields Equation (12–191).

$$\frac{\partial\Omega_{(1,2)}}{\partial z} = (\pm)\frac{1}{2\Omega}\frac{d}{dz}\int_{-\infty}^{\infty} \text{sech}^2 (t - \Omega z)$$
$$\text{sech}^2 (t + \Omega z)\, dt \qquad (12\text{–}191)$$

From Equation (12–191) the inverse group velocity shift can be represented as follows in Equation (12–192).

$$\delta\Omega_{(1,2)} = (\pm)\frac{2}{\Omega}\frac{[2\Omega z \cosh(2\Omega z) - \sinh(2\Omega z)]}{\sinh^3 (2\Omega z)} \qquad (12\text{–}192)$$

The overall time displacement $\delta t_{(1,2)}$ through integration over z is expressed as follows in Equation (12–193).

$$\delta t_{(1,2)} = (\pm)\frac{1}{\Omega^2} \qquad (12\text{–}193)$$

Finally, the three fundamental relationships that can be extrapolated from the above—half channel separation (Ω), maximum frequency shift (δf), and displacement time (δt)—are expressed as follows:

Half channel separation is expressed as follows in Equation (12–194).

$$\Omega = 1.783\,(\Delta f)\tau_s \qquad (12\text{–}194)$$

Maximum frequency shift during collision (δf) is expressed as follows in Equation (12–195).

$$\delta f = \pm \frac{0.105}{(\Delta f)\tau_s^2} \qquad \textbf{(12–195)}$$

Time displacement (δt) is expressed as follows in Equation (12–196).

$$\delta t = \frac{0.1786}{\Delta f^2 \tau_s} \qquad \textbf{(12–196)}$$

Example 12–42

Compute the half channel separation (Ω), maximum frequency shift (δf), and time displacement (δt) for soliton collision, separated by 70 GHz and with pulse widths of 22 ps.

Solution

i) Compute the half channel separation (Ω).

$$\begin{aligned}
\Omega &= 1.783\,(\Delta f)\tau_s \\
&= 1.783\,(70 \times 10^9)\,(22 \times 10^{-12}) \\
&= 2.74
\end{aligned}$$

Therefore, $\Omega = 2.74$.

ii) Compute the maximum frequency shift (δf).

$$\begin{aligned}
\delta f &= \pm \frac{0.105}{(\Delta f)\tau_s^2} \\
&= \pm \frac{0.105}{(70 \times 10^9)\,(22 \times 10^{-12})^2} \\
&= \pm 3 \text{ GHz}
\end{aligned}$$

Therefore, $\delta f = \pm 3$ GHz.

iii) Compute the time displacement (δt).

$$\begin{aligned}
\delta t &= \frac{0.1786}{\Delta f^2 \tau_s} \\
&= \frac{0.1786}{\Delta f^2 \tau_s} \\
&= \frac{0.1786}{(70 \times 10^9)^2 (22 \times 10^{-12})} \\
&= 1.6 \text{ ps}
\end{aligned}$$

Therefore, $\delta t = 1.6$ ps.

TRANSOCEANIC SOLITON TRANSMISSION. Transoceanic optical transmission systems that employ WDM schemes exhibit definite advantages over single channel systems. These advantages are an increase of the system capacity without major changes in the terminal equipment, easy utilization of networking through the implementation of add/drop multiplexing (A/DM), and reliable and secure transmission. Soliton trans-

mission (RZ) was necessary because of the inability of NRZ data formats to achieve transmission rates higher than 5 Gb/s per channel, while soliton transmissions on the order of 10 Gb/s and 20 Gb/s per channel were possible. Furthermore, with soliton transmission, the amplifier spacing can be increased to a maximum length of 200 km, compared to a maximum of 45 km for NRZ transmission. On the negative side, soliton transmission is susceptible to the Gordon-Haus effect and soliton collision induces jitter. Therefore it is imperative that soliton control through proper soliton management be implemented in both frequency and time domains. The two proven techniques used for soliton control are sliding Fabry-Perot filtering and periodic **dispersion slope compensation (DSC).** However, other soliton control techniques, referred to as *in line synchronous modulation* (SM) and *fixed frequency guiding filtering* exist. Both of these techniques are active soliton control techniques. The implementation of active soliton control allows for soliton transmission over great distances. The only disadvantage of active soliton control is that it can only be efficiently implemented in a single channel system. For WDM applications, combined carrier channel synchronization is essential. WDM carrier channels must be periodically demultiplexed, independently synchronized, and finally recombined. Such schemes require the same number of modulators as the channel number in the WDM scheme. Research has shown that soliton transoceanic systems that employ synchronous modulation (SM) techniques for soliton control can achieve transmission rates of about 4×20 Gb/s and amplifier spacing of 480 km.

NON-DSC SOLITON TRANSMISSION. A successful soliton WDM system must synchronize the optical channels within the WDM scheme at all regeneration points. The analysis of soliton transmission utilizing a synchronous WDM scheme and non-DSC is as follows:

The group delay $\tau(\lambda_p)$ at wavelength λ_p per unit length is expressed by Equation (12–197).

$$\begin{aligned}
\tau(\lambda_p) = \tau(\lambda_0)\ &+ 0.5\left(\frac{dD}{d\lambda}\right)_{\lambda_0}(\lambda_p - \lambda_0)^2 \\
&+ 0.17\left(\frac{d^2 D}{d\lambda^2}\right)_{\lambda_0}(\lambda_P - \lambda_0)^3
\end{aligned} \qquad \textbf{(12–197)}$$

where $\tau(\lambda_p)$ is the group delay (ps/nm), D is the fiber dispersion coefficient (ps/nm/km), λ_0 is the zero dispersion wavelength (nm), and λ_p is the wavelength (p) within the WDM scheme. Two channels (λ_1 and λ_P) within the WDM scheme, which are assumed to be synchronous at distance (Z) zero, exhibit a group delay $\delta\tau_{(p1)}$ at distance Z_R (the distance between two in line modulators). Equation (12–198) expresses this group delay.

$$\delta\tau_{(p1)} = [\tau(\lambda_P) - \tau(\lambda_1)]\,Z_R \qquad \textbf{(12–198)}$$

Substituting Equation (12–198) into Equation (12–197) yields Equation (12–199).

$$\delta\tau_{(p1)} = 0.5 \left[D'_0 \left(\Delta\lambda_{p0}^2 - \Delta\lambda_{10}^2 \right) \right.$$

$$\left. + \frac{D''_0}{3} \left(\Delta\lambda_{p0}^3 - \Delta\lambda_{10}^3 \right) \right] Z_R \quad \textbf{(12–199)}$$

For a bit period (T_b), $\delta\tau_{(p1)}$ can be expressed as follows in Equation (12–200).

$$\delta\tau_{(P1)} = P(p) \, T_b \quad \textbf{(12–200)}$$

where $P(p)$ is an integer that represents the number of bit frames over which the two bit streams have passed through each other (collision).

Because the term $\frac{D''_0}{3} \left(\Delta\lambda_{P0}^3 - 2\Delta\lambda_{10}^3 \right)$ is very small, it can be neglected.

Therefore, Equation (12–201) expresses the number of collisions and the bit period.

$$P(p) \, T_b = D'_0 \Delta\lambda_{P1} \left(\Delta\lambda_{P1} + 2\Delta\lambda_0 \right) \left(\frac{Z_R}{2} \right) \quad \textbf{(12–201)}$$

where $\Delta\lambda$ is the channel spacing.

For equal channel spacing, $\Delta\lambda_{P1} = (p-1)\Delta\lambda$, where p is the channel number within the WDM scheme.

Assuming $\Delta\lambda_{\text{syst}} = \sqrt{\dfrac{T_b}{Z_R D'_0}}$, and $\Delta\lambda = x\Delta\lambda_{10}$, where

$\Delta\lambda_{10} = y \, \Delta\lambda_{\text{syst}} = y\sqrt{\dfrac{T_b}{Z_R D'_0}}$, and x and y are spectral positions of channels within the WDM scheme. The number of collisions ($P_{(p)}$) between channel 1 and channel p is then expressed by Equation (12–202).

$$P_{(p)} = (p-1)N + \frac{(p-2)(p-1)M}{2} \quad \textbf{(12–202)}$$

where N is the number of collisions ($P_{(P)}$) between channel 1 and channel 2 and $M = x^2 y^2$. For $p=2$, the value of x and y is expressed as follows in Equations (12–203) and (12–204).

$$x = \frac{2M}{2N-M} \quad \textbf{(12–203)}$$

For $1 \leq M < 2N$

$$y = \frac{2N-M}{2\sqrt{M}} \quad \textbf{(12–204)}$$

The number of soliton collisions between channel p and any other channel is expressed by Equation (12–205).

$$N_{\text{collisions}}^{p,n} = (p-n)\left[N + \frac{(p+n-3)M}{3} \right] \quad \textbf{(12–205)}$$

For channel n, the number of collisions is expressed by Equation (12–206).

$$N(n)_{\text{collisions}} = \frac{2N}{3}n^3 + \left(N - 2M - \frac{MS}{2} \right)n^2$$

$$+ \left[\left(\frac{4M}{3} - N \right)(S+1) + \frac{MS}{6} \right]n$$

$$+ \frac{S(S+1)}{2}\left(N - \frac{4M}{3} + \frac{MS}{3} \right) \quad \textbf{(12–206)}$$

where S is the number of channels in the WDM scheme. The minimum number of soliton collisions can be evaluated in terms of the total number of channels in the WDM scheme as follows:

For an even number of channels, $\left(n = \dfrac{S}{2} \right)$ (Equation (12–207)).

$$N_{\text{collision(min)}}(n) = \frac{S^2}{4}\left[\frac{MS}{2} - M + N \right]_{\text{even}} \quad \textbf{(12–207)}$$

For an odd number of channels, $\left(n = \dfrac{S+1}{2} \right)$ (Equation (12–208)).

$$N_{\text{collision(min)}}(n) = \left(\frac{S^2-1}{4} \right)\left[\frac{MS}{2} - M + N \right]_{\text{odd}} \quad \textbf{(12–208)}$$

For $n = S$ (Equation (12–209)).

$$N_{\text{collision(min)}}(S) = S\left[\frac{MS^2}{3} + S\left(\frac{N}{2} - M \right) \right.$$

$$\left. - \frac{N}{2} + \frac{2M}{3} \right] \quad \textbf{(12–209)}$$

Example 12–43

Calculate the minimum number of soliton collisions in 4-channel and 5-channel WDM systems with no dispersion slope compensation. Assume $M=N=1$.

Solution

i) $S=4$. From Equation (12–207),

$$N_{\text{collision(min)}}(n) = \frac{S^2}{4}\left[\frac{MS}{2} - M + N \right]_{\text{even}}$$

$$= \frac{4^2}{4}\left[\frac{(1)(4)}{2} - 1 + 1 \right]$$

$$= 8$$

Therefore, $N_{\text{collision(min)}} = 8$.

ii) $S=5$. From Equation (12–208).

$$N_{\text{collision(min)}}(n) = \left(\frac{S^2-1}{4} \right)\left[\frac{MS}{2} - M + N \right]_{\text{odd}}$$

$$= \left(\frac{5^2-1}{4} \right)\left[\frac{(1)(5)}{2} - 1 + 1 \right]$$

$$= 15$$

Therefore, $N_{\text{collision(min)}} = 15$.

Likewise, the maximum number of soliton collisions ($N_{\text{collision(max)}}$) is expressed as follows in Equation (12–210).

$$N_{\text{collision(max)}} = S\left[\frac{MS^2}{3} + S\left(\frac{N}{2} - M\right) - \frac{N}{2} + \frac{2M}{3}\right]$$

(12–210)

Example 12–44

Compute the maximum number of soliton collisions in a 4-channel WDM system and for M=N=1,2,3,4.

Solution
For $N=M=1$:

$$\begin{aligned}
N_{\text{collision(max)}} &= S\left[\frac{MS^2}{3} + S\left(\frac{N}{2} - M\right) - \frac{N}{2} + \frac{2M}{3}\right] \\
&= 4\left[\frac{(1)(4^2)}{3} + 4\left(\frac{1}{2} - 1\right) - \frac{1}{2} + \frac{2(1)}{3}\right] \\
&= 4\left[\frac{16}{3} - 2 - \frac{1}{2} + \frac{2}{3}\right] \\
&= 4\,(3.5) \\
&= 14
\end{aligned}$$

Therefore, $N_{\text{collision(max)}} = 14$.

Repeating the above process for different values of N and M yields the data in Table 12–16.

Example 12–45

In a 4-channel, 10 Gb/s per channel WDM soliton transmission system that operates with the following parameters, compute the regeneration spacing (Z_R). Also compute the number of collisions between channel 3 and channel 4.

- $N=M=1$ ($x=2$, $y=1/2$)
- $T_b\left(\dfrac{1}{10\ \text{Gb/s}}\right) = 100\ \text{ps}$
- $\Delta\lambda = 1\ \text{nm}$
- $D'_o = 7 \times 10^{-2}\ \text{ps/nm}^2\text{km}$

TABLE 12–16 Maximum Number of Soliton Collisions

M	N	S	$N_{\text{collision(max)}}$
1	1	4	14
2	2	4	28
3	3	4	42
4	4	4	56
5	5	4	70

Solution

From $\Delta\lambda = y\sqrt{\dfrac{T_b}{Z_R\,D'_o}}$, solve for Z_R.

$$\begin{aligned}
Z_R &= y^2\left(\frac{T_b}{\Delta\lambda^2\,D'_o}\right) \\
&= \left(\frac{1}{2}\right)^2\left(\frac{100\ \text{ps}}{(1\ \text{nm})^2\,(7 \times 10^{-2}\ \text{ps/nm}^2\text{km})}\right) \\
&= (0.25)\left(\frac{100}{7 \times 10^{-2}}\right)\text{km} \\
&= 357\ \text{km}
\end{aligned}$$

Therefore, $Z_R = 357$ km.

From $P_{(p)} = (p-1)\,N + \dfrac{(p-2)\,(p-1)\,M}{2}$

$$\begin{aligned}
P_{(3)} &= (3-1)1 + \frac{(3-2)(3-1)1}{2} \\
&= 3
\end{aligned}$$

Therefore, $P_{(3)} = 3$.

$$\begin{aligned}
P_{(4)} &= (4-1)\,1 + \frac{(4-2)(4-1)1}{2} \\
&= 6
\end{aligned}$$

Therefore, $P_{(4)} = 6$. These results imply that the soliton pulses of channel 3 and channel 4 reach the regenerator point together after crossing reference channel 1 three times and six times, respectively. Because the ratio $\frac{P_{(4)}}{P_{(3)}}$ is two (integer), synchronization can also be achieved at half of the regenerator distance (Z_R).

SOLITON TRANSMISSION WITH DISPERSION SLOPE COMPENSATION. The selection of the regenerator spacing (Z_R), and the modulator position required by the soliton transmission without dispersion slope compensation, can be relaxed somewhat through the incorporation of DSC. Through this method, the dispersion slope (D'_o) is completely compensated for at the end of each regenerator length. A soliton WDM system that employs DSC is analyzed as follows: The group delay $\tau(\lambda_n)$ at wavelength λ_n per unit length is expressed in Equation (12–211).

$$\tau(\lambda_n) = \tau(\lambda_1) + D\,(\lambda_1)\,(\lambda_n - \lambda_1) + \frac{1}{2}\left(\frac{dD}{d\lambda}\right)(\lambda_n - \lambda_1)^2$$

(12–211)

where, $\tau(\lambda_n)$ is the group delay at wavelength λ_n within the WDM scheme, $D\lambda_1$ is the fiber chromatic dispersion at λ_1, and $\dfrac{dD}{d\lambda_1}$ is the dispersion slope (D'_o). WDM synchronization can be expressed as follows in Equation (12–212).

$$\delta\tau_{n-1} = p(n)\,T_b$$

or

$$\delta\tau_{n1} = \frac{Z_R}{2}\Delta\lambda_{n1}\left[\Delta'_o\Delta\lambda_{n1} + 2\Delta\lambda_1\right] \quad (12\text{--}212)$$

Assuming full dispersion slope compensation, D'_o equal to zero, then Equation (12–212) yields Equation (12–213).

$$\delta\tau_{n1} = Z\overline{D}\lambda_{n1} \quad (12\text{--}213)$$

where $\overline{D}$ is the constant chromatic dispersion over the defined wavelength range of λ_{n1}. Equation (12–214) expresses the channel spacing of the above WDM scheme.

$$\Delta\lambda = \frac{T_b}{\overline{D}Z_R} \quad (12\text{--}214)$$

where $\Delta\lambda$ is the WDM channel spacing (nm), $\overline{D}$ is the constant chromatic dispersion (ps/nm/km), Z_R is the modulator spacing (km), and T_b is the bit period (ps).

Example 12–46

Calculate the channel spacing required in a WDM soliton transmission scheme that employs DSC and has the following operating parameters: Chromatic dispersion over the entire wavelength range equal to 0.4 ps/nm/km, a bit period of 100 ps, and modulator spacing of 150 km.

Solution

$$\Delta\lambda = \frac{T_b}{\overline{D}Z_R}$$

$$= \frac{100 \text{ ps}}{(0.4 \text{ ps/nm/km})\,(150 \text{ km})}$$

$$= 1.7 \text{ nm}$$

Therefore, channel spacing $(\Delta\lambda) = 1.7$ nm.

Example 12–47

Compute the bit period (T_b) in a soliton transmission that employs a WDM scheme and operates with the following parameters:

- $\Delta\lambda = 2$ nm
- $Z_R = 200$ km
- $\overline{D} = 0.6$ ps/nm/km

Solution
From Equation (12–214) solve for T_b.

$$T_b = \Delta\lambda(\overline{D}\,Z_R)$$

$$= (2 \text{ nm})\,(0.6 \text{ ps/nm.km})\,(200 \text{ km})$$

$$= 240 \text{ ps}$$

Therefore, $T_b = 240$ ps.

Example 12–48

Calculate the maximum modulator spacing (Z_R) in a WDM soliton transmission that employs DSC and operates with the following parameters:

$\overline{D} = 0.5$ ps/nm-km
$T_b = 80$ ps
$\Delta\lambda = 1.5$ nm

Solution
From Equation (12–214), solve for Z_R.

$$Z_R = \frac{T_b}{\Delta\lambda\overline{D}}$$

$$= \frac{80 \text{ ps}}{(1.5 \text{ nm})\,(0.5 \text{ ps/nm-km})}$$

$$= 106.67 \text{ km}$$

Therefore, $Z_R = 106.67$ km.

In this scheme, the number of soliton collisions can also be expressed by Equation (12–215).

$$P(n) = n-1 \quad (12\text{--}215)$$

Comparing the number of soliton collisions in schemes with and without DSC indicates that the number of collisions in the first case is quadrature compared to a linear of the second case. The number of soliton collisions for any given channel and its adjacent channels can be expressed as follows in Equation (12–216).

$$N_{col}(n) = n^2 - (s+1)\,n + \frac{s(s+1)}{2} \quad (12\text{--}216)$$

where n is any given channel and s is the adjacent channel, for a minimum number of collisions $(N_{col(min)})$. For an odd channel, the minimum number of collisions is calculated by modifying Equation (12–216), yielding Equation (12–217).

$$N_{col}\left(\frac{s+1}{2}\right) = \frac{s^2-1}{4} \quad \text{(odd channel)} \quad (12\text{--}217)$$

Likewise, Equation (12–218) expresses the minimum number of collisions for an even channel.

$$N_{col}\left(\frac{s}{2}\right) = \frac{s^2}{4} \quad \text{(even channel)} \quad (12\text{--}218)$$

The maximum number of collisions $(N_{col(max)})$ is expressed as follows in Equation (12–219).

$$N_{col(max)} = \frac{s(s-1)}{2} \quad (12\text{--}219)$$

Slope dispersion compensation can be accomplished through the utilization of fiber Bragg grading or DCF that exhibits dispersion slope compensation on the order of -15.8 ps/nm^2. Such large dispersion slope compensation limits the usable spectral bandwidth to approximately 2 nm. However, trade-offs between DSC and spectral bandwidth can result in values of anywhere between -0.3 ps/nm^2 and 25 nm of spectral bandwidth. If a fiber Bragg grading is used, the dispersion slope periodicity (Z_D) must satisfy the relationship ($Z_D \leq Z_R$). If DCF is used, the length of the fiber is expressed as follows in Equation (12–220).

$$L_{DCF} = Z_D \left(\frac{D'_{o(DSF)}}{D'_{o(DSF)} - D'_{o(DCF)}} \right) \quad \textbf{(12–220)}$$

where L_{DCF}=dispersion slope compensation fiber length, Z_D=DSC periodicity, $D'_{o(DCF)}$=dispersion slope for DCF fiber, and $D'_{o(DSF)}$=dispersion slope for DSF fiber.

Example 12–49

Calculate the length of a dispersion compensation fiber that employs a DCF fiber with a DSF dispersion slope of 0.07 ps/nm^2, 100 km periodicity, and a DCF dispersion slope of -0.3 ps/nm^2.

Solution

$$\begin{aligned} L_{DSF} &= \frac{Z_D \left(D'_{o(DSF)} \right)}{D'_{o(DSF)} - D'_{o(DCF)}} \\ &= \frac{100 \text{ km}(0.07 \text{ ps/nm}^2)}{0.07 \text{ ps/nm}^2 - (-0.3 \text{ ps/nm}^2)} \\ &= 19 \text{ km} \end{aligned}$$

Therefore, L_{DSF}=19 km.

The Effect of Soliton Collision in a Synchronous WDM System

Because the propagation velocity of solitons is inversely proportional to wavelength, solitons in shorter wavelength channels will travel faster than solitons in longer wavelength channels. Solitons of different channels within the WDM scheme will pass each other at some point (collision). As mentioned, soliton collision is based on frequency perturbations and timing position changes caused by the Kerr effect and FWM. The process of soliton collision is illustrated in Figure (12–116). From Figure 12–116, it seems that at the left-hand side (beginning of the collision process), soliton pulses attract each other, while at the right-hand side they repel each other. This temporal change of soliton propagation velocities is a function of fiber nonlinearities, which are also subject to soliton intensity. If constant fiber nonlinearity is assumed, then the change in propagation velocity at the first half of the collision process will be offset at the second half, and the soliton pulses will arrive together at the next regenerator stage. If a small temporal delay is encountered, it will be equal for both solitons (symmetrical collision).

However, fiber nonlinearities are not constant. They vary during the collision process. This variance (asymmetricity) of fiber nonlinearities during collision results in small (but noticeable) soliton frequency variations, subsequent random timing displacement, and an increase of jitter. The timing shifts are smaller for channels that are located at the center of the WDM scheme and larger at the outer channels. Furthermore, soliton collision resulting from fiber asymmetrical nonlinearities promotes FWM, which is not cancelled out, and which results in random noise generation. This noise is induced to adjacent channels, which causes substantial system performance degradation. An important parameter of soliton collision is the collision length ($L_{collision}$). Collision length is the propagation distance at which two

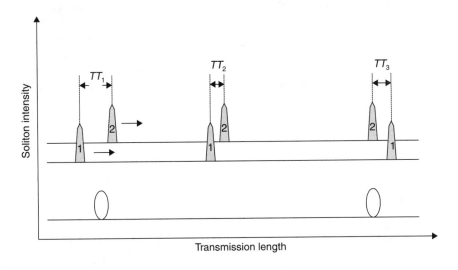

Figure 12–116. Soliton collision (symmetrical).

soliton pulses overlap at half peak power. For two channel (m,n) soliton collision, the collision length $(L_{collision})$ for synchronous WDM without dispersion slope compensation is expressed by Equation (12–221).

$$L_{collision} = \frac{4T_b}{5D'_o\Delta\lambda_{m,n}\left(\Delta\lambda_{m,n} + 2\Delta\lambda_{m,o}\right)} \quad \textbf{(12–221)}$$

where $\Delta\lambda_{mo}$ is the wavelength difference between channel m and reference channel o, D'_o is the dispersion slope, and T_b is the bit period. With dispersion slope compensation, the collision length is expressed by Equation (12–222).

$$L_{collision} = \frac{2T_b}{5\overline{D}\Delta\lambda_{m,n}} \quad \textbf{(12–222)}$$

Example 12–50

Calculate the collision length $(L_{collision})$ for a synchronous WDM system with and without dispersion slope compensation, and operating with the following parameters:

- $T_b = 100$ ps
- $D'_o = 0.07$ ps/nm^2
- $\Delta\lambda_{m,n} = 1$ nm
- $\Delta\lambda_{mo} = 2$ nm
- $\overline{D}_o = 0.1$ ps/nm-km
- $m = 3$
- $n = 1$

Solution

i) Collision length without dispersion slope compensation in km.

$$L_{collision} = \frac{4T_b}{5D'_o\Delta\lambda_{m,n}\left(\Delta\lambda_{m,n} + 2\Delta\lambda_{m,o}\right)}$$

$$= \frac{4(100\,\text{ps})}{[5(0.07\,\text{ps/nm}^2\,\text{km})(1\,\text{nm})](1\,\text{nm} + 2\times 2\,\text{nm})}$$

$$= 228\,\text{km}$$

Therefore, $L_{collision} = 228$ km.

ii) Collision length with dispersion slope compensation in km.

$$L_{collision} = \frac{2T_b}{5\overline{D}\Delta\lambda_{m,n}}$$

$$= \frac{2(100\,\text{ps})}{5(0.1\,\text{ps/nm/km})}$$

$$= 400\,\text{km}$$

Therefore, $L_{collision} = 400$ km.

From Example 12–50, it is evident that employing dispersion slope compensation increases the collision length, with a corresponding decrease of soliton collisions. The total collision

length of any channel within the WDM scheme, in reference to modulator repeater length, is expressed as follows:

For synchronous WDM without DSC, see Equation (12–223).

$$L_{collision}^{tot} = \frac{2}{5}Z_R(s-1) \quad \textbf{(12–223)}$$

where Z_R is the modulator repeater distance (km) and s is the number of channels in the WDM system.

For synchronous WDM with DSC, see Equation (12–224).

$$L_{collision}^{tot} = 2Z_R(s-1) \quad \textbf{(12–224)}$$

Example 12–51

Compute the total soliton collision length $(L_{collision}^{tot})$ for a 4-channel synchronous WDM system with and without dispersion slope compensation and with a modulator spacing of 200 km.

Solution

i) With DSC:

$$L_{collision}^{tot} = 2Z_R(s-1)$$

$$= 2(200\,\text{km})(4-1)$$

$$= 1200\,\text{km}$$

Therefore, $L_{collision}^{tot} = 1200$ km.

ii) non-DSC:

$$L_{collision}^{tot} = \frac{2}{5}Z_R(s-1)$$

$$= \frac{2}{5}(200)(4-1)$$

$$= 240\,\text{km}$$

Therefore, $L_{collision}^{tot} = 240$ km.

From Example 12–51, the advantage of employing DSC is obvious. The longer the collision length, the smaller the number of soliton collisions in a given optical link, and therefore, the less the impact of the generated jitter on system performance.

Application 1

The typical operating characteristics of a transoceanic optical link are listed as follows:

a) time resolution: 0.78 ps

b) frequency resolution: 156 MH

c) input signal: $2^7 - 1$ PRBS

d) amplifier spacing (Z_A): 50 km

e) filter spacing (Z_F): 50 km

f) modulator spacing (Z_R): 350 km

g) modulation depth $\Delta\Phi$: independently controlled

h) intensity depth $I\Phi$: independently controlled

i) fiber attenuation (α): 022 dB/km

j) receiver acceptance window: 0.7 T_b

k) amplifier noise figure (NF): 3.5 dB at 1480 nm

l) channel spacing $\Delta\lambda$: $\dfrac{T_b}{5}$ with filtered WDM

m) F. P. filter periodicity: compatible to $\Delta\lambda$

n) optical power launched into each channel:

$$P_{channel} = P_{soliton}\left(\frac{\alpha Z_A}{1-e^{-\alpha Z_A}}\right), \text{ where } \alpha \text{ is the fiber}$$

attenuation and Z_A is the amplifier spacing.

o) dispersion compensation: fully implemented

THE CLOCK RECOVERY CHANNEL. Important parameters to be considered:

a) SNR: the channel exhibiting the lower SNR must be selected for the modulator driver

b) Number of collisions: for each channel within the WDM scheme

c) The clock recovery circuit: must be driven by the channel that exhibits the lower number of collisions located at the center of the WDM scheme.

CRITICAL SYSTEM CONSIDERATION. Channel synchronicity is subject to the number of soliton collisions.

Q-FACTOR. The Q-factor must be considered, based on the minimum amplitude (Q_A) factor and the timing (Q_t) factor of a specific channel within the WDM system. The Q-factor is expressed in Equation (12–225).

$$Q_t = \frac{T_w}{2\sigma_t} \qquad \textbf{(12–225)}$$

where T_w is the width of the bit and σ_t is the standard deviation ($\sigma_o \rightarrow 0 - bit$ or $\sigma_1 \rightarrow 1 - bit$).

BER. The BER is expressed as follows in Equation (12–226).

$$BER = \frac{1}{(2\pi Q_{min})^{1/2}} e^{-\frac{Q^2_{min}}{2}} \qquad \textbf{(12–226)}$$

For $Q_{min} = 6$:

$$BER = \frac{1}{(2\pi Q_{min})^{1/2}} e^{-\frac{Q^2_{min}}{2}}$$

$$= \frac{1}{(2\pi 6)^{1/2}} e^{-\frac{6^2}{2}}$$

$$= 2.5 \times 10^{-9}$$

Therefore, BER $= 2.5 \times 10^{-9}$.

Application 2

4×20 Gb/s soliton WDM system basic configuration:

a) Channel spacing ($\Delta\lambda$): 1.2 nm

b) Operating wavelengths: Channel 1, $\lambda_1 = 1555.3$ nm; Channel 2, $\lambda_2 = 1556.5$ nm; Channel 3, $\lambda_3 = 1558.9$ nm; Channel 4, $\lambda_4 = 1560.1$ nm

c) Optimization is required for the following: intensity modulation (IM) depth, FP filter bandwidth, and phase modulation index ($\Delta\Phi$).

If the soliton is shifted outside the modulator window, excessive phase modulation (PM) and drift from the optical operating point will result.

d) The optical power launched into each channel must be optimized. That is, it must be equal to soliton average power ($P_{soliton(avg)}$). If the launched power is higher than the soliton average power, then the excessive optical energy will be coupled into the adjacent channels, causing system performance degradation.

The typical operating characteristics of a 4×20 Gb/s non-DSC system are listed in Table 12–17.

TABLE 12–17 Typical Operating Characteristics

Parameters	Symbols		Units
Pulse width	T_w	10	ps
Dispersion management	DM	4	steps
Chromatic dispersion (avg.)	$\overline{D}$	0.178	ps/nm/km
Channel spacing	$\Delta\lambda$	1.2	nm
Amplifier spacing	Z_A	50	km
Maximum number of collisions between repeaters	$N_{collisions}$	13	—
Filter 3dB bandwidth	$B_{W_{-3 dB}}$	0.75	nm
Intensity/phase modulation depth	$IM/\Delta\Phi$	10 dB/6 deg	—

SUMMARY

Chapter 12 was devoted exclusively to the design of fiber optics systems for short and long haul applications. The chapter began with an introduction to the rules and procedures required for optical power budget analysis and continued with a complete optical fiber link design. The very important concepts of dispersion and wave polarization and their effect on system performance were discussed in detail. High speed experimental systems that employ spectral inversion were also introduced. The chapter continued with the presentation of various design techniques for analog optical fiber systems, hybrid analog systems, digital systems, and a number of transoceanic optical fiber systems. Several experimental systems that operate in the Tb/s range were also presented. The chapter concluded with the introduction of the soliton, its theory, and potential for implementation in long distance optical fiber links.

QUESTIONS

Section 12.1

1. Name the basic building blocks of an optical fiber communications link.

2. What is the maximum distance that an optical fiber link can reach, using a multimode (MM) fiber?

3. Define *link power budget*.

4. Name the two most important parameters that affect the power margin of an optical fiber link.

5. Describe the effects of an over power budget and an under power budget on the system performance.

6. List the technological breakthroughs which made the implementation of all optical networks possible.

7. What are the main contributors that limit the full utilization of the fiber's available bandwidth?

8. Draw a block diagram of an IM-DD optical system and explain its operation.

9. Name the main contributors to impairment in optical links that operate below and above 20 Gb/s transmission rates.

10. Name the main factor that contributes to pulse broadening (chirping).

11. With the assistance of a formula, describe how the negative effect of Raman scattering on ISI can be marginalized.

12. Describe the causes of the Kerr effect.

13. Where is the Kerr effect most noticeable?

14. How can SE and ASE be minimized?

15. With the assistance of a formula, describe how the performance of an optical receiver can be evaluated in the presence of ASE and Kerr nonlinearities.

17. Define *dispersion management*.

18. List all the components that contribute to the establishment of the *average dispersion coefficient*.

Section 12.2

19. List the most important assumptions that must be made for power budget calculations based on first order approximation.

20. Name the method that provides the most accurate way for power margin evaluation.

21. Describe the two fiber parameters that determine the level of optical power that can be launched into the fiber.

22. Define *optical receiver sensitivity*.

Section 12.3

23. Name the main applications of 10 Mb/s optical links.

24. What are the most important advantages of using optical fiber links in industrial applications?

25. Describe the operating characteristics of the most preferable transmitter and receiver used in industrial applications.

26. Describe why polymer plastic optical fibers are preferred for industrial applications.

Section 12.4

27. With the assistance of a formula, describe the main contributing factors in the establishment of an optical link's maximum length.

28. Define *optical receiver dynamic range*.

29. What is the result of an optical system that operates outside the receiver dynamic range?

30. List the main components required for the establishment of the minimum link distance.

31. Draw a graph that depicts the relationship between PWD and receiver input optical power, and explain this relationship.

32. Describe *optical power coupling loss*.

33. Express the optical power coupling loss relationship, and comment on the parameters that contribute to this loss.

34. With the assistance of a graph, describe the relationship between transmitter optical output power and junction temperature.

35. Give the formula and explain the relationship between transmitter optical power and operating temperature.

36. Draw a schematic diagram of a series LED driver circuit and explain its operation.

37. Draw a schematic diagram of a parallel LED driver circuit and explain its operation.

38. List all the parameters incorporated in the evaluation of the link losses.

Section 12.5

39. Name the types of dispersion.

40. What factors determine the type of dispersion in optical fiber links?

41. Describe the causes of chromatic dispersion.

42. There are four ways to explain the dispersion effect in optical fiber systems. Name them.

43. Explain why optical fiber systems require dispersion compensation, even though they employ zero dispersion fibers.

44. What are the two critical factors to be considered in the quest for longer transmission distances and higher bit rates?

45. Describe the method by which dispersion compensation is achieved in the transmission section of a fiber optics link.

46. Describe the method of dispersion compensation through the optical fiber.

47. What is the Gires-Tournois interferometer? Explain in detail.

48. Compare the spectral inversion dispersion compensation method with interferometer and fiber methods.

49. What is the most noticeable disadvantage of spectral inversion schemes that use FWM?

50. How can the disadvantage in Question 49 be remedied? Explain in detail.

51. With the assistance of a diagram, explain the polarization sensitive spectral inversion scheme.

52. What is the performance indicator of a spectral inverter?

53. Define *spectral inversion efficiency*.

54. List all the components that contribute to the establishment of spectral inversion efficiency.

55. With the assistance of a graph, explain the relationship between the effective noise figure and spectral inversion fiber length.

56. With the assistance of a graph, explain the relationship between the effective noise figure and pump power.

57. With the assistance of a graph, explain the relationship between the effective noise figure and noise power spectral density.

58. Explain the purpose of inserting an optical band pass filter at the output of the spectral inverter.

59. Draw a block diagram of a polarization insensitive spectral inverter and explain (in detail) its operation.

60. List the noise components of the spectral inverted signal channel in the *x* direction.

Section 12.6

61. Name the limiting factor of optical systems that employ DWDM and operate at 40 Gb/s.

62. Explain the method of GVD and SPM compensation via OPC.

Section 12.11

63. Describe the wave polarization effect in long distance, high density optical systems.

64. What are the causes of different group and phase velocities in single-mode fibers?

65. Define *beat length*.

66. With the assistance of a diagram, define *differential group velocity*.

67. Give and elaborate on the relationship between PMD and long length fibers.

68. Briefly describe the two models that determine PMD in long length fibers.

69. What is the principal state of polarization?

70. Express and elaborate on the relationship between optical span system bit rate and PMD.

Section 12.12

71. Name the main factor that influences the mean dispersion penalty of an LED transmitter coupled into an MM fiber.

72. Give the expression of an LED exit bandwidth.

73. What are the key parameters that determine LED exit bandwidth?

74. Explain why a laser diode coupled into an MM fiber induces only modal dispersion.

75. With the assistance of a formula, define *laser total bandwidth*.

76. Define *chromatic dispersion bandwidth*.

Section 12.14

77. Name the dispersion effect in an optical link when a laser diode is used as the source and coupled into a single-mode fiber.

78. What is the condition by which chromatic dispersion can be theoretically eliminated?

Section 12.15

79. What are the advantages of using optical fiber technology for video transmission?

80. Describe the primary function of an analog optical link.

81. List the main building blocks of an analog optical fiber link and explain its operation.

82. Explain how an AM-VSB signal can be transmitted over a passive optical network.

83. What is the minimum for CNR and the maximum CSO and CTO distortion levels for the system in Question 82?

84. Name the fundamental characteristics of transmitters employed in an analog multichannel optical system.

85. How can CSO and CTO distortions be eliminated in CATV systems that employ external optical modulators?

86. Describe the relationship between CSO, CTO, and modulation depth in a CATV optical system.

87. What are the factors that influence the signal spectrum at the output of a transmitter employed in a multichannel optical system?

88. List all the parameters incorporated into the establishment of CSO distortion.

89. In an AM-VSB video system, name the system parameter that is mostly affected by SPM and SBS.

90. Draw a block diagram of an optical receiver employed in CATV systems and explain its operation.

91. Define *receiver responsivity*.

92. Give the relationship of an ideal responsivity and list the parameters that establish it.

93. What are the parameters that influence the CNR of optical receivers operating in multichannel analog optical systems?

Section 12.16

94. List the main components of a hybrid multichannel analog and digital video optical link.

95. With the assistance of a block diagram, explain the operation of a hybrid multichannel analog and digital video optical link.

96. List the parameters involved in the establishment of the CNR of the analog signal in a hybrid multichannel analog and digital video optical system.

97. Describe the effect of EDFA in a hybrid multichannel analog and digital video optical system.

98. What is the leading performance indicator of the digital signal in a hybrid multichannel analog and digital video optical system?

Section 12.17

99. What was the year that the first transcontinental telegraph cable was installed between North America and Europe?

100. Name the year that the first transatlantic telephone system was installed.

101. Briefly describe the two transmission formats adapted for long distance transmission.

102. With the assistance of a block diagram, describe the operation and performance characteristics of a 60 Gb/s TDM-WDM oceanic optical system.

103. What are the key components that compose the power output control mechanism of an oceanic optical system?

104. Describe the relationship between higher order distortion and waveform distortion.

105. List the main components incorporated into the North America to Europe undersea optical system.

106. Why is the incorporation of a performance monitoring mechanism so important to optical transoceanic systems?

107. Draw a block diagram of an undersea optical repeater and describe its operation.

108. With the assistance of a block diagram, describe the operation of a 20 Gb/s, 8100 km optical soliton transmission.

109. List the main functions of EDFA incorporated into the design of the repeater chain in transoceanic optical systems.

110. What is the serious problem that arises from the use of EDFA in undersea optical systems?

111. What is dispersion mapping? With the assistance of a graph, explain it in detail.

112. Name the parameters that influence the BER of a transoceanic optical system.

113. A digital signal arriving at the input of an optical receiver may fall into four categories. Name these categories.

114. What are the limitations imposed on an oceanic optical system?

115. Which factors define the SNR at the input of the receiver in a transoceanic optical system?

116. How can the repeater span be increased in oceanic optical systems?

Section 12.18

117. Draw the AT&T Bell Labs 1 Tb/s experimental optical system and explain its operation in detail.

118. Name the two methods by which TDM 40 Gb/s optical systems can be implemented.

119. List the components required for 40 Gb/s optical systems.

120. What are the operating characteristics of a modulator required for ETDM optical systems?

Section 12.19

121. With the assistance of a block diagram, describe the operation of the optical clock recovery circuit by the SOA phase detection method.

122. Describe the operation of the MZ interferometer demultiplexer.

123. Draw a block diagram of a 40 Gb/s dispersion management scheme, and briefly describe its operation.

Section 12.20

124. Define *soliton*.

125. Describe, in detail, the *unit of soliton*.

126. With the assistance of a formula, define the *power of soliton*.

127. How are the nonlinear terms of the nonlinear Schroedinger equation cancelled out in soliton transmission? Explain, in detail.

128. What is a path averaging soliton?

129. Explain why soliton transmission is influenced by the Gordon-Haus effect.

130. Define *soliton collision*.

131. With the assistance of a graph, describe soliton collision in detail.

132. Derive the final expression of the time displacement of a soliton collision in an ideal fiber.

133. Name some of the advantages of transoceanic soliton transmission compared to single channel systems.

134. Compare soliton transmission with and without DSC.

135. Describe, in detail, soliton collision in a synchronous WDM optical system.

PROBLEMS

1. In an optical fiber link, if the power generated by the transmitter is 8.5 µW and the receiver sensitivity 0.5 µW, calculate the power margin of the optical link.

2. If an optical fiber link operates with a transmitter power of 32 µW and power margin 10 dB, calculate the required minimum receiver sensitivity.

3. The receiver sensitivity of an optical fiber link is −27.5 dB and the required power margin 3.5 dB. If the optical power launched into a 50/125 fiber is 8 dB, calculate the transmitter power and the link power budget.

4. An optical fiber link is designed to operate at 50 Mb/s. The link incorporates an LED as the transmitter and a PIN photodiode as the receiver with 0.4 A/W sensitivity. If the link safety margin is set at 2.85 dB, and the power budget at 10 dB, calculate the receiver operating power for a transmitter current drain of 120 mA and receiver current drain of 2 µA.

5. Calculate the maximum distance attainable before amplification of an optical link that operates at the 1550 nm wavelength window with a bit rate of 2.5 Gb/s. The link utilizes an SI-SMF with an effective bandwidth of 4 nm.

6. Calculate the maximum chromatic dispersion of an optical fiber link that operates at the 1.25 Gb/s rate. Assume a link distance of 85 km and fiber operating bandwidth of 0.55 nm.

7. An optical fiber link operates at the 1550 nm wavelength window and utilizes an SI-SMF that exhibits a chromatic dispersion of 17 ps/nm-km. If the link distance is 100 km and the fiber bandwidth 0.4 nm, calculate the system bit rate.

8. Compute the chromatic dispersion induced in an optical fiber link by a 120 km fiber that operates at the 1550 nm wavelength window and at a system bit rate of 8 Gb/s.

9. An optical link incorporates a laser diode as its optical source, coupled into an MM fiber that has an operating bandwidth of 800 MHz/km. If the system bit rate is 0.8 Gb/s and link length is 1.2 km, compute the mean dispersion penalty. Assume four splices are required for the link.

10. An optical fiber link incorporates an MLM laser diode as the transmitter, operating at the 1310 nm wavelength window with a spectral line width of 2.5 nm. The laser is coupled into a single-mode fiber that exhibits a dispersion of 3.3 ps/nm-km. If the system bit rate is 1.25 Gb/s and the dispersion penalty is 2 dB, calculate the length of the optical link.

11. An optical fiber network operates with the following parameters:

   i) system bit rate: 200 Mb/s

   ii) length of the link: 60 km

   iii) optical fiber: 8.3/125 SM

   iv) operating wavelength window: 1310 nm

   v) transmitter: 1237J (laser)

   vi) receiver: 1310C

   vii) number of connectors: 8

   viii) rotary slices: 4

   ix) spectral width: 4.4 nm (max)

   x) dispersion penalty: 1 dB

   xi) clock recovery: available

Calculate:

   i) maximum dispersion (D_{max})

   ii) mean link margin (µM)

   iii) sigma link margin (σM)

12. An analog optical fiber link is designed to carry 40 NTSC video channels. If the modulation depth of the link is 6% and the coefficient (a) is 2.8×10^{-3}, compute the CSO distortion.

13. If CSO distribution of an analog optical fiber link designed to carry 60 NTSC video channels is to be maintained at –55.7 dB, calculate the required modulation depth of the link. Assume coefficient (a) of 3.6×10^{-3}.

14. Calculate the CTO distortion of a 40-channel NSTC optical video link that operates with 6.5% modulation depth and coefficient (b) of 1.07×10^{-2}.

15. A 50-channel NTSC optical system operates with an index of modulation of 4.5%. If the modulation depth is to be reduced to 3.5%, calculate the difference between the CSO and CTO distortions for both modulation depths. Assume coefficient (a) of 2.15×10^{-3} and coefficient (b) of 3.95×10^{-3}.

16. An analog fiber optics communications link that utilizes amplitude modulation operates with a modulation depth of 3.5%. If the RIN is set at –160 dB, calculate the receiver CNR.

17. The CNR at the receiver input of an optical fiber communications system is 55.8 dB. If the AM modulation depth of the system is 4%, calculate the RIN.

18. An optical receiver operates in the 1310 nm wavelength window. If the PIN photocurrent is 0.116 mA and the diode quantum efficiency is 82%, calculate the required optical power at the input of the receiver.

19. If the back reflection of a photodiode is specified at –50 dB, calculate the optical power reflected back into the fiber, assuming an optical power at the input of the receiver of 0.15 mW.

20. A receiver employed in an analog multichannel NTSC optical system operates with an input power of 1.2 mW. If the receiver responsivity is specified at 700 V/W and the index of modulation at 4.5%, calculate the voltage at the output of the receiver.

21. If an optical receiver operates with an input power of 1 mW, a modulation depth of 3%, and a specified output voltage of 18 dBm_V, calculate the receiver responsivity.

22. Calculate the required modulation depth of an optical receiver specified with a 850 V/W responsivity, an input optical power of 1.25 mW, and an output voltage of 25 dBm_V.

23. An optical receiver employed in an analog multichannel AM-VSB optical system operates with a modulation depth of 3.4%, a photodiode current of 1 mA, equivalent input noise of 7.5 pA/Hz, RIN of –150 dB/Hz, and bandwidth of 4 MHz. Calculate the CNR at the input of the receiver.

24. A chain repeater composed of fifteen EDFAs is employed in an oceanic optical fiber system. If the chain repeater bandwidth is 4.5 GHZ, the gain is 12.5 dB, and the NF is 5.5 dB, calculate the SNR. Assume 5 mW of optical power launched into the fiber.

25. Calculate the optical power required at the input of the optical fiber of an oceanic optical system that incorporates a chain repeater composed of twelve EDFAs that operate with the following parameters: effective bandwidth of 5 GHz, gain of 10 dB, noise figure of 5.5 dB, and repeater SNR of 50 dB.

26. An optical fiber system employs a single-mode nondispersive fiber. If the system bit rate is 2.5 Gb/s, the operating wavelength is 1550 nm, and the fiber dispersion is 17 ps/nm-km, determine the fiber length. If the system bit rate is increased to 5 Gb/s, calculate the fiber length and compare it with that of a 2.5 Gb/s system.

27. Calculate the soliton characteristic length for a dispersion constant of 0.25 ps/nm-km and FWHM of 20 ps.

28. A soliton system operates with the following parameters:

    i)   wavelength: 1550 nm

    ii) dispersion: 0.22 ps/nm-km

    iii) full width at half maximum: 25 ps

    iv) refractive index average over all polarization states: 2×10^{-16} cm^2 / W

    v)   fiber core effective area: 55 μm^2

    Calculate the soliton peak pulse power.

29. Calculate the standard deviation of a soliton pulse that operates with the following parameters: FWHM of 25 ps, system length of 8 Mm, fiber effective area of 55 μm^2, chromatic dispersion of 0.45 ps/nm-km, fiber attenuation of 0.045 dB/km, amplifier NF of 3 dB, and SE noise of 1.8.

30. Calculate the system BER of a soliton transmission with a quality factor of 6.5.

31. Calculate the standard deviation induced by the acoustic effect of a soliton transmission operating with a dispersion of 0.6 ps/nm-km, a system bit rate of 5 Gb/s, a soliton pulse width of 18 ps, and a transmission distance of 8 Mm.

32. Two soliton pulses in a DWDM optical scheme are traveling through a fiber that exhibits a dispersion coefficient of 0.5 ps/nm-km. If the soliton pulse width is 20 ps and the wavelength difference of 0.4 nm, calculate the collision length.

33. Two soliton pulses of 20 ps pulse width are traveling through a fiber that has a frequency separation of 60 GHz. Calculate the half channel separation, time displacement, and maximum frequency shift.

34. Calculate the number of soliton collisions in a 6-channel WDM optical system without dispersion slope compensation. Assume $M=N=1$.

35. Compute the maximum number of soliton collisions in a 6-channel WDM optical system, assuming $M=N=3$.

36. Calculate the regeneration spacing and the number of collisions between channel 4 and channel 5 in a 6-channel WDM soliton optical system that operates with a bit rate of 8 Gb/s, a channel spacing of 1.2 nm, and a fiber dispersion slope of 6×10^{-2} ps/nm^2 km. Assume $M=N=1$ ($x=2$, $y=0.5$).

37. A WDM soliton transmission system that employs a dispersion slope compensation fiber operates with a 5 Gb/s bit rate, chromatic dispersion over the wavelength range of 0.25 ps/nm/km, and modulator spacing of 200 km. Compute the required channel spacing.

38. A synchronous WDM soliton system operates with the following parameters: $T_b=200$ ps, $D'_o = 0.05$ ps/nm^2, $\Delta\lambda_{m,n} = 1.1$ nm, $\Delta\lambda_{mo} = 2.2$ nm, $\overline{D}_o = 0.15$ ps/nm-km. Calculate the collision length without and with dispersion compensation.

13

NETWORKS

OBJECTIVES

1. Establish the need for optical networks
2. Give a historical perspective of the development of optical networks
3. Define the fundamental concepts of optical networks
4. Describe the difference between point-to-point and multipoint networks
5. List the major building blocks of optical networks
6. Define local area network

7. Describe, in detail, the various types of LANs and LAN standards
8. Explain the operating characteristics of Ethernet and fast Ethernet
9. Describe fiber channel and fiber channel topologies
10. List the main building blocks of an ATM network and explain its operation
11. Describe the purpose and operating characteristics of SONET

KEY TERMS

access group
adaptation layer
arbitrated loop
arithmetic unit interface (AUI)
asynchronous transfer mode (ATM)
attachment unit interface
backbone
bridge
broadband integrated services digital network (B-ISDN)
buffering/framing
bus interface

carrier sensing multiple access with collision detection (CSMA-CD)
cell loss priority (CLP)
cell rate decoupling
client
clock and data recovery
concentrator
convergence sublayer protocol data unit (CS-PDU)
convergent sublayer (CS)
customer premise equipment (CPE)

cyclic redundancy check (CRC)
data capacity
defense advanced research project agency (DARPA)
encoder
Ethernet
fiber channel
fiber to the curb (FTTC)
fiber distributed data interface (FDDI)
fiber to the-home (FTTH)

fiber optics inter-repeater link (FOIRL)
frame multiplexer
full service access network (FSAN)
header error control (HEC)
hub
layer network
linear chain topology
line sublayer
link
local area network (LAN)
medium access unit (MAU)

mesh topology
metropolitan area network (MAN)
microstrip line
multipoint network
multiwavelength optical networking (MONET)
network interface card
network node interface (NNI)
Nippon Telephone and Telegraph (NTT)
operation and maintenance (OAM)
optical carrier (OC)
optical network
optical network unit (ONU)
passive optical networks (PON)
path overhead (POH)
path sublayer

payload type identification (PTI)
peripherals
photonic sublayer
physical layer
physical medium dependent (PMD)
point-to-point
ring topology
segmentation and reassembly protocol data unit (SAR-PDU)
segmentation and reassembly (SAR) sublayer
segment type (ST)
sequence number protection (SNP)
serial data processor
server
star topology
strip line

subnetwork
switched fabric
synchronous digital hierarchies (SDH)
synchronous digital hierarchies-time division multiplexing (SDH-TDM)
synchronous optical network (SONET)
synchronous payload envelope (SPE)
synchronous transport mode (STM)
synchronous transport signal (STS)
token ring topology
transmission convergence (TC)
transmission frame adaptation

transmission frame generation
transport overhead (TOH)
tree topology
user network interface (UNI)
variable bit rate (VBR)
very large scale integrated circuits (VLSI)
virtual channel identification (VCI)
virtual path identification (VPI)
wavelength cross connect
wavelength division multiplexing (WDM)
wavelength division multiplexing-passive optical network (WDM-PON)
wide area network (WAN)
work station

13.1   INTRODUCTION

In the early 1980s and 1990s, the telecommunications industry around the world, and more specifically in the United States, underwent revolutionary changes. In 1982, AT&T was broken up into several independently operated phone companies and was consequently deprived of the privilege of operating inside local access and transport areas. This break up allowed several other companies to fill the newly created vacuum, thus inducing healthy competition. In this way, AT&T lost the monopoly in the telecommunications decision making process in the United States. Because the large number of independent companies, which occupied the local exchange spectrum, required an equal number of high capacity interfaces, the need for standardization became apparent. Another direct result of AT&T's break up was an emerging need for the creation of a **synchronous optical network (SONET).** The termination of AT&T's monopolistic power in the telecommunications industry in the U.S. was followed by similar developments in other countries, such as the proposed privatization of Japan's **Nippon Telephone and Telegraph (NTT),** the privatization of British Telecom, and other movements for the privatization of national telecommunications companies across Europe. However, the implementation of SONET, **synchronous digital hierarchies (SDH),** and **asynchronous transfer mode (ATM)** were possible only through technological advancements which created high speed integrated circuits, **very large scale integrated circuits (VLSI),** high speed microprocessors, and parallel advancements in optical fiber technology. The 1990s also witnessed a phenomenal increase in **data capacity** demands, while Internet and broadband services were reaching critical mass. The increase in capacity per fiber demand from 1978 to 2001 is illustrated in Figure 13–1.

These transmission capacities, which were projected at the beginning of the 21st century, would depend upon the implementation of an all **optical network.** However, in order for optical networks to be fully implemented, breakthroughs in various fields of optical technologies were absolutely essential. These new technologies came into being in the form of optical amplifiers (OA), **wavelength division multiplexing (WDM),** add/drop multiplexers, optical filters, and optical cross connect. Prior to this time, optical networks were limited to long haul, **point-to-point** networks. The new technologies, coupled with advancements in network management software, enabled the implementation of optical networks that had multipoint metropolitan rings, cross connect, customer access, and **local area networks (LANs).** As stated previously, in 1996, transmission capacities in the Tb/s per fiber range were experimentally achieved. These corresponded to the transmission of twenty million digitally encoded voice channels. This potential transmission capability of the optical systems, coupled with advantageous economic and technological considerations such as component and fiber size and a substantial increase in the spacing between optical repeaters, made this the technology of the future. Specifics of the evolutionary process at the beginning of the development of the optical technology and the revolutionary process, which followed in terms of network data transmission capabilities, are listed in the Table 13–1.

Similarities between demands for computer power and computer technology advancement exist: data capacity demands and optical network advancement. In the computer field, the demand for computer power increases by 100% every ten years, while component technology, such as integrated circuits and advancements in microprocessor technology increase by 10% every ten years. Therefore, component technology advancement is unable to meet computer power requirements.

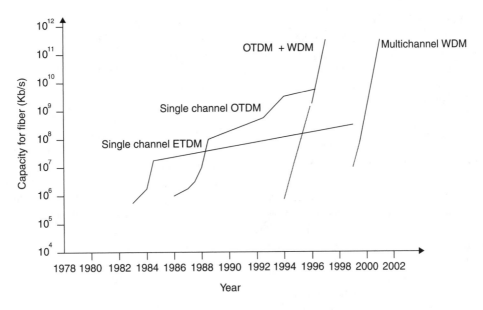

Figure 13–1. Data capacity demands between 1978 and 2002.

However, computer designers have overcome this problem by adopting the concept of parallel architecture. Similarly, in optical networks, the designers adopted the concept of parallel channels. That is, a number of high speed WDM channels are combined in a parallel configuration in order to increase a network's overall carrying capacity, while the data capacity per channel is relatively low, therefore fiber nonlinearities are maintained at acceptable levels. This concept, in conjunction with the recently developed EDFAs not only increased the network capacity but also increased the repeater distance to a length of 10,000 km (undersea networks). The major breakthrough dates in the development of optical networks that employ WDM are as follows:

a) In 1991, an experimental optical network, which incorporated four WDM optical channels (each capable of carrying 1.7 Gb/s of data transmitted a distance of 840 km), was demonstrated. This optical network also incorporated thirteen EDFAs spaced at 70 km intervals.

b) In 1993, Bell Lab engineers working on a project named "the hero" were successful in transmitting 80 Gb/s through eight WDM channels at 10 Gb/s per channel, for a distance of 240 km with a BER of $<10^{-13}$.

c) In 1995, AT&T deployed the next generation light wave network, composed of eight parallel WDM channels, each capable of carrying 2.5 Gb/s of data to a maximum of 20 Gb/s and a distance of 360 km.

d) In 1996, Bell Labs, NTT, and Fujitsu succeeded in their attempts to design optical networks capable of transmitting in the Tb/s range. Bell Lab engineers were able to transmit 1.1 Tb/s through an optical system composed of fifty-five WDM channels (20 Gb/s per channel) at a distance of 55 km, by using NZDSF; NTT succeeded in transmitting 1 Tb/s through ten WDM channels, each capable of carrying 100 Gb/s a distance of 40 km, by using DSFs.

Transmission capacities are limited by fiber nonlinearities. However, the effective window of a silica fiber can be as high as 50 THz. With this available bandwidth, transmission rates of 200 Tb/s can theoretically be achieved through the implementation of QAM techniques that exhibit a theoretical spectral efficiency of 4 b/s/Hz. The employment of such a high

TABLE 13–1 Optical Technology Timeline

Year	System	Wavelength	Fiber	WDM Channels	Voice Channels	Bit Rate per Channel	Repeater Spans
1980	FT3	0.82 μm	MM	1	672	45 Mb/s	7 km
1983	FT3C	0.82 μm	MM	1	1344	90 Mb/s	7 km
1985	FTG-417	1.3 μm	SM	1	6048	417 Mb/s	50 km
1987	FTG-1.7	1.3 μm	SM	1	24,192	1.7 Gb/s	50 km
1989	FTG-1.7	1.3/1.55 μm	SM	2	48,384	1.7 Gb/s	50 km
1992	FT-2000	1.3 μm	SM	1	32,256	2.5 Gb/s	50 km
1992	FT-200	1.3/1.55 μm	SM	2	64,120	2.5 Gb/s	50 km
1995	NGLN	1.55 μm	SM	8	516,000	2.5 Gb/s	360 km
1997	NGLN II	1.55 μm	SM	16	258,000	2.5 Gb/s	360 km
1999	Wave Star 400G	1.55 μm	SM	80	2,580,000	2.5 Gb/s	640 km
1999		1.55 μm	SM	40	5,160,000	10 Gb/s	640 km

FT—fiber technology   MM—multimode   SM—single mode   NGLN—next generation lightwave networks

level digital modulation technique will require a significant increase in the optical power launched into the fiber. However, such an increase of the optical power through the fiber will promote a proportional increase of fiber nonlinearities. Unless the impact of optical power on fiber nonlinearities is solved, such high levels of digital modulation techniques cannot be implemented at this time.

13.2 OPTICAL NETWORKS

The ever-increasing demand for higher capacities necessitated the conception of optical networks. Up to this point, the implementation of optical networks was limited to point-to-point systems, while multipoint and optical cross connect networks remained in the realm of theoretical possibility. The fundamental principle upon which an optical network is based is its ability to share resources with other networks through multiplexing. This, of course, cannot be accomplished with point-to-point systems. Early network technology borrowed multiplexing techniques from traditional communications concepts such as FDM, TDM, and CDMA. However, when the

TDM scheme approached the saturation point, the theoretical possibility of an all optical network was conceived. The breakthrough came in the form of optical amplifiers (EDFAs) and WDM, which enabled the network to add or drop wavelengths at intermediate points and to provide optical amplification at the repeater without the need for electrical/optical conversion. The evolutionary process from point-to-point networks to WDM **multipoint networks** is illustrated in Figure 13–2.

The implementation of wavelength add/drop multiplexing enabled an optical network to add or drop wavelengths at selected points without the necessity of electrical/optical conversion. The implementation of such networks required new families of optical components. A list of components required for the implementation of optical networks is as follows:

- dynamic gain equalizers
- EDFAs
- fibers
- integrated add/drop multiplexers
- tunable lasers/couplers

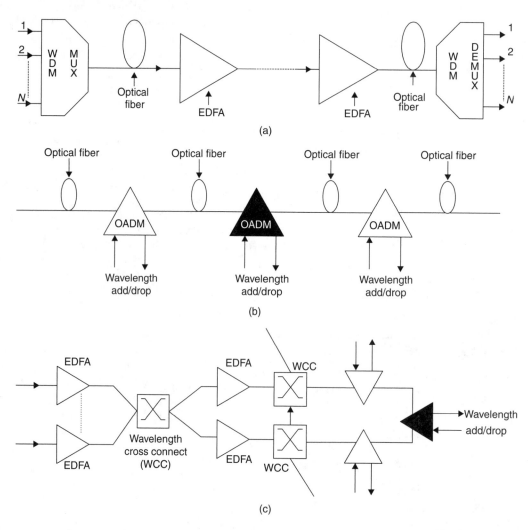

Figure 13–2. a) A WDM point-to-point transport network. b) A WDM multipoint transport network. c) A wavelength add/drop multiplexing cross connect network.

- **wavelength cross connect**
- wavelength division multiplexer cross connect fabrics
- wavelength division multiplexer sources
- wavelength division multiplexer routers
- wavelength monitors
- ultra wideband amplifiers
- optical filters

A block diagram of a basic optical transport network is illustrated in Figure 13–3.

In the early stages of the development, the feasibility of implementing optical networks was confined to the LAN level. However, the drive for optical network implementation beyond the LAN level was proceeding at an accelerated pace. As a result of this effort, the All Optical Network Consortium was formed to coordinate the activities that lead to the development of optical networks. The **Multiwavelength Optical NETworking** (MONET) program succeeded this consortium. The MONET program was sponsored by the **Defense Advanced Research Project Agency (DARPA),** which was composed of companies such as Bellcore, AT&T, Bell Atlantic, Southwestern Bell, and Bell South. The objective of the MONET program was to explore the feasibility of a national end-to-end optical network based on WDM, long reach transmission WDM cross connect, and add/drop multiplexing. The final objective of this program was to bring an all optical network to the customer. The two major forces behind this philosophy were the ever increasing demand for services by the end users led by the Internet, and the progressive cost decrease of optical links. Furthermore, small businesses could also benefit by directly connecting to the core networks through low speed SONET and **synchronous digital hierarchies-time division multiplexing (SDH-TDM).** Further increases of end user demand and fiber network cost reductions promoted the implementation of WDM to the access level. In interoffice applications, fiber optics interfaces have been developed for gigabit **Ethernet.** Such networks, which utilize either VCSELs that operate at the 850 nm wavelength combined with multimode fibers capable of reaching maximum lengths of 250 m, or edge emitter lasers which operate at the 1310 nm wavelength, with single mode-fibers capable of reaching maximum lengths of 5 km. Such systems can be utilized to accommodate 10Gb/s Ethernet formats. Another solution for bringing the fiber to the end user is the implementation of **passive optical networks (PON).** In such a network, an optical fiber connects the central office to sixteen **optical network units (ONU)** through a passive optical splitter. The function of the ONU is to convert the optical signals into electrical signals and to distribute them to thirty-two end users. The ONU, which feeds a number of end users is referred to as **Fiber To The Curb (FTTC),** while the optical network units terminating at the end users are referred to as **Fiber To The Home (FTTH).** To fully implement optical networks at the access level, a number of network operators came together and established the **Full Service Access Network (FSAN)** consortium. The main function of this consortium was to provide a common access platform for network operators. This

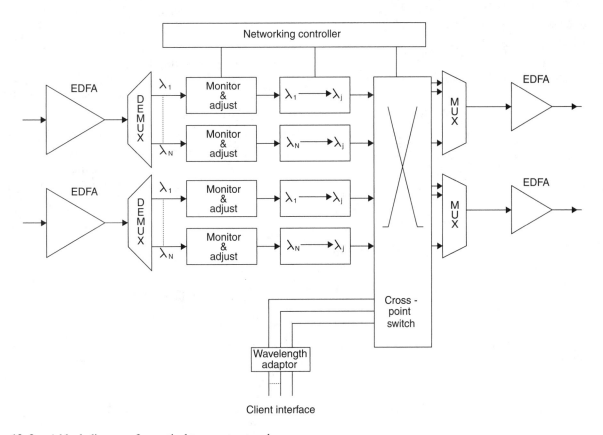

Figure 13–3.   A block diagram of an optical transport network.

platform is based on **wavelength division multiplexing passive optical network (WDM-PON)** systems. Using both a wavelength of 1500 nm for downstream and a 1310 nm wavelength for upstream transmission, 155 Mb/s baseband bidirectional signals formatted in ATM cells can be transmitted through the platform. For CATV, a maximum of 30 video channels can be transmitted through a coaxial cable. However, optical fiber networks can carry a much larger number of video channels of superior quality. For a detailed description of video signal transmission over optical fiber systems, see Chapter 12.

13.3   A REVIEW OF DATA COMMUNICATIONS LINKS

Fundamental Elements of Data Communications Links

Data communications links are designed to transfer information in series form or to establish links between two pieces of electronic equipment such as computers, **servers,** and **peripherals.** However, because the data provided by such equipment is in parallel form, it must first be converted to serial form for transmission. The main building blocks of such a link are as follows:

a) bus interface

b) buffering/framing

c) encoder

d) multiplexer

e) serial data processor

f) driver circuit

g) transmitter (optional)

h) transmitter medium

i) receiver

j) clock and data recovery

k) demultiplexer

l) decoder circuit

A block diagram of an optical fiber data link is illustrated in Figure 13–4.

A FUNCTIONAL DESCRIPTION.

1. **The bus interface:**   The function of the **bus interface** is to recognize and extract the data from the bus, and then transform the data from bus format into data link format. When data is received, the bus interface informs the local bus that data is to be stored in the local bus. It then converts the received data from data link format into local bus format.

2. **Buffering/Framing:**   The function of a **buffering/framing** block is to store the parallel data from the bus in the local memory, to divide the large blocks of data into smaller blocks (frames), and to add more frames (such as error tolerance codes and LAN management information). Finally, it sends all the generated frames sequentially into the encoding circuitry.

3. **Encoding:**   **Encoder** circuitry converts the data from an original format into a prescribed format, which can be recognized by the link.

4. **Multiplexer:**   The multiplexer circuit converts the parallel data into serial data, then applies the serial data to the input of the signal processor.

5. **Serial data processor:**   The **serial data processor** converts the serial data generated by the multiplexer into a form specified by the transmission link.

6. **Driver circuit:**   The driver circuit converts the input signal from a voltage level into the current level required to modulate the optical transmitter, either an LED or a laser diode.

7. **The transmitter (optional):**   The transmitter converts the electrical signal into an optical signal, then launches it into the optical fiber for transmission to the receiver end.

8. **Receiver:**   The receiver's function is to convert the received signal from current levels into voltage levels.

9. **Clock and data recovery:**   **Clock and data recovery** converts the received serial data into the original serial bit stream. The circuit is designed to lock onto the frequency

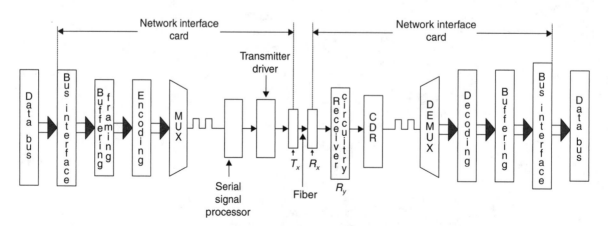

Figure 13–4.   A block diagram of an optical data communications link.

and phase of the incoming signal, to extract the clock signal, and to use it to check the incoming signal at each clock period. In this way, the data rate of the received signal is the same as that of the transmitted signal.

10. **Demultiplexer:** The demultiplexer converts the received serial data into a parallel stream.

11. **Decoding circuit:** The decoding circuit converts the received data into its original form.

The Encoding Process

Through the encoding process, the data signals and the clock signals are combined in a way that enables a timing recovery circuit at the receiver end to regenerate both the clock and the data signals into their original form. The encoding process is essential in order to enhance data link efficiency. The three most common encoding schemes are listed in Table 13–2.

The translation of data from bit (unencoded data) into baud (encoded data) is subject to the encoding scheme that is employed. In the absence of encoding, the number of bits is equal to the number of bauds.

13.4 NETWORKS

Data networks, based on transmission distance and equipment used, are classified as:

a) Local Area Networks (LAN)

b) **Metropolitan Area Networks (MAN)**

c) **Wide Area Networks (WAN)**

Local Area Networks

LANs are networks that are composed of a number of peripherals and other devices shared by several customers, connected and operating together at a distance not exceeding 300 m. The basic components incorporated into a LAN are as follows:

Network operating system: The network operating system is composed of a set of programs used by both the server and the customers on the same network.

Clients: **Clients,** also known as users or nodes, are computers that request the use of files and shared resources within the network.

Servers: Computers that share their hard disks (locally connected resources), peripherals, and communications circuits with clients and other servers are called *servers.*

Network peripherals: Peripheral equipment that incorporate specialized processors to run server software within the network are called *network peripherals.* These

peripherals may be connected directly to the LAN or to a node computer.

Network interface card: A **network interface card (NIC)** (LAN adaptor card) connects the clients, servers, and network peripherals to the LAN through a cable or an optical fiber line. The function of the NIC is to amplify and encode data before transmission.

Backbone: A **backbone** is a high speed network that connects a number of LANs (Figure 13–5).

Hub: A **hub** is a piece of equipment that usually contains four to forty-eight connection points between nodes and the server. In its design, a hub incorporates a repeater for signal regeneration.

Multipoint concentrator: A rack in which a number of hubs are mounted is a multipoint concentrator. **Concentrators** are used to facilitate a large number of nodes.

Bridge: A **bridge** is a store and forward device that separates shared LANs. Through this separation, the number of nodes/segments is reduced, which results in an increase of the bandwidth for each node. LANs connected by a bridge must be of the same protocol, because a bridge is unable to translate between different protocols. A bridge forwards a message it receives from one point of the LAN to another.

Router: In its principal application, a router is similar to a bridge. It is a store and forward device, which is really quite different from a bridge. Routers interpret logical or network addresses and forward messages to the appropriate segment, which is part of the destination network. Routers can perform protocol translations while bridges cannot.

Network Topology

The ITU-T G.805 recommendation does not define network topology, but it does define network **topology** components used to describe the transport network in terms of the topological relationship between sets of points within the same layer network. The literal definition of the term *topology* (Greek: *topologia*) is the study of geometric properties and spacial relations unaffected by the continuous change of space and size of a figure. In the ITU-T G.805 recommendations, four topology components are identified. These are:

1. **layer network**

2. subnetwork

3. link

4. access group

A layer network refers to combined resources that support special characteristic information. The transport processing functions of a layer network are set by **subnetworks,** while

TABLE 13–2 Encoding Schemes

Encoding	Description	Data Rate (Mb/s)	Data Rate (MBd)	Frequency (MHz)	Efficiency
Manchester	Each bit is replaced by two symbols	100	200	100	Low
4B5B	4-bit group is replaced by five symbols	100	125	62.5	High
5B6B	5-bit group is replaced by six symbols	100	120	60	Higher

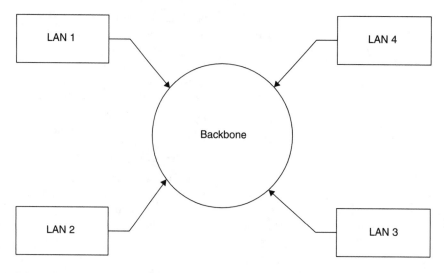

Figure 13–5. A backbone network.

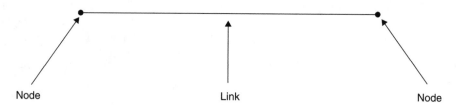

Figure 13–6. Point-to-point topology.

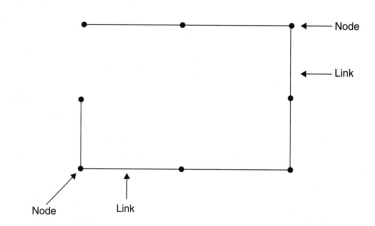

Figure 13–7. Linear chain topology.

transport identities are set by links, which, present fixed connectivity to a specified layer network. In accordance with ITU-T recommendations, G.805 layer networks are divided into two basic categories:

1. transmission media layer networks
2. path layer networks

A transmission media layer network is one which transfers information from a transmission medium layer network access point to one or more path layer networks.

A path layer network transfers information between path layer access points, and is independent of the transmission medium.

Network Topology Types

Eight network topology types exist. These are:

POINT-TO-POINT. A point-to-point topology connects two nodes through a link (Figure 13–6).

LINEAR CHAIN. A **linear chain topology** is composed of a number of point-to-point topologies connected in a series configuration (Figure 13–7). The nodes in a linear chain link are used for cross connect and termination functionality.

STAR TOPOLOGY. The nodes in a **star topology** are connected to a hub node, which provides cross connection.

However, if the hub node is disconnected, star topology connectivity is also lost. Star topology connection is illustrated in Figure 13–8.

Advantages

a) easy to modify cable layouts

b) allows for node expansion

c) provides centralized management

Disadvantages

a) requires longer cable lengths

b) single port network failure

c) expensive

TREE TOPOLOGY. In a **tree topology,** each node must have two or more connections with other nodes (Figure 13–9).

From Figure 13–9, it is evident that tree topology is a combination of linear chain and star topologies. Tree topology's advantages and disadvantages are as follows:

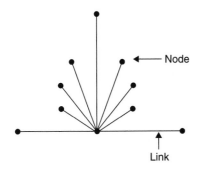

Figure 13–8. Star topology connection.

Advantages

a) allows for uncomplicated network expansion

b) simplified fault isolation

c) network fault local (confined to a single node)

Disadvantages

a) Complete network failure if the primary bus fails

RING TOPOLOGY. In a **ring topology,** each node is connected to two links (Figure 13–10). Therefore, the links form a ring, thus the name *ring* topology. Networks based on the ring topology format allow for route diversity.

Advantages

a) requires shorter cable lengths

b) less expensive

Disadvantages

a) more difficult when network configuration modifications are required

MESH TOPOLOGY. In **mesh topology,** two or more paths exist between at least two nodes. Mesh topology can also be classified as a combination of ring and tree topologies or a combination of all topologies (Figure 13–11).

Mesh topology can provide route diversity in the event that a point-to-point link has failed.

FULLY CONNECTED MESH TOPOLOGY. In a fully connected mesh topology, there is a direct link between all the nodes in the topology. Therefore, there is no need for a cross connect facility (Figure 13–12).

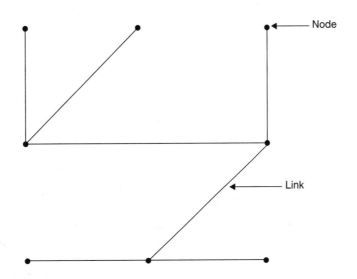

Figure 13–9. Tree topology connection.

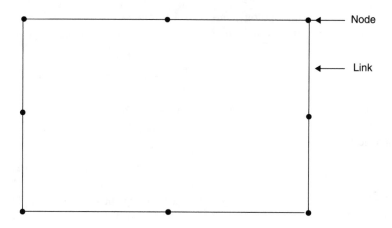

Figure 13–10. Ring topology connection.

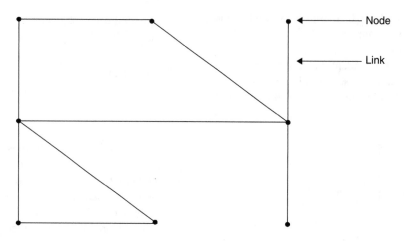

Figure 13–11. Mesh topology.

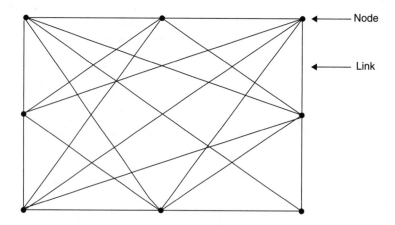

Figure 13–12. Fully connected mesh topology.

From Figure 13–12, it is evident that there are several paths that can provide route diversity in the event that a point-to-point link fails.

HIGHLY CONNECTED MESH TOPOLOGY. In a highly connected mesh topology, two paths exist between each and every other node. Based on this arrangement, the loss of a single link will not disrupt the connectivity of any other links (Figure 13–13).

13.5 NETWORK TRANSPORT ARCHITECTURE

The ITU-T G.805 recommendation defines network transport architecture as "any item used to describe transport network functionality," in contrast to network topology, which is defined as "any item used to describe transport network connectivity."

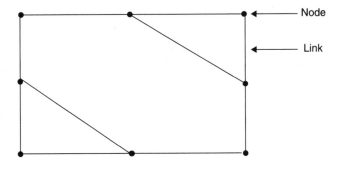

Figure 13–13. Highly connected mesh topology.

Local Area Networks

In LAN architecture, like Ethernet, all nodes communicate via a shared bus, thus allowing only one transmission at a time. In this way, the available bandwidth is shared among all the nodes in the network. However, an increase in the number of nodes will decrease the available bandwidth per node. This problem can be solved by subdividing the main network into a smaller number of LAN network segments through the utilization of routers and bridges. The block diagrams for two LAN networks, one employing a bridge and the other a router, are illustrated in Figure 13–14 and Figure 13–15.

Here, a smaller number of devices are connected to the hub, thus providing reasonably efficient bandwidth per node. The lack of sufficient bandwidth available to each node at the beginning of the development of LAN networks led to the development of switching. Through switching, each node has full access to the available LAN bandwidth. A comparison between a LAN that employs a router and a LAN that employs a bridge is listed in Table 13–3.

The Ethernet

Ethernet is identified by the IEEE 802.3 standard. Since the establishment of the Ethernet in the mid 1980s, it has become the most widely used transmission protocol in the world for LANs. The data rates and encoding schemes for three Ethernet protocols are listed in Table 13–4.

Ethernet is a shared bus network protocol with a **carrier sensing multiple access with collision detection (CSMACD)** access scheme. In an Ethernet configuration, a central hub is used, as defined by the LAN standard (Figure 13–16).

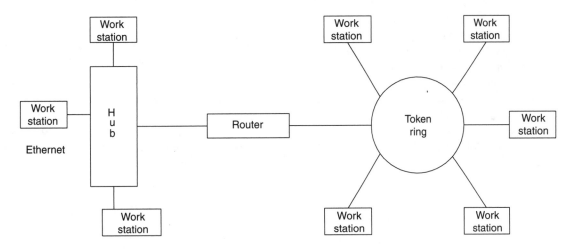

Figure 13–14. A LAN network that employs a router.

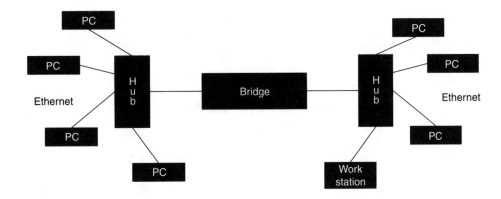

Figure 13–15. A LAN network that employs a bridge.

TABLE 13–3 LANs That Employ a Router and LANs That Employ a Bridge

Switching Support	Router Support
Hardware processing provides higher throughput	Software based route determination and packet processing
Full duplex switched links between nodes	Network node address resolution
Bandwidth management	Network management
Low cost	Protocol conversion and mismatch resolution

TABLE 13–4 Ethernet Data Bit Rates

Standard	Data Rates Mb/s	Data Rates MBd	Encoding Scheme
Ethernet	10	20	Manchester
Fast Ethernet	100	125	4B/5B
Gigabit Ethernet	1000	1250	8B/10B

Ethernet nodes account for at least 90% of all LANs, because of their proven stability, reliability, and adaptability to higher bit rates. From an initial 10 Mb/s, Ethernet has moved to 100 Mb/s, and more recently to 1 Gb/s with the next step being the 10 Gb/s Ethernet. In order for equipment manufacturers to satisfy Ethernet users, they offered 10/100 hub cables for use in both 10 Mb/s and 100 Mb/s connections and have plans underway to provide a 100/1000 hub in the not-so-distant future. However, because Ethernet is not a dedicated network but a standard network, telecommunications users still prefer ATM, capable of simultaneously handling voice, video, and data. Ethernet specifications and IEEE 802.x standards are listed in Table 13–5 (10 Mb/s/20 Mb), Table 13–6 (100 Mb/s/125 MbBd), and Table 13–7 (1000 Mb/s/1250 Mbd).

10Mb/s Ethernet Link

The structure of a 10 Mb/s Ethernet link, illustrated in Figure 13–17, is composed of the Ethernet controller, the **arithmetic unit interface** (AUI), a 10 Base-FL single chip transceiver, an HP transmitter, and an HP receiver connected to a multimode (MM) or single-mode (SM) fiber.

THE ETHERNET CONTROLLER. The Ethernet controller chip performs the following functions:

a) buffering/framing

b) encoding/decoding

c) multiplexing/demultiplexing

d) clock and data recovery

THE ARITHMETIC UNIT INTERFACE. The AUI interfaces either directly with the coaxial cable or through an optical fiber, with a **medium access unit (MAU)** composed of the following subcomponents:

a) 10 Base-FL single chip transceiver. This chip converts data from the AUI into the specific format required by the optical transmitter and receiver. It also incorporates the drive and equalizing circuitry, and monitors the reception, transmission collision, and error functions.

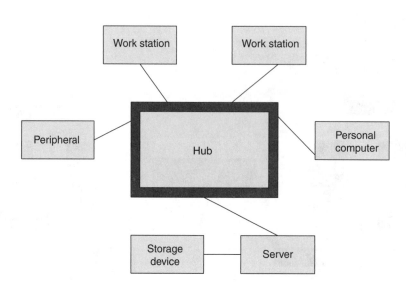

Figure 13–16. Ethernet configuration.

TABLE 13–5   10Mb/s/20MBd Ethernet

Specifications (IEEE 802.3)	Transmission Media	Type of Signal	Link Length (max)
10 Base 2	Coax	Electrical	185 m
10 Base 5	Coax	Electrical	500 m
10 Base T	Twisted wire pair (TP)	Electrical	100 m
10 Base FB[1]	Optical Fiber (MM)	Optical (820 nm)	2 km
10 Base FL FA[1]	Optical Fiber (MM)	Optical (820 nm)	2 km
10 Base FP[2]	Optical Fiber (MM)	Optical (820 nm)	500 m
*FOIRL	Optical Fiber (MM)	Optical (820 nm)	1 km

*FOIRL (fiber optics inter-repeater link)
[1]active repeater: more components required
[2]passive repeater: fewer components required

TABLE 13–6   100Mb/s/125MBd Ethernet

Specifications (IEEE 802.3u)	Transmission Media	Type of Signal	Link Length (max)
100 Base-T	Unshielded twisted pair	Electrical	100 m
100 Base-FB	Optical fiber (MM)	Optical (1300 nm)	2 km
100 Base-FL	Optical fiber (MM)	Optical (1300 nm)	2 km
100 Base-FP	Optical fiber (MM)	Optical (1300 nm)	500 m

TABLE 13–7   1 Gb/s Ethernet Specifications

Specifications (IEEE.802.3z)	Transmission Medium	Type of Signal	Link Length (max)
Gigabit Ethernet	Unshielded twisted pair	Electrical	100 m
Gigabit Ethernet	62.5/125 μm fiber	Optical (850 nm)	220/275 m
Gigabit Ethernet	50/125 μm fiber	Optical (850 nm)	500/550 m
Gigabit Ethernet	50/125 μm fiber	Optical (1300 nm)	550 m
Gigabit Ethernet	62.5/125 μm fiber	Optical (850 nm)	220/275 m
Gigabit Ethernet	9/125 μm fiber	Optical (850 nm)	5 km

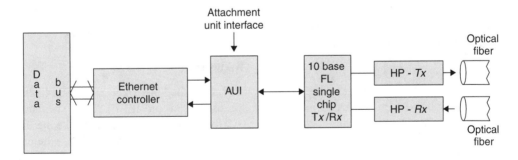

Figure 13–17.   A 10 Mb/s Ethernet link.

b) the transmitter contains the LED

c) the receiver contains the PIN photodiode and transimpedance amplifier

100 Mb/s Ethernet Link

The structure of a 100 Mb/s Ethernet link, illustrated in Figure 13–18, is composed of an Ethernet controller, an AUI, and an optical transceiver. The 100 Mb/s Ethernet controller performs the same function as that of the 10 Mb/s controller. The AUI directly interfaces with the fiber optics transmitter/receiver. The optical transceiver incorporates an LED or laser diode, a photodetector diode, and a driver and quantizer circuitry.

100 Mb/s Ethernet specifications and IEEE 802x standards are listed in Table 13–6.

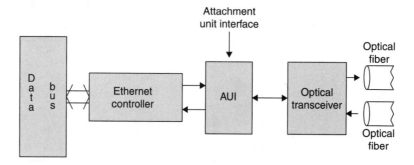

Figure 13–18. A 100 Mb/s Ethernet link.

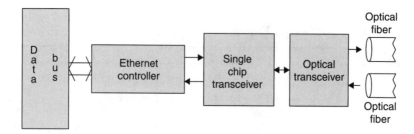

Figure 13–19. A 1 Gb/s Ethernet link.

1 Gb/s Ethernet Link

Gigabit Ethernet is an extension of 10 Mb/s and 100 Mb/s Ethernet, and is also based on IEEE 802.3 standards. Its operating bandwidth is 1 Gb/s, while the bandwidth is shared by all nodes in the network. Gigabit Ethernet allows for high speed operations at both half and full duplex operators. It addresses backward compatibility with 10 Base-T and 100 Base-T topology, and uses the CSMA/CD access method. Gigabit Ethernet can facilitate applications such as high volume data, high resolution video, and medical data, thus becoming the preferred choice for high bandwidth, high speed LAN networks. The structure of a 1 Gb/s Ethernet link, illustrated in Figure 13–19, is composed of the following: the Ethernet controller, a single chip electrical transceiver, and an optical transceiver.

The Gigabit Ethernet controller performs the same function as the 10 Mb/s and 100 Mb/s Ethernet links in Figure 13–17 and Figure 13–18. The single chip interfaces directly with the Ethernet controller of the optical transceiver. It also performs the serializing/deserializing function. The optical transceiver incorporates a VCSEL or an FP laser, a PIN photodiode, a preamplifier driver, and equalizer circuitry. One Gb/s Ethernet specifications and IEEE 802.3z standards are listed in Table 13–7.

An increase in the bandwidth of the optical fiber used as the transmission medium will increase the maximum distance of the network. Utilizing the 1000 Base-LX long wave transceiver and a multimode fiber, a maximum distance of 550 m can be achieved. However, if a single-mode fiber is used (100 BaseSX), a maximum distance of 5 km can be reached. For fibers that operate at lower bandwidths, the achievable distance is set at a 220 m maximum. As shown in Table 13–7, maximum distances for 1 Gb/s Ethernet range from 25 m when using an unshielded twisted pair (UTP), to a maximum distance of 5 km when using 9/125 μm optical fibers operating at the 1300 nm wavelength. However, if the HP transceiver is used, the network distance can be increased to 10 km. The relationship between the Ethernet's physical layer and data rates is listed in Table 13–8.

The evolution of the Ethernet LAN backbone network is illustrated in Figure 13–20.

13.6 LAN STANDARDS

Token Ring

IBM developed the **token ring** with 4 Mb/s and 100 Mb/s speeds, based on IEEE 802.5 standards. Usually, the 16 Mb/s speed is used as the backbone. In a token ring, a node can transmit in the network when it has the token. That is, the token travels around the ring and stops at any node to check for a request for data transmission. If such a request exists at a particular node, the token frame is declared busy, attaches the data and destination information to the node, and continues to travel around the ring. The recipient removes the information and indicates on the token ring frame that the information has been received. The token ring is then declared free. For token ring networks, an active monitor or manager is essential. This manager is provided with the active nodes, which use their serial numbers to determine which one will be the active monitor. The manager or ring monitor initiates the first

TABLE 13–8   Ethernet Physical Layer v. Data Rates

Operating System Interconnections (OSI) Reference Layers	10 Mb/s	100Mb/s/1Gb/s
Physical	Medium dependent interface (MDI)	Medium dependent interface (MDI)
Data link	Physical medium attachment (PMA)	Physical medium dependent (PMD)
Network	Attachment union interface (AMI)	Physical medium attachment (PMA)
Transport	Physical layer signaling	Physical coding sublayer (PCS)
Session	Reconciliation	Gb/s media independent interface
Presentation	Media access control (MAC)	Media access control (MAC)
Application	Logic link control (LLC)	Logic link control (LLC)

three tokens, thus providing equal accesses to the network for all nodes. In a token ring LAN, all nodes are connected in a logical ring configuration, and the information only travels in one direction. This shared LAN configuration ensures that all nodes in the ring have equal access to the LAN network for short or long periods of time. A token ring LAN is illustrated in Figure 13–21.

From Figure 13–21, token ring standards specify that a centrally located hub connected to PCs, **work stations,** peripherals, and the server must manage the LAN.

Token Ring Media Access Unit

A block diagram of a token ring media access unit (MAU) is illustrated in Figure 13–22.

A token ring MAU is composed of the token ring controller chip, the attachment unit interface, the transmitter, and the receiver. The controller chip performs the function of bus interface, multiplexing/demultiplexing, encoding/decoding, buffering, and clock and data recovery. The attachment unit interfaces directly to the coaxial cable or to the optical fiber medium access unit. The transmitter contains the LED and the receiver contains the PIN photodiode and the transimpedance amplifier.

HP, one of the leading manufacturers of components used in the design of optical networks, has developed efficient and inexpensive optical transmitters and receivers such as the HFB0400series intended for both token ring and Ethernet LAN systems. The transmitter and receiver specifications must comply with IEEE.802.5J standards. These standards for LED transmitters are listed in Table 13–9 and the standards for optical links are listed in Table 13–10.

From Table 13–9, the optical power launched into the 62.5/125 µm fiber that has an NA of 0.275 is equal to −12 dBm peak at 60 mA of forward LED current (I_F). However, the optical power drops to −15 dBm (avg) when

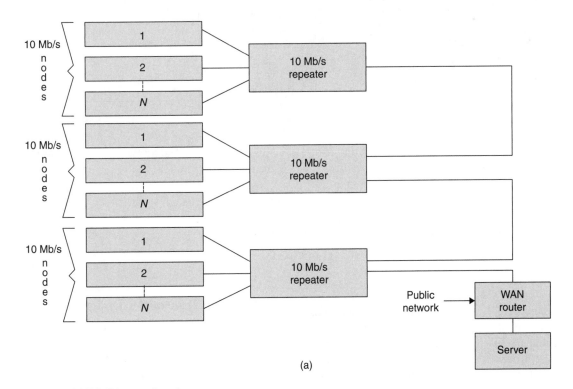

(a)

Figure 13–20.   a) A 10 Mb/s Ethernet shared.

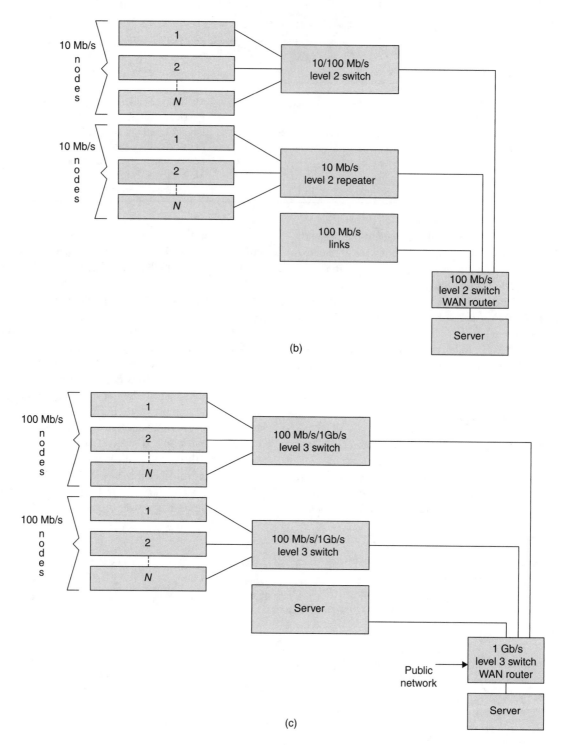

Figure 13–20.   b) A 100 Mb/s Ethernet backbone, c) A 1 Gb/s Ethernet backbone.

the data is Manchester coded (50% duty cycle), and transmitted a distance of 1km. The HFBR-24 × 6 optical receiver contains a PIN photodetector diode and a transimpedance amplifier. Both the HFBR-14 × 4 transmitter and the HFBR-26 × 6 receiver specifications exceed IEEE.805.2J recommendations.

Token Ring Transmitter Design

HP designed two circuits to drive the HFBR 14 × 4 LED transmitter for the token ring network. Both driver circuits comply with the IEEE.802.5J standards listed in Table 13–9 and Table 13–10. The first circuit, a voltage source driver circuit specifically designed for token ring applications, is

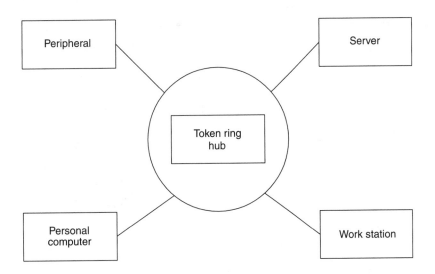

Figure 13–21. A token ring LAN.

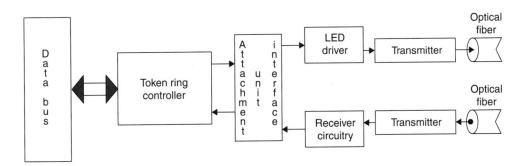

Figure 13–22. A token ring MAU.

TABLE 13–9 Optical Transmitter Specifications (IEEE.802.5J Recommendations)

Parameters	Transmission Bandwidth (8 MBd)	Transsmission Bandwidth (32 MBd)	Symbol	Unit
Optical power launched into fiber over life	–12 to –19	–12 to –19	$Pt_{(ON)}$	dBm (avg)
Extinction ratio	<13 dBm	<13 dBm	$Pt_{(Ext)}$	
Transmitter disabled	–38	–38	$Pt_{(OFF)}$	dBm (avg)
Rise time (optical) max	25	6	t_r	ns
Fall time (optical) max	25	6	t_f	ns
Rise and fall time diff.	12	3		
Symbol width dist.	± 4	± 1.5	OTA	ns

TABLE 13–10 Optical Fiber Link Specifications (IEEE.802.5J Recommendations)

Optical Fiber Length	Receiver 8 MBd	Rise/Fall Time 32 MBd	Optical Power (Received)	Jitter 8 MBd	32 MBd	Eye Opening 8 MBd	32 MBd
10 m	25 ns	6 ns	–12 dBm	9.9 ns$_{(pp)}$	5.8 ns$_{(pp)}$	115 ns	25.5 ns
2 km	60 ns	27 ns	–32 dBm	9.9 ns$_{(pp)}$	9.2 ns$_{(pp)}$	107 ns	22.1 ns

Optical fiber used: 62.5/125 µm

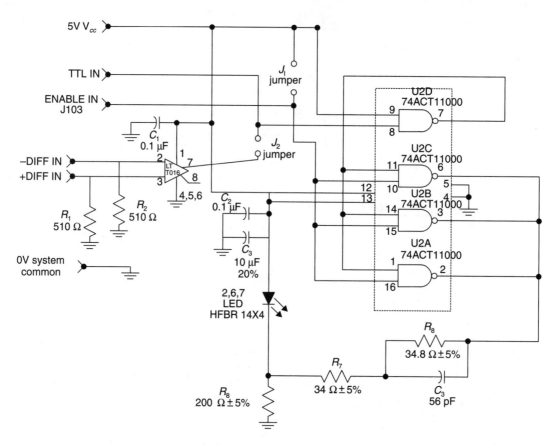

Figure 13–23. Voltage source transmitter for token ring networks.

illustrated in Figure 13–23. The forward current (I_F) of the LED, in accordance with I_F standards, must be maintained at 60 mA for logic-0 at pin 9 of the U4D. However, power supply (V_{cc}), LED forward voltage (V_F) variations, and component tolerances such as R7, R8, and R9 will affect the LED forward current (I_F). R7 and R8 are current limiting resistors that have a combined value of 68.8 Ω connected to the output of the three parallel NAND gates, with a combined output impedance of 1 Ω. Therefore, process variations within the 74ACT11000 chip will not affect the forward current (I_F) of the LED. However, any substantial variations in the LED forward current, resulting from power supply variations, resistor tolerances, and LED forward voltage variations, will result in a proportional variations in the optical power launched into the optical fiber. These variations are within the limits specified by IEEE.805.2J standards.

For proper circuit operation, the values of R7, R8, and R9 are calculated as follows.

LED Driver Circuit (Voltage Source)

Factors to be considered:

B is a constant. For optimization between prebias and LED forward current (I_F), the value of B (empirical constant) is set at 3.97.

N is the number of NAND gates

The required resistor values and the capacitor C_4 value are given by the following expressions.

The value of R_9 is expressed by Equation (13–1).

$$R_9 = \frac{(V_{cc} - V_F)(B + 1)}{I_{F(ON)}} \tag{13–1}$$

The value of R_8 is expressed by Equation (13–2).

$$R_8 = \frac{R_9}{2B} \tag{13–2}$$

The value of R_7 is expressed by Equation (13–3).

$$R_7 = \frac{R_9}{2B} - \frac{3}{N} \tag{13–3}$$

The value of C_4 is expressed by Equation (13–4).

$$C_4 = \frac{2 \times 10^{-9}}{R_8} \tag{13–4}$$

Example 13–1
Calculate R_9, R_7, R_8, and C_4 for the given values:

- V_{cc}=5 V
- V_F=1.5 V
- $I_{F(ON)}$=60 mA
- B=3.97

Solution

i) Calculate the value of R_9. From Equation (13–1),

$$R_9 = \frac{(V_{cc} - V_F)(B + 1)}{I_{F(ON)}}$$

$$= \frac{(5 - 1.5)(3.97 + 1)}{60 \times 10^{-3}}$$

$$= 290\ \Omega$$

Therefore, $R_9 = 290\ \Omega$.

ii) Calculate the value of R_7.

$$R_7 = \frac{R_9}{2B} - \frac{3}{N}$$

$$= \frac{290}{2(3.97)} - \frac{3}{3} = 36.5 - 1 = 35.5$$

Therefore, $R_7 = 35.5\ \Omega$.

iii) Calculate the value of R_8.

$$R_8 = \frac{R_9}{2B} = \frac{290}{2(3.97)} = 36.5\ \Omega$$

Therefore, $R_8 = 36.5\ \Omega$.

iv) Calculate C_4.

$$C_4 = \frac{2 \times 10^{-9}}{R_8} = \frac{2 \times 10^{-9}}{36.5} = 55 \times 10^{-12}$$

Therefore, $C_4 = 55$ pF.

A summary of the voltage source transmitter performance is presented in Table 13–11.

Current Source (Temperature Compensation)

If the optical power launched into the fiber is to maintain a small degree of deviation, a current source circuit will be required to drive the LED (Figure 13–24).

The LED forward current generated by the current source driver circuit is independent of V_{cc} and V_F variations, but it maintains its dependency on resistor tolerance and zener diode (U_3) band gap reference voltage. The current source circuit in Figure 13–24 also provides compensation for LED forward current (I_F) fluctuations resulting from temperature variations. Q_3 and Q_4 exhibit a negative coefficient of -2 mV/°C. With a temperature increase, the quantum efficiency of the LED also decreases, causing a proportional decrease in the optical power generated by the LED. However, this increase in temperature causes the base-to-emitter voltage of Q_3 and Q_4 to increase, reflecting

TABLE 13–11   Voltage Source Transmitter Performance Characteristics

Parameters	Performance (Measured)	Test Conditions
$p_{T(ON)}$	-12 dBm$_{(pk)}$	Logic-0 (transmitter TTL input) $I_F = 60$ mA
$p_{T(OFF)}$	-82.2 dBm$_{(pk)}$	Logic-1 (transmitter TTL input)
Rise time: t_r (LED)	1.3 ns	Input 1 MHz 50% duty cycle
Fall time: t_F (LED)	3.8 ns	Input 1 MHz 50% duty cycle
t_r–t_f	1.77 ns	Input 1 MHz 50% duty cycle
Transmitter jitter	0.023 ns(p-p)	72 MBd D2D2 hex input

a proportional increase in the LED forward current (I_F). This increase in the forward current compensates for the reduction of the LED quantum efficiency, thus maintaining the optical power launched into the fiber at a constant. The design rules for the current source driver circuit in Figure 13–24 are as follows.

Factors to be considered in the design of LED driver circuits:

The LED forward current (I_F) is expressed by Equation (13–5).

$$I_F = \frac{\lfloor \Delta V_{U_3} - V_{BE(Q_3)} \rfloor}{R_5} + \frac{\lfloor \Delta V - V_{BE(Q_4)} \rfloor}{R_6} \quad \textbf{(13–5)}$$

The value of R_3 is expressed by Equation (13–6).

$$R_3 = \frac{V_{OH} - V_{OL}}{I_F} \quad \textbf{(13–6)}$$

The value of C_4 is expressed by Equation (13–7).

$$C_4 = \frac{2 \times 10^{-9}\ s}{R_3} \quad \textbf{(13–7)}$$

Example 13–2

Calculate the LED forward current (I_F), the R_3, and the C_4 of the current source circuit for the given data.

- $\Delta V_{U_3} = 1.24$ V
- $V_{BE(Q_3)} = 0.7$ V
- $V_{BE(Q_4)} = 0.7$ V
- $R_5 = 17.5\ \Omega$
- $R_6 = 17.5\ \Omega$
- $V_{OH} = 5$ V
- $V_{OL} = 0$ V

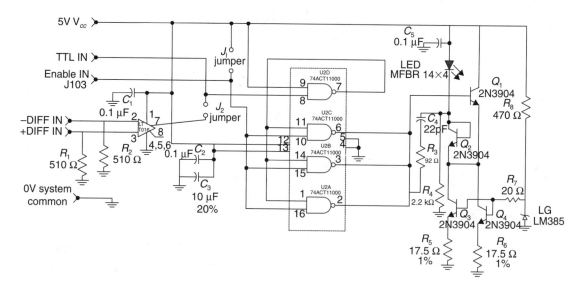

Figure 13–24. A current source transmitter for token ring networks.

Solution

i) Calculate I_F.

$$I_F = \frac{\lfloor \Delta V_{U_3} - V_{BE(Q_3)} \rfloor}{R_5} + \frac{\lfloor \Delta V - V_{BE(Q_4)} \rfloor}{R_6}$$

$$= \frac{(1.24\ \text{V} - 0.7\ \text{V})}{17.5\ \Omega} + \frac{(1.24\ \text{V} - 0.7\ \text{V})}{17.5\ \Omega}$$

$$= 61.7\ \text{mA}$$

Therefore, $I_F = 61.7\ \text{mA}$.

ii) Calculate R_3.

$$R_3 = \frac{V_{OH} - V_{OL}}{I_F}$$

$$= \frac{5\ \text{V} - 0\ \text{V}}{61.7\ \text{mA}}$$

$$= 81\ \Omega$$

Therefore, $R_3 = 81\ \Omega$.

iii) Calculate C_4.

$$C_4 = \frac{2 \times 10^{-9}\ \text{s}}{R_3}$$

$$= \frac{2 \times 10^{-9}}{81\ \Omega}$$

$$= 25 \times 10^{-12}\ \text{F}$$

Therefore, $C_4 = 25\ \text{pF}$.

Token Ring Receiver Design

The optical receiver used in token ring topology, illustrated in Figure 13–25, is composed of an HFBR-24 $\times$ 6 device and an ML4622 equalizer circuit. The ML4622 equalizer also incorporates a **link** monitor function. Through this function, the output data will be inhibited if the optical power drops

below a specified minimum level. The receiver specifications are as follows:

a) sensitivity: –34 dBm at a BER = 10^{-10}

b) BER = 10^{-10}

c) data: 32 MBd (Manchester encoded)

d) jitter: < 7 ns

e) eye opening: > 24 ns

The above specifications were set using a 2 km, 62.5/125 μm optical fiber.

A summary of the receiver performance characteristics when utilizing a 1 m, 62.5/125 μm optical fiber are listed in the Table 13–12.

A summary of the receiver performance characteristics when utilizing a 2 km 62.5/125 μm optical fiber are listed in Table 13–13.

An eye diagram of the receiver output versus clock signal for a long 3.32 km, 62.2/125 fiber, with received optical power of −32 dBm (avg), −14.8 dBm transmitter optical power, and 32 MBd D2D2 hexadecimal data is illustrated in Figure 13–26.

An eye diagram of the receiver output versus clock signal for a short 1 m 62.2/125 fiber, with received optical power of −11.4 dBm (avg) and 32 MBd D2D2 hexadecimal data, is illustrated in Figure 13–27.

Fiber Distributed Data Interface

Fiber distributed data interface (FDDI) is a LAN based on the ANSI standards committee developed in the 1980s. The FDDI was designed for transmission MM and SM fibers. With an available bandwidth of 100 Mb/s, the network can support backbone applications. A block diagram of an FDDI network is illustrated in Figure 13–28. From Figure 13–28, the FDDI network is a token passive network redundant ring

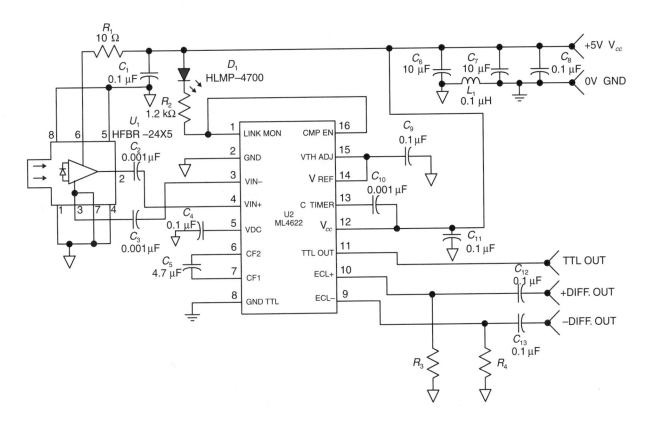

Figure 13–25. A token ring LAN receiver circuit.

TABLE 13–12 Receiver Performance Characteristics (1 m).

Parameters	Measured Performance	Test Conditions
Receiver sensitivity	-36 dBm (avg) BER $= 1 \times 10^{-10}$	32 MBd D2D2 Hex data
Link monitor assert threshold	-34.4 dBm (avg)	32 MBd D2D2 Hex data

TABLE 13–13 Receiver Performance Characteristics (2 km).

Parameters	Measured Performance	Test Conditions
Receiver sensitivity	-34.1 dBm (avg) BER $= 1 \times 10^{-10}$	32 MBd D2D2 Hex data
Link jitter receiver ECL output	6.91 ns (p-p)	$P_r = -32$ dBm (avg) (32 MBd D2D2 Hex data)
Link jitter receiver TTL output	5.52 ns (p-p)	$P_r = -32$ dBm (avg) (32 MBd D2D2 Hex data)

capable of providing high speed connections to a group of slower speed networks. The FDDI 100 Mb/s central backbone is composed of two rings. The outer ring (A) is used for primary network transmission with data flowing only in one direction, while the inner ring (B) is indented to transmit information in the opposite direction. It is also referred to as the redundant ring. In the event that the primary (outer) ring is disconnected, the secondary (redundant) ring will receive the rerouted data, thus maintaining network functionality. FDDI primary and secondary ring arrangement is illustrated in Figure 13–29.

Because the FDDI is a shared LAN network, its bandwidth is divided among all the devices connected to the network. Therefore, the larger the number of devices connected to the network, the less bandwidth per device. In order to increase the number of nodes in an FDDI backbone, concentrators, which allow for the connection of more nodes without reducing the bandwidth per each node, are used. The block diagram of an enhanced FDDI-LAN network is illustrated in Figure 13–30.

In order to avoid the possibility of erroneous connections between ring A and ring B, the standards committee has

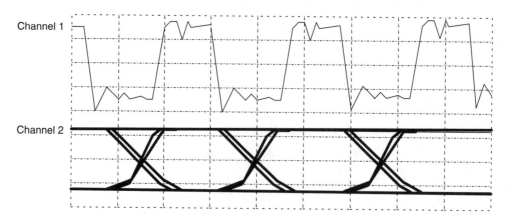

Figure 13–26. An eye diagram of receiver output *v.* clock for a 3.32 km link.

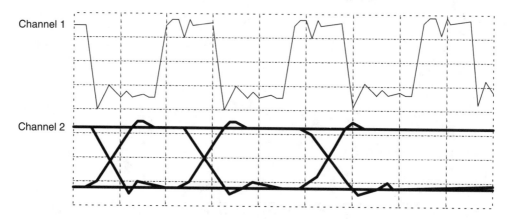

Figure 13–27. An eye diagram of receiver output *v.* clock for a 1 m link.

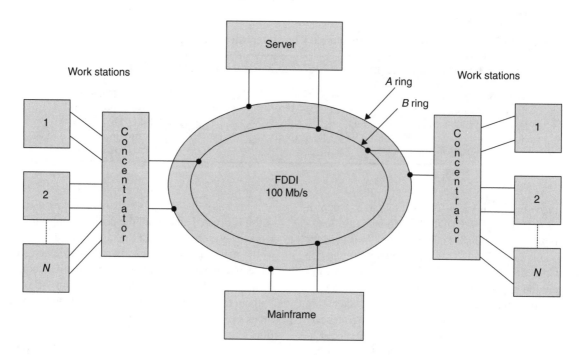

Figure 13–28. A block diagram of an FDDI.

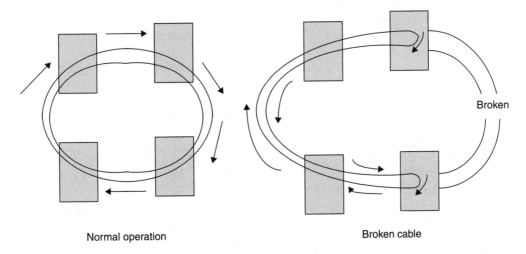

Figure 13–29. FDDI primary and secondary (redundant) rings.

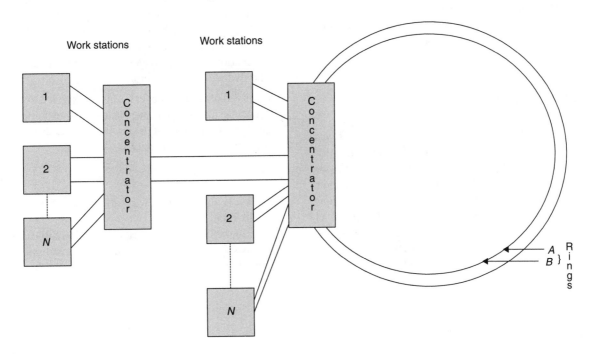

Figure 13–30. Enhanced FDDI LAN ring.

designed a special device with key options defined as follows in Table 13–14.

FDDI Encoding

The FDDI uses a 4B/5B encoding scheme that reflects an efficiency of approximately 80%. That is, for four bits of useful data, five bits are transmitted through the network. A 100 Mb/s useful data stream will require 125 Mb/s in order to be transmitted. Therefore, the FDDI link is a 125 MBd link. FDDI-LAN networks that use a twisted wire pair as the transmission medium can reach transmission distances of

TABLE 13–14   Key Options

Key	Purpose	Location
A	Connect to ring A	Any node connected to the backbone
B	Connect to ring B	Any node connected to the backbone
M	Connected to end device	Non-backbone side of a connection
S	Connect to concentration	End receiver connected to the concentration

100 m, while backbone connections that use optical fibers as the transmission medium can reach distances of 2 km.

FDDI Structure

FDDI links are composed of four layers. These are the:

1. Media access control (MAC), which controls the signal flow to and from the bus interface.

2. Physical layer, which performs the function of buffering/framing multiplexing/demultiplexing, encoding/decoding, and clock and data recovery.

3. Physical media dependent layer (PMD), which controls the transmitter drive circuit. The circuit incorporates a 1300 nm wavelength LED or laser transmitter, a 1300 nm operating wavelength PIN photodetector diode receiver, and the receiver circuit equalizer. A block diagram of an FDDI adapter card is illustrated in Figure 13–31.

4. The station management manages the layer configuration and provides reconfiguration in the event of network discontinuity.

13.7 THE FIBER CHANNEL

The conception and development of the **fiber channel** was driven by the necessity to serve both networks and channels. While networks are software oriented and are designed to perform unforeseen connections and commands, channels are designed to provide direct point-to-point connections between processors and peripherals. The fiber channel was designed to perform both network and channel tasks. The fiber channel standards incorporate three topologies and three classes of services, as illustrated in Figure 13–32.

Point-to-point (direct connect) is used for the transmission of large blocks of data and also as a mass data storage device. **Switched fabric** is used to cluster a number of devices through a switching fabric. An arbitrated loop (FC-AL) is used to connect three or more devices without of fabric. It allows the connection of several devices in a ring, forming a "virtual single device." An arbitrated loop used as a mass storage device is illustrated in Figure 13–33.

Classes of Services

To satisfy the large number of data communications requirements, the fiber channel provides three classes of services.

- Class 1: This is based on a circuit switched or hard switched connection. It is a dedicated link that connects two devices and is capable of providing a continuous and guaranteed delivery of data with acknowledgment of receipt.

- Class 2: This connection is based on frame switched service. It is not a dedicated connection. In frame switching, the switching mechanism reads the frame header code, then makes a decision for the payload destination. It also provides guaranteed delivery with acknowledgment of receipt.

- Class 3 (datagram transfer): This is a service that has no connection, and that allows data to be transmitted very quickly to a number of devices. Class 3 does not provide acknowledgment of receipt.

Efficiency

Fiber channel efficiency for various transmission rates is listed in Table 13–15.

Fiber channel was based on the IBM 8B/10B scheme, and was specifically designed for data transmission over an optical fiber.

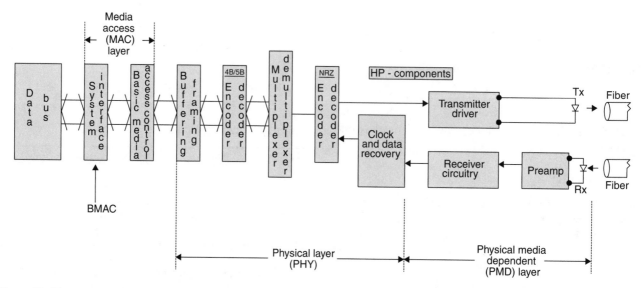

Figure 13–31. A block diagram of an FDDI adapter card.

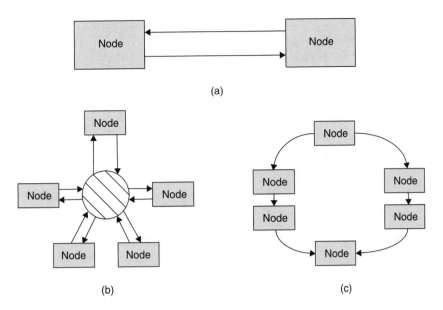

(a)

(b)   (c)

Figure 13–32. Fiber channel topologies: a) point-to-point (direct connect), b) switched fabric, and c) arbitrated loop.

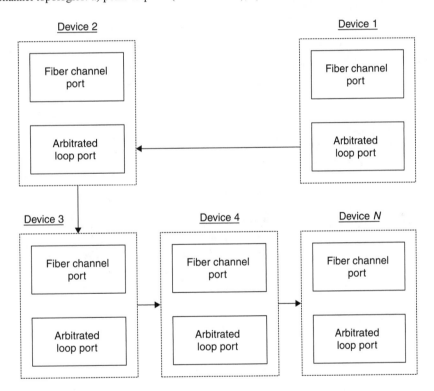

Figure 13–33. An arbitrated loop as a mass storage device.

TABLE 13–15 Fiber Channel Efficiency

Fiber Channel Speed	Serial Data (Mb/s)	Parallel Data (Mb/s)	Serial Signal (MBd)
8th (eighth)	100	12.5	133
4th (quarter)	200	25	266
2nd (half)	400	50	531
1 (full)	800	100	1062.5
2 (double)	1600	200	2125
4 (quad)	3200	400	4250

Fiber Channel Standards

The ANSI X311 committee developed a set of standards for the fiber channel. These standards are as follows:

1. The fiber channel's physical standards (FC-PH) X3.230-1994. This standard is subdivided into five sublevels: FC-0, FC-1, FC-2, FC-3, and FC-4. The fiber channel adapter card is illustrated in Figure 13–34. **FC-0** performs the following functions within the physical layer: serialization/deserialization, clock and data

recovery, and physical media dependent interface. **FC-1** is a transmission protocol that encodes/decodes, utilizing the 8B/10B encoding format. **FC-2** is a signaling protocol that performs the function of buffering framing of the data before transmission. **FC-3** is a common service. Although described in the standards, it has not yet been implemented. **FC-4** is a bus interface. It maps the data from the bus format to the fiber channel format.

2. The Fiber Channel Arbitrated Loop standard (FC-AL) X3.272-199x.

The Fiber Channel Transceiver

Because a relatively large number of transceiver devices for fiber channel applications exist, it is practically impossible to list and describe all of their operational functions. However, it is imperative that one or two transceiver devices from leading optical network manufacturers be described in some detail. One transceiver module designed for fiber channel applications is the HDMP-1536A/46A developed by HP. The HDMP-1536A/46A is an inexpensive transceiver chip intended for use in the design of a high speed fiber channel and an ANSI X3.230-1996 interface. A block diagram of a typical application of the device in a fiber channel protocol is illustrated in Figure 13–35.

General Description

At the transmitter input, the applied 10 bit parallel TTL data is encoded into 8B/10B format and then multiplexed into a high speed serial data stream. The PLL circuit in the transmitter section locks onto the 106.25 MHz clock frequency supplied by the user, then multiplies it by ten to generate the required 1.0625 GHz clock signal for a high speed serial output signal, which can be interfaced directly with the optical module for optical transmission. At the receiver end, the PLL circuit locks onto the incoming 1.0625 GBd serial data, from which it recovers both the data and the clock signals. At the same time, the 8B/10B comma character is recognized and then converted again to a 10 bit parallel data as a TTL output. The receiver section also recovers two 53.125 MHz byte clock signals with a phase difference of 180°, which is used to align the parallel data at the positive going transition. Block diagram of the HDMP-1536A/46A chip is illustrated in Figure 13–36.

Detailed Description of Operation

The HDMP-1546A transceiver is capable of transmitting and receiving 10 bit parallel data over a high speed line in accordance with the fiber channel standards for FC-0

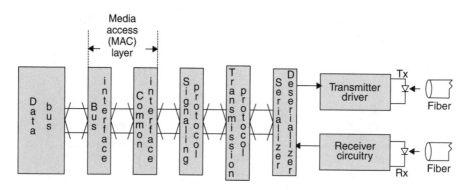

Figure 13–34. A fiber channel adapter card.

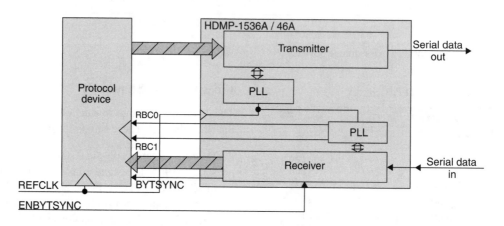

Figure 13–35. A block diagram of a fiber channel application of the HDMP-1536A/46A device.

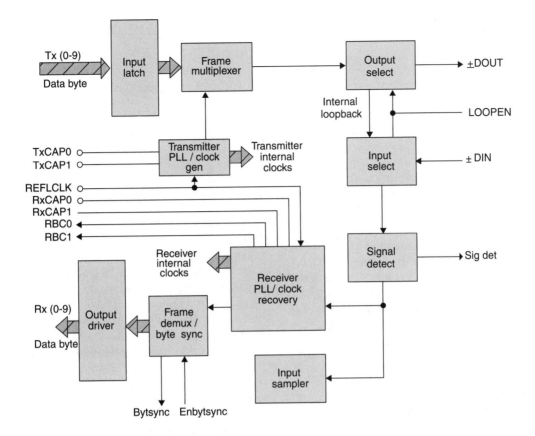

Figure 13–36. A block diagram of the HDMP-1546A transceiver device.

layer specifications. The transceiver incorporates the following:

- TTL parallel inputs/outputs (I/Os)
- High speed PLL circuits (transmitter receiver)
- High speed serial clock and data recovery circuitry
- Parallel-to-serial converters
- Comma character recognition circuitry
- Byte alignment circuitry
- Serial-to-parallel conversion circuitry

The functions are accomplished by the building blocks within the transceiver block. These are:

Transmitter:

a) input latch

b) PLL/clock generator

c) **frame multiplexer**

d) output select

Receiver:

a) input select

b) PLL/clock recovery

c) input sampler

d) frame demultiplexer and byte synchronization

e) output drivers

f) signal detect

The function of each block is as follows.

Transmitter

INPUT LATCH. A 10-bit TTL parallel data is applied at T_0 to T_9 inputs while the reference clock signal (REFCLK), provided by the customer, is used as the transmit byte clock signal. A diagram that illustrates the proper alignment of the data and clock signals is shown in Figure 13–37.

PLL/CLOCK GENERATOR. The PLL and clock generator circuitry generates all clock signals required by the transmitter section and is based on the signal supplied by the user reference clock signal. With a frequency of 106.25 MHz, it is aligned with the incoming parallel data. Furthermore, the same reference clock frequency multiplied by ten (1.0625 GHz) is used as reference frequency for high-speed serial transmission.

FRAME MULTIPLEXER. The frame multiplexer accepts the 10-bit parallel data input and converts it into a high speed (1.0625 GBd) serial data output through an internally generated high speed clock. Processing the input parallel data sequentially to series output generates the high speed serial data.

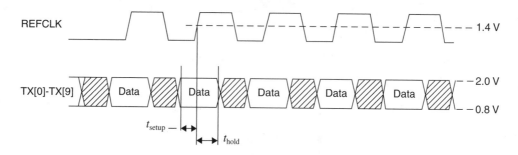

Figure 13–37. Transmitter data and reference clock signal timing diagram.

Output Select

The output select section tests the transmitter's functionality. With the LOOPEN at low state, the transmitter serial data is available at the $\pm D_{out}$ port. With the LOOPEN at high state, $\pm D_{out}$ is held at logic high and the serial high speed data is fed back to the receiver section to be tested. The timing characteristics of the transmitter section are listed in Table 13–16.

A timing diagram of the transmitter latency is illustrated in Figure 13–38.

Receiver

INPUT SELECT. The function of the input select block is to determine whether the incoming data at the $\pm D_{in}$ port or the internal transmitter output data is to be applied to the receiver section. This process is accomplished as follows: If the LOOPEN is at low, the incoming serial data at $\pm D_{in}$ is applied to the receiver input. If the LOOPEN is at high, the internal high speed data is applied at the receiver input for testing.

RECEIVER PLL/CLOCK RECOVERY. The receiver PLL and clock recovery section is designed to lock onto the frequency and phase of the incoming serial data stream and to recover both the byte and bit clocks. The receiver PLL locks automatically onto the incoming serial data stream by continually locking onto the 106.25 MHz clock and phase lock into the high speed serial data. The presence of the input signal is detected by a signal detection circuit, which triggers the phase detection function at the presence of this signal. As soon as bit lock has been achieved, the receiver automatically generates the 1.0625 GHz sampling clock frequency and also recovers the two out of phase (by 180°) 53.125 MHz byte clock frequencies (RBC1/RBC2), which are used to clock the 10-bit parallel output data.

INPUT SAMPLER. The function of the input sampler block is to convert the received serial input signal into a retimed serial bit stream. This signal is then applied at the input of the frame demultiplexer (FRAM-DEMUX and BYTE SYNC) for further processing.

TABLE 13–16   Transmitter Section Timing Characteristics

Parameters	Minimum	Typical	Maximum	Symbol	Unit
Setup time	2	—	—	t_{setup}	ns
Hold time	1.5	—	—	t_{hold}	ns
Transmitter latency	—	3.5	—	$t_{tx.lat.}$	ns

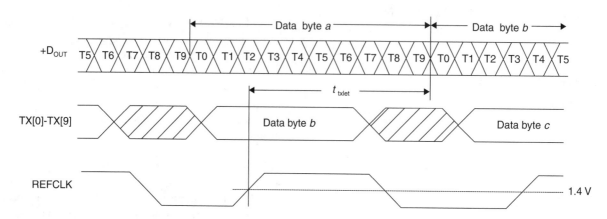

Figure 13–38. Transmitter latency timing diagram.

FRAME DEMULTIPLEXER AND BYTE SYNC. The serial data applied at the input of the frame demultiplexer is converted back into 10-bit parallel data. Furthermore, the same block is designed to recognize the comma character (K28.5) of positive disparity (0011111xxx). Upon successful recognition of the comma character, the frame demultiplexer, in conjunction with the receiver PLL/clock recovery circuit, aligns the receive byte clocks to the parallel data.

OUTPUT DRIVERS. The output drivers convert the 10-bit parallel data generated by the demultiplexer circuit into TTL compatible signals.

SIGNAL DETECT. The main function of the signal detect block is to prevent the generation of random data in the absence of serial data at the input of the receiver $\pm D_{in}$. The signal detect circuitry examines the differential amplitude of the input signal. If this amplitude is very small, it generates a logic low-0 at the SIG_DET output, and forces the receiver outputs (RX 0–9) to logic-1. If the input signal is normal, it generates a logic-1 at the SIN_DET output, and enables the data at the output of the input select to be processed further. A diagram that illustrates the proper alignment between the data and clock signals of the receiver is shown in Figure 13–39.

The timing characteristics of the receiver section are listed in Table 13–17.

A timing diagram of the receiver latency is illustrated in Figure 13–40.

The HDMP-1536A/46A transceiver clock requirements at operating temperatures between 0°C and +70°C and power supply of between 3.15 V and 3.45 V are listed in Table 13–18.

The HDMP-1536A/46A DC electrical specifications at operating temperatures between 0°C and +70°C and power supply of between 3.15 V and 3.45 V are listed in Table 13–19.

The device's absolute maximum ratings are listed in Table 13–20.

The HDMP-1536A/46A AC electrical specifications at operating temperatures between 0°C and +70°C and power supply of between 3.15 V and 3.45 V are listed in Table 13–21.

Design Implementation of the HDMP-1536A/46A Fiber Channel Transceiver

The HDMP-1536A/46A fiber channel transceiver incorporates multiplexing, demultiplexing, clock and data recovery, driver, and equalizer circuitry. In order to exploit the maximum performance from the device, certain rules must be applied during the layout process. Because the TTL lines exhibit rise and fall times of about 1.5 ns and 1.1 ns, respectively, and the high speed lines exhibit rise and fall times of about 0.225 ns, the layout interconnects within the device must be treated as transmission lines. This facilitates impedance matching, which results in a substantial reduction of signal back reflections. Transmission lines can either be **microstrip** or **strip lines.**

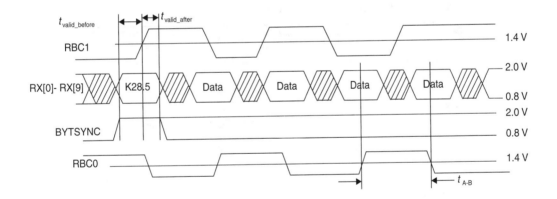

Figure 13–39. Receiver data and reference clock signal timing.

TABLE 13–17 Receiver Section Timing Characteristics

Parameters	Minimum	Typical	Maximum	Symbol	Unit
Bit sync time	—	—	2.5	b_sync	kbits
Time data valid before rising edge of RBC	3	—	—	t_{valid_before}	ns
Time data valid after rising edge of RBC	1.5	—	—	t_{valid_after}	ns
RBC duty cycle	40	—	60	—	%
Rising edge time difference between RBC0 and RBC1	8.9	9.4	9.9	$t_{A–B}$	ns
Receiver latency	—	24.5	—	t_{rxlat}	ns

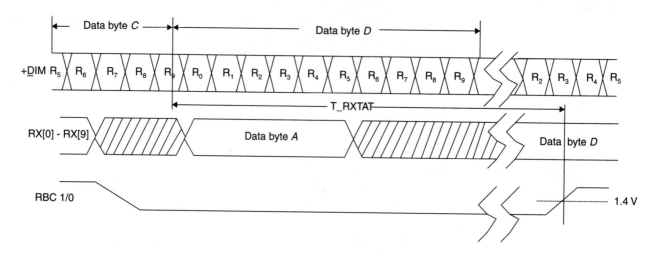

Figure 13–40. Transmitter latency timing.

TABLE 13–18   Transceiver Reference Clock Requirements.

Parameters	Minimum	Typical	Maximum	Symbol	Unit
Frequency (fiber channel)	106.20	106.25	106.30	f	MHz
Frequency tolerance	−100	—	+100	$f_{tol.}$	ppm
Duty cycle symmetry	40	—	60	Symm	—

TABLE 13–19   DC Electrical Specifications

Parameters	Minimum	Typical	Maximum	Symbol	Unit
TTL input high voltage	2	—	V_{cc}	V_{IH}	V
TTL input low voltage	0	—	0.8	V_{IL}	V
TTL output high voltage	2.2	—	V_{cc}	V_{OH}	V
TTL output low voltage	0	—	0.6	V_{OL}	V
Input high current	—	—	40	I_{IH}	μA
Input low current	—	—	−600	I_{IL}	μA
Transceiver V_{cc} supply current	—	205	—	I_{CC}	mA

MICROSTRIP TRANSMISSION LINES. A microstrip transmission line is constructed by placing the transmission line on top of the PC board material and a conducting plate on the bottom (Figure 13–41).

The characteristic impedance (Z_o) of the microstrip transmission line is a factor of the relative dielectric constant of the printed circuit board (ϵ_r), the width (w) of the copper transmission line strip, the thickness (h) of the dielectric, and the thickness of the copper strip (t). However, t becomes significant when its thickness is compatible to h. A very close approximation of the microstrip characteristic impedance (Z_o) can be obtained by Equation (13–7).

TABLE 13–20 Absolute Maximum Ratings

Parameters	Minimum	Maximum	Symbol	Unit
Supply voltage	−0.5	5.0	V_{cc}	V
TTL input voltage	−0.7	$V_{cc}+2.8$	V_{in}	V
Output source current	—	13	I_{out}	mA
Storage temperature	−65	+150	T_{stg}	°C
Junction operating temperature	0	+150	T_j	°C

TABLE 13–21 AC Electrical Specifications

Parameters	Minimum	Typical	Maximum	Symbol	Unit
REFCLK rise time (0.8–2.0 V)	0.7	—	2.4	t_r REFCLK	ns
REFCLK fall time (0.8–2.0 V)	0.7	—	2.4	t_f REFCLK	ns
Input TTL rise time (0.8–2.0 V)	—	2	—	t_r TTL$_{in}$	ns
Input TTL fall time (0.8–2.0 V)		2	—	t_f TTL$_{in}$	ns
Output TTL rise time (0.8–2.0 V)	—	1.5	2.4	t_r TTL$_{out}$	ns
Output TTL fall time (0.8–2.0 V)		1.1	—	t_f TTL$_{out}$	ns

$$Z_o = \frac{60}{(0.475\varepsilon_r + 0.67)^{1/2}} \ln\left[\frac{4h}{0.67(0.8w + t)}\right] (\Omega) \quad \textbf{(13–7a)}$$

or

$$Z_o = \frac{87}{(\varepsilon_r + 1.41)^{1/2}} \ln\left[\frac{5.98h}{(0.8w + t)}\right] (\Omega) \quad \textbf{(13–7b)}$$

where ε_r is the relative dielectric constant of the PC board, Z_o is the characteristic impedance of the transmission line (Ω), h is the thickness of the dielectric (mils), w is the width of the microstrip (mils), and t is the thickness of the microstrip (mils).

The propagation delay (t_{pd}) for the microstrip line is expressed by Equation (13–8).

$$t_{pd} = 1.017(0.475\varepsilon_r + 0.67)^{1/2} \text{ ns/ft} \quad \textbf{(13–8)}$$

Substituting Equation (13–8) into Equation (13–7) yields Equation (13–9).

$$Z_o = \frac{60}{t_{pd}} \ln\left[\frac{4h}{0.67(0.8w + t)}\right] (\Omega) \quad \textbf{(13–9)}$$

The capacitance of the transmission line is expressed by Equation (13–10).

$$C_o = \frac{t_{pd}}{Z_o} \text{ (pF/in)} \quad \textbf{(13–10)}$$

Example 13–2

Calculate the characteristic impedance (Z_o) and the capacitance (C_o) of a microstrip transmission line, given the following data.

- $\varepsilon_r = 4.9$
- $h = 5$ mils
- $w = 10$ mils
- $t = 0.5$ mils

Solution

i) Compute Z_o. From Equation (13–7),

$$Z_o = \frac{60}{(0.475\varepsilon_r + 0.67)^{1/2}} \ln\left[\frac{4h}{0.67(0.8w + t)}\right]$$

$$= \frac{60}{[0.475(4.9) + 0.67]^{1/2}} \ln\left[\frac{4(5)}{0.67[0.8(10) + 0.5]}\right]$$

$$= (34.65)\ln(3.5)$$

$$= 43.5 \ \Omega$$

Therefore, $Z_o = 43.5 \ \Omega$.

ii) Compute C_o. From Equation (13–8), calculate t_{pd}.

$$t_{pd} = 1.017(0.475\varepsilon_r + 0.67)^{1/2}$$

$$= 1.017[0.475(4.9) + 0.67]^{1/2}$$

$$= 1.76 \text{ ns/ft}$$

Transmitter and Receiver Input and Output Identifications

Abbreviations	Definition	Type	Description
BYTSYNC	Byte sync output	O-TTL	Active high output indicating detection of a comma character (0011111XXX)
±DIN	Serial data inputs	HS_IN	High speed inputs. This input serial data is accepted at a low LOOPEN
±DOUT	Serial data output	HS_OUT	High speed output. These outputs are enabled with LOOPEN low
ENBYTSYNC	Enable byte sync input	I-TTL	When high it activates the internal byte sync function for clock synchronization with the comma character
GND	Logic ground	S	Zero volts. Used for internal PECL logic, it must be well isolated from the TTL noisy ground
GND_RXA	Analog ground	S	Zero volts. Provides a clean ground plane for the receiver PLL
GND_RXTTL	TTL receiver ground	S	Zero volts. Provides a clean ground for the TTL outputs of the receiver section
GND_TXA	Analog ground	S	Zero volts. Provides a clean ground plane for the transmitter PLL
GND_TXHS	Ground	S	Zero volts. Provides a clean ground plane for the transmitter high speed outputs
LOOPEN	Loopback enable input	I-TTL	When high, the transmitter serial high speed output is looped back into the receiver input while ±DOUT are held at static-1
RBC0 RBC1	Receive byte clocks	O-TTL	The receiver recovers the two 53.125 MHz clock signals, which are out of phase by 180° and uses these signals to clock the parallel output data at the rising edge
REFCLK	Reference clock and transmit byte clock	I-TTL	106.25 MHz clock provided by the supplier. The transmitter multiplies this clock by ten and uses this frequency for the required internal clock signals
RX (0-9)	Data outputs	O-TTL	10-bit data byte with RX(0) the LSB and RX(9) the MSB
RXCAP0 RXCAP1	Loop filter capacitor	C	0.1 µF capacitor connected across RXCAP0 and RXCAP1 used as loop filter
TX (0-9)	Data inputs	I-TTL	10-bit data byte with TX(0) the LSB and TX(9) the MSB
TXCAP0 TXCAP1	Loop filter capacitor	C	0.1 µF capacitor connected across TXCAP0 and TXCAP1 used as loop filter
Vcc	Logic power supply	S	3.3 V used for the receiver PECL logic
Vcc_RXA	Analog power supply	S	3.3 V used for the PLL circuitry
Vcc_RXTTL	TTL power supply	S	3.3 V used for the receiver TTL output buffers
Vcc_TXA	Analog power supply	S	3.3 V used for the TX PLL circuitry
Vcc_TXECL	High speed ECL supply	S	3.3 V used for the high speed transmission output
Vcc_TXHS	High speed supply	S	3.3 V used for the TX high speed circuitry

or

$$t_{pd} = 1.76\,(1000/12)$$
$$= 146.73 \text{ ps/in}$$

Therefore, $t_{pd} = 146.73$ ps/in.

iii) Calculate C_o. From Equation (13–10),

$$C_o = \frac{t_{pd}}{Z_o} = \frac{147.73 \text{ ps/in}}{43.5 \ \Omega} = 3.4 \text{ pF/in}$$

Therefore, $C_o = 3.4$ pF/in.

In order to minimize back reflections, the total transmission line round trip delay time must be less than the rise or fall time (whichever is smaller) of the driven signal.

STRIP LINE TRANSMISSION LINES. In a strip line transmission line, the signal is transmitted through a line that is confined between the PC board dielectric material and two conducting ground planes (Figure 13–42). As before, the characteristic impedance (Z_o) of the strip line is a function of the relative dielectric constant (ε_r) of the material that composes the PC board, the thickness of the dielectric (h), and the length (b) of the strip line (Equation (13–11)).

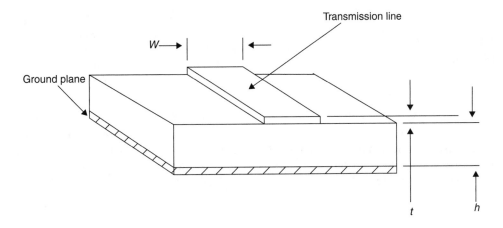

Figure 13–41. A microstrip transmission line.

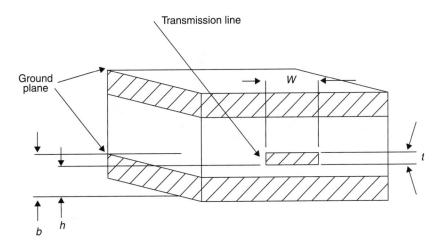

Figure 13–42. A strip line transmission line.

$$Z_o = \frac{60}{\sqrt{\varepsilon_r}} \ln\left[\frac{4b}{0.67\pi(0.8w + t)}\right] \qquad (13\text{--}11)$$

The propagation delay of the strip line is expressed by Equation (13–12).

$$t_{pd} = 1.017\sqrt{\varepsilon_r}\ \text{ns/ft} \qquad (13\text{--}12)$$

Example 13–3

Calculate the characteristic impedance (Z_o) and the propagation delay of a strip line transmission line, given the following data:

- $\varepsilon_r = 4.7$
- $b = 10$ mils
- $w = 4$ mils
- $t = 0.5$ mils

Solution

i) Calculate Z_o. From Equation (13–11),

$$\begin{aligned}
Z_o &= \frac{60}{\sqrt{\varepsilon_r}} \ln\left[\frac{4b}{0.67\pi(0.8w + t)}\right] \\[2mm]
&= \frac{60}{\sqrt{4.7}} \ln\left[\frac{4(10)}{0.67\pi[(0.8)(4) + 0.5]}\right] \\[2mm]
&= (27.67)\ \ln(5.136) \\[2mm]
&= 45.27\ \Omega
\end{aligned}$$

Therefore, $Z_o = 45.27\ \Omega$.

ii) Calculate t_{pd}. From Equation (13–12),

$$\begin{aligned}
t_{pd} &= 1.017\sqrt{\varepsilon_r}\ \text{ns/ft} \\[2mm]
&= 1.017\sqrt{4.7} = 2.2\ \text{ns/ft}
\end{aligned}$$

Therefore, $t_{pd} = 2.2$ ns/ft.

or

$$t_{pd} = 2.2 \text{ ns/ft} \times (1000 \div 12)$$
$$= 183 \text{ ps/in}$$

Therefore, $t_{pd} = 183$ ps/in.

PC BOARD LAYOUT CONSIDERATIONS. Proper layout of the PCB is essential if parasitic inductances and capacitances are to be maintained to a minimum. Control of parasitic elements increases the VSWR of the high speed transmission lines and prevents ringing.

POWER SUPPLY CONSIDERATIONS. The voltage at the power supply pins of the device must be free of noise. To accomplish this, bypass decoupling capacitors must be connected to the appropriate power supply pins. These capacitor values are about 0.1 μF (Figure 13–43).

PHASE LOCK LOOP FILTER CAPACITORS. The PLL circuitry of the receiver section must be protected from picking up stray noise from neighboring components or transmission lines. The values set for these capacitors (CPLLT and CPLLR) are also about 0.1 μF.

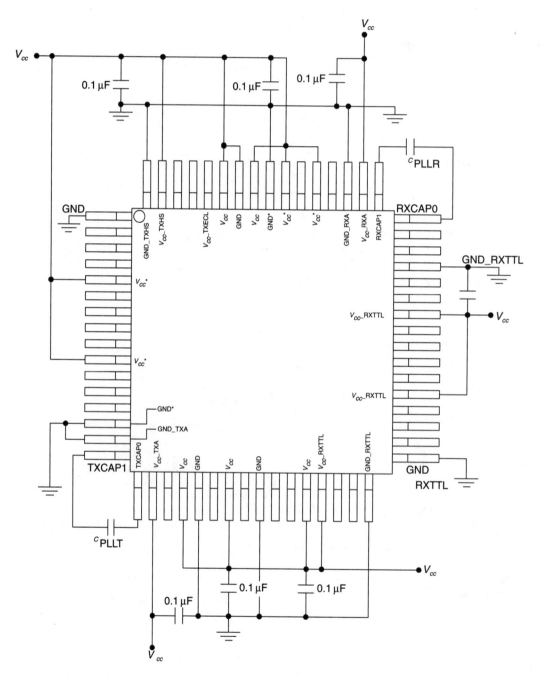

Figure 13–43. Bypass capacitor connections.

INPUT/OUTPUT INTERFACING. In the HDMP-1536A/46A chip, four types of inputs/outputs exist. They are listed in Table 13–22.

From Table 13–22, it is evident that four types of inputs and outputs exist—TTL and high speed. Two high speed connections between the transmitter outputs (+DOUT, −DOUT) and receiver inputs (+DIN, −DIN) are illustrated in Figure 13–44 and Figure 13–45.

The values of R_5 and R_6 are subject to the characteristic impedance of the transmission lines and the transmitter output drives. For example, if Z_o is equal to 50 Ω, then the values of R_5 and R_6 must also be equal to 50 Ω each. The connectors also incorporate a set of series padding resistors (R_3 and R_4) used to dampen load reflections. These resistor values are between 25 Ω and 75 Ω. Figure 13–46 shows that a single differential resistor can be used with C_1 and C_2 of 0.01 μF each, and R_5 equal to 250 Ω.

OPTICAL FIBER INTERFACE. The HDMP-1536A/46A transceiver chip can also be connected to the HFBR-53D3 fiber optics transceiver. In this case, HP has issued the following chip layout guidelines:

HDMP-1536A/46A and HFBR-53D3 Layout Guidelines

a) controlled impedance transmission lines must be used for high speed serial and parallel lines

b) in order to minimize pulse distortion, high speed differential lines must be of equal length and routed together

c) for a much needed reduction of the power supply noise, filters must be connected as close as possible to V_{cc} of the HFBR-53D3 fiber optics transceiver

d) high speed load and source resistors must be kept very close to the device pins

e) charge pump capacitors for transmitter and receiver PLL circuitry must be connected very close to the device pins with very short connections

JITTER TRANSFER. In order to establish jitter transfer for the transmitter PLL circuit, the 106.25 MHz REFCLK

signal is modulated by a sinewave, which is swept between 10 KHz and 7 MHz. In this way, the jitter transfer for a 1.0625 Gb/s high speed output can be established. In essence, the PLL is used as an LPF capable of detecting any noise in the FERCLK signal. Jitter transfer of the HDMP-1536A/46A transceiver is illustrated in Figure 13–46.

The eye diagram for high speed differential output is illustrated in Figure 13–47.

13.8 ASYNCHRONOUS TRANSFER MODE

Introduction

Traditionally, transmission of information voice, video, or data through a LAN requires different methods of transmission from that of WANs, especially when network connectivity is migrating from LANs to MANs to international networks. Furthermore, transmission of voice, video, and data requires separate networks that exhibit distinct transmission characteristics. For example, data transmission can sometimes be transmitted for a considerable length of time without the need for communicating, while other times it needs to communicate information very fast. Furthermore, in contrast to data transmission, voice and video are extremely sensitive to the time and order of arrival at the receiver end. This rather complex and expensive process can be simplified through the implementation of ATM. Because ATM is available with different speeds, it can transmit voice, video, and data simultaneously under one network. ATM technology is very flexible and very powerful. It will soon become the standard for both LAN and WAN networks. At the beginning of the transmission process, the sender requires a path from the network, while defining destination call type and speed of transmission. The sender defines an end-to-end quality of transmission. Furthermore, because ATM is a switch instead of a shared process, it benefits from a higher bandwidth, flexibility of transmission, speed, and defined connection procedures such as fixed cell lengths. A block diagram of the ATM system architecture is illustrated in Figure 13–48.

ATM system architecture, illustrated in Figure 13–48 is composed of the following layers:

a) Adaptation layer (AAL)

b) ATM layer (ATM)

c) Physical layer (PHY)

The adaptation layer divides the incoming data into a 48-byte payload, and the ATM layer adds 5-byte header information into the 48-byte payload, thus producing a 53-byte ATM cell. The 5-byte header is intended to direct the payload to the right connection, and the physical layer defines the network interface.

TABLE 13–22   Input/Output Interfacing

Definitions	Types	Pins
Input TTL (high when open)	I-TTL	Tx (0. . . 9)
Output TTL	O-TTL	RBC0 RBC1 BYTSYNC
High speed output PECL compatible	HS_OUT	+DOUT −DOUT
High speed input PECL compatible	HS_IN	+DIN −DIN

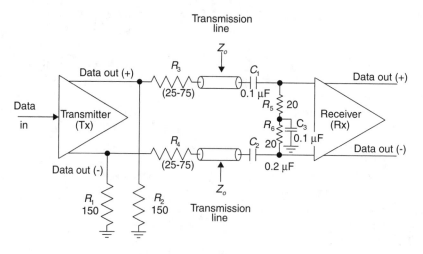

Figure 13–44. High speed transmitter receiver connection.

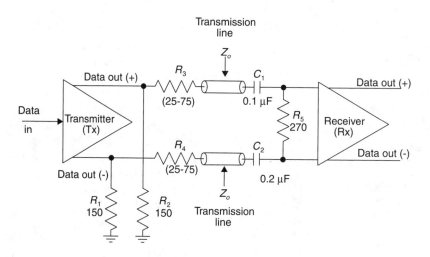

Figure 13–45. High speed transmitter receiver connection with single terminator resistor.

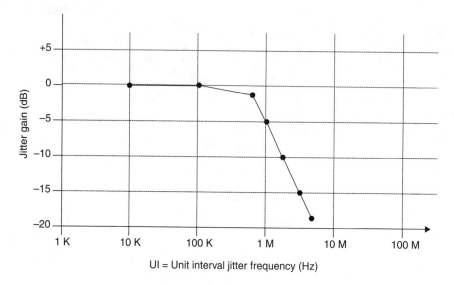

Figure 13–46. HDMP-1536A/46A transmitter jitter transfer.

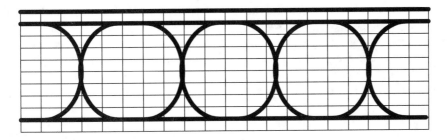

Figure 13–47. Eye diagram for high speed differential output (HDMP-1536A/46A).

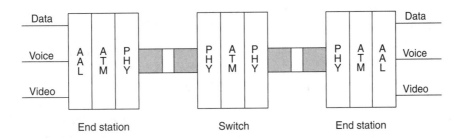

Figure 13–48. ATM system architecture.

Advantages of ATM

ATM allows for simultaneous transmission of voice, video, and data, thus improving efficiency and network manageability. High speed transmission and traffic integration makes ATM a better choice for future network upgrading. ATM is compatible with preexisting physical networks, and can be transported by optical fibers, coaxial cable, or twisted pair lines. Through the implementation of the same technology for applications such as public and private wide area services, network management can be significantly simplified.

Based on the proven flexibility and scalability in terms of number of users, bandwidth availability, and transmission distances, the telecommunications industry is ever closer to ATM universal standardization.

13.9 ATM OR BROADBAND ISDN PROTOCOL

High speed voice, video, and data transmission can be accomplished more effectively through the utilization of **broadband integrated services digital network (B-ISDN)** protocol, commonly known as ATM.

ATM provides the information packet structure. Transmission of the generated packet is accomplished through the physical layer. In an ATM protocol, data intended for transmission is subdivided into blocks of 53 bytes, referred to as ATM cells. These cells are framed and transported through the link, while at the receiver end the same cells are extracted from the frame then reconstructed to the original data form. ATM cells from different databases can be transported

through the same link, with varying bandwidths subject to database requirements. The term *asynchronous* is indicative of the varying bandwidth in the transfer mode. The ATM (B-ISDN) protocol is composed of the user plane, the control plane, and the management plane. The user plane is a layer structure intended to transfer user application information. The control layer is used for connection control, and receives signaling. The management plane is used for layer and plane management and incorporates functions such as monitoring connections and providing ways for information exchange between user and control planes. The user, the control, and the management planes are all layer structures. Layers are defined by the ATM (B-ISDN) protocol as being service specific, designed to meet different application requirements. The function of ATM layer is to transfer the 53 byte cell from the source to the receiver. The physical layer is designed to provide the transport mechanism between the sender and the receiver. The physical layer incorporates functions such as line drivers, connectors, frame/byte alignments, and clock and data recovery. Both ATM and physical layers are service independent. A more detailed description of the layers is as follows:

Transmitter End: The Adaptation Layer

Perhaps the most important feature of the ATM process is that of the **adaptation layer (AAL).** From Figure 13–48 the AAL maps the data provided by the user into the ATM cell, which is composed of 53 bytes, subdivided into the **convergent sublayer (CS)** and the **segmentation and reassembly (SAR)** sublayer.

The CS subdivides user information into smaller segments and adds on a header required by the AAL layer for error detection. The result is referred to as the **convergence sublayer protocol data unit (CS-PDU),** or packet. The segmentation and reassembly sublayer segments the packet into 48 bytes, called **segmentation and reassembly protocol data units (SAR-PDU),** also adding header/trailer information as required by the AAL layer. The SAR-PDU is sent to the ATM layer, where a 5-byte header is added to complete the 53-byte ATM cell. Four types of adaptation layers exist, the AAL-1, AAL-2, AAL-3/4, and AAL-5.

AAL-1. Adaptation layer 1 is used to support constant bit rate (CBR) traffic such as voice and video. The AAL-1 format is illustrated in Figure 13–49.

The AAL-1 (SAR-PDU) is composed of a 47-byte payload and a 1-byte header. The header consists of a 4-bit sequence number (SN) field used to detect incorrectly inserted or lost cells, and a 4-bit **sequence number protection (SNP)** field intended for the protection of the SN field from possible errors.

AAL-2. Adaptation layer 2 is composed of the SAR-PDU header, the SAR-PDU payload, and the SAR-PDU trailer. The header is composed of the 4-bit sequence number (SN) field and the IT field, while the trailer is composed of the **cyclic redundancy check (CRC)** field and the LI field.

AAL-3/4. Adaptation layer 3/4 is used to support **variable bit rate (VBR)** and **available bit rate (ABR)** traffic. The AAL-3/4 format is illustrated in Figure 13–50.

From the 48 bytes available to the AAL-3/4 layer, 4 bytes are used in the SAR-PDU for CRC, error detection, and a **segment type (ST)** field that indicates the start, continuation, and end of a message. Furthermore, 8 bytes are used as length indicators for the CS-PDU payload. However, the utilization of the AAL-3/4 format in a LAN network is bandwidth inefficient. This is because 4 bytes are used as overhead in the SAR-PDU sublayer, thus reducing the available 48-byte payload by 8.3%. Furthermore, if an error occurs during the transmission, it will be difficult to detect at the packet level. However, increasing the bit number of the CRC used as error detection in the packet level improves bandwidth efficiency.

AAL-5. The adaptation layer 5 format, similar to AAL-3/4, supports VBR traffic as well as ABR traffic. The structure of the AAL-5 format is illustrated in Figure 13–51.

AAL-5 contains a maximum payload of 65,536 bytes and an 8-byte trailer that consists of a 2-byte field, a 2-byte control field, and a 32-bit CRC. The 48 bytes available to AAL-5 SAR-PDU are used for payload without the extra addition of header/trailer information.

Receiver End: The ATM Layer

At the receiver end, the SAR sublayer, upon receipt of the ATM cell, removes the header/trailer and sends the remaining data to the CS. The CS checks for errors, then removes the header/trailer information and presents the

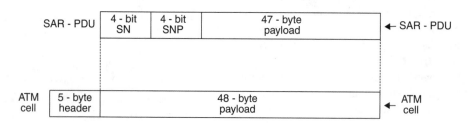

Figure 13–49. Adaptation layer-1 (AAL-1) format.

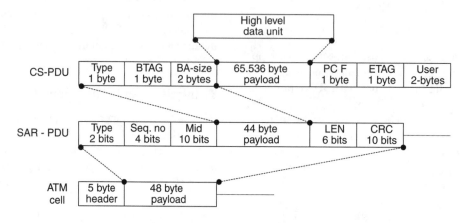

Figure 13–50. Adaptation layer-3/4 (AAL-3/4) format.

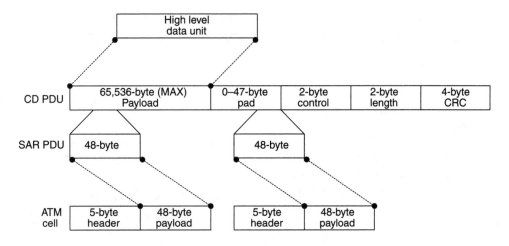

Figure 13–51. Adaptation layer-5 (AAL-5) format.

remaining data, which may or may not be in the original form, to the next.

The function of the ATM layer is to successfully transfer ATM cells from the sender to the receiver. Upon receiving a 48-byte SAR-PDU format from the AAL layer, the ATM adds on a 5-byte header, thus completing the ATM cell. The 5-byte ATM header is formatted in two ways: one is intended for **user network interface (UNI),** and the other is intended for **network node interface (NNI).** Both formats are illustrated in Figure 13–52.

UNI is the interface between the **customer premise equipment (CPE)** and the public network, while NNI is the interface between the different public carriers, for example, interface between two local carriers, a local and a long distance carrier, or two long distance carriers. The basic difference between UNI and NNI lies in the header structure. The UNI adds on a generic flow control (GFC) field, which is used to control the traffic flow at each UNI and schedule the information transfer between various sources. The user UNI header provides 24 bits for data routing, 8 bits for **virtual path identification (VPI),** and 13 bits for **virtual channel identification (VCI).**

The UNI header, composed of 5 bytes, is structured as follows:

Byte 1

a) bits 1–4 allocated for VPI

b) bits 5–8 allocated for GFC

Byte 2

a) bits 1–4 allocated for VCI

b) bits 5–8 allocated for VPI

Byte 3

a) 8 bits allocated for VCI

Byte 4

a) 1 bit allocated for **cell loss priority (CLP)**

b) bits 2–4 allocated for **payload type identification (PTI)**

c) bits 5–8 allocated for VCI

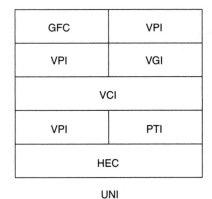

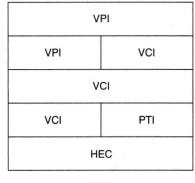

Figure 13–52. ATM header formats.

Byte 5

a) 8 bits allocated for **header error control (HEC)**

The UNI is either private or public. For a private UNI, the frame format, bit rate, line rate, and transmission media are listed in Table 13–23.

For a public UNI, the frame format, bit rate, and transmission media are listed in Table 13–24.

The NNI provides 28 bits for routing, 13 bits for VCI, and 12 bits for VPI. The ATM header also incorporates 3 bits for PTI. The 3 bits are used to differentiate between cells that carry **operation and maintenance (OAM)** information and cells that carry user information. The PTI also provides end of packet (EOP) identification for the AAL-5 layer. One bit in the 4th byte of the header is used as a CLP field. Through this bit, depending on traffic conditions, loss priority cells are identified and consequently discarded, which improves traffic flow. The 5th byte in the ATM cell is used as an HEC and is composed of an 8-bit CRC. Through the HEC, multiple bit errors can be detected while simultaneous correction is performed of single bit errors. At the receiver end, transmission of ATM cell errors can be detected through the HEC and corrective action can be taken. A summary of the NNI cell header is as follows:

Byte 1

a) 8 bits allocated for VPI

Byte 2

a) bits 1–4 allocated for VCI

b) bits 5–8 allocated for VPI

Byte 3

a) 8 bits allocated for VCI

Byte 4

a) 1 bit allocated for CLP

TABLE 13–24 Public UNI

Frame Format	Bit Rate	Transmission Media
DS1	1.544 Mb/s	Twisted pair
DS3	44.736 Mb/s	Coax.
E1	2.048 Mb/s	SMF
E3	34.368 Mb/s	Coax.
J2	6.312 Mb/s	Coax.
NXT1	NX1.544 Mb/s	Twisted pair

b) bits 2–4 allocated for PTI

c) bits 5–8 allocated for VCI

The Physical Layer

The **physical layer** (PHY) U-plane is divided into two sublayers: the **transmission convergence (TC)** and the **physical medium dependent (PMD)** sublayers. The TC sublayer is an independent frame structure. It performs all the functions required to transform a flow of ATM cells into a bit stream that is ready to be recovered by the receiver and converted to the original data. It also provides the detection of cell boundaries, adaptation of ATM layer data rates into line transmission rates, generation and recovery of the frame structure, and insertion and extraction of ATM cells into and from the frame payload. The TC sublayer performs the following functions:

a) **cell rate decoupling**

b) HEC sequence generation/verification

c) cell delineation

d) **transmission frame adaptation**

e) **transmission frame generation/recovery**

A more detailed description of the TC sublayer follows.

TRANSMISSION CONVERGENCE. Physical layer cells include F3 OAM cells that serve as OAM at the transmitter

TABLE 13–23 Private UNI

Frame Format	Bit Rate	Line Rate	Transmission Media
Cell stream	25.6 Mb/s	32 MBd/s	UTP-3
STS-1	51.84 Mb/s	51.84 Mb/s	UTP-3
FDDI	100 Mb/s	125 MBd/s	MMF
STM-1			
STS-3c	155.52 Mb/s	155.52 Mb/s	UTP-5
STM-1			Coax.
STS-3c	155.52 Mb/s	155.52 Mb/s	SMF, MMF
Cell stream	155.52 Mb/s	194.4 MBd/s	STP
			MMF
STM-4			
STS-12	622.08 Mb/s	622.08 Mb/s	SMF, MMF

path section and idle cells that serve as cell rate decoupling. The first four bytes of the F3 OAM physical layer are as follows:

Header Pattern	Byte 1	Byte 2	Byte 3	Byte 4
	000000000	00000000	00000000	00001001

F3 OAM cells must be inserted first, followed by 431 ATM or idle cells, followed by another F3 OAM cell, and so on.

CELL RATE DECOUPLING. The interface structure is composed of a continuous stream of cells. Each stream consists of 53 bytes. In the absence of an F3 OAM cell for transmission, the physical layer will insert idle cells for CRD. The PMD sublayer is application specific (bit rate dependent) and provides the line coding and signal conditioning required for transmission of data over a medium.

13.10 SYNCHRONOUS OPTICAL NETWORKS (SONET)

Introduction

SONET refers to standards developed through the cooperation among Bellcore, the ANSI, and the ITU, which define a set of transmission formats and transmission rates above the 51.840 Mb/s rate. The standards are intended to be used by SONET equipment manufacturers. The need for the development of SONET standards arose from the inability of the SDH to accommodate the rapidly expanding demand for network speeds and services, and to facilitate complex administration and maintenance operations. SONET standards were developed in three phases between 1988 and 1994. In phase one, released in 1988, T1.105 standards define byte interleave, multiplexing formats, line rates, and mapping and monitoring mechanisms. T1.106 defines the standards for long distance transmission utilizing SM fibers. In phase two, SONET standards T1.102-199X, T1.105R1, and T1.117 were released in

1990 and 1991. The T1.102-199X standards specify the electrical signals for STS-1 and STS-3 formats. T1.105 R1 is a revision of T1.105 standards and specifies SONET operating characteristics such as format classification, timing and synchronization, switching protection, layer protocol, and mapping of DS-4 into STS-3. T1.117 standards set the optical parameters for short distance (2 km maximum) optical transmission. Phase three describes the standards for operation, administration, maintenance, and provisioning (OAM&P) set forth by T1.105.01, T1.105.03, T1.105.05, and T1.119. T1.105.01, released in 1994 by ANSI sets the standards for two- and four-fiber bidirectional line switch SONET rings that are capable of failure restoration, which must be accomplished within the time frame of 50 ms. T1.105.03 provides the standards for jitter between SDH and the SONET interface. T1.105 sets the standards for SONET tandem connection overhead, and T1.119 sets the standards for SONET management network equipment.

SONET Hierarchy Speeds

SONET speeds are classified as optical carrier 1 (OC-1) to **optical carrier** 192 (OC-192). Each optical carrier is listed with a corresponding electrical level referred to as a **synchronous transport signal (STS),** a line bit rate, a payload bit rate, and an overhead bit rate. SONET hierarchy speeds are listed in Table 13–25 **(STM-Synchronous transport mode).**

From Table 13–25, the basic bit rate of 51.840 Mb/s is multiplied by the number that specifies the OC, thus generating the corresponding line bit rate. For example, the line bit rate for OC-48 is 51.840 Mb/s, 51.840 Mb/s multiplied by 48 is equal to 2488.320 Mb/s. From all the OCs listed in Table 13–25, only four are supported by equipment manufacturers. These are OC-3 with a line bit rate of 155.520 Mb/s, OC-12 with a line bit rate of 622.080 Mb/s, OC-48 with a line bit rate of 2488.32 Gb/s, and OC-192 with a line bit rate of 9621.504 Gb/s. The building block of the SONET frame is the synchronous transport signal STS-1 composed of 810 bytes transmitted every 125 μs. The SONET STS frame is developed as follows:

TABLE 13–25 SONET Hierarchy Speeds

Optical Level	Electrical Level	Line Rate (Mb/s)	Payload Rate (Mb/s)	Overhead Rate (Mb/s)	SDH Equivalent
OC-1	STS-1	51.840	50.112	1.728	—
OC-3	STS-3	155.520	150.336	5.184	STM-1
OC-9	STS-9	466.560	451.008	15.552	—
OC-12	STS-12	622.080	601.344	20.736	STM-4
OC-18	STS-18	933.120	902.016	31.104	—
OC-24	STS-24	1244.160	1202.688	41.472	—
OC-36	STS-36	1866.240	1804.032	62.208	—
OC-48	STS-48	2488.320	2405.376	82.944	STM-16
OC-96	STS-96	4976.640	4810.752	165.888	—
OC-192	STS-192	9953.280	9621.504	331.776	STM-64

$$\text{SONET frame} = \frac{1}{125 \; \mu s} \; \text{frame} \times 810 \; \text{bytes/frame}$$

$$= 6.48 \times 10^6 \; \text{bytes} \times 8\text{b/s/byte}$$

$$= 51.84 \; \text{Mb/s}$$

Therefore, SONET frame = 51.84 Mb/s (STS-1).

The SONET frame structure is illustrated in Figure 13–53 (**TDH-transport overhead**).

From Figure 13–53, the SONET frame structure STS-1 is composed of 90 byte columns and 9 byte rows. The frame is transmitted row-by-row from 1 to 9. For example, byte-1 located at column 1 and row 1 is transmitted first and byte 810 located at row 9 and column 810 is transmitted last. The time required for the transmission of a complete frame is 125 μs. This provides automatic synchronization for digitized voice signals that require 125 μs for transmission. The SONET frame is formed through a multiplexing process as follows: The segmented user data and the **path overhead (POH)** signals are multiplexed to form the **synchronous payload envelope (SPE),** which is added onto the TOH signal to form the complete SONET frame. The overhead to payload ratio for all SONET speeds is maintained constant at 3.448% (Table 13–26).

From Table 13–26, it is apparent that the overhead to payload ratio for all SONET speeds is maintained constant.

SONET Architecture

SONET is a physical layer composed of four sublayers: the photonic, the section, the line, and the **path sublayers.** A block diagram of SONET architecture is illustrated in Figure 13–54.

THE PHOTONIC (OPTICAL) SUBLAYER. The **photonic sublayer** deals with the operating characteristics of devices such as LEDs and lasers at the transmitter end and PIN photodetector diodes at the receiver end. In essence, the photonic sublayer's main function is to convert electrical signals into optical signals.

THE SECTION SUBLAYER. It refers to signal regeneration procession through the transmission link. At specific length intervals, signal regeneration is required. Furthermore, section sublayers also perform functions such as error monitoring, scrambling, and frame and layer overhead transport.

THE LINE SUBLAYER. The main function of this sublayer is to transport SONET payloads through the physical layer. Some other functions of the line sublayer are multiplexing, synchronization, communication among neighboring lines, and proper maintenance of SONET payloads.

THE PATH SUBLAYER. The path layer is used to transport data between SONET multiplexing equipment. It is also used for mapping SONET frames into a transmission line format while transmission integrity is maintained through the POH. In the data provided by the user (DS-1, DS-2. . .), path overhead bytes are added onto the SPE, which is further processed to the line layer. A block diagram of the SONET overhead sublayers is illustrated in Figure 13–55.

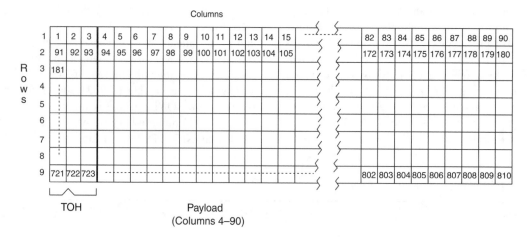

Figure 13–53. SONET frame structure.

TABLE 13–26 Overhead to Payload Ratio

	STS-1	STS-3	STS-12	STS-48	STS-192
Overhead	3	9	36	144	576
Payload	87	261	1044	4,176	16,704
%	3.448	3.448	3.448	3.448	3.448

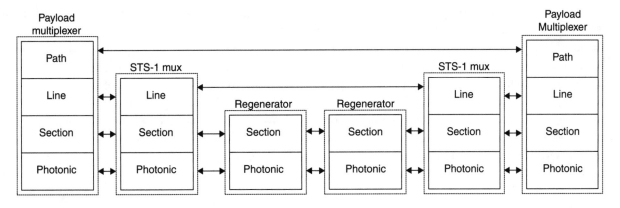

Figure 13–54. Block diagram of SONET architecture.

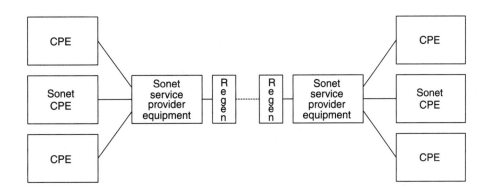

Figure 13–55. A block diagram of SONET architecture.

SUMMARY

In Chapter 13, the fundamental principles of networks were presented. The chapter began with a brief review of data communications and an introduction of optical networks. The chapter continued with the description of network transport architecture, with specific emphasis given to LANs and Ethernet links. The important concept of Gigabit Ethernet was also described in some detail. The chapter continued with a presentation of LAN standards, including the token ring and FDDI, followed by the fiber channel topologies and standards. ATM was presented in detail. Advantages, operating characteristics, and network protocols were discussed, followed by a description of ATM physical layers. The chapter concluded with an introduction of SONET, including SONET hierarchy, speeds, and SONET architecture.

QUESTIONS

Section 13.1

1. What was the cause for the revolution in the communications industry in the 1980s and 1990s?

2. With the assistance of a graph, describe the data capacity demands for the time period between 1978 and 2002.

3. Describe the relationship between data capacity demands and network advancement.

4. List the major breakthrough dates in the development of optical networks.

5. How high can the effective window of a silica fiber go?

6. Based on the available bandwidth, what is the theoretical limit for data transmission of an optical system that utilizes Q-AM?

Section 13.2

7. What is the fundamental operating principle of an optical network?

8. Briefly describe the technological breakthroughs that made the implementation of optical networks possible.

9. Draw a block diagram of WDM, point-to-point transport network and briefly explain its operation.

10. Describe the advantages and disadvantages of a point-to-point and a multipoint transport network.

11. What is the advantage of implementing wavelength add/drop multiplexing?

12. List the components required for the implementation of optical networks.

13. With the assistance of a block diagram, describe the operation of a wavelength add/drop multiplexing cross-connect network.

14. Define *MONET*.

15. Name ways by which fiber can be brought to the end user.

16. Describe the function of an ONU.

17. What is the function of the FSAN consortium?

Section 13.3

18. Define the function of a data communications link.

19. List the main building blocks of a data communications link.

20. Draw a block diagram of a data link, and briefly describe its operation.

21. What is the function of the clock and data recovery circuit?

22. Describe the function of the buffer/framing block.

23. Describe the encoding process.

Section 13.4

24. Define *LAN*.

25. List the main components that compose LAN.

26. What is the function of a network interface card?

27. Draw the diagram of a backbone network and briefly explain its operation.

28. Name the four network topology components.

29. Describe *linear chin topology*.

30. List the advantages and disadvantages of a star topology connection.

31. What are the advantages of tree topology?

32. Name the disadvantages of ring topology.

Section 13.5

33. What does ITU-T recommendation G.805 define?

34. Describe the disadvantage of a shared LAN network.

35. How was the problem in Question 34 remedied?

36. Draw block diagrams of both router and bridge LAN networks and explain the differences and similarities.

37. Define *Ethernet protocol*.

38. List Ethernet and fast Ethernet data rates and encoding schemes.

39. Draw a block diagram of Ethernet configuration, and briefly describe its operation.

40. What are the main functions of the 10 Mb/s Ethernet controller?

41. With the assistance of a block diagram, describe the operation of 100 Mb/s Ethernet.

42. With the assistance of a block diagram, describe the operation of 1 Gb/s Ethernet.

43. Describe the difference between shared Ethernet and backbone Ethernet.

Section 13.6

44. What does the IEEE 802.5 standard define?

45. Draw a block diagram and describe the token ring LAN topology.

46. Describe the function of the token ring MAU.

47. Define *FDDI*.

48. With the assistance of a block diagram, describe the function of an FDDI network.

49. What is the role of the secondary ring in an FDDI scheme?

50. Describe the function and operating characteristics of an enhanced FDDI LAN.

51. Name the four layers of FDDI links.

52. What is the function of the physical media dependent layer?

Section 13.7

53. What were the motivating factors for the development of the fiber channel?

54. Name the three fiber channel topologies.

55. Draw a block diagram of fiber channel topologies and describe their operating characteristics.

56. Briefly describe fiber channel classes of services.

57. Name the fiber channel standards developed by the ANSI X311 committee.

58. Describe, in detail, fiber channel standards.

59. Draw a block diagram of a fiber channel transreceiver module and describe its operation.

60. Describe, in detail, the role of the PLL/clock generator of the transmitter section.

Section 13.8

61. Describe the advantages of ATM in relation to LAN and WAN networks.

62. List the ATM layers that compose its network architecture.

Section 13.9

63. What is an ATM cell?

64. How many cells are there in an ATM protocol?

65. Name the components that define an ATM (B-ISDN) protocol.

66. What is the ATM layer function?

67. Describe the components that constitute an ATM physical layer.

68. What is the adaptation layer?

69. Draw the adaptation layer 1 format and describe its operation.

70. Draw the adaptation layer 5 format and describe its operation.

71. What is the function of the NNI?

72. List the functions performed by the TC sublayer.

Section 13.10

73. What does *synchronous optical networks* refer to?

74. Describe the need for the development of SONET.

75. Define *SONET standards*.

76. Describe, in detail, phase three standard.

77. What is the OC-192 hierarchy speed and its equivalent electrical level?

78. Develop the SONET STS frame and bit rate.

79. How is the SONET frame structured?

80. How long will it take to transmit one SONET frame?

81. What is the relationship between SONET and T1 format?

82. Describe the multiplexing process of a SONET frame.

83. What is the overhead to payload ratio of all SONET speeds?

84. Draw a block diagram of SONET architecture and describe its operation.

GLOSSARY

A

Abberation. Refers to different focal lengths of lenses at different wavelengths

Absolute refractive index. The refractive index of a medium in relation to vacuum

Absolute temperature. The temperature measured at $0°K$

Absorption. The loss of energy of a light beam passing through a medium

Absorption coefficient. The rate at which a characteristic of the medium is lost

Absorption layer. The geometric dimensions of the medium

Absorption spectrum. The optical bandwidth most affected by the absorption layer

Acceptance angle. The maximum angle at which optical power will be accepted at the input of an optical device

Acceptor doping. Type of impurity diffused into a semiconductor material

Acceptors. Impurities diffused into a semiconductor material capable of accepting an electron from the intrinsic semiconductor

Access protocol. A set of procedures required at a specified reference point between a network and a user that allows the user to obtain access to the network

AC coupling. A method of coupling two amplifiers

Acoustic effect. The result of the interaction between an optical wave and an acoustic wave in a medium

Acoustic wave. A sound wave

Acoustic wave photon. The photon generated by the interaction of an acoustic and an optical wave in a medium

Acousto-optics. The study of the interaction of optical and acoustical waves through a medium

Active layer. The layer in a semiconductor device where the intended action takes place

Active layer length. The physical length of the active layer

Active layer thickness. The physical width of the active layer

Add/drop multiplexer. The addition or deletion of a channel in a WDMA optical system

Address. A code that defines the ID of a node on a network

Adjacent channel interference. The electrical or optical power spilled over to a channel by an adjacent channel

Advanced photon source. An accelerator capable of providing the most powerful X-ray beams required for material research

A-Law. A European standard for voice signal digitization

Algorithm. A well defined set of rules applied to the solution of a given problem

Alternate route. A secondary communications path between two terminals, used during failure of the primary route

Aluminum gallium arsenide (AlGaAs). A semiconductor compound used for the fabrication of optical devices

Ambient light. Light present in the space where an optical detector device is operating

Ambient noise. The optical or electronic noise generated by all the noise sources in a given environment

Ambient temperature. The temperature of the medium in which a device operates

American national standards institute (ANSI). A voluntary industry association formed to develop standards without interference from special interest groups

American standards code for information interchange (ASCII). A 7-bit code used in data communications

Ampere's law. It defines the relationship between a magnetic field and an electric current

Amplified spontaneous emission (ASE). Broadband radiation generated by a laser diode

Amplifier. An electronic or optical circuit designed to amplify weak electrical or optical signals

Amplifier front-end. The stage at which the incoming signal is applied for amplification

Amplifier sensitivity. The minimum signal strength required at the input of the amplifier

Amplitude modulated vestigial sideband (AM-VSB). A method of analog modulation

Amplitude shift keying (ASK). A digital modulation method

Analog. A signal that varies continuously with time

AND gate. A logic device performing the AND logic function

Angle of acceptance. The angle at which optical power is coupled into an optical fiber

Angle of convergence. The angle or the rate at which an optical ray descends toward the optical axis

Angle of deviation. The angle through which an optical ray deviates from its original path by reflection or refraction

Angle of incident. The angle formed between the incident optical ray upon a reflective surface and the normal ray to the surface at the point of incident

Angle of reflection. The angle formed between the reflected ray and the normal ray at the point of reflection

Angle of refraction. The angle formed by the refracted optical ray and the normal ray to the surface at the point of refraction

Angular misalignment. An angular deviation from the optimum between optical devices employed in a system for processing optical radiation

Anisophotic source. An optical source with an uneven distribution of the radian energy it generates

Aniso-type. Materials whose properties vary with the propagation of the optical wave

Antireflection coating. A thin layer of coating deposited at the surface of a lens in order to reduce its reflective properties

Anti-Stokes photon. The photon of a particular wavelength generated in material excited by radiation of a different wavelength

Aperture. A hole through which optical radiation is able to pass

Aperture illumination. The field distribution over the aperture containing amplitude, phase, and wave polarization

Aperture ratio. The ratio of the lens aperture to focal length

Application layer. In the OSI reference mode, it refers to programs employed by the user to access the OSI reference mode

Architecture. An overall plan encompassing all the individual components and the functional inter-relationships required to accomplish a set task

Arithmetic unit interface (AUI). The section of a computer designed to perform arithmetic operations

Array waveguide grating (AWG). An optical multiplexing scheme based on optical interference

Asymmetrical digital subscriber line. A method of simultaneous transmission of voice and data over local exchange facilities

Asynchronous communications. In such a system, a start bit and an end bit are sent at the beginning and the end of each data word

Asynchronous transfer mode (ATM). A digital transmission switching format with each cell composed of 5 bytes of header and 48 information bytes

Asynchronous transmission. Data transmission is accomplished character by character and synchronization between transmitter and receiver is achieved by the addition of a pulse at the start and finish of each character

Atom. The smallest unit in an element retaining all the fundamental chemical characteristics of the element

Atomic number. The number of protons in the nucleus of an atom

Atomic weight. The sum total of protons and neutrons in the nucleus of an atom

Attachment unit interface (AUI). The interface between terminal equipment and the medium access unit of a data station

Attenuation. Loss of power during the processing of an optical signal through a passive optical device

Attenuation coefficient. A quantitative measure of a material's internal absorption properties

Automatic gain control (AGC). A circuit incorporated into the design of an amplifier to control the output voltage level

Auto-zero. A function maintaining zero offset voltage levels

Auto-zero function. A feedback amplifier used to eliminate any offset voltage along the traveling path of a signal

Available bit rate (ABR). The information handling capacity of a communications system in bits per second

Avalanche effect. The process whereby photocurrent traveling through a semiconductor junction with sufficient energy is capable of generating additional electron-hole pairs

Avalanche photodetectors (APD). A photodiode capable of amplifying the generated photocurrent when reverse biased

Avalanche photodiode. Electron-hole generation is achieved through photon absorption. At reverse bias close to breakdown voltage, the initially generated electron-hole pairs collide with ions to generate more electron-hole pairs. The process is referred to as avalanche current multiplication

Average power. The pulse energy multiplied by the repetition rate in a pulsed laser diode

Average threshold receiver. An optical receiver consisting of a transmitter using a PCM, a log-normal channel, and an array of photodetectors exhibiting very low BER

Axial vapor-phase deposition. A vapor phase oxidation process employed in the fabrication of graded-index optical fibers

B

Backbone. A high-speed LAN capable of supporting access LANs

Backbone system. A data transmission network designed for high-speed data communications between various regions

Back facet. The mirror at the back of a laser diode

Back facet photodetector. An optical detector used to capture emission of light from the back facet of a laser diode

Background current. The intrinsic current of an electronic or optical device

Back reflection. The optical power reflected back to the origin from the input of a receiver

Backscatter. The deflection of radiation through the scattering process with angles larger than 90° in relation to the initial direction

Backscattering coefficient. The backscattering coefficient is expressed by: $\beta = 4\pi\left(\dfrac{\Phi}{W_i}\right)4\pi r^2\left(\dfrac{W}{W_i}\right)$, where β = Backscattering coefficient, Φ = Unit solid angle, W_r = Power per unit area at a distance r, W_i = Power per unit area

Balanced differential filter. A filter inserted at the output of a preamplifier or at the input of a postamplifier and intended to reduce the noise bandwidth

Band elimination filter. Band elimination or band stop filter designed to pass frequencies above and below a specified band of frequencies

Band gap. The minimum energy required in eV for an electron to transfer from the valence band to the conduction band

Band gap energy. The amount of absorption energy in eV required by an electron to transfer from the valence band to the conduction band

Band gap width. The energy difference between the valence band and the conducting band in a material

Bandpass filter. A filter designed to pass optical signals of specified wavelength without attenuation while blocking heavily attenuating signals outside the specific wavelength

Bandwidth. The range of frequencies in which electronic devices or circuits are specified to operate

Bandwidth length product. The product of the length of an optical fiber and its operating bandwidth

Bandwidth-limited operation. The limitations imposed by the bandwidth on an optical system's performance

Bandwidth polarization dependence. The all-state input signal polarization of a DWDM demultiplexer circuit which also affects the center wavelength of the output channel by promoting a peak-to-peak deviation of the center wavelength

Barrier layer. A layer that is used to create a boundary in the fabrication of optical fibers as a protection against OH-ion diffusion

Barrier potential. The potential difference at the *PN* junction of a semiconductor material

Barrier voltage. The voltage difference at the *PN* junction of a semiconductor material

Baseband. In a communications system, the frequency band through which information is transmitted

Base failure rate. The minimum acceptable failure rate of a component

Baud. A unit of data transmission speed in b/s. Baud rates are as follows: 300, 1200, 2400, 4800, and 9600

B-channel. A 64 Kb/s ISDN user-to-network channel used for both basic and primary rate interface, capable of carrying a voice or data call but not the signaling required for the call

Beam. In optics: A number of converging or diverging optical rays that may be parallel to each other

Beam attenuator. An optical device designed to reduce the flux density of an optical beam through scattering or absorption

Beam diameter. The distance between two opposite points within an optical beam, the power of which is equal to $\dfrac{1}{e^2}P_{pk}$

Beam divergence. Refers to the increase of the diameter of a collimated optical beam

Beam expander. An optical device or a combination of devices used to increase the diameter of an optical beam

Beam optics. A study in the field of optics dealing with waves exhibiting very small angular divergence

Beam splitter. An optical device designed to split a beam of light into two or more beams

Beam width. The angular width of an optical beam reflecting a measure of convergence or divergence of the beam

Bell communications research. An RND organization owned by Bell's regional operating companies

Bending of light. The result of a light beam passing very close to a body with massive gravitational pull, or a beam of light passing through an optical interface at an abnormal angle of incidence

Bend loss. The optical power loss due to the bending of the fiber through which the optical ray is traveling

Bend radius. The minimum radius of the bending curvature that a fiber can sustain before breaking

Bernoulli terms. Mathematical equations defining the changes occurring between potential and kinetic in a wave motion

Bias. DC voltage applied to an active semiconductor device, required to establish the desirable operating point

Bias control. The circuit designed to maintain a constant bias voltage level

Bias-current alarm. A mechanism detecting excess bias current levels

Binary phase-shift keying (BPSK). A digital modulation scheme that uses two phases to represent data. For example, binary −1 is represented by 0° and binary-0 is represented by 180°

Biphase data encoding. An encoding scheme in which the transition from high-to-low and low-to-high takes place in the middle of the pulse

Bipolar. Semiconductor devices utilizing both electrons and holes as the operating current, or a bipolar signal with both positive and negative polarities

Bipolar code sequence. Code words occurring in bipolar form

Birefringence. The separation of an optical beam into two diverging beams referred to as ordinary and extraordinary beams

Birefringent. The optical beam that can be separated into two diverging beams

Birefringent filter. An optical filter that passes light in wide-spaced and well-defined wavelength bands

Bit. Binary character consisting of the value of 0 or the value of 1

Bit error rate (BER). An indicator of performance in a digital and optical communications system, measured by the ratio of the number of bits incorrectly received to the total number of transmitted bits

Bit interleaved parity. A simple parity check mechanism

Bit stream. A series of 0s and 1s

Black body. An ideal body capable of absorbing all the radiation energy impeding upon it

Bohr's atomic model. Describes the fundamental principles of quantum theory

Boltzmann's constant. A constant derived by Boltzmann with an approximate value of 1.38×10^{-23} J/K. This constant was obtained by dividing the universal gas constant with Avogadro's number

Boundary detection. The detection at the beginning or at the end of a pulse

Bragg grating method. Used for the separation of a light beam into two orthogonal polarized light beams

Breakdown current. The current generated by an avalanche photodiode at breakdown voltage

Breakdown voltage. The point at which a further increase of the reverse bias voltage in an avalanche photodiode will result in a current theoretically approaching infinity

Brewster's law. It describes the relationship between the reflected and refracted rays when the incident ray is at a specific angle with the origin

Bridge. A device used to connect two or more LANs of the same type

Brillouin scattering. Spontaneous scattering of the light passing though a medium caused by its interaction with sound waves passing through the same medium simultaneously

Broadband integrated services digital network (B-ISDN). A standard for digital communications incorporating voice, video, and data

Buffer. A device used in image processing to temporarily store data between two processing stages

Buffering/framing. The process of organizing temporarily stored data into fixed units

Bulk scattering. The scattering of light in a medium

Burst. A number of events occurring within a very short period of time

Burst error. A series of errors occurring consecutively in a short period of time in a data transmission system

Bus. Topology assigned to LANs in which all nodes are attached to a single cable

Bus interface. A device connecting all nodes to a single cable

Bus network. A network topology applied to a single communications link that connects three or more terminals together

Butt coupling. The cross-sectional area of the fiber, which is at least equal to the optical emission area of the optical source

Butterworth filter. An electric filter exhibiting flat frequency response

Byte. The number of bits assigned to represent a single character

C

carrier density. The density of majority carriers in a semiconductor material

carrier injection. The process whereby positive and negative charges are driven into a *PN* junction in order for light to be emitted

carrier mobility. The number of electrons and holes moving in opposite directions in a semiconductor material

Carrier number. The number of carriers in a semiconductor material

Carrier recombination. The process through which electrons from the *N*-region and holes from the *P*-region are recombined with the corresponding holes and electrons of these regions

Carrier sensing multiple access with collision detection (CSMA-CD). Ethernet based LAN

Carrier system. The utilization of a number of information-carrying channels over a single communications link. The channels are multiplexed at the transmitter end and demultiplexed at the receiver end

Carrier-to-noise ratio (CNR). The ratio of the signal power to noise power, usually measured at the input of a communications receiver and usually expressed in dB

Carrier velocity. The velocity of the electrons in an *N*-type semiconductor material, or the velocity of the holes in a *P*-type semiconductor material

Carrier wave (CW). A high-frequency continuous wave used to convey information in either analog or digital form

Cascade amplification. A signal amplification process involving more than one amplifier. In such a scheme, the output signal of the first amplifier becomes the input signal of the second, and so on

CATV. Cable TV

Cauchy's equation. It relates mathematically the refractive index of a material and the wavelength of the incident beam

Cavity length. The length between two mirrors in an optical source

Cavity refractive index. The refractive index in the cavity of an optical source

Cell loss priority (CLP). A bit in the 4th byte header used to identify and discard lost priority cells

Cell loss priority field. A priority bit in the ATM header indicating that the header can be discarded if required

Cell rate decoupling. Idle cells in the physical layer of the OSI model

Cells. Blocks of data transmitted in asynchronous transfer mode

Central office (CO). An office of a telephone company that facilitates the switching signals for local telephone circuits

Central processing unit (CPU). The main processing unit in a microprocessor or a switch

CEPT. The European 30-channel multiplexing system

Channel bandwidth-channel baseband. A range of frequencies at which information is transmitted through a carrier

Channel bit rate. The number of bps an optical channel can carry

Channel density. The number of channels handled by an optical fiber at a specified bandwidth

Channel power. The power carried by an optical channel

Channel separation-channel spacing. The frequency or wavelength separating two adjacent optical channels

Channel wavelength accuracy. The degree of deviation of an optical channel from its set limits

Characteristic angle. The specific angle required by a mode to travel through an optical fiber

Character-oriented protocol. A set of rules used for communicating data, which uses special characters such as ETX, SOX, and STX to control the flow of information

Charge pump current. The current required by an optical device to raise atoms from a specific energy level to a higher level while achieving population inversion in the interim levels

Chirp. The level of change of the frequency of an electromagnetic wave

Chirped pulse amplification laser. A laser diode whose optical pulses are chirped before amplification, then compressed again to increase the intensity of the beam without signal distortion

Chirping. A rapid change of the frequency of an electromagnetic wave

Chromatic dispersion. The broadening of a pulse traveling through a fiber, as a result of the wavelength dependence of the velocity of propagation

Chromatic second order (CSO) dispersion. A difference between the peak level of the RF signal and the peak level of the signal generated by the second order modulation product ($f_1 \pm f_2$)

Chromatic third order (CTO) dispersion. A difference between the peak level of the RF signal and the peak level of the signal generated by the third order modulation product ($f_1 + f_2 - f_3$)

Circular polarization. A wave resulting from the combination of two plane waves with a phase difference of 90°

Cladding. A lower refractive index (compared to the core) material surrounding the core; a necessary requirement for the transmission of light through an optical fiber

Cladding ray. An optical ray reflected from the outer surface of the cladding back into the core of the optical fiber

Client. A software program used to establish contact and exchange data with a server

Clock and data recovery. A circuit in the receiver, designed to generate the transmitter clock and to recover the data stream

Clock detect. A circuit in the receiver, designed to generate the clock signal

Clock disable. The function designed to disable the clock signal

Clock driver. The circuit designed to drive the clock signal

Clock enable. The function designed to enable the clock

Clock generator. A PLL circuit generating the required 2.488 GHz clock frequency from a 77.78 MHz reference

Clock jitter. The deviation of the clock frequency from its fixed value

Clock multiplier unit (CMU). The clock in an Ethernet LAN network generating all the required clock frequencies for that network

Clock recovery. The circuit in the receiver, designed to generate the transmitter clock signal

Clock synchronizer. In a receiver, the circuit designed to synchronize the clock frequency with that of the transmitter frequency

Clock synthesizer. The circuit design to synthesize all the required clock frequencies

Code. Symbols used to represent information; for example, ASCII and EBCDIC

Code division multiple access (CDMA). A spread spectrum multiplexing technique allowing a number of users to utilize the frequency spectrum through the use at a specific code

Code division multiplexing (CDM). A spread spectrum multiplexing method allowing a large number of users to share a common carrier through a specifically assigned code for each user

Coder/decoder. A circuit designed to convert an analog signal into a digital signal at the transmitter end, and a digital signal back to analog signal at the receiver end

Coherent. A communications system in which the receiver is synchronized to the transmitter frequency

Coherent communications. A communications system in which the receiver section generates the carrier signal at the exact frequency and phase

Coherent light source. A light source such as that of a laser capable of generating coherent optical radiation

Coherent optical processing. Optical pulses are converted to a phase coherent code sequence by the optical encoder

Coherent radiation. Radiation in which the phase difference between two points in the field of radiation is constant

Collimated radiation. The radiation from an optical source that has all rays parallel to each other (ideal case)

Collision. Result when two notes attempt to transmit at the same time in a CSMA-CD protocol

Collision function. The rate at which two notes attempt to transmit at the same time in a CSMA-CD protocol

Combiner. An optical device capable of combining a number of optical beams into a single beam

Common carrier. The utility that provides communications services to the public upon demand

Common channel signaling. A method using a single signaling channel for use by a number of information carrying channels

Common mode impedance. The impedance between two points in an electronic circuit in reference to ground

Common mode voltage. The voltage difference between two points in an electronic circuit in reference to ground

Common transmit byte clock. The clock circuit generating an 8-bit word and transmitted to all sections of the module

Communications architecture. The sum of the software and hardware used in conjunction to implement a communications system

Composite second order (CSO). A CSO intermodulation product

Composite second order (CSO) distortion. The CSO distortion of a TV signal is expressed as a function of the second order product, a coefficient related to NTSC and modulation depth

Composite third order (CTO). A CTO intermodulation product

Composite third order (CTO) distortion. The CTO distortion is expressed as a function of second order product and a constant related to the 50-channel NTSC signal

Compression ratio. The ratio of the number of bits required to represent a signal to the number of bits required to represent the compressed signal

Compton scattering. In 1923, A. H. Compton observed that a portion of scattered radiation includes longer wavelengths than that of the incident beam

Concave. A hollow, inward curve

Concentrator. A device design to aggregate 100 Mb/s or 1 Gb/s optical layer physical interfaces

Concentric. Mostly referred to as cycles with different radii and a common central point

Conduction band. In semiconductors, the energy band through which electrons have high mobility

Conference of European Postal and Telecomm Administration. The European communications standards committee

Confinement layer. The layer surrounding the active layer in an optical semiconductor device

Connector loss. The optical power loss associated with optical fiber connections and splices

Consultative Committee on International Telephone and Telegraph (CCITT). International committee under the International Telecommunications Union (ITU) assigned the task of developing recommendations associated with all aspects of international communications and the implementation of public data networks

Continuous wave (CW) laser. A laser diode that emits radiation continually instead of in pulses

Convergence. The bending of optical rays toward a common point

Convergence sublayer protocol data unit (CS-PDU). Information in conformity with the specifications of the ATM adaptation layer

Convergent sublayer. The sublayer whose index of refractions bends optical rays toward a common point

Converging lens. The lens used to converge optical rays

Conversion efficiency. The ratio of output energy to pump energy (i.e., pumped laser)

Convolution. In optics, an image enhancing technique mathematically relating a pixel to its immediately surrounding pixels. Through this, the fundamental characteristics of the pixel can be calculated

Coupling efficiency. The ratio of the optical power measured at the output of a directional coupler to that measured at the input

Coupling losses. The optical power loss experienced through the coupling of two optical fibers

Coupling reflections. The optical power reflected back to the source as a result of the coupling of two optical fibers

Critical absorption wavelength. The wavelength at which the absorption of an element exhibits signs of inconsistency

Critical angle. The minimum angle of incident at which total internal reflection is accomplished

Critical aperture. The aperture at which a lens exhibits its best optical characteristics

Cross correlation. A method of improving the SNR by comparing a sampled signal to a known reference signal

Cross phase modulation (XPM). XPM is encountered in DWDM optical systems and is similar to SPM

Crosstalk. The optical power leakage from one channel to another

Crystal. A solid exhibiting a symmetrical and geometrical atomic structure

Crystal lattice. A geometric and symmetrical array of points reflecting the position of the atoms in crystalline form

Current control loop. In an electronic circuit, a feedback mechanism intended to maintain current at a constant level

Current density. The sum total of electrons and holes moving within a semiconductor material under the influence of an external electric field

Current transient. A momentarily increase of the current in a circuit causing damage to the circuit or to other devices within the system

Custom local area signaling service (CLASS). Services such as caller ID provided to telecommunications customers

Customer premise equipment (CPE). Customer premises equipment is owned by the subscriber and is located at the subscriber's location

Cut off wavelength. In a wave guide, the wavelength at which the signal does not propagate

Cyclic redundancy check (CRC). Numerical computation intended to ensure the accurate delivery of data

D

Damping coefficient. A mathematical expression of the exponential decrease of the amplitude of a sinewave

Damping ratio. Defines the rate of decrease of the amplitude of a sinewave

Dark current. In a photodetector diode, the presence of photodetector current in the absence of incident radiation and in the presence of bias voltage

Dark noise. The electronic noise present in a photodetector diode when completely shielded from the outside environment. Also appears in the presence of bias voltage

Database. A collection of information that can be used in a number of applications

Data capacity. The data handling capacity in bps of an optical system

Data compression. Techniques used to compress data to occupy less memory space than normal

Data link. A digital communications system composed of two transreceivers and a transmission medium capable of simultaneous transmission of information in both directions

Data recovery. The data recovered at the receiver end of a communications system

Data set. Synonymous with modem

dBm. Unit of power gain measurement in reference to 1mW

DC biasing. The DC voltage applied to an electronic circuit required to establish its operating characteristics

DC coupling. A direct connection of two stage amplifiers

DC offset. A mechanism adjusting the DC voltage to the required levels

De Broglie wave. A wave related to the motion of a particle with wavelength equal to h/p, where h is Planck's constant and p is the particle's momentum

Decibel. A standard measurement unit of gain or loss

Decision theoretic character recognition. A system by which the incoming character in an optical communications receiver is recognized by comparing it to the number of characters already stored in the memory of the optical receiver

Decoupling capacitor. A capacitor used to perform the decoupling function

Decoupling capacitor array. An array of capacitors used to perform the decoupling function

De-emphasis. The opposite of pre-emphasis

Defense advance research project agency (DARPA). The agency setting and coordinating advanced defense projects

Deflection. The deviation of an optical ray from its expected path

Degradation. The gradual reduction over time of the output signal of an optical apparatus while its input signal is maintained at a constant

Delay distortion. The distortion of an optical signal propagated through a medium caused by different propagation velocities of the various frequencies composing the optical signal

Delay time. The interval between the time an electrical signal modulates an LED and the time the signal has reached ten percent of its maximum at the output of a photodetector

Demultiplexer crosstalk. The spillover of optical power from a channel to an adjacent channel during the demultiplexing process

Demultiplexing. The process whereby a number of signals are separated from a combined signal

Dense wavelength division multiplexing (DWDM). A multiplexing system applied to optical fiber communications in which a large number of closely spaced optical channels carry a large volume of digital information

Depletion area. The area in a semiconductor device that is free of carriers

Depletion region. The region in a *PN* junction that is free of mobile charge carriers

Depolarizer. A device that destroys the polarization of an optical beam by reflecting it in all directions

Detector. A device capable of converting incident optical radiation into a proportional electric current

Device gain. The ability of an electronic or optical device to amplify a signal applied at its input

Diagnostic loopback. A diagnostic signal returned to source

Dielectric. Characteristic of materials that can sustain electric fields with minimum power dispersion, or materials that exhibit properties of an electrical insulator

Dielectric coated grating. A very shallow diffraction grating of a precise dielectric overcoating capable of absorbing polarized incident light of a specific wavelength

Dielectric constant. A numerical indicator of the magnitude of the shift of positive and negative charges within a solid under the influence of an electric field

Dielectric cylindrical waveguide. A cylindrical waveguide fabricated solely by a dielectric material capable of conducting electromagnetic waves

Dielectric medium. The medium through which an optical ray travels

Differential amplifier. An amplifier that provides gain to the difference of the two signals applied at its inverting and noninverting inputs

Differential input. The input of a differential amplifier

Differential input sensitivity. The minimum voltage difference required by a differential amplifier to provide gain

Differential input voltage. The voltage difference between the two inputs of a differential amplifier

Differential mode attenuation. Variations in the level of attenuation of different modes traveling through an optical fiber

Differential mode delay. Also known as multimode group delay, refers to the propagation delays caused by differences in group velocities of the various modes propagating through an optical fiber

Differential mode impedance. The impedance at the input of a differential amplifier

Differential output. The output voltage of a differential amplifier

Differential pair terminator. Termination of a differential pair to its characteristic impedance

Differential quantum efficiency. The slope of a curve plotted as the relationship of the output to the input and indicative of a device's quantum efficiency

Differential signaling. A method using high-speed analog circuitry for transferring multigigabit data through optical or copper wire pair

Differential tapped termination. A method of input/output impedance matching for differential transimpedance amplifiers

Differential transimpedance amplifier. An optical amplifier operating beyond the 1.25 Gb/s range and exhibiting excellent stability, improved dynamic range, and common mode rejection ratio

Diffraction. When an optical beam of light is traveling through a small opening or an opening with rough edges, secondary wavefronts are generated. These wavefronts will interfere with each other as well as with the original wave, creating a number of different diffraction patterns

Diffraction angle. The angle between the incident ray and the diffracted ray

Diffraction efficiency. The ratio of incident flux to diffracted flux

Digital cross-connect. Transmission equipment used to implement a connection through a network manager under the control of the "network operator"

Digital information. A stream of 1s and 0s that can represent data or voice and video signals

Digital loop carrier. Central office equipment located in a remote place providing transmission of voice traffic via multiplexed high-speed trunk interfaces to and from the switch

Digital signal. A signal with a limited number of states by comparison to analog signals, which have an infinite number of states

Digital subscriber line (DSL). A service that transmits digital signals over twisted pair wires at varying speeds; intended for home use

Digitizer. An electronic circuit designed to sample and quantize analog signals

Diode. A semiconductor device composed of a *PN* junction capable of passing current in only one direction

Dipole moment. The product of the distance between two equal in magnitude but opposite charges

Direct band gap devices. Optical semiconductor devices composed of gallium, aluminum, or indium and arsenide materials exhibiting theoretical efficiencies up to 50%

Direct detection. The conversion of an optical pulse detected at the input of an optical receiver directly into an electrical pulse

Direction of propagation. The direction of propagation of an electromagnetic wave defined by the direction of the electric vector component

Direct ray. An optical ray traveling a direct path without refraction or reflection

Direct sequence. A technique applied to a spread spectrum used in CDMA systems in which the bandwidth of the signal is directly modulated by a pseudo noise sequence signal

Direct transmission. Transmission of light directly and without scattering

Discrete. An individual component, passive or active, such as a resistor, a capacitor, or a transistor

Dispersion chromatic dispersion. The separation of an optical beam into various wavelengths. Dispersion in an optical fiber is the result of the propagation of different wavelengths, which compose the optical beam, at different speeds

Dispersion coefficient. Pulse broadening per unit length of optical fibers

Dispersion compensation fiber (DCF). An optical fiber designed to provide dispersion compensation

Dispersion compensation method. A method used in the utilization of optical fibers to offset chromatic dispersion

Dispersion filter. An optical filter employing polarization and interference concepts in order to process monochromatic light

Dispersion flattened single-mode fiber (SMF). An optical filter exhibiting low pulse dispersion over a wide spectral range, thus being suitable for operating at 1300 nm and 1550 nm wavelengths

Dispersion formula. A mathematical expression presenting the index of refraction as a function of the wavelength

Dispersion management. Can be implemented through spectral inversion

Dispersion mapping. The generated dispersion profile in an optical transceiver necessary for the implementation of dispersion compensation

Dispersion penalty. Induced by LED transmitter coupled into an MMF. This dispersion is subject to system bit rate and LED bandwidth

Dispersion shifted fiber (DSF). A single-mode fiber designed for high-speed data transmission. The main advantage of such an optical fiber is the lowering of the dispersion slope at higher wavelengths (1550 nm)

Dispersion slope. The chromatic dispersion in an optical fiber measured in $ps/nm^2 \; km$

Dispersion slope compensation (DSC). The process whereby dispersion slope compensation is implemented in an optical fiber

Dispersion wavelength. The wavelength at which dispersion is measured in an optical fiber

Dispersive power. A measure of the dispersive properties of silica

Distortion. A temporal alteration of the shape of an electronic or optical signal

Distributed feedback Bragg laser. A device with a feedback mechanism used to intensify specific modes within the resonator. With a precise selection of grating space, singe-mode optical oscillations can be obtained

Distributed feedback laser (DFB). A laser capable of generating a single line spectrum under a high data rate of modulation

Distributed modal dispersion. The pulse dispersion encountered when an optical pulse composed of different modes travels through a fiber at different speeds

Distribution trace. The representation of a continuous approximation of all discrete spectral components in the spectrum

Distribution variance. A measure of the spread of the distribution around the average value

Dominant wavelength. The wavelength, defined by a specific color, in an optical spectrum

Donor. The impurity atom capable of inducing an electron into the conduction band of a semiconductor material in order to increase its electrical conductivity

Donor doping. A pentavalent element added to the silicon substrate in order to generate free electrons (donors)

Dopant. The impurity induced into a substance in order to alter the properties of the substance

Doping density. The number of impurity atoms per unit volume in a substance

Drift. A progressive change in time of the signal at the output of a circuit

Drift current. The current generated by the drift of a signal

Drift flatness. The level of drift variations per unit time

Drift mobility. The speed by which a signal drifts

Driver current peaking. The maximum value of the current required for initiating laser action

Driver switching noise. The noise voltage generated by the driving current circuit in a laser diode, the minimum current required for initiating laser action

Duplex. In a data communications system, the simultaneous transmission of data in both directions

Duty cycle. The ratio of the pulse duration to the period, expressed in %

Duty cycle distortion (DCD). The deviation from its 50 percent duty cycle crossing of a square waveform

Dynamic gain. The output/input ratio of an optical amplifier measured at all available voltage levels contained in the signal

Dynamic range. The range of voltages contained in a signal

E

Eccentricity. The displacement of the optical axis from the mechanical axis

Echo cancellation. A technique using digital signal processing in a communications system to prevent echoes generated in the transmitter section from reaching the receiver

ECL buffer. A circuit employing a multi-input differential amplifier and an emitter follower used for impedance matching for combining and amplifying digital signals

Edge detection. The process through which a binary signal is detected at its transition points

Edge detector optical receiver. An improved version of the fixed threshold optical receiver

Edge-emitting LED. Optical power generated from heterogeneous layers. Edge-emitting LEDs exhibit higher coupling efficiency than surface-emitting LEDs

Effective aperture. A percent of the geometric aperture used to collect electromagnetic radiation and guide it into the detector input

Effective gain. The gain provided by a geometric aperture to the collected electromagnetic radiation

Effective noise figure. The logarithmic ratio of the equivalent noise temperature and ambient temperature of a microwave or optical amplifier

Effective numerical aperture. The practical numerical aperture of an optical fiber instead of the theoretical

Efficiency. The ratio of the power applied at the input of a device to that of the power measured at the output of the same device

Elastic scattering. A form of scattering caused by the interaction of the same type of particles and without any loss of kinetic energy

Elastic theory. A mathematical relationship describing the behavior of a solid material when stretched

Elasto-optic effect. A change of the refractive index of the core of an optical fiber resulting from a change in its length caused by external stresses

Electric field. One of the two components of an electromagnetic field

Electric flux density. A measure of the electric field as a function of the equivalent charge per unit area

Electroabsorption modulated laser. Allows for longer transmission distances with a minimum degree of chirping

Electroabsorptive modulator (EAM). An optical device incorporating a CW laser diode and an integrated modulator in a single chip

Electroluminescence. The process of converting electrical energy to optical energy

Electromagnetic compatibility. Standards adapted to maintain the integrity of the electromagnetic spectrum

Electromagnetic force. The force of attraction or repulsion between charged particles

Electromagnetic interference (EMI). An electromagnetic disturbance that disrupts the operation of electrical and electronic equipment

Electromagnetic wave. A wave composed of an electric and a magnetic field perpendicular to each other and both perpendicular to the direction of propagation, traveling with the velocity of light

Electromotive force. The potential difference between two points in an electric field

Electron. A negatively charged particle of an atom with a mass at rest equal to 9.109558×10^{-31} kg, charge $1.6021917 \times 10^{-19}$ C, and a spin quantum number of 1/2

Electron beam. A number of electrons generated and emitted by a single source. They move in the same direction and at the same speed

Electron concentration. The number of electrons per unit volume in a semiconductor material

Electron mass. The mass of an electron measured at rest as 9.109558×10^{-31} kg

Electron velocity. The velocity of an electron traveling through a medium

Electron-volt. The kinetic energy of an electron gained while passing through a potential difference of one volt

Electro-optical efficiency. The conversion efficiency from electric energy to optical energy

Electro-optics. The field concerned with the study of the use of electric fields for the generation, amplification, and processing of optical radiation

Electro-optics modulator. An optical device used to alter the polarization of the optical signal applied at one of its ports in accordance with an electrical signal applied to another port. The resultant optical signal is collected at the output port of the electro-optics modulator

Electrostriction. The process by which the material density of an optical fiber increases through the influence of a transverse acoustic wave

Element. Material, the atoms of which are composed of the same number of electrons, protons, and neutrons

Elliptical polarization. The type of polarization where the electric and magnetic vector components of the wave form an ellipse

Emergent ray. An optical ray leaving a medium in contrast to the incident ray

Emission. The radiation of an optical source

Emission coefficiency. The efficiency through which optical emission is generated

Emission spectrum. The spectrum of an optical signal generated by an emitting optical source

Encoder. The device or circuit performing the function of encoding

Encoding. A process whereby information is converted to digital form

Encryption. A method used for modifying a bit stream so that information appears as a random sequence of numbers

End of life receiver power dissipation. The power dissipation of an optical receiver at the end of its functional life

Energy band. The energy, required by an electron, that allows it to transfer from the valence band to the conducting band

Energy density. The optical energy confined in a medium per unit volume

Environmental factor. The degree by which the environment affects the operation of electronic or optical devices

Epitaxial growth. The process through which a single crystal layer is grown on a crystalline substrate

Epitaxial layer. The growth of a single crystal layer on a crystalline substrate

Equalization. A technique used to compensate for variable signal attenuation within a defined bandwidth

Equilibrium. In a chemical reaction, the forward reaction direction and the reverse reaction direction run all the time

Equivalent inductance. An equivalent inductance representing all the inductances in an electric circuit

Erbium (Erg^{3+}). A stable element composed of six isotopes

Erbium-doped fiber amplifier (EDFA). The main component of an optical amplifier is an optical fiber, the core of which consists of erbium ions diffused at specific levels. The modified fiber core is transparent to the erbium wavelength. When an LED is used as a pump, optical amplification is achieved

Error control. A technique used to detect and correct errors in a data stream

Error-correction code. In digital and fiber optics communications systems, redundant bits are added to the information bits, thus allowing the encoded information to be decoded correctly

Error function. A number defined by a computer program, indicative of optical system performance

Error probability. A statistical probability that an error will occur

Ethernet. A standard for LANs operating at 10 Mb/s, 100 Mb/s, or 1 Gb/s

Excess noise factor. A factor indicating an increase of the shot noise within an avalanche photodiode in relation to a noiseless photomultiplier

Excitation energy. The energy difference of an atom between its ground and quantum states

Excitation potential. The potential energy required for elevating the energy of an atom to the next quantum level

Exit angle. The angle formed between the exit ray of an optical system and the optical axis of that same system

Extinction ratio. The power ratio of a plane polarized optical ray transmitted through a polarizer with its polarization axis parallel to the optical ray and that of a polarizer with its polarization axis perpendicular to the optical ray

Extrinsic. The presence of impurity in optical fibers

Extrinsic absorption. Resulting from the presence of impurities unintentionally injected into the optical fiber mix during the fabrication process

Extrinsic detector. A photodetector diode fabricated by semiconductor materials whose responsivity properties have been modified through the diffusion of certain level of impurities into the basic semiconductor materials

Extrinsic fiber loss. Power loss attributed to the fiber misalignment at a connector or at splicing points

Extrinsic properties. Properties exhibited by semiconductor materials resulting from impurities within their crystalline structure

Extrinsic semiconductors. Semiconductor materials whose crystalline lattice is altered by impurities

Eye diagram. A diagram (displayed in an oscilloscope) that depicts the distortion level of a received bit stream

Eye pattern. A display on an oscilloscope indicating the quality of the received signal in a digital or optical communications system

F

Fabry-Perot cavity. An optical resonator used in the fabrication of laser diodes. Two parallel planes provide the feedback mechanism in a Fabry-Perot laser structure

Fabry-Perot interferometer. It consists of two high reflectivity mirrors, parallel to each other, allowing light to be reflected between them several times. Constructive interference of a specific wavelength light wave can be accomplished through the selection of the appropriate angle in relation to the normal. Fabry-Perot interferometers can be used in such applications as high-resolution spectrometers and laser oscillators

Fabry-Perot laser. A semiconductor laser diode capable of generating a highly directional optical beam

Facet. The cleaved end mirror of the active region of a laser diode

Fall time. The time it takes the output signal of a photodetector diode to fall from 90% to 10%

Faraday's law. A change in the magnetic field in a coil will produce an electromotive force across that coil

Far-field diffraction pattern. The diffraction pattern of an optical source seen from an infinite distance

Federal Communications Commission (FCC). A seven-member panel appointed by the U.S. president with powers to regulate all radio communications within the U.S. and between the U.S. and other countries

Feed. To apply an electronic or optical signal at the input of a passive or active device

Feedback. The transfer portion of a signal from the output of a circuit back to the input

Feedback amplifier. An active circuit maintaining part of the output signal at its input in order to maintain the amplifier's performance characteristics

Feedback circuit. A circuit specifically designed to provide the required feedback mechanism

Fermat's principle. The path that an optical ray takes when traveling through a set of media to render the optical path equal

Fermi-Dirac distribution. A statistical estimation of the electron-hole concentration in the valence and conducting bands of a semiconductor material

Fermi-Dirac function. The probability (mathematical expression) of any given energy state being present in a semiconductor structure at thermal equilibrium

Fermi efficiency. Is defined as the ratio of the band gap energy level to kT

Fermi energy. The energy level determined at absolute zero. At this level, the electrons do not have enough energy to move to the next state

Fermi integral of half-order. A mathematical expression leading to the approximate solution of electron concentration in a semiconductor material

Fiber channel. A computer communications protocol combining both network and channel technologies

Fiber distributed data interface (FDDI). A standard developed by ANSI for data transmissions through fiber optics systems. FDDI is capable of data transmission of 100 Mb/s and above through a LAN incorporating 500 stations in a ring topology with a length of up to 100 km

Fiber effective cross-sectional area. The cross section of the core of a fiber

Fiber grating. A process through which the optical characteristics of a fiber are altered in order to simultaneously transmit a number of optical wavelengths

Fiber index profile. A profile of the refractive index in relation to the distance from the center of the fiber core

Fiber in the loop. Fiber optics transmission used to provide broadband services between the CO and the subscriber

Fiber laser. A laser diode whose lasing mechanism is provided by an optical fiber doped with rare earths

Fiber node. An accessible service pin that is served by fiber and used for telephone system distribution or cable TV

Fiber optic inter-repeater line (FOIRL). IEEE 802.3 standards for fiber optics

Fiber optics. The study of the transmission of information for short or long distances through optical fibers

Fiber optics cable. A cable containing a number of optical fibers

Fiber to home. A network bringing signals through optical fibers to home

Fiber to the curb (FTTC). Fiber optics transmission that provides broadband services beyond the CO to within 100 ft of the customer

Fiber to the home (FTTH). Optical fiber networks terminating at the end users

Fictive temperature. The temperature at which glass is at its thermal equilibrium

Field angular velocity. Defined by ω

Field of view. The maximum area that can be viewed through an optical apparatus

Field pattern. In a given plane, the intensity of emission in relation to direction

Filter ripple. The voltage variations at the filter output frequency response

Fixed threshold detection. An optical detection process with a reduced error probability

Fixed threshold optical receivers. An optical receiver exhibiting low performance at high signal levels

Flux. The luminous power of an optical beam

Flux concentration. The intensity of optical radiation transmitted toward an optical receiver

Flux density. Flux per unit area measured at the direction of propagation

Force of attraction. The force between electrons and protons within the atomic structure

Format. An arrangement of data within a binary stream allowing for identification and control of information

Forward biasing current. The current resulting from a forward voltage applied across a diode

Forward current. The current generated in a *PN* junction based on majority carriers

Forward error correction (FEC). The process through which simultaneous repair of corrupt data can be achieved in a network receiver

Forward voltage. The forward voltage applied across a semiconductor diode

Four wave mixing (FWM). The process whereby optical power is transferred into the adjacent channel in a multichannel optical system

Frame. In digital transmission employing TDM, the bit stream is arranged in fixed units referred to as "frames." These frames are transmitted at 125 μs intervals

Frame byte. An 8-bit word at the start of binary frame

Frame multiplexer. A logic circuit multiplexing frames

Free electron. An electron that has left the outer shell of the atom

Free spectral range. Frequency intervals between adjacent transmission peaks

Frequency division multiplexing (FDM). In electronic communications systems, it refers to the transmission of several signals over a common path. In optical communications systems, it is commonly referred to as wavelength division multiplexing (WDM)

Frequency doubling. The process through which the frequency of an optical beam is doubled

Frequency modulation. An analog modulation technique through which the frequency of the carrier signal is altered by the information signal in accordance to a specified plan

Frequency period. Frequency f (c/s), period $T = \dfrac{1}{f}$

Frequency shift keying (FSK). A digital modulation technique used in the design of low speed modems

Full service access network (FSAN). A consortium of network operators established to provide a common access platform

Full width at half maximum (FWHM). The optical power spectral density at one-half of the peak amplitude

Fundamental mode. In waveguides, the lowest order mode

Fusion splice. Splicing achieved through the application of sufficient heat to a specific location to fuse the two ends of an optical fiber

G

Gain. The ratio of the output optical power to the input optical power, characteristic property of an optical amplifier

Gain-bandwidth product. The product of the gain and frequency of an avalanche photodetector diode

Gain competition. The amplitude increase of the remaining four wavelengths if one of the five wavelengths, generated by a laser diode, degrades or completely collapses

Gain equalizer. An electronic circuit incorporated into an optical amplifier capable of compensating for amplifier gain and improving the SNR

Gain flatness compensation. In an optical amplifier, the mechanism providing compensation for flatness variations outside the set limits

Gallium aluminum arsenide (GaAlAs). A semiconductor alloy used in the fabrication of laser diodes

Gallium arsenide (GaAs) injection laser. A laser diode composed of a *PN* junction confined within a gallium arsenide crystal. Produces a Fabry-Perot resonant cavity. Laser action occurs in the mode exhibiting higher gain

Gaussian beam. Optical beam with Gaussian electric field amplitude distribution

Gaussian optics. The part of optics based on the concept that for very small apertures and field angles, the sin θ in Snell's law can be substituted for by the angle θ

General absorption. The loss of power of an optical beam passing through a medium

Generic flow control field. Priority bits (four bits) in an ATM header that communicate the intention of the switch to implement congestion control to the target end station

Geometric optics. The study of optics based on the notion that light travels in a straight line. Neglects the phenomenon of diffraction, thus recognizing only the wave theory of light

Germanium atom. Composed of thirty-two electrons distributed along four orbiting shells and sub-shells, and thirty-two protons in the nucleus

Gordon-Haus effect. A phenomenon whereby the ASE noise produces random frequency variations which are translated into pulse variations

Graded index. Refers to an optical fiber with a parabolically decreasing refractive index from the core to the cladding. Such an optical fiber exhibits wide operating bandwidth and high efficiency

Gravitational field. The force created around bodies (massive) resulting in the attraction of other bodies

Group delay distortion. The distortion of a pulse traveling through the different modes of an optical fiber

Group velocity. The inverse of the rate of change of the phase constant to angular velocity of a specific mode

Group velocity dispersion (GVD). The distortion of a pulse traveling in an optical fiber as a result of the group velocity

Guard band. The wavelength separating two optical channels

Guided ray. A ray confined to the core of an optical fiber

Guided wave. A wave whose energy is guided through the core of an optical fiber

H

Half bandwidth. In an optical filter, a full width of wavelength at half maximum power (FWHM)

Half duplex. A communications system design to transmit only in one direction

Half power point (HPP). In reference to optical filters, it refers to the wavelength at which the filter output power is one-half of the input. In reference to a laser diode, it refers to the wavelength of the leading or falling edge of the laser pulse where the optical power is half of the maximum

Half width at half-maximum (HWHM). In an optical power, a half width of a wavelength at half maximum power

Hard clad silica fibers (HCSF). A type of optical fiber made up of hard silica

Header error control (HEC). A byte in the ATM cell containing information required for performing error detection on the cell header

Helmholtz constant. Relates the absorption coefficient to the wavelength of an optical wave traveling through a fiber

Heterogeneous. A substance composed of elements with different optical properties

Heterojunction. A junction between a p-type and n-type semiconductor material in a semiconductor device

Heterojunction laser. Laser diodes fabricated with such semiconductor alloys as AlGaAsSb/GaSb and InGaAsP/InP

Heterostructures. An optical device fabricated by growing an epitaxial layer on a semiconductor layer

Hierarchy. Transmission speeds used for multiplexing a successively higher number of circuits

High bit-rate digital subscriber line. The system designed to transport 1.544 Mb/s (T-1) over 12,000 ft of 24-gauge wire or 9000 ft of 26-gauge wire without repeaters

High current sinking. An electronic device capable of sinking relatively high current

High current sourcing. An electronic device capable of sourcing relatively high current

High-frequency distortion. The distortion of the high-frequency component of an electrical signal passing through a passive or active device

High-loss fiber. An optical fiber whose attenuation properties exceed a permissible maximum

Highly-degenerate semiconductors. Semiconductors for which the Fermi level lies inside the energy band

High order mode fiber. An optical fiber, the utilization of which substantially improves the dispersion management process

High pass filter. A filter designed to process the high-frequency components of an optical signal without attenuation

Hold time. The time required to hold the sample of an analog waveform before the next sample

Hole concentration. The concentration of holes in a p-type semiconductor material

Homocentric. Optical rays having the same focal point (parallel optical rays)

Homogeneous. A substance composed of elements with the same optical properties

Homogeneous cladding. The cladding section of an optical fiber composed of a dielectric material of constant refractive index

Homojunction. Junction formed in a common type of semiconductor material

Hub. The center of a LAN in a star topology

Huffman encoding. A data compression method requiring fewer bits to represent frequently occurring characters

Huygen's principle. Each point on the wavefront of a propagating electromagnetic wave is, in itself, a source of a new wave

Huygen's principle of reflection. The application of Huygen's principle to interpret the reflection of light

I

Ideal filter. A filter exhibiting zero attenuation at the pass band and complete attenuation at the outside bands

Illuminance. The luminous flux incident upon a surface per unit area

Immunity (noise). The maximum noise level an electronics circuit can tolerate

Impedance (electr.). The ratio of the voltage to current in a pair of terminals

Impurity ion. An undesirable ion in a crystal lattice

Impurity level. The level of impurity atoms (donors or acceptors) ejected into silicon or germanium, thus converting it to a semiconductor

Incandescence. The radiation of visible light by an appropriately heated source

Incident. Flux incident upon a surface per unit area

Incident ray. An optical ray impeding upon a surface

Incident wave. An optical wave impeding upon a surface

Incoherent. In an incoherent optical system, the received signal is out of phase with the transmitted signal

Incoherent optical processing. Much simpler than the coherent method, the operation is based on light intensity variations of the encoder module

Incoherent scattering. The scattering of an optical wave in an incoherent manner

Index-guided laser. A laser diode, the optical beam of which is contained at the active layer and guided at the output by an optical waveguide of a specified index profile

Index profile. The change of the refractive index as a function of the radius of an optical fiber

Indirect band gap. The classification for Si, Ge, and AlAs semiconductor materials

Indirect modulation. Optical systems where the modulation occurs outside the laser diode

Indium. An element incorporated into the fabrication of semiconductor alloys such as InGaAs, InP, and InGaAsP

Infrared. The portion of the electromagnetic spectrum occupying the 0.75 µm to 1000 µm range

Infrared absorption. The process whereby infrared radiation is absorbed by a crystal when its crystalline lattice vibrates, thus promoting ion motion

Injection efficiency. The ratio of the injected electron to photon generation in an optical semiconductor device

Injection loss. The number of electrons injected into the semiconductor material unable to generate corresponding photons

In-line repeater. An optical amplifier inserted into a long distance optical system to amplify the optical signal along that line

Input buffer. A circuit used at the input or output of an amplifier for impedance matching

Input data pulse shaper. A circuit at the input of a receiver designed to restore corrupt data to its original shape

Input register. A logic circuit designed to store data temporarily for further processing

Input sensitivity. A characteristic operating parameter of an optical receiver defining the minimum electrical or optical signal required for detection

Input signaling. The signaling method at the input of a receiver

Input signal intensity. The intensity of the signal applied to the input of a receiver

Insertion loss. Optical power loss due to the insertion of passive components or splicing into the optical system

Insertion loss variations. The loss of power experienced by a signal passing through a microwave or optical passive device

Inside vapor-phase oxidation. A fabrication method for the construction of low loss optical filters

Institute of Electrical and Electronics Engineers (IEEE). A membership organization of electrical and electronic engineers which has, as one of its main functions, the development of standards applicable to communications industry

Insulator. A material or component resisting the flow of current

Integrated digital loop carrier. Access equipment that extends central office services by connecting a Sonet ring onto the network while providing telephone services to the subscriber

Integrated energy. The integral of instantaneous optical power generated by an optical source whose power output varies over time

Integrated laser. A laser diode fabricated in a single substrate

Integrated optical circuit. An optical circuit composed of passive and active components fabricated into a single substrate

Integrated optics. In optics, similar to that of integrated circuits for electronics. Integrated optics incorporate optical filters, directional couplers, and lenses interconnected by optical waveguides

Integrated services digital networks (ISDN). International standard developed for the simultaneous transmission of voice and data through a communications channel

Intelligent network. A public switching network that provides switching, routing, and control through computer control. Intelligent networks are capable of automatically accommodating new services over wider areas

Intelligent peripheral. Specialized computer services designed to facilitate intelligent networks

Intensity. Flux per unit solid angle

Intensity modulated direct detection (IM-DD). An optical system using multiple subscriber modulation with fixed bias

Intensity modulation. The process through which an electrical signal of a specific frequency varies the intensity of an optical signal generated by an LED or laser diode with the same frequency

Interface. An apparatus inserted between the common boundaries of two systems to ensure proper operation

Interference. A systematic attenuation and reinforcement of the signal amplitude by an undesirable signal

Interference filter. A filter design to control the spectral composition of an optical signal through interference

Interferometer. An optical device used for optical path comparison

Intermediate order mode. In a multimode fiber, the mode occupying the middle range

Intermodal. The spectral interference between two modes in an optical fiber

Intermodal dispersion. Mainly noticed with multimode step-index fibers, it is the result of the different group velocities along the various modes of propagation within the optical fiber

Internal cavity losses. In a laser diode, the optical power loss experienced in the cavity and related to the length of the cavity and absorption coefficient

Internal photoeffect. The internal photoeffect is divided into two categories: a) intrinsic and b) extrinsic. a) Intrinsic: When electrons move from the valence band to the conducting band. b) Extrinsic: When electrons move the valence band to impurity levels

Internal photoelectric effect. The free-electron generation produced in solids by photon absorption

International Telecommunications Union. A telecommunications agency founded by the United Nations in 1965 to provide standardized procedures and practices for the telecommunications industry such as radio regulations and frequency allocation

International Telegraph and Telephone Consultative Committee (CCITT). Consultative Committee for International Telephone and Telegraph

Internet protocol. A standard format of data transmission over the Internet

Intersymbol interference (ISI). The interference produced when power from one bit of information is spilled over to the adjacent bit

Intramodal. A plane wave composed of different frequency components

Intramodal dispersion. The result of the nonlinear change of phase velocity in relation to the wavelength of a plane wave composed of different frequency components propagating through a dielectric medium

Intramodal distortion. In a single mode waveguide, it is the distortion resulting from the group velocity dispersion of the propagating wave through the single mode

Intrinsic. No impurities

Intrinsic absorption. The result of the interaction of free electrons and the operating wavelength within an optical fiber

Intrinsic semiconductor. Silicon or germanium in crystalline form, with the absence of impurities

Inverse square law. A relationship determining the propagated microwave power proportional to the inverse of the square of the distance from the source

Ionizing radiation. The radiation capable of generating ions while traveling through a matter

Ion laser. Stimulation emission of radiation is accomplished between two levels of ion gas

Ion pair. Two particles carrying opposite charges

Irradiation. The bombardment of an object with radiation

Isothermal compressibility coefficient. A coefficient equal to the negative of the inverse isothermal bulk modulus K_T, which in turn relates to the change in volume of a solid substance as a function of the change of the applied pressure

Iso-type. The duplicate of a prototype

J

Jitter. The deviation of an electrical pulse from its original place in time

Johnson noise. The electronic noise generated by thermally agitated electron motion in semiconductor materials

Josephson effect. The phenomenon whereby radiation detectors produce undesirable energy due to photon interaction

Joule (J). Unit of energy or work with a value equal to 1×10^7 ergs

Junction capacitance. The capacitance measured across a PN-junction

Junction diode. A semiconductor device composed of N-type and a P-type semiconductor material. The device is capable of conducting current in only one direction

Junction temperature. The temperature measured across a PN-junction

Junction voltage. The depletion voltage across a PN-junction

K

K characters. A special 8-bit word used in 8B/10B encoders defined by a customer

Kerr effect. An optical fiber nonlinearity caused by the interaction of the field intensity with the fiber refractive index

Kinetic energy. The velocity of motion defined by the mass and velocity of an object

L

Lambert. Unit of luminance

Lambert's cosine law. The flux per unit solid angle away from the transmitting surface in any direction is proportional to the cosine of the angle between the said direction and the normal to the surface

Laser. Light Amplification by Stimulation Emission of Radiation. A device that generates optical beams of various wavelengths

Laser bias control. The bias current required to initiate laser action in a laser diode

Laser bias driver. The driver circuit providing the bias current in a laser diode

Laser current control. A circuit designed to maintain the laser bias current within the specified limits

Laser detector. A device capable of detecting an optical beam generated by a laser diode

Laser efficiency. The ratio of the optical to the electrical conversion

Laser modulator. An external circuit used to modulate the laser diode

Laser pump. It relates to the number of electrons injected in the active region per cm^3

Laser shutdown input. An input at which a voltage is applied to shut down the laser diode when required

Lasing threshold. The minimum excitation power required in a laser diode for maintaining stimulated emission

Lateral wave. Light produced at the interface of two mediums with different refractive indices when an optical ray impedes upon one surface with an angle of incidence that causes almost total internal reflection

Lattice. Mainly referred to as points representing the atoms in a crystal

Launching fiber. A fiber used with an optical source to launch the generated optical power of the source to the next fiber

Layer network. A combination of resources that support special characteristic information

Least significant bit. The lowest order bit in a binary sequence

LED bandwidth. The spectrum around the wavelength at which the LED emits its maximum optical power

LED driver. A circuit driving the LED

LED forward current. The forward current required by the LED to generate optical power

LED forward voltage. The forward voltage required by the LED to generate the forward current

Lens coupling. Allows for maximum coupling efficiency between the output of an LED source and the optical fiber

Lens method of coupling. A method of coupling an optical source to the optical fiber

Level detector. A circuit designed to detect the optical power generated by an optical source

Light. Radiation of electromagnetic form occupying the 400 nm to 750 nm spectrum and visible to the human eye

Light amplifier. A device capable of amplifying the light signal applied at the input without altering its wavelength

Light dispersion. The process whereby a beam of white light can be separated through a prism into individual wavelengths

Light emitting diode (LED). A semiconductor device composed of a *PN* junction capable of generating optical energy when forward biased

Light filter. A passive optical device used to process optical energy of a specific wavelength

Light modulator-optical modulator. A device composed of two input ports and an output port. A continuous optical wave is applied at one port and a modulating signal is applied to the other while the modulated optical signal appears at the output port

Light quantum. Optical systems that incorporate special light detectors capable of detecting light waves of one wavelength and transmitting them in another. Such a system can use only one optical fiber

Light ray. The radii of an optical wave indicating the direction of travel

Light source. A source of visible radiation

Light source power. The electrical power required by the light source to maintain stimulation. This power is provided by specially designed regulated power supplies

Line. A path providing one-way communications between one end user and a switch

Linear chain topology. A series of point-to-point topologies connected in a series configuration

Linearity. In an optical amplifier, the measure of the deviation of the input-to-output voltage ratio

Linear scattering. The phenomenon through which optical energy is transferred from the dominant mode to adjacent modes and is proportional to the optical signal strength injected into the dominant mode

Line protocol. A program designed to control data communications functions such as hand shaking and the movement of data between transmitter and receiver

Line sublayer. The sublayer that transports Sonet payloads through a physical layer

Line termination. An ISDN interface incorporated between the central switch of a telecommunications office and a local loop

Linewidth. Specific range of frequencies or wavelengths over which the optical power is distributed

Link. The sum of two transreceivers and the transmission medium that is necessary to establish efficient communications between two points

Link access protocol-B channel. Standard ISDN protocol defining the data in the two-voice channel 192 Kbps signal

Link layer. In OSI, it refers to synchronization and control over the influence of errors generated within the physical layer

Link power budget. The difference between the optical power launched into the fiber and the power required at the input of the optical receiver to generate a usable electrical signal

Lithium fluoride. A crystal with refracting properties in the ultraviolet, visible, and infrared spectra

Lithium niobate LiNbO$_3$. A ferroelectric material in crystalline form used in the development of optical modulators

Local area network (LAN). High-speed high-capacity computer links used over relatively short distances (a few kilometers) to connect several buildings

Local exchange carrier. The supplier of communications services within a geographic region

Local loop. Part of a telephone network between the switching office and the subscriber

Local oscillator laser. In the receiver section of a synchronous optical communications systems, a laser device used as the local oscillator in order to generate the required transmitter signal

Lock detect. An active-HIGH CMOS signal indicating whether or not the PLL circuit is locked into the reference clock signal

Lock to reference. With a lock to reference low, the PLL circuit at the receiver section locks onto the supplied reference clock signal

Long-wavelength system. Optical systems operating in long wavelengths

Long-wave pass filter. An optical filter designed to pass long wavelength signals and to reject short wavelengths

Loop. A pair of wires connecting the subscriber to the CO of a telephone company

Loopback. A signal sent back to the transmitter section of a data communications system from a point along the transmission path during troubleshooting

Loop bandwidth. The effective operating bandwidth of a loop

Loss compensators. Devices used to compensate for loss of optical power in an optical system

Loss of lock. An output in a multiplexer circuit used to indicate when the clock multiplier unit has lost lock status

Lossy medium. A medium that is capable of total absorption or total scattering of the radiation impeding upon it

Low-loss fiber. The optical fiber that induces a small loss to the optical signal that passes through it. Fiber loss is measured in dB/km

Low order mode. In an optical fiber, the optical ray travels very close to the center of the fiber core

Low-pass filter. A passive or active filter is an electrical or electronic circuit designed to process signals of low frequency and a specific bandwidth and to reject high frequency signals

Luminance. Emitted luminous flux (per unit solid angle) which is projected to a surface (per unit of area) normal to the direction of emission

Luminance factor. Ratio of luminance of an object to a perfect reflector that is identically illuminated

Luminescent fiber. A fiber that emits luminescent radiation when excited by high-energy particles

Luminous efficacy. The ratio of total luminous flux and radial flux

M

Mach Zehnder (MZ) (LiNbO$_3$ optical modulator). The MZ-LiNbO$_3$ optical modulator is composed of two-phase modulators, a 3 dB Y-junction optical splitter, and a 3 dB Y-combiner. It provides ports for an optical carrier wave, a DC-bias control, and an RF signal (data). Its main function is to modulate the optical carrier wave with the RF signal—the complex signal collected at the output port

Mach-Zehnder interferometer ultra-fast nonlinear interferometer (UNI). An extension of Twyman-Green interferometer, the MZ interferometer is composed of two beam splitters and two mirrors. It is used in the design of MZ-LiNbO$_3$ optical modulators

Macrobending. Macroscopic curvatures occurring in optical fibers, results in local axial and wavelength displacement leading to an increase if fiber attenuation

Macrobend loss. Optical fiber loss resulting from macrobending

Magnetic field. The effect produced by magnetic lines in space

Magnetic flux density. The density of the magnetic lines in a magnetic field

Malus's law. Malus states that the intensity of the optical component of a fully polarized optical ray parallel to the analyzer varies with the square of the cosine of the angle between the polarizer and the analyzer

Manchester data pattern (NRZi). The logic-1 is represented by a pulse transition from low to high, and the logic-0 is represented by a pulse transition from high to low

Manchester encoding. A logic transition occurs at the center of each bit. A positive transition indicates a logic-1 and a negative transition indicates a logic-0

Mark. A term used to indicate a binary-1

Mark density. The optical power density of a mark

Master group. CCITT G.232 recommendations for 812 kHz to 2.044 MHz FDM signals

Matched filter. A filter designed with a maximum SNR, capable of separating signals from random noise

Matched transmission line. A transmission line generating no reflective waves

Material dispersion. Dispersion caused by the wavelength dependence of the refractive index of the material composing the optical waveguide

Material scattering. Scattering attributed to the material properties used for the fabrication of the optical waveguide

Maximum system capacity. The capacity, in bps, that an optical system is capable of handling

Maxwell-Boltzman statistics. A relationship establishing the concentration of electrons in the conduction band of a semiconductor

Maxwell's equation. The equation developed by Maxwell, which shows that an oscillating electric charge generates a magnetic field and an electric field perpendicular to each other. Both travel at the velocity of light perpendicular to the direction of propagation

Mean power control. The function of a circuit designed to maintain the bias current of a laser diode within a specified range

Mean spherical intensity. The intensity of an optical source (average value) relative to all directions

Mean spherical luminous intensity. The luminous intensity (average) of a point source

Mean square avalanche gain. The mean square of the primary, background, and dark currents in a PIN optical device

Mean square spectral density. The mean square of the generated photocurrent in a photodiode

Mean wavelength. The center of all the spectral components

Measured center wavelength. The measured wavelength at the center of an optical spectrum

Mechanical splice. It refers to the technique in which splicing is accomplished through mechanical means

Medium. The space or material through which electromagnetic radiation is able to propagate

Medium access unit (MAU). Part of an arithmetic unit interface

Mesh topology. Defined by two or more paths and at least two nodes

Message. A digitally coded information, formatted for transmission

Metropolitan area networks (MAN). An optical fiber or cable backbone interconnecting a number of LANs in a specified area

Microaperture. A slit or a small opening

Microbending. Microscopic curvatures occurring in optical fibers resulting in local axial and wavelength displacement

Microbending loss. Optical loss resulting from microbending, measured in dB

Microstrip line. A microstrip transmission line is constructed by placing the transmission line on top of the PC board material with a conducting plate placed at the bottom

Mie-scattering. Mie-scattering occurs when the size of the physical anomalies within the optical fiber is larger than one-tenth of the diameter of the operating wavelength

Mirror effective length. The length of a quarter wave mirror stack composed of GaAs/AlGaAs

Mirror losses. Defined by front and rear mirror reflectivity

Mirror reflection coefficient. A coefficient indicative of the mirror reflectivity in a resonant cavity enhanced photodetector

Mirror reflectivity. The ability of a mirror to reflect light

Modal dispersion. The result of the interaction of the different spectral components of the optical signal with the dispersive and guidance properties of the fiber; optical distortion due to modal dispersion occurring in a multimode optical fiber

Modal gain. A characteristic of VCSEL lasers defined by the current density at threshold, differential gain, and transparency current density

Modal noise. Mode dependent optical losses are translated as a form of noise. These losses are subject to distribution variations of the radiant power among different modes

Modal number. The number of modes in which the light can travel through a fiber

Modal optical power. The optical power carried by a mode in a fiber

Mode. Propagation characteristic of the light traveling through an optical waveguide. This propagation characteristic is defined by a radiation pattern in the plane perpendicular to the direction of propagation

Mode coupling. The exchange of optical power between different modes

Mode dispersion. Is the phenomenon whereby the modulating electric signal is broadened while traveling through the core of the fiber

Mode filter. An apparatus used to measure the attenuation properties of optical fibers

Mode interference noise. Mode interference causing undesirable optical power output variations

Mode-locked laser. A laser diode that generates high power and very short duration optical pulses (ps) by selectively modulating the optical energy content of each mode

Mode spacing. The average wavelength spacing between individual spectral modes

Modulation. The change of the optical intensity in accordance with the input data signal

Modulation bandwidth. The highest possible frequency at which an LED or laser diode can be driven without inducing signal distortion

Modulator. A circuit performing the function of modulation

Modulator driver. For digital applications, an appropriate modulator driver composed of a differential amplifier is used. This amplifier is designed to supply 100 mA of current for a 2.5 Gbps application and 300 mA for a 10 Gbps application

Molecule. The basic building blocks of matter

Monte Carlo simulation. A statistical method used to evaluate a distributive mode dispersion coefficient

MOSFET solid state relay. Relays designed for MOSFET solid state devices

Most significant bit. The highest order bit in a binary sequence

Motion pictures expert group (MPEG). A standards group that formulates the techniques for video signal digitization and compression

μ-Law. A North American standard for voice digitization

Multifiber cable. A fiber optics cable composed of a number of optical fibers

Multifiber joint. A connector capable of optically joining all the individual fibers within an optical cable

Multilongitudinal mode (MLM). A laser diode with a spectral line defined as the rms of the spectral width

Multimode dispersion. Dispersion of the optical power transmitted through a multimode fiber, subject to fiber modal characteristics

Multimode (MM) fiber. Multimode fibers are capable of carrying the same optical wavelength through different paths corresponding to different arrival times at the end of the fiber

Multimode optical waveguides. An optical waveguide capable of transmitting in more than one mode

Multimode step-index fiber. MM-SI fibers are very similar to SM-SI fibers except for the size of their cores. MM-SI is larger than SM-SI

Multiplexing. The process through which several optical channels are combined into one channel

Multiplication factor. A factor comparing the photon generating current between PIN and APD photodetector diodes

Multipoint network. A complex optical network incorporating add/drop multiplexing and EDFAs

Multiwavelength optical networking (MONET). A consortium monitoring the implementation of a multiwavelength optical network

N

Narrowband ISDN. Encompassing a 2B+D basic interface and a 23B+D primary rate interface with a maximum speed of 1.5 Mb/s

National Association of Broadcasters (NAB). A U.S. lobby representing the radio and TV industry dedicated to the promotion of the industries' interests on a national level

National Television Standards Committee (NTSC). A U.S. Government agency responsible for setting and supervising television standards

Natural frequency. The natural vibrating frequency of an object

Negative dispersion fiber (NDF) Advanced optical fibers manufactured for high-density long-distance optical systems

Network. A system of cables, including optical cables, linking a number of terminals that can communicate with each other

Network access point. An open point in a carrier network used to connect the network to other carriers or to service facilities

Network interface card (NIC). A LAN adaptor card connecting the client servers and network peripherals to the local network through a cable or an optical fiber line

Network layer. In OSI, it refers to switching and routing functions

Network node interface (NNI). An interface between the different public carriers

Network-to-network interface. A standards interface specifying connections between ATM network nodes

Neutron. A subatomic particle carrying no electric charge and with mass almost equal to that of a proton

Newton. Unit of force: The force that will accelerate 1 Kg of mass at an acceleration of 1 m/s^2

Nippon Telephone and Telegraph (NTT). The Japanese telephone and telegraph company, an equivalent to AT&T in the U.S.

Noise bandwidth. The frequency band at which noise is determined

Noise current. The current generated in a semiconductor device or circuit as a result of an undesirable electronic signal

Noise figure. A noise performance characteristic of an electronic or optical receiver

Noise immunity. A level below which an electronic or optical device is unaffected by noise

Noise intensity. The noise intensity measured at a specific bandwidth

Noise jitter. A short term noncumulative variation of the significant instants of a digital signal from its ideal position in time

Noise power. The noise power measured at a specific frequency range (bandwidth)

Noise voltage. The noise voltage measured at a specific bandwidth

Nominal wavelength. The operating wavelength range of an optical device

Nondegenerate semiconductor. The semiconductors that have a Fermi level lies within the band gap and away from the bottom of the conducting band by several kT

Nondispersion shifted fibers. Advanced optical fibers manufactured for high-density long-distance optical systems

Nondispersive shifted single-mode fibers (NDS-SMF). Advanced single mode optical fibers manufactured for high-density long-distance optical systems

Non-Gaussian (sporadic) noise. Noise that does not obey the normal Gaussian distribution function

Nongenerative semiconductors. The Fermi level within the band gap

Nonlinear coefficient. A coefficient indicative to the system or device nonlinearity

Nonlinearity. A characteristic of an optical device relating the disproportional variations of the output signal to the variations of the input signal

Nonlinear loop optical mirror (NLOM). An optical demultiplexer using a birefringence mirror

Nonlinear optical detector. An optical detector capable of recognizing nonlinear optical effects through the utilization of directional beams

Nonlinear optical effect. An effect such as stimulated Raman, that can only be viewed by directional beams (very close to monochromatic)

Nonlinear process. Optical fiber deficiencies negatively affecting DWDM optical systems

Nonlinear scattering. Raman and Brillouin scattering

Non return to zero (NRZ) data pattern. In a non-return to zero (NRZ) data pattern, the logic-1 occupies the entire positive part of the bit cell while the logic-0 occupies the entire negative part of the bit cell

Non-zero dispersion shifted fiber (NZDSF). Advanced optical fiber developed by Corning

No return to zero. A digital signal that is low for zero, high for one, and that does not return to zero between successive pulses

Normalized frequency. A dimensionless quantity, V, given by $V = \dfrac{2\pi a}{\lambda} \sqrt{n_1^2 - n_2^2}$, where, λ=wavelength and n_1, n_2 are refractive indices

n-type semiconductors. Extrinsic semiconductor materials formed by diffusing impurities of pentavalent atoms into their crystalline structure

Nuclear force. An attractive force that acts upon all particles in the nucleus of an atom (for very short distance)

Nucleus. The center of an atom

Numerical aperture (NA). Relates four basic parameters—the angle of acceptance, the refractive indices of air, core, and cladding—of an optical fiber as follows: $n_0 \sin \theta = [n_1^2 - n_2^2]^{1/2}$

Nyquist criteria. For a digitized analog signal to be reproduced without distortion, it must be sampled with a sampling frequency two times the size or larger than the signal bandwidth ($f_s \geqslant 2\,B_W$)

Nyquist frequency. The sampling frequency must be equal or higher than that of the sampled signal

Nyquist theorem. It determines the minimum required clock frequency necessary for sampling a continuous time-varying waveform without the possibility of distortion at the reproduction stage

O

On-off keying (OOK). A simple digital modulation scheme

Operating bandwidth. The frequency range of a device or circuit that is operating effectively

Operating temperature. The temperature of a device or circuit that is operating effectively

Operating wavelength. The wavelength range of a device or circuit that is operating effectively

Operation and maintenance (OAM). Cells carrying operation and maintenance instructions

Operations, Administration, Management and Provisioning. Part of SDH and SONET standards dealing with the administration and management of networks

Optical add/drop multiplexer. Advanced optical multiplexers/demultiplexers employed in DWDM optical systems

Optical amplifier. A circuit designed to amplify photon energy

Optical amplitude modulation (OAM). An optical modulation scheme where the amplitude of the optical carrier changes in accordance with the information signal

Optical analysis. The mathematical analysis of optical systems in order to establish their optical properties

Optical axis. The line passing through the curved centers of lenses

Optical band pass filters (OBPF). A filter design to pass the optical band wavelength

Optical carrier (OC). A unit in the SONET hierarchy in which OC is indicative of optical signals with increments of 51.84 Mbps; that is, OC-1 equal to 51.84 Mbps, OC-3 equal to 155 Mbps, and OC-12 equal to 622 Mbps

Optical circulators. Passive optical components

Optical couplers. Passive optical components

Optical density. Associated with the transmittance of a signal through a medium expressed by the following equation: $D_\lambda = -\log_{10}(\tau_\lambda)$ where τ_λ is the transmittance

Optical driver. Devices that control the transmitted and received optical power in an optical system

Optical efficiency. It defines the rate of a photon to electron generation in a photodetector device

Optical emission. The minority carrier recombination process in both types of materials, under the influence of the externally applied voltage leading to the emission of light

Optical energy. The energy generated by an optical source

Optical energy confinement factor. A critical factor determining the modal gain of a laser diode

Optical extinction ratio. The power ratio of a plane-polarized optical ray transmitted through a polarizer with its polarization axis parallel to the optical ray and that of a polarizer with its polarization axis perpendicular to the optical ray

Optical fall time. The time required by an optical pulse to go from a high level to a low level

Optical fiber loss. The loss of power of an optical ray traveling through a fiber

Optical field intensity. The voltage per meter generated by an optical field

Optical gain. A characteristic of a laser diode expressed as the ratio of the generated optical power and bias current

Optical isolators. Protect optical equipment from damaging ground loops and transient spikes on the data lines

Optical link redundancy. A special link within an optical system intended to maintain normal operation in case of failure of the main link

Optical networks. Networks where data is carried through optical signals

Optical network unit (ONU). The function of the optical network unit is to convert the optical signals into electrical signals and to distribute them to thirty-two end users

Optical overshoot. An unwanted sharp increase of the amplitude of an optical pulse at its rise time

Optical path. An optical path is defined as the product of the distance (d) traveled by a ray through a medium and the refractive index of that medium

Optical path length. The product of the refractive index and the geometric distance of an optical transmission medium with a constant refractive index

Optical peaking. A wavelength at which the power of an optical pulse is at its maximum

Optical phase conjugation (OPC). The same as optical phase inversion

Optical polarization. The process whereby an electromagnetic wave is transformed from three-dimensional waves to two-dimensional waves

Optical power. The power generated by an optical source

Optical preamplifiers. An amplifier at the input of an optical receiver required to amplify a weak optical signal to the level specified by receiver sensitivity

Optical pulse ringing. Small and continuous amplitude variations of an optical pulse

Optical pumping. It refers to the process whereby atoms are elevated to higher energy levels through the absorption of light

Optical receiver dynamic range. The difference between receiver sensitivity and overdrive

Optical reflector tap. Senses the incoming optical power

Optical rise time. The time it takes an optical pulse to rise from 10% to 90% of its maximum value

Optical sensitivity. The minimum optical power required by an optical receiver to be functional

Optical signal processing (OSP). An integrated approach to optical signal processing incorporating both hardware and software products

Optical splitter. Provides polarization alignment

Optical switching. A method whereby each node has full access to the LAN bandwidth

Optical transmitter. A transmitter generating optical power for transmission through an optical fiber

Optical undershoot. An unwanted sharp decrease of the amplitude of an optical pulse at the edge of the pulse

Optical waveguide. An optical device designed to confine and guide optical energy parallel to its optical axis

Optical waveguide termination. An optical fiber termination method intended to prevent reflections back to the optical source

Optics. The study of light

Optoelectronics. Refers to devices designed to convert optical power into electrical signals and electrical signals into optical power

Optoisolator-optocoupler. Composed of an input amplifier, an LED, a photodiode, and an output amplifier. The main function of an optoisolator is to provide matching between incompatible modules within an optical communications system

Orthogonal modal state. The two principal components of an optical wave are perpendicular to each other

Output optical power. The optical power generated by an optical source

Output signal skew. The difference between low to high and high to low propagation delay of a pulse passing through a device

P

Packet energy. Photon energy

Packet switching. A form of data transmission. It refers to data transmission in groups or packets

Packet wave. A wave perceived to be composed of photons

Packing density. The term is applied to both optical fiber bundles and optical fiber system interconnect. In optical fiber bundles, packing density is referred to as the ratio of the cross-sectional area of a single fiber within the bundle to the total cross-sectional area. In optical fiber communications systems, packing density is referred to as the maximum number of optical detectors incorporated into the optical receiver

Parabolic profile. In an optical fiber, the variations of the refractive index of the fiber is a parabolic function of the fiber radius

Parallel to serial converter. A logic circuit performing parallel to serial data conversion

Parametric amplification. A method of optical wave amplification in which an intense coherent pump wave interacts with an optical crystal (nonlinear), resulting in the amplification of two other optical wavelengths

Parametric oscillator. Parametric oscillators generate tunable coherent optical wavelength beams through the insertion of a parametric amplifier into a resonant optical cavity

Parasitic capacitance. Unwanted capacitance encountered in semiconductor devices and transmission lines

Parasitic inductance. Unwanted inductance encountered in transmission lines

Parasitic voltage. Voltage generated as a result of parasitic elements

Parity. In a communications system, a redundant bit is added to the information bit for easy error detection at the receiver end

Parity error. The error occurring in DTE upon receipt of a wrong parity

Parity register. A logic circuit indicating whether a bit error has occurred

Partial coherence theory. In theory, coherent optical radiation is generated by a pure monochromatic point source. In practice, a limited bandwidth optical signal is generated with a non-zero angle, thus partially satisfying the coherent theory of optical radiation

Particle natural vibrating frequency. Characteristic natural vibrating frequency of a particle

Passive optical components. Optical devices capable of responding to the incident optical beams but incapable of either generating or amplifying optical beams

Passive optical network (PON). An optical network using passive components (splitters) for the delivery of optical signals to various users

Path averaging soliton. Occurs in a soliton transmission where the path average values of both dispersion (D) and intensity (I) are maintained equally for all optical spans

Path overhead (POH). Used to maintain transmission integrity in a SONET network

Path sublayer. The network layer that transfers information from a transmission medium network layer access point to one or more path layer networks

Payload type indicator (PTI). A 3-bit field in the ATM cell indicating the type of information carried by the payload

PCM-TDM. A PCM-TDM technique with a bit stream of 64 kb/s

Peak amplitude. Fabry-Perot laser central modal component amplitude

Peak excursion. Allows for the adjustment of the base level of all the discrete spectral components in a spectrum analyzer

Peak function. A single line representing each component of a displayed spectrum

Peak spectral emission. A narrow optical spectrum in which the radiation is at a maximum

Peak wavelength. A wavelength in which the optical intensity is at a maximum

Periodic and random deviation (PARD). An undesirable AC-component on top of the DC-power line of a power supply, relevant to fiber optics applications

Periodic table of elements. A table that lists (in numerical order) the electron number of all natural and man-made elements

Periodic wave. An optical wave in which the radiant energy at each point on the wave repeats itself at equal time intervals

Peripherals. Equipment used to support the main function of a computer

Permeability. In a given medium, the ratio of the magnetic induction to the applied magnetizing force

Permittivity. A constant of proportionality which exists between electric field displacement and electric field intensity

Phase. The segment of a periodic wave measured from a reference point

Phase angle. The measured angle between two vectors representing periodic quantities varying sinusoidally and at the same frequency

Phase array. A demultiplexing method employed in DWDM optical systems

Phase conjugation-spectral inversion. A technique used for dispersion compensation

Phase detector. The main function of the phase detector is to generate the required control signal for the frequency discriminator and also to provide frequency acquisition control

Phase difference. The phase difference between two signals, measured in degrees

Phase lock loop. An electronic circuit composed of a voltage controlled oscillator, a filter, and a phase detector

Phase noise. In a coherent communications system, the phase difference between the transmitted signal and the received signal is referred to as phase noise

Phase shift keying (PSK). A digital modulation method using the phase of a carrier signal to transmit digital information

Phase velocity. The velocity at which a monochromatic wave is traveling through an optical waveguide

Photocarriers. An electro-hole pair in an optical detector

Photoconductive effect. The electric conductivity variation of matter as a function of photon absorption

Photoconductivity. The rate of free-carrier generation; the mobility and carrier lifetime constitute the conductivity of some nonmetallic materials

Photoconductor. A resistor made from a semiconductor material, a polycrystalline film, or a single crystal. Exhibits light sensitivity properties. That is, its resistivity decreases with an increase of the intensity of the impeding light

Photocurrent. The current generated by a photosensitive device resulting from exposure to optical radiation

Photocurrent gain. The current gain parameter of an APD

Photodetection. The function performed by a photodetector converting optical energy to electric current

Photodetector. A device designed to sense optical power and to generate a proportional amount of current

Photodetector noise. The product of the random photocurrent generation at the output of a photodetector as a result of the random nature of the incident photons

Photodiode. A junction semiconductor diode that is sensitive to optical radiation and whose reverse bias current is a function of the optical radiation intensity

Photodiode responsivity. In a photodetector, defined by the ratio of the current generated in the absorption region per unit optical power incident to the region

Photoelastic coefficient. A characteristic parameter of an optical fiber essential for the calculation of Rayleigh scattering

Photoelectric absorption. The conversion of incident optical power into a photocurrent

Photoelectric emission. An electric emission resulting from incident optical radiation

Photoelectron. The generation of an electron from a photon

Photon. A packet of energy

Photon density. Number of photons per unit area

Photon energy. Equal to $E_g = hf$

Photonic sublayer. A convergent sublayer and a segmentation and reassembly sublayer, both composing a physical layer

Photon lifetime. The time required for building or decay of the optical field within a resonant cavity

Physical layer. In OSI, it refers to the activation and deactivation of the physical connections and signal transmission

Physical loop-gain product. An operating characteristic of a PLL circuit

Physical medium dependent (PMD). Provides the line coding and signal conditioning required for the transmission of the data over a medium in a network

Physical optics. The field of study of light as a wave

Pigtail. A passive component, usually a short-length optical fiber, permanently attached to an optical device and intended to couple optical power into the end device and the main transmission fiber

PIN. A semiconductor diode composed of positive doped (*P*-type) semiconductor material and a negative doped (*N*-type) semiconductor material with a large intrinsic region in between (Positive-Intrinsic-Negative)

PIN photodiode. A semiconductor diode composed of positive doped (*P*-type) semiconductor material and a negative doped (*N*-type) semiconductor material with a large intrinsic region in between (Positive-Intrinsic-Negative). The absorbed photons in the intrinsic region generate a proportional number of electron-hole pairs, which are translated into a photocurrent by an externally applied electric field

Planar waveguide. A waveguide fabricated by thin film technology

Planck's constant. Equal to 16.63×10^{-34} J/s 2

Planck's law. This law states that the relationship between electromagnetic radiation and a discrete quanta of energy transfers is proportional to the frequency of the electromagnetic wave

Planck's radiation law. The mathematical expression expressing the black body spectral radian emittance as a function of the emitted wavelength and temperature

Plane polarization. A plane polarized wave is defined as a wave whose amplitude vector of the electric field component always travels in the same direction

Plane waves. In a plane wave, the surfaces of constant phase are parallel planes to the direction of propagation

Plastic-clad silica fiber. A multimode step-index fiber in which the silica composed core is surrounded by a lower refractive index plastic cladding

Plesiochronous digital hierarchy. An original multiplexing hierarchy used in T-1/E-1 and T-3/E-3 systems

PN-junction. The junction formed by a P-type and an N-type semiconductor material

Point light source. An infinitely small source of radiation

Point source. A very small light source

Point-to-point. A communications link connecting only two points

Point-to-point transmission. A type of transmission that carries electronic signals from one point to the next without branching

Polarization controller. An optical device used to provide optical signal polarization from the laser diode to the input of the modulator

Polarization dependent gain (PDG). An impairment to the optical system induced by the in-line optical amplifiers

Polarization dependent loss (PDL). The peak-to-peak output power variations of a specific demultiplexer output channel when the input of the demultiplexer is fully polarized

Polarization hole-burning (PHB). A critical parameter influencing the performance of an undersea optical system

Polarization maintenance fiber (PMF). An optical fiber that maintains the polarization of light traveling through it

Polarization modal dispersion (PMD). A function of the operating wavelength, calculated by the delay time between two orthogonal polarized modes within a single mode fiber

Port. Hardware that allows data to enter and exit from a computer

Positive justification. In a digital multiplexing system, the addition of a bit to the writing clock in the case where the reading clock is faster than the writing clock

Postamplifier. A postamplifier employed in fiber optics systems is composed of a limiting amplifier, a level detector, two ECL buffers, and a circuit providing the disable function

Power amplifier. An amplifier that amplifies both voltage and current

Power dissipation. Power dissipated in the form of heat by a passive or an active component

Power efficiency. The ratio of the output power to the input power ($\eta\%$)

Power loss. The power loss of a signal passing through an optical fiber

Power margin. The optical power above a set minimum at the input of a receiver

Power optical amplifier. An optical amplifier designed to increase the optical power generated by an optical source to a level required by system specifications

Power penalty. The additional transmitter power required to offset connector losses in an optical system

Power supply rejection ratio (PSRR). The ratio of the power supply sensitivity to the circuit closed loop gain under normal operating conditions

Preemphasis. A method applied to frequency modulation intended for maintaining a constant signal level

Presentation layer. In OSI, it refers to code conversion and data formatting

Primary photocurrent. In an APD diode, the sum of photocurrent, dark current, and background current

Primary ring. The path that carries communication on an FDDI

Primary sequence code. This number is indicative of the maximum number of subscribers effectively utilizing a synchronous CDMA optical system

Principle of reversibility. If the direction of the reflected or refracted ray is reversed, each will retrace the optical path

Private branch exchange. A telephone switch connecting twenty or more telephones to each other and to public or private networks

Probability density function (PDF). A mathematical expression of the possible occurrence of an event

Profile dispersion. The optical waveguide dispersion attributed to refractive index profile variations as a function of the operating wavelength

Programmable add/drop multiplexer. An improved version of a standard add/drop multiplexer

Programmable input signal detect. An advanced method of optical input signal detection

Propagation constant. A constant characteristic of the propagating medium

Propagation delay. The time delay a signal encounters propagating through a medium

Propagation distance. The distance traveled by an optical pulse

Protocol. Any agreement that facilitates communication

Protocol data unit. A discrete piece of information appropriately formatted and encapsulated in an ATM payload

Proton. A positively charged particle with an electrical charge equal and opposite to that of an electron $1+1.602 \times 10^{-19}$ C2 and a mass of 1.643 ± 10^{-27} kg

Proton mass. Equal to 1.643×10^{-27} kg

Pseudo-orthogonal code sequence. A number equal to the primary number in a CDMA optical system

P-type material. In a *P*-type material, the majority carriers are holes generated by a dopant with acceptor atoms

P-type semiconductors. Extrinsic semiconductor materials formed by diffusing impurities of trivalent atoms into their crystalline structure

Public switched telephone network (PSTN). A collection of networks providing public telephone switching services

Pulse chirping. The broadening of a pulse through the transmission process

Pulse code modulation. A digital modulation method which converts an analog signal (voice or video) into a digital signal

Pulsed laser. Optical energy emitted from a laser diode in the form of short pulses

Pulse duration. The laser optical output pulse measured between the half power point of the leading edge and the half power point of the falling edge of the optical pulse

Pulse repetition frequency. The number of optical pulses generated by the laser diode per unit time

Pulse shaping. The process by which the shape of the optical pulse generated by the laser diode can be altered through a change in the laser diode supply voltage

Pulse spreading. It refers to a pulse's distortion while propagating through the optical fiber. The degree of the pulse distortion is relevant to material and mode propagation characteristics of the transmission optical fiber

Pulse width. The width of an electrical or optical pulse measured at the –3 dB point

Pulse width adjust (PWA). A mechanism for the automatic adjustment of the width of a pulse

Pulse width distortion (PWD). The distortion induced to the width of a pulse during its transmission through a medium

Pulse width modulation (PWM). A simple digital modulation scheme

Pumping radiation. The level of radiation required for exciting a laser diode to a higher energy level

Q

Q-factor. A number indicative to the transmission performance of an optical system

Quadrature amplitude modulation (QAM). An advanced digital modulation scheme. A digital modulation method whereby the digital information is carried at both the phase and the amplitude of the carrier signal

Quantizer. The process by which a sampled analog signal is digitized

Quantum. The smallest possible amount (quanta) into which the energy of a wave can be divided. This quantum energy is proportional to the frequency of the wave($E = h\triangle$)

Quantum detector. A photodetector device capable of generating electrical charges when electrons in a nonconducting state are converted to electrons in a conducting state through photon energy absorption

Quantum efficiency. The ratio of the number of emitted photons to the number of electrons (bias current) injected into the laser diode

Quantum mechanics. The study of atomic phenomena

Quantum noise. The total accumulated external background noise and internal noise (dark current)

Quantum numbers. There are four quantum numbers. The first quantum number is referred to as the principal quantum number, identified by the letter n; the second is referred to as the orbital quantum number, identified by the letter l; the third quantum number is referred to as the orbital quantum number m; and the fourth quantum number (m_s) describes electron spin direction under the influence of an applied magnetic field

Quantum optics. The study of the interaction of light with the atomic entities of matter

Quantum well (QW) laser. A very high electron density region between AlGaAs and GaAs layers within a laser diode structure, designed to enhance lasing efficiency and to decrease heat

Quaternary. A digital coding method that uses four voltage levels to represent information in an ISDN loop

R

Radial distortion. A magnification variation from the center of the field to any point in the optical field

Radial gradient. A gradient profile with a refractive index variation perpendicular to the optical axis

Radian power. The rate by which radian energy flows

Radiative minority carrier lifetime. The maximum time allowed for minority carriers inside the bandgap before carrier recombination takes place

Raman effect. When vibrating atoms in a crystalline lattice interact with optical waves, the vibrating atoms absorb some of the energy of the optical wave

Raman optical amplifier. It utilizes the principle of Raman scattering for amplification of selective wavelengths through the stimulation emission of radiation

Raman scattering. When a light beam passes through matter, a portion of the beam is scattered in all directions in a random manner. During this process, frequencies of a small portion of the scattered light are removed from the incident beam. These frequencies are the same as those of the vibrating frequencies of the scattering medium. This phenomenon can be used to initiate the stimulated emission of radiation

Rare earth doped fibers. Optical fibers in which rare-earth elements such as erbium, holmium, or neodymium are incorporated into the class-core structure in order to enhance optical fiber performance

Rare earth elements. Rare-earth elements are those with atomic numbers between 57 and 71

Ray. An optical ray is defined as a line normal to the wavefront, it points in the direction of the radian energy flow

Rayleigh scattering. In an optical fiber, Rayleigh scattering is caused by material anomalies with diameters one-tenth or less than the operating wavelength

Reading clock. A synchronization clock used in the demultiplexing process

Receiver sensitivity. The absolute minimum optical power level required at the input of an optical receiver for reliable signal detection

Reference clock. A clock required for the synchronization of the transmit receiver in a communications system

Reference clock jitter. The jitter generated by the reference clock

Reference frequency. The frequency generated by the reference clock

Reflection. The process whereby an optical ray incident upon a surface is reflected from this surface with an angle of reflection equal to the angle of incidence

Refraction. The process whereby an optical ray incident upon a transparent surface separating two media is refracted into the second medium with an angle dictated by the refractive index of this medium

Refractive index. Characteristic of the refractive property of a medium

Refractive index profile. The form and rate of change of refractive index of an optical fiber

Relative intensity noise (RIN). Noise characteristic of a DFB CW laser diode expressed in dB/Hz

Relative permitivity. Permitivity other than that of free space

Reliability factor. LED operational lifetime characteristic based on a model developed to calculate the device maximum operational time between failures

Repeater. A device that connects two LAN segments in order to form a larger segment in OSI

Resonance frequency. In a resonant circuit, the frequency at which the voltage across the circuit is at maximum

Resonant cavity enhanced (RCE). An advanced performance photodiode, fabricated by the combintion of a conventional Schottky photodetector semiconductor device (InAlAs) into a Fabry-Perot resonant cavity

Response time. In photodetector devices, this is the time a carrier takes to cross the depletion region

Responsivity. Responsivity of a PIN photodiode is the ratio of the generated photo current per incident of unit-light power

Return loss. The optical power loss through back reflections

Return to zero (RZ). A data pattern incorporating a third level representing separation between bits

Return to zero data pattern. For logic-1, the pulse remains high for fifty percent of the bit cell then falls back to zero. For logic-0, the pulse remains at zero level for the entire bit cell

Reverse current. The current generated in a *PN*-junction due to the minority carriers

Ring topology. In a ring topology, each node is connected to two links

Rise time. It refers to the time it takes the current generated by a photodetector diode to go from ten percent to ninety percent

Roll off. It refers to a defect at a cleaved end of on optical fiber, shows as a slight blemish of the cleaved edge

Router. In WDM, a device that routes wavelengths to different destinations

S

Sample and hold. The process whereby an analog signal is sampled by a sampling circuit and its value is stored for a specified time before the next sampling. Sample and hold circuits are part of the analog-to-digital conversion process

Saturation current. The turn-on current of a semiconductor diode

Saturation power. The product of the saturation voltage and saturation current of a forward bias semiconductor diode

Saturation velocity. In a photodetector diode, the carrier velocity given as m/s

Scattered light filter. A filter designed to limit undesirable light scattering from reflections from the edges of an optical device

Scattering. The phenomenon whereby optical energy is transferred from the dominant mode to adjacent modes. Scattering is divided into two major categories: linear and nonlinear

Scattering coefficient. A coefficient indicative to a material's scattering properties

Schoedinger nonlinear equation. Expresses the propagation of an optical signal through a fiber in the presence of chromatic dispersion, polarization modal dispersion, fiber losses, and Kerr nonlinearities

Secondary ring. The data path that serves as a substitute in FDDI

Segmentation and reassembly (SAR) sublayer. Segments a packet into 48 bytes called segmentation and reassembly protocol data units (SAR-PDU)

Selective absorption. Occurs when unpolarized optical waves enter an anisotropic material. The material has characteristics that allow it to absorb optical energy in one direction and to transmit optical energy parallel to the crystallographic axis

Self phase modulation (SPM). An undesirable signal modulation caused by refractive index nonlinearities

Sellmeier's equation. Large amplitudes of particle vibrations at resonance will interfere with the traveling wave, thus altering its velocity within a medium

Semiconductor optical amplifier. An optical circuit designed to amplify weak optical signals of a specific wavelength with a minimum degree of distortion

Sequence number protection (SNP). A 4-bit SN intended for the protection of the SN field from possible errors

Server. A computer on a network that provides central data storage, which can be accessed by other computers

Service control point. Central point controlling the database in an intelligent network

Service node. A function of an intelligent network that contains databases for call services and other specialized switching

Service switching point. A term indicating the class 4/5 switch in an intelligent network

Session layer. In OSI it refers to the organization and management of data exchange

Shot noise. Electronic noise generated in a semiconductor material by random variations in the velocity and number of free electrons

Shot noise power. The power generated by shot noise in a semiconductor material

Shunt capacitance. Capacitance formed between adjacent connections

Sigma (σ)**.** The Fabry-Perot mean square of the spectral width, assuming Gaussian distribution

Signaling. Communications between telephone switches for setting up and terminating calls

Signaling system #7. International standards protocol applied to open signaling in a digital public switching network. This protocol is based on a 64 Kbps channel and is used for call control, information transfer, and billing management

Signal level. The optical power measured at a point along an optical fiber

Signal level detect. A mechanism through which the input signal is constantly monitored

Signal power. The measure of the power of an optical signal

Signal to noise ratio (SNR). The ratio of the desirable signal level to the undesirable noise level

Signal transfer point. A function of an intelligent network acting as the transfer point for call signaling and processing

Silicon atom. Composed of fourteen electrons distributed among three orbiting shells and fourteen protons and fourteen neutrons in the nucleus

Single mode. Containing only one mode of operation

Single mode step index fiber (SI-SMF). SI-SM fibers are composed of a uniform cylindrical dielectric core of refractive index n_1, surrounded by a cladding material of refractive index, where, $n_1 > n_2$

Single polarization fibers. Fibers that convey light in only one mode of polarization: step-index fibers with small cores carry light in a single mode

Skew ray. A ray that is propagated through an optical system without being a meridional ray

Snell's law. When an optical ray is incident with an angle θ_1 upon the interface between two dielectric mediums with different indices ($n_1 > n_2$), the following relationship holds true: $\dfrac{\sin \theta_1}{\sin \theta_2} = \dfrac{n_2}{n_1}$

Soliton. An ultra short optical pulse generated by a laser diode exhibiting zero chromatic dispersion while propagating through an optical waveguide

Soliton collision. The process through which a soliton pulse from one channel in a DWDM optical scheme overtakes another pulse of an adjacent channel within the scheme

Soliton laser. A laser diode structure incorporating an external laser cavity capable of generating a soliton optical pulse

Soliton period. The period of a soliton pulse

SONET/SDH. Synchronous Optical NETwork/Synchronous Digital Hierarchy

Source spectral linewidth. The spectral linewidth of an optical source

Space division multiplexing (SDM). In an optical fiber bundle, each fiber carries a single optical channel

Spectral attenuation. The attenuation of a signal traveling through an optical fiber as a function of the signal wavelength

Spectral bandwidth. The bandwidth of an LED measured at half power point

Spectral line width. In LED devices, spectral line width is determined at the half power point of the spectral density in reference to wavelength

Spectral window. A high transmittance narrow optical wavelength slotted between two low transmittance optical wavelengths

Spectrum. A continuous range of frequencies expressed by FFTs

Splice. A junction between two optical fibers

Splicing. A process of jointing two optical fibers

Spontaneous emission. Light generated by electron-hole motion in a semiconductor material

Spread spectrum. A modulation technique that spreads the transmitted signal across a wide band of frequencies by constant frequency shift

Standard deviation. A statistic that defines how tightly all the various examples are clustered around the mean in a set of data

Standard group. CCITT G.232 recommendations for 6 KHz to 108 KHz FDM signals

Standard master group. An FDM scheme that incorporates five super groups, totaling three hundred voice channels and occupying the frequency range between 812 KHz and 2.044 MHz

Standard single mode fiber. A single mode step-index fiber exhibiting zero dispersion at 1310 nm wavelength

Standard super group. CCITT G.232 recommendations for 712 KHz to 552 KHz FDM signals

Standard super master group. CCITT G.232 recommendations for 8.516 MHz to 12.388 MHz FDM signals

Star coupler. Star couplers are passive optical devices incorporating one or more inputs. They are capable of distributing the input optical signal(s) to a larger number of output waveguides in a star configuration. Composed of three or four ports

Start bit. A bit placed in front of a binary word before transmission in an asynchronous data transmission system

Star topology. In star topology, the nodes are connected to a hub node which, in turn, provides cross connection

State of polarization (SOP). Under this principle, if a pulse at the input of the fiber is aligned to one of the principal states, it will appear at the output of the fiber with all its spectral components having the same polarization state

Stefan-Boltzmann law. A relationship governing the total energy (E) emitted by a black body at all wavelengths

Step index fibers. Optical fibers with uniform core refractive index

Step index multimode fiber. A step index fiber with large core carrying light in multiple modes

Step index profile. The graphical profile of the uniform core refractive index and the surrounding cladding of a step index fiber

Stimulated Brillouin scattering (SBS). When a burst of light generated by a laser diode is beamed into a crystal, a light amplification process takes place, resulting in the deformation of the crystal lattice and producing ultrasonic waves as a byproduct

Stimulated emission. Stimulated emission of radiation is the process by which photons are used to generate other photons. This process is applicable in laser diode operations

Stimulated Raman scattering (SRS). SRS is the result of the interaction between the vibrating atoms in a crystalline lattice and the optical wave

Stokes photon. A photon that is transformed into a lower frequency when traveling through a transparent medium

Stop bit. A bit placed at the end of a binary word before transmission in an asynchronous data transmission system

Straight line method. A method used for measuring frequency flatness

Strip line. A strip transmission line is constructed by placing the transmission line on top of the PC board material with a conducting plate placed at the bottom

Submarine cable. An optical cable constructed for underwater operation

Subscriber. A customer of a communications carrier

Subscriber loop. The part of a telephone network connecting the CO and the subscriber

Subscriber loop interface circuit. It provides the battery feed, over voltage protection, ringing, signaling, coding, hybrid, and testing functions in telephony

Surface-emitting laser diode (SEL). SEL diodes emit light in a vertical direction in relationship to their active region. That is, this optical output power is collected from the surface of the semiconductor structure, unlike standard lasers whose optical power output is collected from a cleaved mirror

Surface emitting LED. The optical radiation of the surface LED takes place from the surface of the active layer

Switch. A device or circuit designed to convey electrical or optical signals to different physical paths

Switched fabric. A fiber channel topology

Switched network. A network that routes signals to various physical paths through switching circuitry

Switching current. The current required by a switching device to turn ON or OFF

Switching time. The time required by a switching device to turn ON or OFF

Symmetrical rise/fall times. The rise time equals the fall time

Synchronous CDMA with modified prime sequence. Utilizes unipolar data (1,0). Bit 1 is represented by a code sequence waveform representing the bit address with a length sequence subject to bit period

Synchronous communications. A communications system in which the transmitter and receiver operate with the same clock frequency

Synchronous data link control. A full duplex layer-2 software protocol of the OSI model, a synchronous protocol including a clock signal in the transmitted data

Synchronous detector. Detectors that are sensitive to the phase and frequency (signal) of a particular control signal

Synchronous digital hierarchy (SDH). A synchronous optical network standard

Synchronous optical network (Sonet). An optical fiber communications links interface operating over a single mode fiber at the 1300 nm wavelength window and at 52 Mbps, 155 Mbps, and 622 Mbps transmission rates (a standard for optical fiber communications)

Synchronous transfer point. In an intelligent network, the point at which call control is transferred between the service control point and the service switching point

Synchronous transmission. In a synchronous optical transmission system, the transmitter section also transmits a clock reference signal for data retiming at the receiver end. Synchronous optical communications systems exhibit better performance

Synchronous transport module. The basic building of European synchronous networks with a data rate of 155.52 Mbps

Synchronous transport signal (STS). It refers to the level 1 (STS-1) building block of the North American synchronous network with a data rate of 51.84 Mbps

Systems network architecture (SNA). Set of rules governing interaction between network components in an IBM environment

T

T-1 carrier. A transmission system using TDM multiplexing to carry twenty-four voice channels, each of 64 Kbps to a total transmission of 1.544 Mbps. Also referred to as DS1

TDM-PCM. Twenty-four voice signals digitally multiplexed to 1.455 Mbps

Temperature coefficient. A typical operating characteristic of a DFB laser diode

Temperature factor. An operating parameter critical to the establishment of the maximum LED operational time

Temperature stabilization substrate. One of the two components of the temperature control unit used as a temperature source providing the required internal temperature for the AWG device

Temperature tracking. A mechanism constantly monitoring the temperature control unit

Ternary. A semiconductor device made of three different elements

Thermal runaway effect. The inability of a device to dissipate internally generated thermal energy

Thermal stability. A measure of operating insertion loss variation for each port in an AWG device over the full range of local ambient operating temperatures while the device is thermally controlled

Thermal voltage. Part of the barrier voltage at the *PN*-junction of a semiconductor device

Thermistors. Thermally sensitive resistors with a negative or positive resistance/temperature coefficient

Thermoelectric aperture. A device that converts thermal energy to electric voltage

Third order chromatic dispersion. Same as the Kerr effect (see *Kerr effect*)

Threshold. The minimum level of an input signal required by a detector to generate a response

Threshold current (I_{TH}). The minimum current required by a laser diode to initiate spontaneous emission of radiation at a specified temperature level

Threshold voltage (V_{TH}). The minimum voltage above a set reference voltage required to initiate a change at the output of an electronic circuit (i.e., comparator circuits)

Time constant. The time required by an electronic circuit to react to an input excitation signal by reaching an output level of sixty-three percent of the same input signal

Time division multiple access (TDMA). The multiplexing of several digital signals into a single channel

Time-division multiplexing (TDM). TDM refers to the process whereby several parallel input data channels are combined to form a single output data channel

Time domain CDMA. Every bit of the input binary data sequence is encoded and the resulting waveform, representing the destination address of the binary data, is broadened by a specific factor

Time frame. For PCM signals, equal to 125 μs

Token. A specific bit pattern determining the terminal to be connected to the token ring LAN network

Token passing. A protocol that allows a terminal to transmit in a token ring network

Token ring. A LAN standard (IEEE 802.5) connecting personal computers through coaxial cable

Topology. The physical layout of a network

Total internal reflection. The total reflection of light at an angle of incidence below a critical value

Total power dissipation. Same as power dissipation

Tracking. The process of tracking the frequency and phase of an electrical signal

Tracking error. Indicates the difference between the laser optical power and the monitor photodiode current

Transceiver. A combined transmitter and receiver

Transducer. Performs the conversion of energy from physical quantities to electrical signals. In optics, the representative device that converts optical energy into electrical energy is the photodetector diode

Transfer function. A mathematical expression relating the output signal to the input signal of an electronic device as a function of frequency

Transimpedance amplifier. The main function is to convert the generated photodetector current into a voltage signal of a level that is capable of driving an AGC post-amplifier circuit

Transistor. Active semiconductor devices (bipolar or MOSFET) utilized in linear or switching modes of operation

Transition. Digital signals changing states from high to low or from low to high. In quantum mechanics, a change in energy levels through absorption or emission of photons or particle kinetic energy

Transmission. In fiber optics systems, the conduction of optical energy generated by an externally or internally modulated laser diode through an optical fiber to the input of an optical receiver

Transmission convergence (TC). One of the two sublayers of the physical layer

Transmission efficiency. The ratio of the optical energy detected at the receiver input to the optical energy generated by the optical transmitter

Transmission frame adaptation. One of the five functions performed by the transmission convergence sublayer

Transmission frame generation. One of the five functions performed by the transmission convergence sublayer

Transmission loss. The optical power loss during the transmission of a signal through an optical fiber

Transmission medium. The medium linking the transmitter and the receiver in a communications system

Transmittance. The ratio of the transmitted radian power to the incident radian power

Transmitter. In optical fiber links, the transmitter section is mainly composed of an LED or laser diode, an LED or laser driver circuit, and associated electronic circuits

Transponder. A transreceiver module designed to transmit signals upon receipt of an appropriate interrogation signal

Transport layer. In OSI, it refers to activation and deactivation, transmission synchronization and control, routing, and switching functions

Transport overhead (TOH). A part of the SONET frame

Transverse electric. The electric vector component of an electromagnetic wave perpendicular to the magnetic vector component

Transverse electromagnetic wave. The electric field vector component is perpendicular to the magnetic vector component and both are perpendicular to the direction of propagation

Transverse magnetic. The magnetic vector component of an electromagnetic wave perpendicular to the electric vector component

Transverse mode. Spatial modes perpendicular to a laser diode active region. The fundamental transverse mode is designated as TE_{00} and is Gaussian in nature

Traveling wave tube (TWT). A high power microwave amplifier

Tree topology. In a tree topology, each node must have two or more connections with other nodes

Trunk. The path between two switches conducting only one telephone conversation

Tunable extended vertical cavity laser (TEVCL). TECVLs can be tuned over a wide wavelength range with an almost flat response and a very small level of channel crosstalk

Tunable laser. Laser diodes capable of adjusting the optical power output over a wide (approximately 70 nm) wavelength range

Tunneling. Energy levels required by atomic particles to pass through a barrier. Quantum mechanics laws predict the finite tunneling capabilities of atomic particles

Tunneling current. The current as a result of tunneling

Turn off time improvement. A reduction of the turn off time of a semiconductor device operating in the switch mode

U

Uncooled laser transmitter.   Pioneered by Lucent Technologies, this transmitter is designed to operate in optical fiber systems that are in compliance with SONET and ITU-T SDH formats

Unipolar code sequence.   A binary sequence using either positive or negative signal levels

Unpolarized wave.   Waves that can travel vertically and horizontally

Usage parameter control.   A function that prevents congestion by controlling the data traffic entering the network

User network interface (UNI).   Standard driven protocol defining connections between an ATM network and an ATM user

V

Valence electrons.   The electrons occupying the valence shell. These electrons define the chemical properties of the atom

Valence shell.   The outermost shell in an atomic model

Vapor phase epitaxy (VPE).   A laser diode fabrication process which practically eliminates two of the problems encountered during the fabrication process: control of the physical dimensions and the control of the heterobarrier lattice growth

Variable bit rate (VBR).   Data traffic with variable bit rate

VCO gain.   Voltage controlled oscillator, part of a PLL circuit

Vertical cavity surface emitting laser (VCSEL).   A semiconductor laser in which light oscillates perpendicular to the junction plane and emerges from the surface of the wafer

Very large scale integrated circuits (VLSI).   Integrated circuits with very dense component structure

V-groove.   Etching technique employed in the substrate of optical semiconductor devices for the fabrication of integrated optical modules

Virtual channel identification (VCI).   A 1 to 4 bit word assigned for channel identification in a UNI

Virtual path identification (VPI).   A 5 to 8 bit word assigned for path identification in a UNI

Voice channel.   It consists of a 4 KHz bandwidth, or 64 Kbps bit rate

Voltage level comparator swing ($V\pi$).   The swing voltage ($V\pi$) is defined as the voltage required for shifting the phase of the optical signal in one of the modulator waveguides by 180°

W

Wave.   The form in which the energy of the electromagnetic spectrum propagates

Wave equation.   Applied to electromagnetic waves traveling along the x-axis and perpendicular to the y-axis

Waveform.   The graphical representation of the oscillating variations of a wave as a function of time

Wavefront.   The surface connecting all the equidistant points from the wave source

Wave function.   The amplitude at a point in a wave represented by the wave equation

Waveguide.   A passive device designed to confine and conduct microwave or optical energy to a direction subject to the waveguide physical characteristics

Waveguide dispersion.   Wave distortion while a wave propagates through a waveguide. Waveguide dispersion is a result of the interaction of phase and group velocities of the traveling wave with the waveguide physical properties

Waveguide laser.   A gas laser device encased in a tube and acting as guide to the laser beam

Waveguide scattering.   Optical signal attenuation caused by the scattering of wave propagating through the waveguide. It is attributed to the refractive index profile and geometric variations of the optical waveguide

Wave intensity.   The amount of energy flowing through a unit area per unit time

Wavelength.   The physical distance traveled by an electromagnetic wave during one cycle. The relationship between wavelength and frequency of an electromagnetic wave $\left(\lambda = \dfrac{c}{f}\right)$ where c is the velocity of light in space

Wavelength bandwidth.   A narrow spectrum of the wavelength used for the transmission of information

Wavelength conversion.   A process converting wavelengths in all photonic systems

Wavelength division multiplexing (WDM).   The process whereby multiple optical carriers of different wavelengths utilize the same optical fiber

Wave motion.   Electromagnetic waves propagating through free space. In this motion, the electric field is perpendicular to the magnetic field and both are perpendicular to the direction of propagation

Wave optics.   The study of light, primarily concerned with the nature of waves (physical optics)

Wave particle duality.   Light can be considered both a wave and a discrete particle

Wave theory. The study of electromagnetic waves as they propagate through a medium

Weak force. Gravitational force

White noise. A random noise with a spectral density exhibiting a very small degree of frequency dependence over a specified frequency range

Wien displacement law. A relationship between the peak energy distribution at wavelength λ and the temperature T of black body radiation

X

x **axis.** The horizontal axis in a Cartesian coordinate system, or the reference axis in a quartz crystal

Y

y **axis.** The vertical axis in a Cartesian coordinate system, perpendicular to the *x*-axis, or perpendicular to the parallel and opposite faces in a crystal structure

Z

Zero dispersion wavelength. An optical wavelength exhibiting zero dispersion while traveling through an optical fiber

Zero order. The point in an interference pattern where all the optical paths are equal to zero

Zero order filter. The removal of the zero order in a Fourier spectrum distribution

ABBREVIATIONS

A

AA	Acceptance Angle
AAL	ATM Adaptive Layer
AC	Angle of Convergence
ACI	Adjacent Channel Interference
ADM	Add/Drop Multiplexer
ADSL	Asymmetrical Digital Subscriber Line
ALOHA	Advanced Laser Optical Hazards Analysis
AM	Amplitude Modulation
ANSI	American National Standards Institute
AO	Acousto Optics
AP	Access Protocol
APS	Advance Photon Source
ARI	Absolute Refractive Index
AS	Absorption Spectrum
ASCII	American Standards Code for Information Interchange
ASE	Amplifier Spontaneous Emission
AT	Asynchronous Transmission
ATM	Asynchronous Transfer Mode
AVPD	Axial Vapor Phase Deposition
AWG	Array Waveguide Grating
AWGN	Additive White Gaussian Noise

B

BB	Base Band
BD	Beam Divergence
BELLCORE	BELL COmmunications REsearch
BER	Bit Error Rate
BFSK	Binary Frequency Shift Keying
BIP	Bit Interleaved Parity

B-ISDN	Broadband Integrated Services Digital Networks
BISYNC	BInary SYNchronous Communications
BPDU	Burst Protocol Data Unit
BPF	Band Pass Filter
BPS	Bits Per Second
BPSK	Binary Phase Shift Keying
BR	Burst Error
BSC	Back Scattering Coefficient
BW	BeamWidth

C

CA	Critical Angle
CAD	Computer Aided Design
CAW	Critical Absorption Wavelength
CC	Coherent Communications
CCITT	Consultative Committee of the International Telephone & Telegraph
CCS	Common Channel Signaling
CDMA	Code Division Multiple Access
CEPT	Conference of European Postal and Telecom administration
C/I	Carrier to Interference ratio
CLASS	Custom Local Area Signaling Service
CLP	Cell Loss Priority field
CLS	Coherent Light Source
CMOS	Complementary Metal Oxide Semiconductors
CNR	Carrier to Noise Ratio
CO	Central Office
CODEC	COder/DECoder
COP	Carrier Oriented Protocol
CPAL	Chirp Pulse Amplification Laser

CPE	Customer Premise Equipment
CPFSK	Constant Phase Frequency Shift Keying
CPS	Characters Per Second
CPU	Central Processing Unit
CSA	Canadian Standards Association
CW	Continuous Wave

D

DCC	Digital Cross Connect
DCW	Direct Cylindrical Waveguide
DEMUX	DEMUltipleXing
DFBL	Distributed Feedback Bragg Laser
DFSMF	Dispersion Flattened Single-Mode Fiber
DLC	Digital Loop Carrier
DMA	Differential Mode Attenuation
DPN	Digital Packet Network
DPSK	Differential Phase Shift Keying
DQE	Differential Quantum Efficiency
DQPSK	Digital Quadrature Phase Shift Keying
DS	Direct Sequence
DS0	Digital Signal, level 0
DS1	Digital Signal, level 1
DS2	Digital Signal, level 2
DS3	Digital Signal, level 3
DSF	Dispersion Shifted Fiber
DSI	Digital Signal Interpolation
DSL	Digital Subscriber Line
DSP	Digital Signal Processing
DSSS	Direct Sequence Spread Spectrum
DTCR	Decision Theoretical Character Recognition
DTE	Data Terminal Equipment
DWDM	Dense Wavelength Division Multiplexing

E

EDFA	Erbium Doped Fiber Amplifier
EDSL	Extended Digital Subscriber Line
EIA	Electronic Industries Association
ENOB	Effective Number Of Bits
ET	Exchange Termination
ETSI	European Telecommunications Standards Institute

F

FCC	Federal Communications Commission
FDDI	Fiber Distributed Data Interface
FDF	Fermi-Dirac Function
FDM	Frequency Division Multiplexing
FDMA	Frequency Division Multiple Access
FDX	Full DupleX
FEC	Forward Error Correction
FFDP	Far Field Diffraction Pattern
FH	Frequency Hopping
FHSS	Frequency Hopping Spread Spectrum
FIFO	First-In First-Out memory
FITL	Fiber In The Loop
FM	Frequency Modulation
FOC	Fiber Optics Cable
FOV	Field Of View
FPI	Fabry-Perot Interferometer
FSK	Frequency Shift Keying
FTTC	Fiber To The Curb
FWHM	Full Width at Half Maximum

G

GaAlAs	Gallium Aluminum Arsenide
GaAs	Gallium Arsenide
GBP	Gain Bandwidth Product
Ge	Germanium
GFC	Generic Flow Control
GFSK	Gaussian Frequency Shift Keying
GW	Guided Waves

H

HBw	Half Bandwidth
HD	Half Duplex
HDLC	High level Data Link Control
HDSL	High speed Digital Subscriber Loop
HDX	Half DupleX
HEC	Header Error Control
HLF	High Loss Fiber
HPF	High Pass Filter
HPP	Half Power Point
HSSI	High Speed Serial Interface

I

IDLC	Integrated Digital Loop Carrier
IEC	IntErchange Carrier
IEEE	Institute of Electrical and Electronic Engineers
IF	Intermediate Frequency
IMD	Inter Modulation Distortion
IN	Intelligent Network
IP	Intelligent Peripheral
IP3	Third order Intercept Point
IPE	Internal Photoelectric Effect
IR	Incident Ray
ISDN	Integrated Services Digital Network
ISI	InterSymbol Interference
ITU	International Telecommunications Union
IVPO	Inside Vapor Phase Oxidization

J

JPEG	Joint Photographic Experts Group

L

LAN	Local Area Network
LAPB	Link Access Protocol B channel
LAPD	Link Access Protocol D channel
LEC	Local Exchange Carrier
LED	Light Emitting Diode
LiNbO$_3$	Lithium Niobate
LNA	Low Noise Amplifier
LNB	Low Noise Block
LO	Local Oscillator
LP	Line Protocol
LPC	Linear Predictive Coding
LPF	Low Pass Filter
LSB	Least Significant Bit
LT	Line Termination

M

M$_{1,3}$	Multiplexer DS1 to DS3
MAN	Metropolitan Area Network
MBw	Modulation Bandwidth
MD	Modal Dispersion

MF	Multi Frequency
MLL	Mode Lock Laser
MMD	Multi-Mode Dispersion
MMOW	Multi-Mode Optical Waveguide
MPEG	Motion Picture Expert Group
MQW	Multi-Quantum Wavelength
MSB	Most Significant Bit
MSK	Minimum Shift Keying
MSLI	Mean Spherical Luminous Intensity
MUX	MUltipleXer

N

NAP	Network Access Point
NFS	Network File System
NIC	Network Interface Card/controller
N-ISDN	Narrow band ISDN
NL	Network Layer
NNI	Network to Network Interface
NT	Network Termination
NTIA	National Telecommunications and Information Administration
NTSC	National Television Standards Committee
NTT	Nippon Telephone and Telegraph

O

OA	Optical Axis
OAM	Operations Administration Maintenance
OAMP	Operations Administration Maintenance and Provisioning
OC-n	Optical Carrier
OD	Optical Density
OE	OptoElectronics
OPL	Optical Path Length
OQPSK	Offset Quadrature Phase Shift Keying
OSI	Open System Interconnect
OW	Optical Waveguide
OWT	Optical Waveguide Terminator

P

PBX	Private Branch eXchange
PC	Personal Computer
PCI	Peripheral Component Interconnect

PCM	Pulse Code Modulation
PCS	Personal Communications Service
PCSF	Plastic Clad Silica Fiber
PDH	Plesiochronous Digital Hierarchy
PDMA	Phase Division Multiple Access
PDN	Public Data Network
PDS	Premises Distribution System
PDU	Protocol Data Unit
PE	Parity Error
PIN	Positive Intrinsic Negative
PL	Physical Layer
PLCP	Physical Layer Convergence Protocol
PLL	Phase Lock Loop
PMF	Polarization Maintenance Fibers
PN	Pseudo random Noise
POC	Passive Optical Components
PON	Passive Optical Network
PPM	Pulse Position Modulation
PRF	Pulse Repetition Frequency
PSE	Peak Spectral Emission
PSK	Phase Shift Keying
PSTN	Public Switch Telephone Network
PTI	Payload Type Indicator

Q

QAM	Quadrature Amplitude Modulation
QD	Quantum Detector
QE	Quantum Efficiency
QFSK	Quadrature Frequency Shift Keying
QM	Quantum Mechanics
QN	Quantum Noise
QO	Quantum Optics
QW	Quantum Well

R

RAM	Random Access Memory
RCE	Resonant Cavity Enhanced detection quantum efficiency
REE	Rare Earth Elements
RF	Radio Frequency
RIP	Refractive Index Profile
RS	Raman Scattering
RSSI	Receive Signal Strength Indicator

| RTU | Right To Use |
| RZ | Return to Zero |

S

SAP	Service Access Point
SAW	Surface Acoustic Wave
SBS	Stimulated Brillouin Scattering
SC	Synchronous Communications
SCP	Service Control Point
SDH	Synchronous Digital Hierarchy
SDLC	Synchronous Data Link Control
SDM	Space Division Multiplexing
SE	Stimulated Emission
SELD	Surface Emitting Laser Diode
S/H	Sample and Hold
SI	Step Index fiber
SIMMF	Step Index Multi-Mode Fiber
SL	Subscriber Loop
SLIC	Subscriber Line Interface Circuit
SN	Service Node
SNA	System Network Architecture
SNR	Signal to Noise Ratio
SOH	Section OverHead
SONET	Synchronous Optical NETwork
SPF	Single Polarization Fiber
SPM	Self Phase Modulation
SQ	Signal Quality
SS	Spread Spectrum
SS7	Signaling System #7
S-SMF	Standard Single Mode Fiber
SSP	Service Switching Point
STM	Synchronous Transport Module
STP	Signal Transfer Point
STS	Synchronous Transport Signal
SW	Spectral Window

T

TCM	Time Compression Multiplexing
TDM	Time Division Multiplexing
TDMA	Time Division Multiple Access
TE	Terminal Equipment
TEVCL	Tunable Extended Vertical Cavity Laser

TF	Transfer Function
TIR	Total Internal Reflection
TL	Transport Layer
TM	Transverse Mode
TR	Technical Requirement
TTC	Telecommunications Technology Committee
TXE	Transmit Enable

U

UDI	Unrestricted Digital Information
UNI	User Network Interface
UPC	User Parameter Control
UPC	Universal Personal Communications
UPS	Uninterrupted Power Supply

V

VCO	Voltage Control Oscillator
VCSEL	Vertical Cavity Surface Emitting Laser
VDSL	Very high bit rate Digital Subscriber Line
VSELPC	Vector-Sum Excited Linear Predictive Coding

W

WDM	Wavelength Division Multiplexing
WF	Wave Function
WGD	Wave Guide Dispersion
WGS	WaveGuide Scattering
WLAN	Wireless Local Area Network
WN	White Noise
WO	Wave Optics

X

XPM	Cross Phase Modulation

Z

ZDW	Zero Dispersion Wavelength
ZO	Zero Order
ZOF	Zero Order Filter

REFERENCES

Adam, L., Simova, E., Kavehrad, M. 1995. "Experimental Optical CDMA System Based on Spectral Amplitude Encoding of Noncoherent Broadband Sources," *Spie* Vol. 2614 (p. 122).

Adams, L. E., Kintzer, E.S. 1995. "Characterization of an all-optical clock recovery figure eight laser for systems applications." *Spie* Vol. 2614 (p. 2).

Agilent Technologies. "Designing With the HDMP/1536A and HDMP/1546A Fiber Channel Transceivers." *Application Note.*

Agilent Technologies. Hardman, Dennis. "Wide Area Network Analysis and Troubleshooting." 1994.

Agilent Technologies. Unverrich, Rod. "Voice Over IP—Understanding H.323." 1994.

Antoniades, N., Wei, X., Stern, T. F., Pathak, B., Yang, E. S. 1995. "Use of Subcarrier Multiplexing/Multiple Access for Multipoint Connections in All-Optical Networks." *Spie* Vol. 2614 (p. 218).

Applied Micro Circuits Corporation. "OC-48 Application Note With Amazon/S3044/83/40 and Sumitomo Fiber Optic." Applied Micro Circuits Corporation. January 2000 "S3019 With 2 X 9 Sumitomo Fiber Optics Application Note." March 1998.

Applied Micro Circuits Corporation. "S3019 With 1 X 9 HP Fiber Optics Application Note." April 1998.

Applied Micro Circuits Corporation. "S3019 With 2 X 9 Sumitomo Fiber Optics Application Note." March 1998.

Applied Micro Circuits Corporation. "S3037 With 1 X 9 HP Fiber Optics and S1201 Congo Application Note." October 1999.

AT&T Microelectronics. "Lithium Niobate Intensity (Amplitude) Modulator," *Technical Note.* March 1995.

AT&T Microelectronics. "The Relationship Between Chirp and Voltage for the AT&T Mach-Zehnder Lithium Niobate Modulators," *Technical Note.* October 1995.

Atatüre, M., Sergienko, A. V., Saleh, B. E. A., Teich, M. C. "Dispersion-Independent High-Visibility Quantum Interference in Ultrafast Parametric Down-Conversion." *Physical Review Letters*, Vol. 84, No. 4, January 24, 2000.

Bell Laboratories & Lucent Technologies; Brinkman, W. F., Lang, D. V. "Physics and the Communications Industry." 1997.

Bell Labs. Al-Salameh, D. Y., Fatehi, M. T., Gartner, W. J., Lumish, S., Nelson, B. L., Raychaudhuri, K. K. "Optical Networking." *Bell Labs Technical Journal*, January/March 1998.

Boche, B., Müller, Böhm, R., Tränkle, G., Weimann, G. "Monolithic Integration of GaAs-AlGaAs Quantum-Well Lasers With Directional Couplers Using Vertical Coupling of Light." *IEEE Photonics Technology Letters*, Vol. 8, No. 12, December 1996.

Boston Optical Fiber, Inc. Ilyashenko, V., Berman, E. "Graded Index Plastic Optical Fibers: Quo Vadis." 1999.

Boucouvalas, A. C., 1995. "Asymmetry of Free Space Optical. Links." *Spie* Vol. 2614 (p.69).

Corning Incorporated. Dowdell, E. A. "High Data Rate Networks, The Latest Fiber Technologies for Long-Haul." 1999.

Corning Incorporated. Hluck, Laura. "Optical Fibers for High Capacity Dense Wavelength-Division Multiplexed Systems."

Corning Incorporated. Whitman, J. "Polarization Mode Dispersion (PMD): What is the statistical method for determining Link PMD and why is it important?" March 1999.

Department of Electrical and Computer Engineering, Department of Physics, and Center for Photonics Research, Boston University. Herzog, W. D., Singh, R., Moustakas, T. D., Goldberg, B. B., Ünlü, M. S. "Photoluminescence microscopy of InGaN quantum wells." September 1996.

Department of Physics and Astronomy, University of New Mexico. Finley, Daniel. "Circular polarization states for light, and quarter-wave plates." January 23, 2001.

Faulkner, D., James, K., Cook, A., Quayle, A., 1995. "Advances in Fiber Access Systems Design and Application." *Spie* Vol. 2614 (p. 313).

Fiber Optics, Infineon Technologies. "V23814-K1306-M230 Parallel Optical Link: PAROLI^{TM}T$_X$DC/MUX-DEC, V23815-K1306-M230 Parallel Optical Link: PAROLI^{TM} T$_X$ DC/DEMUX-DEC." May 2000.

Fiber Optics, Infineon Technologies. "V23814-U1306-M130 Parallel Optical Link: PAROLI^{TM}T$_X$AC 1.6 Gbits/s, V23815-U1306-M130 Parallel Optical Link: PAROLI^{TM} T$_X$ AC 1.6 Gbits/s." September 1999.

Garcia, M.M., Uttamchanani, D. "Influence of the dynamic response of the Fabry-Perot filter on the performance of an OFDM-DD networks." Lightwave Technology, Vol. 15 (10) (pp. 1778–1783). August 1997.

Ghassemlooy, Z., Reyher, R. U., Kaluarachchi, E. D., Simmonds, A. J., 1995. "Digital pulse interval and width modulation for optical fiber communications." *Spie*: Vol. 2614 (p.60).

Hewlett Packard. "Low Cost Fiber-Optic Links for Digital Applications up to 155 MBd." *Application Bulletin 78.* 1997.

Hewlett Packard in conjunction with AMCC. "SONET/SDH OC-48 Transceiver Reference Design." *Application Note 1172.* 1999.

Hewlett Packard. "Characterization Report on 155 Mb/s Single Mode Fiber Transceiver for ATM, SONET OC-3/SDH STM-1." From *Application Note 1141.* 1998.

Hewlett Packard. "Characterization Report on 622 Mb/s Single Mode Fiber Transceiver for ATM, SONET OC-12/SDH STM-4 (I4.1)." From *Application Note 1132*. 1998.

Hewlett Packard. "Characterization Report for 2488 Mb/s Single Mode Fiber Transceiver for ATM, SONET OC-48/SDH STM-16." From *Application Note 1167*. 1999.

Hewlett Packard. "Characterization Report for MT-RJ Duplex Single Mode Transceiver." From *Application Notes 1181&1180*. 1999.

Hewlett Packard. "Characterization Report for Small Form Factor MT-RJ Fiber Optic Transceivers for Fiber Channel." From *Application Note 1182*. 1999.

Hewlett Packard. "Characterization Report for Small Form Factor MT-RJ Fiber Optic Transceivers for Gigabit Ethernet." From *Application Note 1183*. 1999.

Hewlett Packard. "Characterization Report on 1.25 Gb/s 1300 nm Laser Transceiver for 100-SM-LC-L Fiber Channel." From *Application Note 1165*. 1998.

Hewlett Packard. "Complete Fiber-Optic Solutions for IEEE 802.3 FOIRL, 10Base-FB, and 10Base-FL." From *Application Note 1038, 1998*.

Hewlett Packard. "Dense Wavelength-Division Multiplexing Systems: An Overview."

Hewlett Packard. "Fiber Optic Technical Training Manual." 1998.

Hewlett Packard. "Fiber Optic Transmitter and Receiver Data Links for 155 MBd." *Technical Data*. 1996.

Hewlett Packard. "GBIC Module Design Guidelines for Gigabit Ethernet and Fiber Channel Data Communication Applications." *Application Note 1158*. 1998.

Hewlett Packard. "Inexpensive 20 to 160 MBd Fiber-Optic Solutions for Industrial, Medical, Telecom, and Proprietary Data Communication Applications." *Application Note 1123*. 1998.

Hewlett Packard. "Interfacing to PECL Optical Transceivers." *Application Note 1173*. 1999.

Hewlett Packard. "SDX/FDX Characterization Report." *Application Note 1093*. 1996.

Hewlett Packard. "SERCOS Fiber Optic Transmitters and Receiver." *Technical Data*.

Hewlett Packard. "Understanding CDMA Measurements for Base Stations and Their Components." *Application Note 1311*. 1998.

Hewlett Packard. "W-CDMA, EDGE and Bluetooth Technology-Developing next-generation communications with the HP 89400 Series." 1999.

Hewlett Packard. 1.25 Gb Multimode and Single Mode Small Form Factor (SFF) Transceivers." *Application Note 1184*. 1999.

Hewlett Packard. Perry, E., Ramanathan, S. "Experiences from Monitoring a Hybrid Fiber-Coaxial Broadband Access Network." 1998.

Hewlett Packard. Proprietary Data Communication Applications."*Application Note 1122*. 1998.

Hewlett Packard. *Technical Data*. 1999.

"High-Performance IR emitter and IR PIN photodiode in subminiature SMT package."

IEEE. García, J. S., Galindo, A., Iribas, M. L. "Polarization Mode Dispersion Power Penalty; Influence of Rise/Fall Times, Receiver Q and Amplifier Noise." *Photonics Technology Letters*, Vol. 8, No. 12. December 1996.

IEEE. Goel, A., Shevgaonkar, R. K. "Wide Band Dispersion Compensating Optical Fiber." *Photonics Technology Letters*, Vol. 8, No. 12. December 1996.

IEEE. Howerton, M. M., Moeller, R. P., Gopalakrishnan, G. K., Burns, W. K. "Low-Biased Fiber-Optic Link for Microwave Downconversion." *Photonics Technology Letters*, Vol. 8, No. 12. December 1996.

IEEE. Jacobs, S. A., Refi, J. J., Fangmann, R. E. "Statistical estimation of PMD coefficients for system design." *Electronics Letters*, Vol. 33, No. 7. March 27 1997.

IEEE. Jian, Benjamin B. "Etched Corner Reflector Array Lasers: A Detailed Study." *Photonics Technology Letters*, Vol. 8, No. 12. December 1996.

IEEE. Kiyan, R., Kim, S. K., Kim, B. Y. "Bidirectional Single-Mode Erbium-Doped Fiber-Ring Laser." *Photonics Technology Letters*, Vol. 8, No. 12. December 1996.

IEEE. Lin, G., Yen, S. T., Lee, C. P. "Extremely Small Vertical Far-Field Angle of InGaAs-AlGaAs Quantum-Well Lasers with Specially Designed Cladding Structure," *Photonics Technology Letters*, Vol. 8, No. 12. December 1996.

IEEE. Okugawa, T., Hotate, K. "Synthesis of Arbitrary Shapes of Optical Coherence Function Using Phase Modulation." *Photonics Technology Letters*, Vol. 8, No. 12. December 1996.

IEEE. Patel, N.S., Rauschenbach, K.A., Hall, K.L. "40-Gbits/s Demultiplexing Using an Ultrafast Nonlinear Interferometer (UNI)," *Photonics Technology Letters*, Vol. 8, No. 12. December 1996.

IEEE. Salehi, J. A. "Code Division Multiple-Access Techniques in Optical Fiber Networks-Part I: Fundamental Principles." *Transactions on Communications*, Vol. 37, No. 8. August 1989.

IEEE. Teshima, M., Koga, M.. "100-GHz-Spaced 8-Channel Frequency Control of DBR Lasers for Virtual Wavelength Path Cross-Connect System." *Photonics Technology Letters*, Vol. 8, No. 12. December 1996.

IEEE. Yuan, Y., Zhang, X., Bhattacharya, P. "Low Photocurrent GaAs-Al$_{0.3}$Ga$_{0.7}$As Multiple-Quantum-Well Modulators with Selective Erbium Doping." *Photonics Technology Letters*, Vol. 8, No. 12. December 1996.

Infineon Technologies. "5V V23806-A84-C6 Single Mode 155 Mbd ATM/SDH/SONET 2x9 Transceiver with Rx Monitor." *Fiber Optics*. May 2000.

Infineon Technologies. Leininger, Lars. "Interfacing PAROLI® and LVPECL signals." *Appnote 80. Fiber Optics*, August 2000.

Ivankovski, Yuval; Mendlovic, David. "High-rate-long-distance fiber-optic communication based on advanced modulation techniques." *Applied Optics*, Vol.38, No. 26. September 10, 1999.

Iversen, K., Hampicke, D. "Comparison and classification of all-optical CDMA systems for future telecommunication networks." *Spie* Vol. 2614. (p. 110). 1995.

Jackson, K. W., Mathis, T. D., Patel, P. D., Santana, M. R., Thomas, P. M. "Advances in Cable Design." Lucent Technologies: *Optical Fiber Telecommunications*, Vol. IIIA, 1997.

Jan, P. J., Sergienko, A.V., Jost, B.M., Saleh, B.E.A., Teich, M.C., "Dispersion in femtosecond entangled two-photon interference." *A Physical Review*, Vol. 59, No. 3. March 1999.

Jenkins, Francis A., Harvey E. White. 1957. *Fundamentals of Optics*. 3rd Ed. New York: McGraw Hill.

Kacehrad, M., Zaccarin., "Optical code-division-multiplexed systems based on spectral encoding of noncoherent sources." *J. Lightw. Techn.*, Vol. 13, No.3 (p. 534545). March 1995.

Kaiser Optical Systems, Inc. "Virtual Raman Tutorial." 2000.

Kajita, Mikihiro; Kasahara, Kenichi; Kim, T. J., Neilson, D.T., Ogura, I., Redmond, I., Schenfeld, E., "Wavelength-division multiplexing free-space optical interconnect networks for massively parallel processing systems." *Applied Optics*, Vol. 37, No. 17 (p.10). June 1998.

Kapon, Eli. 1998. "Semiconductor Lasers II:Materials and Structures." Academic Press.

Keller, S., Keller, B. P., Kapolnek, D., Abare, A. C., Masui, H., Coldren, L. A., Mishra, U. K., and Baars, S. P. D. "Growth and characterization of bulk InGaN films and quantum wells" *Applied Physics. Letters*, Vol. 68, (p. 3147). 1996.

Kitayama, K., "Novel Spatial Spread Spectrum Based Fiber Optic CDMA Networks for Image Transmission," *J. Select. Areas Commun.*, Vol.12, No.4. May 1994.

Kwait, P. G., Mattle, K., Weinfurter, H., Zeilinger, A. "New High-Intensity Source of Polarization-Entangled Photon Pairs." *Physical Review Letters*, Vol. 75, No.4. December 11, 1995.

Kwong, W. C., Prucnal, P. R. "Ultrafast all-optical code-division multiple-access (CDMA) fiber-optic networks." Computer Networks and ISDN Systems. Elsevier Science B.V. 1994.

Laboratory of Metrology. "Physics of Brillouin scattering." *Activities*. 1998.

Lucent Technologies. "10 Gbits/s and 20 Gbits/s Lithium Niobate SLIM-PAC Electro-Optic Modulator." *Advance Data Sheet*. September 1998.

Lucent Technologies. "10 Gbits/s Lithium Niobate Electro-Optic Modulator." *Data Sheet*. August 1999.

Lucent Technologies. "127A/B/C InGaAs Avalanche Photodetectors." *Data Sheet*. September 1998.

Lucent Technologies. "128-Type Long-Wavelength PIN Photodetector." *Advance Data Sheet*. February 1996.

Lucent Technologies. "131-Type Long-Wavelength PIN Photodetector." *Data Sheet*. April 1998.

Lucent Technologies. "1725-Type Gain Block Erbium-Doped Fiber Amplifier." *Preliminary Data Sheet*. February 2000.

Lucent Technologies. "2.5 GHz Clock Synthesizer, 16:1 Data Multiplexer." *Advance Data Sheet*. September 1999.

Lucent Technologies. "263-Type 0.98 m Pump Laser Module with Fiber Grating." *Data Sheet*. March 2000.

Lucent Technologies. "40 Gbits/s Lithium Niobate Electro-Optic Modulator." *Product Definition Sheet*. August 1999.

Lucent Technologies. "40-Channel Arrayed Waveguide Grating Multiplexer/Demultiplexer." *Preliminary Data Sheet*. January 2000.

Lucent Technologies. "Arrayed Waveguide Grating Multiplexer/Demultiplexer." January 2000.

Lucent Technologies. "Arrayed Waveguide Grating Temperature Controller." *Technical Note*. November 1999.

Lucent Technologies. "Bell Labs Technology: Trends & Developments." 1998.

Lucent Technologies. "D171-Type *FastLight*™ PIN Photodetectors." *Advance Data Sheet*. January 1999.

Lucent Technologies. "D2570, D2526, D2555 Wavelength-Selected Laser 2000 Direct Modulated Isolated DFB Laser Module." *Data Sheet*. September 1998.

Lucent Technologies. "E2500-Type 2.5 Gbits/s Electroabsorption Modulated Isolated Laser Module(EM-ILM) for Ultralong-Reach Applications (>600 km)." *Data Sheet*. September 1998.

Lucent Technologies. "E2560/E2580-Type 10 Gbits/s EML Modules." *Advance Data Sheet*. January 1999.

Lucent Technologies. "E2560/E2580-Type 10 Gbits/s EML Modules." *Advance Data Sheet*. January 1999.

Lucent Technologies. "E2580 EML with Integral Driver IC: Pin Definition and Operation." *Application Note*. May 2000.

Lucent Technologies. "Electroabsorptive Modulated Laser (EML): Setup and Optimization." *Technical Note*. May 2000.

Lucent Technologies. "Extended Band (L-Band) 1735-Type Gain Block Erbium-Doped Fiber Amplifier." *Preliminary Data Sheet*. January 2000.

Lucent Technologies. "High-Power D254xPx Laser 2000 Isolated DFB Laser Module with Polarization-Maintaining Fiber." *Advance Data Sheet*. September 1999.

Lucent Technologies. "High-Speed x170A Long-Wavelength PIN Photodetector." *Advance Data Sheet*. July 1996.

Lucent Technologies. "LG 1600KXH Clock and Data Regenerator." *Data Sheet*. June 1999.

Lucent Technologies. "LG1626DXC Modulator Driver," *Data Sheet*. February 1999.

Lucent Technologies. "LG1626DXC Modulator Driver." *Data Sheet*. February 1999.

Lucent Technologies. "LG1628AXA SONET/SDH 2.488 Gbits/s Transimpedance Amplifier." *Preliminary Data Sheet*. January 1998.

Lucent Technologies. "Low-Cost, High-Voltage APD Bias Circuit with Temperature Compensation." *Application Note*. January 1999.

Lucent Technologies. "Lucent Technologies unveils new optoelectronic components for high-speed communication systems." February 24, 1998.

Lucent Technologies. "Microelectronics Group News Announcement." February 27, 1996.

Lucent Technologies. "NetLight™ 1417G4A and 1417H4A ATM/SONET/SDH Transceivers." *Data Sheet*. July 1998.

Lucent Technologies. "NetLight™ 1417G5 and 1417H5-Type ATM/SONET/SDH Transceivers with Clock Recovery." *Data Sheet*. June 1999.

Lucent Technologies. "NetLight™ 1417J4A 1300 nm Laser 2.5 Gbits/s SpeedBlaster™ Transceiver." *Advance Data Sheet*. July 1998.

Lucent Technologies. "NetLight™ 1417J4A 1300 nm Laser Gigabit Transceivers." *Data Sheet*. July 1999.

Lucent Technologies. "NetLight™ 2417J4A 1300 nm Laser Gigabit Transceivers." *Data Sheet*. July 1999.

Lucent Technologies. "Networks and Communications Documentation."

Lucent Technologies. "New DFB Isolated Laser Modules For 1999." *Microelectronics group*. September 1999.

Lucent Technologies. "Optical Amplifier Platform, 1724-Type Erbium-Doped Fiber Amplifier (W Series)." *Data Sheet*. January 2000.

Lucent Technologies. "Selecting Lucent Fiber-Optic Transmitters and Receivers for SONET/SDH Applications." *Application Note*. July 1996.

Lucent Technologies. "Six companies sign multi-source agreement for next-generation fiber-optic transceivers." *Optoelectronics Group*. February 1999.

Lucent Technologies. "SLM-16 Synchronous Line Multiplexer." *Optical Networking*. 1999.

Lucent Technologies. "System Analysis, Component Selection, and Testing Considerations for 1310 nm Analog Fiber-Optic CATV Applications." *Application Note*. June 1998.

Lucent Technologies. "TMOD0110G 10 Gbits/s Modulator Driver." *Preliminary Data Sheet*. November 1999.

Lucent Technologies. "TRCV012G5 2.5 Gbits/s and TRCV012G7 2.7 Gbits/s Limiting Amplifier, Clock Recovery, 1:16 Data Demultiplexer." *Advance Data Sheet*. February 2000.

Lucent Technologies. "TRCV012G5 2.5 Gbits/s Limiting Amplifier, Clock Recovery, 1:16 Data Demultiplexer." *Advance Data Sheet*. September 1999.

Lucent Technologies. "TTIA0110G 10 Gbits/s Transimpedance Amplifier." *Preliminary Data Sheet*. October 1999.

Lucent Technologies. "Using the Lithium Niobate Modulator: Electro-Optical and Mechanical Connections." *Technical Note*. April 1998.

Lucent Technologies. "Wavelengh-Selected D2525P Laser 2000 Isolated DFB Laser Module with PMF." *Advance Data Sheet*. September 1998.

Lucent Technologies. "WaveStar™ ADM4/1." *Optical Networking*. 2000.

Lucent Technologies. "x2821C Dual-Output Lithium Niobate Modulator and x2821CA Integrated Phase Dual-Output Lithium Niobate Modulator." *Advance Data Sheet*. April 1998.

Lucent Technologies. DiGiovanni, D. J., Jablonowski, D. P., Yan, M. F. "Advances in Fiber Design and Processing." *Optical Fiber Telecommunications*, Vol. IIIA. 1997.

Lucent Technologies. Fan, C., Kunz, J. P. "Terrestrial Amplified Lightwave System Design." 1997.

Lucent Technologies. Forghieri, F., Tkach, R. W., Chraplyvy, A. R. "Fiber Nonlinearities and Their Impact on Transmission Systems." *Optical Fiber Telecommunications*, Vol. IIIA. 1997.

Marquis, D., Castagnozzi, D. M., Hemenway, B. R. "Description of All-Optical Network Testbed and Applications." *Spie* Vol. 2614 (p. 44). 1995.

Mendis, F. V. C., Haldar, M. K., Wang, J. "New method for reducing distortion in directly modulated lasers in subcarrier multiplexed system." *Spie* Vol. 2614 (p. 236). 1995.

Microcosm. "Complete, Pure-CMOS IC for 10Base-T to 10Base-FL Ethernet Conversion." Preliminary Information: MC4664/9_D.

Nakamura, M., Kitayama, K. "System performances of optical space code-division multiple-access-based fiber-optic two-dimensional parallel data link." *Applied Optics*, Vol. 37, No. 14. May 10, 1998.

Nakamura, S., Mukai, T., Senoh, M., Nagahama, S. *Journal of Applied Physics*, Vol. 74 (p. 3911). 1993.

Nakamura, S., Mukai, T., Senoh, M., Nagahama, S., Iwasa, N. "InGaN/InGaN superlattices grown on GaN films." *Journal of Applied Physics*, Vol. 74 (p. 6). September 15, 1993.

Nakamura, S., Senoh, M., Nagahama, S., Iwasa, N., Yamada, T., Matsushita, T., Kiyoku, H., and Sugimoto, Y. " InGaN multi-quantum-well structure laser diodes grown on substrates." *Applied Physics Letters*, Vol. 68 (p. 2105). 1996.

Nakamura, S., Senoh, M., Nagahama, S., Iwasa, N., Yamada, T., Matsushita, T., Kiyoku, H., and Sugimoto, Y. "Characteristics of InGaN multi-quantum-well-structure laser diodes." *Applied Physics Letters*, Vol. 68 (p. 3269). 1996.

National Radio Astronomy Observatory. Emerson, Darrel. "Elliptical Polarization in the Ionosphere." February 1998.

Nortel Networks. "AB52 155Mb/s Transimpedance Amplifier." March 1999.

Nortel Networks. "AC03 Automatic Gain Control Amplifier." March 1999.

Nortel Networks. "AC30 Multiplexer and Laser Diode Driver." May 1999.

Nortel Networks. "YA18 2.5 Gb/s Clock and Data Recovery Circuit." *Data Sheet*. February 1999.

Nortel Networks. "YA19: 2.5 Gbits/s 16:1 Multiplexer and Clock Generator." February 2000.

Nortel Networks. "YA20 2.5 Gb/s 1:16 Demultiplexer." *Data Sheet*. 1999.

Onat, B. M., Ünlü, M. S., "Polarization Sensing with Resonant Cavity Enhanced Photodetectors." January 28, 1997.

Özbay, E., Islam, M. S., Onat, B., Gökkavas, M., Aytür, O., Tuttle, G., Towe, E., Henderson, R. H., Ünlü, M. S. "Fabrication of High-Speed Resonant Cavity Enhanced Schottky Photodiodes." Http:// Photon.bu.edu/selim/papers/ptl-97/PTL.html.

Philips Semiconductors. Geurts, J. M. M. "Fiber optic transceiver demo board STM16 OM5801." *Application Note AN96051*. August 1, 1996.

Philips. "OQ2536HP SDH/SONET STM16/OC48 demultiplexer." *Data Sheet*. March 10, 1998.

Philips. "OQ2541HP; OQ2541U SDH/SONET data and clock recovery unit STM1/4/16 OC3/12/48 GE." *Data Sheet*. May 27, 1999.

Philips. "SA5211 Transimpedance amplifier (180 MHz)." *Data Sheet*. October 07, 1998.

Philips. "SA5212A Transimpedance amplifier (140 MHz)." *Data Sheet*. October 07, 1998.

Philips. "SA5217 Postamplifier with link status indicator." *Data Sheet*. October 07, 1998.

Philips. "SA5223 Wide dynamic range AGC transimpedance amplifier (150 MHz)." *Data Sheet*. October 24, 1995.

Philips. "SA5224 FDDI fiber optic postamplifier." *Data Sheet*. October 07, 1998.

Philips. "SA5225 Fiber optic postamplifier." *Data Sheet*. October 07, 1998.

Philips. "TDA1300T; TDA1300TT Photodetector amplifiers and laser supplies." *Data Sheet*. July 15, 1997.

Philips. "TZA3004HL SDH/SONET data and clock recovery unit STM1/4 OC3/12." *Data Sheet*. February 09, 1998.

Philips. "TZA3019 2.5 Gbits/s dual postamplifier with level detectors and 2×2 switch." *Data Sheet*. April 10, 2000.

Philips. "TZA3023 SDH/SONET STM1/OC12 transimpedance amplifier." *Data Sheet*. March 29, 2000.

Philips. "TZA3033 SDH/SONET STM1/OC3 postamplifier." *Data Sheet*. November 03, 1999.

Philips. "TZA3033 SDH/SONET STM1/OC3 transimpedance amplifier."*Data Sheet*. July 08, 1998.

Philips. "TZA3043; TZA3043B Gigabit Ethernet/Fiber Channel transimpedance amplifier." *Data Sheet*. March 28, 2000.

Photcoef. "Photon Interaction Coefficients of the Elements." August 6, 2001.

Photonic Integration Research, Inc. "Mach-Zehnder Interferometer (MZ) FDM Module." December 5, 1999.

Photonic Integration Research, Inc. "Normal Band Arrayed-Waveguide Grating (AWG) DWDM Module." September 22, 1999.

Photonic Integration Research, Inc. "Single Mode 1 × n and 2 × n Splitter Modules." December 07, 1999.

Photonic Integration Research, Inc. "Single Mode Thermo-Optic Switch Module." March 01, 1999.

Raytheon Commercial Electronics. "User's Manual for the Distributed Fiber Optic Controller (DFOC)." August 26, 1998.

Raytheon Commercial Electronics. "User's Manual for the Enhanced Standard Modular (SMX) Fiber Optic Transceiver." August 26, 1998.

RSOFT Corp. "Direct Modulated Laser." *LinkSIM Version 1.0 Beta 1*, Chapter 5:LinkSIM Models, p. 35.

Sargis, P. D., Haigh, R. E., McCammon, K.G. "Dispersion-reduction technique using subcarrier multiplexing." *Spie* Vol. 2614 (p. 244). 1995.

SDL Integrated Optics Ltd. "10 Gb/s Family."

SDL Integrated Optics Ltd. "2.5 Gb/s Family."

SDL Integrated Optics Ltd. "Application Notes: 2.5 Gb/s and 10 Gb/s External Modulators."

SDL Integrated Optics Ltd. "Application Notes: Introduction to Modulators."

Siemens Semiconductor Group. "Fiber Optic Transceiver Product Selection." *Appnote 79*. November 1998.

Silva, M.D. "Interpreting CDMA Measurements." *Wireless Design Online*. June, 1998

Spark Notes: "Introduction to Optics."

Sribar, J., Puksec, J. D. *Physics of Semiconductor Devices-Solved Problems with Theory: Vol.1.*

Sumitomo Electric. "Technical Specification for 2.5Gbps Fiber Optic Transceiver Module (SDM7128-XC)." November 1998.

Sumitomo Electric. "Technical Specification for Optical Transceiver Module (SCM7111-XC)." April 2000.

Sumitomo Electric. "Technical Specification for Optical Transceiver Module (SCM7311-XC)." April 2000.

Sun. Ünlü, Selim. "RCE Optical Logic and Systems."October 8, 1995

Sun. Ünlü, Selim; "RCE Optical Logic and Systems." October 8, 1995

Sun. Ünlü, Selim; Strit, Samne."Resonant Cavity Enhanced (RCE) Photonic Devices." October 8, 1995.

Systran Corp. "Fiber Channel Network Technology Applied to Advanced DSP Systems." 1994.

Texas Instruments. "TNETE2201A 1.25-Gigabit Ethernet Transceiver." March 1999.

Tulane University. Nelson, S. A. "Interference Phenomena, Compensation, and Optic Sign." January 17, 2001.

Wang, J., Qiao, C. "An Evaluation of the Time Domain Approach for Crosstalk Free Communication in Photonic Switches." *Spie* Vol. 2614 (p. 204). 1995.

Wickramasinghe, W. R., Ghassemlooy, Z., Chao, L., "Pulse time modulation for subcarrier multiplexed systems." *Spie* Vol. 2614 (p. 229). 1995.

Wilson, B., "Digital pulse interval modulation for fibre transmission." *Spie* Vol. 2614 (p.53). 1995.

Wolfram Research: Resource Library. "Blackbody Radiation."

Ziemann, O., Iversen, K. "On optical CDMA based on spectral encoding with integrated optical devices." *Spie* Vol. 2614 (p.142). 1995.

INDEX